'The smallest
point is a world in in itself'
Leibniz, on the fundamental particle of Reality:
The fractal point=world of space-time.

$\cdot$

i

GM_i

$[GM]_{st}$

(A GST of MSTs)

A General Systems Theory of
Multiple Worlds of Space-Times

THE 4TH PARADIGM OF SCIENCE: SOCIAL NETWORKS

(The Whys of the Universe)

Luis Sancho

The 4th Paradigm of science: Social Networks
A General Systems Theory of Multiple Spaces and times.

iUniverse books may be ordered through booksellers or by contacting:

iUniverse
1663 Liberty Drive
Bloomington, IN 47403
www.iuniverse.com
1-800-Authors (1-800-288-4677)

ISBN: 978-1-4620-4529-7 (sc)

Printed in the United States of America

iUniverse rev. date: 09/20/2011

INTRODUCTION:

The 4th Paradigm of science: Social Networks.

0

CREATION

The Cycles of Time Arrows

i

FRACTAL UNIVERSES

Space-Time Points: Worlds of Time Arrows

A.

ASTRO-PHYSICS

The 3 scales of the Universe.

B.

BIOLOGY

The Plan of Evolution.

C.

CIVILIZATIONS

The FMI Complex vs. History

FOREWORD: THE FRACTAL, ORGANIC, DUAL UNIVERSE

The 4th paradigm in a nutshell: The Universe is an organic system of energy and information.

All what you see in reality responds to a dual structure: Entities of Nature are complementary systems of simplex networks of energy and information, which combine to form a complex system that switches and merges both states. In the process the system might reproduce an entity self-similar to itself. And this is the definition of an organism, reason why we affirm that reality is dual and organic.

In Physics all what exists are particles of information associated to energetic fields that move them, and only both together exist (Complementarity Principle), creating a physical entity, which can decouple creating self-similar particles.

In Biology cells have a DNA nucleus of information, surrounded by cellular cytoplasm that provides energy to the nucleus; or an informative head moved by an energetic body. And only both together can exist, creating a biological organism. There are no bodies without heads and red cells without nuclei die faster than any others. And cells and organisms can reproduce self-similar forms.

Finally, in the economic ecosystem money acts as the informative language that gives orders to the physical economy, which reproduces all types of serial machines.

We formalize all Dual systems and events of reality born of complementary networks of energy and information with an Equation we call the 'Fractal Generator of Reality': $\sum E \Leftrightarrow \prod i$. Since we claim this equation resumes all the events and forms of the Universe. It is the key equation of organicism, the philosophy of science called General Systems Sciences that considers the Universe and all its parts an organic system generated by feed-back, iterative exchanges, $\Leftrightarrow$, of energy (E) and information (I) between both type of networks, $(\sum, \prod)$. *The symbol $\sum$(E) means that body systems are loose sums of cellular herds that only relate to their neighbors. $\prod$(I) means that informative systems form complex networks in which each element connects with all others, multiplying its 'axons'. For example, 5 informative elements have 5x5=$\prod$5 axons. Thus, information networks co-exist in two scales. They are more complex than energy systems and dominate them. So informative particles, heads and money dominate respectively energetic fields of forces, bodies and the physical economy. The Generator Equation of Reality will allow us to derive a set of rules from where to extract the properties and laws of all disciplines of science, each one specialized in the study of the entities of energy and information of a certain scale of reality, from the smallest complementary entities studied by quantum theory to the biggest one, the Universe, a system of galaxies related by networks of dark energy and informative, gravitational forces. We can, departing from those concepts, define also the Universe and all its parts:

'The Universe is a fractal, organic system, which reproduces complementary, social networks of energy and information that self-organize themselves in bigger, self-similar scales'.

We mean by energy, motion; that is, displacement in space, which is a relational concept, first defined by Leibniz: The sum of all motions, simplified by the eye-mind as a continuous, static extension creates our perception of space. And we mean by information, dimensional, cyclical form. And the sum of all cyclical forms-in-action, in-form-ations, which our mind perceives as cycles of clock-time, creates relational, 'fractal' time, also defined first by Leibniz, one of the fathers of the 4th paradigm[1]. Energy without motion or form without energy is not perceivable. Thus, systems to exist need a minimal quantity of both or they cease to exist - they die. Or as Taoist, the closest, classic philosophy to this book put it: 'All yin-information has a drop of yang-Energy; all Yang has a drop of yin'.

Further on, because all systems respond to the same laws, we can recognize the shapes of a network dominant in energy by its form that will be lineal, planar, extended in space, as the line is the shortest distance between two points; hence the shape with more motion. And we can recognize the shapes of information because they are smaller in space, broken and cyclical, as the cycle or sphere is the shape that stores more form in lesser space; hence the perfect form of information.

So DNA nuclei, particles, heads and coins of money, made of gold, the best informative atom of the Universe, are cyclical and small. While fields of forces, moving bodies and weapons, made of iron, the most energetic atom of the Universe, are lineal and big. Finally, real physical, biological and mechanical systems are those who transform back and forth energy and information, between their two poles: $E \Leftrightarrow I$.

It follows also that cyclical information, frequency and time cycles are related. On the other side, expansive energy, motion and space, which stores energy in the vacuum are also related. So energy is fractal space and information fractal time – facts that will allow to fusion the jargon of many sciences.

Thus, reality is a creative process in which infinite forms constantly imprint surfaces of energy with fractal patterns, defining the dominant arrow of time in the Universe: information, which constantly grows the complexity of its forms from a relative past into the future. Thus, we define for any system the arrow of information from the perspective of the ages or dimensions of time, past, present and future: All complementary systems warp the spatial energy of the Universe into dimensional form, reaching a limit of warping, as they move from a *past* age of 'pure energy with minimal form' or *young age* of the system, into a relative *present*, mature age of balance between energy and form, or classic, *reproductive age* of the system, in which they often decouple=iterate=reproduce a self-similar form (hence creating a dynamic present, as the repeated form seems to leave reality unchanged) to end in a future 3rd age of maximal form and minimal energy. Then the form explodes back to the past in minimal time, unfolding all its warped form into pure energy, hence dying. And this is the 2nd arrow of time, entropy – a fast motion of time towards a past without form. So we can unfold the Generator equation in 3 ± 1 ages to define the cycle of life and death of all systems between birth and extinction:

$$\infty I \times 0 E \ (seed) \rightarrow Max.E \times Min.I \ (youth) \rightarrow E=I \ (reproduction) \rightarrow Max.I \times Min.E \ (3^{rd} \ age) < \infty E \times 0I \ (death)$$

Death implies that reproduction is needed for the Universe to conserve its logic forms. The previous equation represents the evolution of all systems. In Cosmology it represents the 3 solutions to the Equations of Einstein, from the big-bang to the steady state to the big crunch, in a repetitive Universe of infinite cycles. In Physics represents the 3 states of matter: the energetic, gaseous state, the balanced liquid state and the informative, solid state. In Biology it represents the 3 ages of life, and so on.

Those 3 ages of the Generator Equation can co-exist synchronically as 'subspecies' that form a ternary system with lineal energy limbs, spherical heads of information and elliptic reproductive bodies that combine both. So we can represent also living systems: as E(limbs)<E=I>Bodies>I(Head).

The equation of the ages of life also defines more clearly the meaning of social networks and the fractal, scalar structure of the Universe: Self-similar, self-reproductive forms chain each other into informative networks without energy, (seeds), which 'emerge' in a 'higher' scale of existence, going through the 3 ages of life. Then they dissolve the information acquired in that higher scale, dying back into their lower plane of existence. So a seed of genetic information of the relative cellular scale, i-1, emerges as a fetus out of the womb after self-reproducing and organizing its form, in the scale of biological organisms, i, where it lives and dies after a 3rd age of dwindling energy, dissolving back to its lower cellular scale, i-1.

So we can define a parameter of information, I, as a scalar arrow of growing complexity: The simplest i=1 'universal constants' of spatial energy and temporal information, (magnetic and electric constants) combine to form Planck constants, h, also called 'actions' of energy and time. They will reproduce on the pure vacuum without form, creating light waves, social networks of h-actions; which evolve further warping into cyclical forms, called particles. Those particles will reproduce and evolve socially into atoms, with an informative pole (I), the quark, with more formal mass and an energetic pole, (E) the electron, with more spatial motion=energy. Both exchange, ⇔, flows of electromagnetic energy and informative gravitation. Thus mass and gravitation are the informative forces and particles that interact with light and electrons, the energetic, spatial ones; creating together the next scale of atoms.

So a pattern of social evolution in growing scales of complexity emerges: particles become complementary atoms of energy and information, reproduced in the big-bang and the interior of stars. Then atoms form molecular networks, creating the chemical scale. Then molecules reproduce and self-organize into organic, cellular systems or inorganic, celestial bodies. Next cells reproduce and self-organize into organisms that reproduce and self-organize into societies; while celestial bodies self-organize into galaxies that self-organize into Universes.

All those facts must be formalized with new tools of mathematics and Logic. The mathematics of a fractal, organic, self-generating Universe are Non-Euclidean geometries whose 5 postulates we complete going beyond the work of Riemann, who defined a point with 'volume' through which many parallels can cross. Hence a point must be a fractal that grows in size as we come closer to it, allowing more parallels to go through. Such fractal point with breadth, with form and motion, with energy, becomes the new unit (new 1st postulate), of Fractal, Non-Euclidean mathematics, (of which all other geometries are simplifications), in which lines are waves (as points have motion and form) and planes are social networks. A fractal point is the 'organic unit' of a certain i-scale of the Universe that from a higher plane of existence appears as an Euclidean point without breadth, but as we can closer it becomes bigger and shows its form and motion. So we humans see in the limit of our perception, quarks and electrons as fractal points without parts and in the galactic realm we see with naked eyes stars and galaxies as fractal points without parts; but as we come closer those fractal points should acquire volume. And all of them will follow the Mathematical Laws of Fractal, non-Euclidean geometry.

In logic we also have to upgrade Aristotelian Logic to understand a Universe with two arrows of time, one of energy and other of information, of relative past and future, which combine to create the dominant arrow of 'present=existence'. We call this logic, which adds the arrow of information, i-logic. It explains the meaning of past, present and future, the 3 dimensions of time, and how they interact to form reality. Two are the essential equations of the 3 dimensions of time: past->present ->future, which gives us a diachronic analysis of reality, through the 3 ages of time. And Past (energy system) x Future (informative system) = Present organism, which defines the systems we perceive in space.

It follows that what we call a mind is a structure that perceives dimensions of time and forms in space, reordering them according to those causal equations, to generate 'a virtual image' of reality. Or in other words the mind is a perceptor of logic structures of space (present systems) that flow in time from past to future, as they live and from future to past as they die. So the Universe is immortal, a zero sum of all the flows of life towards the future and all the flows of death to the past that create an eternal present.

First the book will introduce the Philosophy of Organicism, the Generator Equation and its 2 arrows of time. Then it upgrades the logic and mathematics of science. Once we have those 3 formal tools, organicism, Duality and Non-e Mathematics we shall apply them to each science showing how they can resolve most of the unanswered questions of each discipline. Since the r=evolution that implies to upgrade Euclidean and Aristotelian Logic and the mechanist philosophies of the Founding fathers of science is a jump in complexity in our understanding of the Universe that has not taken place since the quantum and relativity revolution of the XX century, we consider this and similar books of other pioneers to initiate a new paradigm of science. The power of the 4th paradigm will be shown precisely in the solutions provided to many standing questions of science and the synoptic capacity of its theoretical tools, which coming from a simple equation and its rules of manipulation, derived from i-logic and Non-euclidan, Fractal Geometry can deduce all other scientific equations and represent all forms of reality.

The difficulties of a change in paradigm in science and its implications for society.

The advances of a Unification Theory of all sciences brought about by the discovery of a similar set of laws that can describe all systems of the Universe means a change of scientific paradigm, from the mechanist philosophy of the founding fathers of science to organicism; from the simple, static Euclidean mathematics of classic science to the complex, dynamic Non-Euclidean, Fractal points explained in this book – the new mathematical unit of the 4th paradigm that substitutes the 'point without breadth' of Euclid; and finally from the Aristotelian, unicausal logic of classic science to the Dual Logic of systems in which there are two arrows of time, energy and information, which converge in all sytems to create the present reality. Unfortunately a change of paradigm of such depth will take time to be accepted by scholars. And it is a fact, painful to recognize that in the last decades fractal theory, systems sciences and organicism, the pillars of this model have garnished little attention among mainstream science, which keeps exploring the previous models of a continuous space-time in physics, a mechanist philosophy of the Universe and a single time arrow (energy or entropy) despite is growing difficulties to resolve many facts of science that this book will explain easily.

The general reasons of the natural resistance of scholars to change a scientific paradigm were masterly described by Thomas Kuhn in its work on sociology of science[2]: Most things we know, Kuhn explains, are learned by 'memetic' repetition not by reasoning. Routine, personal prestige and established knowledge imprinted by repetition, generation after generation makes very difficult for a new paradigm to displace what scholars with authority have learned. Authority is always more important than truth for us to 'believe', *even in science*. And authority is obtained not by innovation but by learning and teaching in achademia what it is already know. The rat bites its tail.

Further on, the 4th, organic paradigm collides with what Kuhn calls the industrial interests of 'big science', which are by definition 'mechanical', based on the use of machines to understand the Universe. Yet if the Universe is a dual system of networks of energy and information, as organisms are, this means that man not the machine is the measure of all things – *even machines,* which are simple systems of energy (weapons, transports) or information (chips=brains, cameras=eyes, mobiles=ears), that copy the functions of our own organs; put together now into organic robots.

Thus the 4th paradigm makes of Humanism and its disciplines, Biology and Humanities the queen sciences. It also means the revaluation of the languages of the human mind, his verbal ethics and sensorial experience, expressed through spatial art and temporal words as important forms of knowledge. It also means in a Universe made of evolving, social networks that socialism not capitalism is the natural theory of economics in which man not technology matters more. Hence it recommends as the new goal of our societies a welfare state that satisfies human necessities not a supply economy to the service of companies, re=producers of machines. In brief, the 4th paradigm challenges the 3 dogmas of our industrial civilization, capitalism, mechanism and nationalism; since its social, organic laws, show that socialism, organicism and humanism (the treatment of all humans as members of the same species) are more complex and truthful to those laws. This means that our civilization has taken the wrong path by abandoning the worship of the tree of life for the tree of technology and the goal of evolving the human, social organism we call history for the goal of evolving the machines of the economic ecosystem. This, as we shall see in the last chapters, has grave implications for our survival as a species.

The future of the 4th paradigm. Knots of thought and growing proofs.

On the long term, though, the perspectives for the 4th paradigm are bright. Since as Kuhn explains, time works in favour of a new paradigm, as the old one becomes stuck with no further solutions to its unresolved problems and the new paradigm moves from the phase of the pioneers to the massive entrance of new researchers, which in the field of fractals and social networks occurred in the 2000s.

In that regard, the experimental proofs of the fractal, self-organizing Nature of reality and all its parts are now definitive in Cosmology, after the Sloan mapping has found that even at the greater scales of the Universe galaxies also self-organize themselves in fractal networks. On many other sciences proofs keep mounting: The New Scientist published in 2008 the first article on Fractal Evolution and affirmed that the Sloan's mapping has made the Big-bang obsolete. The crisis of e-money of 2008 was also predicted with chronological exactitude in our pioneer books on biological history, back in the 90s with our models of organic machines, within the fractal, long, medium and short, 800-80-8 waves of economic and historic evolution. Later we shall analyze some other the predictions published in those older works[3], now proved across many sciences, and advance further solutions to key questions still unresolved by the previous paradigm of 'mechanist' science. Thus, I believe the time is ripe for this re-edition of a General Systems Model of Multiple, Fractal Spaces-times which is the philosophy of science and General Theory that science needs to enter a more mature, interdisciplinary stage[3].

I imagine now the reader rising his eyebrows at this point, given the ambitious plan set for this work – it is this book a joke of a crackpot? No, it is not. In my defense I will bring the words of Mr. Mandelbrot - the last recognized master of many disciplines that illuminated so many new fields of science - explaining the difficulties this kind off approach has to break into the scholar world:

'When they told me my work was impossible to understand, i charged again and gave new explanations tirelessly. If I had been inconsistent or shy, or too arrogant, or my book had been unreadable it is quite probable that the discovery of fractals had never been known. But there are many researchers who are shy, arrogant and illegible (this I recognize to be defects of this writer). What happens then with their work? Nothing. Plus our discipline requires swimming between several disciplines, which always irritates scholars. Today as researchers become more specialized, a discovery as this would be impossible. For that reason none of my students tries to imitate me. All of them have chosen a specific discipline to apply the general theory of fractals.'

In brief a General Systems Theory has to fight the belief among scientists – even among physicists and system scientists that try to find such theory – that a Theory of the Whole is imposible to find, either because it does not exist or because humans do not have enough intellectual capacity to achieve it.

Yet the intention of this book is to develop a general theory of the fractal, organic Universe *that harmonizes and fusions all other theories of science, advancing them*, once we explain the logic and mathematical principles common to the fractal points of all disciplines. We will do so, as Mandelbrot put it, 'charging again' with 3 increasing degrees of complexity, which go deeper into the concepts explained in the previous paragraph: Introduction; 0I - logic and mathematical laws of all systems; and ABC – a detailed scientific analysis of all the species of all the scales known in the Universe.

Unfortunatelly, paraphrasing Mandelbrot this book adds to that 'arrogance' of trying to explain it all, the shortcomings of a 'shy' scholar that hardly participates in Congresses of Science and prefers to self-publish his books not to deal with editorial reviews; which added to the fact English is not my native tongue and the new logic, mathematical and philosophical concepts of General Systems might be unfamiliar to the reader makes his work somewhat 'illegible'. It thus might happen 'nothing' with it.

But I disagree with Mandelbrot's assertion that 'fractals would have never been discovered'. Even when an innovator is shy, arrogant, illegible and lives outside the centers of mainstream science, as Copernicus or Leibniz or Mendel did in the past, science is always a collective effort and sooner or latter the subconscious collective of human knowledge catches up. Since in the future the Universe will still be here and will still obey the same laws that those pioneers discovered first, for others to find them.

It is the only fact I can bring in my defense to encourage you to go through the difficulties of reading this book: You read the future of science and if you go through all its hurdles, your mind will be upgraded into a deeper, more satisfying explanation of the Universe and all its parts, including yourself.

If we were to apply the fractal, organic theory of this book to the new paradigm of science, knowledge also follows the general pattern of all self-organized systems: a pioneer, or first 'generator' cell, a knot of human thought, makes a synopsis of prior knowledge and gives a giant leap on the understanding of 'the thoughts of god', which latter are expanded by a fractal wave of self-similar minds, specialists in each discipline. Then again all those 'details' on the thoughts of God are summarized by a new knot of thought, which includes them in a wider, interdisciplinary view that will in turn become expanded by specialists of all sciences: knot of interdisciplinary thought < wave of detailed specialists > collapse in new knot of interdisciplinary thought < wave of specialists... and so on.

Thus a series of antecedents in classic western thought (Plato, Aristotle, Leibniz and Riemann, being the most remarkable), collapsed in the first fractal knots of General Systems thought, Bertalanffy in the 50s and Mandelbrot in the 70s, who defined and expanded the concepts of the fractal, organic paradigm to all disciplines. Then several specialists illuminated them with detailed studies of the fractal Universe on physics (Nottale, causal triangulation), chemistry (Mehaute) and other disciplines in the 80s and 90s[1].

Further on, since as structuralists put it, each writer is merely known by his works, we must properly talk of 'books' that are knots of thought rather than of people. In that sense this book, first published in Spain in the 90s[3], is a knot of thought that completes the insights on a General Systems Theory of the Western tradition, from Plato to Mandelbrot going a step beyond, thanks to the wealth of discoveries of those specialists into a more complex, interdisciplinary attempt to lay down the formalisms and laws of all fractal, organic systems and sciences. Hopefully, now that it is available to the English world (most of the pioneers of the fractal paradigm, as those mentioned before and this author belong to the French, Latin European culture, a fact which has hindered its distribution in the English-speaking globalized world), it will inspire a new wave of specialists who will take General Systems Sciences to its completion with more detailed studies. As such, the book has a minimal quantity of logic and mathematical formalisms, needed for deductive rigor in its description of space and a higher level of philosophical analysis. So any specialist of any discipline can understand it. Yet it is also written for polymaths, who are interested in the great questions previous paradigms failed to respond: What we are, why we exist, what is our destiny in the Universe, as self-similar parts to the whole.

Though if Mandelbrot is right, we will get no readers. It might happen 'nothing'. Yet even in that case, it was a satisfaction to write it to peer again, as my countryman Picasso put it, 'into the window of the absolute'. You are welcome to accompany me.

[1] 'Timeus' with the myth of the cave and 'Monadology' with its definition of a fractal world are the 2 key pioneering works of classic western tradition, while the 'I ching' is the key work of the Eastern tradition. 'Fractals' by Mandelbrot in 1975 is the first key work of the new paradigm that expanded into a series of disciplinary masterpieces, among them 'L'espace-time brisse' by Mehaute, which proved the existence of an arrow of information and 'microcosms' by Nottale, which is the first formulation of a fractal theory of quantum physics with deep insights in the fractal, invariant morphology of the Universe. Finally, the fractal, self-organizing nature of space-time was resolved by causal triangulation (Ambjorn, Jurkiewicz and Loss), who showed the evolution of space-time from bidimensional systems (holographic principle) into the Einsteinian, 4-dimensional Universe.

[2] 'The nature of scientific revolutions', Thomas Kuhn.

[3] 'The codes of the Universe' c.94, 'Radiations of Space-time', c.97, 'A Theory of Unification' and 'Bio-History, Bio-economics' c. 97, published by Bookmasters, are my pioneer books of the 90s on a General Systems Theory of multiple, fractal space-times, available at www.unificationtheory.com & www.economicstruth.com. 'Los Ciclos del Tiempo', 'Los Ciclos de la Economia y la Historia', Arabera, 97, Spain are the Spanish versions.

INTRODUCTION

THE 4TH PARADIGM OF SCIENCE: SOCIAL NETWORKS

The Universe is a system made of social networks of complementary energy and information, which

grow in scales of complexity from the simplest particles to the Universe - a network of informative galaxies and dark energy. In all scales of reality species are social, cellular systems, constructed with self-similar energetic and informative networks (fields and particles in physics; bodies and heads in biology), the whys that cause their existential cycles. In the graph: social dwellings of life superorganisms - insects and humans.

1. The organic whys of the Universe; Systems sciences. The 4th paradigm of knowledge.

A simplex explanation of this work considers the meaning of knowledge and what questions it must answer to exhaust the study of a certain subject. In journalism we ask what, who, how, when and why.

The evolution of science implies big shifts in the paradigms, languages and philosophical dogmas we use to understand the Universe. In the evolution of science we observed first 'what' (experience) and then asked 'who'. It was the mythic age of science, the first paradigm of knowledge – when an anthropomorphic being, often a god was the cause of all events.

Then the Greeks used reason to ask the what (experience) and how of things, its causes and consequences. It was the 2nd paradigm of knowledge: logic thought.

The 3rd paradigm started by Galileo with his use of machines (clocks and telescopes), responded to when, measuring space distances and time frequencies in great detail with a single space-time system. The ticks of the heart, the stomach, the moon, the atom and the clock are different, but to measure them we needed a unit of time and so we equalized all rhythms with a clock, and to compare the spatial trajectories of those cycles we needed a 'background of space', so we put together all the broken spaces of reality into a joined puzzle, which we called Cartesian space-time. The error of a single spacetime came when we forgot those simplifications and considered that the abstract space-time continuum of Descartes used to measure all other spaces and times was the real space-time.

The culmination of this process of mechanical measure came with quantum theory, which refined the measures of the cyclical trajectories of particles in the microscopic world and General Relativity, which refined them in the cosmological realm by correcting the deformations of those rhythms of time and distances of space caused by the limits of speed of our light-based Universe.

In philosophical terms, the paradigm of measure meant the birth of mechanism, the fundamental philosophy of our world today: the machine - no longer man, an organism - became the 'measure of all things'. This was a simplification, as today we realize that machines merely imitate our organs of energy and information with networks of metal-atoms (so a crane is an energetic arm of metal and a chip an informative brain of metal), which now we fusion into 'organic' robots. And so the change of paradigm from the Greek, Aristotelian and Asian tradition of organicism to mechanism is only a hiatus on a richer, more complex understanding of the whys of the Universe.

Mechanism changed also the language of understanding of the Universe, from Aristotelian Logic to mathematical Platonism, since mathematics was the language used by machines to measure the Universe of time and space with clocks and telescopes; while logic was the language embedded in the syntax of words, which measure time with causal verbs that describe the logic relationships between its 3 dimensions of past, present and future. So in terms of philosophy of science, mechanism meant a pendulum law that changed the paradigm from Aristotle (organic, temporal causality) to Plato (mechanical, spatial geometry). This was a wrong choice, because a truly inclusive theory of reality has to put together both languages and approaches as we shall do in this work: the geometric how & instrumental when matched by a temporal why, which must be by definition a causal, temporal process. The 3rd paradigm obsessed by spatial measure was not very kin of such inquire, as Feynman famously put it: 'the why is the only thing a physicist never asks'. And yet the why has always been a fundamental question of knowledge.

That why should respond to the existence of a program of 'evolution', creation and extinction of the deterministic reality that we see all around us, which always gives birth and extinguishes the same entities, repeating their forms once and again. What is the purpose of the Universe and all its repeated parts? Why they have those forms and follow always a life and death cycle?

Thus, scientists, not satisfied with the limits of the 3rd paradigm of measure kept asking the why, which could not be a personal God (the who of the 1st paradigm), neither the machine, the instrument of measure of the 3rd paradigm ('God is a clocker' said Kepler, because he used clocks to measure it and 'God speaks mathematics' said Galileo, because those machines translated the events of the Universe into mathematical data).

According to *Deism* the whys of existence are due to a personal being, external to the Universe that makes it all happen and cares for humans more than for the rest of His work. According to *Mechanism*, this is due to the self-similarity between the Universe and the primitive machines we humans construct to observe it. Mechanism though needs 'someone' to make the machine, which is not self-generated; so it is similar to deism, reason why the founding fathers of science, all pious believers, adopted it as a proof of the existence of God, which had given man self-similar properties – the capacity to make machines to the image and likeness of the Universe. The problem with those 2 approaches, which in fact are the same is obvious: a personal God is an anthropomorphic, subjective myth and science must be objective; while a mechanical view of the Universe still needs an internal, self-sustained process of growth, creation and synchronization caused by an external God that made and rewinds the clock - as Leibniz clearly stated in his critique of Newton[nt.0.4]. Scientists today are unaware that mechanist theories are in fact deist theories, reason why Kepler and Newton, pious believers, liked them; since they were a metaphor of their self-centered, anthropomorphic religious beliefs: If man created machines because we were made to the image and likeness of God, God had created the ultimate machine, the Universe.

Unfortunately such theories cannot satisfy the rational, objective, self-sufficient nature of science, as they require external myths to work. This leaves us only with a 3rd objective, scientific possibility, organicism - a more complex theory that comes of age in this work: if the why is neither God nor the machine, there is only a 3rd option in between - the organism, which is an intermediate concept; since an organism is self-sufficient as God is, yet it is part of Nature as the machine is. Such organicism means, as Aristotle, the father of this alternative to Platonic, mechanical science, well understood, a return to Logic - the understanding of the causal processes that move and transform organic systems from past to future – albeit a logic more complex than the single, unicausal logic of Aristotle. It also means, as he famously affirmed that we 'are all Gods' in an Aristotelian sense (unmoved focuses of informative perception that control and move the bodies of energy around them). That is, all what exists is structured with 2 complementary systems, one of information that gauges reality (particles and heads of physical and biological entities) and one of energy that moves them (fields of forces and bodies). Thus the answer to the why can be resumed in a word '*networks*' that defines the multiple realities we see around us:

'*All what exists are organic, fractal systems, made of self-reproductive, complementary, social, topological networks of energy (bodies and fields) and information (heads and particles)*': $\sum Se < X > \prod Ti$

Such networks can be described not only with the logic of organisms but also mathematically, because as Plato put it, 'numbers are forms', meaning that a set of self-similar entities are a number and so mathematics, the science of numbers, is by definition a social science. Further on those numbers create social networks with form: so 1 is the point, 2 is the line, 3 is the network and 4 the square and so on. Thus each number defines in geometry certain topological forms that favor certain flows of energy and information between the points of the network.

Reality is never static, because geometry becomes topology; planes and forms become networks and distances become flows between points, which can be described with feed-back equations, $E \Leftrightarrow I$, no longer with equalities.

Biology is in fact the central science of the 4th paradigm of social networks, since organic systems are traditionally studied by Biological Sciences. Thus, the 4th paradigm or 'why' of the Universe, based in organicism, expands many of the laws of Biology to all other sciences. It also adds many laws of Physics, topology and new mathematical discoveries (fractal and non-Euclidean geometries), creating a corpus of knowledge that applies to all realities, formalized by 2 relatively new disciplines, General Systems Sciences and Duality:

Duality studies the self-similar laws of all complementary networks of energy and information: Physical entities are made with fields of energetic forces and particles of information; biological entities have heads of neuronal information and bodies of energy. Gender is the duality of female, informative, reproductive beings and male, energetic, spatial ones.

Those systems are made of 'cellular parts' that create social networks, which evolve becoming wholes; fractal units of a larger whole in a bigger scale of space-time. And so General systems studies how networks evolve socially from parts into wholes, called 'systems', units of bigger social wholes.

Reality evolves in scales of size under the same organic laws of dual, complementarity networks of energy and information that become units of new wholes: The Universe evolves from particles into atoms into molecules, cells and solar systems, organisms, societies, planets, galaxies and Universes under those self-similar laws of social networks - the why of reality.

Thus Systems Science completes our search for the why of the events and forms of Universal entities.

Recap: The 1st paradigm of science responded to the what and who of events, with myths and anthropomorphism; the 2nd paradigm to the how; the 3rd paradigm to the when, measuring space distances and time frequencies in great detail; the 4th paradigm responds to the why of the forms of space and cycles of time that all entities of the Universe follow. That why is organic, social: the Universe is made of complementary energetic and informative, cellular networks that evolve and combine together, creating reproductive networks. And such complex systems create, new bigger, fractal networks units of a larger plane of space-time, evolving reality in scales of size that follow the same organic laws, from particles that become atoms that become molecules, cells, organisms, societies, planets, galaxies and Universes. Duality and social systems become thus the key disciplines of the 4th paradigm - the why of reality. The Universe is neither born of a personal god (1st paradigm of knowledge) or a mechanism (3rd paradigm of science) but an organic, complex system of networks of energy and information, whose self-similar parts, made to the likeness of the whole, evolve socially from its simple particles to the more complex organisms and cosmological entities.

2. Types of social networks and time arrows: The tenants of organicism.

Let us then consider a definition of a network to fully grasp the structure of all systems:

A simplex system or *network i*s a group of 'cellular elements' of energy (a herd with minimal communication between its cells that associate only to those that surround it) or information (a network proper, joined by multiple connections between each cell and all the other members of the network).

Simplex systems are unstable, since reality requires 2 *complementary networks*, one with motion=energy and other with in/form/ation to create perceivable actions, exi. Motion without form cannot be perceived and form without motion either, as it cannot transmit its information. So for example, an informative machine requires a human to activate it. And the informative network dominates despite being smaller in space, because it is better connected.

Further on Natural systems are self-reproductive because a system requires a 3rd reproductive network or enzyme to ensure the survival of its in/form/ation. As all systems wear and break up dying with its repetition of cycles of time. So a system that cannot be reproduced becomes extinguished.

Thus, dual networks tend to evolve and reproduce new points through exchanges of energy and information that imprint the energy network of a body/field with the form of its particle or DNA code, repeating the whole system. The result is the creation of 3rd reproductive network/system. So most systems start as energy networks that acquire form and evolve into a complex dual network of information, associated to an energy system and finally both mix to create reproductive systems. Even fundamental particles, electrons and quarks, reproduce when absorbing enough energy and evolve socially into networks. And this means that the Universe is an organic system of networks and so are its parts, despite anthropomorphic myths that deny the self-similarity of all entities. *We call such ternary systems, complex systems or superorganisms:*

For example a human is a system made of 3 physiological networks: the energetic, digestive network; the reproductive blood network and the informative nervous network, with outlets to the exterior called senses. Those 3 networks provide the cells with energy, information and reproductive hormones. And the result is a human being, himself a cell of a more complex superorganism – the economic ecosystem, which can also be described as a complex network with an informative system, money, an energetic system, the military system and a reproductive system of 'company-mothers' of machines. We call this network of 'memes of metal' (money or informative metal; weapons or energetic metal and machines or organic metal that imitates our living organs, from arm=cranes to chip=brains), the eco(nomic)system or Financial-Military-Industrial System.

Thus complementary networks and complex systems are everywhere in the Universe:

In quantum physics, they explain the complementarity principle: all particles of information are associated to a field of energy.

In biology they explain the duality of cell body/DNA informative nucleus, or body/head that all biological systems have to exist.

In sociology they explain duality of the informative classes that create the languages of information that rule societies, money, and verbal laws and the social body of people who earn money and obey it.

In economics they explain the duality of machines of energy and information; and the duality of the financial, informative economy that organizes the physical body of the economy.

Recap.. All entities are social networks. Even the smallest atom, is made of a network of simpler particles. We distinguish 3 types of social systems:
   -Simplex waves of energy (herds) or information (networks).
   - Complementary systems of energy and information with 2 networks, an informative particle and an energetic limb/field. Complementary systems create actions of energy and information, exi, the unit of the universe (Complementary principle of quantum physics, simplest viruses).
   - But most systems in the Universe are complex systems made with 3 networks: energetic limbs, informative heads/particles and a reproductive network that ensures the survival of the whole system beyond the discontinuity of death. Thus even those complementary systems that do not reproduce internally are reproduced by an external enzymatic process, which acts as the 3rd network.

3. Space-Time Cycles, Knots and networks: The Generator Equation: Se⇔Ti.

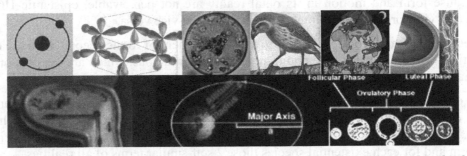

Times are cyclical and multiple; each event has a frequency that carries information and defines a closed cycle or trajectory, which breaks space into an inner, 'fractal' region and an external world. The result is the creation of multiple spaces with motion=energy and multiple clock-cycles with information, carried in the form and frequency of those cycles. The age of metric science measured all those space-time cycles by equalizing them with the rhythms of a single clock, putting them together into a continuum space-time graph. The deformations this caused was the loss of motion in the 'energetic space' that underlies reality (light-space and gravitational space) and the loss of form in the cycles of time drawn with that energy. The parallelism between space and energy or lineal motion and time, form and cyclical motion was lost; and the organic, complex, moving Universe became an abstract, static reality. Further on, the metric age adopted a philosophy of science based in the new machines used to measure those space-time cycles (clocks and telescopes, today evolved into computers and cameras), called mechanism that equalled reality to a machine. So Einstein affirmed: 'Time is what a clock measures'. But as we measure more time cycles and perceive more forms of space, a new paradigm is needed to account for that variety of 'multiple spaces and times'. Such fractal theory of spaces and times introduces the discoveries of the sciences of systems and information. Its main thesis is the existence of multiple cycles of space-times, which have form=information and motion=energy, the 2 properties of all entities, which are 'complementary systems' with a network that moves the system and one that gauges information: In physics all systems are dual: particles of information and fields of energy (complementary principle). In biology all systems are ternary networks with energetic limbs, reproductive bodies and heads of information. Thus we define reality as a system of dual & ternary networks with motion & form. We call this Theory of General Systems=Networks of Multiple Points of Space-time, the 4th paradigm of science as it unifies all other sciences, finding the self-similar properties of all its systems, which are precisely the properties of spatial energy=motion and temporal clocks of information, the 2 substances of which all beings are made.

In the study of the ontal side of reality, as opposed to the epistemic point of view (the object vs. the subject), the reader must always be aware that 'anthropomorphism' and 'utilitarianism' reduces our definitions of Universal truths to human uses. In that regard the units of information in Nature are not the human bytes invented by Shannon to study the transport of information in machines, but the cyclical, clock-like trajectories of all entities, which have form; hence carry the information of reality. When we look around all what we observe are 'forms' with motion. Yet forms in action are in-form-a©tions, and both together, energy and form, construct reality. Form stores information in its dimensional warping. So dimensional information is the key to understand the whys of reality, which can be considered a game of forms-in-action, forms with movement.

Motion without form is meaningless, as Hemingway well understood in his famous advice to Dietrich: 'don't confuse movement with action'. The unit of reality is not motion but action, which carries both motion, synonymous of energy and dimensional form synonymous of informative, cyclical clocks of time. And for that reason, Planck, the founder of quantum physics, clearly stated that all is made of actions of energy=motion and time=formal clocks. However pure motion, which in the physical realm is represented by the gravitational membrane of 'non-local action' (infinite motion) is not perceivable. Neither, it is pure form, as it does not emit any signal that the observer can pick up. This was already understood by the Taoists, who said that all yin=form had a minimal drop of yang=motion. So because form and motion in its ontal reality are not perceivable, epistemic Humans use terms in which the 'yin=form has a minimal yang=motion' and vice versa.

Speed, energy and work, which study the displacement of forms become then the closest human units of motion; while clocks of time and dimensional informations, which are cyclical or curved forms with minimal motion become the commonest human units of form.

So energy is synonymous of motion and information of form, the 2 essential parameters of all *actions of the Universe, which are the substance of what exi=sts.* Yet because epistemic humans are analytic minds, they have used an infinite number of different jargons to define for each specific class of motions and form and for each existential species those 2 self-similar terms of all realities.

In physics the terms for the 2 initial principles are: energy=entropy= lineal speed=space & information=form=time clocks=cyclic motions. The simplest scales of reality are dominant in energy= motion =entropy forces, reason why physicists believe it is the only arrow that creates the future and have disregarded the study of form, information, even in their discipline. Indeed, most of the information of the Universe is carried by bidimensional and tridimensional, physical clocks, masses and charges, which are vortices of space-time of the quantum, electromagnetic and cosmic, gravitational scale that carry in their clock-like frequency, $E=Mc^2+E=hv->M=kv$, most information of the Universe[A2].

Yet physicists still make ceteris paribus analysis of reality with a single arrow of time, entropy, which is the cause of many errors of modern physics such as the evaporation of information in black holes, which in fact are the systems that reproduce the mass of the Universe or the concept of mass as an external property to the particle, given by collisions with the flawed, never seen Higgs. Unfortunately, a single time arrow is not enough to explain the Universe. Since as Cheng-Tzu put it: from one (energy) comes two (information), from 2 comes 3 (reproduction) and from 3 the infinite beings. Yet from a simple one complexity cannot arouse. Thus the existence and proper definition of the second primary arrow of time, information is the key to understand the complexity of the organic Universe.

Information, *form, is one of the two parameters of reality with the same importance than energy, motion.* This fact is an intuitive truth of reality: whenever we look around we observe forms *with motion* or motions *with form.* And so any theory of reality must account not only as physicists do, with the motion of things, but also with the meaning of its forms. Darwin was a first step in that direction, thanks to his recognition that life was a process of evolution of form in living beings. In Physics Einstein gave a first step when he affirmed that 'time curves space into masses', which carry most of the information or form of the Universe. But physicists have been very slow to recognize the existence of an arrow of information in the Universe to the point they call it 'negantropy', the negation of entropy. This is due to the origin of their science, which has always been the study of the motion of things, since Galileo started their discipline studying the trajectory of cannonballs.

Physical energy also called entropy is lineal, expansive motion unlike forms made of implosive, cyclic time-clocks, masses and charges. They carry most of the form or 'information' of reality, defined as:

$$I^d, \text{ where D is the number of dimensions of the system.}$$

This means that the definition of information in the field of computers as the frequency carried by a lineal stream of mathematical data (Shannon) represents a minimal part of the information of the Universe. Information in Nature is carried by all types of dimensional 'forms' and languages: in physics information is carried by the forms of waves or the rotational frequency of masses and charges; in biology by the dimensional warping of DNA and protein molecules; in social sciences by the form of words and art; and so on. Dimensional form is in fact one of the two essential variables that creates reality; along with 'energy', also a wider notion in Nature similar to the concept of lineal motion, speed or expansive entropy. For that reason we consider energy=lineal motion=entropy and information=cyclical form, the two components of all systems of reality.

Yet only when we combine lineal, 'energetic motions' that move and occupy 'space' and cyclical forms, clocks of time that carry information – masses and charges - we obtain an action of reality:

In physics informative time cycles=clocks break and mold the forces and *energy of space* into particles. The energy of vacuum space warps into broken cyclical particles that act as clock of times - charges (quantum scale) and masses (cosmological space), which carry in their cyclical motions most of the information of the Universe and attract, as hurricanes do with its imploding motions, whatever floats in the space-time membrane (quantum or cosmological) in which they act.

In Biology sequences of genetic code become imprinted and form carbohydrates.

That duality of time arrows - energy, stored in the spatial vacuum, and information, stored in the cyclical clocks of charges and masses - defines a new Fundamental Principle of Science:

'All what exists are herds of spatial energy that warp and unwarp themselves in creative and destructive cycles into networks of dimensional in/form/ation: $\sum E \Leftrightarrow \prod I$ *'.*

The Duality of Energy and Information substitutes the law of conservation of energy; and it is formalized into the 'Generator, feedback equation' of the Universe, $E \Leftrightarrow I$, or $E x I = K$, from where we will deduce all the other equations of science.

All what you see is the existence of ∞ cycles of transformation of energy into form that create time frequencies as they are performed by all entities of reality in all its scales of size. There are ∞ of such exchanges; hence there are ∞ time cycles and ∞ vital spaces that interact through those feed-back equations, which represent a complementary system in space with a body/force and particle/head or a space-time cycle of exchange of energy and information in time, a flow of 'existence'.

In this dynamic reality equations are not static 'equalities', X=Y, but feedback exchanges and transformations between those 2 variables - lineal, expansive, entropic motions and cyclical, implosive clocks of time, described with Feedback equations that can be reduced to the 'Generator' equation of all space-time cycles: $Se \Leftrightarrow Ti$ (dynamic exchange) or $E x Ti = K$ (complementary system).

This classic equation of quantum physics (exi=k) and relativity (ext=k) finds now its why as clocks of time and cycles of information mean the same; and becomes in the 4th paradigm of system sciences what the space-time field equations of Einstein were to the 3rd paradigm of physical measure – the departing, unification equation of all forces/bodies of energy and particles/heads of information; the mandala, which originates it all through iterations of space-time cycles and its combinations into networks that become whole units of a new fractal scale of space-time. Thus the duality law that substitutes the law of conservation of energy is formalized by the 'Generator, feedback equation' of the Universe from where we deduce all the other equations of science, including the equations of physics.

For example, E=M (Planck's notation) becomes now a simple transformation E=M(i): as mass carries the information of the Universe in its cosmological scale and in its dynamic form, $E \Leftrightarrow M(i)$, explains easily how mass becomes energy' uncoiling its informative vortex and vice versa: $E = Mc^2$ transforms lineal energy into cyclical motion=mass by merely coiling motion into cyclical mass. Thus particles are rotational vortices with dimensional form that carry information, as complex life-molecules do, according to their dimensional warping.

The properties of the arrow of form *are opposite to those of the arrow of lineal, expansive energy. Information seems still because it creates complex bidimensional or tridimensional patterns of* cyclical motion, which allows the reproduction of self-similar form; while the key property of energy is motion and lack of form (linearity). And so the Universe is a series of $E \Leftrightarrow I$ beats that generate reality, as entities switch between their 'particle' and 'force' state. For example, Mr. Mehaute[2] proved that even in

16

classic entropy processes such as chemical reactions, when a system lacks energy 'time doesn't stop': the system doesn't die but it starts to produce fractal patterns that store new information, as it happens when you 'warp' a bidimensional sheet of paper in a new 'dimension of form'. In life also when a system becomes still (for example, a larva forms a chrysalis) the evolution of information accelerates.

So motion and from create the duality of stop=creation of information and go=energetic motion, self-similar to that of a movie film, which stops so light can focus and enlarge the information of the picture and then it moves. Information requires stillness to 'perceive' or 'construct' networks and systems with form. So all what exists moves and stops, to focus and gauge information, then it moves again. And those stops and goes are the basic beats that balance motion and form in the Universe.

All those balances between energy and information were denied in the age of monism, which accepted only lineal inertia= entropy, and denied cyclical inertia, information. There is an obvious reason for such earlier monism – simplification is easier to understand and facilitates calculus and measure with machines. Indeed, a Universe constructed with two elements, energy and information, is more complex than one created with a single system of energy and a single cycle of time – that of a clock simplified into lineal time in a Cartesian plane. But the Universe is a system of multiple, formal entities moving and tracing cycles - multiple clocks of time and multiple spaces - reason why Leibniz not Newton was right with his model of multiple, relational times and spaces, properly formalized in this work. In this new paradigm fixed space and solid particles are no longer relevant, as we define space in terms of expansive motions and particles, masses and charges in terms of cyclical, implosive motions. Solid substances and static space become then a maya of the senses, since even vacuum space carries energy and the pictures we observe of charges and masses show vibrating, cyclical frequencies, which carry most of the information of the Universe. So the big why is to give an order, purpose and causality to the complex interaction of all the motions with form of the Universe, all its informations.

In that regard, the 4th paradigm improves both the empirical, mathematical solutions of science but especially the logic concepts that the paradigm of metric measure, obsessed with digital machines, brushed aside and now regain importance. In this work the explanation of the meaning of words like time, space, dimensions, information, energy, motion, mass, universal constant, society, organism, machine, network, wave, light, point, etc. which were not key questions of the 3rd paradigm, will occupy many pages; as information is not only carried by numbers but also by logic words, better suited to describe qualities and properties ill-translated by numbers. And only, when those concepts are fully understood in its complexity we will be able to explain the laws of time and space, systems and networks and provide detailed examples and solutions to those sciences.

Recap. The Universe is made of forms with motion that trace space-time cycles. In physics energy is synonymous of spatial, expansive, lineal motion or entropy and form is synonymous of cyclical clocks of time, masses and charges, which carry in their frequency the information of reality. Form dominates motion and b*oth combine creating actions, exi, the complementary unit of all physical realities.*

Those 'time forms and spatial, energetic cycles' are described with feedback equations, through which entities exchange back and forth flows of energy and information. They define the new law of physics, the Principle of conservation of motion and form, 'all is energy that transform back and forth into information: e⇔I, Se x Ti=K' from where all other laws of physics are deduced.

4. The 4 Time cycles/arrows: energy feeding, forming, reproducing and evolving socially.

What are networks made of? They are knots of time cycles: the flows of polar communication that webs them become the strings knotted into cells of the networks. They are not made of substances but motions, events; which express the program or game of construction and destruction of social networks.

Time is motion and change and what we observe are motions with a purpose: to construct networks. We are always in construction. We express that organic, social program of the Universe with the concept of an arrow of time, or 'will of the universe', which defines the type of events that will happen into the future as a result of the activity of those networks. In fact all those systems deploy 4 arrows or actions or dimensions of space-time: energetic, informative, reproductive and social actions. They are equivalent to the drives of a living being and suffice to explain all 'biological' events of all entities of existence. We call the 2 primary arrows, energy or motion and form or information, simplex arrows, e and I; and the 2 secondary arrows born of the combination of the previous ones - reproduction, exi, and the social organization of a herd of self-reproduced exi beings, $\sum$, into networks – complex or secondary arrows.

How tight the organization of those networks is - the quality of bonding - will determine if the herd is energy in motion or rotating information. Networks must achieve a degree of smoothness in its axons of information and acquire a level of warping in 3 dimensional space that are by essence different than herds of motion cells, which follow a lineal path and flatten its form to minimal resistance to motion.

Bodies and limbs move; networks of neuronal information map reality and control a field of energy that moves them. We use different terms for *energy cells* and *neuronal networks*, since organisms are a game of 2 types of networks, a neuronal, warped dimensional form and a planar moving, energy herd. And so we can define formally any network system with an equation that represents the 4 elements of an organic system, its cells, $\sum$, energy networks, e, informative networks, I and the exchanges of energy and information that reproduce the system, exi or $e \Leftrightarrow I$ (spatial, temporal analyses): $\sum e \Leftrightarrow I$. We call the previous network the Generator feedback equation of systems of energy and information.

Thus, the recognition of 4 arrows/dimensions of space-time is a key difference between the organic paradigm and the 3rd paradigm of mechanist measure, which only recognizes an arrow of time, motion or energy or entropy, all self-similar concepts discovered in the analysis of energetic machines (weapons, used by Galileo to study motion and steam machines used to study entropy and heat in the XIX C.). Those earlier scientists, departing from those studies defined a mechanical world based in entropy. But now that we have constructed informative and organic machines (robots) made of complementary networks and social systems of machines; even the world of technological science is ripe to accept a more complex view on the arrows of time.

All scientific paradigms define first the fundamental units of time and space, which in General System sciences are space-time points made of cycles of time with spatial motions, which wherever the specific type of point-species we describe, gather in lines=waves that gather in planes, defined with numbers=sets of points as networks of points that become wholes. Thus in the 4th paradigm each point/species will have a volume of vital space made of cyclical, temporal motions, and it can be described as a social knot or network, including humans, stars, atoms, organisms and any other system.

And we can analyze them all mathematically, establishing the 'fractal, generator equation' of all systems of the Universe across a minimal number of 3 scales, 3 ages and 3 topological regions. Let us then use those elements to define the generator equation of the Universe, in its more complex, exhaustive formulation.

Since Time and space express the widest Laws of both, Philosophy and Science, in their search for an explanation to the Future, a Theory of multiple Times and spaces is both, a Philosophical, Religious Theory of Reality and a Scientific Epistemology of Nature. For that reason, even if this work uses the experimental method and its logic and mathematical languages to describe those arrows, we can easily translate the classic jargon of philosophy and religion to the modern jargon of Multiple Spaces-Times. Then each time arrow becomes an essential *Will* or *Why* of the Universe, explained in classic religions through Gods or anthropomorphic 'Avatars', each of which represented one Will or Arrow of Time.

The concept of a causal arrow of time or *Will* of the Universe exists since the beginning of knowledge, albeit expressed with different languages. Religion and philosophy has used the verbal language, to study those arrows. In ancient times the concept was 'impersonated' by God as the 'seer' or 'will of times' (Saint Augustine) that in its different manifestations represented those arrows.

In modern times, science used mathematics and logic to study them, albeit in its simplest forms (Euclidean mathematics and Aristotelian logic), which we will upgrade in order to understand the more complex, Non-Euclidean, fractal geometry and Non-Aristotelian, simultaneous logic of those Arrows. Since modern science lacks the needed 'linguistic tools' to grasp the interrelationships between the 4 main arrows of time, reason why most scientific theories use only one arrow to explain the Universe. Though being reality multidimensional more complex philosophies of science have arised:

'Monism', sponsored by physicists, which considers only 1 time arrow, energy, also called entropy.

'Duality', sponsored by Eastern philosophies in the past and Complexity Theorists in the present, which consider two types of changes: changes in the information or 'form' of beings and changes in their energy or motion.

'Trinity' sponsored by all kind of philosophical and Religious doctrines, which considers time to have 3 ages, past, present and future, youth, maturity and old age, renewed by death.

'4-D': More complex theories of change of biological or religious origin, consider 2 other arrows of time: 'reproduction' (the constant repetition of beings that happen all over the Universe) and 'love' or 'eusocial evolution' (the coming together of individuals into groups).

We call such model a General Systems Theory of Multiple Spaces and Times or MST theory, since its two main theses are:

- The existence of multiple entities, whose motions in space trace multiple cycles of energy and information, multiple time clocks that break that space in ∞ pieces..

- And the dual nature of those entities; since all entities in the Universe are 'complementary systems' of energy and information, from particles of information and fields of energy (quantum physics) to reproductive bodies and heads of information (biological systems).

Let us then consider the new mathematical foundations of this model, based in the nature of *fractal information, which was not fully understood in the past because its mathematical formulation through dimensional form and fractal repetition appeared at the end of the XX century.* Indeed, if the 3rd paradigm extended the discovery of molecular entropy to all scales of systems of reality, the understanding of information creates the fundamental beat of the Universe, the feedback rhythm, e⇔I, from where all other equations, from the life (>I) & death (<E) cycle, to the dual membranes of the Universe (Gravitational, attractive masses > I Vs. Entropic, electromagnetism <E), to the processes of reproduction (ExI=k) will be deduced.

Recap: There are 2 simplex actions, feeding on energy and gauging information, and 2 complex actions combinations of the simplex ones: reproducing, exi, and evolving socially, with self-similar reproduced cells into a whole. Those 4 arrows or wills of time are the 4 dimensions of all space-time beings.

Historically 2 types of theories on Time Arrows dominated human cultures. *Monist* theories consider the Universe to be caused by a single causal arrow, called 'God' in religions and 'Entropy or energy' in Physics. *Dualist* theories consider the existence of 2 arrows, energy, the destructive arrow, called Shiva or Yang, and information, the creative arrow, called Vishnu and Yin, the creator of Live. Only a few, complex philosophers (Leibniz), disciplines of science (Biology, Theory of Evolution, System sciences) and religions (Taoism, Buddhism, Zurvanism, Christianity), have also understood the two more complex arrows of future time: the arrow of reproduction of energy and information (exi) and the arrow of social evolution or arrow of love.

General Systems sciences fusions all those disciplines, unifying the Time jargons of Philosophy, Religion and Science and provides 3 levels of time arrows, dual theories of energy and information, which suffice to explain most physical systems; 4-dimensional theories which add the arrows of reproduction and social evolution, needed to explain most biological phenomena and finally 9-dimensional theories of 9 arrows/cycles of space-time that exhaust all possible events and cycles of the most complex species of reality, the human super-organisms studied by social sciences and the Universe at large. To represent those arrows the algebra of multiple spaces-times defines a Unification equation of all arrows, events and morphologies of space-time, whose exhaustive and particular analysis reveals both the thoughts of god and its details of each species, which we call the Generator equation of space-time cycles: $\sum Se \Leftrightarrow \prod Ti$.

5. Non-Euclidean points: knots of time wills, the Souls of the Game of Existence.

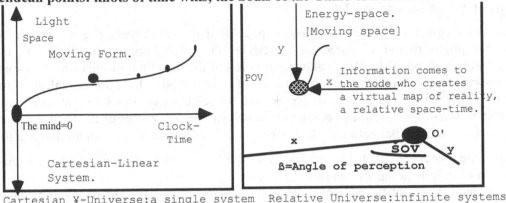

Points with parts are knots of time cycles that become the unit of all networks. All what you see are forms with motion, forms-in-action, in-form-ations. We call all those forms with motion Non-Euclidean points, (mathematical jargon), Space-time cycles (physical jargon) or complementary networks (biological jargon) the new units of the physical, mathematical and biological formalism of Systems sciences. They define a new mathematical and logical language that advances the previous A-ristotelian and E-uclidean logic to better map out the Universe, called i-logic geometry (as i represents information and comes after A and E). An i- point or space time cycle (1ˢᵗ postulate of i-geometry) has form and motion. Any entity of the Universe is an i-point, as all of them have form and motion. 2 i-points will form a wave (2ⁿᵈ postulate) that communicates energy=motion and form=information between them. Thus all lines are in fact waves with motion. Self-similar points then will form a network by exchanging energy and information along certain paths. Thus all planes are networks (4ᵗʰ postulate). And each of those points crossed by flows of energy and information that shape its internal form (5ᵗʰ postulate) will be able to move and gauge the information of the external Universe, becoming a 'relative mind', a perceptive element of an intelligent Universe made of the sum of infinite such points. Finally the 3ʳᵈ postulate of i-logic geometry will define according to the self-similarity between those points the type of networks they create when they socialize with other points. And so the application of those 5 postulates and its laws to any system of self-similar beings, from atoms to humans to galaxies, give us the whys of reality, which is game of points that socialize into bigger wholes, points of a new scale: particles become points of atoms that become points of molecules, cells, organisms and societies in the biological realm, or parts of celestial bodies, galaxies and Universes in the physical world. In the past science has focused mostly on the study of motion and energy and its measure with mechanical instruments. The 4ᵗʰ paradigm of science reveals the why of the motions and forms of reality by adding to those studies the understanding of the properties and laws of Information. Since only both together, the arrow of 'energy' or 'entropy' dominant in motion and the arrow of evolution of form, or 'information' explain the fundamental unit of reality: a feed-back cycle that transforms back and forth energy into information. The Universe is made of an infinite number of such cycles of exchange of energy and information between entities that shape the primary substance of reality: 'cycles of time' that occupy a surface of space. Time is cyclical and multiple. Each event has a frequency that carries information, defining a cycle or closed trajectory that breaks space into an inner, 'fractal' region and an external world. The result is the creation of multiple, 'vital' spaces and clock-cycles. Those cycles become chained in knots and organized in networks of knots, which share flows of energy and information, creating the complex systems and scales of reality.

What are the units of those networks? Can we find, as we have done for the wholes, unified under the concept of networks, a unifying principle to explain the simplest of the simplest parts? Indeed we can. Since the units of a network are points, but a very special point, one with form or information and motion or energy, called a non-Euclidean point. *So the Units of reality are forms with motions, which we call in physical terms, space-time cycles.*

The creation of a systems sciences formalism is possible thanks to the advance of 3 new disciplines that become the pillars of the 4th paradigm of science: theory of information, of 'form', the other substance of reality 'forgotten' by the founding physicists of the paradigm of measure, obsessed by the measure of energy and its motions; the mathematics of information, which are fractal, non-Euclidean geometries, only fully developed in the last decade; and the science that uses information to explain the Universe - system sciences also called complexity, which studies all species of the Universe as systems made of networks of self-similar entities, which are responsible for its actions with motion and form.

In the graph, we observe a key difference between the 3rd mechanist single space-time continuum and the 4th paradigm of infinite points of measure:

In the 3rd paradigm, scientific rods of measure are limited to the human world and scale, which Descartes precisely defined in a book he called the 'world', warning that they were the rods of the human mind. Yet they have ever since being used to describe all the other points, without realizing that they were only the characteristics of the space-time designed by our eye with the parameters of light-space, (the electromagnetic vacuum in which we exist with its 3 light-caused, perpendicular Euclidean coordinates and its c- speed). Once scientists started to make measures and inventing machines to help them in that task they forgot the admonitions of Descartes, who told them that perhaps a devil had put that world in his mind: 'I think therefore I am' was indeed the meaning of it all: our mind is the best metronome, which creates a 'meaningful mapping' of reality with light – but it cannot measure or penetrate in other scales of spacetime or opaque points of different substances around us.

In any case Cartesian space (Galilean Relativity) and c speed (Einstein relativity) became the absolute rod for all other measures. This allowed the development of a 'referential world', similar to the way the visual mind perceives time and space, but this frame of reference made with light pixels is not the only possible mapping of all broken spaces and time cycles of the Universe. There should be other species with other rods of measure, which perceive spaces we don't see (smelling spaces which dogs use to make their mind, gravitational spaces which seem to guide the actions of black holes, etc.).

This creates a new method of knowledge, proper of the 4th paradigm that we call the informative or linguistic method: all languages have a content of truth that gives knowledge. So if the total truth of a being stored in itself gives a probability of truth equal to 1, any external point of view gives us a lesser amount of knowledge that increases when we include more perspectives and languages.

Since all kind of languages and senses give us images of reality, the linguistic method states a higher probability of truth is achieved with several perspectives and linguistic descriptions of a certain system; as each point of view will create a self-similar image of reality. And so by integrating them all we obtain a complex, kaleidoscopic image of the whole. Reason why most systems have redundant, parallel systems and biological organisms have bilateral eyes, ears, etc.

Perceptive points are difficult to accept by mechanist science that measures and digitalizes reality only with numbers, because it is a philosophy that defies the linguistic, anthropomorphic ego-centrism of mankind, which admits only his languages and perceived spaces as real.

Let us then expand Non-Euclidean geometries with the definition of a fractal, Non-Euclidean point.

In the graph, because a Cartesian graph (left) is an abstract continuum of lineal time, based in the single rhythm of a mechanical clock and a continuous space of points with no 'breath' (Euclidean definition of a point), ever since it became the canonical representation of space-time science has believed in a single, absolute time and a single absolute space, with a single point of view, that of the human mind in the origin of coordinates. Yet the Cartesian plane only represents the continuous space-time of 'the membrane of light' it represents with its 3 perpendicular coordinates, equivalent to the height=electric field, width=magnetic field and length=wave speed of light. Since *light is the space we see and so its 3 coordinates are self-similar to those of the Cartesian, human mind made with light.*

The real Universe has ∞points of view that gauge information in ∞scales of space-time. Those points of view by definition have volume to fit ∞ parallels - flows of energy and information that go through them. Each of those parallels is in fact a cyclical, curved trajectory (Einstein's definition of a parallel); and so we can consider each Non-Euclidean point of view a knot of time cycles with form and motion.

The new 5 postulates of a Non-Euclidean, fractal geometry of Multiple Spaces and Times are:

1st Postulate: *'A fractal point is a world with an inner content of information that creates its 3 internal, topologic, organic dimensions and a content of energy that traces its external motions=time arrows'*

2nd Postulate: *'A line is a wave of fractal points.'*

3rd Postulate: *'2 fractal points are self-similar when their external, spatial perimeter or their inner information is equal. Similar points form organic networks by sharing their energy and information. Dissimilar points ab=use each other in Darwinian relationships'*

4th Postulate: *'A plane is a network that joins points through waves of energy and information.'*

5th Postulate: *'Non-Euclidean points perceive energy and information: A fractal point has inner apertures to the world, through which multiple waves of energy and information can cross.'*

The 5 postulates of Non-Euclidean geometry define a point with parts, a line as a wave and a plane as an organic network of points.

The first postulate defines a point with form and energy. Points are not 'points with no breath' (Euclid), an abstract definition that created the simplified concept of a space-time continuum but 'fractal points', which grow in size when we come closer to them. And so stars are points in the sky that become huge as we come closer to them. And a microscope discovers an entire world in the minimal size of a cell. Einstein offered a partial solution to this conundrum considering that from our scale of size, those Non-E points seem to curve the information of the universe that fluxed on it. But Leibniz had given a better answer: 'the smallest point is a world in itself'. This is the meaning of a fractal point, whose internal parts respond to the 3 topologies of a 4-dimensional Universe: a hyperbolic, informative system; an energetic, planar or spherical membrane and a cyclical, toroidal region with reproductive organs. *And we shall find those 3 topological regions in all points-species of the Universe.*

It is now clear that non-Euclidean points are points of view that gauge information and move their complementary body with a will, seeking for more energy and information and reproduction, seeking for an organic will to repeat their cycles and motions. And this will gives the point the category of mind, or Atman or soul, a perceptive point of view that absorbs information and moves its body accordingly after 'thinking' the information and 'bending' it subjectively to its will.

That we have no free will but we have will is the deepest consequence of stating that all points of view search for its 3+1 dimensions of existence, its energy, informative, reproductive and social arrows.

To do so the point must perceive and gauge. If it is a particle it will gauge and move to capture energy (2nd quantum number) or information (spin number), while going around its business of self-reproducing its wave. Points are moved by an ego, a desire to acquire more energy and form which they better achieve in networks, with the flexibility of herd hunting and network thinking. But in each scale of reality there is a rational point of view calculating its strategies to get more energy and form.

Non-Euclidean points are the souls, atmans, monads of the Game of Existence, the knots with a living will, which they display in their action, exi. Leibniz failed to realize they would communicate. Buddhist with the concept of Atmans are closer: knots of relationships with other i-points. Such multiple, perceptive Universe merely extends the nature of being human, a self-perceptive point always feeding, perceiving, reproducing and socializing, which is also what life does, to all the entities of the Universe. All is life. All shows the will of life. All keeps reproducing self-similar points of view, points of order. A point is any entity of the Universe. In a more detailed analysis those parts turn out to be self-similar in geometrical terms in all of them, defined by the three canonical topologies of a four-dimensional Universe, which describe an informative, hyperbolic region or 'head', a toroidal, 'reproductive body' and lineal or planar energetic limbs and membranes, common to all the points of the Universe. The 2nd postulate explains the interaction between two points connected by a wave of communication or 'line'; the 3rd postulate explains the type of interactions between 2 points according to their relative equality, which will bring them together into a social network or dissimilarity, which will make them not interact or enter into a Darwinian relationship in which a point absorbs the energy of the other. Self-similarity is *required to start an organic process of eusocial evolution; or else systems that do not understand their information use each other as energy, in Darwinian hunting processes or ignore their paths.*

The 4th postulate defines the creation of networks made of systems of points across multiple scales of space-time. It defines spaces as networks of points, interconnected by flows of energy and information: All in the Universe are thus complementary systems made of networks of non-Euclidean points.

Finally, the 5th postulate that explains the processes of absorption of waves of energy and information that the point *gauges to act-react with the Universe.*

Thus, we define time cycles, forms with motion, its multiple knots of time cycles and networks with a new geometry of space-time, Non-Euclidean geometry, born of the completion of the work of Einstein in physics and Riemann, Lobachevski in the field of mathematics. Since now all points have form and motion, they have breath; and since now all lines have form and motion they are waves; and since now all planes are networks, they are discontinuous planes. Thus, the new formalism of space-time redefines also the geometry of the Universe in dynamic, discontinuous terms.

Networks are made of relative points with form and motion, which can be defined as time cycles – trajectories traced by an entity in search of energy, information, reproduction or social interaction. Those time cycles enclose a surface of vital space or energy and so they define a certain space-time. Any entity will however trace many cycles in search of its organic motions to feed, gauge, reproduce and evolve socially into bigger networks. So each entity will be a knot of multiple time cycles.

Those knots of time cycles are what we call a fractal Non-Euclidean point. The basic unit of the new paradigm is thus mathematical – the point – but a very special point, a fractal point which grows in detail and information the closer we come to it, till it becomes, as Leibniz put it, a world in itself with internal complementary networks=organs of energy and information (fields and particles in physics; limbs, body and head in biology).

In the formalism of Systems sciences we call them Non-Euclidean points, since they are points through which infinite flows of energy (parallels) and information can cross. We also represent them with the symbol 'i', which is the next vowel to the A-ristotelian, E-uclidean paradigm, the symbol of information and a visual image of the 2 components of the point, the informative, cyclical part, o, and the lineal, energetic one, |, which we call the body and head of a biological Point; or the field and particle of a physical point (Principle of Complementarity).

The universe is made of networks of such i-points, and each network is what we call a world or discontinuous space-time, st, unit of a bigger fractal network, a new st-point in itself. So a network of particles becomes an atom, which becomes an st-point of a molecule and so on till creating the Universe.

Even humans can be studied as Non-E points, in which each head is indeed a spherical point that communicates energy and information with other humans, forming social networks.

The anthropomorphic reader might think that humans are different from the rest of points of the Universe, but it is a fact that all points obey in their actions and communicative flows within a network the same laws: humans and electrons behave the same when they move through slits or in herds; the geometries of social groups are also the same; and the ultimate purpose of those points, to feed on energy and information, whatever kind, is also the same in all networks of the Universe. Thus the laws of networks become the social, organic, reproductive why of all beings of the Universe: the Universe reproduces information and organize forms socially into networks. From magnetic and electric fields, made of magnetic and electric constants that mix and reproduce a light wave to fundamental particles, quarks and electrons that absorb energy and reproduce new quarks and electrons, to energetic males and informative females that reproduce together, all in the Universe can be described with the formalism of networks, connects the why with how and when of the 3rd paradigm of metric measure that has analyzed those networks, its motions and forms in detail. Why those i-points with an ego-driven will that desires selfishly more energy and information and self-reproduction collaborate in herds, becoming parts of a whole has to do with the complexity by which the arrow of social evolution impose systems to its parts. Parts become enslaved by their dependency to the higher energy and information provided by networks.

Recap. Advances in sciences always depart from the evolution of mathematics and logic, the languages of space and time of the human mind. In the 4th paradigm the 5 postulates of i-logic geometry based in the concept of a fractal point, achieve that evolution: Leibniz, Einstein and Riemann, who in the XIX c. realized that 'through a point infinite parallels can cross' (5th postulate of non-Euclidean geometry) are the points of departure. 'Non-Euclidean points' with form and motion, made of knots of time cycles, are the final elements needed to understand the why of the Universe. They socialize into networks that become points of a higher scale, which reproduce and organize new networks; and so the Universe keep growing in fractal scales, from particles that organize networks and become atoms that organize networks and become molecules that organize networks and become cells, that become organisms, that become planetary societies; while planets and stars form gravitational networks that become galaxies, organized by dark matter into Universal networks. Thus each Non-E, fractal point is a world in itself - a topological point with a volume of energy and information that relate it to other points through waves, which carry energy and information and create networks that warp into bigger points.

All what exists are cycles of space-time gathered in knots called Non-Euclidean points, gathered in complementary networks of energy and information; which become the physical systems of reality. Thus the 4th paradigm uses the new mathematics of non-Euclidean geometry, topology and fractals and the new logic of multiple time cycles, to define a new fundamental particle of space-time, the fractal point - a world of space-time in itself - an entity made with 3 networks of energy, information and reproduction, which constantly try to feed, inform and reproduce the point. Those Non-Euclidean points of energy and information evolve into networks, forming organic systems. Each system becomes then a unit-point of a bigger network, determining the dominant arrow of creation: social evolution in growing scales of self-similar forms.

The Universe is a game of chained time cycles that form knots and networks, which carry the information of the Universe. The evolution of parts into social wholes with form is the why of it all.

All those points perceive a limited reality decoded as information, extracted from the flows of energy its relative perspective and informative parts can observe. All points of view in the Universe move and gauge the information about the limited reality they perceive, creating with those exchanges of information and energy a mapping of reality. And so only the sum of all those st-points creates the absolute Space-Time of the Cartesian plane, whose rods of measure and time clocks are those of the human 'point' - our eye ball, who sees light and measures space with a rod of light and see space with the 3 perpendicular dimensions of light (the magnetic, electric and c-speed fields).

6. Ternary method: laws of topology and Plan of Evolution.

The 3 networks of Non-Euclidean points, which are knots of time cycles; their mathematical and logical formalisms; their symmetries in time and space, its ternary topologies (since there are only 3 topologies in a 4-dimensional Universe, which correspond to energetic, planar networks; informative, hyperbolic ones and toroidal, reproductive systems) and the study of its Feedback, generator equation, which formalizes their polar exchanges all with the tools of fractal, non-Euclidean geometry, form the

4th paradigm of science define also a ternary plan of creation and evolution that all entities of the Universe follow.

We shall briefly introduce those 'ternary' laws, derived from the existence of 3 type of social networks, since they will be used to classify all biological species, and unfold its the plan of evolution:

-The restrictions of topology, which limits the number of possible speciations and forms in all type of organisms to 3 unique topologies of 4-Dimensional space, *which correspond to those ternary networks* – energetic, planar topologies; hyperbolic, informative ones and reproductive, cyclical forms. This means that species can be classified in ternary speciations. For example, the gorilla (energetic ape), the chimpanzee (balanced, reproductive) and the homo (informative), latter decoupled in 3 dominant subspecies of races and sexual genders (with the gay gender, mixture of the energetic male and informative woman), and so on.

-The restriction of time events: those 3 topologies of networks correspond to the 3 Horizons/ages of all species: the energetic youth, reproductive maturity and informative 3rd age. So for example, men evolved in 3 phases, from the energetic Australopithecus, to the Homo Erectus, which came out of Africa and reproduced all over the world, to the Homo Sapiens, the informative species.

- The final social phase that integrates the 3 topologies and its parallel ages: when the 3 networks are created or the 3 subspecies are formed, they often integrate into a super-organism. So the 3 networks of memes of metal, informative money, energetic weapons and re=produced machines, now create a global superorganism, the FMI complex. The 3 networks of cells created life super-organisms. The 3 networks of insects, the reproductive queen, informative worker and energetic soldiers create the ant super-organism; and so on.

- The existence of 3 scales, as all systems are made of cellular points (lower scale) that form complementary networks (middle scale) and become organic units of an ecosystem (upper scale.)

Such topological/temporal 'plan' or 'why' of reality caused by the existence of 3 social networks is common to all systems:

Each entity of the Universe has at least 2 networks, one able to measure information and one able to absorb energy=move.

The complementarity principle thus becomes no longer an isolated law of physics ('all physical entities are fields of forces of energy guided by particles of information') but rather a Universal law.

Mathematically we express it with the feedback generator equation of energy and form, which defines existence itself: $E \times I$, $E \Leftrightarrow I$, and the law of conservation of energy and information: 'all what exists is energy that trans/forms itself back and forth into information'.

$E \times I$ or $E \Leftrightarrow I$, in dynamic terms becomes the fundamental structure and event of reality - a dual network of energy cells or body/field with an informative network of neuronal cells (head) or particles, which constantly exchange energy/motion and information in a complementary manner, and in this process they self-reproduce, $\Leftrightarrow$, $\times$, their motions, creating new combinations of energy and form.

Thus, according to the ternary principle knowledge starts in monism – the study of single networks. Duality is the next level that studies 'complementary beings' made of networks of energy and information. Such systems are made of lineal limbs/forces and an informative head/particle on the other.

Most systems have an energetic, informative & reproductive system. Such ternary systems follow the ternary principle - the 3 symmetries of fractal spaces-times: the 3 ages in time of all systems, dominated in each age by a network (the energetic youth, reproductive maturity and informative, 3rd age), the 3 topologies in space that define those networks (hyperbolic, informative center; planar or spherical, energetic limbs and membranes and toroid, cyclic, reproductive organs); and co-exist 3 scales of existence – the cellular, organic and social scale in which the organism becomes a unit of a bigger scale.

Organicism. The 4 arrows/dimensions of time: Feeding, gauging, reproducing and socializing.

Thus each spatial organ/topology has a temporal function: beings are made of lineal forms/motions /limbs, cyclical heads/motions/ informative systems and toroidal, elliptic reproductive bodies. And this is so because the line is the shortest distance between two points and the sphere, the form that stores more information and the ellipse combines both. So from viruses with a head of information and a lineal limb to semen, trees, light with a photon and a transversal tail, fishes with a head in front and a moving body or trees with a lineal trunk and a ball of roots and leaves on top, we find some basic topologies with vital functions that bring us the biggest of all whys: forms are topologies and topologies have motion, which expresses an organic function – either a simple energy/information system, with a head that gauges and a field/limb that moves, or in the most complex beings, an intermediate cyclical toroidal, elliptical body that reproduces the system. Those are the ultimate whys of reality: gauging heads/particles; lineal limbs/fields that move; elliptical bodies that reproduce the system and social networks that put together multiple st-points into bigger structures that will again repeat the same game:

'All topological Non-Euclidean space-time points are knots of time cycles, which trace motions that gauge information, feed on energy, reproduce or evolve socially the point into networks of points that become knots of a higher, fractal space-time plane.'

The most remarkable success of a theory of Multiple Spaces and times based in the 5 postulates of Non-Euclidean geometry, the definition of a point as a knot of time arrows, a plane as a network and an organic system and a ternary network of energy, information and reproduction, is the completion of a Mathematical Theory of evolution of form, of in/form/ation, which applies to all sciences and will be used in this work to classify the main physical entities of the universe.

Its basic tenant is that we live in a 4th dimensional Universe in which forms are restricted to the limited topologies of space and the logic ages of time found for all systems[1]. Those restrictions of topology limit forms of all systems to the 3 unique topologies of a 4-Dimensional space: the energetic plane, the informative hyperbole and the cyclical toroid that combines energy and information in reproductive events.

In time those 3 topologies correspond to the 3 Horizons/ages of all species: energetic youth, reproductive maturity and informative, 3rd age. It is the main symmetry between spatial form and time events: All networks belong to the 3 topologies of the Universe, developed in a sequential order from the energetic age to the informative 3rd age, with a reproductive intermediate phase. And those 3 topologies and ages define the life/death cycle.

Finally the arrow of social, network evolution creates 3 scales: entities are constantly increasing its information and social evolution from points into networks and ecosystems. Thus the 3rd set of key laws of system sciences is the study of the parts and the wholes and how information and energy is shared or distributed among them: From those facts stem among many other laws, the laws of genetics in biology and the laws of physical states.

And all those laws put together are expressed in the existence of 4 drives of life existence or time arrows, common to all beings, which absorb energy with its energetic bodies/fields, gauge information with its particle/heads, reproduce whenever they have enough energy its networks and evolve socially into bigger systems.

Thus the triad of essential system laws is the law of the 3 topologies, of the 3 ages and the law of social evolution, which gives birth to the 3+1 arrows or wills or whys of reality.

Thus the understanding of reality with the 3 types of networks of complex systems starts with monism – the study of single networks. Duality is the next level that studies 'complementary beings' made of networks of energy and information. Such systems are made of lineal limbs/forces and an informative head/particle on the other. Yet most systems follow the ternary principle and have an energetic, informative & reproductive system. Such ternary systems follow the 3 symmetries of the complex Universe: the 3 ages in time of the system, dominated each one by a network (the energetic youth, reproductive maturity and informative, 3rd age), the 3 topologies in space that define those networks (hyperbolic, informative center; planar or spherical, energetic limbs and membranes and toroid, cyclical, reproductive organs); and co-exist 3 scales of existence – the cellular, organic and social scale, in which the organism becomes a unit of the next scale of reality. So the study of systems through its dualities and its 3 symmetries of spatial form, temporal ages and scalar size, is the essence of the ternary method that we shall follow in this work to classify all physical species and study its evolution.

Those ternary ages and topologies, and dual laws and symmetries between the spatial motions and temporal, informative states of all physical systems, studied under the formalisms of non-Euclidean postulates, ternary topologies and fractal, scalar geometries are the mathematical foundations of a theory of multiple spaces and times, extended through several scales of reality, whose origin can be traced to the work of Leibniz on relational, multiple time cycles and vital spaces, but now truly acquire thanks to systems sciences all its solving power. Since we classify all space forms and time events in 3 topologies that put together in social networks, create organic systems.

Thus the duality of energy and information and the ternary method is the key system used to resolve and reorganize the elements of physics, from the 3 ages of creation of matter (energetic gas, reproductive liquid and informative solid) to the study of particles and its social forms (the 3 families of increasing mass-information, the duality proton/neutron, quark/electron, strong/weak forces, gravitational/electromagnetic membrane; the 3 ages of the Universe equivalent to the 3 solutions to Einstein's equations – which are phases of evolution in time of the Universe that goes from a big-bang age, through a steady state into a Godelian vortex. While the 3 topologies of all 4-dimensional systems that structure the 'parts' of any physical entity as a 'Non-Euclidean point'[1] will allow us to model black holes, despite not being able to observe them. Let us then study the ternary method in more detail.

Organic functions and motions in time of spatial forms: The topological structure of non-E points.

To understand what all the Non-Euclidean points with its motions and mental worlds, maps of the Universe have in common, we need to consider a mathematical structure for the Universe, which includes all its possible worlds and systems, based in a type of space more general than 'metric spaces' called 'topological spaces', *spaces* whose fundamental properties are no longer 'distances' and 'sizes', which are relative, but properties like 'adjacency', 'formal curvature or information', 'self-similarity' (instead of equality), 'complementarity', 'function', 'geodesics', etc.

Change or motion are the classic definition of time traced back to Aristotle, who also said there were two types of change or motion, the motion in the location of beings studied by physics and the motion of the form of beings (the life/death cycle, the evolution of form) studied by biology. And so this new/old definition of time, which gives us whys to the change of reality substitute the definition of the 3rd paradigm: 'time is what a clock measures'. And it gives us the organic whys to all motions, which are either energetic motions, of lineal topologies or informative motions with more form, cyclical, clock-type topologies. And its combinations, reproductive motions that create new curves and forms.

Thus, in a model of relational, multiple discontinuous spaces we can also describe network systems with general properties given by their topology and the functions and cycles those topologies perform.

What all those different spaces of reality, from the chair to the man to the galaxy have in common is the fact that despite their different size and form, when we do measures – the length of their limbs, legs or the surface of its external membrane, or the density of its different parts - all of them have parts that structure a whole; so all of them are *systems*. And all of them are a combination of the 3 possible and unique topological spaces of a 4 dimensional universe; that is, all of them combine *the hyperbolic, informative topology, the spherical, energetic topology and the cyclical, reproductive, toroidal topology* that mathematicians use in different combinations to map out all beings of reality.

We give to each of those 3 topologies a function, since the experimental method shows that information is accumulated in hyperbolic, broken structures with complex forms (such as your brain, a black hole or a computer); on the other hand tiled membranes made of small planes - triangles, squares and hexagons - are the most resistant 'energetic systems' while reproductive processes are cyclical, feedback equations that 'repeat' a certain form. Thus a science of formal, topological structures in space and its parallel functions in time -a 'system sciences' - is more general than a science of measure. Since it studies the properties of those 3 type of functions/forms/networks assembled through messings allowed by its dark spaces between networks and describe with them all species of the Universe.

And this type of structure, which combines in most physical systems two networks, one of information (particle, head) an one of energy (limb, field), often mixed in biological systems within a 3rd reproductive network (body) turns out to be the Universal structure of all efficient, long-lasting entities of physical and biological space, the fundamental particle of systems sciences and complexity: *a complementary entity of energy and information.* Thus we establish a new way to look at the universe with topological spaces and networks instead of metric spaces and continuous planes– a universe of infinite fractal scales of networks that become parts of bigger wholes.

Finally we analyse how all those different spaces are connected to form the puzzle of a total universe, and to that aim again topology with its concept of open and closed balls and the laws of thermodynamics with its analysis of exoergic and endoergic systems, come handy to understand how those fractals, broken, discontinuous spaces share certain common regions, membranes and bridges through which energy and information can flow in one or other direction to peg them together in stable, bigger forms.

Thus the structural analysis of systems, networks and topological spaces is a wider way to see the order, structure and meaning. Yet it does not eliminate the old method of analysing metric spaces and make measures; because once we have that wider view, each different self-similar species will be differentiated by its precise 'metric'. For example we shall establish with the new laws a topological category of beings called 'cars' according to the assembly of its parts into a whole, made of certain combined materials that define a series of specific topological structures and functions: the car will then be an energetic system, where a human with informative height and hyperbolic brain, assembled to it will play the informative role. And all cars will have always 4 wheels as that is the most efficient structure to create a parallel 4-square plane of minimal friction to move over another plane –the road (since a network of 4 points or square and one of 6 points or hexagon are the most stable bidimensional structure), and so on. Yet to differentiate a Mercedes from a Renault, we need metric measures – the Mercedes will be longer and wider, and so on. So what we are creating is a new superstructure over all the other disciplines and detailed analyses of all the species of the Universe that unifies them from a higher point of view – a goal of philosophy of science that can be traced to the 1st man who looked at the Universe and felt an integrated part of it.

The difference between both types of spaces is clear: metric spaces must be equal in distance and size to be the 'same', and so we cannot go beyond each detailed description to create homologies that give us the wider whys of reality. Topological spaces however are self-similar based in very general properties, which allows classifying many forms as a single type of topological function and concluding that all functions of reality respond to the same 3 topologies in space and functions in time, needed to describe the entire 4-dimensional Universe. Since in topological forms 'distances' and sizes are relative; so what is conserved is the overall 'lineal, planar shape' of energetic functions, (limbs, weapons, proteins); the cyclical, toroidal shape of reproductive cycles, which combine energy and information to recreate a form, (body cycles; mitochondrial cycles, factory cycles), and finally the hyperbolic, broken, convex, warped, cyclical forms of informative systems, (brains, cameras, eyes, languages, black holes.)

The 4 arrows of space-time have a clear order in time:

- Birth: We are born as a seed of information that:

-Max. E: Goes through an energetic youth of fast motions, in which the energetic limbs and external sensorial membrane dominates.

-E=I: Balances energy and information in a mature, reproductive age, in which the internal, topological, reproductive system dominates.

- Max. I: Till it warps all its energy into form, in the 3rd age in which the central, hyperbolic, informative brain/particle singularity is dominant. This age ends in:

- Death: Exploding back into energy, erasing its information in the process of death.

And so this order from energy to information or life E->I and then its local reversal of time arrows from information to energy or death, I->E, becomes the fundamental cycle of existence of all species of reality, from human beings, to matter (energy=gas state, liquid= reproductive state, solid=informative state), to the Universe itself from the big bang to the steady state to the Gödel's vortex, described by the 3 solutions to Einstein equations, which must be ordered in time.

Thus after making a topological, spatial analysis of any entity, the 4th paradigm makes a temporal, causal analysis of its life/death cycles, *since both are related as each topology dominates each age of the species or individual form.*

The 3 fractal scales of space-time.

Those ages start in a simpler st-1 space-time scale as an 1st cellular seed of information, the singularity of the Universe or the plasma state of matter, then the system reproduces and evolves socially till surfacing in the st-plane of existence, which we consider the main life cycle of the system, in which it will herd with self-similar systems of its st+1 social scale performing social tasks with other self-similar organisms of the st+1 social scale, till it dies and returns back to the st-1 cellular level. And so existence is also a travel through 3 st±1 scales of reality. And we do a ternary analysis of the 3 fractal scales of space-time in which all systems exist. They are the atomic, matter and cosmic state of physical entities, in which atoms gather into gas, liquid and solids that form part of cosmic entities, planets and stars.

They are the cellular, organic and social or ecosystemic scale in biological entities. And we do a ternary analysis of the 3 fractal scales of space-time in which all systems exist.

The study of systems through those 3 symmetries of time ages, topological spaces and fractal scales is the essence of the ternary method that we follow in to classify all biologic species:

When we put together those Biologic laws (Genetics, Evolution, Physiology, Organicism), the 4 drives of existence or will of life that coincide with the 4 arrows of time developed in our work on General Systems, proper of all species of the Universe – feeding on energy, gauging information, reproducing and evolving into social, multicellular organisms and ecosystems – the laws of topology, fractal and non-Euclidean geometry, the 3 horizons or ages of increasing information of all evolving systems; and the laws of social evolution that create from individual species, whole superorganisms of a bigger scale of reality; suddenly a clear, impersonal, efficient plan of evolution appears:

As it happens in physics that 'freezes' certain species through the big-bang evolution of matter, all species of reality will be forms that must be specialized as energetic, reproductive, social or informative systems. Thus we have a restricted template of topological forms and types of networks and its combinations in which we can classify all the species of Biology (as we did in our lessons on physical species with all the particles, atomic, molecular and cosmic bodies from the big-bang to the big-crunch). This plan of social, topological evolution will define the process that brought the simplest atoms of life,

CHNO into the most complex of all beings, man, and beyond, the social structures of humanity, its economic and cultural super-organisms (Gods and civilizations), which obey the same laws.

Yet the existence of a plan of evolution does not validate the religious creationist's groups, since the plan is impersonal, mathematical, topological and organic, *as it applies to all systems, including physical systems,* acting as a guidance, a template which all systems follow, whose memorial bytes, recorded in the genetic, lower, cellular scale of each reality code in detail each particular species.

Recap. The new method of knowledge of the 4th paradigm is the linguistic, ternary method, which accepts all languages of knowledge and uses the ternary structures of fractal points in space (its 3 topological regions), in time (its 3 ages of evolution) and its structure across 3 st±1 planes to exhaust our knowledge of any system. There are 3 types of systems, simplex systems of energy or information, complementary systems of energy and information and complex systems which add a feedback reproductive system: $e \Leftrightarrow i$. They dominate in 3 ages of time, energetic youth, reproductive maturity and informative, 3rd age, in which a species travels through 3 scales of self-organization, born as a st-1 seed that reproduces and self-organizes, surfacing as an individual form in the st-plane, which will perform certain tasks in an st+1 ecosystem.

Organicism and its mathematical units, fractal points that gather into social networks called topological spaces substitute the restricted concepts of Euclidean points and continuous space-times, explaining why all those cycles exist: Since we can reduce all forms and systems of reality to the only 3 topologies of a 4-dimensional universe: the energetic plane, the informative hyperbole and the cyclical toroid that combines energy and information in reproductive events. Thus self-similar, simple topologies of energy and information create complementary systems, which evolve further into organic systems, made of reproductive, energetic and informative networks. Thus we classify all space forms and time events in 3 topologies that put together in social networks, create organic systems guided by 4 time arrows: energy feeding, information gauging, reproduction and social evolution. Those 4 categories are the so-called drives of living beings. Motion, form and its iterative repetitions, organized in social networks, are the 4 events or arrows of future of all systems in all scales of reality - the game of existence or thoughts of god (paraphrasing Einstein). Thus, there is a 'Universal Plan' with an existential finality: to create organic systems, departing from energy bites and information bytes evolved 1st into social networks, then into complementary systems and finally into organisms: particles become atoms that become molecules that become cells, organisms, planets, galaxies and Universes. It is the 4th organic why that completes the adventure of science and this work explores in all its consequences, departing from the simplest feed-back cycles/equations of time-space events. In that regard, the most remarkable success of duality and General systems sciences is the completion of Evolutionary Theory, whose evolutions of in/form/ation is limited by the laws of networks and Non-Euclidean geometries, restricted to the 3 topologies of 4D space and the 3 ages/horizons of all systems and species, each one dominated by a topology of energy (youth), reproduction (Maturity) and information (3rd age)[1]. We thus add to genetics and Evolution, a 3rd leg of laws developed in our work on General Systems and Duality.

7. Philosophy & Praxis of the new paradigm: organicist solutions to old questions.

In the philosophical perspective, by recognizing two arrows of time, and its reproductive combinations, systems science returns to the concept of organicism, whereas a machine is just a simplex network system of energy or information (or in the case of robots and company-mothers, an evolving organism) and it denies the 2 alternatives of 'deism' and 'mechanism'. Thus the 4th paradigm implies a philosophy of man and the Universe, more proper of Eastern traditions than Western, anthropomorphic, self-centered thought: Organicism substitutes mechanism; duality of energy and information substitutes monism (a single arrow of entropy or energy), form becomes more important than motion, the social network more important than the individual and life a better model than the machine. It also has deep implications for our praxis of science and policy, since on one side puts man again as the measure of all things – the more complex organism of information known to man – it also vindicates the importance of biological and social sciences over physical ones and it ushers a warning to mankind, which is obsessed by the technological evolution of machines, forgetting the need to evolve humanity at social level and to take care of the life ecosystem developing a sustainable economy. Since it turns out that machines – organisms of metal – follow their own path of evolution as an independent species, which we men merely assembly, following the laws of evolution of all organic systems. Hence, we are now building, after making energetic bodies of machines in the XIX c. and informative heads (chip-brains, mobile-ears and camera-eyes) in the XX century, organic, robotic machines in the XXI c., which are competing with us as a new species in labor and war fields and could easily displace us as the top predator species of this planet by the end of this century. Thus the 4th paradigm is not only a theoretical advance in our knowledge of the Universe but also as science has always been, a tool to improve the future of mankind, which faces in this century challenges originated by the 3rd paradigm and its worship of machines of measures, largely ignored by the cult to the machine the 3rd paradigm caused.

The Universe is an organic fractal of ∞ self-similar points of view, entities that form social networks, which evolve into complementary organisms of energy and information, whose 'evidence' sometimes is not so easy to find, because each specific information is invisible to all other systems that do not understand its coded language or perceive its supportive, energetic force. Those limits of the mechanist method, all too evident today, when 76% of reality is a network of dark energy and matter we cannot perceive, are breached by using the general laws of all networks and systems, which fill the gaps left by the experimental method, as Mendeleyev did, describing the unknown elements of the Periodic Table by using the properties of its self-similar neighbors. Thus the 4th paradigm can overcome some of the limits of the experimental method with the tools of self-similarity and the use of 4 organic whys or drives of existence that reorder the wealth of data of the age of mechanist measure and go beyond its limits.

Thus 'complex organicism' goes beyond the dual philosophies of Aristotle and Plato and fusions them together, using the last discoveries of mathematics (fractals and non-Euclidean points) and system sciences (the study of networks, its units, time cycles described with feedback equations, its social knots and topological forms). As such it is a philosophy of science aimed to substitute the anthropomorphism of religion and the limits of inquire of mechanist science, mature enough to give the answers to the whys that neither religion or physical measure could give; and can be summarized in a sentence: *the creation of organic, social, self-reproductive, complementary networks with energy=motion and in/form/ation that become cells of higher wholes, evolving from the simplest particle to the entire Universe.*

This new philosophy of science has a name: system sciences, the science of networks of information. System sciences were founded at the death of Einstein in the Macy's congress 50 years ago under a new principle of philosophy of science that was meant to substitute mechanism, called organicism, as it was understood that all what existed was a system or network of self-similar beings and those networks were the principle that structured all organic systems. It also added a 2nd arrow of time or 'primary substance' in the study of the Universe – information, which being fractal, discontinuous and non-differentiable had not been properly mathematized in the previous paradigm.

We can study the specific informative particles and energetic fields of forces of the physical scales of the Universe, as classic physics does. But since they represent information and energy, we will also apply the common laws and properties of all informations and energies of reality, studied by Duality and General systems Sciences. This approach resolves many problems of physics.

You might think the ternary method is a very simple game to be able to explain it all. And indeed it is. 'God is simple but not malicious' said Einstein, the intellectual summit of the 3rd paradigm. The 4th paradigm, as all of them did before, starts with simple principles, but the devil is in the details. Each change of paradigm in science, while simplifying the ultimate equations of reality – in this case till arriving to the generator equation of space-time cycles, Se$\Leftrightarrow$Ti - means also a jump in the complexity of both the details and the logic principles of reality. Such is the case of the logic principles of networks, because its 'units' - non-Euclidean points, knots of dynamic time cycles - are far more complex than the simplex, static Euclidean points with 'no breath' proper of the age of metric measure: its structure is fractal, discontinuous, leaving dark spaces in the holes of the networks; and its logic is based in 2x2 arrows of time from past to future, energy=motion and from future to past, information and its 2 complex, reproductive and social combinations; hence it is paradoxical, unlike the simplex Aristotelian causality of the models of reality constructed by the 3rd paradigm of metric measure based only in the arrow of entropy and motion, proper of physical studies.

Why this road has not been taken before is obvious: the 3 mathematical foundations of it are the 3 most recent branches of mathematics, only developed in the XX century, topology by Poincare, Fractal mathematics, developed by Mandelbrot (1973, fractals) and the completion of Non-Euclidean geometry by this author[1], which I introduced at the 50th anniversary of the foundation of Systems Sciences at the International System Sciences Congress at Sonoma.. So in the same manner that quantum and relativity required the solution of the 5th non-Euclidean postulate (Riemann) and the understanding of Hilbert spaces in the XIX C., the new paradigm foreseen by Plato and Leibniz and acknowledged by Einstein and Poincare, happens once its mathematical structure is formalized.

Complex Physics. Multiple spaces-times applied to physics.

The importance of the ternary method of topological time-space arrows – the ultimate why of reality - is obvious: It does not only provide responses to most questions unanswered in previous paradigms *but it also provides a limit to the possible realities of truth that eliminates* mathematical fantasies and false theories and a method of self-similarity to reinforce the certainty of our discoveries.

Once we determine *what* we want to study (experimental evidence), we must observe its *why and how* – its topological structure and causal logic (how or who), which the combined use of time arrows and topologic spaces clarifies. Those 2 systems of truth, experiment and theory, limit with the *experimental what*, and the *why and how of topological time arrows,* what things are certain and what are myths, linguistic fantasies, unreal events or false interpretations. Finally we can study the details, analyzing with clocks and instruments of science, the metric properties of all systems, knowing its logic hows.

In that regard, the laws of the new paradigm should follow the key proofs of veracity in science: simplicity (as we have found a generator equation for all events and forms of reality), higher reach (as multiple spaces times solves with the same laws questions in all disciplines) and experimental evidence (its capacity to resolve those final questions, which we shall show now). How far a 'Theory of multiple spaces & times' will go to fulfil the claim of being the space-time Theory of the XXI C., completing the structure of science will depend on its capacity to provide solutions to unresolved questions of science and predictions testable by experimental results (Popper). Yet the odds are good, since many of those solutions and predictions done as earlier as the 90s when the first texts on Multiple Space-times were published, had already being tested - an encouraging sign for its future. To show that avenue we shall finish this introduction enunciating many of those solutions, which become trivial with the new tools of the 4th paradigm and will be explained in detail in other pages of this work.

Simplex physics is the study of the universe with a single arrow, energy or entropy or motion. The spatialization of time, of information is its main consequence. Complex physics is the study of the Universe with 2 arrows, energy and information and its complex combinations, reproductive process and social processes. Of course, the difference is enormous. It is like moving in a wheel or in 4 wheels. All becomes resolved in complex physics as all the questions have now 4 elements to find a solution.

The advantage of having a higher understanding of the two fundamental parameters of the Universe, space and time and its 2 human languages of perception, geometry and logic, becomes immediately recognizable when we try to resolve the long-standing questions of all disciplines of science, many of which become trivial consequences of those advances.

This happens also in physics, as we can answer questions, which for decades have engaged physicists, as we have new laws to define what is possible and what is a mere mathematical fantasy. For example:

Questions answered by the 3 causal ages of Time events.

-Why there are 3 self-similar families of particles of increasing mass?

Answer: they are the 3 ages/horizons/evolutions of all forms of space/time.

- *Why there are 3 solutions to the equations of space-time of Einstein?* They are not 3 parallel types of Universe, but the 3 ages of our Universe, the energetic big-bang (Friedman solution), the mature, steady state (Einstein solution) and the informative, cyclical big crunch (Gödel's solution).

- *Why there are more particles than antiparticles if both have the same probability of being formed?* There are no more particles than antiparticles, but being antiparticles the inverse arrow of death and dissolution, they last much shorter in time and so as we see less people dying than living, we perceive less antiparticles than particles.

- *Why there are 3±st states of matter?* They are the 3±st ages of all space-time creations between 3 relative planes of existence: plasma (st-1 birth), gas (energetic state), liquid (reproductive, balanced state), solid (informative state), Bose condensate (st+1 emergence as a more evolved form).

...by the existence of 2 time arrows, entropy and information.

- *Why information is bidimensional?* It is the Universe a holography? Informative & energetic systems are bidimensional, made of fractal points with a very small 'height' dimension, which combine in cyclical 4-dimensional patterns to create the 'volume' of all complementary systems.

- *What is mass?* A vortex of space-time, which carries most of the information of the Universe. Since information is proportional to the number of dimensions of a form.

- *Why particles have different masses?* Because depending on the speed of rotation of a mass-vortex, like a hurricane they attract more (faster turning quarks) or less (slower turning electrons) or nothing (open, lineal forces, light and gravitation).

... by the existence of 2 fractal space-times: the gravitational and electromagnetic membrane.

- *Why gravitation is so weak?* Because we exist and perceive the light-space membrane and gravitation is a force of the gravitational membrane we don't perceive.

- *Can we unify charges and masses as Einstein wanted?* Yes; both can be unified NOT with quantum equations, but as 2 informative vortex of 2 fractal membranes of different size, with a simple vortex equation, U.C. x $m^2 = r^3$ x w^2 once we translate the electron to the jargon of gravitational vortices.

- *What is the weak force and why it breaks the space symmetry?* Because it is not a force of space but an event in time that transforms particles between both membranes. And time is not symmetric: motion to the past is different than to the future.

- *Why the Higgs is not found?* Because it does not exist; it is a particle only useful if the weak force were a spatial force.

- *Why the Universe expands and yet there is a balance between its dark energy and mass?* The Universe does not expand, since space-time is discontinuous. So vacuum expands between galaxies, but galaxies contract vacuum into masses and the total effect is in balance.

- *What is dark energy?* Transversal gravitational waves - the energetic, expansive arrow of the gravitational membrane. *What is dark matter?* Quark condensates.

- *Do black holes evaporate?* And if so what is the solution to the information paradox? They don't because they are topological open balls, doors between both membranes and the event horizon that evaporates *is* in our light-membrane. So they evaporate us.

- *It is the inflationary big bang the origin of it all?* No; the Universe has infinite fractal scales in which there are fractal big-bangs, which are the death and release of energy of a previous informative particle/singularity/Universe. So the beta decay (neutronic big-bang), a super-nova (star's big-bang), a quasar (galactic big-bang) and the big-bang (of a universe, cell of a hyper-universe) are just relative deaths. And so on.

We shall elaborate those answers latter in this work in the sections dedicated to physical systems. We can in fact answer *all the questions* unresolved by physicists, as trivia questions, deduced directly from the discontinuous space-time topologies of Non-Euclidean geometry and the multiple causality of time arrows. In the previous small sample, we have simply answered the 'catalogue' of fundamental questions about physics that the most prestigious research center of physics, CERN, has established as the key research program for XXI century physicists. Such is the power of the new formalism of fractal spaces and multiple time arrows. But in this lectures we want to go further than a trivial quiz and set up the foundations of XXI physics; so future researchers can complete in all its details our understanding of the Universe. And for that reason we need a more formal approach, establishing the principles of discontinuous space and multiple time applied to physics, correcting the errors caused by the use of a single space continuum and a single arrow of time and a single clock to measure it; and finally once the principles and corrections are met, to 'paint' the complex Universe as it is in all its splendour.

Complex Biology.

In the field of biology multiple spaces-times brings first a sense of respect to the whole science. Since the 4 simplex and complex arrows of time, energy feeding, information, reproduction and social evolution turn out to be the 4 drives that define life; they expand the concept of a living Universe to all systems of reality and apply many laws of biology to other systems. It does also solve the big questions that have always wondered the mind of man:

-*What is life?* The expression of those 4 arrows in complex ternary systems of light atoms: carbon (reproductive atom), nitrogen (informative atom) and oxygen (energetic atom).

- *It is life unique?* No, all systems follow the same arrows.

-*Why we live and die?* The life and death cycle is the main cycle of dual networks of energy and information, defined by a causal order: we are born as a seed of information, st-1, which reproduces and evolves socially till surfacing as a complex social organism (foetus), which go through 3 ages:

- An energetic youth, Max. E x Min I; a reproductive age, E=I and an informative, 3rd age, Max. I x Min.E, in which the informative, dominant system consumes the remaining energy.

Those 3 ages are the 3 partial equations of the cycle of energy and information E⇔I, which all complementary systems of the Universe follow and ends in a local time reversal, Max.I -> Max. E, a 'big-bang' that erases all information, called death.

- *There is a plan of evolution?* Besides the genetic model and the Darwinian fight between species, a 3rd element guides the process of evolution, the restricted number of topological combinations and the ternary principle that constructs systems with energetic, informative and reproductive topologies. Thus all biological systems tend to build efficient organisms under those rules: all living beings have informative heads on top of energetic fields they command; all of them have lineal or planer limbs to process energy, and on top reproductive bodies and on top informative heads. Evolution follows a pattern of speciation, according to which systems decouple into more energetic, reproductive and informative ternary species. Further on, the fundamental stop and go, information-motion, E<->I, rhythm of all systems creates a reproductive radiation <->punctuated evolution rhythm in all species.

- *Are organisms mere expressions of genes?* No, systems transform flows of energy into information *in all its relative planes of organization.* Thus not only the genetic parts but the wholes, which in organisms are its nervous systems, should codify some key systems of a living organism.

- *What is palingenesis?* In a process of multiple time arrows, there are several causal chains between the 4 main arrows of time, which create sequential processes of evolution, such as the one explained above. Palingenesis is a memorial process that following those sequential chains develops at an accelerated time rhythm a life species.

- Why there is altruism in life species. It is an expression of the 4th arrow of organic evolution where species can be considered whole organisms, in which each individual is a cell. In the same manner we can study with the 3 ages of life such superorganisms and observe that species appear in a young, first horizon as top predator, energetic species, which grow into reproductive radiations, evolve into informative, tall species and then further evolve into eusocial forms (insects, humans) or become extinct by a new radiation of a fitter species (dinosaurs eliminated by superorganisms of small mammals; primitive insects eliminated by eusocial ants; life species extinct by eusocial humans).

What is the common morphology of all life systems? At cellular level the proteins are the lineal, energetic elements, the RNA, are the active, reproductive, balanced forms and the DNA act as informative storage. In the multicellular system, the protein-rich skin and digestive systems are the energetic topology; the brain is the hyperbolic informative system and the blood the hormonal reproductive one that sets the cellular clocks. Those 3 systems have apertures to the world through the senses. So we are an expression of the 3 physiological networks whose will make us search for energy information and reproduction.

Social sciences

Finally, the same laws of multiple planes of existence and multiple arrows, or drives/wills apply to us, human individuals as parts of bigger social organisms, religions and civilizations. So we can answer also many questions of social sciences.

- What is a human being? An entity which exists between 3 relative eusocial scales of reality, the cellular, individual and social plane.

- What is the reason of human actions? Our actions are expressions of the interaction of our 'arrows of time' at cellular and social level; and as such can be described as a complex system of energetic, informative, reproductive and eusocial actions, coded by two 'informative fields, the genetic, biological field of cellular existence and the 'memetic' cultural field of social existence. The interaction of genes and memes program our actions, who search for energy, information, reproduction and social evolution of its cells (actions coded by genes) and as individuals, guided by memes, seek for the same arrows of its social organisms.

-What is a religion? An eusocial organism, whose texts of revelation act as DNA codes in organisms, creating simultaneous actions in all believers that share energy and information through the memes of love, of which the religion is its expression. Thus regardless of the text, in the same manner species with different DNAs are able to create multicellular organisms, the purpose of religions, cultures and legal codes is to create such simultaneous organisms.

- What are machines? Organs of energy (weapons, transport) and information (audio-visual machines) that enhance our energetic and informative capacities.

- What is the Economy? In the 3 past centuries the expression of a new type of superorganism of humans and symbiotic machines that evolved in 3 ages: the first age we made machine bodies (steam, British age), the 2nd age we made machine 'hearts' (German age of electro-chemical engines) the 3rd age we made machines heads (chips-brains, camera-eyes, mobile-ears), during the electronic age of America. Now we fusion all those components into organic robots.

- What are the laws of the economy? The same laws of all complementary systems in which a digital language of information, money guides a physical economy of machines, through a global 'nervous /informative system', the stock market and financial system. As such it obeys the laws of all complementary systems.

And so on. Since indeed, what we learn on this new model is the self-similarity of all topological systems, which follow a general plan of evolution or order of the 4 main arrows of time:

Information seed ->energetic growth ->reproductive radiation->Informative evolution and reorganization into a social system, unit of a new higher plane of existence.

Thus, the evolution of the global superorganism of mankind and machines in which we exist can be modelled also with the topological laws and time arrows of multiple spacetimes.

Today the excessive use of machines of measure, overdeveloped by the Industrial R=evolution have obscured the importance of the logic whys and the synthetic approach it provides to science; while the use of computer models increasingly creates another aberration – the belief that anything, which is mathematical and can be modelled in the 'virtual reality' of a computer is truth. This makes a*ny theory truth, because it can be proved by definition that any mathematical theory can be represented in a computer, which is a mathematical machine; as any verbal fiction can be written in a novel, which is a verbal 'machine'.* And given the amount of information the computer provides in 3-dimensions, such virtual models might seem more real than reality itself, even if they are quixotic fictions created by the combined imagination of the computer and the scientist. Thus, the 4th paradigm returns research to reality as it is; not as the metrics of 'Touring machines' resolve and represent it.

Yet the reader must be aware that such paradigm implies a completely new philosophy of man and the Universe, more proper of Eastern traditions than Western, anthropomorphic, self-centered thought. Organicism substitutes mechanism; duality of energy and information substitutes monism (a single arrow of entropy or energy), form becomes more important than motion, the social network more important than the individual and life a better model than the machine. It also has deep implications for our praxis of science and policy, since on one side it puts man again as the measure of all things – the more complex organism of information known to man – and so it vindicates the importance of biological and social sciences over physical ones, ushering a warning to mankind, which is obsessed by the technological evolution of machines, forgetting the need to evolve humanity at social level and to take care of the life ecosystem developing a sustainable economy. Since it turns out that machines – organisms of metal – follow their own path of evolution as an independent species, which we men merely assembly, following the laws of evolution of all organic systems. Hence, we are now building, after making energetic bodies of machines in the XIX c. and informative heads (chip-brains, mobile-ears and camera-eyes) in the XX century, organic, robotic machines in the XXI c., which are competing with us as a new species in labor and war fields and could easily displace us as the top predator species of this planet by the end of this century.

The 4th paradigm is thus not only a theoretical advance in our knowledge of the Universe but also as science has always been, a tool to improve the future of mankind, which faces in this century challenges originated by the 3rd paradigm and its worship of machines of measures, largely ignored by the cult to the machine the 3rd paradigm caused.

The content of this work, which tries to give an overview of the philosophical and mathematical foundations of the 4th paradigm and its application to the 3 main bodies of human sciences, physics, biology and sociology is divided accordingly in 5 lessons which I gave at ISSS's congresses at Tokyo, Madison and Waterloo as the International Chair of the Science of Duality[1]:

Lesson 1: The Philosophy of General Systems Sciences.

The first lecture explains the structure of the Universe made of self-similar networks of energy and information, which are the 2 components of all the complementary beings that 'exi=st', constantly evolving into more complex systemic networks, which become units of bigger wholes, building the scales of the fractal Universe, from the simplest particles that become parts of atoms that become parts of molecules, parts of cells, planets and stars parts of galaxies, parts of Universes.

Lesson 2: Fractal Universes.

We formalize the philosophy of General Systems Sciences with the advances of fractal and Non-Euclidean geometry. We resolve the 5 postulates of Non-Euclidean geometry, which define all beings as 'points with parts', able to process energy and information; all lines as flows of energy and information between points; all planes as networks of n-points connected by energy and information flows and departing from those structures we deduce the mathematical laws that all complex systems made of multiple networks and extended through different scales of size, follow. We illustrate those laws with examples, and use this formalism in the following lessons to resolve and complete the metric models of physics, biology and economics, which now find their why in the structure of networks and systems, their 3 ages of evolution, its 3 basic topologies and the laws that all networks follow derived from the 5 postulates of i-logic geometry.

Lesson 3: Complex Physics, the arrow of Einstein.

Physics drags errors derived from its definition of space as a continuum and the future of time, born out of a single arrow of entropy and energy. We introduce the model of duality with the arrows of energy and information and the model of multiple spaces, to define a complex Universe made of the interaction of 2 different membranes of spatial energy, the light-membrane and the gravitational membrane, and 2 arrows of time,, the arrow of entropy and electromagnetic energy and the 'arrow of Einstein', the cyclical, broken vortices of mass and gravitation that attract and in/form the physical Universe.

Lesson 4: Duality in Biology.

We study Biological structures and the evolution and differentiation of species in systems specialized in energy and information, such as gender or body/head systems with the laws of Non-Euclidean topology and the two arrows of energy and information. Since evolution follows the 3 'ages' or horizons of energetic, young species, mature, reproductive radiations and informative, 3rd age of speciation in time and its 3 parallel topologies in space – the hyperbolic, informative topology of brains and DNA nuclei, the toroidal, reproductive topology of body organs and the lineal topology of energetic limbs. Thus we unveil a guided plan of evolution that completes the work of Darwin and Mendel.

Lesson 5. The superorganisms of history and economics.

We complete the Introduction to a Dualist model of system sciences with the study of the 2 superorganisms that are being created on planet Earth: The superorganism of human beings that we call history and its smaller super-organisms, (civilizations) and the FMI complex, the superorganisms of money, weapons and machines of metal, called the economy, which develop in a mutual process of symbiosis and competence.

The application of duality and systems theory to economics unveils a biological model of evolution of machines of energy and information with a series of 72 years cycles of evolution of energy machines, information machines and now robots, with extraordinary consequences for the future of mankind. Since economics is a biological discipline, in as much as machines its main produce are systems that imitate our human organs of energy and information. This is needed to realize that life and humanity are not different from any other species of reality, except in its informative complexity – which however brings, according to the balance between both parameters, exi=k, an 'energetic weakness' that we do not recognize but easily jeopardizes our existence, both as individuals and as a species. In that regard, for mankind is essential to understand the unity of all systems for several reasons:

- Because we will be able to inscribe life and mankind in a continuous morphological plan of evolution, which stretches from the first molecules of life to man and beyond to the machines we are evolving under the same laws that brought us into existence.

- Because those laws determine an arrow of social evolution in all species, which has been denied by science for too long but it is the why and will of the Universe: all systems evolve into social networks; including man who has evolved into societies and should evolve into a global, sustainable planet instead of committing suicide in tribal wars.

- Because life is a super-organism in process of extinction and only understanding the common laws of all species, humanity could manage Earth in a sustainable manner to survive the growing competence and process of obsolescence we are suffering at the hands of our 'son' species– intelligent machines.

Recap. The why of the Universe is a social, organic, reproductive why: the Universe reproduces information and organizes forms socially into networks: From magnetic and electric fields, made of magnetic and electric constants that mix and reproduce a light wave to fundamental particles, quarks and electrons that absorb energy and reproduce new quarks and electrons, to energetic males and informative females that reproduce together, all in the Universe can be described with that simple scheme of things, which can be as detailed as needed to connect it with the 3rd paradigm of metric measure of those networks, its motions and forms. All those systems, their mathematic and logic formalisms, the ternary plan of creation and evolution, they follow; their symmetries in time and space, its ternary topologies (since there are only 3 topologies in a 4-dimensional Universe, which correspond to energetic, planar networks; informative, hyperbolic ones and toroidal, reproductive systems) and the study of its Feedback, generator equation, which formalizes them all with the tools of fractal, non-Euclidean geometry, form the 4th paradigm or why of science.

A model of multiple spaces and times has applications in all sciences, illuminating the why of all time events and spatial organisms. Its ternary method, and limited topologies, ages and scales, proper of complex networks also limits the type of truths and theories possible in the organic Universe. . This introduction to the meaning of it all is the beginning of the 3rd age of science, no longer the age of myths or the age of measure but the age of meaning, in which all systems of the Universe, including man, can be described with the same laws. This higher scaffolding or philosophy of science can illuminate and resolve all the questions unanswered with more detailed systems of knowledge based in metric spaces and a single time arrow. Yet the 4th paradigm is not only a theoretical advance in our knowledge of the Universe but also as science has always been, a tool to improve the future of mankind, which faces in this century challenges originated by the 3rd paradigm and its worship of machines of measures, largely ignored by the cult to the machine the 3rd paradigm caused.

Notes:

[1] Congress of Cancun, Sonoma, Tokyo, Madison and Waterloo (2005-2010) where I developed in 10 lectures the theory of multiple spaces and times or 'General Systems Theory' and its applications to all sciences. The complete model was published in Spain in the 90s under the title 'Los ciclos del tiempo' Editorial Arabera.

[2] 'Broken Spaces-Times'; Alan de Mehaute.

CREATION

Cycles of Time Arrows:
Se⇔Ti

'To be, or not to be: that is the question:
Whether 'tis nobler in the mind to suffer
The slings and arrows of outrageous fortune,
Or to take arms against a sea of troubles.
And by opposing end them? To die'
Hamlet, W. Shakespeare,
On the arrows of exi=stence.

The path that can be traced is not the eternal cycle:
the game of yang (energy) and yin (information)
that combine to reproduce 10.000 waves of existences.'

Tao-Te-King, Lao-Tse,
On the function of exi=stence that all i-points trace.

God is the seer, the creator of Time
Saint Augustine on the Creator of Existences

INDEX:

I. THE MANY CHANGES OF MULTIPLE SPACES&TIMES

1. The expansion of space and time concepts: A Systems theory of Multiple Spaces & times.

In science advances on the understanding of space-time set the basis for a new flourishing of all branches of knowledge, a *new paradigm of science,* as all what exists happens in time and space. In that regard, in terms of philosophy of science, a General Systems Theory of Multiple Spaces and Times breaks the space-time continuum into multiple spaces and times, responding to the why of the forms of space and cycles of time that all entities of the Universe follow.

Those paradigms start always with the work of a few authors and take time to be accepted – so the reader should not be surprised that despite centuries of insights in a model of multiple spaces and times, only now in the XXI century, we have formalized such a model.

Indeed, if we were to consider the history of previous paradigms, there were always a few authors and landmark books on time and space that set in motion the evolution of those paradigms and yet took a long time to become mainstream:

- The transition from the mythic age of knowledge in which the Gods, entities outside time and space controlled those parameters at will to the age of reason, was signified by two of such books, the 'Elements' of Euclid that defined a static, geometric space of 3 dimensions and the 'Logic' of Aristotle, which considered a single arrow of temporal causality, A->B for all temporal events. But Aristotle was ignored till the Middle Ages and in fact 90% of his work is lost; while Euclid became a household name… in the renaissance.

The new age of metric space was born also in the XV C. with Copernicus, but it was not seriously analysed till the XVII C. when Descartes in his book the 'World' established a system of lineal light-coordinates of time and space, departing from the human point of view placed in the center of coordinates and Galileo wrote his 'Dialogues' giving us the first mathematical definition of time in terms of 'space location' and 'motion', v=s/t. He perfected the first machines that measure ever since the rhythms of time and space (the clock and the telescope, today evolved into the computer and the camera). And those were the most influential books of… the next century… long after Maese Galileo had died at home under house arrest by the inquisition of thought of the age – the most respected astronomers of the age working for the Vatican Church.

In the XVIII C., Leibniz gave us his infinitesimal calculus to measure with precision small distances and changes in the motion of beings, and the concept of relational, multiple times – the first insight into the 4th paradigm. But Leibniz was too advanced for the age and so his rival Newton, who set the first laws that generalized the measures of the trajectories and rhythms in time and space of cosmic bodies, plotted in a single, continuum Cartesian space and measured with a single clock carried the day. Leibniz's concept was clearly superior from the point of view of the why of the Universe, but Newton's rods of time (the clock that equalized all time cycles and compare them) and space (the continuous, Cartesian plane and its reflective telescope), improved upon Galilean scientific measures and was adopted universally. Such was the awe his work caused that Pope exclaimed: 'And god said, Let Newton be! and all was light'; while the true genius Leibniz died abandoned by his prince and future king of England, for whom he had toiled 30 years writing… a genealogy.

In the XIX c. it was the time of other predecessor of the concept of time as form, Darwin, whose 'Origin of species' would define time as the evolution of information, or form of living beings. He opened the most talked about debate of the century… but he is still ignored in half of the schools of America who prefer to believe in a Book of History of the Bronze Age… So happens to Bertalanffy, who completed his work with the study of the processes of social evolution that create complex systems departing from self-similar cells.

In the XX C. it was the turn of Einstein with his Theory of Relativity and his use of Non-Euclidean Geometry, which few people understood as it is obvious by the insistence, 100 years later, on theories that deny his principle of equivalence between cyclical acceleration and mass, as a vortex of space-time that carries the information of the Universe - Mr. Higgs particle – and by the use of only a Non-Euclidean postulate to define the geometry of space-time.

Mr. Einstein followed in his research a key tenant of the scientific method, the Principle of Correspondence, according to which a new theory of science must include the results of all previous theories as particular cases of the wider theory. For that reason, he called his Theory of Physical Time, Relativity, since it was an extension of the original Galilean Relativity - a more detailed study of the definition of time in terms of space and motion ($v=s/t$). But he used a new rod of measure; instead of the clock, the speed of light and in this manner he further refined the measure of 'metric spaces'. Thus, his work exhausted the measure of time as a dimension of spatial, energetic change in the motion and location of beings.

But Einstein did more to foster our understanding of time by realizing that there were multiple times with different speeds in the Universe and time was a curved geometry, similar to the cycles of a clock. Thus he spatialized the concept of time as morphology of space, saying that 'time curves space into masses'. And indeed, in physics clocks of time are the cyclical trajectories of masses and charges – motions that change the position and energy of the Universe. Thus with Einstein the study of time as change in space, which Galileo initiated, based in the formula $v=s/t$, became exhausted.

We however have defined in complex physics a second arrow of time=change, information and so there is still an entire new field to explore – *the change in the information of physical beings*. Indeed time is any process of change - not only change in motion and location of space. Time is also change in the form or morphology of beings defined by Darwin in Biology and several authors of the evolutionary school of History and Economics in social sciences (Butler, who applied the laws of evolution to machines; Schumpeter, who study those technological cycles; Vico, who saw cyclical ages of war and darkness, followed by times of enlightenment; Marx, who saw history, evolving in cycles, further studied by Kondratieff, and Spengler, who analysed cultural change with the morphological laws of biology.) All those theories of time changed were based not in the analysis of spatial, entropic, energetic changes in the motion of beings, but in the analysis of morphological, informative change, making certain the first insights of Aristotle that divided the sciences of change and time in physical studies of translational change and biological studies of morphological change.

Thus, according to the constant advance of our understanding of time and the Principle of Equivalence, in the XXI century we should expect a new book whose theory fusions the discoveries of Mr. Einstein in the XX century and Mr. Darwin in the XIX century - the concept of biological time, as change in the information of beings and the concept of physical space-time, as change in the spatial location, formal trajectory and speed of beings.

It follows that such theory must be a theory of multiple modes of time-change, and since each temporal change affects a vital space (Biology) or a physical space-time, it must be also a theory of multiple types of spaces. This is the theory explained in this work – a theory of multiple spaces and times that unifies biological, sociological and physical modes of time-change.

Such a book has to be written, from the perspective of the science that tries to fusion Biology and Physics - the science of 'Duality', which puts together the arrow of energy and the arrow of information and can advance our understanding of space and time into the XXI century, by fusioning the work of Einstein and Darwin into a comprehensive theory of all types of spaces and time-changes.

Einstein said that Leibniz was right and there were multiple times and hence multiple spaces but to create such complex model of reality required starting from the beginning, 'scrapping' all what science had found since Galileo. So he did not try. We shall try in this work without 'scrapping' but merely adapting the previous knowledge to the new view, according to the Principle of Correspondence. Yet to do so we have to correct first the error of an entropy-only Universe, the dogma that prevents physicists from creating a proper theory of multiple time cycles and arrows that break space-energy into multiple vital spaces – the physical theory of this work – *a general systems theory of multiple spaces and times.*

Physics is the science of Energy since it has constructed all the weapons and machines of energy of our civilization and this has established a bias towards energy only theories that handicaps its evolution and Complex physics breaks. In detail, the error of a single entropy arrow for all physical phenomena is due to the extension of the laws of electromagnetic forces to the field of gravitation, attractive forces that 'in/form' reality, despite the obvious fact that both forces belong to two different scales of reality – the microcosmic membrane of quantum electromagnetism and the macrocosmic realm of mass.

The duality of 2 universal membranes and the particles and forces that transfer energy and information between them defines a Universe of Multiple, Fractal space-times. In such a model the Universe extends through 2 main scales of size, the quantum and gravitational scales, with man sandwiched between them. Yet forces are constricted to each membrane and scale in which we always find a system of energetic forces and a rotational particle of information, according to the duality of lineal and cyclical motions. This explains why there is no gravitation in the quantum scale. In the quantum electromagnetic scale charges are the vortices of an electromagnetic membrane of space-time made of 'light', whose limits of energy (c-speed) and information (T=0 k), set the limits of what we perceive. Yet there is a bigger scale of cosmological masses and gravitational forces.

And further on, there is a 3rd scale, which participates of both forces – the human scale; and a 4th smaller scale in the internal nucleus of atoms – the realm of strong and weak forces. And so in each of those scales we find again informative particles and energetic forces.

In the past, there was not a clear theoretical scaffolding in which to fit and order all those forces and particles – on the contrary, the dogma that space-time was unique and continuous obliged physicists to try to unify them all instead of consider their dualities, self-similarities and fractal scales as we do in a Theory of Multiple Spaces and Times. Now we will be able to unify those scales with simple fractal equations that show charges and masses to be merely the vortices of information of the quantum and cosmological scales. This proof of the fractal nature of space-time and the informative, vortex like nature of charges and masses, seeked for by Einstein and other physicists of the XX century, ushers simplex physics into the new paradigm of duality in which the Universe is in perpetual motion and lineal energy and cyclical form are in constant balance, exi=k (quantum physics) and ExT=k (Relativity).

The age of metric science measured all the space-time cycles of the Universe by equalizing them with the rhythms of a single clock, putting them together into a continuum space-time graph. This age adopted a philosophy of science based in the new machines used to measure those cycles (clocks and telescopes, today evolved into computers and cameras), called mechanism, from Kepler, who said 'God is a clocker' to Einstein who affirmed 'Time is what a clock measures'. But as we measure more cycles and perceive more forms of space, a new paradigm is needed to account for that variety of 'multiple, self-similar spaces and times 'enclosed by those cycles' that break into fractal patterns of form the Universe. Such fractal theory of spaces and times introduces the discoveries of the sciences of complex systems and information; and the recognition that information, *form, is one of the two parameters of reality with the same importance than energy, motion.*

In that regard, the denial of an arrow of information is an error of physics that other scientists must overcome without fear to fully grasp the meaning of it all. The error, in essence, stems from the expansion of a local 'asymmetry of time', the 2nd law of entropy, obtained in the study of steam machines and molecular heat, which naturally increases its disorder, to a Universal law (global asymmetry), generalized without any proof to all other sciences, scales of reality and forces of the Universe – *that wrongly affirms the entire Universe is in a process of increasing disorder and it will die.* To expand a local phenomenon (the disorder of gaseous molecules) into a global one is very common among humans, which like to feel the center of the Universe. So we believe that our Gods created the Universe and our theories are Universal, but in the case of the 2nd law of entropy it has brought an array of errors and shut down the advance of science as a whole for more than a century, by preventing the full development of a science of information in equal footing with the science of entropy. Yes, there are many electromagnetic processes that show entropy – the increase of disorder and the expansion of space. But one of the two fundamental forces of the Universe, gravitation, creates order and form, by bringing together, by informing masses. So gravitation balances electromagnetism and creates cyclical form, information. This was proved when analysing atoms, in which due to the gravitational force of the nucleus, electrons turned around in cyclical, exact patterns. Soon it was established that in quantum processes the two directions of time, which in multiple space-time theory (MST) equal to the arrows of energy and information, are completely equivalent. And yet the 2nd law that defines an entropy-only Universe was not revised, but a series of Machiavellian arguments were devised to convert gravitation into an 'entropy-only' force (from the evaporation of black holes to the esoteric, probabilistic arguments and abstract definitions of in/form/ation, no longer as form – classic philosophy). Instead of accepting the obvious: *that there is besides the arrow of energy and lineal inertia,* expansion and chaos, an arrow of informative creation, gravitation and cyclical inertia. The same can be said of life. Life is based in certain points of cyclical, formal nature (DNAs, brains) that store or gauge information and create order around them. The same happens in the Universe at large where we have found large masses called black holes, which order and warp the energy of vacuum into rotational galaxies with increasing information towards its center. And so the expansion of space between galaxies is balanced by its implosion as form in galaxies, which explains w*hy we humans, who exist in a galaxy are informative species that keep creating complex information on planet Earth, itself an informative 'vortex' of evolution.*

Today scientists of measure scorn philosophical and logical analysis of causality in time because it cannot be easily put in numbers. But once we realize that we are 'made of time cycles', knots of time cycles and networks of knots of time cycles, the spiritual, intelligent, formal universe kicks in and wipes out with a stroke all the dogmas of mechanism. We can now fusion philosophy and science answering the fundamental question, 'why we exist'. 'What is existence' cannot be revealed from the simplex point of view of a mechanical world. Since the fact that we are made of actions in time, which are motions with form that leave a trace on space but are essentially time=motion, and the fact that those actions have a social finality – to create more complex networks, chaining knots of actions into systems – is completely at odds with a mechanist, fixed, solid, senseless Universe. Motion in Time, social evolution, creation of networks are those organic concepts are now the staple of reality; since motion brings will and sensation. We exist, thus to become part of networks, to sense flows of energy and gauge information. Existence justifies itself. Yet this intelligent, active, temporal, informative Universe can be described with the formalism of logic and mathematics because its fundamental unit, a cycle can be explained with 'feed-back' equations, used in system sciences to explain the back and forth interaction between two poles or elements of an equation. Se$\Leftrightarrow$Ti where Se is a component of spatial energy, a motion element, body or field and Ti is a cycle of time that carries information, particle or head, becomes the syntactic, logic, minimal byte of reality, the new principle of transformation of energy into form: 'energy never dies but transforms back and forth into information'. And so all what you see is the existence of infinite cycles of transformation of energy into form, which create 'time frequencies' as they are performed by all the entities of reality in all its scales of size. There are infinite of such exchanges; hence there are infinite time cycles and infinite vital spaces that interact through those feed-back equations, which represent a complementary system in space with a body/force and particle/head or a space-time cycle of exchange of energy and information in time, a flow of 'existence'.

Physical time-clocks are many as each cyclical mass or charge is a small clock in itself; whose time-cycle varies according to the length of the cycle: the Earth's translational cycle is a time-year and within the planet each vital being has its circadian clocks and each particle with a different speed of rotation is a different clock – being the general law that the smaller in space an entity is the faster its time cycles are, and so space and time together become balanced: e x t=k. And so a set of key laws of Complex, Dual Physics is the study of the synchronicities and symbiotic chains established between the different scales of time cycles of a given physical system and the balances between its arrows of energy and information, which explain the harmony and order of the Universe, its species and scales of reality and the existence of many quantitative parameters for particles and forces that communicate both membranes, 'breaking its symmetry' (neutrinos and electro-weak forces).

What system sciences and duality does is to synthesize all jargons of form and motion (time and space in physics; information and energy in biology; mind and body in philosophy) to distil the properties of the ultimate realities of the Universe, energy and form (Duality), and the complex systems and social networks that entities with energy and form create (General systems sciences). Thus, *the 4th paradigm of science fusions the work of Biology and Physics by considering not only the arrow of lineal time and entropy, but also the arrow of information and the transformations between both*, which create the cyclical beats of time that we observe in nature, defining a new foundational law of science: *'All what exists are complementary systems of reproductive energy and information that transform into each other 'ad eternal': E ⇔ I'*. And it goes beyond metric models of a single space continuum, a notch higher into *the more general topological and Non-Euclidean spaces,* whose self-similarities will allow us to extract wider laws of reality, *the final whys of the events and geometries we observe.*

Recap. Time and Space are the final elements of reality. As humans evolve their understanding, new scientific paradigms are found. Time is synonymous of information. Energy is synonymous of motion. Form dominates motion. The Universe is made of 'time-space cycles' described with feed-back equations, through which entities exchange back and forth flows of energy and information.

The Universe has also a scalar structure, broken in fractal membranes of space-time of different size: the cosmological scale of gravitational, lineal forces and cyclical masses and the quantum scale of electromagnetism and charges, which are the informative units of the quantum scale. We exist sandwiched between both scales.

2. Time is Change.

Science is the search for answers about the 'future' - hence directly concerned with time=change as its primary subject of analysis. Scientists learn what will be the position, energy and form of certain species in the future – how the species will change – according to self-similar events that happened in the past; and they call those regularities scientific laws. Thus, scientific laws can be considered the extrapolation into the future of cyclical regularities found in the past, on the modes of change of every entity of the Universe.

Scientific Laws make Time and the causality that relates past and future events the most important theme of philosophy of science – a discipline, which is, despite its relative obscurity, the summit of scientific thought. In that regard, the 1st fact we must re-address to evolve our philosophy of science is a wider understanding of time=change - since there is a philosophical meaning of time as change, different from the mathematical concept of time as 'what the clock measures' - *A. Einstein[0].*

To differentiate those 2 types of times - the wider, philosophical concept and the restricted, physical one - we shall call the philosophical concept, time=change, and the restricted mathematical concept used in physics, clock-time.

In philosophy of science, time is the perception of *any* type of change, most likely a change in the form of beings (biological time, as in evolution or in the life/death cycle) or a change in the motion of beings (physical time, measured by clocks as in Galileo's formula: V=s/t). So Time as Change studies together all time-related changes in all disciplines from Biology to Physics. In physics the instrument physicists use to measure time, a clock, becomes essential to model all times-changes similar to the clock's rotational frequency, a 'cyclical geometry' of space. Time then becomes synonymous of cyclical trajectories – a geometry of space – and the Universe becomes a game of clock-like motions – the description of all those trajectories:

'God waited 6000 years to find an intelligence like His, to understand His clock-work', *Kepler[1]*.

This geometrical concept of Time, which appears with the use of clocks and advances with the work of Galileo and Kepler, is properly formalized in the work of Einstein in which time becomes associated to the spatial geometry of 'clock-time cycles' - the curved 'geodesics' of Relativity physics. We talk thereafter of the geometries of space-time. Thus the physical concept of clock-time is of great interest to understand the cyclical paths and trajectories we find elsewhere in the Universe, mainly orbital bodies and orbital particles, cyclical masses and charges. So Einstein says that Time is a geometry and 'time curves space' into cyclical, clock-like masses. Indeed, all those particles behave like small, cyclical clocks, which can be modeled comparing them with our mechanical clocks. And so in Relativity Physics we get a very detailed analysis of a specific type of change – the clocklike motions of physical particles that change their position in space.

Fair enough, but this is a restricted analysis of *all* the times= changes of the Universe, which are studied by many sciences besides physics, all of which use the concept of time. Thus, while it is important to study physical motions=changes in space, it is absurd to think -as so many ill-informed people do- that time is 'just' a geometry of space, the '4 dimension of space'. Since this restricted concept of time-change doesn't study other types of changes, like those in the form of beings (biological change). This is a grave defect because we, humans, are biological beings, who experience time mainly in a biological form (through changes in our morphology, from life to death), *not* as particles do, mainly by moving at high speeds. We never move at light speed as the particles studied by Einstein did, and so most findings of Relativity about time are irrelevant to human 'time', which is mostly about morphological, informative change. This was already understood by Aristotle, who talked of 2 types of time-changes, informative/morphological change, studied by Biology and physical, translational change, studied by Physics. And that duality between 'changes in the form, the in-form-ation of beings', and changes in the 'energy, the motion of beings', still stands, defining 'energetic change' and 'informative change' as two primary modes or *Arrows* of time-change.

Thus, we need a philosophy of time=change that goes beyond the study of the geometrical, spatial motions and orbital cycles of physical particles to include at least biological/informative time-change. And to do so, it is needed to incorporate verbal, logic concepts of change, as Theory of Evolution, the main theory of Time in Biology, does.

Let us then forget for the 'time being' the restricted, physical concept of clock-time, defined by Galileo as v=s/t, a measure of the change in the motion of beings, and think about the intuitive, verbal, wider concept of time as change. Time is then defined as *the perception of change*. And the science that studies the causality of all types of time-changes is called logic, which tries to order those different types of temporal change.

Unfortunately the dominance of clock-time - the use of a single language and instrument to measure all the types of time-change of the Universe - has restricted in the last centuries the understanding of the logic of time, reason why we need to disengage from the dictatorship of clock-measure and upgrade our time analysis with the language of logic, the supreme language of all modes of time. And therefore *a language that includes also the language of mathematics, in the same manner time-change includes clock-time*. Indeed, if mathematics derives from logic – a fact proved by Frege and Gödel[2] in the XIX and XX centuries, and by every computer which uses logic circuits to do mathematical calculus – we will prove in this work that clock-time is a partial case of time-change.

Recap: Time is change. Physics is dedicated to measure particles and forces in space and time with the restricted concept of 'time-clock' – the study of change in the cyclical motions of beings with an instrument called 'the clock' and a single language, mathematics. Logic is dedicated to the wider analysis of all types of time-changes, with all types of instruments and languages. It follows that the Logic definition of Time as change is wider and more important for a General Theory of Time than the Physical definition of time-clock.

3. Cyclical Time arrows and scientific laws.

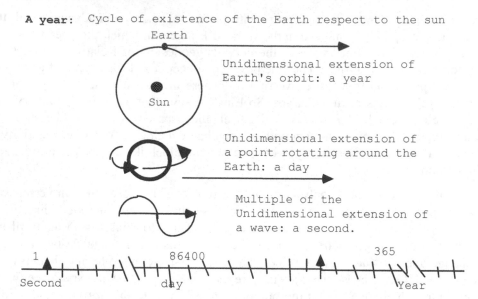

A year: Cycle of existence of the Earth respect to the sun

Earth

Unidimensional extension of Earth's orbit: a year

Sun

Unidimensional extension of a point rotating around the Earth: a day

Multiple of the Unidimensional extension of a wave: a second.

1   86400   365

Second   day   Year

In the Universe there are infinite discontinuous time cycles, in which a certain entity displays a cyclical motion, even if we equalize them, comparing their rhythms with those of the time cycle of a mechanical clock. A theory of multiple space-times goes beyond a quantitative measure of those cycles, explaining the causal, logic 'whys' that drive entities to trace them: Those space-Time cycles are geodesics traced by the ∞ species of the Universe, which in search for energy and information imprint a surface of space leaving behind the patterns of in-form-ation and space-time cycles, we perceive in Nature. Thus each entity of reality is a knot of space-time cycles, whose purpose is given by their existential will – their desire to gauge information, feed on energy, reproduce their form and evolve socially – the 4 fundamental 'arrows' that become the why of those cycles. We talk of a fundamental unit or event of reality, the space-time cycle traced by any entity that constantly uses its energy to reproduce the same formal trajectories in search for energetic, informative, reproductive or social events that create social chains with other cyclical, temporal trajectories. And so we call a knot of time cycles a non-Euclidean point.

The basic unit of the new paradigm is also mathematical – the point – but a very special point, a fractal point which grows in detail and information the closer we come to it, till it becomes, as Leibniz put it, a world in itself. It is in fact a Non-Euclidean point (which can be anything from atoms to cells to human heads), a point through which infinite flows of energy and information can cross. We also represent it with the symbol 'i', which is the next vowel to the A-ristotelian, E-uclidean paradigm, the symbol of information and a visual image of the 2 components of the point, the informative, cyclical part, o, and the lineal, energetic one, |. The universe is made of networks of such i-points, and each network is what we call a world or discontinuous space-time, st, unit of a bigger fractal network, a st-point in itself. So a network of particles becomes an atom, which becomes a unit st-point of a molecule and so on till creating network Universes. Consider humans as such points, in which each head is indeed a spherical point that communicates constantly energy and information with its own body and the external Universe. Such complementary points are made of two networks, one of energy and one of information, which we call roughly the body and head of the Points.

Because the unit of time-change is an 'event', the way to do time science with logic is to classify all observed events into a series of basic types of change, proper of each species of reality – what we call 'laws of science'. Those laws show in detail the changes of some entities under certain events, which will allow us to predict in the future how self-similar entities will behave.

A more generalizing way to do Time science is to relate self-similar events that happen in many different species and try to extract in this manner more general events or Laws of Science.

We can search then for wider and wider Laws, which collect more and more events of change, to define more 'fundamental categories' of Time-Changes in reality (even if we lose some precision in that search of wider types of change). In this manner we will find the widest types of Change. Each of those types of change will relate an enormous amount of similar events in all the species of the Universe, till extracting an absolute type of event/law of science – called in philosophy of science an arrow of time - that will help us to understand the Future Types of Change of the whole Universe.

In that regard, the difference between a 'philosophy of science' and a 'detailed scientific analysis' - a paper about certain events of the Universe - is clear: While both draw from experience and search for general laws, the approach is the opposite. In a philosophy of science the search is for the most general laws and events that happen in the Universe, the big whys from where to deduce the smaller whys. In day-to-day practical science, those questions are not required as long as we have a particular law to define a specific type of event; so we lose generalization but we win precision. Unfortunately today we have little research on the ultimate questions, since what we have evolved more are our instruments of analysis of the details of reality *not* the human mind that synthesizes that knowledge. Such synthesis is achieved only when we find the widest of all types of change – the final arrows of time or generic types of change that become the fundamental events of Logic, the science of causality in time.

Causal arrows of time are important; because if we are able to define the key arrows of causality or modes of time-change, we can forecast an entity's future when we observe a certain event A that will trigger a certain result B. This is the ultimate goal of all sciences: The chemist needs to know how a chemical reaction will develop, according to a certain type of A->B event already observed in the past; the astronomer knows where the sun will be, according to the orbital cycles it displayed in its past time. So a Theory of Multiple Time Arrows act as a general guide to classify the actions of beings; but its fundamental importance lays in its capacity to explain for the 'first time' the scientific why of reality – not only the how of those space-time trajectories provided by clock-time, but the reasons why events happen, why birds feed, planets turn in orbits, space-time has dimensions, waves collapse into particles, etc. Since all those motions and events of change will be ascribed to a General Arrow of Time. In this way, *the why of the Universe becomes reduced to explain 'why' those arrows of time exist*, what are their properties, the relationship between them, and what their existence tells us about the purpose of the Universe as a whole and the purpose of all its parts, which obey the direction of future signaled by those Time Arrows.

Recap: The Unit of Time-change is an Event. Science is the search for generic, causal processes of time-change - called scientific laws - that explain a big number of events and can be applied to many similar species. Those causal relationships will be then repeated when the initial conditions are met, allowing mankind to understand the future events of those species submitted to a law of science. 'Time arrows' are the most general of those scientific laws and hence the 'units' for a General Philosophy of Science. There are 3x3 time arrows in the Universe.

4. The relative truths of different complex time theories.

When the question of how many types of Time Arrows exist is put forward, the easiest way to go is to think there is only a fundamental type of change/arrow that includes all other types of changes or 'events' in the Universe.

Physicists do this when they talk of 'entropy' - the constant increase of energy/motion; hence of disorder observed in many physical entities of the Universe - as the single arrow of time. Latter we shall see this simplification is due to the simplifying limits imposed on physical time studies by the only-use of clocks and classic, Euclidean mathematics, which are not very good tools to perceive the 'form' or 'in/form/ation' of beings.

Single, Monist theorists of time are also common in Religions that consider an entity called 'God (to be) the final cause without further cause', (Saint Thomas[3]).

But these kinds of theories are meaningless and full of errors and simplifications, since as Parmenides proved from 1 not even motion can arise. Or as Einstein said, 'the Universe is simple but not the simplest'. And we shall show many of the errors of physics caused by single-entropy only theories.

Complex, Dualist theories, from Taoism to Biology, consider at least 2 arrows or 'Gods=Wills of Times', energy=yang and information=yin, as the ultimate events-causes of change. In Hinduism the arrow of energy is impersonated by God Shiva and the arrow of information by Vishnu - in Taoism by energy =yang and yin= information. While in modern times physicists translate the energy arrow with the concept of entropy and biologists study the arrow of information as the origin of life.

So Dualist theories are good enough to analyze multiple phenomena and according to the Principle of Equivalence that validates a theory of science even if it is not complete, as long as it can explain a set of events of reality, we consider dualism, as the 2 initial 'elements' from where more complex theories can depart, and call those 2 arrows the simplex arrows of time.

Finally, the Bible considers also the existence of the reproductive arrow (And God said 'grow and multiply') and the social arrow ('Love each other as I have loved you'), which can also be found in many other religions. So, for example, Buddhism talks of Love and Taoism says that from the 'combinations of yin and yang, 10.000 beings are reproduced'. In that sense, religion is the study of God, the creator of Times, the mind of the Universe, which we shall now explain in more detail with the Generator Equation of Times, the mystique God of Science.

God is a biological principle or rather we are, as Aristotle said, all Gods. We are all self-similar parts of the whole, with its same behavior and form, with the same game of existence, biological beings, super-organisms that play a game, and the game is probably what we should call God from here on - *the game of life and death*; the cycles that create a being and destroy it and the process of creation and destruction in any scale from the infinitesimal to the infinite; the game we describe in this work.

Thus, there are different theories about the number of arrows of time that exist in the Universe and different disciplines which study different arrows, and also different languages (verbal, visual and mathematical languages) to describe all those arrows of times-changes. Yet an exhaustive analysis of all Time theories of religion, science and philosophy shows that we can classify most events in time within 4 categories of time events - *energy, information, reproduction and social evolution* – which become the 4 'whys' that explain all the cycles, behavior, motions and events of reality.

Recap. The arrows of time, energy feeding, information gauging and reproduction are common to all species and the essence of the 'will of times', the ultimate meaning of God, the mind of the Universe.

II. SIMPLEX ARROWS: ENERGY ⇔ INFORMATION. THE BEAT OF THE UNIVERSE

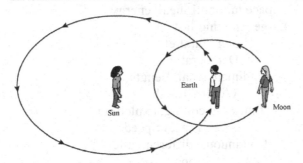

Why the Earth moves if human beings see it still? The answer to the Galileo's paradox is fundamental to understand the duality of all systems that have energy and information, motion and form. Yet our relative point of perception will observe only one of both states of the being.

5. E pur si muove, e pur no muove: Galilean Paradox.

What is reality, static form or motion? The answer is both. Yet this duality, origin of an enormous number of phenomena in the Universe, from the Complementary Principle of Quantum Physics to the reasons why you can see a movie, is still misunderstood by monist science… We call it the paradox of Galileo because Galileo rightly found the Earth was moving when it is perceived as still, but he forgot to ask the right question: why we see the Earth quiet when it is moving? Galileo didn't realize of that contradiction, inaugurating the Philosophy of science sponsored by physicists, called 'Naïve Realism'. This paradox never answered by science is the first question we must explain to fully grasp the meaning of reality. Because if all is motion, if all is events in time, we exist in time NOT in space. Space is submissive to time. We exist in time-space, where the dimensions of space are just 'motions of the energy of vacuum, perceived as static forms'. So in cosmology the expansion of the energy of vacuum is perceived as galactic motions. Once the existence of motion in every entity of reality becomes clear to the observer, a r=evolution in our way of perceiving reality, like nothing that has happened since Galileo, takes place. Since motion - no longer substance - is the primordial content of all realities. All what exists is a motion in time. Yet we perceive some of those motions as static space, since we perceive simultaneously a minimal 'duration' of time, making 'still photographs' of the length of those times cycles and motions as if they were trajectories that occupy space. So we see cars in motion as lines of space in a night picture and the Saturn's orbital planetoids as cyclical rings.

Perception puts together lineal and cyclical motions into a series of static forms that extract motion from universal entities *to fit their information into the limited 'space' of the mind.*

Recap: All what seems static in a 'naïve' perception of reality has motion, when observed in detail. Hence we exist in a Universe of 'motions, of events in time', not of 'substances that occupy space'.

6. Inversion Laws of energy and information arrows.

Aristotle defined 2 types of time=change in the Universe, translational change in space, which physicists study, and biological, morphological change in the information of beings (evolution and life/death cycles). Those 2 types of change are the 'arrows of energy and information' of a Universe made of 'time motions', in which the 'forms of space' are merely 'still photographs' of those flows of times. All what exists are those two states or elements, motion and form, which are equivalent to the abstract concepts of time and space. Yet since, according to the Galilean paradox, all forms have motions, we exist in a Universe of infinite motions that are often perceived as static forms of space. And so the true differentiation of reality must be made between lineal motions that occupy extended space, and we call lineal energies, forces or 'motions'; and cyclical motions that implode space, creating dimensional form, which we shall call cyclical in/form/ations or time-clocks.

Informative time, Ti	Vs.	Energetic Space, ΣS
Time-clocks:Information	Vs.	Space motion:Lineal energy
Small, still	Vs.	Large, moving fast
Tall, Perpendicular	Vs.	Long, Parallel,
Hierarchical	Vs.	Democratic
Bidimensional Height	Vs.	Bidimensional, Length,
Repetition	Vs.	Width-Growth
Cyclical, Rotation, imploding	Vs.	Lineal, uncoiling, exploding
Informative Frequency	Vs.	Lineal Speed
Broken form	Vs.	Continuous, differentiable
Intelligent, perceptive	Vs.	Strong, fast.
Social, organic, creative	Vs.	Darwinian, destructive behavior
Future, evolved predator	Vs,	Past, energetic victim
Life arrow	Vs.	Death arrow.
Waves of space	Vs.	Particles of Time
Female, yin principle	Vs.	Male, yang principle
Masses, charges	Vs.	Forces, fields
Heads & senses	Vs.	Limbs & Bodies.

Thus, the 2 primary arrows of future in the Universe are the creation of energy=lineal motion and information=cyclical motions with more dimensional form. Yet the properties of those 2 geometrical motions are inverted. *Those Inversion Laws/Properties are the fundamental Laws that related both simplex arrows.* Let us consider some of them:

To create cyclical information we warp lineal energy into bi-dimensional form and vice versa - we can create energy by uncoiling bidimensional clocks of information, expanding its shape into lineal space. This dual event can be generalized to 3-dimensional spheres, which 'explode' into bidimensional 'sheets' of lineal space-energy ($E=Mc^2$). And vice versa: the formless, lineal energy of gravitational and electromagnetic forces that fill space can be trans/formed into forms with a cyclical clock-like shape, called a mass or a charge.

Information is form stored in patterns that can be memorized, reproduced and imprinted. And since form is dimensional, the information a system carries is proportional to the number of dimensions it has:

$$Information: v^D$$

This distinction is important because humans defined Information (Shannon) in its simplest dimensional form – a lineal wave with a frequency V – useful for the simple 'computer machines' we use to store information with only two digits, 0 and 1; yet the Universe stores dimensional information in more complex volumes (bidimensional vortices of mass; 3-D electronic systems; 3 dimensional DNA, hyperbolic structures, convoluted brains etc.); such as each dimension stores a level of information and only adding all the scales of form, we can calculate the total information of a system

Thus, we can study those 2 types of events or Time arrows, according to the Duality of the Galilean paradox, as changes of motions or changes of forms, from static lines into cyclical, broken shapes. Yet the mathematical tools we use to describe those 2 types of Time Arrows differ: Since lineal energy is continuous and so we use differential equations to describe energy; while cyclical, closed information is discontinuous, breaking space-time into an inner and outer region - so we need to use discontinuous, fractal mathematics discovered in the past decades to study the creation of form – reason why most physical studies ignore information, since the mathematical description of its properties was not available till recent.

In the verbal age of Science, when men used words to describe the Universe, philosophers called the 2 arrows, Yang-energy, synonymous of male forms (since men have a lot of energy, constantly move, control space and have lineal forms) and yin, related to cyclical time and its patterns of information, which became the female principle (since women are more perceptive, informative, memorial, ordered, made of cyclical curves). Taoists said that yin=form, in/form/ation, which they represented with cycles or fractal lines, - -, and yang=energy, which they represented with a continuous line, ___, combine into yin-yang forms, and further on into ternary sets, which they represented with trigrams (I Ching), giving birth to all the possible forms and events of the Universe. In Hinduism the 'wills of time' were represented by Shiva, God of death and energy vs. Vishnu, God of life and creation.

Duality has always been present in philosophy. Parmenides proved beyond logic fault that from one only one could be created, while 2 could combine themselves into infinite new forms. Heraclites said that 2 elements are necessary to generate by iteration and combination all other forms of the Universe. On the other hand, dialectic philosophers, from Heraclites to Hegel, considered that from a thesis and an antithesis arise all synthetic forms.

- In the modern age, when mankind switched from verbal, temporal languages to mathematical, spatial numbers, in order to describe with higher accuracy the Universe in Space, scientists stressed that formal, geometrical duality of reality. Desargues, a French mathematician, found in the XVII century that all curves and functions could be extracted from the combination of 2 Generator forms, a cycle and a line (that together create a cone, from where those curves, called conics, could be extracted). The father of Western Science, Descartes, also said that the Universe was ultimately made of 'res extensa' (lineal space represented by his Cartesian graph) and vortices of form (charges and masses), similar to time clocks. String Theorists reached the same conclusion in modern science, postulating that all particles are made of lineal and cyclical strings, described by the Beta function. And Einstein in his General Relativity Theory deduced that the Universe was a flat surface of energetic, spatial forces that time curved into cyclical masses, the 2 fractal units of the physical Cosmos. Thus, time and space are geometrical shapes closely related to each other, since cycles of Informative Time bend any plane of Energetic Space, creating the geometrical duality of the physical world. Moreover, those 2 geometries are irreducible to monism, as the ancient problem of squaring the cycle proves.

- In any logic language 2 symbols suffice to represent any being of the Universe. All languages have a syntax based in 2 parameters, one of information and one of relative energy that combine through a 3rd active principle, an operandi or verb, that merges them. We talk of a *ternary Universal Grammar:*

A (Informative subject) <Operandi/verb> B (energy/object)

This ternary, Universal grammar applies to all languages and codes able to represent the time cycles of reality: from the code of colors (red that represents energy, blue, the color of information and green, the reproductive color) to verbal languages, where an informative subject relates to an object through an exchange of energy and information, described by a verb, shaping the genetic structure of all human languages (Chomsky[8]). It also happens in mathematics, where $f(x)= g(y)$ is the universal equation that summarizes all the others.

- Computer Logic is also based in 2 elements: the symbols of the cycle and the line, 0 and |, related by an algorithm. Those digital symbols are not ordered in space as geometric elements, but sequentially, in time, since logic is a time language. Yet they can model any mathematical, spatial relationship, showing the primacy of time over space.

- In Physics, cyclical and lineal movement, gravitational vortices and lineal electromagnetism, cyclical particles and lineal forces, are the only 2 shapes necessary to explain the material Universe. And such duality is irreducible to monism, as the failure of all Unification Theories of forces show.

- If we grow in scale into the biological world, the morphological invariance of both substances is maintained: Time has evolved masses of undifferentiated cells into complex living organisms, which are also made of cyclical heads that store and process Informative Time and lineal limbs of energy. Another less pronounced energy/information duality is between reproductive body, which absorbs energy to reproduce the cycles of the organism and delivers part of it to the limbs and the informative head:

The head is a cyclical, spherical form, and so are the informative senses, all of which accumulate in the head. The head is 'small', 'perceptive', informative. It sits on top of the body, on the dimension of height. It is 'broken, discontinuous' (as it holds more cellular, neuronal parts that the rest of the body'). The head is hierarchical and dominates the body, imposing its directions of future as they guide their body towards energy and information fields. The head is still. It is in metaphysical terms, the Aristotelian, unmoved 'relative God' that controls the movements of the body.

Its senses also show the same duality: the main informative senses are broken in dual elements (left, right eyes and ears) - while the energetic sense, the mouth that feeds on energy is 'bigger' and continuous (a single one). The eyes, the most perfect of those senses are in fact a perfect sphere. And they process 'bidimensional information' (which latter both eyes mix to create the illusion of 3-dimensionality).

On the other hand the body is a plane, dominant in lineal or elliptic, 'reproductive' forms, bigger than the head, on the bottom of the body-head system. The body moves and processes energy. It has however hardly any sensorial elements, as it gets the information from the body. Its detached elements, the limbs are even more lineal, and only become broken, discontinuous on the fingers, which are the sub-elements that process information. Within the body, the most lineal elements are those who process directly energy: the guts and the lungs (in the brain the cerebellum that controls movement is also a lineal network of neurons that would extend over a meter in length).

Further on biological dual systems are irreducible to monism, since a headless organism cannot process information and survive. Even in the social, mental scale of existence, men have always represented energy with lineal arrows of expansive movement, and time with cyclical clocks of rotating, imploding movement. Thus in our plane of space/time, we find 2 invariant elements, formally homologous to the energetic line/movement and the cyclical, temporal rotating forms we have found in all other scales.

- Finally, in Economics, machines and lineal weapons of metal that release energy are valued by cyclical bytes of metal-information (money), 'the brain of the economy', through prices, in a dual, physical and financial economic structure so far irreducible to monism, as the failure of planned economies proved. And when we consider energy and information - machines that men create to enhance their own energy and information capacities - they maintain the invariant forms of energy and information: spherical cameras are informative, and energetic, moving machines are lineal planes.

Thus, the only logic able to explain all forms and species of the Universe is a self-generating dualism, caused by the existence of 2 fundamental invariant, geometrical shapes of energy and information that repeat and emerge in every scale of space/time of the Universe - the lineal wave of energy or shortest movement between 2 points and the cycle, rotating spin or sphere, which stores the maximal volume of information in a minimal volume. They are the most perfect, inverse shapes of energy and information, which repeat in all the species of the Universe:

- All systems that create inner form, information, from DNA nuclei to eyes to brains have similar cyclical or spherical forms because a sphere stores the maximal amount of information in minimal space. And their function is creative, in/form/ative.

- On the other hand, all moving bodies that process energy are lineal, because a line is the fastest, shortest distance between 2 points. So fast felines, F1 cars, light beams and rockets are all lineal in form. Further on, the function of systems that absorb energy is often destructive: for example, felines are predators that erase the information of its preys and missiles are weapons that erase human beings.

- While in any scale of reality, complex, 'organic systems' combine both forms into Ti=k=Se balanced systems. So the 2 commonest forms of Natural fractals are the tree, with an informative head sitting on top of a line/limb of energy and/or reproductive cycle (from virus to sperm, from humans to plants, from robots to missiles); and the spiral, which can easily mutate between a cyclical state, coiling onto itself and a lineal, moving state, as an uncoiled snake (from particles that become waves, to worms, to galaxies that mutate from bars into spirals). Dual organic systems are everywhere: informative particles and fields of energy create physical beings, while all biologic species are made with reproductive bodies and informative heads.

Recap: The 2 simplest arrows of time, creation of energy and information have opposite properties: From a logic perspective, Energy and information have a paradoxical, dual, causality: $E \Leftrightarrow I$, which means to create one we must destroy the other. Thus all events are dual: a creation means a parallel destruction event.

7. Some basic morphologies of temporal energy.

We can consider for each science and plane of existence some basic forms, which imitate the morphology of energy and information; and therefore are very common in the Universe:

- *Lines*: Energy element. I.e.: Swords, limbs, rays.

- *Planes:* Social group of lines. I.e.: Stellar planes, solar planes, mobile platforms.

- *Cycles:* Informative unit. I.e.: coins, cameras,

- *Spheres*: Social groups of informative cycles. I.e. animal heads with a smallish sphere - the eye, the center that processes the biggest quantity of information - and a bigger sphere, the brain also specialized in handling information.

Those forms are the commonest of the Universe. Yet those elements tend to be complementary parts of an organic system. So they give birth to a 3rd element, whose form combines both, the reproductive body, which is ellipsoidal or become combined giving birth to the commonest systems with both, a lineal and a spherical part:

- *Trees, combinations dominant in energy*: They have a lineal trunk, with 2 fractal ends, one specialized in energy absorption (planar leaves); and the other, a smaller 'head' – a network of broken roots that absorb the chemical elements of the system.

- S*pirals, where the dominant part is information.* They often fluctuate between a lineal and a cyclical state. When the spiral cycles inwards, it is transforming energy into information. When the spiral opens its arms it is expelling energy, often for movement purposes. So the RNA, the worm or the snake coils to sleep and uncoils to move. Sometimes spirals have 2 elements: a central zone of maximal information (nucleus), and an external zone of maximal energy (lineal body). Such is the case of spermatozoids or spiral galaxies with a central black hole of gravitational information and 2 arms of energetic stars.

Finally we can talk of *Social groups* that gather those 'non-Euclidean points' in networks of energy and information that repeat the same forms in a bigger scale. The simpler of those social groups acquire 3 morphologies of energy, information or its combination:

- *Rings*. Informative Disks and spheres gather into rings when they come together as groups. So happens with molecules such as the carbon benzene; with groups of men commenting on any type of information, from informative money (stock-market rings), to social parties; to religious circles (Muslims, Indians), in search of the perception of the network of mind, called God.

- *Strings:* Energetic, lineal systems made with rings or points tied one after another in a queue.

- *Coiled Springs:* Balanced forms that combine individual elements dominant in energy and elements dominant in information. They often tie a series of spirals through a social dimension developed in the Z coordinates. The best-known case is the DNA spiral - an association of informative rings called Nucleic Acids, joined by a lineal chain of energetic sugars and Phosphoric acids.

Those simple forms, repeated and combined ad infinitum, allow the creation of very complex macro-organic systems made of micro-organic systems.

For example, in physics and biology, orbitals and bacteria repeat once and again those forms:

So there are 3 basic species of social bacteria, called coccus, bacillus and spirillus. Where, *Spirillus are Spiral forms; Bacillus are Trees and coccus are cyclical forms.*

While the electronic orbitals that turn around the nucleus of an atom are either:

S orbitals, which are spheres. *Or P orbitals,* which are lineal *forms...*that gather in social molecules repeating again the two basic shapes of information and energy:

π *orbitals,* which are cycles. *Or σ orbitals,* which are Trees.

And so we can keep building new scales of forms, combination of those simplex forms, perfectly suited for their functions: Lines and spheres are perfect forms of energy and information, and spirals and ellipses are the perfect reproductive forms that combine both.

Temporal in-form-ation, like its name says, is a measure of inner form, which we perceive when there are discontinuities in a certain bidimensional surface - a clock, a page, a computer screen - either because the static form is broken into informative patterns (--) or its motions makes a sudden peak (>), bouncing in an action-reaction cycle. Universal entities are constantly generating informative dimensions or erasing them back into extended space, with lesser form. Imagine a hand, in mythic terms the hand of God, wrinkling a paper, the ultimate energy of vacuum space, till it crunches into pure cyclical form without motion - to explode again in a big bang of energy that erases information. Such is the universal rhythm that happens in all the parts and entities of reality.

The interaction of the arrow of expansive, lineal, spatial, energy, described by Thermodynamics and the clock-like arrow of cyclical information, caused by gravitation in physical space of by the fractal geometry of any life process that generates information in Biology, *defines the main beat/cycle of the Universe, its fundamental particle:* a system of space-time, constantly absorbing energy from its environment, creating and destroying form, expanding and imploding.

Look around, observe the cycles of existence, legs opening and closing as you move; eyes winking, mouths eating and closing on the energy they will re=form; the beats of the heart; the wings of the bird; night and day; stars of energy (white holes) and black holes of information; big-bangs and big-crunches; the will of the Tao; Chang, the function of existence; bodies and minds; hardware and software; 0 and |; males and females. Since all those dual, complementary systems that follow the essential beat of existence, also merge together their energy and form, e x i, to reproduce and maintain the immortality of their evolved information, which is the ultimate 'reality' the Universe maintains 'ad eternal'. The simple fact that species, which do not reproduce exhaust their energy and end their cycles, becoming extinct explains the overwhelming presence of reproductive events in all systems of the universe. Since all is motion reproduction exists by the mere fact that a motion is constantly reproducing its path. All what we see are paths, trajectories of entities in search of energy and information to reproduce themselves. The how of those paths and reactions is what scientists study in detail. Their whys is what a philosophy of science aims to provide.

Recap. From a geometric perspective both arrows have inverse forms: Energy expands and creates formless space. Yet space has lineal shape, since the line fills the maximal distance with minimal volume. While information implodes, creating cyclical forms, 'clocks of time' that store the maximal quantity of information in lesser space.

III. FRACTAL POINTS: THE FUNDAMENTAL PARTICLE

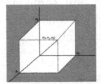

The fundamental particle of the Universe is not a physical form but a logic particle: a knot of time arrows, which in any scale of reality, from physical particles (quantum knots of energy and information) to biology (knots=networks that absorb energy, information and reproduce and evolve into bigger knots) act under a single mandate: to maximize those time arrows, a fact that we formalize with an equation,
the function of Existence: Max. $\sum exi$; which is the fundamental function of both, logic and mathematical languages. In the graph, all such points of view, will define a system of relative perpendicular coordinates, through which it will enact its time arrows, departing from a central knot of information.

8. Geometry of Multiple Spaces-Times: fractal creation.

Reality is made of entities that are knots of Time Arrows constantly tying themselves up with other knots of Time Arrows, forming networks in different scales of existence, evolving as complementary organisms of energy and information. And the question that science asks next is how to formalize that game of existence, its knots of Time Arrows or 'entities of reality', its fluxes of energy and information, its reproductive flows, herds and motions; its complementary networks of energy and form that create super-organisms. And the answer is, as all models of science with its primary languages of space, mathematics, and time, causal logic, albeit more complex than the logic developed by the Greeks to explain a simpler Universe. In this new Non-Euclidean topology, planes are networks of knots of Time Arrows; and those knots of Time Arrows, the beings of reality are 'Non-Euclidean points' with a volume, or 'organs' that transform back and forth energy into information, creating reality. All what you see becomes then a game of knots of Time Arrows, or 'Non-Euclidean fractal points', connecting themselves to other knots, forming complex planes, networks of points of two types, networks of energetic points (herds in motion) and networks of informative points (still networks). In any scale of reality those networks of points, which are knots of Time Arrows take place. The simplest scales of atoms can be described as such networks, but also the human scale. Consider a meeting: a series of human heads, moved by a lineal limb will start to share energy and form by producing waves of smaller 'particles' (sounds), and acquire a cyclical geometry as they create an event of information.

The Universe is a self-similar reality of infinite processes of creation and dissolution of networks made of knots=points of arrows of energy and information; and the best instruments to analyse it are the languages of the mind. In that sense, humans, before clocks reduced our conception of time to a single arrow and language, were guided by psychological time, which is our inner perception of our time cycles and drives of existence - our desire to perceive, inform, reproduce and evolve socially. So we lived according to our cycles of energy (feeding hours), reproduction (family cycles), and social evolution (religious and cultural activities). And followed the cycles of the seasons of this planet, according to which all living species calculated its reproductive cycles.

Recap: The universe is a game of creation and destruction of networks of fractal knots of time cycles=st-points.

9. Geometry & logic:2 languages of Spaces & Times

We describe a carrot with images made by human painters or machines, with numbers, with smells or touching it as a worm does. But only the carrot/being has all the truth and information there is on itself. Any sense or informative language gives us partial information on the species it analyses, its cycles of time and vital space; since all species perceive the Universe through the intermediate syntax of a language. Thus, to evolve human knowledge we improve the syntax of the verbal, temporal and mathematical, spatial, languages we use to describe the Universe.

Time arrows are not new. Its discovery can be traced back to the work of Eastern and Greek philosophers (Lao Tse, Buddha, Plato, Aristotle), which conceived a Universe (eastern philosophy[1]) made of two Time Arrows, energy and form that combined in reproductive acts, in which time was synonymous of motion (Aristotle's definition), and motion was proof of organicism ('when things move they will be alive' Aristotle).

Yet the ancients failed to formalize properly those Time Arrows. In the West, perhaps due to an excess of detail and the desire to prove all facts of science as dogmas without possible error, the choice has been always for simpler models of reality, even if they were inexact, in order to facilitate calculus and axiomatic proof. Indeed, monism by limiting the study of Time Arrows to the easiest of them all – continuous, energetic, 'spatial', evident events – seems more truth, but it arises from a false, limiting postulate. How this has been solved in western science is clear: all sciences have postulates without proof which are selected a priori, but are by no means certain -only simplifications of the whole truths and causes of reality. This 'righteous' mode of dealing with science proper of the Western psyche explains why from Aristotle to quantum physicists a single arrow of causality is preferred: monism, the belief in a single time arrow of energy seems truth because it is very simple; but the postulate of a single arrow is false. So the truths acquired with such simple 'camera' focused on reality are limited by the syntax of the mathematical and logic languages we use to see reality. Euclidean geometry seems truth because it 'forgets' all the inner content of information of a point, but it has been proved ad nauseam wrong. Aristotelian logic with a single cause seems truth, but it is indeed like a camera, which only works in black and white. What Multiple Spaces-Times Theory offers is a camera with all the colors of reality, by evolving the time/logic and spatial geometry we use to mirror the data of the Universe. Thus what this work tries to do is to upgrade the languages of the mind, the a priori categories[4] where the data of science must be fit to obtain theoretical 'images' that make sense. Truth in that regard is product of both, the information we obtain and the syntax in which we fit that information. This chapter will be dedicated to evolve that syntax and provide mankind with a better mirror/camera in which fit information. This is ultimately as far as humans can go in its search of truth. Since only the Universe has all the information in itself, the absolute truth, truth in human terms will be always the product of the best linguistic syntax, the best logic and mathematical mirrors, the best camera of the mind, and the best data, the biggest quantity of experimental facts fitted in that mirror.

Let us consider what kind of landscapes those 2 languages, spatial mathematics and temporal logic, see, and what escapes their vision within the total information of reality.

Unfortunately, it is impossible to define with the limited knowledge of Euclidean Geometry and Aristotelian logic developed by the Greeks, the formalisms of a dual Universe. *We need a new logic and a more complex geometry, better mind tools that have been developed only at the end of the XX century.* In the East, this lack of a rigorous formulation of logic and geometry prevented Chinese and Indian philosophers - aware of the existence of multiple arrows - to explore the complex geometries and multiple logic Arrows, which were explained instead in mystique texts with metaphors and philosophical parables. So while the west lost 'meaning' for precision, the east understood the 'game of existence' but it did not resolve the details of the 'thoughts of god'. Since in essence, all what you see is a manifestation of the game of 'exi=stence' and its limits, which we could call the game of extinction. Indeed, 'reality' is a simple game in which beings 'appear' and become knots of Time Arrows and then dissolve, unknot their form and disappear. And this happens to everything, from galaxies to atoms, with self-similar patterns caused by the common nature of all those forms. So the big question, no longer asked by western science, which merely collects machine-enhanced pictures of reality, is to explain the rules of the game of existence and extinction, which can be summarized as follows: All what exists are complementary systems that reproduce energy and information (fields/ particles, body/ heads), made of networks of self-similar 'entities', which as such can be studied with numbers, since a number is precisely a collection of entities so equal that we do not need to distinguish them as pears or cars but merely say '3'. Numbers, thus by the mere fact of existence means that the Universe has social properties, makes herds, waves and organisms. And so numbers are an excellent language to explain the properties of the 4th arrow of time, organic evolution. Those numbers however exist in 'space' creating 'geometries', some of which are better than others to enhance the chances of existence, since they allow better absorption of information (the point) or energy (the triangle, whose lineal, structural strength and single form – you cannot join 3 points except forming a triangle – fits this role), or are fit to reproduce (the couple or 2, which can easily combine its code) or are perfect units for social evolution (the 4, which is the first number with several possible join configurations, the cross, the quadrangle, the 'zigzag' snake, able to perform different functions). This very simple example of 'organic geometry', which shows the relationship between the 4 first numbers and the 4 main arrows of time, might seem 'magic' to dogmatic, abstract mathematicians, as it is indeed related to earlier Pythagorean schools both in Greece and China[3]. What modern System Sciences has done is to merge the 'right intuitions' of the East with the western tradition of exhaustive analysis, under the experimental method. And so the eastern intuitions about e/i geometries can be tested with experimental proofs of the geometrical configuration of stable atoms, which indeed are *always more stable when they have a number of particles, multiple of 4, or in crystallography, in which only those configurations that follow the geometrical properties that enhance the 4 arrows of an atom exist in the Universe,* as we have shown in our work on chemical compounds[4]. Since those common laws of Multiple Spaces-Times apply to all the scales of reality.

The first conclusion of this initial analysis of 'Time-space' geometry is clear: mathematics *is* one of the two languages of 'god', the mind of the Universe, because it is basically the language that explains the behavior and paths traced by the Time Arrows of its entities. Yet mathematics is *not* the only language of reality because it is not adequate to show the logic, causal chains of those arrows. And so we need a second language, logic, and again we need to evolve the present state of logic – Aristotelian unicausality – to define the 'temporal' properties of reality. Thus the formalism of Multiple Spaces-Times tries to evolve the logic and geometry of the west, departing from the philosophical concepts of the East, following in the steps of Spinoza, Leibniz and the masters of evolutionary and organic theories of the Universe (Darwin, Butler, Spencer, Spengler, Bertalanffy), which Duality, Complexity, System Sciences, Fractal and Non-Euclidean mathematics are completing now.

The key to grasp the scientific description of the 'game of existence' is to evolve our understanding of the 2 'languages' in which existence in Time-space is perceived, the main language of space, geometry, and the main language of time, causal logic. Since classic Geometry is used only to describe fixed forms of space without taking into account its motions in time. There are however 2 new types of geometry, Non-Euclidean and fractal geometry, which include the properties of Time-space described here – its motions, discontinuities and structure in multiple layers of 'fractal, self-similar forms' that create a complex scale of discontinuous space-times. Thus, what we do is to fusion and complete those branches of modern mathematics, in order to have the proper tools to study Time Arrows in space. Thus we describe a Universe made of 2 types of formal motions, energy and information, which create the topological geometries of reproductive bodies/fields and informative particles/heads, which in turn shape the entities of reality. Because those topologies have motion, they exist in time, leaving behind 'traces' in space. And because those motions have a causal purpose, they have an order and synchronicity that chains those traces into complex forms.

Recap: The evolution of knowledge requires not only the gathering of experimental information about the species of the Universe but a better linguistic mirror of logical and mathematical languages to achieve a higher truth: a more accurate image of reality.

How can we explain a complex world of intertwined geometrical and logical patterns with both mathematical and logic languages at the same time? Thanks to the use of 2 new forms of geometry which mimic the properties of Time-space: Non-Euclidean Geometries, which have motion and fractal equations, which are self-reproductive, cyclical systems that repeat and self-organize complex networks of 'numbers' - groups of self-similar forms.

10. Fractal, Non-Euclidean Points of view.

Mathematics, as a language that represents reality with simplified symbols, has a limited capacity to carry information. Its symbols, geometric points and numbers simplify and integrate the fractal, discontinuous reality into a single space-time continuum, the Cartesian Space/Time graph, made of points without breath. However the points of a Cartesian plane or the numbers of an equation are only a linguistic representation of a complex Universe made of discontinuous points with an 'internal content of space-time'. In the real world, we are all pieces made of fractal cellular points that occupy spaces, move and last a certain time. When we translate those space-time systems into Euclidean, abstract, mathematical 'numbers', we make them mere points of geometry void of all content. But when we look in detail at the real beings of the Universe, all points/number have inner energetic and informative volume, as the fractal geometry of the Universe *suddenly increases the detail of the cell, atom or far away star into a complex complementary entity.* So we propose a new Geometrical Unit - the fractal, Non-Euclidean point with space-time parts, which Einstein partially used to describe gravitational space-time. Yet Einstein missed the 'fractal interpretation' of Non-Euclidean geometry we shall bring here, as Fractal structures extending in several planes of space-time were unknown till the 1970s. So Einstein did not interpret those points, which had volume, because infinite parallels of 'forces of energy and information' could cross them, as points, which when enlarged could fit those parallels, but as points in which parallels 'curved' converging into the point. This however is not meaningful, because if such is the case parallels which are by definition 'straight lines', stop being parallels. So we must consider that what Einstein proved using Non-Euclidean points to explain the structure of space-time is its fractal nature: points seem *not* to have breath and fit only a parallel, but when we enlarge the point, we see it is in fact self-similar to much bigger points, as when we enlarge a fractal we see in fact self-similar structures to the macro-structures we see with the naked eye. That is in essence the meaning of Fractal Non-Euclidean geometry: a geometry of multiple 'membranes of space-time' that grow in size, detail and content when we come closer to them, becoming 'Non-Euclidean, fractal points' with breath and a content of energy and information that defines them.

Einstein found that gravitational Space-Time did not follow the 5th Euclidean Postulate, which says:

Through a point external to a line there is only 1 parallel

Euclid affirmed that through a point external to a parallel only another parallel line could be traced, since the point didn't have a volume that could be crossed by more lines:

Abstract, continuous, one-dimensional point:

. —————

Instead Einstein found that the space-time of the Universe followed a Non-Euclidean 5th Postulate:

A point external to a line is crossed by ∞ parallel forces.

Real, discontinuous, n-dimensional points: =========== o

This means that a real point has an inner space-time volume through which many parallels cross. Since reality follows that Non-Euclidean 5th postulate, all points have a volume when we enlarge them, as cells grow when we look at them with a microscope. Then it is easy to fit many parallels in any of those points. Such organic points are like the stars in the sky. If you look at them with the naked eye they are points without breadth, but when you come closer to them, they grow. Then as they grow, they can have infinite parallels within them. Since they become spheres, which are points with breadth - with space-time parts. *So space-time is not a 'curved continuum' as Einstein interpreted it, but a fractal discontinuous.* The maths are the same, the interpretation of reality changes, adapting it to what experimentally we see: a cell-like point enlarges and fits multiple flows of energy and information, and yet it has a point-like nucleus, which enlarges and has DNA information, which seems a lineal strain that enlarge as has many point-like atoms, which enlarge and fit flows of forces, and so on. So *each point is in fact a 3-dimensional point, and if we go to the next scale, a 3x3=9 dimensional point and so on. Yet those dimensions are the so-called fractal dimensions, which are not 'extended to infinity' but only within the size of the point.* In Euclidean geometry, a point has no volume, no dimension, but string theorists say that even the smallest points of the Universe, cyclical strings, have inner dimensions that we observe when we come closer to them. That is the essence of a fractal point: *To be a fractal world, a space-time in itself.*

'Any Non-Euclidean point is a fractal space-time with a minimal of 3 internal, topological, spatial dimensions and an external time motion in the st+1 ecosystem in which it exists'

This simple law is the most important law of the 4th paradigm, foreseen by Leibniz in his Monadology, the foundation of the mathematical model of Multiple spaces-times that completes the 5 Postulates of non-Euclidean geometry and gives us the tools necessary to create a complex new logic and new mathematical model of the Universe, easy to connect through topology with the laws of the previous paradigm of a single metric space-time continuum.

Further on those points must be described always in 4 dimensions, with motion. This should have been obvious, but abstract mathematics simplify entities into numbers and static forms, and organic motion properties disappear. Yet we still say 'San Francisco is at 8 hours from LA', because we mean that journey is a combination of the motion of a car and the spatial distance. Thus we measure reality in Time-space, not only in space as Euclidean maths do. Thus, in the same way Saturn's rings stop being planes without volume when we come closer and observe them as fractal points, called planetoids; Non-Euclidean points acquire both motion and volume when we approach to them. *In words of Klein, a sphere is not a continuous static form, but a group of points in cyclical movement. So in the same way the Saturn's rings are a group of planetoids, a Klein space - the space-time that fills a point has motion - it is the sum of a series of cycles[5].*

Einstein didn't go further, adapting the other 4 Euclidean postulates to the new Geometrical unit: a fractal point with volume. Only then we will be able to define the 2 planes of physical forces, the plane of gravitation and electromagnetism, or any system in which several planes of space-time co-exist together (as in a human being extended from atomic to social planes of cyclical existence). In all those systems planes are made with cellular points, Riemannian spheres with volume that form lines, which are waves between points that exchange energy and information and planes, which are organs of self-similar points that process energy or information in parallel networks. Thus the 5 Postulates of Non-E Geometry vitalize the Universe as a series of networks of energy and information of self-similar cellular points. Since the line and the plane acquire volume and become self-similar to the commonest forms of the Universe, the wave and the network of points with a 3-D volume.

This simple fact explains one of the most important discoveries of modern physics, the Holographic principle, according to which information might be bidimensional, as in the screen of a computer or the page of a book. Now bidimensionality no longer becomes 'magic' since the 3rd dimension is the relative size of the 'fractal point-particle'. Thus bidimensional sheets of information do have a minimal 3rd Dimension; the inner content of the point, which in a relative universe of infinite sizes seems to us a particle-point without volume, as we don't see either the volume of a sheet of paper or a pixel.

All in the Universe are thus complementary systems made of networks. Now this might sound absurd to the anthropomorphic reader that thinks humans are different from the rest of points of the Universe, but it is a fact that those points obey in their actions and communicative flows within a network the same laws: humans and electrons behave the same when they move through slits or in herds, the geometries of social groups are also the same and the ultimate purpose of those points, to feed on energy and information, whatever kind, is also the same in all networks of the Universe. And so that group of laws of networks becomes a primary why for all beings of the Universe.

A fact the leads us to the final element needed to understand the why of the Universe: 'non-Euclidean points' organize networks that become points of a higher scale, which reproduce and organize new networks; and so the Universe keep growing in fractal scales, from particles that organize networks and become atoms that organize networks and become molecules that organize networks and become cells, that become organisms, that become planetary societies and planets and stars form gravitational networks that become galaxies, organized by dark matter into Universal networks.

Because all entities have motion reproduction is merely the repetition of a motion with form. Because each entity has 4 time arrows, all of them trace multiple trajectories in search of those arrows, hence they realize multiple time cycles.

For example, a physical particle traces energetic cycles described by the principal quantum number; organizes itself socially, an act described by the magnetic number, feeds on energy, changing the spatial extension of its trajectory, a fact explained by the secondary quantum number and gauges information, a fact described by its spin number. And those 4 numbers define it as a Non-Euclidean quantum knot of complementary energy and information. A human also feeds on energy and information with body and head, reproduces through multiple social cycles and evolves into societies. Our actions are more complex but essentially the same of those of any particle.

So the unit of reality is a space-time cycle, and many of them create a knot of time cycles or entity of reality, which will be reproduced by repeating those formal cycles with motion in other region of space-time; and many of those knots of time cycles, which are self-similar, since they are born from the reproduction of a first form, come together with self-similar beings into networks. Some of those networks are spatially extended with a lot of motion (fields in physics, bodies in biology) and some are very tight, formal, with a lot of in/form/ation, (particles and heads or nuclei in physical and biological jargons). Both types of networks together then create a complementary organism, which is fitter to handle both energy and form; hence it survives better, it 'exists'.

Today scientists of measure scorn philosophical and logical analysis of causality in time because it cannot be easily put in numbers. But *numbers are only one of the languages of information in the Universe, and many of its properties of bio-logic nature are better described with logic words.* In that regard, we can now fusion philosophy and science answering the fundamental question, 'why we exist'; since once we realize that we are 'made of time cycles', knots of time cycles and networks of knots of time cycles, an intelligent, informative, eternal universe of motions and wills of existence makes the dogmas of deism and mechanism, childish myths. 'What is existence' cannot be revealed from the simplex point of view of a mechanical world, which cannot explain the fact that we are made of motions with form that leave a trace on space but are essentially actions in time that have a social finality – to create more complex networks, chaining knots of actions into systems. This social will of every point and entity of the Universe is completely at odds with a mechanist, fixed, solid, senseless, dumb Universe. Motion in Time and social evolution are concepts that require the capacity to gauge information and interact with other self-similar points to create those organic networks. Further on, dual networks tend to evolve and reproduce new points through exchanges of energy and information. The result is the creation of 3rd network/system: the reproductive network. And so most systems of the Universe are organic, ternary systems made of points (which can be anything from atoms to cells to human heads) organized in 3 networks. We exist as organic networks, to sense flows of energy and gauge information. Existence justifies itself.

Those Non-Euclidean points crossed by infinite parallels are able to gauge information, which implies a perceptive, intelligent Universe in all its fractal, self-similar scales of reality – a world in which even the smallest atoms can act-react to the environment, 'aperceiving' light and gravitational forces. Aristotle and Leibniz, the 2 foremost predecessors of the 4th paradigm of biological whys distinguished conscious perception from vegetative and mechanical perception. It means perception has degrees of complexity. So the simplex particles of the Universe act-react in a mechanical way; yet they still gauge information, reason why quantum physicists called their theories gauge theories, and they still have 2 complementary networks of energy and information, reason why quantum physics is based in such complementary principle.

Yet this intelligent, active, temporal, informative Universe can be described with the formalism of logic and mathematics because its fundamental unit – a spacetime cycle - can be explained with 'feedback' equations, used in system sciences to explain the back and forth interaction between two poles or elements of an equation. Se⇔Ti where Se is a component of spatial energy, a motion element, body or field and Ti is a cycle of time that carries information, particle or head, becomes the syntactic, logic, minimal unit of reality.

These simple first elements of reality - points with volume that exchange energy and information, creating waves of energy with form (no longer lines - 2nd postulate), according to a set of laws based in their self-similarity, (no longer equality - 3rd postulate), that makes them evolve into different topological networks (no longer planes - 4th postulate) - make mathematics an organic language able to describe the logic of creation of all systems of the Universe made of infinite fractal, organic networks intersecting and creating when we put them together under those laws and topological restrictions, the puzzle of reality that the simpler, 3rd paradigm called the space-time continuum and becomes now a General System of Multiple, Fractal Spaces with vital energy and Time cycles with information, the two substances of which all beings are made.

And so with those 3 scales of 'existence': time cycles, knots of time cycles and networks of knots of time cycles (Non-E Points) we can explain all the 'actions' and systems of reality made of those cycles, knots and networks; and describe a complex Universe that exists 'in time' more than in fixed space, since it has always motion; it is also dynamic, made of cyclical, feed-back equations whose causal relationships, forms and trajectories are the essence and purpose of existence. We thus consider a more complex analysis of time arrows, beyond the duality of energy and information, which combine creating a reproductive arrow, exi, and further on socialize, $\sum$exi, creating networks. And so the universe has also an organic will: to create networks of self-reproductive points of energy and information.

Yet the most astounding property of those points is to be points of view, points with will, which perform actions with the purpose mechanical or not, but probably felt in all scales as the inner freedom of the point, of obtaining energy, information reproduction and social evolution. The 4 wills or whys of the Universe are indeed embedded in the postulates of i-logic geometry. The point to exist has to be complementary, to feed and gauge energy and information and to last beyond its wearing it has to form part of bigger social networks or reproduce itself to last beyond death. This simple program self-selects those species that reproduce and evolve socially even if that contradicts the primary individual arrows of the point. Thus the engine of the contradictions of behaviour of points is that tug of war between the Galilean paradox of all points which gauge bigger his nose than Andromeda but need to hunt in herds and control the forms of the Universe with self-similar minds, joined in networks, this eternal duality of freedom vs. order, individual ego vs. collective spirit.

To exist is to act with motion and form, trying to achieve the 'arrows of time' or will of the Universe – feeding your energy network, absorbing information for your informative network, reproduce your system and in doing so, starting an external process of social evolution with self-similar entities to yourself. Those processes can be described with mathematics but we have to accept an intelligent, perceptive, fractal, self-similar Universe of infinite points of view gauging reality in a mechanical, vegetative or conscious way to explain why it happens. Their mathematical description stems from the duality between geometric form and logical function (hylomorphism). Thus, the postulates of fractal i-logic geometry define also the basic arrows= cycles/dimensions of the Universe: the 1st and 5th postulate define a point as a system whose inner parts are able to transform and emit energy and information, e>i<e; the 2nd postulate defines an exi wave of communication that reproduces energy and form between 2 fractal points; the 4th postulate defines the social evolution of a herd that creates a fractal plane - a

network with dark spaces; and the 5th postulate explains a point mapping reality, as it absorbs energy and transforms it into information through its small apertures to the Universe. Since even a minimal quark, as Einstein affirms, should be crossed by a relative ∞ number of strong forces.

It *is the organic will of all systems that search for those 4 arrows what makes the quark to exist as a knot of such flows of time arrows: a physical particle traces energetic cycles described by the principal quantum number;* organizes itself socially, an act described by the magnetic number, feeds on energy, changing the spatial extension of its trajectory, a fact explained by the secondary quantum number and gauges information, a fact described by its spin number. And those 4 numbers define it as a Non-Euclidean quantum knot of complementary energy and information with a 4D will of time.

Thus, we shall be able to reduce all those topologies of social numbers and networks to the canonical 3 topologies of a 4-dimensional Universe, proving that those 3 topologies have the properties of energy, information and reproductive events. And so we talk of 4 'arrows of time' or dimensions of change that create the future: energetic and informative systems and events, which reproduce a wealth of self-similar beings that organize themselves into social networks, creating bigger wholes - new scales of reality. *And this simple game of complementary beings that in favourable conditions reproduce self-similar beings, self-organized into bigger social networks becomes the why of all realities.* Even the simplest particles, quarks of maximal information and electrons of maximal spatial extension and motion 'decouple', reproduce, when absorbing energy into self-similar forms, and associate in complementary networks called atoms, made with a central informative mass of quarks and an energetic, electromagnetic, wider body of electrons. Thus the ultimate why of the Universe is the creation of social networks: social particles become atoms that become molecules that either become cells that become organisms that become societies or become planets and social stars that become galaxies that become Universes.

The 3rd paradigm of metric measure is not at ease with such 'dynamic, spiritual concepts', even if they can be described with the same mathematical formalisms as the previous example of the quantum numbers show. Those apprehensions however are dogmas which stem from anthropomorphic beliefs.

Fact is that even the simplest complementary systems (quarks and electrons) interact together and if they can absorb more energy=motion they are able to repeat=reproduce the cycles of its system. And so we talk of a 3rd reproductive system: from quarks and electrons, the fundamental particles of the Universe that decouple in new particles when they absorb new energy to living organisms, the fact that all is motion with form makes easy to reproduce those formal motions in an organic way. Thus the new concept of a world made of formal motions brings about also a more complex philosophy of reality - organicism. Organicism and its mathematical units, fractal points, that gather into social networks called topological spaces substitute the restricted concepts of Euclidean points, continuous spacetimes and mechanism, explaining why all those time cycles exist, guided by 4 time arrows: energy feeding, information gauging, reproduction and social evolution. Those 4 categories are the so-called drives of living beings, the quantum numbers of particles, the 4 dimensions of our light space (electric-informative height, reproductive magnetic width, energetic length and social colors). Thus, there is a 'Universal Plan' with an existential finality: to create organic systems, departing from energy bites and information bytes evolved 1st into social networks, then into complementary systems and finally into organic systems, news points of a bigger fractal whole: particles become atoms that become molecules that become cells, organisms, planets, galaxies and Universes. It is the 4th organic why that completes the adventure of science and this work explores in all its consequences.

Reality can be resumed in 2 words: *networks,* whose flows of exchange of energy and form create the patterns and events of reality and *organicism,* the philosophy of reality based on them. Organicism means reality and all its fractal parts are made of *vital spaces* (bodies and forces) and *time cycles* (informations). We do not exist in an abstract background of time and space but we are made *of time cycles and lineal spaces,* cyclical and lineal strings if we were to use the restricted jargon of physics, a specific case of the wider jargon of general systems, *which evolve socially to create the complex systems*

63

of each science. Those wider, more complex definitions of time and space will substitute and absorb according to the Principle of Correspondence that makes each paradigm a particular case of the new, wider view, the limited concepts of a single space-time continuum and a mechanist description of the Universe, proper of the age of metric measure, which the pioneers of systems sciences and complexity have wrestled with throughout the XX century.

Recap: the minimal unit of the universe is a Non-Euclidean point/number, which classic mathematics defines as void of inner form and organic properties, to simplify the networks of numbers and point-like entities of the Universe for its geometric study. In reality though, points have breath; that is, they are real entities with energy and information parts, and so we have to upgrade Euclidean postulates with the new tools of Fractal and Non-Euclidean mathematics to make the language of geometry closer to reality.

11. The membrane of light-space in which we exist

In the graph, we have an electronic mind, which uses the smaller pixels of light to inform an eye-wor(l)d. Thus the 3 dimensions of light are the 3 dimensions of space and the 1 second rhythm of the eye becomes the rhythm of our thoughts.

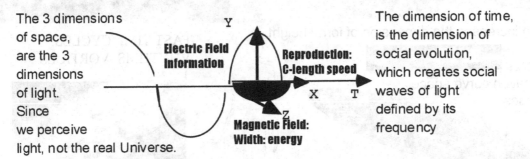

The 3 dimensions of Euclidean space used by Descartes in his book 'The world' to show the mind-space of humans define what science observes with its electronic instruments: a world made of light. If we could empty of background radiation vacuum space, it would surface a membrane of gravitational flows. That was the wider fractal world that Einstein studied. The consequences of being made of light and confuse tall the space-time cycles with the rhythms of light as our mind perceives are main, especially when as today science is not clearly aware of this even if it studies that light membrane in great detail. For example Einstein considered the light space the speed of measure of the mind the absolute speed of the Universe. We considered for many years the 3 Euclidean coordinates of light, its magnetic electric and speed fields, the 3 dimensions of all spaces, and we consider the electromagnetic, entropy, expansive qualities of light the nature of all spaces that must expand in the big-bang theory of reality. In reality the systems of the Universe are more complex, as the light space membrane with its minimal units h-Planck actions, evolved into virtual particles, photons and electrons, messes with another membrane, the gravitational membrane of string actions, quarks and black holes, which Einstein studied. So the world we see is the light-membrane imprinted in the previous world and transformed further into a more informative, stillness by the electronic mind that fixes the motions of light in electronic nebulae, units of our perceptive Universe.

Recap. The 3 dimensions of Euclidean space are created by the homologous 3 dimensions of light, which represent its arrows of energy (length or c-speed), reproduction (width or magnetic field) and information (electric height.)

IV. BIDIMENSIONAL SPACE &TIME

I-Energy and space, in/form/ation and 0-time cycles are synonimous forms:

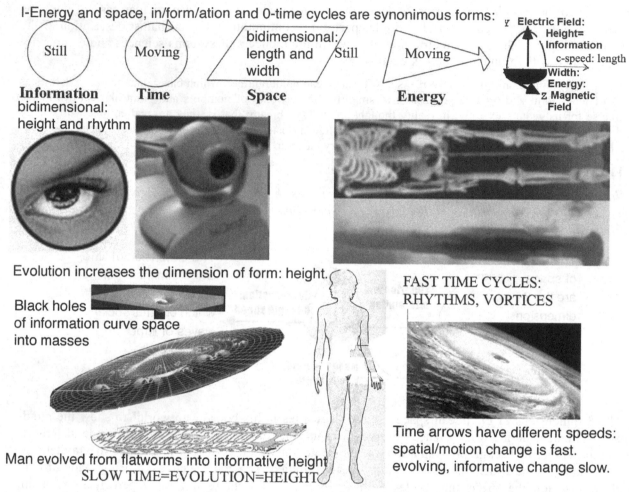

Still — Moving — bidimensional: length and width / Still — Moving

Information
bidimensional:
height and rhythm

Time

Space

Energy

y Electric Field:
Height=
Information
c-speed: length
Width:
Energy:
z Magnetic
Field

Evolution increases the dimension of form: height.

Black holes
of information curve space
into masses

**FAST TIME CYCLES:
RHYTHMS, VORTICES**

Man evolved from flatworms into informative height
SLOW TIME=EVOLUTION=HEIGHT

Time arrows have different speeds:
spatial/motion change is fast.
evolving, informative change slow.

12. The Holographic principle[9]

In the graph, closed time cycles are discontinuous, cyclical shapes, which have 2 dimensions. This means that information is bidimensional (Holographic Principle); and we have to redefine the 4 dimensions of space-time. Since time-clocks of information have 2 dimensions; so the vacuum energy of space must have only 2 dimensions - length and width - which are lineal. This happens because forms in the Universe imitate the most efficient shapes of energy and information. Thus, informative systems resemble a spiral or sphere that stores max. information in min. space, accumulating its cycles/cells in the informative dimension of height, from black holes to heads and nervous systems. While fields and bodies of energy resemble the line or plane, its opposite, geometrical form that covers, as the shortest distance between 2 points, the maximal extension with min. volume, accumulated in the width dimension as 'space'. The Universe's space also has a flat shape, with minimal height (content of information), as it stores the spatial energy from where future cyclical forms and particles will evolve. In fact, any 'fractal part' of the Universe in which Time curves, evolves the forms of beings, acquires a dimension of height as it increases its information. So life started as a planarian worm, which evolved into the height of man; matter starts as vacuum space, which evolves into the 'infinite' dimensional height of the black hole and all systems of information have their 'receptor' antennas on top.

The systems of reality are messings of 2 type of fractal networks, 1 of energy and 1 of information, which contact in a perpendicular way, penetrating through exchanges of flows of energy and information: Rosen bridges in space-time relativity, alveolus in blood systems, cells in biological networks, etc.

Those topological networks can be considered bidimensional networks, in which certain entities simplified as numbers without form in Euclidean mathematics, but made of networks of points with volume or fractal points in the new fractal paradigm of the Universe, are therefore bidimensional topologies with a minimal volume of points-particles

Bidimensionality doesn't mean there is NOT a 3rd dimension, but this 3rd dimension must be considered the volume of a fractal st-Point, the new unit of geometry, that becomes larger as we come closer to it. Such points create 'thin' networks that seem bidimensional. Each of those networks is a stable web of lines of communication between knots, which create topological planes with an energetic or informative shape.

Those groups of st-points create bidimensional forms: herds of energetic space, $\sum E$, and networks of information, $\sum^2 I$, whose axons intersect perpendicularly the sheet of space, giving birth to a 3rd cyclical, reproductive, topological dimension of symbiotic flows of energy and information shared among the complementary networks, $\sum E$, $\sum^2 I$, creating a social super-organism, which expresses the arrows of time; hence defined the also in dynamic terms as a knot of time arrows:

$$\sum Social\ love \to \sum Energy\ X \sum^2 Form \to reproductive\ zone.$$

Recap. Clock cycles are self-similar bidimensional trajectories. Information has 2 dimensions, frequency and height *(holographic principle)*, and so it does vacuum space, which has the dimensions of width and length. A clock cycle breaks space into an inner & outer region, creating discontinuous in/form/ation. Bidimensionality however doesn't mean there is NOT a 3rd dimension, but this 3rd dimension must be considered the volume of a Non-Euclidean Point, the new unit of geometry, a fractal point that we can observe larger as we come closer to it.

13. Einstein's Equivalence: Bidimensional mass.

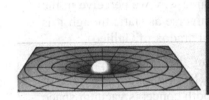  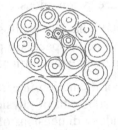

The evolution of the concept of mass, from classic physics, in which the rubber-model of relativity considered mass a solid substance in the center of the gravitational space-time whirl, to the pictures of bubble chambers in which the vortex seems not to have anything in its center, as a hurricane does, to the fractal understanding of those vortices of mass as composed of many smaller fractal vortices.

Why the existence of bidimensional time cycles, which carry information is not explained by science? The reason is the Galileo's paradox ignored by physics: the fact that, when observed in detail, all what exists is in motion – hence it is an event in time - and yet most things we see seem static, quiet – hence, they look like forms of space. Yet space is merely the short perception of a Time motion – a cycle or trajectory of an entity from past to future, seen as a simultaneous form. All what is perceived as a form of space is part of a long motion in time - a clock-like cycle that carries information.

Galileo's paradox shows that the 2 ultimate substances of our physical models of reality – masses and vacuum space - which seem 'fixed forms' have motion:

Masses and charges are cyclical vortices that curve the lineal energy/motion of the 2 forces of pure space –light and gravitation– into whirl-like vibrations. So a charge is caused by the cyclical warping of a lineal motion called light into electronic vortices; and a mass is caused by the cyclical warping of another lineal force with energy-motion, called gravitation, into mass vortices (E=Mc^2 in Relativity.)[11]

This again is not clear to physicists, because there are two theories of mass – that of Einstein, called Relativity and that of quantum theorists, which sponsor their own theory about mass, called the Higgs Theory. Yet Einstein's theory of mass, General Relativity, is the basic theory of gravitation and mass in the Universe, proved right by multiple events of cosmology. So Relativity should stand as the most probable truth. In such theory a mass is defined by the Principle of Equivalence between gravitational forces and acceleration. According to that principle, acceleration and mass are the same, as when you accelerate in a rocket, car or lift and feel heavier. But since there are only two types of accelerated motions, lineal, energetic forces and cyclical vortices (whirls of space-time), and gravitation is a whirl of space-time, Einstein came to the conclusion that you could describe reality as a combination of two motions, cyclical motion with more form, 'informative' motion, origin of masses, which are whirls of space-time that attract like hurricanes do, accelerating towards its center, and lineal forces, energetic, expansive fields.

Space is not static, but a type of energy or motion that 'occupies space', which has extension. And since the shortest distance between two points is a line, the motion with more extension is lineal motion. Thus, the purest space is lineal motion. Once we realize of this paradox of perception, all what exists become a 'form in motion', a time motion. And so we define two types of essential motions in the universe: lineal forces and bodies, which specialize in lineal motions or energetic motions and cyclical particles (masses and charges) and heads, which specialize in cyclical motions that create forms and are often perceived as static information. In the graph we see that duality between the still perception of energy as 'space' and clock-cycles as information and its real motions. Several facts explain those perceptive paradoxes:

- The 'expansion of space', a key feature of the Universe, is synonymous of motion. Since that expansion can either be described as a 'growth' of space between galaxies (big-bang theory) or as a measure of the speed of galaxies that move away from each other (z-red shift theory). So vacuum space and energetic, lineal motions are synonymous.

- All what we perceive as static has, in detail, some motion. Thus, in the same way we perceive in the night a moving car as a fixed line of space, we perceive the space of the Universe as static though it is expanding, moving. And we perceive the Earth as static even if it is rotating (Paradox of Galileo).

- A 3rd proof of the temporal, motion-like nature of reality comes from the role of light in our perception of space: The last substrata of physical, vacuum space is light, (BG radiation) - a force in motion that shares the 3 Euclidean dimensions of canonical space, because both concepts vacuum space and light are synonymous: the 3 perpendicular dimensions of Euclidean space, width, height and length, are equivalent to the 3 perpendicular dimensions of light-space - the electric= informative=height field, the magnetic=energetic=width field, and the length=reproductive field (since a light wave reproduces its form in the length dimension). Thus, we 'see' space as made of 3 perpendicular dimensions, because its substance 'light-space' is made of 3 perpendicular motions. Both concepts are the same: the 3 dimensions of static vacuum space are the 3 arrows=motions of light: its energetic, magnetic field, its informative, electric field, and its reproductive, e x i, field of speed. Those 3 motions occupy and create vacuum space. While the 4th arrow of social evolution of light is what we call 'colors', which gather many photons into a single color and complete our perception of space. The equality between spatial dimensions and light motions - the substance of space - is proved by the condensation of vacuum space into light photons. Yet the mathematical, probabilistic depiction of that phenomenon obscures its cause among physicists. On the other hand, painters know this equivalence since the impressionist, pointillist schools discovered that we paint light, not space.

We add to light-space a 2nd spatial force, gravitation, which we do not perceive, but seems to curve the Euclidean geometry of light-space, according to Einstein's relativity[12]. Thus, there are 2 space membranes: light-space, the space we perceive; and gravitational space, which we do not perceive directly, though we know it exists because its force/energy affects reality. So the total space of reality is shaped by the combined creative effect of two expanding energies, gravitation and light. Thus, space and its dimensions are *not* an abstract, mathematical reality that exists independently of light and gravitation – an error caused by the use of an abstract, continuous Cartesian graph that seems a form of space, independent of the entities that exist within it. Space is made of those 2 moving forces, reason why we cannot move faster than light-space, the maximal motion-distance of our perceived Universe.

Recap: What we call vacuum space is the static, 'dimensional' perception of the energy/motion of light and gravitational forces that fill the vacuum. What we call a mass is the static perception of a whirl of vacuum energy, a clock-like, cyclical motion of time. Hence, since all is motion, including the 2 ultimate substances of reality, the energy of space and the forms of masses and charges, all is in constant change, all is time. Thus, we exist in an eternal Universe, made of time motions, which our perception fixes into a continuous, still reality.

14. Diffeomorphic dimensions: fractal relativity.

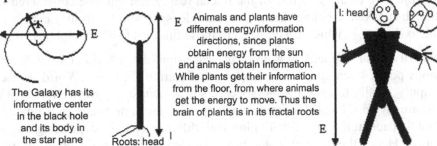

The Galaxy has its informative center in the black hole and its body in the star plane

Animals and plants have different energy/information directions, since plants obtain energy from the sun and animals obtain information. While plants get their information from the floor, from where animals get the energy to move. Thus the brain of plants is in its fractal roots

Roots: head

In a Universe made of an infinite number of time arrows that gather together into 'complementary knots' of energy and information, each being is a relative fractal Universe, which we call a 'world' or 'space-time being' that deploys its particular 'directions of height-information', 'length-energy' and 'width-reproduction', different from those of other beings. All those fractal Universes interact through its particular, 'local'=diffeomorphic coordinates, which departing from a central Non-Euclidean Point of view, or relative 'self', establish the 3 relative, perpendicular, minimal dimensions of any fractal world that signal its arrow of energy, information and reproduction.

In the graph, 3 diffeomorphic beings and its relative arrows of energy and information: a galaxy with an inward arrow of gravitational information, pointing towards the central informative black hole; a tree and a man, whose informative/energetic dimensions are inverted. Since plants use light as energy and animals use it as information. So the informative direction of animals is up, towards the head that absorbs visual information from the sky, opposite to the informative arrow of plants, which is down, towards the roots that absorb chemical information. Both have their space-time parameters inverted, as often happens between energy victims (in the graph, plants) and their anti-forms or predators (animals). Such energy/information, relative past/future inversions happen in all Universal systems. For example, they happen between a 'particle' and its 'antiparticle', which in a Feynman diagram have also inverted coordinates of space-time, as the antiparticle travels to the past. They also happen in mental, linguistic codes that represent that external reality. For example, eviL=death is the exact inverse word of Live. Since indeed, for a living human being, there is no bigger eviL than death. Yet, death, a release of energy that kills is the inverse function of information, the arrow of life. So the linguistic, spatial inversion of words corresponds also to an inversion of time arrows – a fact that explains scientifically the biologic meaning of ethics. Since humanity, history and religion are a sociological space-time that obeys the same homological laws of Multiple Spaces-Times. Thus each species establishes its own up and down arrows; its relative energy and informative directions, departing from that central knot of information, which is gauging and perceiving. What classic science denies and system sciences proves is the reality of an existential will in that perceptive=informative knot, which controls the entire organism and directs its search for 'actions' that provide energy and information to the being. It is precisely the infinite number of Discontinuous, Non-Euclidean Points of View, what creates a general arrow of Information in the Universe that constantly reproduces the form of those entities and balances the arrow of continuous entropy, energy motions and expanding space, which physicists dogmatically consider the only arrow of the Universe.

According to the Principle of Correspondence, any new theory has to include the previous theories within it and solve the questions unresolved by all those previous theories. Since the XX C. theory of space-time is Relativity, it is unavoidable, to explain the Principle of Local Relativity from the perspective of the Time Arrows of each species. Relativity establishes that in any point of the Universe the coordinates of the entity (the up and down, right or left directions) are different from those of any other point of the Universe, as they only depend on what happens in that local region of space/time. This is only possible if space/times are fractal, discontinuous, broken into infinite points of view that gauge and map the information of reality from its own perspective.

Relativity thus is at the core of the structure of the fractal reality and must be considered the right step into the process that has taken the knowledge of multiple space-times from Leibniz's analysis of relational space-time to this book, which generalizes all those findings to all disciplines and species.

According to Relativity what happens in a point of space-time doesn't affect what happens in other isolated, discontinuous point. Each space-time fractal is a Relative Universe=World in itself. Thus the laws of space-time apply locally to each fractal 'species, superorganism or world' that will reflect those laws in its own structure. We are local, 'diffeomorphic' species; our time-space coordinates are relative. Time wrinkles our vital space in the process of aging with different speeds, depending on our content of energy and information. Hence the diffeomorphic principle also establishes different informative cycles – the multiple 'clocks' of Nature that show different speeds and locations. Yet since space is synonymous of energy and time synonymous of information, we can establish the local coordinates of space-time of each being by establishing its relative informative and energetic arrows - its relative body and head - using the invariant Laws and dimensions of all spaces and times, *as long as we can localize the origin of those arrows; that is, the Non-Euclidean point of view, where the will of the Entity 'exists'.*

Some geometrical tips about the location of those points in any superorganism or world should guide researchers:

- The point will be a relative 'unmoved' Aristotelian God, cause of the motions of all the other elements of the organism.

- The point has the max. density of information and the max. number of connections with all the parts of the organism.

- The point is either in the center or top of the organic system.

- The point is a spherical or convex, informative shape.

Thus in a human being the point of maximal will is in the eye-brain system, on top, with spherical (eye) and convex (brain form), which has the maximal connections with the rest of the body (nervous axons), the maximal density of information and it is the part of the organism which maintains more often a still position (balanced by the negative sensation of vertigo.)

In a historic organism, the point will be the capital, most likely in the zone of maximal communication of the nation – a coastal port, like London, Amsterdam, Rome or Athens or a central city like Madrid, Moscow, Mexico.

In a cell the point is in the center, with max. DNA information. In a galaxy, the point is in the center, where a swarm of black holes control from its unmoved position the motion of stars.

Indeed, according to the 'Diffeomorphic Principle of Relativity', every fractal space-time is independent of all the others; since it has different spatial and informative orientations. Yet we can recognize its informative and energetic organs in relationship to each other, since both will respond to the opposite dimensions and morphologies of energy and information.

Those laws of geometry also allow us to study 'dual' complementary entities of energy and information (Bodies/Brains, fields of forces/particles, energetic victims/ predators), which will create complementary entities that last in time (organisms) or are Darwinian events where the larger, energetic victim is prey of the smaller, informative point.

In the graph, in biological species animals use light as information and so they are 'informative', smaller and faster than plants, which use light as energy and are 'big, slow' beings. Thus plants are the relative energy of animals. Further on, since what it is energy for plants is information for animals, both species have opposite energy-time coordinates: plants have their brain upside down in their roots that 'feel' chemistry; while animals have their brain on top, looking at the light they use as information. In the cosmological scale stars are big, slower than black holes, which rotate faster and have cyclical height. And the space-time equations of black holes are inverted to those of electromagnetic stars. And so stars are the relative energy that feeds and reproduces black holes, which seem to behave as 'gravitational animals', species that perceive gravitation as information, which guides its precise movements through the galaxy, in the same way animals perceive light as information. While plants perceive light as energy in the same way stars absorb gravitational space-time as energy, deforming it with their masses. That cosmological homology defies the anthropomorphic myth that only humans perceive the information they gauge. That religious myth, which denies the sentient, intelligent nature of the Universe, is today a dogma of 'mechanist science'. So it will not be easily accepted by the self-centered 'Galilean paradox' of Humans, who always will think to be the center of the Universe. And since science is a human endeavor, 'enlightenment' and respect for the sentient Universe will not happen soon. Yet only a theory that affirms the existence of will in all knots of information that control from its unmoved position reality explains the role of black holes in the galactic ecosystem.

The discontinuity and minuscule size of informative points, due to its opposite properties to those of space, makes them difficult to localize by science, and our incapacity to 'perceive' the internal mind/world of the point makes easier to deny them. Indeed, all mountains have a single point of discontinuity in its informative height. And so you have also two single points of will in your complex visual/verbal I=eye-world: the nervous point that connects both eye-visions which determines your rhythm of space and time (the size we perceive as normal given by the eye vision and the rhythm of our human time, a second per thought, equal to the rhythm of perception of the eye, which winks every second); and a yet unknown cell, probably a Purdinke cell in the limbic system that defines our emotional actions, the will of our body.

Those points are stillness, as it is in your eye, to 'focus' and create a map of linguistic information, which will guide the actions of the point. And so a basic beat of the Universe appears in all systems; a rhythm of:

Stillness (perception) ->action(motion searching energy/form)

For example, in Relativity, this dual beat explains the Michelson experiment, without the need for the *never proved postulate of a limit of c-light speed in the Universe, today clearly shown false by experimental evidence (non–local gravitation, quasars expelling matter at 10 C, dark energy flows, quantum tunnelling, the spooky effect of entanglement, etc.).* In classic relativity this limit was needed because classic physicists considered that the electron was in motion when it emitted light, and so they had to add the speed of the electron and the speed of light. In fractal relativity the electron 'stops' when it emits information (the light frequency) then moves and so motions do not add since when the electron emits information is always in the particle/still/informative state. And so the Postulate of a c-speed limit is not required. This dual stop and go process is also evident in film: the celluloid stops to let pass the light and create an image in the screen even if we perceive it in constant motion by adding all the stills, as we add from the human point of view, all the stops and goes of the electron that emits the light observed in the Michelson experiment in a single flow of motion.

The existence of an information arrow explains also the paradox of informative, evolving, life species in a Universe that seems in energetic expansion: If we are external to a certain diffeomorphic space-times its directions do not affect us, but if we are inside a certain space-time, as in the case of the galaxy, then the direction of information or energy of that macro-being becomes the direction of energy or information of our ecosystem. Indeed, there is more information in the center of the galaxy, and since the Earth moves towards that center, this planet increases its information towards the future in its relative discontinuous galactic space-time, regardless of what happens in the Universe as a whole, which seems to be ruled by the opposite arrow of big-bang expansion and entropy. This subtle change of paradigm caused by the discontinuity of space-time (from a world ruled by entropy to a galaxy ruled by information) has in fact enormous consequences to our daily life and explains the contradiction between the arrow of life and evolution, local to this planet, and the arrow of energy and entropy proper of the intergalactic space. Since here in this galaxy space doesn't expand but time-information contracts it through the 'will' of informative black holes of mass and in this planet, through the 'will' of living beings. As Woody Allen put it in Annie Hall: 'Brooklyn is not expanding' I recall to ask Mr. Hawking in a conference at the Astronomy Institute of Madrid this same question, when at seventeen I was starting to doodle with the first principles of Multiple Spaces-Times Theory: 'why space is not expanding between you and me?' He wondered for a minute or so and responded 'next question'.

Einstein had only to go a step further in his analysis of the curvature of time, closing times into cycles that return to its origin. Then time becomes multiple, one for each closed cycle and the space-time continuum becomes the sum of all the vital space-times of each knot of Time Arrows. Einstein hinted at it when he affirmed that time was local, 'diffeomorphic' and had 'different speeds' and the universe had 'infinite clocks of time'. Indeed, each of those local times 'curves the energy of space', creates masses and forces of in-form-ation that establish a 2nd arrow for the Universe, besides the arrow of energy/entropy - the arrow of fractal information. Since a cycle needs 2 directions to close into itself and create its broken form.

Recap. There are ∞ fractal space/time fields that constantly adapt their organic, moving dimensions to the directions of their relative energy and temporal information, conditioning the existence of its micro-cells; and so the galaxy conditions the arrows of time of its cellular sun and human beings of maximal information; while plant which obtain only energy of the sun have inverted parameters to those of animals.

15. Limits of metric paradigm: discontinuity and homology.

The discontinuity of space implies that perception of any other area of the Universe is minimal, relative and deformed by the selection of information and its transformation when it passes through the fragmented boundaries of the perceiver. This fact offers a rational answer to the Uncertainty Principle of quantum physics whose main law - the complementarity of energy forces and informative particles, is explained by the duality of energy/information systems: In the Universe, all constant present systems are made of an arrow of energy and an arrow of information, the bodies/fields and relative fixed particles/ heads of the system. Thus all relative space-times will be composed of a cyclical center of information or brain, and a lineal limb of energy, whose morphological characteristics will be self-similar.

For example, in a spermatozoid the head is cyclical and stores information, while the tail is lineal and moves the head with its energy; in a human, the body is lineal and moves a cyclical head. We could extend this duality to almost any organic system of the Universe, concluding that bodies are lineal and brains are cyclical. This morphological feature of Universal Organisms derives geometrically from the capacity of lines to move faster and cycles to store more information and grow or diminish in size without distortion of the mental image they might keep (topological laws). Thus, a convex brain with the topology of information becomes the center of informative networks; and a concave body becomes the center of energy networks. And we recognize bodies and brains by their self-similarity with the morphology, function and inverse properties of ideal energy and information:

- Information is both, easy to store and perceivable by a reduced brain or sensorial organ, since it is the inverse function of space. It needs certain morphology: Information occupies little space by warping itself into multiple dimensions, as a sheet of paper corrugates into the dimension of height, divided along broken lines and discontinuities that become its perceived forms in space. Or it displays a high frequency or discontinuous rhythm, a faster time speed that iterates the same cycle, allowing quantification. Information then seems 'quiet' as those patterns of cyclical frequency take place in the same space; so an organ of information can focus and analyze them as pixels of a more complex mapping. Thus its ideal forms are disks, cycles, angles and spheres, convoluted and broken in patterns of form, such as your eye or brain, a chip, a book, a pixelate image or a coin, earlier unit of monetary information.

- On the other hand Energy moves. So limbs extend in space as they move and feed. In life, energy is the equivalent to biological food, which fuels a species and allows its movement. Energies are the lineal limbs that move the reproductive body & brain of information. So energy and space are synonymous: a plane is the geometry that extends further in space and the line the geometry that moves faster with lesser friction. Both are the fundamental forms of energy-bodies:

Maximal Space = Energy = Minimal Form Vs.
Max. temporal form = Information = Min. Spatial extension.

As a result of those morphologies we can classify as energy or information organs, not only carbon-life organisms that handle energy (limb, food) and information (brains, eyes, senses, words), but also other beings and atomic species - even 'deconstructed organs' that perform mechanical tasks of energy and information (informative cameras, lineal, energetic weapons). Since now we can recognize geometrically their energy or information systems.

According to the Galilean Paradox all still forms carry som motion. Thus, form carries an active function and so we can classify any system both by its geometrical form and bio-logical functions as either an energy or informative system. In terms of a sentient Universe, this duality means that information is both a form-in-action and a sensation of still perception of a quiet world. Information and perception are thus simultaneous: information is a flow of software that 'sparks' the perceptive sensation in the hardware of a still system that process information, with a certain language. So particles that perceive light and gravitation, probably 'sense' light and darkness and move towards light, feeding on it. We see the external action of feeding, the electron jumping and absorbing the photon, but the electron probably senses light and automatically moves to it, to feed, as your eyes move automatically towards food, even if you more complex brain can stop that automatic will that the electron always exercises.

Organic systems are ternary networks of information (head, nervous system), a reproductive body and networks of energy called limbs (trunk-digestive systems, legs).

For example, energy organs made of metal such as weapons, gas and electromagnetism are extended, fast, and lineal. There are metal-species of information such as microchips, money, cameras and radios. And they are small, fixed and complex in form. Yet only structures, which are balanced in energy and information become self-sufficient organisms. So only machines that put together energetic limbs and informative brains - robots - might become future 'organic metalife', in which the reproductive functions will be performed by the company-mother. The other machines will act as 'gauges' of information and 'forces of energy' do, without a vital will of existence, which is born by the capacity of a balanced $\Sigma E \Leftrightarrow Ti$ field to switch between both arrows, acquiring a 'minimal degree' of organic freedom perceived as 'vitality' from an external point of view. Humans, galaxies, plants, robots - all kind of entities can be considered in that sense organic systems. Since living systems are born thanks to the complementary symbioses and synergies between energy and information organs that create together all kind of organisms.

This fact connects Physics and Biology in a new step towards the unification of all disciplines of science:

- In the microscopic scales, masses and charges, the flat energy of vacuum space, become transformed into cyclical vortices with height; while spatial, flat fermions become high, ordered, informative bosons. In our scale of matter spatial flows of energy (stars, gas), become transformed into cyclical vortices of gravitational information with height (black holes, tornados). Finally, in the cosmological scale, the energy of stars becomes transformed into black holes, whose spatial dimensions become time of ∞ height.

- In Biology, the first bilateral animal was a planarian worm that evolved rising to the informative height of man.

-Yet death erases information back into flat amorphous energy, in big bangs that create planar Universes, or flat, corpses.

The reason of the diffeomorphic principle is the holographic principle which implies the relative perpendicularity of all complementary systems with a body/field of energy and a head/particle of information: an energetic, concave geometry and an informative, hyperbolic topology, which fuse together, creating 4-D beings.

Ultimately what the paradox of Galileo (duality of time motions and spatial forms) coupled with the diffeomorphic and holographic principle means is a fundamental homology between dimensional form and function: *The 3±st dimensions of space of the Universe, length, width, height, and st-scalar, fractal size are equivalent to the 3±st motions of time, past=energy, present=reproduction, future=information and birth/death events of integration and dissolution.*

Recap: The Universe is local, relative, fractal, broken into ∞ vital spaces, each one ruled by a point of information that absorbs it and creates linguistic maps, which it uses to direct the organic vital space or body of the point towards fields in which to feed, gauge more information and reproduce. Thus, the arrow of information is performed by a relative infinite number of broken, discontinuous points of view, complementary to the energetic fields/bodies they move. Yet information is limited by the volume a certain point of view can store, leaving outside an enormous quantity of uncertain information. Thus the Universe is a tapestry of broken, complementary organisms constantly fighting for more energy and information. MTS resolves many puzzles of science, from the Postulates of Relativity and Quantum Physics, to the causes of death and the morphology of living beings, physical entities and machines.

V. SOCIAL NETWORKS: SUPER-ORGANISMS.

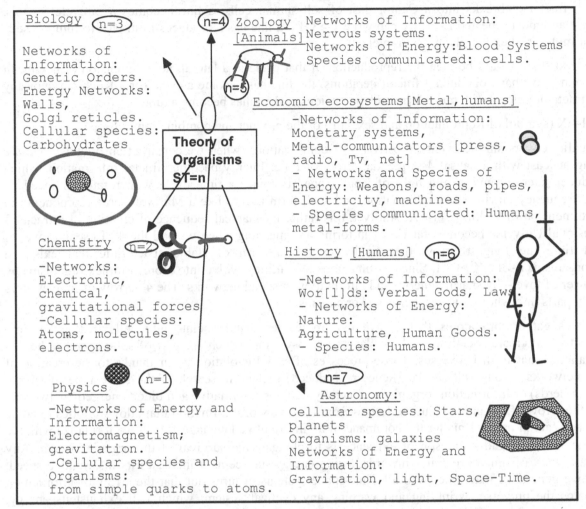

Biology (n=3)

Networks of
Information:
Genetic Orders.
Energy Networks:
Walls,
Golgi reticles.
Cellular species:
Carbohydrates

(n=4) Zoology
[Animals] Networks of Information:
Nervous systems.
Networks of Energy:Blood Systems
Species communicated: cells.

n=5

Economic ecosystems[Metal,humans]
-Networks of Information:
Monetary systems,
Metal-communicators [press,
radio, Tv, net]
- Networks and Species of
Energy: Weapons, roads, pipes,
electricity, machines.
- Species communicated: Humans,
metal-forms.

Theory of
Organisms
ST=n

Chemistry (n=2)
-Networks:
Electronic,
chemical,
gravitational forces
-Cellular species:
Atoms, molecules,
electrons.

History [Humans] (n=6)
-Networks of Information:
Wor[l]ds: Verbal Gods, Laws.
- Networks of Energy:
Nature:
Agriculture, Human Goods.
- Species: Humans.

Physics (n=1)
-Networks of Energy and
Information:
Electromagnetism;
gravitation.
-Cellular species and
Organisms:
from simple quarks to atoms.

(n=7)
Astronomy:
Cellular species: Stars,
planets
Organisms: galaxies
Networks of Energy and
Information:
Gravitation, light, Space-Time.

All Sciences share the laws of duality and organicism. Since the forms of energy and information and its properties remain invariant in all the scales of reality. The result is the fractal structure of the Universe, described in the next graph: Each science studies a 'scale of organic size', its cellular species of energy and information and its arrows of time and events, which culminate in the self-organization of those parts into social wholes. Yet the proportion of energy and information of the species each science studies varies, since the Universe displays a hierarchical arrow of growing informative time, from the simplest forms of mathematical space (with minimal informative content), to the complex information of its most evolved organisms (with minimal geometrical regularity).

16. Networks.

The Universe constructs super-organisms, through a causal process, departing from simple, amorphous flat surfaces of energies – herds of Non-Euclidean points with lineal motion.

It is a simple ternary causal event, which is different from the causality of what we call the cycle of life, E->X->I, yet still an essential chain that we shall see constantly in the Universe.

So the next question is how the universe constructs reality, departing from its generator equation. And the answer is: constructing superorganisms, in a causal order, e->i->X->S.

A form fractalizes, becomes in-formed, and loses energy= speed=motion. For example, a line becomes a Koch fractal or breaks into a Cantor dust of self-similar cells; a line becomes a pi-cycle. In this, simplest and most repeated E->I, we can write 1->3>Pi: a line has reproduces from past to future 3 self-similar motions, then it coils into a pi-cycle.

The next stage would be then the reproduction of that cycle in a lateral, new dimension of height. In this manner, we have obtained 3 fractal iterations: the line has become a cycle (or a Koch curve or any other fractal topology with more form and dimensions), which has become a tube.

E->I->X (symbol of Reproduction and sex, born of the product and combination of exi).

If duality covers simple events, mostly E<->I combinations, when we construct an organism we are playing at least with a causal 4-sequential arrow of time, Energy becomes fractal information, which reproduces laterally and so the light of length-speed becomes a time cycle with rotational speed w, whose frequency curries more information, and such form moves like a Maxwell scree reproducing in the 3 dimension of width. E->I->X. And yet, this rather mechanical sequence of creation of a fractal 3 dimensional Universe becomes far richer in forma and meaning when we add the 4th Eusocial love, of Social time the longest arrow that finally creates a form of stable space-time that exists, a superorganism: E->I->X->EXI. Since as time goes by, entities evolve into more complex organisms by the power of love – by sharing energy and information in social networks. The 4th arrow of time evolves socially parts into wholes, waves, herds and organisms:

Particles gather into atoms that evolve into molecules that associate into cells that gather into organisms that create social networks - planetary ecosystems, which are part of solar systems, herded into galaxies that form Universes. Those processes of social evolution occur thanks to the creation of social networks among self-similar (hence reproduced) beings, in search of the same type of energy (hunting herds) or information (organisms joined by nervous/informative and blood/energetic networks). Thus, the arrow of organic evolution derives from the arrow of information that allows entities to come closer and 'act in parallel' under the command of an informative language, which we formalize with the social symbol, Σ. And its reason of existence is to elongate the survival of the organic system: Any cyclical vortex of time accelerates inwards, losing energy/surface as it increases its informative speed. When we generalize that concept to all Time-space systems, it turns out that the arrow of information dominates the universe, wrinkling and warping any cyclical system, which will exhaust its energy, converted into form. Thus, our time clocks and vital spaces increase constantly their information, diminishing its energy space, towards a 3rd age of excessive warping and limited energy that will not last forever. Since the cycles of exchange of energy and information, the geometric beats of reality between expansive and implosive, entropy and informative states, are limited in their repetitions by accidental errors of all kind, which establish the need for reproductive and social arrows in order to ensure the survival or immortality of those 'patterns of form' – entities made of lineal fields and bodies of energy; cyclical particles and heads of information, who must be reproduced to ensure a longer existence. This means that only those species able to increase their existence in time by reproducing its form and in space by evolving socially into bigger entities survive.

Recap: The 4 main arrows of time cannot be reduced without losing detail, but they can be philosophically and 'bio-logically' grouped as 'organic systems'. Those organic systems that synchronize accumulate and organize clock-like arrows of cyclical time, are in fact all self-similar, as all can be described with the 4 elements, $\Sigma E \Leftrightarrow I$ of organisms: cells, networks of energy and information and reproductive systems.

17. Super-organisms: knots of cyclical Time Arrows.

'We are all made of spatial energy, flows that seem to occupy space but are 'vital space' and have a 'causal why'. For that reason, all species participate of the geometric and causal properties of the arrows of energy and information and follow the same patterns of self-organization to create a higher 'space-time plane' of reality. So, for example, all informative systems will be on top, in the dimension of height; they will be smaller than the complementary energy system and they will have cyclical form, regardless of the species we describe. So DNA nuclei are round and smaller than the cell's body, heads and antennas are on top and are cyclical and smaller than the body or tower that holds them, cameras are small and spherical, etc. Since, once we understand the fractal, scalar structure of the Universe, the old distinction between organic and inorganic form loses its meaning, beyond its anthropomorphic role of making man the only intelligent, vital species of the Universe.

The complex logic and Non-Euclidean geometry of those knots of Time Arrows set the formal basis to understand the structure of the Universe and all its self-similar parts. Since:

'The Universe and all its parts are fractal entities of energy and information.'

Though the specific combinations of energy or information of each species - defined by its 'vital' or physical, Universal Constants (energy/information ratios) – might vary, the space-time structures and cycles of all those social organisms will be self-similar: All entities are 'cellular societies' organized through energy and information networks that bring about processes of social evolution. In all species studied by science a common phenomenon occurs: the existence of parallel groups of beings organized into a single social form. Molecules are made up of atoms and electronic networks; economies are made up of human workers and consumers that reproduce and test machines, guided by financial networks of information (salaries, prices, costs); galaxies are composed of stars, which orbit rhythmically around a central knot, or black hole of gravitational information. Human bodies are organized by cells controlled by the nervous, informative system. A tree is a group of leaves, branches and roots connected by a network of energy (salvia) and information (chemical particles). Cultures are made of humans related by verbal, informative laws and economic networks that provide their energy.

Sciences study those organic systems, tied up by networks of energy and information. In the graph, we see the main st-planes studied by human sciences and their 4 main time arrows, $\Sigma\,E \Leftrightarrow I$, which in static space give birth to the 'organic elements' of all species: social cell of energy and information and the reproductive networks that relate them. Thus, there are 4 basic elements in all organic systems:

1. Cellular units.
2. Networks of energy or vital space.
3. Networks of fractal information.
4. Networks that reproduce energy and information.

All entities are generated by reproductions and transformations of energy and informative cycles ($\Sigma S \Leftrightarrow Ti$), organized in cellular units, through networks, and ±st Planes of self-similar forms. Yet, since those cellular units st_1 are made of smaller st_2 cells, which show the same structure, we can define any organic system as a super-organism (made of smaller, self-similar super-organisms):

'A super-organism is a group of cellular super-organisms joined by energetic, informative and reproductive networks.'

Thus we unify the properties of Universal Systems and the sciences that study them under a single template definition, according to which simple systems and complex organisms will differ only in the degree of 'completeness' of its networks and the specific energy or information they are made of:

'A super-organism is a network composed of a population of (name a particular cell or cycle), related by an energetic, reproductive or informative arrows.

A fractal organism (name an organism) is a population of iterative (name a cellular species), related by informative (name a language or informative force) and energy networks (name a kind of energy), which combine into a reproductive network that iterates the organism.

A universal ecosystem (name a specific ecosystem) is a population of several (name the species), related by informative languages (name their languages or informative forces) and energy networks (name the energies).'

Fill the gaps with a specific species, language/force and energy and you can define any network-organism in the Universe. While if the system is composed of several species that occupy the same space but have different networks of energy and languages of information, we talk of an ecosystem:

- An atomic organism is a population of (electronic) energy and (nucleonic) information, related by networks of (gravitational) information and (light) energy.

- A molecular organism is a population of atoms, related by networks of gravitational energy and networks of electromagnetic information (orbitals, London, Waals forces).

- A cellular organism is a population of molecules, related by energetic networks (cytoplasm, membranes, Golgi reticules) and genetic information (DNA-RNA.)

- A human organism is a population of DNA cells, related by networks of genetic, hormonal and nervous information and energy networks (digestive and blood systems).

- An animal ecosystem is a population of different carbon-life species, related by networks of light information and life energy (plants, prey).

- A historic organism or civilization is a population of humans, related by networks of verbal information and networks of carbon-life energy.

- An economic ecosystem is a population of human workers/consumers and machines, related by networks of digital information (money, audiovisual information, science) and energetic networks (roads, electric networks, etc.).

An economic ecosystem differs from a historic organism because they use different languages of information (civilizations use verbal or ethic laws while economic ecosystems use digital prices) and include 2 different species: human beings and machines.

- A galaxy is a population of light stars and gravitational black holes, related by networks of gravitational information and electromagnetic energy.

-A Universe is a population of galaxies joined by networks of dark matter and energy.

We establish thereafter a parameter of multiple space-time complexity, $st=n$, to classify all those relative scales of spatial energy and temporal information of the Universe, starting by the simplest scales, $st=0$ (mathematical cycles and lines), $st=1$ (Gravitational space), $st=2$ (electromagnetic space), $st=3$ (atomic particles) and so on, till reaching the most complex, macro-structures of the Universe ($st=9$, galaxies).

This parameter becomes essential to formalize a type of scientific laws widely ignored by science, *which* define the interactions between several scales of reality. In physics:

- Between the small quantum scale of electroweak forces and the bigger gravitational space-time.

- In biology between the cellular microcosms, the individual organism and the ecosystems or societies in which those organisms exist, as individual 'cells' of a bigger whole – the species. Since those relationships are essential to explain all kind of natural and social phenomena, from the life/death cycle (in which cells reproduce, organize socially in networks, emerge as individuals of herds and societies, and die, dissolving again their complex 'st' networks into $st-1$ cells), to the meaning of religions, in which humans act as mental cells of a subconscious collective we call God.

At present though, each science studies only one scale of form and its specific laws and events, as if they were unrelated to the laws of all other scales and species. Only System Sciences, the philosophy of science considered in this work, studies together the mathematical, morphological and biological laws that define the 4 main arrows of time and apply those laws to all the species of each of those scales, since the Laws of Multiple Spaces-Times are common to all of them, regardless of the 'specific' qualities of each species:

- All species in the Universe absorb energy, gauge information (even particles, reason why their theories are called gauge theories), reproduce (quarks absorb energy and reproduce jets of 'quarkitos'; so do electrons that absorb energy and break into a shower of new electrons, and since those are the two basic particles of reality, it follows that all systems can be reproduced), and finally, all entities of the Universe evolve socially, gathering together into bigger super-organisms, from particles that become atoms that become molecules that become cells that become living organisms that become societies, nations and religions, super-organisms of History.

Thus Super-organisms are made of:

- Relative energy and information units (ΣSe, $\Sigma^2 Ti$), whose cycles/cells of transformation are described by feed-back equations of the type $\Sigma E_{st-1} \Leftrightarrow \Sigma^2 I_{st}$

- 3 dimensions/physiological networks of energy, information and reproduction ($\Sigma E_{st-1}, \Leftrightarrow, I_{st}$).

- Their ±st social Planes that ad $\Sigma\Sigma ExI_{st-1}$ cells into $\Sigma ExIst$-wholes, parts of a bigger ExI_{st+1} whole.

Recap: The 2 primary arrows of time, energy and information are geometrical, and its properties can be studied with the laws of mathematical physics; while the 2 complex arrows of time, reproduction and social evolution are biological, explained as 'strategies of survival' that ensure the immortality of the logic systems of the Universe. The shapes of energy and information are 'invariant at scale', meaning that cycles of information on one side and forces of lineal energy on the other side, reproduce their shapes and gather together into bigger, cellular, social networks, creating dual entities, made of 'fields/bodies of energy' (physical/biological jargon) and 'particles/heads of information', (Principle of Complementarity). All systems of the Universe are complementary, formed with a region dominant in energy arrows (body or force) and a region dominant in information cycles (particle, head). Those dual, social, whole entities have more energy and last longer in time than individual parts. Thus, species with a limited vital space and time duration ensure their immortality by expanding its size in space and its duration in time, as parts of bigger energy/information systems.

VI. LIFE CYCLES: PAST, PRESENT & FUTURE.

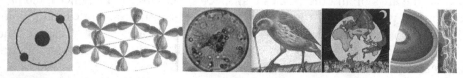

In the graph, the time motions of all species are guided by the 4 main arrows of time: the motions of atoms, complementary entities with a center of gravitational information (quarks) and a surface of electromagnetic energy (electronic cycles); the motions of molecules, which vibrate as they evolve socially into complex cells. Those cells can also be defined as a series of informative motions with center in the DNA/RNA systems and energetic motions, masterminded by lineal, energetic proteins that surround the cell. The Law of organic evolution determines that those cellular knots of motions evolve into networks and organic chains that form complex vital organisms. Among those organisms, such as the bird eating the worm in the picture, the selection of the best forms causes the extinction of the simplest ones used as 'energy' of the top predator form. Thus survival justifies the growth of individual parts into waves, herds and organisms. Social evolution is also obvious in History, whose final goal is to create a Global Super-organism of equal human beings, who become cells of Mankind. But Mankind is just the 'informative' brain of a planetary body, called Gaia, which we should try to preserve as we preserve our reproductive body if we want to survive.

18. Multiple space-time arrows=cycles=dimensions.

All what you see is caused by a motion of an entity through space-time. This motion can be perceived as a dimensional form, if we consider the entire trajectory of the entity (a world-line in the jargon of metric spaces), like the lines of a car in the night, taken at slow motion. It can also be considered an action born of the need of the entity for energy, information or more complex needs, born of the combination of the previous 2; and then we shall call it an arrow of time, or tendency of future caused by the action of the being. Or it can be seen as a repetitive cycle, in as much as species switch between the different arrows and needs, so cyclically they return to certain trajectories and actions – feeding, informing, reproducing and evolving socially.

What we shall do in this work is use those 4 arrows to explain most phenomena, and then refine the analysis of the details of the Universe by differentiating some of those arrows in new dualities. For example, social evolution is different among energetic 'species' which form only waves with minimal communication and informative particles, which form networks. Finally in the Universe processes of social evolution are multiple and fractal so atoms become molecules that become cells that become organisms that become societies and so we have to consider those fractal scales of reality and define a fractal or transcendental arrow. This means we divide the social arrow in 3 different sub-arrows of time, for a total of 6.

On the other hand the arrow of reproduction can reproduce a simple action or 'minimal cellular' event that combines energy and information, exi, which is used in physics; or it can reproduce a far more complex social organism till recreating the entity, $e \Leftrightarrow I$, which is the arrow of reproduction proper and as a result of this, when multiple reproductions happen in time, it gives birth to a series of generations between the birth and extinction of a species, and so we can divide in 3 sub-arrows the arrow of reproduction, for a total of 8.

Finally, if we take the entire Universe or any of its worlds or ecosystems as a whole, it is obvious that we can consider this final teleological goal of creation that puts together all the parts of a set into a whole, the 9th and final arrow, which defines the absolute space-time, EXI, of reality and exhausts our analysis. So the total arrows and dimensions of the whole Universe are 9 - the maximal arrows of time we shall use here. But in this introductory course we use basically the 4 biological arrows that suffice for a detailed analysis of reality: the arrow of energy feeding, informative perception, reproduction and social evolution.

Consider what you do as a human being with your lifetime, and try to put your events and actions within those 4 categories: All family events relate to reproduction. All feeding events relate to energy. All social events relate to eusocial evolution. Your work events relate to the accumulation of money, the 'language of economical reproduction' that allows you to acquire the energetic and informative goods your reproductive body and informative head require. When you drive you are in an event that converts energy into motion. When you read this book or watch TV or surf the Internet, you are accumulating 'information', albeit expressed in different languages, made with different 'formal symbols'. At the end, you will not find any action that does not become explained by those 4 arrows. And since you are in fact, as a human being, the most complex 'form' of the Universe, it is easy to deduce that all other simpler species also perform events related to those 4 arrows.

For example, Physical matter & space can be described with those 4 arrows: We know that the simplest physical beings – particles - gauge information (so quantum theories are called gauge theories), absorb energy (called forces), reproduce into self-similar particles (so quarks and electrons produce self-similar quarks and electrons when you give them energy), and evolve socially into atoms, molecules and celestial bodies. So the two limits of simplicity (atoms) and complexity (human beings) we know, do follow the main Postulate of the Theory of Multiple Spaces-Times – that the Universe can be explained departing from the study of the properties of those 4 main arrows of time.

This self-evident truth however is hidden to science due to the religious traditions 'subconsciously' embedded in the *work* of the founding fathers of science, mostly pious physicists of the Jewish-Christian tradition, for whom there was a single, monotheist God. And so, when a Philosophy of science was devised by those 'founding fathers', it became natural to seek for a single arrow of time that matched the Christian beliefs of Kepler, Galileo, Descartes and Newton[4]. The choice of arrow was thereafter defined by the nature of Physics: since earlier physical studies were based in clock-measures of the motions in space of physical particles, expansive motion=energy= entropy has become the single arrow of modern physics. Yet entropy-only models ignore the arrow of information, of morphological change that defines life. In that regard, there are in science two theories of time, which this work fusions and upgrades into the more complex knowledge about mathematical and logic languages and complex natural phenomena (fractals, chaos theory, black holes, cyclical time, information sciences, etc.) proper of the XXI C:

- Theory of Evolution, the Biological Theory of time developed by Darwin, based in causality and logic time, according to which Time evolves the in/formation, the form of species. Since the causality of Time is based in the extinction of the less perfect morphologies of the Universe, whose form is destroyed and used as energy for the future reproduction of the best forms

- And Relativity Theory, in which the analogy of Time-clocks is taken to its ultimate consequences, defining times as a 'curved geometry of space', self-similar to that of clocks.

This work unifies both Time theories into a single scientific theory of time=change, by expanding the Einsteinian concept of a curved geometry of time events, which are 'cyclical in nature', as the time cycles of the clock, to biological sciences; and the concept of the survival of the fittest 'forms of time', proper of biology to physical particles, where species with maximal momentum survive.

Thus the ultimately surprise of the new paradigm is that you live in a vital universe because the 4 arrows just described are the 4 properties biologists call the drives of life. Indeed, the Universe is made of infinite number vital, fractal, topological space-times that live, exist through those arrows.

Yet a Multiple Spaces-Times Theory goes beyond Evolution Theory, which studies the Time Arrows of Information and Reproduction and Physics, which studies the Time Arrow of Energy, to include also the Time Arrow of Eusocial Evolution, especially relevant to understand Religion and History.

Recap. There is an homology between the concepts of a cycle of time, a dimension of space and an arrow of space-time or action, exi, which becomes the unit of exi=stence of all beings.

. In this introductory course we shall use basically the 4 biological arrows that suffice for a detailed analysis of reality: the arrow of energy feeding, informative perception, reproduction and social evolution.

19. Cyclical times, discontinuous spaces.

If we were to resume this model of multiple spaces and times in a couple of pages, we could say that it takes the alternative theory of cyclical, relational time, and multiple vital spaces, envisioned by Eastern Philosophers, the Greeks and Leibniz to its ultimate consequences, revealing the laws of time cycles, applying them to find the cyclical patterns all species of all sciences (since a science indeed must have predictive capacity by finding the cycles of the entities it studies and projecting them into the future), and establishing the rules of engagement between discontinuous 'fractal spaces' which co-exist in the same Universe but have different densities and forms of energy an information, and so need to establish communicative flows of those substances between them to maintain the harmony we observe in the whole reality.

Each of us is therefore a vital space that performs and lasts a series of time cycles. We are not in an abstract background of time and space (Cartesian-Newtonian paradigm) and the Universe has not only a single universal time-clock with a rhythm similar to the ones we humans use (the second of our mind thoughts, eye blinks and mechanisms). Neither there are only a space-time continuum where all can be perceived, but at least to space-times of universal reach, the light-space we inhabit and the gravitational pace we do not see, and within the light space we live on, there are infinite 'complementary atomic forms that web evolved particles of the mass membrane (quarks) and particles of the light membrane (electrons), in complex networks and organisms, which are themselves 'broken pieces of space that perform specific time cycles'. Man is one of the most evolved forms of the Universe, but it is not different to the rest of them: we are a knot of biological time cycles that occupy a vital space.

And this simple scheme of things, which has always even known to philosophers but put aside by scientists because of the difficulty of measuring reality from infinite points of view, and infinite time rhythms must be ordered to become scientific. Let us tackle this most important question from the perspective of space, time and its combinations.

A verbal, logic definition of the 4 arrows will be as follows:

Energy = Motion; In/form/ation=Form; Reproduction=Repetition; Social Evolution=Network.

The energy arrow is a change in the motion or translation of beings; the information arrow is a change in the morphology of beings; the arrow of reproduction is the repetition or iteration of an entity in other region of space-time and the arrow of social evolution is the process of evolution of individual, self-similar species (hence reproduced/repeated species), into complex herds, waves and organisms, thanks to their creation of social networks in which those entities share energy and information. So the entities of the Universe move, change form, repeat themselves and evolve into *social networks*.

A Universe made of lineal and cyclical motions is eternal, with ∞ parts in perpetual motion, which show 2 essential properties:

- Time arrows/events are cyclical and discontinuous. What happens in the past will happen in the future again, once the initial conditions are met. And so we can map out 'frequencies' of events that repeat themselves in a cyclical manner: Time cycles are patterns that happen 'from time to time', with a 'frequency', v, which is the essential mathematical parameter to study Time Arrows/cycles. This brings also the geometry and concept of clock-times, which are instruments that mimic those 2 properties. In the same manner we eat from time to time in certain places (our favorite restaurant or dinner room), clocks come from time to time to the same place of space. Yet clocks only show the how and when, the geometry and frequency of those cycles. They do *not* explain why events repeat (we eat because we

81

obey an arrow of time called 'energy'; we make love because we obey an arrow of time called 'reproduction' so we need to copulate and so on.)

The discontinuity of time events causes the discontinuity of vital spaces. Since a cyclical trajectory encloses a 'space within it' and outside it. So time events become cyclical trajectories that isolate 'vital spaces', creating space-times of curved nature, which physicists describe with Theory of Relativity.

- Time arrows/events are causal. Time events have an order, A->B, which means event A must happen to trigger event B. We have to be hungry (lack energy) to need more food...

- Time arrows/events are self-similar. Time events happen in self-similar fashion in many different species. All life species for example feed on energy, and so do all particles that absorb energetic forces. And this is the reason why we consider 'energy' one of the fundamental Time Arrows of the Universe that its entire species absorb. Those general 'arrows of change' become in reality specific types of change of species. So science can use those arrows to understand the general meaning of each specific A->B event of each discipline.

- Time arrows/events are multiple, since each species has different types of change, and even though we generalize them with the concept of Time Arrows, those changes do happen individually. There are infinite time-clocks in the Universe, each one with a specific form of change or 'trajectory/motion', and/or different duration/frequency. Thus, absolute time is the sum of all the causal, deterministic paths of events that create the reality of each entity of the Universe.

All those repetitive cycles and changes put together is what we call time, but in many civilizations is called 'Times'. This 'relational concept of times', clearly explained by Leibniz, differs though from the single clock-time concept of Newton[10], which might be easier to use for calculus (reason why it is accepted in the praxis of science) but will not make us advance much in the philosophical understanding of times-changes.

The existence of those 4 arrows/cyclical actions, which a certain entity of time/space performs, arises from a bio-logical analysis of the fundamental fact of existence: we are made of quanta of spatial energy and temporal information. This implies that the existence of all beings has a limit of duration in time or death and a limit of size in space or individual discontinuity. So all beings need to absorb fractal energy for its reproductive body and information for its mind to continue their existence in time/space. Further on, beings reproduce in order to surpass their limits in time, repeating their form in another region of fractal space/time, overcoming in this way the temporal limit of their existence, the cycle of death. Finally, beings associate to other similar beings to surpass their limits of fractal, spatial size, growing in macro-organisms that are simultaneously a sum of individual cells and actions put together into an Organic Whole that acts as a unique being.

Yet the being will never be eternal and at a certain point its time-space will collapse after performing a sum of those survival cycles. We call the sum of all the energetic, informative, reproductive and social cycles of existence of a being, its generational cycle.

So from electronic cycles that exchange energy and information, described with quantum numbers, to man who acts seeking those cycles, feeding, learning, loving and creating societies, the 'will to create dimensions of existence' is embedded in any cyclical Space Time field that tries to overcome its fractal limits, creating $3\pm st_i$ physiological dimensions. The 'existence' of those cycles is in itself a tautology. Since a being that doesn't perform those cycles becomes extinct and hence no longer 'exist'. Thus all what 'exists' accomplishes those $3\pm\sum i$ cycles/dimensions, including a light photon or a crystal, which repeats its form, associates in molecules and processes energy into information, creating an image within its internal, informative center or vibrates emitting energy. Since all exists to cycle and cycles to exist.

Recap. Time events are cyclical, happening 'from time to time', when an entity absorbs energy or information with its field/ body or particle/head. The Universe is a tapestry of fractal, vital spaces, imprinted by cycles of temporal information, which create infinite beings. Those beings reproduce, combining those arrows and then associate in larger social groups. They are the 4 arrows/cycles of time that the 3rd paradigm of metric spaces pegged together into a single space-time continuum.

20. The beats of the Universe: E⇔I

The simplest of those cycles of course is the clock: the needle comes always to the same point and closes a cycle. But the clock is a metric cycle of time, as the point is always in the same metric distance from the needle and the structure doesn't move. In a topological space, a time cycle is not concerned with distances and continuity of the cycle but the entity just has to return to a self-similar topological form, which has one of the 3 functions mentioned before – energetic, informative and reproductive topologies/functions.

So a key law of the 4th paradigm is an old law of science:

Each form of topological space performs a time function/arrow

This law becomes the key to understand those cycles. For example, when a lion comes to a topology of energy called water, and drinks, he is completing a cycle of energy feeding, by intersecting with the plane of water and absorbing it, and that is a time cycle even if he doesn't return to the same water well and doesn't return with the same metric distance, as a clock does in a metric space. What the lion accomplishes is a time cycle in a topological space, in which what matters is no longer the form but the *function of energy feeding.*

Thus we establish a new concept of a space-time cycle, one in which function supersedes metric form, and consider that there is a common 'arrow of time' called energy feeding, which all Complementary entities of energy and information achieve. Suddenly we realize that all entities feed on energy because they need to move and they all have a reproductive body/field of energy which must be replenished or else it will be spent. And this overall function of all systems, feeding on energy becomes the first arrow/cycle of time of the Universe.

Yet all those complementary forms do have also an information network that requires perceiving the Universal tapestry. So in fact, even before the entity feeds it will have to perceive, gauge and calculate information. And this becomes the second arrow of time, of all complementary beings. And again we realize that even physical particles gauge information, and certainly move with it field of forces. So do biological beings that process information with DNA nuclei and heads, and process energy with their cellular bodies. And so now we have 2 arrows of time – a concept that means tendencies of the future - energy and information, and two types of cycles to accomplish them. But unlike the classic concept of science of a single continuous time arrow (the arrow of energy or entropy), because we have now two arrows, time must be discontinuous, as the entity must go from energy cycles to information cycles and this establish the universal, fundamental beat of reality: all entities of the Universe go through a process of motion (energy process) and stillness (perceptive informative process). Since we realize that to perceive and map information you have to 'measure' in stillness. Here it comes the meaning of all those metric spaces of classic science. They describe the arrow of information and measure; while all those analysis of motions and speed describe the arrow of energy.

Thus we define a universal beat of existence, E⇔I, motion, stop, motion, stop, energy, information. Like a movie in which the frame moves and stops and illuminates in stillness creating a form of information, all entities of the universe have stop and go rhythms, day motions and night sleep, and when they move they process energy and when they are still they process information.

In this manner the generator equation of spatial energies and temporal cycles of information appears, E⇔I.

Thus a new field of science of enormous richness is the algebra of time, which studies the previous equation and all its possible beats from where all events of reality will emerge. This equation is to systems sciences and complexity what the Unification equation of Physics is to physicists, a sub discipline of the wider view we bring here - since we shall be able to deduce all the equations and species of all the sciences of mankind from that simple first beat.

For example, all physical systems are complementary, made of a field of energy (e) and a particle of information (i), which constantly switch between one and the other state: e⇔i.

All cells have a nucleus of informative DNA and a body that absorbs energy and reproduces, and so do all biological organisms with an informative head and a body that absorbs energy and reproduces, i⇔e. And the difference between both is that in physical systems energy=motion dominates and in biological systems, information dominates.

We could say and 'God say move and stop, feed and perceive with your body/field and particle head'; and in this manner he set in motion the game of infinite forms of existence.

Let us then try to classify all those space-time cycles, breaking them 'apart' from the equalized time of the clock that metric spaces have used today to measure all of them.

Recap. All what exists is a complementary system of energy and information that switches between both states: e⇔i.

21. Dimensions/arrows/cycles of Existence in space-time.

The first astounding conclusion of a careful study of all structures, forms, cycles and events of Nature is that we can reduce all of them to a dualist->4-dimensional scale, according to which all what exists is created by energy and information, the simplex arrows of space-time. Those 2 arrows in a first layer of complexity combine in social events, and reproductive events, creating the 4 'drives' of all biological beings, since biologists define life as entities that feed on energy, gauge information, reproduce and evolve socially.

Thus from 2 simplex, energy and information arrows come 2 complex, reproductive and social arrows, which give us the 4 'dimensions' of space-time.

We represent them with a generator equation, $\Sigma E \Leftrightarrow \Sigma^2 I$, where we add 2 symbols of social evolution, Σ a sum of the cells of the 'body' and Σ^2 a multiplicative network. Since informative systems form networks that relate all cells among them, multiplying its social power, and ⇔ the symbol of feedback cycles between energy and information systems that reproduce the entities of the Universe. The equation represents also any complementary system of energy and information in the Universe, and depending on the operandi we use to substitute ⇔ we can represent one or other system or event.

This simple scheme of 4 arrows, dimensions, cycles or whys of the Universe suffices to understand the nature of reality. And so in most models of multiple time-spaces we use them to explain the events and forms of the Universe.

Recap. There are 4 whys in the actions of all beings, whose cycles create the space-time trajectories that we perceive as space-time: energy cycles, informative cycles and their combination, reproductive and social cycles.

22. Energy->Form->Reproduction->Social Evolution

The main arrows of time are energy, information, reproduction and social evolution. We cannot unify further those 'classes of events' without losing 'key information' about them, as Physics or monist religions do, with their obsession with a single 'God' and a single 'Clock-time' to measure all types of times-change. Instead, we can study in detail those arrows, its properties, cyclical periods and the relationships between them. And the first logic question to make on this complex philosophy of science/time is: Can we use Logic Time – the existence of causality; of a successive order in the path of events that go from A to B to C, etc. - to classify those arrows, which include all other phenomena into themselves?

What was first the egg or the chicken is a key question of science. The answer is intuitive – the egg; the seed of information, the perceiver point of view, head or particle of information that gauges reality to decide its energetic motions with a purpose. The Universe, not only man, is intelligent, and so are all its parts. Each mathematical point of view, each knot of Time Arrows, from the simplest atom to the mind, decides first intelligently the 'time arrow' or purpose of its motion in its still, perceptive, informative state and then it moves its mass or body/limbs, where it is the energy to feed itself, or in more evolved beings, where to reproduce or join a social network.

Thus, there is a causal chain that produces the more complex Time Arrows, reproduction and social evolution, from the simplex ones, energy and information:

Information (subject) absorbs Energy to move in order to Reproduce or Evolve Socially->IERS

This logic order is the most important causal chain of reality, responsible, for the cycle of life and death, the laws of evolution and the eternity of the Universe.

In all universal actions, we must start with a perceiver, I, the subject of all actions, of all sentences. This perceiver - the unmoved one of Aristotle, the Atman, the Soul, the Monad, the Non-Euclidean Point of View of Time Theory - is a knot of Time Arrows, or motions with a will, of wishes and memories. And the first thing it does, automatically, is to perceive flows of energy and create a mapping of the Universe, in its particle-head, then it will move with energy, space, extension, the ultimate substance of the vacuum in search for one of its arrows; it will untangle a knot of memories looking for a replication, I->E->R. The subject will spend its energy/motion, tracing geodesics, world lines, curved trajectories, all of them mappings in a complex plane of a mere circle. In this manner the subjects become 'forms-in-action', informations. And once we have an entity of energy and form, with momentum, with energy and time, with a cyclical motion, a purpose, once Schopenhauer's Idea becomes Will [14] in search of the immortal repetition of its forms and motions we exist in an event of reproduction that will produce waves of self-similar forms…

It is reproduction, the purpose of existence? No, reproduction is only the beginning of the final arrow of the fractal Universe, which is to evolve individuals into societies that last longer in time, have bigger vital spaces, and hence survive better.

The 4 time arrows imply that the substrata of reality is Time-space, not space-time, since the 3 dimensions of time, past=lineal energy, present=wave /repetition, future=form /cycle are motions; and the existence of causal relationships and social laws that create complex systems of motions, introduce an entire new category of knowledge in the human mind: that of the laws of time-space, the morphological relationships between its transcendental forms and the relationships between the different fractal scales of reality.

Recap: The 2 simplest Time arrows of are the creation of energy and information. The *simplest events* of time seem to be energetic feeding. To create 'form' we need first 'formless' energy. Yet p*rior to all motions and transformations of energy there* must be a perceiver *or point of view, the fundamental logic particle of the Universe, either* a particle of informatio*n - a biological head or any other 'linguistic processor' of information, which defines a path of acti*on – a motion in search of energy to reproduce itself. It is the game of Existen*ce.* On the other hand, to reproduce an entity of energy and information we need both arrows, exi, which combine to give 'birth' to the infinite species of the Universe. Once reproduction happens, the Universe shows a 4th organic arrow of time that evolves socially self-similar, reproduced individuals into waves, herds and organisms. The 4 main arrows of time together define a game of creation of energy and information entities, which reproduce and evolve socially, emerging once and again into more complex forms: lineal and cyclical strings evolve together into particles and forces, which evolve together into atoms made of quarks and electrons, which evolve together into molecules, cells, organisms, societies, planetary systems, galaxies and Universes. For that reason we can establish a causal arrow of Time between the 4 arrows, E->I->R->S.

23. The 4 cyclical, dimensional actions of existence.

Any organism or species between conception in its inferior plane of seminal form, (st-1) and its death back to the same plane of cellular existence, exists through those cycles in a social st-plane in which he performs them. Which in Cyclical Time terms means that it performs a fractal, a social sum $(+\Sigma e \times \Pi i)$ of energetic (E), reproductive ($\Leftrightarrow$) and informative (I) cycles. The study of reality is the analysis of those space-time cycles that we can formalize, departing from the space-time field equation, $\Sigma Se \Leftrightarrow \Sigma^2 Ti$, defining any cyclical action as a partial event of that cyclical equation:

- E<=I: Energy, feeding cycles/dimensions transform information into energy or negative arrow of entropy and direction towards the relative fractal past. They dominate the young 1st age of an Exi field.

-E=>I: Informative cycles/dimensions form the positive arrow of entropy and future that transforms energy into information. All forms process spatial energy into temporal information, from atoms to black holes, which either acquire form or perceive reality. So all beings become 'informed'. They dominate in the old, 3rd age of the Se x Ti field.

Informative and energetic events/forms are inverse arrows that co-exist together, shaping a dual cycle, which sometimes occurs simultaneously, as in a Darwinian event of hunting, when a predator energizes itself, killing the information of the victim. And sometimes 'stretch in the 2 directions of time', from past to future in the process of life and from future to past in the process of death. So we write: *Victim's Death: energy bang = Life: Informative feeding by a Top Predator*

Thus, the existence of many beings can be explained as a dual fluctuation of a space-time field that first evolves, increasing its information, as the entity 'lives' and then devolves, decreasing its information as the entity 'dies', closing in this manner a 'space-time cycle', easily described with the space-time equation. Thus, we 'travel to the past' when we lose information or devolve our form, as when we die, or when we go to a 'relative past culture' that has less information. And vice versa, we travel towards the relative future when we evolve and increase our information.

- E$\Leftrightarrow$I: e$\Leftrightarrow$i: The reproductive cycle/dimension, a mixed, internal and external, dual direction that repeats the energy and information of a being in another place of space/time and requires, therefore, the co-existence of the other 2 cycles, and dominate the mature age of each field. The reproductive cycle sometimes is a social, dual cycle (sexual reproduction) and sometimes is an internal, individual cycle. Yet in all cases is the fundamental cycle of existence, which ensures a certain degree of 'fractal immortality'. Since those cycles exist to preserve the being through the repetition= reproduction of its logic form in other place of the Universe. Thus species inform themselves to accumulate energy in order to reproduce their form and cheat its fractal nature, surviving in time by creating reproductive events that happen in all kind of entities. Reproduction becomes the why and will of the fractal Universe: all existential beings perform informative and energetic cycles transforming those 2 substances into replicas of their own morphologies.

- Σ or Σ^2: The being performs many of its existential cycles as external social cycles, integrated in a herd, Σ of parallel body cells or an informative network, Σ^2 displayed in two scales (cells and axons), that inform, energize and reproduce together, simultaneously as a group. Since social behavior allows the being to increase its form in space as part of the whole. And so it is also essential for spatial survival, since a bigger form displays a stronger IXE action and survives better than a small one. A fact that we formalize with the fractal symbols, Σ & Σ^2 the algebraic parameter that defines all actions as a sum of smaller fractal action-reaction cycles: $\Sigma Ei \Leftrightarrow \Sigma^2 Ti$. We thus talk of a series of cycles of social evolution, which create hierarchical scales of growing spatial size and organic complexity (max.ExI) that gather atoms into molecules, molecules into cells, cells into organisms and organisms into societies or disorders waves and organisms into fractal cells and particles. It is proper of all ages, studied by each science with different jargons. So in Physics we talk of the wave-particle duality that gathers quanta in particles and explodes them in waves and in Biology of networks that order cells into physiological systems, etc.

- $E=>i+I<=e$: The sum of all those cycles gives origin to the life cycle, the cycle of the total existence of the being, from its birth to its death; which according to the ternary principle is divided in 3 ages defined by the sequential dominance of the previous cycles, as the existential species or organism will perform more energetic cycles in its first youth horizon, (max.$\sum Ei$), more reproductive cycles in its maturity, (max $\sum Re$) and it will accumulate more information in its old age, prior to extinction (max. $\sum^2 i$).

In all fractal, cyclical actions can be subdivided in 3 ages:

- Max. E: When we 'look' to absorb visual information or we 'eat' to absorb energy, in the first age, in the beginning of the cycle, we maintain the eyes or the mouth opened, with max. energy=extension.

E=I: Then we close and open the mouth or the eyes, blinking and chewing in an intermediate alternating movement that reproduces the cycle many times.

- Max. I: Finally, we swallow, closing the mouth to 'reform' the food in the stomach into our proper 'in-form-ation' or we think with the brain, and extract the Max. informative perception of the image, in the third, implosive, informative age of the cycle. So:

Existential cycle=$\sum$ energy cycles + reproductive cycles+ informative cycles

Thus, the existential cycle, between birth and extinction, is the fractal sum, $\sum$, of its energetic, informative and reproductive cycles, often performed in social groups (hunting herds, waves, couples), in which energy and information mutate into each other ($\sum E \Leftrightarrow \sum^2 I$).

Recap: The Universe and any of its systems composed of multiple parts form a deterministic, interconnected whole of infinite harmonies between all its cellular cycles, chained by their symbiotic nature, such as the fastest informative cycles gather in ternary and decametric scales, connected to the frequency of the energetic cycles, which are connected to the frequency of reproductive cycles, which are connected to the frequency of the social cycles: I->E->R->S. Since an informative cycle is required for the entity to orientate itself towards a field of energy, and both energy and information together in great quantities/cycles are required to create a reproductive cycle; and finally multiple, self-similar reproductive entities are needed to create a super-organism. The same hierarchical chain happens if we observe those entities in space, such as the pixel of information is minimal, making informative systems smaller than bodies and bites of energy. And each of them is smaller than the total organism composed of both, which is smaller than the eusocial system of which the organism is a single cell.

24. The life/death cycle and the generational cycle.

The formalism of the 2 simplest arrows of time, E->I and I->E, show that the properties of Information and Energy, of Form and Motion, are reversed:

- Energy moves; it is extended, lineal, big and simple, without form. Since the line is the shortest distance between 2 points. Information seems to us still - a pattern of form, which chains different cycles into shapes. It is small, cyclical and discontinuous; since the circle is the figure that stores more form in a lesser perimeter. Yet in dynamic terms, we must think of expansive big-bangs, extensions and implosive in/formation; 2 intuitive substances that Descartes considered the primary of reality (res extensa=space and vortices of mass /information). Once this geometrical duality is clear, we can deduce how the causal order of those Time Arrows creates in all space-time beings the life/death cycle:

Energy warps into Information (Life Ages) + Information explodes back into Energy = Death.

- We are born as energetic, moving children that warp into a 3rd, old age of information.

- Matter is born as a plasma or gas, extended in space, which evolves into in/form/ative solids (atoms, black holes), through an intermediate exi, liquid phase.

So all creations of cyclical particles (masses and charges), all states of matter and all processes of aging can be explained with the two primary arrows: e->i, i->e and an intermediate state, which combines both, the reproductive, mature age of life, the liquid, most complex form of matter.

- Then after a 3rd age of information matter or life explodes, dies, extending in space, reversing the arrows of time: I->E.

Thus the energy and informative arrow together define an immortal Universe made of ∞ bites of energy and bytes of information in cyclical trans-form-ation: a young surface of Energetic Space reproduces in/form/ation till it becomes old and wrinkled. Then it erases its form back into energy in the inverse process of death, completing an existential life/death cycle, which we formalize as the 3 phases of the generator cycle of the Universe: *Max.E (youth), exi, Max.I (old age).*

All beings that exist in time go through 3±st ages that correspond to the 3±st arrows of time.

-St-1->St-1->St: We are born as a seed of information that surfaces into the new plane of existence after a process of self-reproduction and organization.

- *Max. E_{st}:* Youth: The new born grows its energy-limbs.

- $E_{st}=I_{st}$. Maturity: After evolving its reproductive systems, the life being will reproduce, combining energy and information.

- Max. I_{st}. 3rd age: The life being acquires more information, warping and exhausting our energy as we implode our form.

- St->St-1: Death: Finally without energy, the warped in/form /ative st-organism dissolves back into st-1 cells; and further on into the st-2 scale of molecular form from where they evolved.

Yet the life-death cycle is not only a travel through 3 ages of increasing information till death reverses locally time and explodes information into energy, but also a travel between 3 planes of space-time existence, as a seed of st-1 information reproduces and evolves socially, surfacing into a 'higher' st-plane of existence and then reverses through death the social process, dissolving back into its cellular stage. In fact, the process normally takes, when studied in more detail two jumps on planes of space-time organization and what we see as 'life' with its 3 ages is only the 'surface' of the iceberg of the complete cycle. In the graph, we observe those processes of creation and extinction of super-organisms departing from forms of a lower scale of existence in self-similar systems of different scales, from top to bottom:

-St-2 atoms become St-1 DNA molecules that transcend into st-cells, which go through the 3 ages/cycles of growth, reproduction and information till its final apoptosis, when they die returning to its st-1 molecular and st-2 atomic scale.

-St-2 DNA becomes seminal St-1 cells that transcend into st-human beings, which go through the cycles of growth, reproduction and information till their final death, when they descend back into St-1 cells and St-2 amino acids used by insects to evolve.

- St-2 seminal cells become st-1 prophets, which create a code of information, the Book of Revelation which is memorized as a meme by all believers, creating a civilization, which goes through the ages of growth, reproduction and informative, baroque art/thought (since art is the mind of civilizations) and finally become extinguished in an age of war.

The life death cycle is *clearly the consequence of 2 networks messing, in whicch the informative network exhaust the energy of the body herd of cells, warping them and committing selfish suicide. Without the dominance of the informative network the 3 ages of life would not happen.* Death is then the just return of the selfish ego-driven, information remembering, warped network a zero to the ∞ universe:

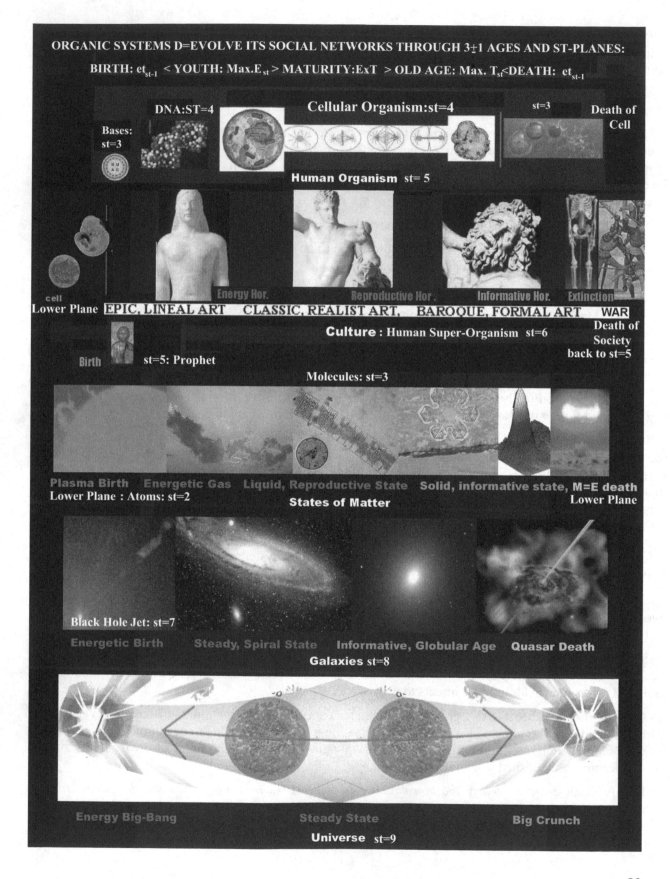

ORGANIC SYSTEMS D=EVOLVE ITS SOCIAL NETWORKS THROUGH 3±1 AGES AND ST-PLANES:

BIRTH: et_{st-1} < YOUTH: $Max.E_{st}$ > MATURITY:ExT > OLD AGE: Max. T_{st} < DEATH: et_{st-1}

DNA:ST=4   Cellular Organism:st=4   st=3   Death of Cell

Bases: st=3

Human Organism st= 5

cell   Energy Hor.   Reproductive Hor.   Informative Hor.   Extinction

Lower Plane   EPIC, LINEAL ART   CLASSIC, REALIST ART,   BAROQUE, FORMAL ART   WAR

Culture : Human Super-Organism st=6   Death of Society back to st=5

Birth   st=5: Prophet

Molecules: st=3

Plasma Birth   Energetic Gas   Liquid, Reproductive State   Solid, informative state, M=E death

Lower Plane : Atoms: st=2   States of Matter   Lower Plane

Black Hole Jet: st=7

Energetic Birth   Steady, Spiral State   Informative, Globular Age   Quasar Death

Galaxies st=8

Energy Big-Bang   Steady State   Big Crunch

Universe st=9

89

In the graph, a human being goes from an energetic youth into an informative 3rd age to become erased back into energy. In the Universe those 3 phases of existence in space-time are the 3 solutions to Einstein's equation, which physicists erroneously think to be parallel Universes in space, when they are phases of the same universe in time: the Universe evolves from an energetic big bang into the steady, mature universe of Einstein that will end into an informative, imploding, curved, vortex-like big-crunch Universe - Gödel's solution that will explode again into a big-bang, which is simultaneously the death of a previous Universe and the birth of a new one.

The same process is followed in the lower part of the graph in physical systems:

-St-2 plasma particles become st-1 atoms, which evolve into social states of matter, st-energetic gas, st-reproductive liquids and st-informative solids that further evolve as st+1 bosons into a new social plane of existence or dissolve back into st-1 atomic, ionic plasma and st-2 radiation ($E=Mc^2$).

- The same process in a cosmic membrane creates from a st-2 seminal nebula of atomic gas, st-1 stars, which further evolve through 3 ages from gas, to liquid to solid black holes that organize a herd of stars into st-galaxies, which die, exploding into quasars and reverting into intergalactic dust (st-2).

-So finally, we can consider also the Universe and its big-bang process a self-similar evolution from an initial st-2 quark gluon soup that produces all kind of st-1 systems of dark matter, st-galaxies and st+1 networks of galaxies, giving birth to the Universe that will go through the 3 ages/solutions of Einstein's equations, from a young, expansive big-bang into a steady-state and a final solid big-crunch, giving birth to a hyper-black hole that will explode back into st-2 quarks, restarting the process.

The reproductive age that renews the being into an offspring of self-similar beings defines the generational cycle, which is also a fractal dimension; that is, a dimension with a time-space limit, as all species have a finite number of self-reproductive generations, after which the reproductive system becomes 'tired' and fails. So life has telomere clocks in its genes; and light has a mean life of 10^{10} years, after which its self-repetitive wave becomes tired and red-shifts back to dark energy.

Recap. The generational cycle is a finite cycle with a limit set by internal clocks: Energy and information arrows are reversed, since to create information we have to destroy energy and to create energy we must destroy information. Thus, the causality of the Universe is 'dualist': I->E (energy creation) + E->I (informative creation) - a fact, which explains the causal cycle of life (E->I) and death (I->E). Thus the 3 dimensions or ages of time are: past, the energetic, young age; present, the age of repetition= reproduction in which the being doesn't seem to change and future or age of information.

25. An overview of the complex 9-dimensional world.

We conclude that most applications of 'real mathematics' are studies of the networks and species created by the Generator Equation:

'All what exists are organic, fractal systems, made of self-reproductive, complementary, social, topological networks of energy (bodies and fields) and information (heads and particles)': $\sum Se<X>\prod Ti$

The previous principle from where all events and forms of space-time arouse is a creative dynamic principle, reason why geometry becomes a ceteris paribus analysis without motion of a topology; planes and forms become networks and distances become flows between points, which can be described with the generator equation, $E \Leftrightarrow I$, with *different degrees of complexity.*

In this work we reject simple monist theorists, which introduce errors. Instead we accept 3 degrees of complex analysis: dualist studies of energy/information systems useful in physics, the simplest systems; 4-dimensional analysis with the synthetic arrows of energy, information, reproduction and social evolution; and the most exhaustive model of multiple time arrows and fractal spaces that divides those arrows in different modes of social evolution and reproduction. We study most systems as dual space-time systems, guided by its 4 dimensional arrows information, energy, reproduction and social evolution, which have 3 topological regions that dominate each of its 3 ages of time, as the species develops from its st-1 initial seed into an organism that performs functions in the st+1 system. And that will be enough to understand the general nature of any species, event or form of the Universe; but from time to time we will provide a more comprehensive vision to exhaust the model with the 9 arrows of space-time. Since if we want to map out all the types of events in space-time with the maximal number of arrows of time, we have to use 3x3 main arrows/cycles/dimensions of time-change in reality, which will be enough to explain *all* the events of *all* the species of the Universe, including the Universe itself. Thus in the exhaustive model we differentiate 2 reproductive arrows:

- exi: The physical arrow of reproduction called an 'action', that combines energy and form, *exi,* becoming the unit of physical events.

- $e \Leftrightarrow i$, The biological arrow that combines informative heads and reproductive bodies, which constantly interact through feed-back cycles that exchange energy and information, creating and reproducing organisms, the essential particle of biological space-times.

- We also differentiate two types of social evolution:

- Σe: Herd or wave evolution in which each entity only relates to those that surround it.

-$\Sigma^2 i$: Network evolution of informative bytes in which each entity relates to all other entities of the group, hence creating i^2 'axons', - the main elements of the network.

- Finally we add the fractal arrows of multiple space-times, ignored by the 3rd paradigm that uses a single space-time:

- $\int \partial$; The Generational, temporal life cycle integrates a self-reproductive cell, which multiplies its numbers, into a super-organism and disintegrates it in the process of death. It can be extended in time as one self-similar form repeats itself generation after generation between birth and extinction.

- $\Sigma, \prod$: The ecosystemic, spatial arrow that integrates all those systems in space and time, in bigger Σ-herds and $\prod$-networks $\Sigma (\Sigma E \Leftrightarrow -\Sigma^2 i) \Leftrightarrow \prod (\Sigma e \Leftrightarrow -\Sigma^2 I)$

- st=n, X^n: The fractal, transcendental arrow of multiple spaces that creates super-organisms made of smaller organisms, evolved across many fractal planes of existence, from particles, to atoms, to molecules, cells, organisms and planets, stars & galaxies till reaching for st=10, the Universe and beyond the absolute reality, whose number of minimal cells will be X^∞:

$$(\Sigma(\Sigma E \Leftrightarrow \Sigma^2 i) \Leftrightarrow \prod(\Sigma e \Leftrightarrow \Sigma^2 I))^{st=10,\infty}$$

It is the Fractal Generator, feedback equation of the Universe and all its exi=stences, whose partial equations and/or synoptic analysis connects the 4th paradigm with the 3rd paradigm of metric measure. Since we can in fact treat both members, integrated across all species of space and all species of Time for St=1, as the Absolute, Single space-time continuum of the metric method.

The different operations born out of that equation can be very complex but in essence all of them are reduced to growth of herds, diminution and balanced exchange, <=>, to integration and disintegration of networks and pixels, reproductive multiplication or division, emergence as a whole or fractioning as a part; and so from mathematics we move into physics, mathematics with motion and biology dimensional topologies with increasing form.

If we want to map out all the types of events in space-time, or modes of change of reality into a maximal number of arrows of time, we shall use 3x3 main arrows/cycles/dimensions of time-change in reality, which explain *all* the events of *all* the species of the Universe, including the Universe itself.

We will not, of course, be able to enunciate in detail all the events of reality, since only the 'Universe stores all the information about itself'[5] (Haldane), but we invite the reader who understands in depth this work, to find a single event that cannot fit on those 3x3 arrows. Since according to Popper, the proper way to prove the truth of any scientific theory is *not* to enunciate exhaustively all the events it explains – an impossible task, as events will happen in the past, in the present and in the future, and in locations we cannot observe – but to search for 'exceptions' that invalidate the theory. Mr. Popper puts the example of the black swan, which invalidated the theory that all swans were white after many centuries of thinking this was a regularity or scientific law about swans. A single black swan was enough to invalidate the 'white swan' theory. Yet for 20 years I have found *not* a single exception to the Theory of Multiple Spaces-Times. Perhaps there are more 'dimensions/arrows' of space-time than those 3x3 arrows we have developed to explain an enormous number of *Events* and species of the Universe. So I challenge the scientific community to find them and/or any event, which cannot be explained with the energetic, informative, reproductive and social arrows used in this work to resolve the 'why' of existence in time.

The previous equation though widens the number of functions and themes of complex mathematics. An interesting branch explored by Cantor deals with the concept of infinity, which now is always local, restricted to the size of the maximal ecosystemic network, which the observer studies. Infinite infinities are impossible in a Universe in which space-time membranes made of multiple networks of knots of time cycles called non-Euclidean points, are limited by its discontinuities. So infinities tend to be decametric scales of magnitude.

Recap. All the life and death cycles of existence create a virtual fluctuation between past and future that makes the universe an eternal present, Absolute time is the equalization of all those cycles and arrows of all entities that change, with a single time-cycle, that of the clock – an abstraction we must ignore if we want to go deeper into the 'why' of those arrows.

9 arrows of timespace exhaust all Universal events. They are the arrows of energy, information and its complex combinations, the different arrows of social evolution, reproduction and organization in complex, fractal planes of existence, from the simplest atom to the Universe itself.

26. The immortal Universe, sum of all cycles.

So the fundamental arrow/cycle that summons them all is the cycle of life and death, which makes us 'transcend' from a seminal seed reproduced into self-similar cells into individuals. While we die, when cells separate from each other and return to the inferior 'plane of existence' from where they departed. Because we are all dust of space-time that creates and dissolves ∞ 'existences'.

Big-bangs that create motion/energy (lineal forces) are self-similar to biologic death processes that dissolve information, expanding and erasing a warped form that explodes into its cellular parts (atomic big-bang, biological death). And they can be explained with the second arrow, i->e, the arrow of creation of energy, showing the astounding homology in terms of Time Arrows of all entities of reality. Because the generator equation of the Universe, $\Sigma E \Leftrightarrow I$, has only 2 initial elements, energy and form, there can only be 3 'limits' to that function, Max. Energy or youth, e=i, an intermediate age of balance in which both parameters equalize, allowing its combination and reproduction in other form of space-time and Max. Information or 3rd age. Then the information of the organism explodes and its social network dissolves into death, as the arrows of time reverse their order: E>(life)... I <E (death)

Why e->i, the warping of energy into form, proper of the life arrow comes first, and the second process, i->e, the creation of energy from form, and all processes of death come second?

Because to die (i->E) an entity must live first (e->i) (biological proof); to create form we must start at least with a line... of energy (mathematical proof), which bends into cycles or breaks into fractal form. Ultimately to create its particles/forms, the Universe started from an extended formless vacuum (physical form). So the main causal order of the arrows of time is: e->i.

Further on to reproduce, species must copy their energy and information in other part of space-time, creating a replica of the original. So reproduction is an arrow of time that combines two simple arrows, energy and form, which therefore come first:

E x i = Reproduction.

Thus reproduction happens once a system has formed its energy and has a minimal content of both, energy and form – reason why systems reproduce a seed of pure information, (max.i), which lives a first youth of growing limbs of energy, a second, mature age of reproduction after adolescence, when the species repeats itself and a 3rd age of warping of form, or age of information; after which they die, completing the life-death cycle, the main causal cycle of time. So, we can consider the existence of 3 ages in life:

Max. Energetic Growth (youth)-> e x i; e=i (reproduction) -> Max i (3rd age)

Even if we can now mathematize for the first time, the 3 ages of life, thanks to the formalism introduced by the arrows of time, those 3 ages have been known always to mankind, reason why in the classic age of Religions those 3 ages/arrows were described by Taoism 'the combination of yin and yang reproduces 10.000 beings' in the East and Zurvanism (the philosophical version of Zoroastrism and the most extended religion in the West from 500 BC till Islam), where Zurvan[6] the 'God of Infinite Time' has 3 'avatars', the energetic youth, the age of pleasure and the age of knowledge. Yet of more interest is to relate those 3 arrows with the 3 classic dimensions of 'time', a relative past or energetic youth, a relative future or informative age and a 'repetitive' present, when we reproduce ourselves. Each of us goes through the same life-death cycle, as we wrinkle our young energy into form to explode back into death. Such is the geometric nature of existence in time-space.

Thus, we order the causal arrows of change in 3 'dimensions', past, present and future, which are the 3 logic dimensions of time, embedded in the logic of verbal thought, the biological language humans use to describe time events. Since when something doesn't change, we say that time doesn't seem to pass. But if we are just repeating the same event/form, we might think we are seeing the same reality that doesn't change. So we feel we are in an eternal present.

Finally we feel intuitively that the future as a place with more information, so when we travel into a country like Japan based in the industries of information, we think we are in the future. So we talk of an arrow of future information or life, an arrow of past energy or death, and a present arrow of repetition-reproduction. Yet since e->i and i->e are inverse in properties, and one is the meaning of life and the other is the meaning of death, and all lives end in deaths, past and future balance each other into an eternal present: 'The separation between past, present and future is an illusion' said Einstein[7].

What Einstein meant becomes now clear, with the understanding of the 3 ages of times, which for each 'knot of Time Arrows', for each entity of the Universe, represents its relative past, present and future. Your past is your youth; your future is your old age. And so if you are in the future of your exi=stential journey, you still will co-exist with the past of your sons, their youth. Further on, since there are species that have evolved further their organism, they are a relative future knot of times. A human is a relative future knot of Time Arrows that co-exists with a worm, our relative past. Thus, not only we exist in a universe of motions, of Time Arrows, but we co-exist with all kind of relative past and future forms.

Since 'times arrows' from future to past (energy) and past to future (form) are infinite, causal events that take place in infinite different beings.

The error of a single time is due to the massive use of time-clocks to measure time, by equalizing all the rhythms of the Universe with the single second/minute/hour standard rhythm of human time clocks, yet the tic-tac frequency of 'other time cycles' is different. Your heart has a slightly faster tic-tac of 70 beats; and a milipulsar of a millisecond and so on. All those systems have an E->I->E->I beat, the fundamental beat of the Universe, which you find in infinite systems of energy and form: legs extended in energy position, warped in form; mouths and lungs opening to absorb energy, imploding when they take in the energy and start the process of converting it into new forms, and so on.

All causal chains of events depart from a young age of energy that warps the form of the species till energy is exhausted, all is warped information and the entity explodes in a big-bang. For example, galaxies start as extended nebulae and evolve into informative masses - black holes of maximal information and minimal extension - human beings also have an energetic youth and end up warped into a 3rd age of information. So do all types of matter that evolve from energetic plasma into a balanced liquid into a solid form of minimal motion.

The immortality of the Universe is therefore the immortality of the properties of those Time Arrows – the geometrical shapes of energy and information; the causal chains of time events those shapes create; the constant reproduction of the basic platonic forms of reality; its association in herds, waves and organisms. Or in mystique terms: 'the forms that can be seen are not the immortal forms', *Lao-Tse.*

Absolute space-time is a simplification of those arrows - the sum of all complementary entities of spatial, extended energy and all time clocks of information in the Universe. Whereas absolute space is the sum of all the vital spaces occupied by those entities and absolute time is the sum of all the events of energy and information taken place in that Universe.

And we must consider that the overall absolute time of the Universe never moves to the past or to the future but the sum of all the cycles of information or life cycles and energy or death cycles, creates an eternal present space-time, sparkled with infinite existences (life death cycles) fluctuating between the past and the future and the past, living and dying…

Recap. For reason of calculus is easier to create theories of time that use a single clock-rhythm but a philosophy of science is more concerned with principles than detailed analysis. So we cannot forget, as physicists often do, the fact that time is cyclical, *not* lineal and hence discontinuous - since a circle always closes and breaks space into an inner and outer form. In other words, the Universe is a mass of discontinuous, vital spaces, whose cyclical motions chain to other motions and cycles, creating complex organisms and particles. All those motions and rhythms form complex space-time beings, in which a region of dominant cyclical motions - informative particle or head - associates with a region of lineal motions - force or body - that moves it, creating a Complementary energy/information being - the 'Fundamental, Logic particle' of the Temporal Universe.

94

VII. TOPOLOGICAL EVOLUTION.

CONCEPTION AS A BLACK HOLE OF MIN. SIZE = MAX. INFORMATION

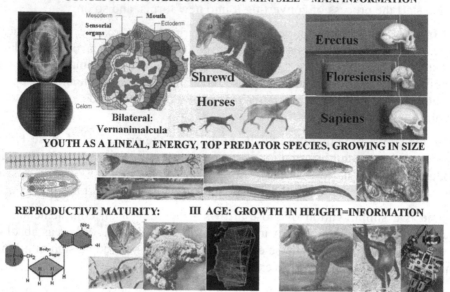

YOUTH AS A LINEAL, ENERGY, TOP PREDATOR SPECIES, GROWING IN SIZE

REPRODUCTIVE MATURITY: III AGE: GROWTH IN HEIGHT=INFORMATION

Evolution follows the arrow of growing information, increasing both, the dimension of height and the social organization of individuals into herds and complex organisms, through a common language of information.

27. The Evolution of Information.

In the graph, the understanding of the arrows time, its dimensions, causal order, and the way in which they evolve individuals into social organisms, can also resolve and advance the science of Biology and evolution. Since living beings are also knots of time cycles of energy feeding, information gauging, reproduction and social evolution, perceived as still organisms, whose purpose is to evolve individual 'cells' of the same species into more complex beings. Thus chemical life cycles chain each other into complex systems called cells that evolved into bigger scales, till creating the biological cycles /clocks of information of human beings, the most complex informative species of the Universe. Further on, the duality of cycles/particles/heads of information, complementary to energies/lineal motions/forces/bodies also applies to biology, where bodies and heads create living systems.

How life evolves is disputed between evolutionists, who believe in a selective chaos and creationists, who think there is a design. Time Theory shows that neither is wrong or right. There is not a personal God that creates us - an ego trip of religious fundamentalists - but there are limits to the chaotic forms of evolution, set by the fact that the morphologies of energy and information are fixed by the efficiency of energetic lines and informative cycles, by the need of complementary designs, by the existential arrow of social evolution and reproduction that favors survival; and ultimately by the mathematical topologies and dimensions of 'fractal information' and spatial energy, which define the outcome of evolutionary tendencies. Indeed, watch the drawings of different life beings made of informative nuclei/heads and reproductive body. Informative systems are always smaller than bodies. They are always on 'top' of bodies, perceiving more from the advantage point of view of the dimension of height. Further on, since information is bidimensional, informative systems are bidimensional (only a few animals have 3-dimensional vision). A book, a screen, a pulpit, an antenna, emits information from its top. A head, a camera, an electric field is on top of a body of energy.

This dominant arrow of information is *dimensional*; so species increase its height dimension, from planarians to human beings, a tall species of information. It is *topological*, so they evolve cyclical forms or heads on top of lineal forms or bodies. And it is *eusocial*, so they evolve from individual cells into social organisms through a common language of information (insects, human beings). Those dimensional, topological and eusocial laws complete Darwin's Theory of evolution:

Ultimately, each species made of self-similar individuals is also a cellular super-organism, which follows the same causal arrows of organisms, with an increase of information towards its '3rd age'. Those 3 ages of 'life' in cellular species become then the 3 'horizons' of evolution of species, which start as energetic, planar top predators (first shark fishes, first worms), with minimal size. Soon the species goes through a young, energetic age of growth in spatial size. Then it follows a II horizon of speciation and massive reproduction of the species (radiation age); and finally the species goes through an informative age of evolution in height and brainpower. Then, either a new, more evolved species appears, causing the death of the previous species, which becomes extinct or dwindles in numbers; or the social evolution of the individuals of the species reaches such a degree of integration, through a common language of information that a macro-organism emerges. So ants successfully create superorganisms called anthills becoming the most successful animal on Earth, whose total life mass is equivalent to the life mass of humans, which also organize themselves into super-organisms of History, our Gods and civilizations. The success of those 2 species shows that the ultimate survival strategy in the Universe is that of social evolution. Since the total mass of 'ants' and 'humans' are the biggest 'living masses' of the planet, as the most complex social superorganisms of Earth.

Recap: Life species can be considered as superorganisms made of individuals=cells, which evolve through 3 horizons=ages, following all the geometrical and logical laws of space-time systems, as life systems are also complementary systems made of reproductive bodies and cyclical heads of information.

28. The fifth dimension of time: generational cycles.

Species become extinct or preys of trophic pyramids, if they don't evolve informatively in social super-organisms, since as species their generational existence is also fractal, finite: all species have a genetic clock, the telomeres, which make them degenerate after a set number of reproductions. It happens to cells that become cancerous or suffer apoptosis after their telomere clock stops; it happens to organisms that degenerate, especially in groups with little genetic diversifications; it happens in light that gets tired after 10^{10} years light of wave reproduction; as it is the best explanation of the cycle of light/dark energy; it happens to others particles whose mean life is often confused as in the case of light with its total sum of generational, particle/antiparticle, wave/particle cycles of existence; and it happens to cultures and civilizations, which have a decametric cycle, becoming destroyed every 800 years after 10 human generations. Yet those who evolve socially as 'organic cells' of a bigger organisms, seems to survive the generational cycle, since an organic cells needs less exi force and it is protected from the environment by the bigger superorganism it lives in.

Recap. The 5th dimension of time is the generational clock-cycle.

29. The existential force. Strategies of survival.

We conclude that evolution is a guided evolution of form: Species evolve towards higher information. And those who don't increase their capacity to understand the language of information of its ecosystem with better brains and better social systems become extinct in its relative universe. In the graphs we trace existential curves of species that survive better thanks to their informative arrow:

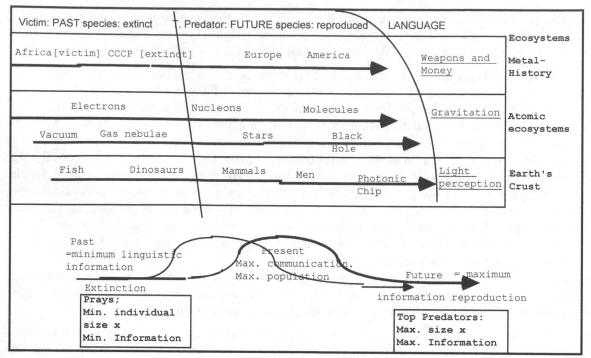

In the graph, the Universe favors Top Predators, which are species with the maximal 'existential force', exi (particle with more momentum in physics, M(i)xV(e); top predator in biology). Those species survive better as they perform better actions and so 'fill up' the present Universe. Yet that simple equation is maximized when i=e, favoring a balance between the field and the particle, the body and head ('mens sana in corpore sanum'). The outcome of that simple law is the massive reproduction into the future of the top predator entity, which becomes for that reason, the 'species that survives into the future'; while the species with less energy and information become extinct. However information is always relative to the bigger ecosystem in which the species exists, whose language of information defines that ecosystem. Thus in the galaxy, where the dominant language is gravitation, the black hole dominates the system; in this planet where the language is light, the species with better eyes (squids, then amphibian-reptiles, then mammals, especially felines with night vision and apes with 3-dimensional vision, and finally computers that will 'talk' with images dominate. Finally in the economic ecosystem made of humans and machines, the country with better memes of metal, better weapons (metal-energy), better money (metal-information) and better machines dominates and reproduces its technology worldwide.*

Energy and information motions become the generating, primordial cellular units of all species, including humans. Any system of reality is complementary. It has both -a lineal limb/force of energy and a processor of information or cyclical 'head/particle'- which co-exist and evolve together. So the Law of Complementarity relates both inverse arrows in a static field/organism, creating complementary systems, composed by systems that process energy or 'fields/bodies' (physics/biology) and systems that process information or 'particles/heads' (physics/biology). How can we measure the efficiency of the 2 parts of the system together? Through the existential force.

The existence of multiple knots of Time Arrows implies that in the Universe of multiple time cycles, the words past, present and future, are also relative, quantized to each individual, who lives his past/youth or age of energy, present/maturity or age of reproduction and future/old age of information before dying, dissolving i=ts exi=stence. Thus we need to analyze the relationships between entities which are in a relative past and interact with others that are in a relative future.

The simplest case is a family relationship between a grand-father, a man of the future with the wisdom of knowing in advance the cycles of time that his offspring will live, and its younger, less informed species: the young learns from the elder knowledge of previous cycles that will happen to the 'past being' into his future.

If we draw the example from different species, however the encounter will be Darwinian and only species which share the same information evolve together. Then the old, more evolved species with higher information will dissolve the simpler species into its relative past, as victim: there will be a Darwinian fight and the more evolved time knot from the future will kill and feed on the simpler form. And that is why we can establish that the relative future is guided by the arrow of information and social evolution, which creates better species. Since in a contact between past and future species, the future species survives, due to its better energy and information, its better existential force. It is thus possible to study from a mere temporal, evolutionary perspective the outcome of encounters between different 'knots of time'.

When we consider the 2 'simple' arrows of time that create the future, energy and information, as we do in Duality, all simplifies and explains itself, like it happened in earlier astronomy when Copernicus put the sun in the center. And one of the things it simplifies is the meaning of survival, which can be resumed in a simple equation: Max E x I = Survival. This function E x I, is the existential force, which explains that a species with the best energy body/force and informative particle/head will survive.

The existential force is equivalent to a particle's momentum, Max MV, which determines the survival form in material collisions. In biology the species with better body and brains survives, which made lions and humans the top predator species of this planet... till we invented weapons and computers, energetic and informative species, made of metal, which now we fusion into robots, potentially the new top predator species of this planet. In the galaxy the highest existential force or momentum is that of a black hole. And it is proved in the Galaxy that the collision of a black hole of any size with any type of star converts the star into a black hole.

An important characteristic of the Existential force is that it favors a relative balance between the energy/body/force and informative/particle/head of the system: the more self-similar are both in quantity of cells and the higher the number of its networks/connections, the more synchronic they act together, and the higher the Existential force of the system is. This fact is the origin of the classic, harmonic, 'beautiful', mature state of most forms of existence that are in balance between both parameters, in a 'mature', steady state age.

Recap: The Universe selects the species with maximal energy and information: Max. E x I ->E=I. It will be a biological top predator or a particle with maximal momentum.

VIII. REPRODUCTIVE ARROW: FRACTAL NETWORKS

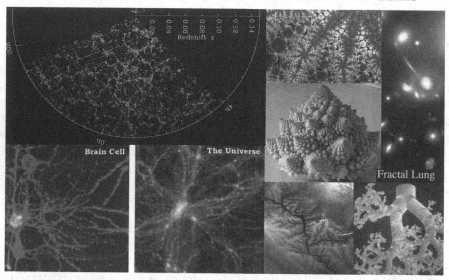

In the graph, several experimental proofs of the fractal, organic structure of all beings of reality: On the left, the Sloan fractal map of galaxies and its comparison with a neuronal cell; on the middle, an ice fractal, a plant and a river; on the right, a cluster of galaxies and a lung. Fractal and non-Euclidean mathematics have evolved our understanding of information in topological terms, defining a new Fundamental particle of the logic/geometrical Universe, the Non-Euclidean Point, which unlike classic Euclidean points, has inner parts – an informative center and a field of energy that moves it. Such points constantly communicate energy and information with self-similar points, through non-Euclidean lines (waves which share the energy and form of those points), shaping together 'Non-Euclidean space-time planes', the network-entities of the graphs, which all structures of nature constantly reproduce organizing a canvas of simpler, energetic 'cellular' motions. From the simplest particles, quarks and electrons that absorb energy and reproduce new particles to the most complex informative species, human beings, reality is made of bytes of information and bites of energy, evolved in complex, social networks through fractal scales, from atoms to molecules, to cells to organisms, planets and galaxies. We can mathematize those processes using non-Euclidean topologies and generator fractal equations that iterate and self-organize in networks those points of nature - an ice geometry, a DNA code, a cellular structure, a physical particle. Thus, when we fuse the mathematical and organic understanding of reality, the Universe appears as a fractal of energy and information, made of self-similar parts, which constantly reproduce their forms. And the quest of the fractal paradigm is to find an equation able to define all those systems - the 'Fractal Generator of the Universe': $E \Leftrightarrow Ti$

30. The why of science: the topological, fractal paradigm.

All what you see can be reduced to numbers, because numbers are 'social groups'. In the 3rd paradigm of metric measure those social groups were described merely in its external properties. System sciences analyses its internal, topological structures, as societies of Non-Euclidean Points (points with form and motion), whose exchanges of energy and information create complex networks of self-similar beings. Can we reduce all those topologies and organic functions of network spaces to a few postulates as the previous Euclidean Geometry did?

Yes, we can, but for that we must complete the work started by Riemann and Lobachevski, who discovered the 5th postulate of Non-Euclidean geometry and affirmed that through a point infinite parallels could cross. For a century and a half no mathematician seemed to have realized that if the 5th postulate has changed, we had to adapt the other 4 postulates to the new definition of a 'point with breadth', through which parallels cross. This point is the fractal point, which grows in size and scale, becoming a topologic network of smaller points, as we come closer to it. The Universe is indeed absolutely relative because 'each point is a world in itself' (Leibniz).

Yet geometric restrictions that allow only 3 types of topological forms in a 4-dimensional Universe -the informative, hyperbolic topology, the energetic, planer sphere, and the cyclical reproductive toroid - reduces all the internal parts of a fractal point to 3+1 functions or arrows of time, energy, information, reproduction and social evolution that conform the 'program' or why of the Universe explained in detail in the twin lecture of this work[1].

Those 3 functions become the functions of 3 specialized fractal networks of points that construct the systems of reality, from your organism, made of 3 physiological networks (the energetic, digestive network, the reproductive, blood network and the informative nervous network), to the galaxy, organized by a network of electromagnetic energy, gravitational information, and a series of herds of stars, non-Euclidean points that reproduce the atoms of the galaxy.

The principle of transformation of energy into form: 'energy never dies but transforms back and forth into information; Exi=K, E⇔Ti' represents the two parts of energy and form of those non-Euclidean point, the new mathematical unit of the 4th paradigm, where points become 'topological balls', and the Universe a geometry of multiple fractal scales of space-time, which grow in size and information when we come closer to them. Each of those points have a different rotational clock, rhythm or 'speed of time' and a different size, *which in each scale is used by those particles to gauge information and measure distances and energies and motions, as each of those points interact with other points of its relative space-time world.*

Thus the key to understand the structure of organisms are the 4 new postulates of non-Euclidean geometry based in the concept of a fractal point. Since we have also to account for the other key property of the spaces we see around us: each of them when observed in detail is made of multiple, smaller self-similar parts, we call generically 'cells'. So even the smallest point, when seen in detail, becomes a fractal point with multiple parts -a non-Euclidean point in which the parts assembly themselves into wholes by connecting themselves through networks, sharing flows of energy and information, and then become parts of even bigger wholes (so particles become parts of atoms parts of molecules parts of cells and planets parts of organisms and galaxies, parts of societies and Universe, in biological and physical spaces). Yet those networks leave dark spaces between their connections where other networks can occupy a place. So thanks to those dark spaces, we find that the system mixes two types of networks – hyperbolic informative networks and energetic, planar or spherical networks (as a sphere has in small surfaces the properties of a plane).

Reality is constructed with 3 levels of increasing complexity: In the first level any trajectory that returns to a point creates a closed form or time cycle, a clock, the unit or minimal event of Reality. Then many time cycles, chained to each other with certain synchronicities create a knot of time cycles that fills a vital space and becomes an entity of reality. We will describe those knots of time cycles as Non-Euclidean points, whose topological parts and laws become the mathematical foundation of *Multiple Spaces and Times*. Finally many knots of time cycles, exchanging flows of spatial energy and temporal information create complementary, social networks, which in its more complex forms are able to reproduce the system and create 'super-organisms', wholes made of parts that are wholes of an inferior scale of reality.

Since non-Euclidean Planes are networks with 'dark spaces' not perceived by its points that only see the flows of the network; those 'holes' allow the creation of complex systems with many networks, webbed within those dark spaces.

Thus the why of the Universe is organic, social: the Universe is made of complementary energetic and informative networks of 'Non-Euclidean points' (points with form and motion) that evolve and combine together, creating reproductive networks. And such complex systems create, new bigger, fractal networks units of a larger plane of space-time, evolving reality in scales of size that follow the same organic laws - from particles that become atoms that become molecules, cells, organisms, societies, planets, galaxies and Universes.

'Each point of the Universe is a world in itself 'said Leibniz. What he meant was understood in the XX century when we discovered the fractal structure of the Universe. Indeed, each point of the Universe, when we come closer to it, becomes a super-organism, made of cells with self-similar functions of time, repeated in all scales where those fractal points display the same logic behavior.

The Universe is a fractal of Temporal, informative cycles that are constantly generated and imprinted in reality. It is a fractal of 4 self-repetitive motions, energy, form, reproduction and social evolution, constantly happening around us in bigger and smaller scales, in bigger and smaller cycles, all type of beings feeding, reproducing, energizing and informing themselves. And because those actions are geometrical, they leave a certain trace which betrays the purpose of the action: a lineal path of feeding, a vibration back and forth of reproduction, a cyclical, implosive, informative process, and so on.

The Universe creates superorganisms by creating first fractal complex informative cells from flows of simpler energy, reproducing them and self-organizing them in 2 parallel, complementary networks, one of energy and one of invisible, faster, thinner information that controls the entire system. We often see only with the scientific method one single of those 2 networks, as gravitation and since most of our human existence magnetism appeared as invisible to us. And it is a general rule that both space-times, the one of energy and information, have different Universal Constants which are ratios, e/I and i/e and exi, which define the arrows of energy, information and vectorial reproduction of any organic dual space/time plane of the Universe. Where the space-network of energy and the informative-network create a holographic organism as those we see all around us in reality.

The best mathematical and logic description of all those qualities that structure reality -self-repetition in smaller and bigger scales, motions with geometrical form, feed-back action-reaction processes, absolute relativity of size - are provided by Non-Euclidean and fractal geometries, the geometry of information, reason why the 2nd part of this book will develop the new formalism of the fractal paradigm, poised to substitute the quantum paradigm and resolve all its questions in physics.

The result of course is that we see fractal structures created by the paths of knots of time, following its arrows of exi=stence in all systems of the Universe. This is what we see, the how of those Time Arrows: fractal paths of existence.

Thus, we shall now finish this first section dedicated to the 'metaphysics' of the fractal paradigm', the why of those paths, with a brief analysis of the fractal structures caused by self-reproductive knots of time, as they follow their paths of existence. The next chapters will be dedicated to a thorough analysis of the mathematical and logical laws that define the paths of Time Arrows, and how they build step by step those fractal networks we call 'Nature'.

Since all what exists studied in detail is a cellular network of self-similar beings sharing energy and information and reproducing and evolving in social networks.

Recap. The Universe is a fractal of information that reproduces, by imprinting and breaking vital spaces into complex forms. Each fractal knot of Time Arrows performs in his search for exi=stence a series of paths and reproductive and social acts whose final result is the creation of a fractal networks. All what we see are fractal networks caused by those knots of time.

31. The Universe is a fractal of knots of Time Arrows.

It is now clear what reality is: a series of knots of Time Arrows, which latter we will formalize as 'Non-Euclidean points'. Such points are in constant communication with self-similar points, reproducing their form and evolving socially into networks. The reproduction of form takes place in a fractal manner, by producing 'seeds' in a microcosmic, 'lower scale of existence', which latter grow. The networks are built with flows of energy and information that tie those points. And the result of those simple, yet repetitive processes of reproduction and social evolution is the creation of a world of 'herds', 'cellular organisms' and networks, extending through multiple scales of size in which the forms of energy and information and the structures and laws that build super-organisms remain invariant.

In physics, scientists talk of the invariance of motions discovered and generalized by Galileo and Einstein (Relativity). Yet we must ad in the fractal paradigm two new invariances, the invariance of scale – the same Time Arrows take place in all scales; and the invariance of topological form – the same shapes of energy and information emerge and repeat themselves in all scales (though, and this is essential for the reader to learn how to compare them, they are topological forms, where the 'distances' and sizes are relative, so what is conserved is the overall 'lineal, planar shape' of energetic functions, such as lineal limbs, lineal weapons, planar membranes; the cyclical, toroidal shape of reproductive cycles, which combine energy and information to recreate a form, such as those of your body organs or the cycles of an assembly factory or a mitochondria, and finally the hyperbolic, convex, warped, cyclical forms of informative organs, such as a brain, a camera or an eye or a black hole.)

Those 3 invariances create a new 'paradigm' of science, no longer built on the belief of a single arrow of time, a single space-time continuum and a single clock-speed to measure it. The result of course is a more complex reality, but that was to be expected. We, humans are just one of the many games of existence of that fractal Universe, and it should not be expected that we are, regardless of our anthropomorphic myths, different from the rest, and able to fully understand in the diminutive brain all what exists, unless we simplify it.

What this work tries to do is to return to a minimal degree of complexity to be able to understand the fractal knots of times, which imprint the energy of reality with its information, constantly reproducing their form and playing the game of existence, constantly achieving the immortality of form.

The Universe is made of 'time fractals' that participate both of the properties of mathematical fractals and living organisms. Yet organic fractals are more complex than mathematical fractals, because they have motions. And so the reader should not confuse both terms.

Fractals are mathematical or organic forms which iterate or reproduce their information in self-similar forms that extend through several scales of size and information which we shall call 'i'-scales.

They are the most accurate models to depict natural phenomena related to informative growth and reproduction, the essence of life organisms:

Arrows of Reproduction & Information: ->Δi

All fractals originated by a Generator Equation (mathematical jargon) or mother cell, (biologic jargon) iterate=reproduce a series of self-similar fractals, called a set in mathematics or a family in biology:

Mother/Generator Cell: ->Σst-1 fractal family/set

The paradox of fractals is that the smaller the fractal scale is the more information it accumulates and the more detail it has. How this is possible? Because information is form-in-action, form with motion, and so the logic paths that systems use to process information are completed faster in lesser space. Further on, in physical systems the cyclical speed and frequency of a vortex with form (a mass or an eddy) increases when the vortex is smaller (VoxRo=k). So the increase of 'cyclical speed' diminishes geometrically with size. This result can be generalized to any system of knots of time. So the metabolism of small animals is faster than that of big ones. The cycles and clocks of atoms are faster than those of our scale, faster than those of the cosmological scale. The same can be said of geometric fractals: a Mandelbrot fractal hold more information in its smaller scales, as we enlarge them.

However, there is a fundamental difference between the logic, linguistic fractals of the mind and the physical Universe: While static, ideal, mental mathematical forms can be of any shape, as movement and friction don't test their efficiency, in the Darwinian Universe only forms of Nature whose properties allow the being to process efficiently energy and information, survive. Thus, mathematics is an ideal language that can create any form, but in the Universe forms have to be efficient in their motion/function and parallel form. Thus, in Physics and Biology we add movement to the mental shapes of geometry, which enormously reduces the possible shapes. Movement acts then as a 'reality check', an Occam's razor that eliminates all complex forms, which won't be able to move properly under friction, stress, turbulences and other obstacles proper of the real world. A Mandelbrot fractal might be very beautiful. But it wouldn't be able to move very fast, and its attached microforms probably would break with an air stream. So there are no such fractals in the Universe. Instead, we find physical entities made of cycles and lines in movement, for reasons of biological efficiency. Since a lineal movement is the shortest, most efficient path between 2 points. On the other hand, information is cyclical, because a sphere accumulates the maximal quantity of form in minimal space. While a cyclical rotation is the movement with less friction and the only one, apart from the line of energy, that maintains the self-similarity of any form, even in trans-form-ations of size. A fact that explains, in terms of survival efficiency, the morphological invariance of energy and information shapes at scale.

Since Mathematical fractals lack, as pure forms of mental information, the energetic dimension of movement and growth, their fractal sets are attached to the Mother- cell's boundary, where the self-similar forms nest, as in the case of the well-known Mandelbrot set. On the other hand organic fractals not only iterate their information as mathematical and mental fractals do, but also make it grow. Species are not mathematical but organic fractals, whereas the word organic refers to the non-mathematical properties of existential entities, which are better described with visual and logic/verbal languages. Those Biological theories of Time (Evolution theory, Theory of Organisms) include also the logic of survival that extinguishes the less fit – hence a relative past species - and reproduces the fittest, hence the future species; and it includes the visual languages that give movement to the forms of energy and information described by mathematics.

Nature's fractals combine the arrow of energy and the arrow of information in an organic way.

They differ from geometric fractals in their dynamism and capacity to absorb energy from their environment. They not only iterate their form in microcosmic scales, but feed their offspring of information till they grow into full replicas of themselves. It is precisely the combination of both arrows what makes the behavior of an organism more complex and varied than the behavior of a mathematical fractal. So in the study of natural fractals, we have to add to the informative arrow - whose laws we deduced from the study of mathematical fractals and informative seeds - the energy arrow of entropy that thermodynamics has analyzed:

Energy, young arrow: $e_{->\infty} x\ i->_0$ X $e_{->0} x\ i->_\infty$: *Information, old arrow*
For any Σ number of cycles: $\Sigma E\ x\ \Sigma i = $ *Organic fractal* $\Sigma E <+> \Sigma T$ *Organic cycle*

The understanding of fractal processes provides a more detailed, scientific analysis of those processes. In that sense, unable to define information with the same rigor used to define the arrow of energy and entropy physicists have ignored for centuries events connected to the arrow of information (chaos, turbulences, super-conductive and super-fluid phenomena, etc.). Physical events and particles are also subject to both arrows, as everything else in the Universe. Thus we define abstract equations (as the back and forth transformation of energy into mass, $E=mc^2$) in terms of process of creation and destruction of information.

A single language cannot describe all the information that a certain being has within itself. So mathematical languages are also synoptic, and reduce the total information of the being. And that is the case of a fractal equation, which reduces the entity to its fundamental essence. In that regard, the Theory of Time Arrows is not Pythagorean, as Quantum Theories often are, in the sense that it does not consider the Universe to be made of mathematical entities, but merely affirms that among the linguistic mirrors that can map out the Universe, mathematics is undoubtedly the key language to represent the trajectories, cycles and motions of its species.

Because all languages observe the Universe, humans 'talk' those arrows of time with logic, verbal, visual, and mathematical languages. Since the syntax of verbal thought with its ternary structure and of visual forms with is line-cycle-wave structure and red-green-primary colors describe entities constructed with those arrows.

Recap. The Universe is based in 3 types of invariance, invariance of motions (Galilean relativity, now applied to the invariance of the 4 arrows/motions of time), invariance of scale (self-repetition of the same processes and arrows in all scales of size) and invariance of topological form (self-repetition of the energy and information shapes of reality. The 3 invariances define a Universe better described with the mathematical laws of fractals geometries, based in the existence of a generator equation that constantly reproduces information, creating networks of self-similar points, which structure all realities we see. The difference between a mathematical and an organic, Universal fractal is evident: organic fractals do have motion and their processes of reproduction and evolution are far more complex than those of mathematical fractals. So we must depart from a narrow-minded mathematical-only approach to study those fractal systems and consider mathematics one of the main but not the only language to describe its organic properties.

32. The 3 elements of all fractal beings.

We see a static, simple Universe in our relative human 'plane of existence'. Yet reality is organic, extended from microcosmic to macrocosmic 'dimensions or scales', and dynamic, as all particles and forces have movement-energy and form, in-form-ation; and all of them are structured into networks, which share their energy and information. Those 3 elements: bites *and bytes of energy and information; networks and scales* define the structure of all natural fractals, including the Universe.

They can be formalized with the Generator Equation we use to formalize the arrows of energy, information, reproduction and social evolution. Since *we defined Universal fractals as 'fractals of Time Arrows', hence both concepts a fractal Universe and a Universe made of knots of self-repetitive Time Arrows are the same.* We observe different degrees of complexity in a chemical or mathematical fractal, in a fractal organism or in the fractal Universe; but those 3 essential elements repeat themselves, to the point that scientists use similar terminology for all of them:

- The elementary, iterative *unit-cells* of a fractal, which according to the duality of Universal arrows, will consist on relative bites *of energy and bytes of information. Those bytes and bites are not static forms, but arrows that exchange energy and information; flows and cycles that the organism repeats constantly, generating self-similar forms.* So we can define a simple iterative, cyclical equation with 2 terms, the 2 arrows of the Universe, which can be considered the generative cycle of all Natural Fractals:

$$i: Fractal\ Cycle:\ e \Leftrightarrow i\ (dynamic\ state) + exi\ (static,\ formal\ state)$$

- *The networks or systems of energy and information* that organize those cells into bigger structures.

How feed-back cycles of energy and information grow into complex organisms, made of many cycles?

-In time through processes of discontinuous reproduction and evolution, which create similar action-reaction cycles, cells and organisms in other parts of the Universe. Most natural fractals have a dual network structure that attracts the micro-cells of the organic fractal by distributing the 2 arrows of energy and information to them. For example, in a human fractal, the nervous system delivers bytes of information and the blood system delivers bites of energy, control together its cellular units. In the Universe the same roles are played by the informative, attractive, gravitational force, and the electromagnetic energy forces.

-In Space through the organization of cellular networks that deliver energy and information to each cell-cycle. Those cell-cycles attach to the networks *to process constantly new energy and information in their iterating cycles*, creating bigger organic systems, made of multiple, similar forms.

When both iterative processes are combined, the fractal units of the Universe *grow in energy through space and evolve its informative complexity in time,* giving birth to the infinite combination of those fractal arrows that we call species, systems or organisms.

The equation includes an Σ-term to signify the gathering of individual exi units into networks and webs:

Σi: Fractal networks: Σ (e⇔i) or Σ (exi).

It still looks a simple equation to generate the Universe in its infinite variety. But the Complexity that arises from such simple scheme is enormous when we introduce the 3rd fundamental concept of fractal structures: Dimensionality. Since indeed, space-time cycles can exist theoretically in any number of dimensions. So we should consider in more detail the 3rd element of fractals: the fractal dimensions, planes or scales of existence invisible to human perception through which the fractal exists. They are:

- i: *The relative st-scales or dimensions* across which fractals extend its organization. We use for the 3 relative scales of size that define most Nature's fractals, the symbol i for information.

In humans there are *3:* the cellular, individual and social scales; as most entities do not exchange energy and information beyond its upper and lower organic level. For example, humans, as individuals belong to a relative st-scale, and are aware of social tasks (our st+1 plane of existence) and cellular sickness (our st-1 plane), but the events that affect the galaxy, the macrocosms to which our planet belongs (st+2) or affect our molecules, the lower microcosms that make up our cells (st-2), hardly affect our existence. Thus depending on the detail and scale at which we observe most fractals we perceive a fractal cycle (minimal size and detail), a fractal cell or a fractal organism (max. detail and size):

Natural Fractal= ΣΣst-1 Cycles > Σst-Cells->st+1 Organism.

How many elements in space make up a fractal, social network? The most obvious structure is a 'tetrarkys', already considered by Pythagoras to be the magic number of the most perfect social networks. The reason is that a tetrarkys can be written as a system with 3x3+st elements. Imagine then such tetrarkys as a triangle with 3 vortices, each with 3 elements that will perform the arrows of energy, information and reproduction, while the central st+1 element that communicates all others will perform the arrow of social evolution, representing the entire fractal as a unit of the next st-scale. Thus a tetrarkys is a highly efficient form, found in all systems of nature, even in human, social systems: 9 soldiers and a caporal make a platoon and 9 caporals and a captain a division.

Recap: All fractals have 3 elements that define them: Cellular units, networks of energy and information that organize those cellular units, and st-scales in which a self-similar but not equal fractal structure emerges, once and again with the same energetic and informative complementary forms.

33. Fractal Reproduction of information

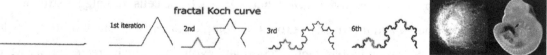

A mathematical fractal like the Koch curve is an energy line that triples its in-form-ative bytes every iterative cycle. But since it doesn't grow in length, it exhausts its energy. An organic fractal also starts as a single cell that clones its in-form-ation. But since it can absorb energy, it can iterate billions of times. In both cases what defines a fractal is its reproduction of in-form-ation that bends energy into form, creating the arrow of futures. So the abundance of fractals proves the existence of an arrow of information in the Universe. We are all fractals, Cantor dust of energy and information: we fractalize so much that finally we lose energy, volume and return to the dust from where we departed in the process of death. For that reason, the only solution to death by excess of information is fractal reproduction of a younger micro-form, when the being has enough energy to replicate in a discontinuous zone of space-time, ensuring the immortality of its logic in-form-ation.

The evolution of Time-space requires giving motion to Euclidean geometry. This is done by fusioning fractal mathematics and Non-Euclidean Mathematics. Mandelbrot in the 70s considered the fact that self-repetitive forms diminish and grow in scale maintaining its forms invariant. The why of this grow ultimately responds to the arrows of reproduction and social evolution: A fractal can be seen as a reproductive process, which is common to all generator equations of information of the Universe.

Fractals can also explain in more detail the reason of aging and the life cycle. The 2 arrows of time follow a natural order from yang=young energy into yin=old information, which defines the existential cycle of most beings in the Universe. For example, the life and death cycle can be explained as an iterative process of reproduction of information that degenerates into aging, when the organism has exhausted all its energy in a 3rd age in which instead of reproducing, it wrinkles and fractalizes its cells.

Recap. An organic fractal always iterates its form over new energy not to become fractalized. Yet after a young age, even natural fractals start to lack energy end up becoming wrinkled. In fact, all forms in Nature including the Universe in the future big crunch, suffer some aging process, guided by the time arrow of information.

34. Fractal reproduction in a discontinuous Universe.

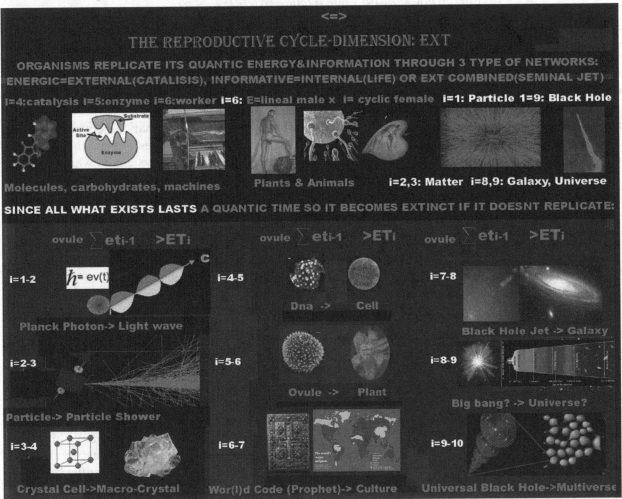

In the graphs, fractal systems are self-generated by seeds of its cellular scale, which encode all its information, compressed in very limited space (an equation, a DNA code, a Unification equation that generates the energy and forms of the Universe). Then the seed replicates its form in 'biological radiations' that multiply the fractal cells of the system, which finally *follow the plan of network evolution of all systems,* recreating the original species. The process follows a series of 'beats' that change the arrow of time of the system from reproductive, energetic waves to informative, self-organizing phases in a pattern proved mathematically for chemical systems by Mehaute, which showed in the 80s that for time to continue when a system was not releasing energy; it had to create fractal information, *the 2nd arrow of time in the Universe.* Thus, the Universe never stops its E->I->E rhythms.

When entropy ceases is because the feed-back equation changed side. In Nature those micro-structures detach themselves from the being and absorb energy in the environment growing in size, self-replicating. So the fractal transmission of form is the most important event taken place in the Universe. From genetic information transferred between living beings through a fractal process of palingenesis to the creation of mental ideas that we convert into bigger machines, fractal replication between different planes of fractal space-time with finite dimensions, is the essence of reality.

Fractal, transcendental reproduction iterates cells and reorganizes them in decametric scales of growing social complexity till reaching a limit number of 10^{10-11} cellular units. The organization of those scales follows a simple law:

Energy flows from st_{-1} into a future st-scale: $E_{st-1} -> I_{st.}$

Information flows from st into a simpler st_{-1} scale: $I_{st} > E_{st-1.}$

Reproduction happens in the same st-scale $\Sigma S_{st} <= I_{st}$

Since those planes are networks existing with different space-time constants, disconnected from the other planes, how a given space-time scale, the cellular or atomic scale, relates to its higher st-plane? Riemann hinted at such scalar structure of space with his analysis of Riemannian surfaces, polynomial planes which communicate through a narrow path that transfers energy and information between them. Those paths between light and gravitational space are in Relativity 'Rosen Bridges' - black holes and other massive regions of the Universe. Yet the concept can generalize to any dual system, in which there are asymmetric flows of energy and information, according to a fundamental law of fractal space-time:

Information flows from the macro-plane of relative future to the micro-points of simpler energy and energy flows from the micro-points, st-1 from past to the relative, more evolved future macro-points, from the microcosms to the macrocosms.

Thus there is an asymmetric, hierarchical single flow of information from a relative future, the higher st+1 plane, to a relative past, st-1 plane and vice versa, a fact which explains the lack of parity of 'temporal, informative flows' (weak forces in physics, hierarchical social classes in sociology, etc.)

Thus the order between those scales is hierarchical. Species in different planes of space-time interact within each other transferring energy and information. Yet the bigger structures control and order the smaller pieces - its parts that become wholes through the arrow of life or disintegrate those wholes back into parts through the process of death. Such interaction is possible because in those different Planes, from microcosms to macrocosms, energy and information have self-similar forms that make the shapes of atoms similar to the forms of galaxies and stars. And follow self-similar laws of harmony. In terms of hierarchical order, we observe that the bigger an entity is in space, the longer it lasts in time, $\Sigma S \Leftrightarrow Ti$. That is, the amount of time and space of the different structures of the universe is in balance. While in terms of hierarchy the bigger scale orders and transfers information to the smaller scale that transfers mainly energy to the bigger form. Those 2 arrows of time, energy and information imply that the bigger scale or network – for example, the nervous system in man – controls the lower scale of cells, which in turn feeds the upper scale.

So we describe Temporal, Organic fractals with 2 terms of increasing complexity: a fractal system or network, which is incomplete, since it doesn't reproduce internally and might not be able to self-organize itself, but requires the collaboration of an external agent; and a (super)-organism in which the arrow of reproduction is internal, iterating the initial fractal units of energy and information into 2 networks that finally combine its energy and information creating the 3rd reproductive network that perpetuates the self-sufficient species.

In biology such processes of self-organization and reproduction of parts into wholes are called Palingenesis – a term, which complex sciences that use both the jargon of physics and biology, extracting self-similar laws that apply to all those scales, uses also to describe physical processes of creation of matter and social processes of creation of cultures. Since all social and physical systems display complex, palingenetic behavior, from physical particles to organisms, from minds to nations. All of them constantly reproduce seminal information - called actions of energy and time in the quantum world, offspring in living organisms, ideas in the linguistic mind, technology and culture in human societies. And in all those cases, the seminal information grows into a species self-similar to the mother. So forces reproduce particles, cells become organisms and nations create self-similar cultures, called colonies. All of them are organic fractals whose different planes (the cellular, individual and social planes) are connected by flows of energy and information: Life reproduces in a smaller space-time plane (st_{-1}) seminal seeds that grow into replicas of original st-mother; human existence is a travel between 3 Planes of relative existence, the cellular Plane (st_{-1}), the organic, individual Plane (st) and the social Plane in which a human can be considered a cell of a social organism, a nation, religion or civilization (st_{+1}). But electrons (st) also produce smaller photons (st_{-1}) that evolve into new electrons, which become cellular units of electric flows (st_{+1}); while galactic black holes (st), made of billions of social quarks (st_{-1}) emit flows of matter that evolve into new galaxies, cells of Universes (st_{+1}). Thus, Social Palingenesis explains also why in all the scales of the Universe, the geometry of energy and information remains invariant; since a st_{+1} Plane is born by the social evolution of energy bytes and information bytes of the inferior Plane, which repeat its self-similar forms in all the multiple, self-similar relative 'planes of existence' of the universe, from the atom to the galaxy, from the cell to the factory. Thus, social palingenesis is not exclusive of living organisms. All systems display those complex, social Planes. The 2 simplest particles of the Universe self-reproduce and socialize: quarks absorb energy and reproduce smaller quarks. Electrons produce a seminal light ray that becomes a new electron. So happens to the heavier form we know: the black hole, which reproduces its form when crossing through a star. Even the Universe is a space-time fractal, which might have reproduced its 'cellular' particles in the big/bang.

Recap. All systems of the Universe reproduce through a fractal, palingenetic process, in which the system produces a seed of information that reproduces and evolves socially till surfacing in the same plane of existence that the parental form.

35. Complex, reproductive motions.

In the Universe there is a simplex cycle, $e \Leftrightarrow I$, that converts lineal motions into cyclical motions, and can at best be made to exist in 3 dimensions (spherical motions) of space and 3 of time (expansive youth, or big-bang, steady state and implosive 3rd age). Yet complex motions - acts of reproduction and social evolution - are different from simplex motions, in as much as they take place between at least 2 planes of different space-time. This difference between the simplex and complex arrows of time creates jumps between space-time planes:

The most complex phenomenon of the Universe is the fractal jump in space-time. Both bidimensional cycles of time, and planes of space are fractal and so their changes and displacements are also fractal. And since time is geometric, bidimensional, as space is, it has also 2 directions of motion.

So the oldest fractal question of philosophy is: how can continuous movement exist in a fractal world filled with discontinuities? Zenon, a Greek philosopher, put the example of Achilles running against a Turtle: Achilles will never reach the turtle because there are infinite fractal spaces it has to cross.

Parmenides gave him the answer: continuous movement doesn't exist in a discreet Universe; it is a mirage of the senses. Since a wave is not really moving, but reproducing information over the potential energy of the vacuum in a sequential, fractal jump, through a series of wave-lengths, drawn one after another, which appear to the senses as a moving wave. Movement is always a discontinuous displacement either in space or in time, in any of its 2 directions, towards the past or the future. The simplest analogy is a television screen where new images are created constantly without hardly any cost of energy, because they are virtual images created by illuminating 3 colors that are already potentially in the screen. In the case of a wave, the relative energy imprinted by the logic form might vary, from the vacuum energy of a light wave to the placental energy of a mother, yet the process is always the same: a form imprints another region of space, changing its spatial position and temporal morphology. *So speed is synonymous of reproduction, the main cyclical action of the Universe. Thus we can explain all physical events in terms of organicism. Since even movement - the external, mechanical change physicists study - turns out to be also an organic cycle of reproduction.* We distinguish, according to their quantity of energy and information 'translated' in the process from a place of space-time to another place, 3 fractal, reproductive jumps:

- *E=I: Space-time reproduction* dominant in spatial displacement that implies a minimal movement of information that fluctuates from a future to a past form and returns back to the future. For example, when an electric impulse arrives to the end of a nerve, it becomes translated into chemical information and then it is transferred chemically through the synapses to the other cell, where it becomes translated back into electric information. That process implies a 'spatial, fractal jump' from a neuron to the next neuron, but also a fractal jump in time from a level of informative complexity, the electric language to the lower level of chemical languages, which is the relative, past informative system of life. The informative essence of the process implies a transcription of form from each discontinuous point to the next point but also a movement in space, since 'all yang has a drop of yin', all is temporal energy, also spatial and temporal reproduction requires a minimal dual movement in time and space.

-*Max. E: Spatial, wave reproduction* of a wave a particle that jumps from one place to the next place, reproducing a minimal temporal form. It is very common in physical waves, defined by its speed of reproduction, $V=S/T$. Depending on the density of a form, its speed of reproduction will vary. So the most massive forms are slower than those whose informations are minimal and can imprint the energy of vacuum very fast. Those forms that hardly move however evolve a lot in time:

- *Max. i: Temporal, palingenetic reproduction,* when an organism produces a past, smaller, simpler seed that reproduces the being, evolving faster towards the future. So a foetus is 'born' when it comes out of the body of the mother to the next discontinuous region of the Universe, making a minimal spatial jump and a huge evolutionary, temporal jump.

Recap. There are 3 types of reproduction: self-reproduction by a complementary system that mixes its energy and information, without concourse of external agents; spatial reproduction, or motion of a wave that imprints a simple form of energy as it translates in space and enzymatic reproduction when an external agent assemblies the parts into a whole.

36. Reproduction of physical, biologic and mental fractals.

In graph, cosmological, chemical, biological and mechanical fractals during their reproduction cycles.

All fractals iterate=reproduce by absorbing energy transformed into fractal information by a Mother cell or external enzyme- as it happens with carbohydrates or machines iterated by 'enzy=men'.

Unlike mathematical fractals, made of fixed, spatial forms, Nature's fractals are made of temporal, iterative cycles, in which energy is transformed into information and growth, as the fractal family becomes detached from the mother cells. Organic fractals are not fixed, geometrical forms that merely repeat in spatial scales, void of life and temporal change. They are *temporal beings, whose bio-logic cycles and functions* reproduce at different scales and time intervals. Since not only space has a fractal structure. Time is also discrete, made of cyclic trajectories that repeat themselves, as it happens with the cycles of a mechanical clock. It is precisely the repetition of the organic cycles of energy feeding and informative reproduction, what makes natural organisms fractal structures. The Universe is made of such temporal, organic fractals that reproduce into multiple clonic forms that latter re-organize themselves, giving birth to macro-forms similar to the original 'cell'.

The most obvious natural fractals are those created by the iterative, repetitive nature of 3-D space. They are similar to mathematical fractals where geometrical form dominates.

For example, an ice crystal *is a chemical fractal* that jets up triangular, tall spikes, in its relative dimension of height, growing into a fractal; a *salt grain* is a cubic crystal that reproduces its form as it grows in scale and recomposes itself into huge cubes. While a hurricane fractalizes air streams into a vortex of wind cycles that grow also in height, the fractal dimension of information. Those physical fractals already iterate its form, despite the simplicity of its fundamental elements; since all natural fractals reproduce the shape of the 'mother-cell' into multiple clonic forms that latter re-organize themselves, giving birth to macro-species similar to the original 'cell'.

In Biology, fractals are used routinely to explain how the iteration of simplex growth processes gives birth to complex botanic shapes: Branches, leaves and roots are all easily created by the iteration of an initial mother-form. A DNA molecule, which iterates its form both in micro-fragments (amino acids) and at macro-scale (recreating a cell). While a tree repeatedly fractalizes its form in 3 branches, ever slimmer, with less energy but ever increasing the information of the tree.

A *human being is also a temporal fractal*, made of fractal cells, whose organic functions repeat themselves at individual macro-scale in a longer time-interval: the feeding, reproductive, informative (sensorial), and social behavior of human beings are a macro- repetition of the same cycles that take place in cells. Cells feed on energy, absorb information, reproduce themselves and form social groups, as human beings do, at a smaller, faster scale. The fractal sum of those cellular cycles creates a human being. And then, the fractal sum of human cycles at a bigger scale creates socio-biological organisms, called nations and civilizations. A human is an organic, bio-logic, temporal fractal, extended in 3 scales, the cellular, organic and social scale.

Recap. The arrow of information and entropy together describe all kind of natural fractals, not only the well-known biological fractals but also social and physical systems that display complex, fragmented borders, (from physical particles to organisms, from minds to nations) and constantly re=produce information (actions in the quantum world, offspring in living organisms, ideas in the linguistic mind, technology and culture in human societies). They are also organic fractals, since once we understand their structure in multiple scales connected by flows of energy and information, the old distinction between organic and inorganic matter loses its meaning, beyond its anthropomorphic role to make man the center of the Universe.

37. Linguistic fractals. Paradox of inflationary information.

Minds can also be modeled as fractal systems that replicate the forms and movements of the Universe in a smaller scale, through the syntax and semantics of a certain language. Indeed, modern science concentrates in the study of the parallelisms between the hardware of the mind, the neuronal brain, and its software, the depiction of the Universe made with visual and verbal languages. But what truly matters is the bio-logical parallelism between those mental structures made of syntactic and semantic, fractal units (numbers, geometrical forms, phonemes and sentences), and the external Universe they map out. Since organisms survive thanks to their accurate, mental perception that orientates them in that bigger Universe. So fractal minds are a fundamental tool of evolution: The mind acts as a virtual, fractal mirror, allowing the being to interact efficiently with the Universe, absorbing energy and information from it to replicate its form. This means that we can study minds as mental fractals similar to its 2 main languages of space-time; mathematics and logic.

For that reason, mental and geometric fractals are both defined as entities whose informative boundary tends to ∞ while the energy/space of those boundaries tends to zero:

$$\delta e/\delta t\text{->}0 \text{ x } \delta i/\delta t\text{->}\infty\text{: } \textit{Mental Fractal.}$$

Thus the information of a mathematical fractal is found in the boundary. It happens also in black holes, where information is attached at the boundary in a bidimensional space (holographic principle, which we proved in a simple manner as result of the 2 dimensions of a cycle of time that carries information and now we can observe from a more complex perspective). It is also the case of the human mind, an i=eye+wor(l)d, created by the eye's retina and the electromagnetic waves of the cortex brain. Thus the informative parts of organisms, what we call 'senses', are also external, attached to the boundary and one of the fundamental functions of Nature's fractal arrow is to create senses, systems of perception of fractalized information. Minds and senses are similar to mathematical fractals. It should not come as a surprise; since mathematics after all is a mind's language, and minds fix most cycles of Natural fractals into still images that don't capture the vital energy of the Universe.

Mathematical and real fractals share the properties of information, inverse to energy. Information increase when the fractal becomes smaller in its surface. This paradox, that the smallest regions of space hold the maximal quantity of time: Max. I =Min. E is however essential to full grasp the Universe.

Languages hold more information than the forms they describe. Information is bidimensional and can be warped and cut. This essential property, that information is in the void, the cut, not the substance, as it will *be first perceived by a mind in order to exist, and so it is not real till the perceiver observes it (quantum paradoxes), is* however the reason why the smaller regions, in which voids multiply are more conductive to information frequencies that long ones. This paradox already understood by Gödel simplifies the distinction between mathematical physics and reality. Information is inflationary.

The Universe is not all the possible mathematical models. As Riemann, the key mathematician of the XIX C. explained, we cannot consider all possible spaces real, because the physical universe has selected only one geometry[6], to which we shall add – it selects the most efficient geometry that maximizes the 4 main arrows of time of species and helps the organic, Complementary, exi, bio-logic forms of reality to survive. Unfortunately neither mathematicians nor physicists today accept restrictions to their imagination, so they invent 'multiverses, cosmic strings, multiple, infinite dimensions' without 'upgrading' the concept of a continuous space into the reality of fractal discontinuous scales. The confusion of languages with reality is one of the main 'errors' of science we must resolve in order to improve the 'logic/mathematical mirror' the mind uses to perceive reality.

This distinction often lost to physicists in modern times, clearly expressed by Riemann in his 'foundations of geometry' and further proved by Gödel implies that not all possible mathematical spaces and mathematical particles exist. The space of the Universe is far more restrictive in form and dimensions than the space of mathematicians. Why this happens can be explained 'bio-logically' by the fact that the inverse function of existence, extinction, eliminates all forms which are not efficient even if it creates them a priori – in the same manner all mutations in Nature, which are not efficient disappears. While our informative mind keeps producing imaginary models of geometry, which are pumped up by the scholar who seeks to convince the world his geometry is also right. For example, in the field of physics, multi-universes, super symmetries, entropy in black holes, etc. are mathematical theories with no equivalent in the world because they ignore the arrow of information in mass and the need for no more than 2x2 dimensions to construct the Universe.

While the Copenhagen interpretation of quantum physics in which an organic wave of photons, parts of whole electrons, guided by the 4th arrow of organic evolution is considered a herd of 'probabilities' is wrong. As in the case of the Mandelbrot set that doesn't exist in nature, unable to resist energetic motion, in all those cases the 4 main time arrows of reality and its need to be performed by all entities that wish to exist, limit the information that becomes real.

All this requires understanding 2 of the basic properties of information – the fact that 'fractal information' is constantly created by the motions of reality and the paradox between energy with maximal spatial extension but minimal information vs. information with minimal energy and extension. In brief, information increases, as in fractals, when we diminish in size. So brains are smaller than bodies but have more information. And a language has maximal information packed in minimal space. A language in fact has more information than a huge surface of vacuum space, which makes theories more abundant than the forms we find in the real Universe. Information is inflationary and so are physical theories with bizarre space/times that reality never will select.

Mathematics is, as all languages, an informative, 'fractal', whose self-similar forms/theories multiply in excess ($\Sigma Se_{->0}$; $Ti_{->\infty}$). Thus, if certain sentences of verbal thought can be imaginary fictions, despite its beauty, so happens to many 'quixotic' geometrical forms, which do not exist. The Universe selects only efficient geometries, based in forms that can move and perform the arrows of time in the real Universe. So for example, the most beautiful fractal form of the mathematical Universe, the Mandelbrot set, does not exist in reality, because all its small self-similar attachments would easily break if they were moving as reality does.

Hence the importance of having a geometric theory of space-time that describes accurately the forms of the physical Universe. This was the key discovery of Einstein. He realized that Euclidean, lineal geometry, used to describe ideal space, could not describe the cyclical paths created in gravitational space-time. So he switched in his description of reality to a space called Non-Euclidean, where its minimal fractal unit, the point, could be crossed by multiple gravitational curves and hence, unlike the abstract point of Euclid, it required inner volume, inner form and energy. This point, *which has always in Nature a region that can absorb /gauge information and a lineal or cyclical, rotational motion*, is what we call a Point of View= Measure or Non-Euclidean point of view, the relative unit of Non-Euclidean mathematics.

Philosophers of science have wondered often if mathematics is merely a construction of the mind or an image of reality and why they are all pervading in the description of any 'system' of self-similar forms in the Universe. The answer is both: languages are construction of the mind that reflects the external universe's information in a small space, hence they must reduce the total information to fit the mental mirror, trying also to make an image the closest possible to reality. This is what logic languages do with time and mathematical languages with space; because the Universe is logic-mathematical, albeit of a type of logic and mathematics more complex than the abstractions without motion devised by the Greeks 2300 years ago. Indeed, the same concept applies to the new logic of time, which is based in multiple causality, either a dual 'biological causality' in A<->B events in which an element A is predator and the other B victim; hence B is considered by A mere energy when B considers itself an informative entity worth to exist (so A is both A and B, victim and predator, energy for the predator information for the I-self of the victim); or 'multiple, spherical causality', when multiple agents create a certain network or form. So for example a real circle is not drawn in abstract by a single pen, but happens when multiple self-similar points of view converge and form a cycle.

Ultimately we are all made of spatial energy and temporal information, whose synoptic descriptions are achieved with 2 languages, mathematics and logic. And so the search for knowledge by the human mind is not only the recollection of experimental data, but its description with the languages of mathematics and logic. Hence even more important than the collection of data is the creation of a mental container – those 2 languages – efficient and evolved enough to fit properly all the data of reality we observe. Yet since the revolution of thought represented by Evolution Theory in Biology and quantum /relativity in physics, there is a gap between the behavior of universal entities and the linguistic syntax of mathematics and logic, hardly evolved since 2300 years ago, when Euclid and Aristotle founded those 2 sciences. Thus, we shall in this work summarize the new advances and laws of spatial geometry and temporal logic that are needed to describe a complex Universe of multiple, fractal space-times in which points are Complementary systems – entities that gauge and feed on energy with an informative and energetic center (body/brain, field/particle system). What still stands is a simplification of reality to 'fit' it into the mind. So while we are not concerned with the specific energy and information the point of view process (the details that don't fit), we cannot longer hold truth the syntactic errors of Euclidean mathematics (points, lines and planes without breadth, width or height), and Aristotelian Logic (single causal events of the type A->B).

We call the evolved syntax of those 2 space-time languages, i-logic geometry, or simply i-logic geometry (as i represents 'information' and it is the next vowel after the A-ristotelian and E-uclidean languages used till now. Yet it also represents a 'complementary system' with a cyclical, informative 'head', O, and a lineal, energetic 'limb-body' |, which are also symbols used in 'complex algebra' to represent energetic and informative functions)). Thus, we affirm that the Universe of multiple space-times follows *the complex logic of information or* 'i-logic, fractal geometry' pioneered in the XIX and XX C. by Darwin, Riemann, Planck, Einstein and Mandelbrot whose discoveries were the first steps of a 'new paradigm of science', which this work completes, upgrading philosophy of science (mechanism and monism), stuck in Aristotelian, Euclidean theories since the XVII century. So we shall talk often of the 'i-logic, fractal paradigm' as opposed to the 'A-logic, Euclidean paradigm' of a mechanical world of single 'ceteris paribus' causes and entities which are supposed to exist in a single, continuous, abstract, non-moving space.

The second part of this work after this brief introduction to 'metaphysics', the description of the ultimate meaning of existence – the accomplishment of the 4 main arrows of time – will be dedicated to the development of those new formalisms of mathematics and logic needed to make detailed, rigorous analysis of the events and forms of natural, organic fractals.

In that regard, since all languages are fractals of the entire Universe perceived in a slower, shorter, memorial pattern, a mirror of information with less dimensions of 'motion' and 'form' than the original. So these texts could be also considered a fractal, linguistic mirror of all Realities.

Recap: Information is inflationary. So are informative languages, reasons why we must differentiate fictions from theories that truly describe the universe. Multiple Spaces-Times improve our logic and mathematical tools to describe the Universe. This is done advanced the Aristotelian logic and Euclidean geometry into the next vowel of the alphabet: i-logic geometry, the logic and geometry of information.

38. Man is a fractal organism.

Man as a fractal is made of cellular units and networks of energy and information (nervous and blood systems), whose cycles and functions extend through 3 main st-scales: the cellular, individual and social scale. A human being described as a fractal will be composed of:

- *DNA cells*, the Mother cells of the human fractal, whose genetic functions and forms repeat themselves trillions of times, composed of informative units (the DNA genes) and energetic ones (amino acids, bricks of genetic reproduction).

- A palingenetic process of reproduction and differentiation of those cells, which latter gather into energetic, reproductive and informative *networks*: the digestive, blood and nervous systems that create a human.

- The scales that group DNA molecules into cells, which gather themselves into organs, which gather themselves into physiological systems, put together into a human being.

Those scales break further in ternary and decametric systems: man has 10 organic systems, 10 fractal fingers, 10 glial cells feed a neuron, etc.

Of all the possible fractals that can be described with those 3 elements I share the fascination of Kant for what is inside the mind of men and what is above it, on the skies. So we will dedicate most of the pages of this work to study those 2 fractals, man and the Universe.

Recap: Humans are fractals of energy and information defined by the 3 elements of all fractals, cells, scales and reproduction.

39. The Universe is a Fractal organism.

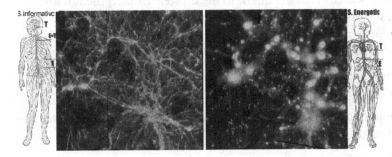

The Universe and man described as fractals, whose cellular units are blood and nervous energy/ information networks and informative masses (MACHOs, quark stars?) and light galaxies that shape 2 networks of radiant energy and dark, informative matter. Each point of those 2 networks (compared to the similar networks of a human organism) represents a galaxy.

The Universe can be described as a fractal, based in the existence of the 3 elements that define all Natural fractals:

- The Universe has 2 networks, one of electromagnetic energy and one of gravitational information (black holes, dark matter) that controls the electromagnetic world. It has also informative vortices (masses and charges) and lineal forces.

- Its fundamental particles, quarks of mass/information and electrons/photons of electromagnetic energy can absorb energy, gauge information and reproduce self-similar particles.

- The structure of those particles and forces is scalar: lineal forces of electromagnetic energy and cyclical particles of gravitational information grow in scale and size, from the tiniest lineal and cyclical strings, which are so small that we cannot perceive them, to the atomic scale of lineal forces (light, gravitation) and cyclical particles (electrons and quarks), to the cosmic scale of gravitational and electromagnetic fields, caused by spherical, celestial bodies (stars, black holes and planets), creating the macrocosms we perceive.

The galaxy has a dual network of informative, black hole masses and electromagnetic stars, which are the cells of the informative and energetic networks of the galaxy, join by flows of gravitational and electromagnetic space. And in the same manner, if we take each galaxy as a st-point, we can map out the Universe as a dual network of dark, gravitational information (left side of the picture), which joins the central black holes of the galaxy and electromagnetic energy, which joins the electromagnetic planes of the galaxies (a point in the right side of the graph).

Since systems extend in '4-dimensional space' as a series of networks webbed among them, and in the hierarchical scales of size as a series of parts that become wholes, units/parts of new wholes and in time as a sequence dominated by one of those 3 networks, which will grow from the lower cellular scale into the social scale and die back into the cellular scale. And also we can consider that the Universe will go

114

through 3 ages that correspond to the 3 solutions of Einstein's spacetime equation and start as a relative 'seminal cell' in the big-bang and finally warp into the big-crunch of maximal form.

Bio-chemical and geometrical fractals have been known for a century. But only recently we have studied *the physical Universe as a fractal*, born out of the dynamic combinations of those 2 arrows of time. One of its most remarkable successes is the description of Einstein's space-time continuum as an iteration of some basic fractal geometries (Nottale). Thus Einstein's sentence, 'Time curves space', means in fractal terms that the arrow of informative time curves the vacuum energy of the Universe, fractalizing it into fractal particles, charges and masses, which are fractal, spiral vortices of electromagnetic and gravitational information. Moreover, if we order the sizes of the Universe in a *decametric scale*, we find that at certain intervals the same forms and functions appear once and again. In the 60s, the American filmmaker Eames made a documentary called 'The Power of 10', in which that iterative repetition of cyclical, spiral forms appeared from the quark to the cluster. In fact, the homology of form and function between those scales is so remarkable that the same equations describe the tiny magnetic field of an electron or a star, made of trillions of particles. Thus, since that decametric scale is the most common scale considered for the entire Universe, we can define the Universe in fractal terms:

'The Universe is a fractal super-organism made of particles of information and forces of energy gathered in networks, whose formal cycles repeat in decametric scales'.

While the fractal Universe and the continuous model of the big-bang use the same experimental information, the final outlook is very different. Since the fractal Universe is:

A) Eternal. B) Cyclical. C) ∞ in energy x information scales.

Indeed, even if there were a process of organic reproduction, of a big-bang like radiation of forms and particles over a surface of void energy, this will be subject to the duality of arrows of time, of life and death cycles, of big crunches (life, creation of information) and big bangs (death, creation of energy). So B->A, and since the Universe is cyclical, it is eternal. And so, in a fractal Universe the metaphysical question is: do Universes become 'microscopic black holes', quarks of a bigger, fractal Hyper-Universe? A detailed study of the physical constants and parameters of those particles seem indeed to foresee a fractal Universe, which is not only infinite in its relative spatial size, but in the number of scales/planes of organic evolution, in such a manner that each galaxy could be the atom of a bigger Universe, as it happens in fact in the theoretical models of relativity, (Einstein-Walker model of a Universe made of galaxies, interpreted as 'hydrogen atoms'[13].)

What is the absolute form of the entire Universe? If we observe the previous graph, we realize that the known Universe shows the clear form of an electromagnetic wave, with a wider base shaped by the 'gravito-magnetic' field of dark matter... So we might consider it again a 'point' of a bigger network.

Recap: The Universe is a fractal of energy and information that extends in 10 perceived self-similar scales of Euclidean networks in which the 3 invariances of motions, form and scales are kept. But since in the fractal model the parameters of atoms and galaxies are self-similar, the Universe could be infinite in its scales.

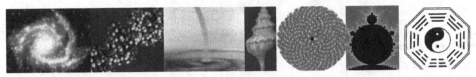

In the graph, a galaxy, a DNA molecule, a tornado, a shell, a sunflower, and 2 logic, mental mirrors of the Universe - a Mandelbrot set and a yin-yang, eightfold life cycle - are all made of shapes that combine the most efficient morphologies of energy and information: the line or shortest distance between 2 points and the cycle that stores the max. quantity of information in lesser space. Both combine to reproduce the infinite beings of reality. It is the game of existence we all play.

40. A Universe of actions of energy and information.

Once we have established the existence of a Universe of motions in time and forms of space, whose organic parts are guided by the 'will' of the 4 main arrows of time, we can respond to the fundamental questions about man and the Universe - why we are here, what we are. And the answer is obvious: since we are all complementary beings, made of energy and information, we exist to combine, absorb and reproduce energy and information, exi; what we constantly do through actions that imprint our form and deliver it, carried by a flow of energy= motion. We make actions of energy and temporal information. So we often say 'I don't have time and energy to do this'.

This simple expression, exi, is therefore the unit of reality, an action, specific of each species, which will imprint the external world with its form and energy, as many times as it does. Yet actions are also fractal, discontinuous, quantic 'steps'. And so we define for each system, including the biggest systems of reality (the gravitational and light-space membranes we inhabit), a minimal 'existential unit', or action that defines the system:

$$Se \times Ti = K = Action\ of\ existence.$$

In all systems we shall find such actions: the h-Planck constant is the minimal action of the light-membrane; in the gravitational world a Lambda-string is its minimal action. When we humans live, we constantly act with 'energy' an time, performing cycles of existence, whose minimal actions is a 'thought', an act of perception (I think therefore I am), which takes a second of time, our existential beat. In Physical systems, actions can also be translated as momentum ($m(Ti) \times V(Se)$); in biological systems, exi will define the 'existential force' of a top predator. In the economic world an action is a minimal quantity of a Company, the Free Citizen of the Economic ecosystem that re=produces machines, money and weapons and structure the financial-military-Industrial complex in which we live. In a wave of light it is the minimal step, as the wave reproduces its form over the surface of magnetic and electric quanta (constants of vacuum). There are in that sense many ways to define an action but in most cases, as its formula indicates, it will be an action of reproduction that imprints a form on a surface of energy; and therefore also the minimal unit of 'form' display but any existential system.

41. Mandate of existence: Maximize your Time Arrows.

The meaning of existence is a tautology in itself: we 'exi=st' to inform ourselves in order to absorb energy with the aim of reproducing our organism in other zone of space-time, avoiding in this manner 'extinction', the inverse function of existence - since we are only 'parts' of the total Universe with a limited quantity of vital energy/space and a limited duration of time. So to ensure the survival of our in/form/ation we must either reproduce in other zones of space-time or increase our quantity of vital energy and our duration in time, by becoming a part of a bigger entity, a wave, herd or organism.

So our actions become 'bricks', units, frequencies of those 4 main arrows of time. And while all 'yin=information has a bit of yang=energy' and vice versa, we can distinguish those actions by the quantity of energy or information they deploy, within the limits of an action, whose total product is constant:

- Max. I x Min e: *Informative action*
- Max E x Min i: *Energetic action*
- E=I: *Reproductive action*
- $\sum$ e x $\sum$2i: *Social action*

The sum of all those actions is what all beings do as existential beings whatever type of existence, the entity carries on. All this can be formalized in a simple Mandate, which all the languages, equations and actions of all particles, beings and entities of reality follow: maximize your time arrows, maximize your function of existence:

$$Max. \sum exi$$

Thus, if the Universe can be defined externally, objectively, as a geometry of two inverse motions, energy and information; from the internal perspective of the 'will' or 'why' of those motions of time, the Universe is a game of 'exi=stence', where 'to grow=evolve and multiply' is the meaning of it all.

Some scientists think that this mandate can only expressed in the language of particles and forces, machines and spatial forms, geometry=mathematics. But this is by no means truth. Indeed, all languages of all species, which each particle or entity of reality use to guide its behavior in the Universe express the mandate to its species. As Pythagorism affirms, physical entities are mathematical entities in as much as they use the language of mathematics to gauge energy and information and maximize their exi=stence. This is self-evident. We can describe the actions of physical particles only with mathematics because that is the language which they possess. Probably all particles are like small computers and vice versa. We are about to be able to make quantum computers using the simplest physical particles. But you need then the 'errors' of the Galilean=Ego paradox of Physicists – aka reductionism – not to go further than that, and affirm that the Universe is only mathematics, and that the mathematical actions of particles that gauge constantly information and energy, have no will behind, no purpose, no goal. This of course means that nothing else matter, nor Biology, nor life, nor Humanity and History, because when we enter into those other scales and entities, mathematics is not enough to describe the behavior of humans and living beings. Yet, *with 3x3 time arrows and the mandate of existence, we can still explain all the actions and forms of biology and history.* So again, according to the Principle of Correspondence, we can consider that mathematics as the language of physical particles and physics, as the science that studies only those particles of information and its forces of energy, are part of a bigger Science of Sciences, the science of Multiple Spaces-Times.

Consider indeed, Biological beings. They have a geometrical, topological form, which we briefly described in previous paragraphs and will analyze in more detail. Essentially, any biological being is a complementary being, made of networks of cells that try to maximize the simplex arrows of time: perception of information, with their heads; and absorption of energy with their body. They also try to maximize their complex arrows of time, which are less common (yet still exist in physical particles): the arrow of reproduction and the arrow of social evolution that gathers cells into organisms. Physical particles, quarks and electrons, also reproduce when they absorb a lot of energy and evolve in social networks, molecules, planets and stars; yet their dominant arrows are energy and information, the simplest ones, given their minimal 'volume of existence'. And so E$\Leftrightarrow$I, the fundamental beat of reality will suffice to explain most physical actions. In Biology though, the dominant arrow is reproduction, Max. exi; and so we shall understand finally the ultimate meaning of theory of Evolution encoded in Mr. Darwin's dictum: A struggle for existence follows from the constant reproduction of species in an environment of limited resources.

The way this fight for existence between species takes place is what Darwin described. The Biological networks or organs of each species (nervous=informative system, digestive= energetic system and hormonal=blood, reproductive system) define its physiology, the key science of medicine and biology at the scale of organisms. Yet, because entities of reality spread over several fractal scales of size, the language that living beings 'speak' to reproduce and create a self-similar entity is the genetic code. And so we have a new language, genetics, which encodes the function of existence of living beings, trying to maximize its reproduction, energy and information absorption and social evolution.

Finally, in the scale of human beings, the main language that expresses the function of existence is NOT mathematics, neither genetics, which encodes our biological life as individuals, since we are guided in the external, social world, by the arrow of social evolution and the memetic, cultural objects we use to organize societies. Thus, again the language has changed: humans encode their wantings and will for energy, information, social evolution and reproduction in verbal mandates, which are either philosophical, logical, religious or legal codes. So Genesis said 'and God told man, grow and multiply', which again means Max. ExI, reproduce, grow your energy and information. And Jesus said 'love each other as I have loved you', which means share your energy and information with other self-similar beings of your species or culture (in more reduced, tribal religions, which have not evolved verbally to understand that we all humans belong to the same species) so you can create an efficient social organisms, nation, God, civilization, all parts of the whole: mankind.

It is thus obvious that to maximize the function of exi=stence and its actions has self-similar expressions in all languages of the Fractal Universe – reason why indeed, in English the verbal expression, to exist and the equation, exi=st is self-similar.

The physical description of the mandate gives origin to the type of physical Laws, which describe the geometrical forms generated by those cycles of time in physical entities.

The verbal, historic description gives origin to the type of memes, Books of Revelation or Codes of Laws that are handed from generation to generation and develop human societies, which are either Religious bodies of God, the temple of the spirit, the 'word that became man and inhabited among us', Saint John 1.1, (God means then the subconscious collective of a social organism, expressed through the verbal mandates of its prophets that tell the believers to maximize their collective existence through the arrow of social love).

The result is what you see around you, the game of existence of infinite points of view, trying to gauge more information and energy, to maximize its reproductive and social existence, forming knots of time arrows at individual and collective level in all scales of reality.

Recap. The Universe is a game played by infinite actors, which try in all scales of reality to maximize their energy and information in order to multiply and share that energy and information with self-similar particles, forming networks and superorganisms. The function of maximal existence can be expressed in all languages, Max. e x I is its logic and mathematical expression, 'grow and multiply'' its most successful verbal expression (Bible). The will of times is also expressed in verbal mandates by Biologists (theory of evolution as a struggle for existence) and explains our existence as human beings at all levels.

42. Tug-of-war between the selfish and selfless arrows.

Now we come to a philosophical question that might seem trivial when considering the automated, topological paths of existence performed by the actions of the simplest physical beings, but has an enormous importance when addressing the will and actions of human beings – the 'Maslow' pyramid of hierarchies between the 4 wills of existence. Maslow studied that pyramid between the needs of man as a biological being, yet since the 4 arrows coincide with the 4 drives of living beings it is easy to notice how those 4 arrows and hierarchy are common to all other species. In detail study we can observe that hierarchy to be equal to the main chain of causality between the 4 arrows: we need first to perceive to locate energy to feed ourselves, and only when we have enough energy we can reproduce, being the arrow of social evolution, the least desired, because it also conflicts with the first arrow of perception from the self-centered perspective of our 'Galilean Paradox'. And so the 4th arrow of 'belonging' to a complex, social organism is often imposed from outside, and yet it happens all the time as complex social organisms are stronger than individual cells and hence survive.

Thus we can consider that the strongest will of the pyramid is our self-centered perception of the Universe that makes us feel superior to all other beings – the center of reality, since 'Every point of view measures reality from its local perspective, confusing the entire Universe with its informative, still mapping or 'mind' that makes him feel the center of the world, and ignore all other minds, languages and perspectives he doesn't observe'.

We, as all entities of reality are indeed, self-centered, arrogant Galilean Paradoxes that think to be infinite when we are just an infinitesimal fractal point absorbing a very reduced quantity of the total in/form/ations of reality with a specific 'rod of measure' of space and time, which constructs that mapping but will differ from other rods of measure and 'time speeds' of processing information of all other species of reality. So humans measure/perceive light and use the rod of absolute light speed to measure space and the frequency of a second – the blink of an eye, the heart beat and the speed of human thought – to measure time.

Yet there might be other beings who measure space with gravitational rods we do not perceive, with atomic or molecular rods as we do with the sense of taste and smell, etc.

Even the use of an absolute light speed could be changed by a rod that would make the 'frequency of information' of light absolute and then change the speed of the light waves according to the amplitude of those waves and have an absolutely different mapping of reality, which will certainly make sense, bring new perspectives a seem the absolute universe to those who 'measure' reality in that manner.

All those reasons imply that a topological model of reality is absolutely relative: from an existential point of view, regardless of the importance humans ascribe to their 'intelligence'=mind perception, to their point of view, and to their ego and Galilean Paradox that make us think that our nose is bigger than the Andromeda Galaxy, *we are nothing on the objective sum of all things, and we have no privilege neither we are treated with special difference by the game.* And so because, each and all points of view are 'meaningless' for the entire game, only those points of view that submit the Galilean, self-centered paradox to the 4th will of eusocial love, and sharing of energy and information with other members of the same species to create a bigger superorganism do survive on the long term.

It is the basic tension human beings experience in real life, between our selfish ego-driven desires and biological, individual will and the need to give part of those desires for the common good of our societies and ultimately of the human kind. Indeed, the most successful nation on Earth, China, is based on the submission of the individual will to the collective superorganism. 'Unfortunately in the case of the human-kind, the natural, final arrow of eusocial evolution that should have converted the entire species into a global superorganism able to survive in balance with our energy body – Gaia –has failed, and we have not evolved from the social level of tribal history, divided in nations, created by hordes of the most selfish, self-centered, arrogant type of humans, the warrior who imposes his will because otherwise he can kill and eliminate from existence those who oppose them.

Indeed, regardless of the paraphernalia of 'memes', bytes of cultural information that praise the job of the military, it is self-evident that the biggest risk of extinction of mankind is war and the evolution of weapons of mass-destruction, driven by the failure of egocentric tribes to evolve into a global humanity. And the Universe always punishes those failures with extinction.

Thus we can consider that the simplex arrows of time, informative perception and energy feeding, in which we also ignore the 'existence' of the victim consumed as energy, after eliminating their tasteless 'head of information' are the selfish wills of existence. And the two complex arrows of reproduction, which implies to give for free energy and information to our offspring and social evolution, which implies to give up our simplex wills for the common good of the society are the selfless arrows. And there is a tension between both kinds of arrows in all systems of existence.

Recap. The fundamental will of all points of view is the Galilean, self-centered paradox or will of subjective perception. Yet the Universe favors species, whose individuals are selfless and offer their energy and information to reproduce a self-similar replica of themselves or to build complex super-organisms of which they are only a cell.

43. The synchronicities of knots of time arrows.

Each active organism is a complex chain of cyclical actions performed to achieve the 4 main arrows of time. Yet that chaos of action is ordered by causal processes, synchronized clocks and energy/information networks that harmonize all those seemingly chaotic cycles of actions into a dance of existential beats, interdependent and in permanent symbiosis - being the key to those harmonies among different Time Arrows the concept of 'time-speed', the speed at which a certain cycle of time closes its trajectory into a '*loop*' or *event* (spatial/temporal description) that will be repeated with a frequency that defines the regularity of the time cycle and its speed of perception.

For example, the subjective cycle of time perception in man is the cycle of informative perception determined by the wink of the eye which opens to absorb energy and closes when it processes it as information every second. This rhythm is self-similar to the rhythm of the heart, which is approximately of a beat per second, when blood is expanded as energy and imploded carrying the released chemical information of the body. So, since we have a thought per second and a beat per second, the second is in fact the 'clock-time' of the human brain and the mechanical instruments we use to compare all the cycles of time of the Universe with our own cycles.

Each species and cycle of time has a different frequency, which depends on the 'size' of the species we study, and the causality of those arrows, such as:

Max. Energy-extension=Min. Speed-Frequency of Time cycles:
Max. Speed/frequency: Informative cycle->Energetic ->Reproductive ->Social cycle: Min. Frequency.

For example, the cycle of perception of the human eye – each second –determines the beat of the heart/breathing cycle of energy for the body; the cycle of rotation of the Sun-Earth system – a day – determines the Energy cycle of the planetary surface, whose microscopic, life-species adjust their energetic /formative, awaken/sleep cycle to that of the bigger organism. And vice versa, the smaller microscopic cells of the body reproduce each day, according to that awaken/sleep cycle. There are in fact a series of General Laws of synchronicity between those Time Arrows/cycles that apply to any entity or complex system of the Universe, which extends through several scales of spatial size and informative complexity.

All the actions of all species can in fact be described by 'beats' of existence, partial sub-equations of the Generator Equation, which define rhythmical chains of action. In an individual those beats of existence become the life rhythms of each of our organs, from the eye that feeds on light and then informs the brain, which produces a visual thought every second (the blink time of the eye), (i->e->i) to the night/day, dreaming/acting rhythm (Release of chemical information->dreaming->release of energy->acting), etc.

A 2nd series of ternary rhythms occur as organisms switch between the 3 physiological networks of hormonal, reproductive activity; nervous, informative activity and feeding, energetic activity (Re->in->en).

Those rhythms happen also in the social scale of existence as a single organism, cell of our civilizations. In society those rhythms shape the ternary bio-rhythms of our daily life; the 8 hours of work (obtaining our social energy, money), 8 hours of play (dedicated to family reproduction), and 8 hours of sleep (when our organism repairs its form). Moreover those rhythms of our daily life are synchronous to the rhythms of the global organism, Gaia; and self-similar rhythms are found in the activity of animals and plants.

Since in objective terms humans are created by the same temporal arrows and display the same properties than the rest of the entities of the Universe. A human being is the same kind of entity than 'an atom' – a complex chain of time cycles of energy, information, reproduction and social evolution. The difference is not of quality but of quantity and hence of complexity: Humans are knots of an enormous amount of time cycles whose spatial topologies, actions and motions configure our existence.

On the other extreme of simplicity, an atomic particle (quark, electron or photon) is the simpler species, we can study with those arrows, in terms of size and complexity of its 'EIRS' cycles, compared to a human being. Since a particle can be described with just 4 quantum numbers that resume the 4 kinds of 'actions' particles constantly perform:

- Particles gauge information (reason why quantum theory is a 'gauge theory'). That is, they calculate the distances with other atoms and act-react to exchanges of energy and information with them.

- Particles feed on energy, absorbing electromagnetic and gravitational forces and other particles.

- Particles decouple, repeating themselves in other regions of space-time (they reproduce by iterating its particles, quarks, electrons and photons).

- Finally particles evolve together into complex social structures, called atoms and molecules, which also feed on energy, gauge information, reproduce and evolve socially.

All those 'motions', quantum numbers, molecular vibrations, decouplings and physical events can be re-considered as manifestations of the 4 main arrows of time, chained in synchronic patterns and cycles, cyclical and lineal geometries, whose abstract definition conform the Laws of quantum physics.

The description physicists do of those arrows is an abstract, mathematical, mechanical description, which corresponds to the limits of the mechanical, monist, 'religious' philosophy of science of our founding fathers[4], which did not require an 'internal', self-organizing will to describe the Universe, since it was understood that the 'will', the 'why' of all the clock-like motions of the Universe was 'God', an entity that seems to have appeared to a pastor of the bronze age.

System sciences reasons those ultimate questions from the perspective of science and its mathematical and logic languages, albeit with a degree of complexity superior to that provided by monist, reductionist, mechanist, classic scientific philosophies of reality. Thus we substitute the 'will of a personal God' or any other anthropic principle with the 4 objective 'arrows of causality' or Time Arrows that all particles and entities of 'existence' follow, which can be described objectively with mathematical and logical equations, and finds in the subjective will for survival its sufficient reason.

Further on, all quantum entities obey the Complementarity Principle. They show 2 states or structures that co-exist simultaneously: a field of energetic forces that moves a particle of information, which gauges the space-time that surrounds the particle. And we cannot distinguish or perceive both entities: the force/body and head/particle together. In other words, physical entities are made of 'motion fields' and 'gauging particles'. They constantly move, stop and gauge information, move, gauge... So because they have 2 components, one specialized in energetic motions (the field of forces) and one specialized in informative gauging (the particle, mass or charge), all their actions are either 'energetic actions', informative gauging, the combinations of both type of actions (decouplings, which produce self-similar particles) or more complex actions in which the particle, atom or molecule share smaller particles, called bosons, creating social networks, made of fields of forces and particles of information.

If we widen the concept of energy as lineal, expansive motion, and information as a cyclical, in/formative motion, we realize that the biological realm also follows a Principle of Complementarity between reproductive bodies and informative heads, which co-exist together in living beings. And bodies are responsible for lineal motions, while heads map out the Universe, creating images, reflected in the mind of the external motions of reality, imitated by small cyclical motions within the brain and its neurons. So again we observe a dual structure, body/head, with energetic/informative properties that move and gauge, feed on energy and perceive information. And in more complex events, reproduce the biological entity or use languages of information to create complex social structures.

All what exists is a 'fractal knot of energy and information motions', whose why is 'to absorb more energy and information' from the perspective of the entity or 'point of view' that gauge and moves in the Universe. Imagine a universe of infinite 'species' made of 'lineal motions', 'forces or bodies', upon which certain structures of information, called 'cyclical motions', particles or heads, are sustained. Reality is a fractal sum of those 2 motions, energy and information, which combine to create an infinite number of entities that gauge information, feed on energy, reproduce their form and evolve into social structures till reaching the complex Universe.

We cannot describe human actions with simple equations because humans can gauge many types of information, feed in many forms of different energy, choose many different couples to court, love and reproduce, and make very complex choices in our decisions on how to relate socially to other human beings. But at the end of the journey, amazingly enough we observe that all what we humans do is also gauging information, absorbing energy, reproducing and evolve into social systems. So those 4 types of events or causal arrows of time are all what we humans do; even if we use a complex, 'ambiguous' language – words -to describe those actions.

To exist, to be in this Universe means to absorb energy, gauge information, reproduce our form and evolve socially by acts of communication with other beings. Those 'actions' are considered mechanical in the physical world and actions of will in the human world. Yet they are described in both worlds with the same '2 elements': energy and clocks of time. So Planck proved that all physical forces are made of h-quanta, whose parameters are 'energy and time', while we say in layman terms that 'we do not have enough time and energy to do this' – meaning we are made of a limited quantity of vital space, our energy/body, and time-clocks of information – our minds.

There is no difference between 'us', human beings and 'particles' that also have a limited energy and time, which they spend performing the same game of 4 types of actions - the Game of Existence. We all play the same game, regardless of what plays it, an atom, or a human being.

Religions will tell you that the 'will' of man is a property of 'our soul', which is supposed to exist according to the tradition of our 'pious' founding father, Mr. Descartes, in some non-local space-time[4]. Science will tell you that the 'will of man' is the 'why' we don't ask, or at best it will be born of a chaotic series of circumstances and physical events that come together to 'create life'. While biologists will affirm that the will of man has the same drives of existence (gauging, feeding, reproducing and evolving socially) that all forms of life have, but *only* forms of life have.

System sciences depart from anthropomorphic traditions, validating instead the ideas of Eastern philosophies (Taoism, Buddhism, Zurvanism), upgraded with the scientific method, showing that the Universe is 'organic', not mechanical, because all its particles share the '4 drives of existence' of biological beings, whose 'why' is self-evident in a world made of perpetual motions. Since if we all are made of motions, life is embedded in the existence of those motions that repeat themselves by the mere 'action of moving': Reproduction becomes natural to the existence itself of cyclical and lineal motions that repeat their forms as they trace their trajectories.

If we leave aside metaphysical, subjective questions on the 'nature of consciousness' and 'will', we could affirm that all entities of the Universe have both, a 'will of existence', shown in its constant pursuit of those 4 arrows, and an objective nature, in as much as those 4 arrows or wills can be proved objectively by the events that atoms, humans, animals or any other entity perform. It is though necessary to understand in which scale of reality a certain region of Time-space 'acts'. For example, a chair, made

122

with human formal imagination is not organic neither it shows any of the 4 main arrows of time unless it is attached to a human being (so it helps humans to think- process information - in a 'still', formal position). Yet at the spatial scale of atomic wood, the chair shows those arrows and in the temporal distance when the chair was wood, it had also at macroscopic level those vital arrows of time.

What is then reality – the ultimate meaning of those time arrows? At first sight, it seems there is no meaning at all, but merely the eternal beat of existence, energy becoming form, becoming energy… and the sensations related to those motions, which might be shared by all realities – pain/pressure and pleasure/release. A closer view on both arrows and the complementarity of all physical and biological entities, which achieve immortality through reproduction of their own form – from light imprinting its waves on the vacuum, to quarks jetting bundles of 'quarkitos', to humans 'growing and multiplying' all over the Earth - seemingly makes the 3rd existential arrow of reproduction, the meaning of it all, and the Universe a constant orgasm of repetitive, present forms. But then, we realize that reproduction in itself is just a step towards self-organization, as self-similar cells that can 'decode' their common language of information come together into stronger, bigger waves, herds and organisms, which survive better, due to their higher $\sum ExI$, existential force. So the answer to the meaning of it all is clear: the Universe has a dominant arrow, the creation of 'fractal superorganisms', species which are self-similar to their parts and emerge through processes of evolution into bigger wholes. Since the 4 main arrows of time and its discontinuous cycles come together, creating vital knots of motions that energize, gauge information, reproduce and finally evolve socially into herds and organisms: I->E->R->S. In other words, a fractal infinite Universe of energy and information struggles to create 'more of it' by expanding inwards and outwards its relative scales of organic size; since a Universe made of motions has no limits of size and form, as it becomes ever more organized in bigger and smaller social herds and organisms: I->E->R->S.

Yet even if social evolution is the dominant arrow of the Universe, it is also the less common, more distanced in the frequency of its events, since all other arrows must exist 'a priori' for an event of social evolution - the final causal arrow of reality - to take place, thanks to a common language that carries information among self-similar cellular individuals.

The highest frequency of events always corresponds to informative events, particles and heads gauging and thinking, mapping reality with its hardware, as they run inside the software of a force, whose frequency carries informations decoded and transferred into such mappings. So the atom figures out first in its inner quark-systems how to move, searching for a flow of energy or its relocation in harmony with the other atoms of its web, while an office clerk thinks first and then writes or makes a call related to those informative thoughts - an act of measure precedes an act of motion.

Accordingly to those frequencies and the generic law, Min.E= Max.I, the light pixels of informative mappings are smaller and faster to allow faster events/cycles of perception than the bytes of energy that feed the body - amino acids, which are smaller than the seminal cells that carry the information of reproductive acts, which are smaller than the units of social evolution (other humans). So happens to the frequency of perception (each second), faster than that of feeding, which happens more often than the frequency of sexual reproduction, which is more common than our acts of love - social acts of evolution that 'collapse' a wave of individuals into a particle/organism, emerging as a whole action, born of the self-organization of its parts, such as a religious mass or a voting action, or a marriage, all acts that create social groups. Yet, the ultimate why of any chain of causal Time Arrows must be found in the goal of 'survival', the final drive of all entities of the Universe, maximized by its 2 complex arrows, reproduction and social evolution that evolves cellular entities into macrocosmic wholes, which become cellular entities of a higher 'space-time plane'.

What we are then can be expressed in more poetic terms: knots of existence, flows of arrows of time, mirrors of information, bodies in action, 'wills of the penis'[14], loving cells of bigger social units, exi=stential flows, 'vanitas, vanitatum et omnia vanitates'; motions of time in perpetual conflict.

The rules of engagement of all those knots of time, 'points of view' that gauge information, seek for energy, reproduce and evolve socially, creating 'ad eternal' the infinite scales upwards and inwards, is the ultimate knowledge of the game of existence, which system sciences study. This holistic view is a mystique revealing existence of the unity of it all, which few self-centered humans accept. Scientists rejected it in the past, because it had not been properly formalized. But one of the advantages of the new formalism of Non-Euclidean, fractal, logic, causal chains of Time Arrows/events, is that it renders a mathematical analysis of those arrows of times are its generic laws, hence allowing their application to the study of the species of any scientific *network of self-similar points in an open flow of communication.* Thus, the range and power of the formalism of Multiple Spaces-Times will not become evident till we develop the logic of multiple causes and the non-Euclidean geometry of flows of communication between the fundamental particles of the Universe, its non-Euclidean, Fractal points[15]. A consequence of those postulates is in fact the definition of a soul as a non-Euclidean point of view:

'Every point of view=mind feels the center of the Universe, but it is only an infinitesimal Non-Euclidean point that stores a limited mapping of reality with its informative languages.'

We are all Atmans, informative CPUs, souls that gauge in different languages reality, to construct a mental mapping that caters to our point of view and arrows of time. So we create our 'perceived Universe' – a fact known from Descartes to Schopenhauer, but lost to 'naïve, realist' physicists. All *of them thinking clocks are the only point that matters to measure time, and the human mind, the only ego that knots energy into form, since that is the only perspective they perceive.* That is why most of history, humans thought the Earth to be the center of reality; and still think they are the only intelligent species.

All is memorial because cyclical inertia, the repetition of the cycles of existence of any form, is common to all space-time fields. This is a tautology: a species that forgets, that does not repeat a fractal cycle of existence becomes extinct and its cycles of existence disappear. Thus, a being has to remember and repeat its fractal cycles to last in time. That is the existential game of any being: to feed on energy and use its genetic, memorial information to transform it through reproduction into new cyclical actions.

Yet those actions are encoded in 'fractal equations of information' which are logical chains that constantly repeat themselves, becoming memorial patterns. Those patterns have different names in science: genes in biology, memes in history, Universal constants, which are proportions between physical energy and information in physics, etc. Yet what all those fractal equations and e/i proportions become are cyclical, self-repetitive events that respond to an arrow of time.

Recap: All what exi=st is a fractal knot of cycles of energy, information, reproduction and social evolution. There is not a difference of quality among universal entities – only a difference in the quantity and complexity of those cyclical arrows that 'inform' each species of reality. The ultimate meaning of a Universe made of time motions not of spatial forms is closer to a Philosophical tradition that ranges from Plato and Buddha to Leibniz and Schopenhauer: reality is a game of 'fractal mirrors', souls=Atmans= quantum knots of information that gauge reality to absorb energy in order to repeat=iterate=reproduce their form. Since we are all fractal parts of the whole with a limited duration in time and a limited quantity of space. And so only those species, which reproduce their 'logic arrows' of time, their information, in other zone of space survive. Thus the game of existence is in essence a game of repetition of forms, a game of exi=stences.

44. Generational cycle. The 5th dimension/arrow of time.

The previous, simplified analysis of the synchronicities between time cycles raises again the existence of more dimensions of space-time beyond the classic, 'individual' cycles of energy feeding, informative perception, reproduction and social evolution. We already showed the existence of a generational cycle in 'reproductive waves' of physical energy, such as light, whose motion is just the sum of all the reproductive cycles of the wave; and the same can be said of the complex particle/wave generational cycle, equivalent to a complete life-death cycle, as the physical entity switches between its energy and information states, $E \Leftrightarrow i$. In the case of the body those generational cycles of cells are fundamental to understand the synchronicities of the body, as all its cells except the neuronal cells that dominate its informative system have a limited numbers of reproductions, setting up a generational cycle. And the same can be said of the species of life in this planet.

Thus, there is a 5th dimension of time or generational cycle, which creates in physical entities a lineal dimension of speed or width in the light and gravitational membranes of space-time between which humans are sandwiched; and sets a telomere limit to the number of reproductions of living species, both at cellular and multicellular level. And since such dimension is a 'complex of the reproductive, complex dimension', which is in itself a complex of the energetic, simplex dimension (as all acts of reproduction require a surface of vital space to imprint; so for example women cannot reproduce without a 17% of body-fat volume and particles only reproduce in highly energetic environments), we write:

$$\Sigma\,Reproductive\,arrows\text{-}\!>\!Generational\,arrow.$$

It defines the existence of a new arrow or dimension of time. We have not considered though such arrow in this book for simplicity and because ultimately in the fractal universe, *the number of dimensions when we keep adding complex planes made of networks of st-points, which restart again new macro-energetic, informative, social and reproductive cycles is infinite*. And ultimately because we live in a 4-dimensional Universe, in which the 'generational cycle', except for simple, short-lived entities as those who create the 2 Universal membranes of gravitation and light, are non-perceivable except for long spans of time.

Recap. The sum of reproductive cycles gives birth to the generational cycle, which is also a finite arrow, as all forms have a limited number of generations, after which the reproductive systems fail.

45. The transcendental arrow: Emergence.

In the same manner that the sum of multiple reproductive cycles gives birth to the generational arrow, the product of multiple networks of informative beings gives birth to the transcendental arrow, such as:

$$\prod\Sigma^{2}\,informative\,particles\!:\prod Social\,networks\!:Transcendental\,arrow.$$

Thus a new dimension of time is the social evolution of organisms into bigger super-organisms, scale after scale, from particles to atoms to molecules to cells to organisms, to superorganisms to the entire Universe with its infinite scales.

Though each organism is when observed in detail a super-organism, if we were to consider a rigorous distinction between both, the difference would be of our detail of perception: a super-organism observes each of its cells as an organism of smaller 'cellular element' and an organic view stops in the analysis of its cells. A fact, which leads to a metaphysical question: it is transcendental arrow infinite in scales; or it is as all other dimensions a finite dimension? Unfortunately an argument on the number of scales of the Universe will always be theoretical, because unlike the other dimensions we cannot observe the absolute totality of space extensions and informative, scales created through time durations to have any experimental evidence of such limit. The *Transcendental* dimension, the dimension of emergence, studied by the 'science of emergence' or transcendental arrow, as it is often not perceived, especially for the individuals who become part of the whole; believers that form part of an eusocial human organism through the arrow of love; physicists, who don't see the super-organisms of galaxies, etc.

The difference between the transcendental arrow and the organic arrow is subtle. Unlike organic systems mediated by the 4th arrow, the transcendental arrow is far less brutal than the stick and carrot system by which the informative, nervous systems of organisms dominate the cells of its body; in as much as it implies a high degree of homogeneity and collaboration between the parts that transcend.

In that regard, we could consider that the social arrow forms 'herds', waves and bodies, while the transcendental arrow is the arrow that puts together complex, informative particles together, and while the arrow of organic evolution would be based in the concept of 'safety', the transcendental dimension will be based in the concept of love, as the particles know each other more intimately, and in certain cases can even become fusioned in a bosonic structure, in which all occupy the same space.

Thus the difference between the Generational and Transcendental arrows and its fractal units, the arrows of Reproduction and Organic Evolution is minimal and that self-similarity brings about some philosophical, metaphysical conclusions:

-As in a succession that converges with less difference between terms, the *convergence* of time arrows seems to indicate that there are no more 'dimensions of reality' beyond the *Transcendental* arrow.

-The *Transcendental* and generational arrow seems to act, 'reinforce' each other, as the generational limit obliges species to evolve into eusocial organisms if they want to survive, hence 'transcendence' is the ultimate mechanism of survival in the Universe. In the same manner the Simplex and Complex arrows are intertwined as one cannot exist without the other. Such is the harmony of all the systems of an interconnected Universe.

- The more complex arrows of time seem to create more perfect worlds. If we assign to each arrow a sensation and/or its negative lack of it, blindness vs. perception =information; hunger vs. taste=energy feeding, pleasure= reproduction; safety=organic evolution; eternity=generational cycle and love= transcendental cycle, and postulate that *all entities of existence are moved to action by the existence of sensorial fields associated to its arrows;* we reach some philosophical conclusions:

- The simplex arrows are automated and reinforced by the duality of negative/positive sensations, where the negative sensation is so powerful that it obliges the entity to act without delay to achieve the arrow. We can then image the simplex physical particles whose basic fields is a dual energetic /informative system to be far more automated than the next scale of beings, biological beings dominated by the reproductive and organic arrows.

- Those 2 arrows are what Aristotle called 'vegetative wills' and are reinforced by the objective laws of survival, as those species who do not reproduce or evolve into bigger organisms tend to perish and become food of larger biological systems. Thus, we observe in those 2 arrows and systems a higher degree of freedom, as species can live without reproducing or acting in social groups, but the penalty a 'posteriori', extinction is equally deterministic.

- Finally the generational cycle and the arrow of transcendental evolution *seem not to be imposed from within the system but from outside and despite of it.* Indeed, the generational limits of existence are not positive and certainly not liked by the species, which is subject to it. A solution is found when we realize that the generational limit as well as the organic arrow that creates herds does not affect to the informative cells and its networks; since indeed, black holes are immortal, as the informative elements of the universe and neuronal cells live all the existence of the organism. And so it becomes obvious that the limits of existence of those arrows are imposed by the neuronal, informative networks of the system.

And so the transcendental arrow of eusocial love is *the summit of all the arrows of existence, unique to informative networks, whose members know each other deeply, share energy and information in higher degree, command totally the other elements of the organism, last forever, as long as the organism exist and have a higher* degree of freedom that all other species. Reason why there is not clear systems of reinforcement. It is the summit of existence, free and eternal, within the limits of the ecosystem or super-organism in which those cells exist.

Recap. The transcendental arrow makes individual organisms transcend into super-organisms, creating the scales of the Universe.

46. The function of existence of the universe.

We humans have a clear limit in our perception of those scales, given by a series of parameters proper of radiant matter. We can't see beyond quarks and galaxies; we can't absorb information coming from faster than light forces (non-local gravitation) or perceive ordered vibrations below 0 K temperature. It is a metaphysical question to wonder if those are the hard limits of reality; in which case the smallest point, the quark, won't be as Leibniz put it, a world in itself, but a hard limit: the simplest rotational movement or 'spin' of the Universe; or they are, as I believe, only the limits of human perception. Then all what exists is truly relative in its scale, and the atom is an object self-similar to a galaxy, which would reveal infinite detail if we could approach the ultra-diminutive Planck scale. A fact, which translated into fractal notation, gives us the simplest formalism for all Universes, sum of all potential cyclical forms of all its ∞ fractals:

$$\prod (\Sigma e <=> \Sigma^2 i)^{n=\infty} \quad \textit{Universal function of Existence}$$

The previous equation resumes all the possible infinite combinations of cycles of times, made of n-points in communication, that create all type of networks of energy and form, which become complementary organisms, and start a function of existence. So first energy appears, energy becomes form, energy and form reproduce and create social organisms with the reproduced cells.

We can observe the universal equation also as a game that grows causal dimensions: from lineal energy (1) comes from (2) and both combine to reproduce (3), self-similar waves of being. This simple description of the game of the Universe, E ->EI->ExI->$\sum$exi, is already present in Taoism (from 1 comes 2 from 2 comes 3 from 3 the infinite beings, Cheng Tzu). It resumes the constant, dimensional creation and extinction of reality (as finally the fractalization of so many cells, exi, dissolves the form or the form explodes and dies in a big bang).

The Cycles of the Universe are thus self-repetitive; sine as Taoists understood, the universe you see is not its function of existence, which is expressed in that synoptic equation. The processes of creation and destruction of planes of existence however are too complex to study them as a whole without doing some basic distinctions. To start with there are two different type of arrows, simplex, E$\Leftrightarrow$I processes, more proper of physical sciences (as in changes of state by increase of energy, from solid to liquid to gas or vice versa; or as changes from lineal energetic motions into cyclical vortices, as in E=Mc^2), are easier to analyze as they happen in the same place of existence. But we, men co-exist between the quantum and galactic scale of physical objects and between the cellular and social scale of biological beings. And so most of the events we must consider are events that take place between 3 co-planes of existence. In physical events between the quantum, human and gravitational planes and in biological events between the cellular, human and social plane.

The human mind is an I=Eye+wor(l)d who selects perceptively from the infinite number of forms of the Universe only those that matter to us. Thus, we do not see beyond the galactic, gravitational plane; though there are hints about a network structure of Galaxies, a big-bang model, which in organic terms would mean the birth of a first cell-particle that reproduced the entire cosmos and a model of fractal space-times with infinite scales of size that this work represents. Since the classic model of space-time continuum only considers a single light-space, our existential membrane with limits of energy and information (c-speed, 0 Kelvin).

Yet both in human and cosmic organisms only multiple space-time scales explains those cycles of life and death, which happen as a 'travel between 3 scales of size, the cellular semen, human individual and social super-organism of history. In the same manner cosmology needs to grasp the fact that physical space is organized in 2 planes, the electroweak membrane we exist in and the gravitational space membrane of masses, and gravitational forces, which we do not perceive. Yet it is needed to resolve the meaning of dark matter, dark energy, gravitational non-local action, the death of light into gravitational space, its formation of a background radiation, the actions taken place in black holes, etc.

Recap: A Non-Euclidean space or plane, defined by Riemann in the XIX C, is a network space - a space created by a network of self-similar points, related by constant flows of energy and information, whose properties are defined by the degree of 'homogeneity' or self-similarity of those points. So your body is a discontinuous spacetime, a small world in itself, constructed with 3 clear 'cellular spaces', the nervous space, an informative network, the 'energetic space' – your digestive and breathing networks – and a reproductive space, the blood and hormonal networks and organs. We shall see then reality in organic terms as a series of 'network spaces', which will have all the properties of Non-Euclidean spaces (motion, curvature, dark spaces in the 'holes' left by the network, etc.).

47. The laws of creation: ternary principles.

Some dual, creative analyses forgotten by AE sciences that we consider in MST Theory: the Universe as a spatial system, machines as evolutionary beings, and civilizations as organisms. The game of creation must be understood through its 'dual symmetries' and ternary networks that allowed its finite reproduction of fractal forms in the infinity of continuous space

The most satisfying function of existence is reproduction, the Generator Arrow, the maker of the game and all its species - the purest sensations of them all.

The Generator Feedback equation of spatial energy and temporal information is a reproducer, which embodies all the combinations of the game in itself, the creative algorithm of reality, which Touring once looked for with its simplest example a touring machine. But the creator potentially is more than a touring machine, it is what the touring machine observes, reality itself.

The rest of beings are just following the combinations of that grand design.

The active creator is limited by the parts of the whole that will assembly. So we shall call it with a more humble name, the assembler

With the syntactic combinations of the i-logic generator, the creator will always end repeating a form of finite time that had existed in ∞ space.

The proof that we are all repetitions is clear: space is infinite because it has not dark spaces, but it is a continuous number of closed and open topologies assembled in triangles, squares, hexagonal, polygonal and hexagonal, planes and spheres, with no darkness.

Information however is always limited and time always ends: it is a fluctuation of existence over that ∞ energy.

All existences have possibly existed in the infinite fractal space of a finite number of times cycles.

We all will be assembled again as an i-logic form of space-time in some other self-repetitive fractal printed in the infinite energy of the trophic pyramid of existence that gave you birth.

Let us then operate the creator: Its simplest combinations will be complementary systems of energy and information that take advantage of the inverse properties of both:

The dualities of Creation.

In time the generator combines past, energy forms and future informations, generated by its universal, feedback cycles, and its arrows that imprint and create the chains of forms of present existence. Creation is structured by a series of dualities that merge creating an eternal, fluctuating present that appears and disappears as cyclical action-reaction systems cancel each other, making its total sum zero:

- Time and space have opposite morphologies and functions, which cancel each other:

- Informative, temporal singularities are convex, implosive and energetic membranes are concave, explosive. Thus a flat plane of temporal energy fluctuates into 2 virtual forms whose total value is null. For example, particles and antiparticles are born of a vacuum plane without form but they keep their inverse CPT parity. Thus the sum of its spatial form (P), its temporal arrows (T) and its dual, organic charge cancel.

- The duality of energy and information explains also *sexual differentiation:* females are specialized 'time beings' dominant in cyclical forms, memory, information, temporal verbs and perceptive languages. While males are specialized 'energy beings' with bigger, lineal forms, dominant in spatial tasks. Yet both can be further differentiated into an informative cyclical head, and a lineal body. So, in graph 3.1 we draw a man with "temporal cycles" that represent his informative organs, the head and senses; and energy lines that represent the body and his members. Yet, despite the simplicity of that design, he is recognizable as a man.

- Duality causes the creation and reproduction of space-time fields through the process of palingenesis, by which a certain form of relative future, a father, emits a relative form of past, with lesser evolution, a seminal seed, that then evolves very fast towards the future, till becoming again a present form parallel to the father.

- Duality exists in biologic organisms evolved by the dual influence of macro-ecosystems and micro-genes.

- Duality is the cause of informative perception, as flows of spatial forces become transformed in a point of relative time or particle, in which they 'merge' into a 'bosonic', accumulative image of information that represents reality.

- Duality also applies to behaviour: there are Darwinian acts between different beings that destroy each other vs. social, evolutionary acts between equal beings that evolve together, sharing energy and information through common networks. Thus all organic life ends up cancelled by a predator. It is a key duality as it responds to the 2 arrows of space-time, the arrow of symbiotic order, of information and the arrow of energetic entropy of destruction defined by the 3rd postulate of illogic geometry.

- Finally Duality implies that all forms that evolve in time through 3 horizons, then organize in space those 3 horizons as the 3 regions of an i-point.

The number of events we can describe in all sciences departing from duality is enormous. Yet in as much as temporal information is dominant in living beings, illogic time is more important in biological and sociological sciences, explaining phenomena such as reproduction, perception, life and death, organic structures, etc. While, physical and cosmological particles can be described better with the use of Non-Euclidean, spatial geometry, as particles are dominant in spatial energy. We will consider in the next chapters the most important biological applications of illogic time: the duality of Darwinian and symbiotic behaviour; the way in which perception occurs; the concept of a top predator species, as information selects species with better brains; the palingenetic reproduction of biological forms which implies a dual travel in time back and forth from the future to the past…

But duality is only the beginning. Most systems evolve, self-combine and create finally a 3rd system, a reproductive one, becoming autonomous, without the need of 'assemblers' and 'enzymes'. And then the game becomes richer in variations, because it becomes guided by ternary symmetries.

We understand those organic, self-reproductive systems according to the *Ternary Principle* studying the 3 *temporal ages* and functions of a MST field and then putting them together as the 3 *geometric, organic, spatial* regions of i- point. Since all MST field require 3 elements to exist: a Ti, informative element; an Es, energetic, spatial form, and an intermediate 'present, simultaneous region', dual flux of temporal energy that merges them into a whole. The result is an Se-Ti rhythm of evolution and reproduction of forms in time that become latter reorganized as i-points in space, creating dual organic, real forms of temporal energy. For example, the Universe first created fundamental particles, temporal quarks and spatial electrons that recombined into ST atoms; the body reproduces cells that latter evolve, becoming organs of energy and information, etc. Thus MST theory shows the complementarity between the ternary horizons of species that once have evolved energetic and informative particles interact creating an intermediate ST zone, shaping a new i- organism:

- Max. E: The external membrane and energetic network of the system that performs *energy cycles*, transforming information into energy, appears first: it is the cell's fat membrane, the stars of the galaxy, the endodermic cells of the future digestive systems….

-*Max. i. Then it will appear the informative quanta of the future informative network and 'brain' of the* system that perform *informative cycles*, transforming energy into information: they are the cells' informative nucleotides, the black holes of the spiral galaxies, the brain.

- Σexi: Finally the interaction of energy and informative systems creates *an intermediate, reproductive region that* combine energy and information: so protein membranes and nucleotide acids create the cell; stars and black holes create galaxies; the energetic endoderm and the informative ectoderm create the middle mesoderm region, each one the blue print of the future energetic, digestive, informative, nervous and blood, reproductive systems.

In formal terms we write the process as a decoupling of an initial *exi form* that differentiates into a more energetic membrane (Max.e) and higher informative quanta (Max.i), that interact, creating the intermediate region, Σexi, shaping in this manner a new, $\Sigma E \Leftrightarrow \Sigma^2 I$, MST field equation – the ultimate definition of any species. Since we are all self-repetitive MST field equations

The creator, MST equation, $\Sigma E \Leftrightarrow \Sigma^2 I$, represents both: a temporal event between 2 relative 'points', an E-point with higher content of energy or relative past form, and an i-point with higher content of information or relative future form, that communicate implosive information (>) and explosive energy (<), through a wave of temporal energy, $\Leftrightarrow$; *creating a ternary, multicausal, simultaneous structure of spatial present.*

Multifunctionality: 3 st-points functions. Ternary Principle:

Creation happens due to the 'diversification' of any Space-time field in 3 subspecies in space or 3 ages in time; e, exi, i.

It is the ternary principle once and again written in the book of Nature, caused by the fact that there are only 3 elementary forms in the Universe, energy, information and a combination of both. Thus, events, species and space-time fields, both in time and space, have 3 elements: *3 ages, 3 horizons, 3 dimensions or 3 physiological networks*, whose functions correspond to the arrow of energy, information and reproduction that create the Universe. The Ternary principle is the origin of an evolutionary, impersonal plan of creation that diversifies species in all scales of reality into 3 forms, (Max. ΣE, Max. Ti, ΣE=Ti), from the 3 families of masses to 3 the types of Universes. It allows organizing all biological species in a tree of ternary horizons and differentiations of energetic, informative and balanced organisms, which co-exist in 3 st-scales of existence:

'Any form can be subdivided ad eternal in new ternary forms.'

For example, the human body can be subdivided into 3 networks; then the digestive network can be subdivided into the stomach, liver and intestine system that can be subdivided into the colon, small, and large intestine, which can be subdivided into the left, top and right side, etc. The Ternary principle implies that the 3 spatial dimensions of any space-time field perform 3 temporal functions:

'Any species, which is part of an ecosystem or an organism, maximizes its survival developing 3 functions as an energetic, reproductive and informative system for the higher scale'

I.e.: a cilia act as energetic limbs that move the cell, sensorial antennae that inform, and they evolved as centrioles that help to reproduce it. Hormones are also multifunctional.

And so on. This first rule of creation developed either in space as a topology of 4-dimensional reality with 3 elements, hyperbolic ring, toroidal cycle and spherical plane, or in time as a game of energy that warps into information reproducing along the way, is your limit.

So the creator descends a notch more into self-appreciation since he is nothing but a contemplator of a game which has created before. That is he is created by the game and as assembler he is so determined to do one of the known-known combinations that, indeed, there is nothing new under the sun.

The laws of the game of creation are in themselves an entire sub-discipline of multiple space-times theory, studied in detail in my files. In essence the process is self-similar in all scales:

A species will differentiate in ternary sizes (its cellular, organic and social size); it will differentiate in ternary topologies (an species dominant in energy, one in information and one in reproduction); in ternary ages (a neoteny species aborted in a palingenetic phase; one mature species and one with an excess of information, *which will become cell of a new scale of evolution, a superorganism communicated by the language); and* in ternary functions, symbiotic to the higher st+1 organism in which the entity exists (as an energetic part, an informative part and a reproductive part or else the organism would not 'tolerate' the presence of the microcosmic species with no function).

Further on all those creative strategies of survival become more complex, when we consider its combinations: species with several functions, often in several scales; complementary, sexual specialization in an energetic, male entity and an informative, female entity; ternary structures that form complete topologies; open balls that act as doors between membranes, without center and membrane; a*nd combinatory varieties of 'sexual species' or 'complementary' species, in which the energy/informative components, each diversify in 3 topologies, scales, ages, etc.* Those combinations further enlarge the number of subspecies, though many of them, especially 'anti-species' in which the lesser informative 'male' plays an informative role and the lesser 'energetic' woman an energetic role (for example an antiparticle, with an energetic electron in its informative center and an informative proton in its energetic membrane), will not be stable and will not survive. This brings another essential law of creation, the inflationary nature of information checked by the laws of survival that extinguish unsuccessful species and tends to reduce the explosive age of creation to 3 'basic ternary differentiated' types that survive, while other transitional or non-balanced combinations disappear:

'The Universe creates an inflationary number of 'forms' which are then reduced by natural selection to ternary species and balanced, complementary, dual systems'.

Those rules allow the entity to play an interconnected, synchronized role as part of a bigger organism and/or ecosystem, in which it will play its dynamic, causal chains/arrows of existence and life cycles:

Max.I(seed)->Max.E(youth)->Max.Reproduction (Maturity)->Max.information (3rd Age)->Death(I->E)

But Life cycles can be be immortal, if instead of dying after its 3rd Age the entity evolves socially as cell of a higher system communicated with its informative language (E_{st-1}). Then the species will 'transcend' into a complex super-organism. Some systems might even attempt a feed-back cycle of immortality (Physical particles, simple biological jelly fishes:

I-> energy youth ->reproduction <-Energy Youth...
Energy (wave) -> Information (Particle) ->Energy (wave)

And so, with those simple rules we can classify as we shall do in our studies of physical, biological and sociological species of the Universe all the entities of reality and its events.

Recap: The ternary principle explains the creation and diversification fractal super-organisms both in time ages, scalar planes and network-spaces. Space is infinite time is not, time games are less than space. Because the volume of space is bigger than the volume of time all forms have been repeated. And so all is repeated again.

X. SYSTEMS SCIENCES

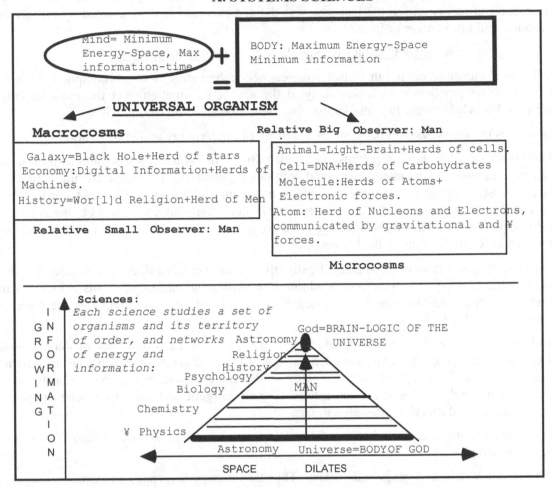

There are infinite Fractal worlds, 'st-points' where 'st=n' is the scalar parameter that defines both, the size of the relative Universe and the relative science that studies it, specialized in a certain scale. For example, Chemistry studies fractal molecules; Biology studies fractal cells and History fractal, human societies, divided by spatial borders and cultural, informative discontinuities. Thus it is possible to describe mathematical, physical, biological and mental species with the same forms and functions of fractal structures; and create a Science of Sciences, dedicated to the study of all fractal structures in the Universe, and the description of its species with those common functions and forms.

48. Unification of sciences: General Systems Theory.

We considered the 6th dimension of time to be the constant growth in complexity of all systems that evolve into new scales, from simple particles till forming the Universe in physical space and human societies in biological space. And so there should be a science that studies that final dimension of existence. And indeed, it exists It is Complexity, more properly called Systems Sciences, was founded at the death of Einstein, in the Macy's congress, where the most advanced scientists of the age met to find new avenues to the 'dead end' in which quantum theorists have placed the evolution of science, with their definition of a Universe guided by a single arrow of energy, found in the restricted analysis of heat, and later expanded to all universal phenomena. In this manner the idea that the Universe has a single arrow, expansive entropy had become dogma, without taking into account the cyclical motions of electronic particles and the existence of the arrow of informative gravitation - an attractive force, which therefore informs and creates order. Since as Einstein put it gravitation is an arrow of 'Time (that) bends

132

space into mass'. Yet in the Macy congress it was understood that information played also a fundamental role in the creation of reality and so the Universe was defined as a dual system, created by two arrows, energy or motion and information or form.

Duality & Complexity: Simple and Complex Time arrows

The best-known science to come out of that congress was cybernetics in which computer science is based, because of the practical uses it has today in the age of information. Yet the most important sciences for pure knowledge were two less-known disciplines, Duality and Complexity:

- Duality studies the Universe with the 2 'simplex' arrows of time, energy and in/form/ation[16].

Scientific Duality has important precedents in Eastern cultures. In Taoism yin translates as information and yang as energy, and the Universe is explained with those 2 elements. In Europe Leibniz introduced Taoism to the West, establishing it as the foundation of German philosophy. There, Hegel translated philosophically Duality into Dialectics - the concept that from a thesis (the equivalent of energy) and an antithesis (the equivalent of information, with inverse properties to those of energy) a synthesis arises, by merging both in 'Complementary, dual systems'.

So even if Duality was not understood scientifically till Systems Sciences filled it with scientific data, it has been always a 'Theory of It All', in the tradition of religions and philosophy. This is logic – since humans have understood Duality with different degrees of detail and different jargons by the mere fact that they observe a dualist Universe.

What modern Duality has done is to upgrade those concepts with the discoveries of mathematical, biological and physical laws, which show the Universe as a Fractal System of energy and information. In that regard, the central concept of modern duality is also the Principle of Complementarity, which states that all systems in the Universe are constructed with an energetic and informative element, whose geometrical properties and functions are self-similar:

Energetic systems store energy and move. They occupy more space and are lineal (the shortest distance between two points)

Informative systems store and gauge information. Thus they occupy less space and tend to be cyclical (the form that stores more information per unit of volume).

This dual complementarity is the main Law of quantum physics (principle of complementarity between a lineal field of energy and a cyclical, particle of Information). It is the principle that explains the structure of all living organisms (from animals with cyclical heads and lineal limbs; to plants with energetic trunks and planar leaves, which store and process energy, and quantized, fractal roots that absorb chemical information; to cells with a central nucleus that stores the DNA/RNA information of the system). It explains biological reproduction (since males have lineal limbs, specialized in the collection of energy and females, with curved bodies, reproduce our information). It is the principle of Computer science, based in the duality of symbols, 0 and |.

The Universe has two inverse ways of creating reality: it creates either energy by erasing form, as when a quasar explodes, $E=Mc^2$, expelling 'dark energy' and flows of expansive space; or when any being dies, dissolving its form. Or it creates information by warping energy, as when energy becomes mass, $M=E/c^2$ (in fact the first equation that Einstein found), since 'Time bends (vacuum) space into mass' (Einstein) or 'Time evolves the form of beings' (Darwin). And so all what exists fluctuates between the arrow of information, of life, of creation of particles with form, E->I, and the arrow of death, of erasing of form into energy, of big-bangs, I->E.

Both together, in their transformations, E⇔I, generate the cycles of creation of form (life cycles) and destruction (death cycles), which define the existence of its parts.

Thus Duality advances a step further in the great insights of Eastern and German Philosophy, opening new avenues of thought and explaining the why of many phenomena that the use of the single arrow of quantum physics cannot resolve.

- But Duality only studies simultaneously the energy and form of what we call a 'plane of existence', or 'scale of the fractal Universe' (Religious/philosophical vs. mathematical jargons). And when we consider reality, it is obvious that the Universe is structured in 'layers' of entities of growing size and complexity – from particles to atoms to molecules, cells, organisms, societies, planetary systems and galaxies. In each of those layers we find Dual species: quarks of massive information and electrons of energetic space, which form atoms; Nitrogen rings that store information and phosphate chains, which form DNA molecules; Nuclei that store information and cellular bodies that store energy; heads of information and bodies of energy; audiovisual networks of information and economic networks that distribute energy, forming together nations. And this pattern of social evolution and emergence into new, more complex systems that create a bigger plane of existence also occurs in physical space, as molecules become planets and stars that organize themselves into galaxies. So we must add to the 2 Simple arrows of time, energy and information, 2 Complex arrows of time: the creation of self-similar species or reproduction and the social organization of those self-similar, cellular species into networks and more complex systems. And we find those 2 universal arrows in both, physical and biological systems. Even the simplest particles of the Universe, quarks and electrons reproduce into new quarks and electrons when they absorb energy, as biological beings do, and organize themselves into complex cellular systems in atoms that have many quarks and electrons. In that regard, Complexity studies how reality is created by multiple layers of reality, which from the simplest strings and particles evolve into molecules, cells, organisms, planets, stars and galaxies of increasing size and complexity.

So the arrows of energy and form are not enough to understand the complex Universe, which exists through a series of 'planes of increasing complexity and self-organization', starting from the first theoretical duality of 'lineal strings and cyclical strings', till arriving to the final duality of black holes of gravitational information and stars of electromagnetic energy (Galactic duality).

Such networks gather an enormous number of individual energy and information cycles/cells of space-time into a single whole or social organism deployed across multiple planes of space-time. The study of those scalar, network structures is the field of Complexity Theory, the sub discipline of System Sciences, which departing from Duality, analyzes how lines and cycles of energy and information reproduce, evolve, self-organize and emerge as a bigger system, in a 'higher', ever more complex space -time plane, creating in this manner complex systems, made of simpler, 'cellular parts'.

Thus Systems sciences require a second discipline, besides Duality, to study complex systems and their self-organization thanks to the arrows of reproduction and social evolution, which create networks of energy and information that become wholes, units themselves of a new plane of existence:

System sciences: Duality (analyzes how the beats of energy & information, E ⇔ I, create the future) +

Complexity (studies how complex, reproductive, exi & eusocial Σ exi, events create the future).

Thus Complexity studies the 2 'complex' arrows of time, derived from the combinations of energy and form, exi, or 'reproduction', which creates self-similar 'cellular beings' in a given plane or scale of reality and 'social evolution', which creates complex, tight networks made of those reproduced exi parts, that become together a single unit of a new, higher, bigger, more complex plane of existence. Those 2 'complex' arrows of time derived from the Simplex ones (energy and form), create the 4 'dimensions' or 'arrows' of time needed to explain the Universe, and its understanding through the formalism of the Generator Equation of Multiple Spaces-Times, Σ exi, represents an enormous leap for all sciences, which prior to the Systemic Paradigm worked only with the arrow of entropy and energy proper of XIX century physics. We ad now to that 'arrow of energy', the 'canvas', the paint - the information in which that energy transforms itself, and the painter, since we contend that the finality of the Universe, the

134

ultimate arrow of 'future time' is the creation of social organisms through the use of a common language of information - hence the creation of history, the superorganism of mankind, the most complex form of the Universe. We prove all this with the experimental method and the logic and mathematical formalisms derived from the generator equation, Σexi, whose symbols define the 4 main arrows of time:

e or Se, which is the symbol of spatial energy, the energy-motion, which is stored in vacuum space, from where all 'forms' of information are created.

I or Ti, temporal information, which is the dominant arrow of future, as time bends energy, trans/forming it into particles and 'forms' of information.

E x I is the formalism of reproduction. Since all systems reproduce their reproductive bodies/fields and informative particles/heads and the networks, X, that relate both systems.

Σexi describes the Social evolution of self-similar cells, exi, into herds and organisms.

Thus the Generator Equation formalizes the main cycles, laws and events generated by the principle, 'energy never dies but transforms itself into information', which applies to any form or motion, in any scale of reality. Since the generator equation of the fractal Universe is just a complex expansion of the main equation of classic science, 'energy never dies but transforms itself', to include information. Yet when those arrows of time develop all its 'repetitions', feed-back iterations and combinations, as it does a simple fractal generator, like the Mandelbrot fractal, all the possible cyclical events/forms of transformation of energy into information appear in the Universe. In that sense, the study of the generator, fractal equation of the Universe could be to XXI century science, what the study of the final equation of Entropy, the arrow of time studied by physicists, $E=Mc^2$, was to XX century science. Yet instead of generating pure energy as the equation of entropy has done, the fractal generator creates life, which makes its study much more rewarding for mankind, - *a life form.*

Recap: A true Theory of Unification, as systems science is, should explain all what exists, including humans - why we are here, our role in the Universe, what *truly matters to mankind.* This is what System sciences do by using the 4 main arrows of time. Since all Space-time beings are causal networks of cyclical time knots, created by multiple chains of organic, e⇔i cycles, and its complex combinations, reproduction and social evolution, where each 'time arrow event' becomes a flow of energy and information that converges into the space-time of the knot. Those knots can be formalized with cyclical, fractal space-time parameters in terms of frequencies and dimensions. Such perception of the World through Time Arrows was natural to Eastern philosophy. System Sciences and its 2 main disciplines, Duality and complexity study those Time Arrows with the scientific method.

49. The Philosophy of Science of Multiple Spaces-Times.

We are times of existence, cycles of time motions with a will and a purpose, to carry about a causal arrow of time. Those causal arrow exist, they are there. Why, who made them, are they eternal as the game seems to support? This we don't argue. We just know that certain causal chains exist and design a game, which is biological and perfect.

In the graph, depending on the species and plane of space-time each science studies, it will analyze species with different proportions of space & time. The science that studies the simplest scales of maximal content of energy and space, Physics, is for that reason tendencially lineal and based in spatial geometry and its self-similar science of algebra; but has been unable to understand time beyond its spatial use to describe movement in space (Galilean and Einsteinian definition of time as a parameter of speed, s=vt=ct. Yet far more important to the human kind is the science of biology that explains how those simplest morphologies have mutated, evolved and changed, bending the original space into complex morphologies of 'time'. Biology is however a complex science, made of two sub disciplines 'bio', the study of life and its wills of reproduction, information and feeding, and logic, the science of time, which studies the interaction of those 3 wills/arrows of existence and its cycles. In that sense, if there is a primary science of God, it would be Biology and specifically Logic, as God, the mind of the

Universe that creates its forms, is indeed, time, Tao, the sum of all the temporal cycles of existences of all its beings and its common laws. Mathematics and logic are the foundation of any combined study of space and time, and their postulates can be applied to any relative ±st scale of the Universe, where the same invariant geometries and logic functions will be repeated. For that reason, the combined postulates of mathematics and Logic, which we call in complexity the Postulates of i-logic Geometry, resume the space and time laws that rule the events and actions of all superorganisms and its 4 arrows of existence.

Indeed, the Universe is a super-organism of space and time, extended in multiple scales, self-similar in form and function. Despite the infinite potential combinations of spatial bytes and time bytes that give origin to the super-organisms of the Universe, all of them, are made - paraphrasing the Bible, 'to the image and likeness of God', the fractal super-organism of spatial energy and temporal information, which self-organizes bytes of information and bites of energy in 3 dimensional networks, creating all kind of entities in all the scales of the Universe. Thus Metaphysics, the science of 'God', is the study of the Dual laws of Space/Time and its 3±st arrows/dimensions that give origin to the fractal Organisms of the Universe. While each specific science studies a particular scale of the Universe and its super-organisms.

In that regard, a fundamental new law of the 'fractal paradigm' of self-similar beings is the recognition of the homology between the self-similar basic 'bricks' that structure the quantum world (the atom), the biological world (the cell), the economic ecosystem (the factory) and the astronomical scale of reality (the galaxy). Indeed, we shall constantly bring to the reader the remarkable self-similarities between atoms and galaxies (proved by the Unification Equation of charges and masses, which finds protons and black holes self-similar; galaxies and cells (both with an informative center of DNA and black holes, which controls the position of its 'star and mitochondria factories); economic ecosystems (nations) and biological organisms with its nervous/informative=audiovisual networks and blood, economical, reproductive networks, etc.

Those self-similarities which were always taken as a metaphor must be regarded as a reality that springs from the limited number of combinations of the 3 canonical topologies of the Universe, its 4 arrows of bio-logic time and the 3 scales of construction of 'fractal super-organisms'; so atoms, cells, factories and galaxies are the first scale of physical, biological, economical and universal structures; where the basic cycles, synchronicities, topologies and functions of existence are more clearly 'drawn'. They will give birth to a second scale of molecules, organs, company-mothers and galactic clusters, which signal the transition to the final scale of physical systems and biological organisms, economical nations and universes, in which complexity and differentiation of form is maximal, *even if the topological structures and causal arrows are maintained.*

Of all those 'higher' systems the more interesting super-organisms are its 2 known limits of energy and form:

- The Physical Universe, studied by the science of Astrophysics, *the largest in space,* based in the duality of Temporal Particles and Spatial forces, extended through 3 main scales of size, the Planck scale of Gravitational forces, the human scale of Electromagnetism and the cosmological scale of stars, galaxies and Worm Holes.

- And the Human World, studied by Biological and Historical sciences, *the most complex in information,* which extends also in 3x3 planes, inscribed within the wider Universe: the cellular, individual and social scale of Gods and civilizations.

Thus, as a human being, I share with Kant the fascination for 'what is inside the human mind and above it' without limit.

All those sciences can be study with the laws of Time arrows and multiple, fractal space-time planes, which are parallel relative worlds, of different scalar size; each one studied by a science specialized in a certain scale. For example, Chemistry studies the st=5 plane of molecules; Biology studies the st6 plane of cells; History the plane of human societies, divided by spatial borders and cultural, informative discontinuities, etc. While Duality is the Science of Sciences, dedicated to the study of the common laws of space and time shared by all those mathematical, physical, biological and mental super-organisms, wholes made of parts which display the same forms and functions of the whole; all of them defined by the $3\pm st$ dimensional cycles/arrows of feeding, information reproduction and social evolution they perform during its existence. Duality studies the homologous, invariant laws at scale of energy and information that all the species of the Universe follow. While Systems Theory studies the growth and organization of those bites of energy and bytes of information into more complex organisms and new st-planes. Since the Universe is indeed 'an organism with a body and a soul, called Logos'[17].

Systems theory merely culminates the search for a Unification theory of reality, based in such organic principles. Indeed, departing from the first monist theories of time (anthropomorphic religions and time-clock in physics), there has been a progress towards a more complex analysis of Time Arrows that culminates in the XX century with Biology and System sciences, which put together all those arrows in order to explain what Biologists call the 'drives of existence of living beings', and System Sciences now expands to all other species. Biologists said that living organisms perceive information, feed on energy, reproduce and evolve into herds and organisms. So the creation of energy, the gauging of information, the reproduction of species and their social evolution into herds and organisms are the 4 types of changes or fundamental arrows of time in the Universe. Reason why in complexity, we say that the Universe is *an organic fractal of energy and information'* in which all its parts are self-similar super-organisms made of cellular parts, which are connected by networks of energy and information, and perform cycles of feeding (on energy), gauging (information), reproduction and social evolution.

System sciences relate all those processes with a common jargon, taken from biology and physics, which applies to any process taken place between different scales of reality, proving that those 4 main arrows of time are common to all the species of the Universe, including physical species. And so, by using those 4 main arrows of time to study the events of all sciences, we can rebuild the science of philosophy as a philosophy of science and finally understand what time is, what is the meaning of existence in the space-time Universe, why we live and die (the entire process of time changes we experience), and other classic themes of philosophy, which the older, restricted, monist theories of time – mainly physics – have not answered.

In System Sciences we dissent from Quantum Monism, which considers the Universe made with a single arrow of expansive space (entropy or energy) and defines mass with a quantum particle and a new Universal field never observed before (the Higgs). Instead, we back the established work of Einstein's Relativity and its Principle of Equivalence between masses and cyclical acceleration, which defines a mass as a cyclical vortex of gravitational space-time, the in/formative arrow of the physical Universe. Moreover information is considered the dominant arrow of creation of the future, since it is the passing of time, what 'forms' the energy of vacuum space. 'Time bends space into masses' said Einstein. The same can be said of Biological species, since 'time evolves the morphology of living beings' (Darwin). In that regard, this work can be also considered an alternative to the dominant philosophy of science and the many bizarre interpretations of the quantum paradoxes, which Multiple Spaces-Times resolve from an organic, dualist perspective – so for example, the collapse of waves of quanta into tight particles, when confronted with an electronic flow (the observer's microscope), which so many puzzles has caused among philosophers of quantum science, becomes the same self-evident process that makes fishes come together into tighter herds when a shark comes against them: the event is a strategy of survival based in social geometries. For the same reason, we see the same formal patterns in the path of particles crossing through a slit or humans evacuating a theater: in both planes of existence a group of self-similar particles will form the best geometrical flow to cross the slit/door in minimal time. Yet all those self-similar processes cannot be understood without the invariance at scale of the forms and properties of energy and information, and requires more arrows than entropy.

The arrow of energy, the 'canvas' of reality does *not* create the Universe, but merely destroys information in entropy processes of death, such as the big-bang or the destruction of a living organism. So we need to add to the mix, the creative 'paint', the information arrow, explained both with the mathematics evolved in the XX century, (fractals and Non-Euclidean geometries) and causal words that describe those processes with its '3 verbal dimensions', past, synonymous of energy; present, synonymous of reproduction=repetition of the same beings and future, synonymous of evolution.

Yet information is the dominant arrow of future of most universal processes, which are guided by the bio-logic laws of eusocial evolution that gather the individuals of the same species into organisms and in human history were expressed by the mandates of religions of love. Thus, the arrow of information is dominant in Religion and Philosophy, needed to explain the meaning of God, Man and the Universe, beyond the simplest scales of reality dominated by energy and matter, which physics study.

Because all fractals have inner microscopic cells with form, every fractal 'point of the Universe is a small world in itself' (Leibniz), which displays the same elements and cycles than the whole. Since all of them are made of bites and bytes of energy and information, gathered in networks through several scales of relative size. Those 3 specific parameters - the type of energy and information the fractal system uses, the relative size of the fractal and the complexity of its cellular networks, might vary - but the general laws and properties of the organic fractal remain; since they are connected to the properties of the fundamental bites and bytes of energy and information of the Universe.

Traditionally scientists called such Science of Sciences, *General Systems Science*, which we systematize in this book, defining it as *the Science that studies the logic laws and fractal similarities that rule all the organic species and worlds of the Universe*. We aim to define the main fractal forms and functions of the Universe, studying with them all the species of classic science across its main scales of spatial size and formal evolution. We will do so in incremental degrees of complexity, mimicking the fractal structure of the Universe, as we increase in each iteration of this short prologue the depth of our understanding of the 3 main elements of fractal structures, making in the 3rd Iteration a detailed analysis of the fractals of the Universe with the 2 arrows of energy and information.

Recap: Reality is structured in hierarchical scales of growing form and diminishing extension, according to the reversed properties of energy and information (Max.E=Min.I) described by the laws of Non-Euclidean, Fractal Geometry and Non-Aristotelian, multi-causal logic. Each science studies a scale/plane of space-time with self-similar forms. Those st-planes vary in space size and time duration, according to Universal Constants that related those parameters, yet all obey the invariant morphologies of space-time. The study of those general laws of any species of energy and information is the realm of General Systems Theory. Each science studies a scale of self-similar forms in the fractal Universe. It follows that since those laws are self-similar in all scales, yet the detail of our observation is maximal in our plane of human existence, closer to us, while the microscopic Plane of Physics is subject to Uncertainty Laws, Biology and social sciences are the most important science to understand Man - the measure of all things (contrary to common thought that places higher value in microcosmic sciences subject to Uncertainty of perception (Physics).

50. The informative, linguistic, cyclical, topologic method.

What is the difference between the classic method of knowledge of the previous paradigm of a single space-time and the new paradigm of multiple times and spaces?

In the old method we performed measures with machines, reducing them to a single lineal time and a single space graph, from where 'regularities' on those measures were taken to create wider laws of reality, forecasting the repetition of those regularities in the future, which became the limit of our inquire. We didn't know however why many of those regularities formed because time was considered lineal and so it contradicted the cyclical patterns of those events. Regarding our analysis in space of the forms of beings we also described and measured the organs and systems of species, but most likely we could not connect those measures in space to the events in time. Only in biology which accepts cyclical time and the natural arrows of its species, there was a harmony between knowledge in space and knowledge in time. Yet in physics, where ½ of reality (the gravitational world) is not perceived or even acknowledged as a different scale of space-time, since it contradicts the dogma of the continuum; the difficulty to measure dark spaces and dark energies, which occupy 97% of reality, the uncertainty of quantum measures and the use of s single clock of time, which fails to recognize they cyclical patterns of particles as inertial motions that shape an arrow of information in the Universe, which balances the arrow of entropy, all those errors of the lineal paradigm of a single time arrow and space makes impossible to resolve its pending questions.

In the new paradigm those questions become trivial answers under the new method of inquire; since we know why there are cyclical regularities and can fit events as actions that fulfil a certain time arrow. Indeed, because time arrows are discontinuous, given the fact there are multiple arrows for each entity of reality, which switches between those arrows; events happen in a discontinuous, cyclical manner. For example, we humans switch between feeding, perceiving, reproducing etc. So the first thing we do in the new paradigm when we observe a certain experimental regularity in the behaviour of a species is to search for the arrow of time accomplished by that species. Once we have ascribed all the events of a certain form to its time arrows, we can then search for the synchronicities between those different arrows. For example, in the quantum world we ascribe the 4 elemental arrows to the 4 quantum numbers, and then we find its synchronicities. This is already known by quantum physicists, but *the difference is that now we know the why of those numbers, before we only knew they existed.* In space, a self-similar process is required: we observe an entity and describe its forms, but since we know those forms are topological shapes, we can ascribe them to the 3 topologies of a 4-dimensional universe, find therefore not only the form but also the interrelated structure and functions of each form. And so we can now compare each form in space with a function and arrow of time and harmonize both, our spatial analysis and temporal knowledge of the events those organs perform.

Even if there are scales of the system we don't perceive, we have laws that explain the structure of all systems in 3 networks/planes, one of energy, information and reproduction. Thus, as Mendeleyev did with the atomic table, we can 'fill' the gaps by self-similarity with other systems. This is especially useful in cosmology where the gravitational membrane of dark energy and quark matter is invisible but since it is parallel in structure to the quantum membrane of electromagnetism we know by heart, we can model it as a self-similar membrane and complete the standard model of physics, proving our topological and causal analysis with indirect proofs of gravitational events that 'surface' in our world.

Those methods complete in depth the meaning of particles, physical events and cosmological structures; but perhaps where they render more astounding results is in the ideological, anthropomorphic, abstract analysis of economic ecosystems and human societies. And this brings the final knowledge provided by an organicist model of the Universe: all those events and patterns, arrows of time and structures we find in all systems are in fact mere expressions of a teleological goal of the Universe, the creation of super-organism, the arrow of social evolution that guides all species who feed in energy and perceive information in order to reproduce and when reproduced in enough numbers, self-organize themselves in more complex complementary systems of reproductive energy and information. And this becomes evident in the study of civilizations, which are superorganisms of history and markets, which are economic ecosystems in which machines and human beings are in a dual relationship of symbiosis and competence (in labor and war fields). So we might say that if the metric paradigm was searching reality from the bottom up and never reached the summit of the why, in the new paradigm we converge from the bottom (the experimental facts, whats and whens) and the top (the logical and topological whys and hows), merging them into a deep, exhaustive answers of all the sides of a being, including man and the economy.

For example, in social sciences, the less developed till today, we find first certain cyclical regularities of the species we study, civilizations (superorganisms of human beings) and company-mothers (superorganisms of machines). Those cycles with an 800 and 80 years regularity turn out to be cycles of evolution of the memes of eusocial love that took mankind from the individual family to the tribe, the village, the city-state, the nation and the religious civilization and the memes of metal (money, weapons and machines), which also evolve in informative complexity and social organization till creating the planetary super-organism ruled by the informative language of money that we call the global market. Moreover, we find that the 800 year cycle of civilizations is related to changes in the energy fields of the planet (draught and hot and cold weather changes) as the glaciation cycle is related to the punctuated evolution and biological radiation (reproduction) of life species. So we find synchronicities between the parts and the whole of the planet.

We can further explain our actions as expressions of the drives of life, our desire for energy, information, social evolution and reproduction, and the symbiosis with machines as direct effects of those needs that machines, which are organs of energy and information, further enhance. Yet we find also Darwinian relationships with machines of energy, weapons, that can kill our body and informative systems of metal (money, TVs, computers) that substitute the values of our verbal language and atrophy and hypnotize our brains. So suddenly the lineal concept of history as technological progress is no longer positive but has negative and positive sides, good and bad fruits, ages of war in which the negative goods of metal dominate and ages of prosperity. And this brings us the final result of a proper science of history not an *ideology* or culture that always favors the machine of measure over man as the measure of all things: we can create a proper policy that takes into account all the elements of the social world, including those 'hidden' effects that were ignore in the previous paradigm in which the finality of history and science, the creation of machines and measures make with them, becomes secondary to the understanding and evolution of man, the most informative species and new summit of our renewed, deeper vision of the organic Universe where information is the meaning of it all.

This also means that in the new paradigm all languages of information are equally valid and so words also carry meaning and values from a human perspective that cannot be ignored. Ethical questions thus become scientific as they express the arrow of eusocial love and evolution of mankind into a single superorganism. And so now there is no dystopia between knowing more the organic paradigm and making a world better for mankind, the measure of the new paradigm.

So the new method fusions the 'thoughts of God' and its details, with a clear causal change of inquire:

-Experimental facts x Laws of multiple times and spaces-> Mapping of the time arrows and topological structures of being-> Metric measure of all its regularities in time and space->Prediction of future events->Man as the measure of all things->Praxis and Policy of science which shall either prevent the event and/or forbid the species to limit its damage to our biological existence or promote the entity, industry and event if it enhances the natural cycles of man.

This causal chain adds an ethical policy to all forms of scientific research is specially needed today, when the religion of the machine as the measure of all things make even those tools that can destroy the humankind (nuclear devices and weapons, organic robots that compete with us in war and labor fields), always positive because the meaning of knowledge is the instrument we use to make digital measures and so the 'message is the instrument' not it use to improve our lives and reveal the why of things, now bring about by the mind of man and all our languages of information, not only numbers.

Recap. In the organic paradigm we know both the details provided by the experimental method and the thoughts of god, provided by the arrows of time and the topologies of space. So we can complete knowledge departing from the bottom and the top describing the form and function of all beings, its cyclical events and its relationship with all other structures of reality.

Notes.

[0] ` Zeit ist das, was man an der Uhr abliest.'

[1] `Harmonices Mundi' (1618).

[2] In an attempt to realize Leibniz's ideas for a language of thought and rational calculus, Frege developed a logic notation as the foundation of mathematical reasoning. Though this notation was first outlined in his Begriffsschrift (1879), the most mature statement of Frege's system is 'Grundgesetze der Arithmetik' (1893/1903).

In 1931 Kurt Gödel demonstrated in his paper On Formally Undecidable Propositions that within any given branch of mathematics, there would always be some propositions that couldn't be proven either true or false using the rules and axioms ... of that mathematical branch itself. You might be able to prove every conceivable statement about numbers within a system by going outside the system in order to come up with new rules and axioms, but by doing so you'll only create a larger system with its own unprovable statements. The implication is that all logical system of any complexity are, by definition, incomplete; each of them contains, at any given time, more true statements than it can possibly prove according to its own defining set of rules.

[3] Summa Theologica.

[4] The influence of religious beliefs in the ideas of the first philosophers of science is hardly recognized, except in the work of Thomas Kuhn. For example, Newton dedicated most of his work-hours to alchemy and Biblical studies, and believed God sent to him comets as personal messages. Only such beliefs explain the obvious contradictions between reason and the mechanist, monist philosophies of reality sponsored still today by many scientists. Since a mechanical version of the Universe as a series of clock-like cycles, which particles and entities do not command, but obey blindly, requires a creator - the 'clock-maker' of Newton and Kepler, who puts the cosmos on track - as Leibniz, precursor of rational organic theories explains in a letter to Clarke: 'Sir Issac Newton and his followers also have a very odd opinion concerning the work of God. According to them God Almighty needs to wind up his watch from time to time'; probably referring to Newton's Optiks, p. 402, in which Newton, aware of certain irregularities in the orbital paths, latter resolved by Einstein, affirms that those regularities will increase 'till the system wants a reformation'. Yet rational science should exclude any external, mythic agent to put it on track.

[5] 'The universe - said also J. B. S. Haldane - is not only queerer than we suppose, but queerer than we can suppose.' One of the queerest things about it is that its properties at the very largest scales--galactic super clusters--are very intimately related to those at the very smallest--subatomic particles. The big and the small are related; because they are indeed, self-similar fractal structures, whose energetic and informative properties have transcended through multiple scales of size, emerging again in the macrocosms.

[6] Mircea Eliade in 'History of Religious Beliefs' shows how pervading was in classic pre-Christian cultures the concept of a God of Time, (Zurvan in Zoroastrism), who lives through 3 ages, and the belief in a Universe made of Time, in which space is a Maya of the senses (Vedas, Buddhism).

[7] Letter to Besso's wife.

[8] 'Syntactic structures' 1957.

[9] The Holographic principle is explained in physics only for the limited case of the surface of black holes. It is however a general feature of all systems of information, which this author has used in all disciplines – from biological analysis of cellular functions to the study of form in art theory and bidimensional painting.

[10] Letters to Clarke (1715-16).

[11] A mathematical proof of its equivalence with information, extracted from Einstein's equations is:

$$E = mc^2 + E \times T = k \rightarrow Mc^2 = k/T \rightarrow M = (k/c^2) x (1/T) \rightarrow M = K \times v$$

Where v is the frequency of a mass as a vortex of space-time.

[12] Einstein could only apply the 5th postulate of Non-Euclidean geometry to describe gravitational space-time. Today, thanks to the discovery of the other 4 postulates of Non-Euclidean Geometry, which I introduced to the world of science in the 50th anniversary of Complex Sciences at the Sonoma Congress, the complex structure of Space, created by the interaction of Euclidean light-space and Non-Euclidean Gravitational space, can be resolved to explain many pending questions of astrophysics – from the nature of dark matter to big bang theory.

[13] Relativists model the Universe with Einstein-Walker equations in which each galaxy is treated as a hydrogen atom. This theoretical trick that facilitates the calculus of relativity shows the self-similarity between a hydrogen atom and a Galaxy of the upper scale. Scientists also use models of electronic nebulae to describe the behavior of stars around the central black hole; and some have tried models with high-density photons. Yet self-similarity is not identity. So quantum cosmology, which uses quantum equations to describe the macrocosms as if it was identical to the quantum world, makes no sense. Instead we must observe the self-similarities of those scales, caused by the emergence of fractal parts into wholes. Can we know if those scales of relative size are infinite? Not really, because if there is a higher fractal scale and galaxies are self-similar to atoms, the extension of the cosmos will be so vast that we shall never find its limits. Indeed, in our Universe there

are trillions of atoms. So even if there are trillions of self-similar atomic galaxies, the maximal perception we can have is of a few billions, within the total cosmos.

[14] Schopenhauer's philosophy culminates 3 millennia of western tradition in dualist models of the Universe, with his analysis of a reality composed of ideas – the informative mappings of the human mind - and the will – the satisfaction of the human arrows of time, our desire for energy, form and reproduction. While the idea has its site in the brain, the will has its point of maximal force, according to Schopenhauer, in the penis, whose pure desire of reproduction, guides in this philosophy of action, as the ultimate goal of existence, the world of the mind. Thus, for Schopenhauer – and we agree – the arrow of reproduction dominates the arrow of information.

[15] In his thesis, 'On the Hypotheses which Lie at the Foundations of Geometry', 1854, Riemann defined space in terms of motions and self-similarity, setting the basis for the advances of Fractal, Non-Euclidean geometries, formalized in the 70s and 90s by Mandelbrot ('Fractals', 1976) and completed by this author with his definition of the fractal point and the 4 remaining non-Euclidean postulates ('Radiations of space-time', 94; ISSS, Sonoma, 2006.)

[16] This work resumes 20 years of research in System sciences and its multiple disciplines: non-Euclidean geometry, fractal mathematics, duality, complex logic, complex biology, complex physics, complex history, etc. It is part of the 9 conferences which I will give at the International Systems Societies, during my tenure of the SIG of duality in the congresses of Cancun, Sonoma, Tokyo, Madison and Waterloo (2005-2010), to complete an overview of what I call the '3rd age of science' when Man will learn he exi=sts within a self-similar, organic, eternal Universe of 2 fractal, scalar, self-generative, i-logic motions - energy & information: $\sum e \Leftrightarrow i$.

Modern scientific Duality is a relative new science, whose formalisms and main laws were systematized in my book 'The cycles of Time', c. 94 (Spanish edition). Yet prior to this work, there has been an enormous quantity of laws, data and 'similarities' between all the species of the Universe (called isomorphisms and other technical names in our science) found during 50 years of studies by System and complexity scientists. Santa Fe Institute of Complexity and Len Troncale from Pomona University have done important work classifying those isomorphisms. While an array of specialists in system sciences and complexity, from Rossler to Capra, from Prigogine to Miller, and the founding fathers at Macy's, Bertalanffy and Norbert Wiener, provided many of the theories and logic and mathematical insights required for the consistency of a Theory of Multiple Spaces-Times. The complete development of such theory would require though the rewriting of the entire 'Encyclopedia of human knowledge', since we contend all can be explained with those 4 drives of 'exi=stence'.

[17] Timeus, Plato. The organicist paradigm has always been understood by the highest minds of mankind. It was evident for Plato and Aristotle. It was also foreseen by Spinoza and Leibniz in its Protea. But it has always been censored by the mechanist view. Leibniz's master book, to put an example, was not translated to English till this century.

i.

FRACTAL UNIVERSES
SPACE-TIME POINTS

1. Mathematical and Logic Languages. The syntax of Algebras and equations.

2. Geometry: Non-Euclidean Postulates.

3. 1st Postulate: Points with Parts: The 3 Fractal Topologies of Super-organisms.

4. 5th Postulate: i-logic Points of View.

5. 2nd Postulate: Waves of Energy and Information.

6. 3rd Postulate: Self-similarity.

7. 4th Postulate: Planes of Existence.

8. Organic, Social Classes.

9. Knots of Times: Synchronicity.

10. Palingenesis: Travel between Space-Time Planes.

11. Exi=stential Algebra: $\sum Se<X>\prod Ti$

I. LOGIC AND MATHEMATICAL LANGUAGES.

1. The nature of mathematics and logic.

The first concepts and ideas of mankind tend to be simple, intuitive and truth. The forest is clear and the details of the trees do not hide the whole. So the oldest philosophy of the Universe, that of the Chinese agricultural Neolithic about a world made of yin=information and yang=energy, which combined to recreate the infinite 'waves of existences' (Ch'ang), is still the most accurate philosophy of reality, now forgotten with the arrival of so many scientific details that the overview of the forest is lost.

The same happens with the understanding of time, which all earlier civilizations considered cyclical and causal, synonymous of change; and so they study all different types of change, establishing repetitive rhythms of change and causal relationships, creating the science of Logic (Aristotle).

The Greeks also defined mathematics as the language of space and so Geometry became the foundational science of mathematics; and it remained so till the XVII century, to the point that Al-Jorizim, the founder of modern mathematics (Algebra and Algorithms are words derived from his names and books, translated by Middle Age Spanish mathematicians and then spread during the renaissance to Europe), proved all its quadratic theorems with geometrical methods.

Mathematics is defined in encyclopaedias as the science of sequential numbers (algebra) and static space (geometry). Till the XIX century, the science of sequential numbers or 'mathematics of time' used only Aristotelian Causality with its single time arrow to order them. On the other hand, the 'mathematics of space' considered only a lineal, continuum space defined by Euclid. Both sciences were fusioned by Descartes (analytic geometry) with his Cartesian Plane; and we can thereafter talk of mathematics of spacetime, albeit in its simplest conception: a single, static space, and a lineal, unicausal time.

It would be latter in the XIX and XX centuries (with the exception of the insights of Leibniz[1]), thanks to the work of Frege, Boole, Riemann, Einstein, Poincare, Cantor, Bourbaki, Mandelbrot[1] and Computers, which this work advances a step further, when Space and Time – hence Logic and Geometry – were first mixed, but without a full formalism that made possible to extract all the laws of spacetime and grasp in depth the meaning of those languages, due to the limits of the Cartesian single space-time plane. A theory of multiple spaces-times thus has to refound the key concepts of mathematics, the plane, which now is a multiple plane in which each point creates its own frame of reference, interacting with other points and planes through exchanges of energy=motion and form that re-establishes the balances between their distances and topographies; and the number is now a network of self-similar points that have at least a formal bidimensionality. Mathematics includes several sciencces in one: the science of numbers can be stretched to consider the science of networks as each number is a point of a network with a geometrical form. Numbers and geometry are thus essential sides of the same coin. On the other hand geometry is mainly topology, related to causal transformations of a complementary networks of energy and form, whose 'numerical' cells go through growths and extinctions through life, described with differential equations. Those equations thus can be reduced to cases of the generator equation, Se⇔Ti, which in this manner resumes all other differential equations. And indeed, when we use equations to explain repetitive events of nature we are normally mapping out an exchange or transformation of energy into information, which in a generic way is represented by the exi function.

Thus topological networks, differential equations that map out the life/death cycle of an event with its growth and diminutions of energy and information, exchanges of energy and form between networks and membranes and causal chains that repeat those events till creating very complex systems, are the key mathematical operations of 2 sciences, logic and mathematics, in which the most general laws of networks are written. Those Mathematics of multiple space-times have illustrious precedents:

In the XIX and XX century mathematicians developed theories of multiple spaces and defined spaces with motion (Klein), and spaces as networks of self-similar points (Riemann) but maintained 4 of the 5 postulates of Euclid, the single Cartesian spacetime continuum and a single causal arrow of lineal time in its theory of numbers. Thus a theory of multiple space-time arrows requires the evolution of the science of space (geometry), completing the 5 postulates of Non-Euclidean Geometry and also the evolution of the science of temporal, sequential numbers, expanding its laws of causality. This was partially done with the use of the Complex Plane, since real numbers roughly correspond to the properties of continuous energy and complex numbers to the property of discontinuous, informative time. But its dual nature as an expression of the different properties of space and time has never been fully understood even if Physicists use the Complex plane in modern physics.

In that regard there are two possible representations in a plane of complex, informative numbers:

- They could be placed in the inverse, negative side of the graph of Cartesian coordinates as the arrow of information - to better represent the duality of energy and form ($I \Leftrightarrow E$). Then many results obtained with complex numbers become intuitive, as it is the case of the metric of Special relativity, where the time factor has a negative side $-c^2t^2$ meaning indeed that time-information contract space and 'rests' to the positive arrow of Einstein's equations.

- Or they can be placed, as it is customary in the perpendicular Y dimension. Since a dimension of a mathematical plane becomes a sequential order of numbers, then the Complex plane becomes a bidimensional mathematical representation of the 2 simplex arrows of time, energy and form; while a 3D complex plane represents 3 time arrows where $exi=z$ represents the reproductive arrow.

Finally a fractal, multiple space-time built with a series of complex planes (a task started by Riemann with his description of polynomials as a series of planes of space, stacked one after another), is the appropriate form of defining the 4th arrow of time.

From a philosophical perspective the study of multiple sequential time (algebra of Multiple Spaces-Times) and multiple, fractal spaces (5 non-E postulates) improves the accuracy of mathematics as an image of the reality of multiple space-times and explains its ultimate meaning, as the science of space (geometry) and time (numbers): *Space is mathematical, geometrical; time is logical, sequential, numerical and the conjunction of both is organic; reason why complexity and system sciences, who observe the Universe from an organic perspective are the sciences of sciences.*

We cannot treat all the aspects of complex mathematics in this introduction. We shall therefore focus on. its 2 key elements: the geometry of numbers wth form and its postulates that describe how the 4 arrows or wills of time construct networks of fractal points and the algebra of the generator equation, which describes that geometrical construction from an analytical, numerical, differential perspective.

Recap: Geometry is the language of spatial form; Logic is the language of time events. The algebra of the generator equation of space-time cycles and the non-euclidean geometry of fractal points fusions both creating a model of multiple, sequential dimensions of numbers and multiple planes of space closer to a description of reality and its organic systems. This is done by resolving the 5 non-Euclidean postulates of Geometry, which describe knots of time arrows as points, and by observing the self-similarity of mathematical equations, which describe groups of knots of time arrows as numbers and the equations of Multiple Spaces-Times derived of the Generator equation of the Universe or Principle of conservation of energy and information. Thus, the 4th paradigm upgrades the postulates of Non-Euclidean Geometry and the foundation of Logic to include the causal relationships between time arrows and explain better a universe of multiple spaces and a logic to of Multiple Time cycles and arrows of time cycles.

II. NON-EUCLIDEAN POSTULATES

2. The 5 Postulates of Non-Euclidean, i-logic Geometry.

According to the Correspondence Principle a new scientific Theory has to include all previous ones. So happens with Multiple Timespace Theory, which includes as specific cases all other simpler geometries of space-times, by completing the 5 Non-Euclidean postulates of geometry and adding the concept of a fractal point developed in XX C. fractal geometry. The result is a geometry that includes all other geometries; since the incomplete non-Euclidean geometry discovered in the XIX c. (5th postulate: a point can include many parallels) already included the previous Euclidean postulates based in points through which only one parallel can be traced. Those new postulates describe with more accuracy all type of space-times belonging to all kind of sciences, from the physical space/time of atoms (described by quantum Physics) to that of galaxies (which Einstein described in his General Theory, using the 5th Non-E postulate); from the organic space/time of cells and species to the historic space-times of nations.

Thus, according to the Correspondence Principle, we establish a hierarchy of completeness among all those geometries that stretches from the simplest Euclidean Postulates to the i-logic geometry of fractal space-time, which includes all others as particular cases; and it is in itself a partial image of the total information about space-time we call reality - always bigger than any 'linguistic' image we made of it with the languages of the mind that reduce the total information of reality to fit it in the brain:

Euclidean Mathematics (Light- Space)> Classic Non-Euclidean geometry> (Gravitational Plane) > Fractal space-times (all scales) > Total Reality [2]

We consider Euclidean geometry, the natural geometry of a single, continuous space-time, made by the 'eye' only perception of light-space and its 3 perpendicular dimensions (height/electric field, width/ magnetic field, and length/c-speed). But that space is not all the spaces of reality. It misses at least a second fractal scale, gravitational scale and *all the complex, vital spaces of organisms of information and matter*. In Non-Euclidean mathematics, the new unit of 'fractal space-time' is the 3-dimensional Fractal Point of view, whose main property is to have in/form/ation - an inner content, which in a Universe made of multiple scales of size, we perceive as we come closer to the point and observe it in detail. Those fractal points with volume are described by the 5 Postulates of Non-Euclidean Geometry, which adapt to fractal, discontinuous space and cyclical time, the classic 5 Euclidean postulates.

The first postulate explains a fractal point, as a point with parts. In a more detailed analysis those parts turn out to be self-similar in geometrical terms, in all systems, due to the specific lineal vs. cyclical, concave vs. convex geometries of energy and information and its 'planar combinations'. In its more complex versions those geometries are defined by the three canonical topologies of a four-dimensional Universe, which describe the informative 'head', reproductive 'body' and energetic limbs and membranes of all the superorganisms of the Universe. The 2nd, the interaction between two points connected by a wave of communication or 'line', the 3rd, the type of 'biological' interactions between 2 points according to their relative equality, and finally the 4th, the complex structure of a system of points across multiple organic space-time planes, such as those who create a human being.

Thus, the more evolved Fractal, i-logic Geometry of space/times includes all previous geometries as simplified cases of the more complex structure of the Universe in multiple scales. We call it a fractal, i-logic Geometry because some postulates can be interpreted in geometric terms, as the definition of the essential elements of any geometry of space/time (the point, 1st postulate; the line, 2nd postulate and the plane, 4th postulates) and some in logic terms, as the description of the different causal chain of events that can happen when st-points enter in communication -3rd, 5th and 4th postulates:

146

Further on, according to the duality between geometric form and logical function, those postulates of fractal i-logic geometry define also the basic arrows= cycles/dimensions of the Universe: the 1st and 5th postulate define a point as a system whose inner parts are able to transform and emit energy and information, e>i<e; the 2nd postulate defines an exi wave of communication that reproduces energy and form between 2 fractal points; and the 4th postulate defines the social evolution of a herd that creates a fractal plane, a network with dark spaces; and the 5th postulate explains a point in its iterative actions as it absorbs energy and transforms it into information through its small apertures to the Universe. Since even a minimal quark of mass, as Einstein affirms should be crossed by ∞ parallels.

The importance of the 'linguistic laws' of i-logic geometry resides in the fact that all systems of the Universe follow in space the laws of geometry and in time, the laws of logic, so the enhanced linguistic mind of science obtained with those laws can be used, as this author has done in multiple works to resolve long-standing questions in all sciences, which can further be ordered in a pyramid of increasing information and diminishing energy: Physics studies the simplest forms with maximal energies. But if we consider the 2nd arrow of 'creation' of reality, that of form, of information, the summits of science are biology, which studies the most complex forms of information (life) and Sociology/Religion, which goes a step further into the analysis on how the arrow of eusocial love creates Human Societies according to the Organic laws of the Universe (eastern religions).

To be able to consider biologic behavior (social evolution) and 'mental perception' (informative gauging), as factors dependant on geometrical qualities is an astounding break-through for science, which will be developed in its details in this work, to show an enormous number of experimental proofs. Indeed, fractal, i-logic Geometry achieves a long-sought goal, which Plato, Spinoza and Leibniz tried to realize, but failed, with the use of simpler Euclidean geometry: to unify the logic and mathematical principles of the Universe. Since the 3rd and 5th Non-Euclidean postulates are logic, the 1st and 2nd are geometric and the 4th, the nature of a topological plane or network of non-E points is both logic and geometrical, as it defines the fundamental logic-geometric structure of the Universe: the superorganism.

The logical and mathematical conclusions we shall arrive in this chapter are the same we achieved in the previous one with a more descriptive, less formal analysis: the Universe's final goal is to create 'topological, organic planes, super-organisms' – the 4th, more complex 'geometrical postulate/form of reality' – departing from simpler forms, points and cycles. We shall try to get to that conclusion step by step, first by describing the geometrical postulates, then by fusioning them with fractal mathematics and finally by studying the most fascinating of all of them, the 3rd postulate, which proves that behavior among the points of the Universe is based in the degree of equality or self-similarity between their inner form that determines if those forms communicate symbiotically or destroy each other, thus defining the duality of eusocial evolution as a group of a number of self-similar beings or destructive extinction.

Some, first-hand consequences of those postulates are:

- *1st Postulate:* We correct Euclidean geometry, increasing and relativizing the dimensions of a point that fits inner energy and information. Since all points have either 1 motion, a time dimension, when we consider them from the perspective of the upper st+1 Plane for which they are in the limit of invisibility (what quantum scientists call a point-particle) *but still has a motion or 'function' in that upper ecosystem, st+1 in which it exists*; and 3 dimensions/networks/functions when we consider them from the inner perspective of the point. This is the case even in the smallest planes of theoretical strings, made of points with parts, with volume – since we require 3x3s+1t theoretical inner dimensions to describe them - a paradox that can only be resolved if we consider the to be 'strings' of fractal points with inner, fractal dimensions.

- *2nd postulate:* When we observe a one-dimensional line as a form with inner parts it becomes then a 4-dimensional wave made of cyclical points with motion. Hence in quantum theory we say that any particle in motion has associated a wave. *Thus the 2nd postulate resolves the wave/particle duality,* as all lines are now waves traced by a point with inner volume. Further on, since all lines have volume, they carry information and so all forces can in fact act both as a source of energy and as a language of information - as physical experiments prove. Since when we observe a ray of light in detail it becomes a 4-dimensional wave with electric height and magnetic width, often exchanging flows of energy and information in action-reaction processes of communication between bigger points.

- *3rd Postulate:* Equality is no longer only external, shown in the spatial perimeter of any geometrical form (3rd Euclidean Postulate) but also internal and further on it is never absolute but relative, since we cannot perceive the entire inner form of a point – hence the strategies of behavior such as camouflage. Forms are self-similar to each other, which defines different relationships between organic points, according to their degree of self-similarity. The 3rd postulate is thus the key to explain the behavior of particles as the degree of self-similarity increases the degree of communication between beings. Some of the most common behaviors and 'events derived from this postulate are:

1) Reproductive functions in case of maximal self-similarity or complementarity in energy and form; $ei\text{->}\Sigma ei$ or Max E x min. I (male)= Min. e x Max I (female).

2) Social evolution, when points share a common language of information, $i=i\text{ -> }2i$.

3) Darwinian devolution when forms are so different that cannot understand each other's information and feed into each other: $i \neq i$. In such cases if those 2 entities meet they will start a process of 'struggle for existence', trying to absorb each other's energy (when E=E) or simply will not communicate (when $E{\neq}E$, since then there is neither a common information to evolve socially nor a common energy to feed on). Yet because any point absorbs only a relative quantity of information from reality, self-similarity is relative and it can be faked for purposes of hunting, allowing biological games, such as camouflage and capture, or sociological memes that invent racial differences, allowing the exploitation of a group by another.

The geometric complexity of the 3rd Postulate is caused by the topological forms created by any event that entangles Multiple Spaces-Times. Since it describes the paths and forms of dual systems, which connect points: Self-similarity implies parallel motions in herds; since equal entities will maintain a parallel distance to allow informative communication *without interfering with the reproductive body of each point.* Darwinian behavior implies perpendicular confrontations, *to penetrate and absorb the energy of the other point.* Finally, *absolute, inner and outer self-similarity* brings bosonic states, which happen more often to simpler species like quarks and particles that can form a bosonic condensate as they do in black holes, *where the proximity of the points is maximized. And indeed, the same phenomenon between cells with the same inner information /DNA originates the 'collapse' of waves into tighter organisms.*

The 3rd Non-Euclidean postulate is implicit in the work of Lobachevski and Riemann who defined spaces with the properties of self-similarity (Riemann's homogeneity), which determines its closeness (Lobachevski's adjacency).

- *4th postulate:* Now because a plane has an inner volume – that of its points - it is a cellular, organic topography, a network of self-similar points. And because networks of points of energy and information are complementary, often we find systems with 2 complementary networks that form ever more complex geometries – based in the geometrical dualities of lineal energy and cyclical information – with

148

the results we observe in nature: *the creation of an enormous number of complementary systems which are, as we shall see latter, all of them self-similar in its geometries and functions.* Thus, the types of Non-Euclidean planes of space-time range from the simplest Euclidean planes to the more complex organisms with a volume given by the relative point/beings that form its space-time networks.

Thus, a plane becomes a real topography made of points with volume, extended as a cellular surface. We can observe its surface as a bidimensional membrane of information (for example your skin, or the screen of a computer made of pixels, or the sheet of this work). Or we can consider the 3-dimensional inner structure of its points and then it becomes a network with inner motions, as those points will form a lattice in which they communicate lineal flows of energy and information that maintain the lattice pegged. Often 2 topological planes of energy and form combine to create a 4-dimensional organism. Such is the most common structure of the Universe, a 4-dimensional World, which is a Universe in itself, made of self-similar cells or networks of points that constantly exchanges energy and information within the ecosystem in which it exists.

- *5th postulate:* It defines points as informative knots or linguistic eyes - minds of information that absorb a flux of forces used by the point to perceive a relative world. A non-Euclidean point corresponds then to our concept of a relative mind that gauges the information reality with a certain force, similar to the concept of a monad in Leibniz relativistic space-time. In words of Einstein: a point of space is a fixed frame of reference.

Thus, Non-Euclidean mathematics fuses the logic and geometry of the fractal Universe, greatly improving our understanding of Reality even in terms of mechanist measure. Since mathematical solutions to problems with several points of view are impossible to find in continuous space (i.e. 3-body problem in gravitation), given the fact that a network of infinite points of view is local and relative and each point is a focal knot that acts from its perspective. Thus, the absolute truth of a system is the sum of all its points of view, which influence each other. Yet even if we cannot calculate precisely with mathematics, systems with more than 2 bodies, since those systems are organic, hierarchical, made of networks with attractor points, fractal structures and self-similar paths, the new mathematics of attractors, fractals, scales and Non-Euclidean systems, refine greatly our analysis. In essence, indeed, we observe that 'networks' integrate parts/points into wholes, which then 'act' as a single point. So in the complex models of i-logic geometry we can tackle many problems by defining sets of points as 'wholes' of a 'higher space-time'.

But the most rewarding consequence of the new language of the mind is the 'deep' understanding of the meaning of many experimental facts about the species of the Universe till now merely explained as hows. For example, the previous principle of local measure, where each point is a relative center of the Universe, is called in relativity the diffeomorphic principle, which now becomes explained as a partial case of the wider law we called the 'Galilean paradox'; the duality particle of information/wave of energy becomes a specific case of the application to physics of the duality of energy and information found in all systems and so on. The expansion of the laws of quantum theory (complementarity principle) and Relativity (relativity of scale, local measure, etc.) to other sciences and the organic principles of the 3rd and 4th postulate to physical particles is therefore the consequence of those postulates. Yet it requires the understanding of the new, i-logic, organic laws of the Universe and its networks, because E-mathematics has clear limits to extract all the information of the Universe, given the fact that it syntax includes a priori errors and simplifying postulates (single space-time continuum of points without parts, etc.). Thus, when the event described is complex, performed by a great number of points/variables you enter into non-lineal systems, which require topological descriptions (chaotic attractors, fractal non-differentiable equations and Non-Euclidean mathematics), and the i-logic laws of organic networks and systems – a better syntax in which to fit experimental evidence, especially in phenomena of informative nature (since only formless energy is continuous and resembles the models built with a single arrow of energy and a single plane of space-time). So while classic physical systems calculate accurately the energetic, continuous properties of the Universe an overview on how multiple points of view emerge into wholes requires organic laws. Of course, this search of whys also applies to the understanding of mathematics. For example, the previous postulates resolve the long-standing question of what is the nature of the 3rd and 5th postulates that seemed redundant (as the 5th describes also properties of a point like the first does, and the 3rd seem to describe a non-geometrical property). They are no longer redundant, but they are more concerned with causal logic and time than spatial geometry in its purest forms (points, lines and planes.)

There is also self-similarity between the fractal postulates of i-logic geometry (since the 5th is geometrically self-similar to the 1st, as both are concerned with points) and the 4 dimensional time paths/arrows of the universe. This is not casual since all languages of space-times depict in self-similar ways the 4D Universe. Thus if the 1st and 5th postulates define *a gauging point of information* as the fundamental unit of the Universe, the 2nd postulate defines a line or flow of communication of energy and information between 2 points, which *reproduces* part of the information of the 'generator' point across a surface of space; the 3rd postulate defines those points, which are not similar as *energetic* substances that will be absorbed by the points. Yet if those points are self-similar they will gather through the arrow of eusocial love, creating according to the 4th postulate a network of space/time, a new organic plane of existence. So the 4 arrows of space-time are explained by the 4 postulates of i-logic geometry.

Recap. The study of the geometric paths of Time Arrows with the tools of mathematics, explains the whys of mathematical laws: Geometry and topology acquire now its 'why': topology explains in detail the 3 canonical parts of all Points of view, while the i-logic postulates of geometry explain the universal structures that those points of view create.

In organic terms, the 5 postulates of fractal Geometry describe how points become parts of social webs, which self-organize fractal planes made of networks of points, which emerge as cellular units of a higher fractal space-time or new superorganism... Thus, according to the Principles of Correspondence and Relativity, proper of physical reality, those different geometries are relative descriptions of the same fundamental structure of the Universe: the point with parts and its more complex social forms, lines and planes. The fractal generator of the Universe is a logic equation that represents the main interactions between the arrows of time. Yet all languages mirror that logic equation in its syntax, since all minds gauge and represent the cycles of the Universe with languages of perception.

III. 1ST POSTULATE: POINTS WITH PARTS: 3 FRACTAL, ORGANIC TOPOLOGIES

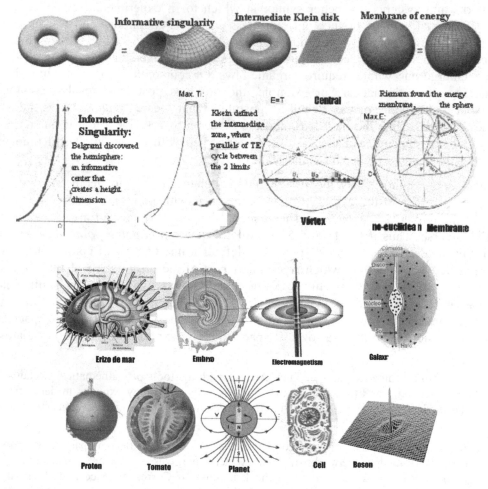

3. Topological Spaces. The why of geometrical forms.

Reality is made of fractal points, which are knots of Time Arrows, able to perform energetic, informative and reproductive functions. As complex as one of those points-entities might be when observed in detail, *any fractal point is made of 3 regions whose geometry responds to the topological forms of a 4-DimensionalUniverse, the convex plane, the torus and the sphere.*

The Universe comes down to two bidimensional elements, energy and information, and its 4-dimensional combinations. Thus all entities can be described as wholes made of 3 internal parts whose geometrical properties maximize their energetic, informative and reproductive functions:

- Max Ti: an inner, dual center, corresponding to convex topologies (left), made with 2 cyclical forms. It is the dominant informative topology of any fractal organism, described by Belgrami in the XIX c. as a conical form with 'height', with negative curvature.

- ⇔: A middle, reproductive zone, described by Klein as a disk of quanta in cyclical motion that communicate energy and information between the inner and outer zones.

- Max. E: An outer membrane of energy, described by Riemann's spherical geometry.

When we see fractal points far away we describe them as points with breath, with the tools of Euclidean geometry since the 'inner space' shrinks to a point and so the 'bulk' or curvature of space-time shrinks to a plane. Yet, when we come closer to them, they grow into points with volume. The volume of those Fractal, Non-Euclidean points can thereafter be studied with the 3 types of canonical, Non-Euclidean geometries or topologies of a 4-Dimensional Universe – the Universe we live in. Those 3 topologies make up the 3 regions of the point, which correspond each one to the 3 essential arrows/functions of any species: the external, energetic membrane; the central, informative brain and its reproductive combination, exi.

It follows that the first part we observe in a point is the external membrane, which without detail seems to have a continuous, energetic appearance. But on close view, we observe most external membranes store and/or absorb information (Holographic principle) due to its fractal geometry. This can be generalized to any membrane which shows a 'bidimensional surface' that acquires more form, more 'fractal steps' when we come closer to it. Thus, the Holographic principle, which physicists know in the restricted field of black hole theory but can't explain why exists, is both explained and extended to any bidimensional membrane of information. We find bidimensional, warped, fractal membranes that store information not only in black holes, but also in the development of organic senses, departing from the exoderm (external membrane of the fetus), in the seminal cells that reproduce life, formed as an outgrowth on the body surface (genital systems) or in the complex forms pegged to the surface/skin of the Mandelbrot and Julia sets, the best known mathematical fractals.

This external, energetic membrane has the topology of continuous surfaces called *the Riemann sphere*, which is the external surface of any point - the skin and limbs and any of its multiple self-similar entities, some of which are drawn at the bottom of the graphic. In any system of the Universe, the membrane acts as the energetic, external surface of the informative, point through which the point absorbs energy from the external Universe, to process it into information.

- The toroid, which is the body or reproductive region (seen as a cyclical path of space-time), which fills up the space-time between the nuclei and the external membrane. It is the zone where the reproductive organs of the system exist, and where the information of the system is born.

- Energy and information systems have inverse properties: energy is expansive, external, more extended, and information is implosive and smaller. Thus we find the informative head or system either on top of the reproductive body, with a spherical and smaller size or in the center, and its topologies will correspond to those of maximal form; the so-called *hyperbolic topology*. In the upper graph, it is the double ring, or convex, hyperbolic surface, with maximal form, or informative center of any entity of the Universe, the point in which the fractal reproduction of information reaches its zenith. It is the same form than the toroid but reproduced into a higher content of information, by doubling the initial form.

- Finally between both topologies of energy and information, the point will have a middle region of exchange of energy and form, made of cycles that go back and forth between those regions, which correspond to *the 3rd canonical topology*. It is the torus, plane or Klein disk - a curved region of energy & information quanta, with cyclical motions, confined by 2 limits, an external, spherical membrane of max. energy and an internal informative nucleus. It must be noticed that according to Klein, the topologist that studied better this type of surface, the toroid is NOT really a fixed form of space, but we must consider those cycles' motions and add the parameter of speed. Thus Klein introduces the Paradox of Galileo to describe the Non-Euclidean geometries of the Universe as we have done in this book. Cycles are mere static perceptions of motions and we must always consider distances as space-time distances. So we say: London is at 4 minutes distance because we consider distance and speed together.

This is ultimately the meaning of time in physics, a measure of the speed of motion as a way to gauge space-distances: v=s/t.

In the graph, we observe several non-Euclidean points created by those 3 canonical topologies that can adopt multiple forms by deformation, but suffice to construct all the shapes of our Universe. Indeed, a topology is deformable. So an external membrane, which corresponds to the topology of a sphere, can become any shape, as long as it is not torn up, to enclose a reproductive and informative zone. So your skin is in topology a sphere, which encloses the complex forms of your reproductive organs. Those organs are Klein cycles of great complexity that exchange energy and information. Since those 3 topologies suffice to describe any 4-Dimensional form, it follows that the Universe is merely a puzzle of energetic, informative and reproductive parts, associated in 'numbers', groups and all kind of entities as those shown in the previous graph, from different sciences, which are created with those 3 topologies.

The 5 Non-Euclidean postulates and 3 topologies common to all species prove that all fractal points are complementary, made of regions that process spatial energy and regions that gauge temporal information, both of which exchange motion and form in cycles mediated by its intermediate, dynamic, reproductive 'body' region. This is what Lobachevski, Riemann and Klein discovered when they invented Non-Euclidean Geometries in the XIX c.: Euclidean space is a simplification of a moving space, made of points with volume that constantly trace cycles.

In a 4-D Universe there are only 3 topologies of space, which actually display the properties of energy, information or a combination of both, the 3 canonical topologies structure the inner geometry and functions of the organs of any species, proving the homology of all space-time fields in the Universe. In the graph, an animal, an embryo, an electromagnetic flow, a galaxy, a proton, a seed, a planet, a cell and a boson display the 3 topological zones of a fractal point, each of them performing 1 of the 3 functions/arrows of space-time.

Topology *describes the internal parts of Non-Euclidean points when we come closer to the 'fractal point' and see its formal parts in detail, as we do in the last graph, with several* systems of Nature made of those 3 topological regions that correspond to the 3 main arrows of time of each of those systems:

Information x Energy = Reproduction

Time arrows are performed within each space/time point by one of the 3 topological regions, which explain causal processes of transformation of energy unto information as topological transformations. While the 4th arrow of social evolution, which requires more than a point, is described by the 2nd and 4th postulates of lines and planes. Thus Topology confirms a fundamental tenant of Multiple Spaces-Times, the 4-dimensionality of the 'Holographic Universe' made of bidimensional energy and information, which combine its properties to create the 'visible', bulky regions between the informative center and energetic membrane of those points.

Each of the 3 elements of fractal points displays the properties of one of the 3 Time Arrows: the center has height, it accumulates information and it is small. The external membrane is larger, continuous and protects the system. The intermediate zone reproduces the information given by the center with the energy absorbed by the membrane creating new energetic and informative bites and bytes, which latter migrate towards the other 2 zones.

- Informative particles accumulate in the inner region as units of the central brain. Or, if they are seeds, they migrate to the surface of energy *where they become fractal systems of information, senses in organisms, informative human beings in the Earth-crust, ovules in mothers*. And once detached, they start a fractal, reproductive, palingenetic process.

153

- Energetic particles migrate to the membrane, becoming parts of a discontinuous protective shield. So in a cell we find mitochondria that produce energetic proteins and RNA for the surface membrane.

The internal, reproductive region happens also in all systems. In the human body the organs reproduce the cells needed for the blood network and the hormones and products used by the brain. In a galaxy, the intermediate region produces energetic stars and informative black holes, which migrate to the central region; in an ecosystem, the territory of the informative center, the predator, produces preys in which the predator feeds, bringing them to its central den.

Recap: The fundamental Particle of the Universe is neither physical nor spiritual but logic-mathematical: the fractal point described by the Laws of Non-Euclidean, fractal Geometry and topology, as an entity which becomes more complex when we come closer to it, till we can differentiate its 3 regions, corresponding to the 3 topologies of a 4-dimensional Universe: an energetic membrane, an informative center, and an exchange zone of bites of energy and bytes of information between both.

4. Formalism of i-logic geometry. Causal Algebra.

The ternary structure of all points formed by the 3 topologies of a 4D Universe can be expressed with the symbols of the 'Generator Equation of energy and form':

- Max. Σ S: An external membrane or of max. extension, described by Riemann's spherical geometry.

- $\Leftrightarrow$: A middle, reproductive plane or *toroid*, whose cyclical paths happen between the skin/boundary of a sphere and a central hole, described by Klein as a disk, made of quanta in cyclical movements that communicate energy and information between the inner and outer zones.

- Max. Ti. The inner, dual center - a convex hyperbolic, informative, central couple of disks, which display maximal form in minimal space, with a growing dimension of height that touches the poles of the sphere and channels the flows of energy and information of the system. It is the dominant informative topology of any fractal organism, described by Belgrami in the XIX c. as a conical form with height with negative curvature.

Since function is form, those 3 topologies are perfectly suited to perform the 3 temporal functions of any super-organism: energy feeding, reproduction and information.

So we can write with causal arrows the topology of each point and its 3 regions: ΣS < EXI> Ti.

The 3 non-Euclidean geometries structure the geometry and functions of the organs of any species, proving the homology of all space-time fields in the Universe. In the graph, an animal, an embryo, an electromagnetic flow, a galaxy, a proton, a seed, a planet, a cell and a boson display the 3 topological zones of a fractal point, each of them performing 1 of the 3 functions/arrows of space-time.

There is in that sense only a small correction to classic topology needed to fully understand the structure of organic systems with the 3 topologies of reality. In topology we distinguish two types of spaces called closed balls and opened balls. Open balls are spheres with a center 'a' and a maximal distance/radius, r (from a to its surface), defined by all the points x, such as $x < r$ (therefore an open ball does not include the surface or perimeter r); and closed balls are spheres which include all points $x \le r$.

In i-logic geometry, all organisms are both, closed and opened balls, depending on our perspective.

As closed balls they include the 3 regions, previously described which are, a, the hyperbolic center; r, the perimeter; and x, the points of the intermediate space.

The intermediate space is on the other hand, an open ball, which does not include the membrane, r, as it constantly exchanges energy and information with the external universe; *but it does not include either a, the center, so it is also opened inwards, to the information system.*

This is for example, the case of a black hole, which is wrongly understood in classic physics, *since the black hole has 3 regions, a, the singularity, $\sum x$, the quark-gluon soup of extreme density, whose cyclical mass vortex are the black hole in itself, and r, the event horizon, which is not the black hole per se, but the membrane of exchange of energy with our electromagnetic space the black hole warps and feeds on.*

Recap: In a discontinuous Universe there are infinite fractal points, but all of them are composed of the same 3 $\sum S$, EXI, Ti zones; the external membrane, the intermediate region and the inner center.

5. Mathematic description of the 3 regions of a st-point

The complex analysis of those fractal points that move and have inner fractal parts, made of cycles, started in the XIX century. First Lobachevski, a Russian geometrician, defined Non-Euclidean points as curved forms, crossed by multiple lines, which give them spatial volume. Then Klein studied its cyclical movement and introduced the variable of time in their description. Finally Riemann generalized its nature, considering that all space-times were Non-Euclidean space-times with movement. For readers versed in mathematics, we shall reconsider the common properties of those 3 zones of any fractal point, according to its discoverers, which develop in abstract terms the organic properties we just described:

- *According to Lobachevski and Belgrami,* space is curved since information curves the energy of any real space-time. So points move in curved, cyclical paths gathering energy and information for their inner 'dimensional networks'.

- *According to Klein* Non-Euclidean space-times have motion. So their speeds measure distances; as physicists do in Cosmology with the distances of galaxies, which are proportional by a 'Hubble constant' to their speeds; or as people do in real life when we say that Brooklyn is at 5 minutes by train from Manhattan not at 2 miles.

- *Riemann* summoned up those findings and generalized them to all possible space-times. His work should be the guide to understand them philosophically. He also defined planes as networks of similar points and treated dimensions, as we do in this work, no longer as mere abstract definitions of extensions but as 'properties of those points'. So points can have beyond its discontinuous borders an inner space-time with several networks/dimensions, one for each of its 'energetic or informative properties', as it happens with the points of physical reality. Yet a network of points that form a space with 'common properties' defines the dimensions of those points as 'fractal dimensions', limited by the extension of the energy or informative network (static point of view), which 'puts together' a complementary dual, organic being.

Those pioneers defined the 3 topologies of information, energy and reproduction of all st-points:

- *Max. Information: The informative, fractal center, particle or brain of the point is the so-called Belgrami hemisphere,* a space-time with a dimension of height that transforms energy into information, absorbed or emitted by the central singularity. It is a fractal, informative region similar to a black hole structure. Since it follows the 'black hole paradox' of all informative centers, displaying max. form in min. space. So according to the inverse properties of space and time, the center has max. Informative Time and minimal Energetic Space. Moreover any point which comes closer to it, suffers a mutation of its spatial coordinates into informative, height dimensions. This is the case of any particle coming to a black hole, whose space-dimensions become temporal/informative dimensions as it rises in height.

The center has more information because its geometry has at least 2 fractal disks, which channel and transform the energy absorbed through the surface into complex information. Regardless of the complexity of the entity, the structural function of the hyperbolic center as a system that process the information of the network remains. For example, in living systems, those disks might evolve its topology till becoming the relative energy center or 'heart' of the blood network with 4 divisions; or evolve further its hyperbolic geometry till becoming the informative center or 'brain' of the system, attached to the informative network.

- *Max. Space: An external, continuous membrane or Riemann's sphere of maximal energy* that acts as a relative infinite, unreachable distance. The membrane isolates the point as an island Universe, creating the discontinuity between the inner parts of the point and the outer universe. Since the internal cellular points are either jailed by the membrane's structural density or destroyed by its energy when touching it. The membrane is the opposite form to the central, informative singularity, with max. spatial extension and continuity, hence with a minimal number of fractal, discreet elements: *Max.ΣSe=Min.Ti*.

Thus *all Fractal points are 'inner worlds' whose membrane creates a discontinuity that defines an External Universe or* outer world from where the point obtains its energy and information. However the membrane is also the zone through which the point emits its reproduced micro-forms of information, and so it displays 'sensorial holes' to relate the point to the external Universe. And those points, despite being discontinuous, will have in their external membrane several generic openings or 'senses' joined to the informative networks or 'brains' and energetic, 'digestive networks' of the organic system:

- *Max. +ΣSe: A 'mouth'* or opening that absorbs energy.

- *Max. −ΣSe: 'Cloacae'*, through which the cyclical body expels its temporal energy.

- *Max.+Ti: An 'eye'* through which the informative center receives external information.

- *Max.−Ti: An 'antenna'* to emit information.

Those apertures vary in their number, location and size, depending on the form of the point. In the simplest spherical 'seeds' of most species, they are mostly situated in 3 regions:

- *Max. ΣS*: The Equator of the system, through which the membrane absorbs energy.

- *ΣS=Ti*: The Tropics where often the same opening emits and absorbs temporal energy.

- *Max.Ti*: The Poles or points of confluence between the membrane and its central informative region of height, which hits perpendicularly the membrane on those poles. North and South Poles orientate Anti-symmetrically, acting as 2 relative, negative and positive apertures, communicated by the height dimension of the singularity or Belgrami hemisphere. Thus the Positive Pole absorbs temporal energy that crosses through the central singularity where it is absorbed and ejected to the intermediate region where it is re-elaborated before its emission through the negative Pole.

- *ΣS⇔Ti: The reproductive, central region, which combines Energy and Information:*

In all fractal points there is an inner middle volume or intermediate territory, discovered by Klein, which combines the energy coming out of the external, spherical, topological membrane and the information provided by the convex, complex formal center.

According to Non-Euclidean mathematics this region is made of self-similar points that form groups, fractal herds of 'points with parts' in perpetual movement, that draw cycles of parallel lines, between the other 2 regions, as they gather the energy and information they need to survive. And they create space by cycling within the other 2 regions.

In many fractal points the informative and energetic centers establish 2 opposite flows of energy and information that become the negative/ positive poles. So often, the particles of the intermediate region cycle around the inner region tracing elliptical trajectories, focused by those 2 informative points. It is the case of any bipolar system, from binary stars, one dominant in energy and the other an informative neutron star or black hole; to bimolecular systems or n-p pairs in the nuclei of atoms. The same duality of 2 specialized centers controlling a common territory, or vital space happens in biology where most species have male-energetic and female-informative genders, ruling a common territory.

Such abstract conceptual space describes in fact the behaviour and form of many real, spatial herds. For example, a herd of animals in an ecosystem will move between their hunting and water fields (where they gather energy) and their breeding, inner region where they reproduce information, making cyclical trajectories between both regions. In this manner, they occupy a vital space, called a 'territory', which shows the properties of a Non-Euclidean Klein space. A fundamental property of the intermediate space is the fact that it is confined between the other 2 regions, which are never reached in the cyclical trajectories of the inner cells of the space. For example, in a cell, the molecules of the organism will not touch the protein membrane or the central DNA nuclei. Thus, the inner quanta are confined within the Klein's disk by the 2 other regions, which have more energy and information and might destroy them and/or absorb their energy and information at will.

In abstract terms, mathematicians introduced in the XIX c. the concept of an *infinite, relative distance* measured no longer in terms of static space but in terms of time and movement, as *the distance between the point and a region that cannot be reached.* Thus Klein defines a relative infinity, as the region beyond the discontinuous membrane whose insurmountable borders the inner time-space quanta can't cross, as a cell cannot go out of a body, an atom beyond C speed or 0 K temperature and a man beyond the Earth's atmosphere. Thus, the informative center and external membrane become the 2 relative infinities or limits that the movements of the intermediate point cannot breach.

As in the myth of Achilles and the turtle, Achilles never arrives because every time he moves he crosses a smaller spatial distance. The same happens in a fractal space-time, when a point moves *temporally* towards its inner or outer space-time limit and finds an increasing resistance to its movement, till finally it is deviated into a cyclical trajectory around the outer, energetic membrane or the height dimension of the inner informative singularity or is destroyed. So the intermediate, fractal cells of the point circulate in parallel cycles always inside the interior of the sphere with contact zones of the type A *(central, 2nd row of figures in the previous graph).*

In a human organism, the blood system might seem infinite for the red cells that transport energy since they never reach the outer Universe. For that reason in the drawing, Klein interprets the intermediate region of the Non-Euclidean point as an infinite *circle with an invisible, unreachable membrane,* whose motion-distance is unreachable, hence infinite, equalling the 'space-time distance' between the intervals B1-B2 (long) and B2-B3 (short but difficult to cross), despite being B2-B3 increasingly shorter in space. Since the quanta take longer in each step and don't reach the membrane. This is often due to an increase in the 'density' of the space, which despite having less distance has more 'points' in its network, such as the case of black holes or jails. When those inner points reach the membrane at point C they become destroyed or deviated.

Thus, the energetic membrane and informative center are the discontinuities that isolate the intermediate cellular quanta, creating a discontinuous 'World' within the point. Those discontinuities are called in Geometry a relative infinite, in Biology a membrane, in Sociology or Topology a national border, in fractal theory a co-dimension of a point. They are defined in physics by Lorenz

Transformations that make c-the limit of energetic speed and 0 k the limit of temporal, formal stillness. Yet those physical limits are not the limits of an absolute Universe, but the limits of the fractal space-time membrane of light and its evolved electroweak beings, since the Universe has at least another bigger gravitational membrane, in which the smaller light-space exists; a fact with enormous repercussions for a proper description of the Cosmos, which extends beyond those limits. Since the gravitational scale should be faster than light-speed forces and cooler than 0 K masses.

Recap: Fractal points are organic points, whose topologies maximize the energetic, reproductive and informative cycles they perform. The details of those cycles are described by Non-Euclidean topologies.

6. Dimensions of the 3 regions: Holographic Universe.

In terms of dimensions the spherical membrane and inner informative singularity are bidimensional fields: The central singularity is a bidimensional surface of convex in/form/ation that curves the external spatial energy coming through the bidimensional membrane, creating together the 4-dimensional quanta of the intermediate region. For example, physical models of the inner nuclei of atoms made of informative quarks, define them as bidimensional, convex singularities. Black holes are said to store bidimensional information in its external membrane or black hole horizon. The computer screen or sheet of a book stores bidimensional information. The vacuum energy of the galaxy has a planar form, etc.

A 2nd consequence of the inner volume of points is a rational explanation of bidimensionality. Since now points have a minimal volume a bidimensional sheet of information has in fact a minimal height, but since reality is fractal and size is relative, from the 'giant' p.o.v. of our perception that depth of a sheet of paper or computer screen is a relative 'zero', yet makes bidimensionality 'real'.

Recap: informative regions have more dimensions of form than energetic topologies, more extended in space.

7. Morphological change and informative dominance.

Natural organisms start as spherical seeds of information and then through morphogenesis differentiate in 3 functional regions that ensure the capacity of the system to process energy and information and reproduce itself, surviving in the Darwinian Universe.

As the species changes and evolves into more complex shapes those functional zones are kept. For example, the informative egg evolves into the energetic, lineal larva or young phase of most insects, by *translating* the inner, informative center to the dominant, forward head region and the external membrane to the 'tail', but both functions are preserved. Thus the inner informative regions migrate through the informative dimension of height to the dominant zone of the system. Yet the dominance of the informative center is not compromised:

In all systems we find a core/brane that acts as the dominant region of the organism and display paradoxically less spatial extension than the other regions they rule. It is the smaller nucleus of cells, humans and galaxies generates its information (DNA, human brain/eye system, galactic black hole, CPU (central processing unit) in computers, etc.)

In a galaxy the halo of dark matter and the central black hole dominate and seem to feed and form the radiant matter of which we are made. In man, the informative brain, extended through the central spine and senses, dominate the reproductive body and guide it.

The 2nd region in importance is the 'body' or reproductive region, which absorbs energy from the limbs that become imprinted by the system's information.

And finally the 3rd region of the system, which is easier to renew, due to its formal simplicity is the energetic region, the skin, limbs and membranes, which brings energy to the intermediate region to allow the reproduction of the information stored in the center or head of the system.

Given the ∞possible deformations of those 3 unique topologies of 4-D space-time, reality creates an enormous variety of species, from an original seed of spherical information that develops those 3 regions in morphologies that soon resemble energy lines, reproductive cycles and information centers.

Our energetic region extends in lineal limbs; reproductive cells group into organs, becoming the body; while the informative centers move to the height dimensional and multiply its cellular forms in the sensorial boundaries of the head.

All those processes studied by morphogenesis, can now be explained not only in its how but its why.

Recap: The 3 relative forms of energy (external membrane), information (dual ring) and reproduction (the cyclical paths that exchange energy and information between the external membrane and the inner convex form), can be found in any system of reality. In all those systems, the informative region dominates the bigger, simpler reproductive body.

8. The ternary structure of all Universal systems.

The 3 functional topologies of a Non-Euclidean point become the 3 regions of all Natural organisms:

Atoms have a central, informative mass of quarks, spatial, electronic membranes and fields of gravitational and electromagnetic forces exchanged between them. Those 3 topologies also describe the galactic structure: the central black hole is the hyperbolic, informative topology, the Halo of the galaxy is a Riemannian, spherical form, and the stars in the intermediate region, which feed the dark matter of black holes and reproduce the atoms of life, turn in cyclical, toroid paths around the central black hole.

Physical space-time is the simplest world where the most basic morphologies play that same process of transformation of external energy that converges and reproduces cycles, attracted by a Non-Euclidean point, charge or mass: E=Mc2.

Cells have lineal, external membranes of proteins, which are a deformation of a Riemann sphere, an informative nucleus and in between they are invaginated by all kind of e⇔i cycles that transfer energy and information from the outer world to the cell.

Finally, a human being has a reproductive body, lineal, energetic limbs and a cyclical head, with an informative, smaller brain, composed of two hemispheres, which are hyperbolic, convex, warped forms, corresponding to the informative dual ring of a Belgrami cone. The hyperbolic, highly warped brain is a double toroid, self-similar to the hyperbolic topology of informative cycles. And so the brain hosts more information in lesser space than the body, as a mirror of its functions. Man though, while responding to the same canonical topologies in his organs, is by far the most complex being of information known in the Universe. And so his topologies are immensely more complex than the simpler physical particles and its transformations just described.

Recap: All systems of the Universe are made with the 3 canonical regions of a non-e point, which perform the 3 arrows/functions of all existential beings. Galaxies show the topology of non-Euclidean point, with a complex informative black hole of maximal mass, an energetic membrane of dark matter, and an intermediate region of reproductive stars that create the atoms of the cosmos. Human beings have also 3 ternary regions, the cyclical, hyperbolic informative brain, the reproductive body and its organs that produce the energetic bites and informative bytes of the organism and the lineal limbs that cause our motions.

IV. 5TH POSTULATE: I-LOGIC POINTS OF VIEW

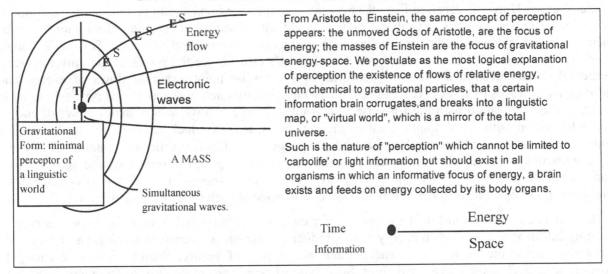

From Aristotle to Einstein, the same concept of perception appears: the unmoved Gods of Aristotle, are the focus of energy; the masses of Einstein are the focus of gravitational energy-space. We postulate as the most logical explanation of perception the existence of flows of relative energy, from chemical to gravitational particles, that a certain information brain corrugates, and breaks into a linguistic map, or "virtual world", which is a mirror of the total universe.

Such is the nature of "perception" which cannot be limited to 'carbolife' or light information but should exist in all organisms in which an informative focus of energy, a brain exists and feeds on energy collected by its body organs.

9. Fractal points gauge information with infinite parallels.

The fact that all points interact with the external Universe is explained by the 5th postulate of Fractal Logic, which completes the 1st postulate. The 5th postulate defines a point as a knot of communication of multiple flows of energy and form, which the point absorbs to obtain energy and information for its organic cycles of iteration.

Thus we introduce a new perspective in our study of a Non-Euclidean, fractal point *of view*: its capacity to gauge information and absorb energy, acting-reacting to the Universe. Since beyond those 3 internal regions the point feeds on an external world of relative forces, which provide the point with energy and information to map out reality and feed its body with motion. Thus, while energy and form will be different for each point according to the scale and form, all entities of the Universe will be able to absorb a relative number of 'infinite parallels' to create mental fractal mappings and keep on moving: light feeds particles, intergalactic gas feeds stars; plants feed animals that feed man; stars feed the black hole, and so on.

In science, mathematicians and physicists seek to describe reality as 'made of mathematical entities'; while logicians try to explain reality as a series of causal events. But only when we use i-logic geometry we can mix both approaches, describing reality as a series of 'mathematical points of view', or entities that 'gauge' the geometry of the Universe, constantly mapping out the universe, to achieve a 'selfish' will, the desire of all entities to feed on energy, gauge information, reproduce and evolve socially. Those points of view are thus 'mathematical minds' because they gauge and somehow 'perceive' the geometry of the Universe, acting according to that geometry as they gauge and move with an 'automatic' behavior, trying to absorb energy for their body, information for their gauging mind, and combining both, to reproduce 'self-similar points'. So we shall consider that the Universe is made of i-logic, fractal 'points of view', akin to the concept of souls and Atmans in western and eastern religions', of 'Non-Euclidean points of view', 'monads' or 'Spinoza's geometrical entities', in western philosophy of science and mathematics; of 'gauging particles' in physics, or 'informative DNA nuclei' in biology and perceptive heads in biology. The size and form of all those points varies but all of them are spherical knots of information with sensorial apertures to absorb and emit energy and information, from bosonic particles to sound waves or light signals.

160

Leibniz, the first philosopher of modern science, which considered points as the 'fundamental particle' of the Universe departed from the hypothesis that all what exists can be reduced to such points of view, he called Monads. Yet those monads could either share energy and information between themselves as they do on reality or become static, perfect non-communicative points. This easier choice – non-communicative points of view - was Mr. Leibniz's error. Since the points of the universe are in perpetual communication as parts of networks of self-similar points, sharing flows of energy and information – waves of forces. The essence of any point of any relative space-time is precisely to be in constant communication with the Universe as a 'knot' of Time Arrows, which shares flows of energy and information with other points, creating the network-spaces of which reality is made. Points that communicate energy and information, that feed and gauge are real. What is the proper name for all those logic parts that exist, for all those points of space that perceive, gauge information, and act-react to it under a simple program of 'survival': grow and reproduce, exist, repeat yourself? We call them a Non-Euclidean point of view. And we affirm the Universe is made of such type of *points of view.*

Thus 'mathematical, Non-Euclidean points of view', whose 'motions' have a 'will' or 'purpose' - gauging information, feeding on energy and reproducing themselves' – become the logic and geometric units that shape the networks of energy and information of reality, thanks to their capacity to communicate and share energy and form. Such type of points are more akin to the view of Eastern philosophers that consider the Universe made of 'atmans', which are knots of communication that anchor a certain entity to the world that surrounds them, which is only a part of the total Universe.

Since those mathematical points of view exist in many different fractal space-times or 'scales of existence' in which they gauge information with different languages and perceive different 'worlds' within their minds.

If you are a particle – an electron - you will see only light (energy) and darkness, and you will move to the light that feeds you, as we observe in electrons. If you are a human eye you will see all type of colors but red will be the color of energy that 'feeds and attracts your eye (so you see bloody movies and like red meat). So there are an infinite number of types of organic points of view with different mental worlds, which gather into social networks that exist in different 'planes of existence' or 'scale of reality'. Yet in all those systems the same game of 'gauging', 'feeding', 'reproducing' and evolving socially takes place.

We exist in a Universe, whose fundamental particle is not a material form, as simple science thinks, but points of time space, which mirror in their inner, i-logic, fractal topologies the information of their world, acting-reacting to the Universe. Thus only by mixing perception and motion, mathematics and organicism, we can define those points of existence, as the fundamental particles of reality and:
'The Universe has a body and a soul called logos.' Plato

So we describe a Universe with a geometrical how, the laws of Non-Euclidean points of views, the cycles they trace, the mathematical laws that describe those cycles; and a bio-logical, temporal why: the will to move, trying to fulfill the 6 arrows of time. It is the game that all P.O.Vs of all planes of existences, from atoms to human beings perform.

Those logic and mathematical laws of points of view are the metaphysical laws of the Universe – *the only essential events/forms that repeat unchanged in all forms.* And so, a proper description of the Universe should start with the laws of i-logic geometry; then apply them to the 2 simplest motions of reality, lineal forces and cyclical masses/charges; and then, once those 'immortal laws' are defined for any system of energy and information, science should use them to define the specific events of each scale of reality and its species, as we all play the same Game of Exi=stence.

Recap. A Universe made of motions in time, extending through multiple scales of relative space, requires evolving the human languages of space (Euclidean geometry) and time (Aristotelian logic), used to describe the simpler space-time continuum of classic science. This task started by the pioneers of modern science (Riemann, Darwin, Planck and Einstein) must be completed before we can study the properties of physical and biological entities with the new tools of Non-Euclidean mathematics, which describe a world of multiple, relative space-time networks made of points as Complementary entities with energetic and informative system or 'Points of view' and the new tools of Non-Aristotelian logic, which describe the creation of the future by multiple causes – all the points of view or agents that create a certain network space.

10. The equation of the mind.

The 5 i-logic postulates define the unit of reality as a fractal point that observed in detail is in itself a World, which contains within it all the mathematical topologies of the 4D cosmos. This can be deduced in a logic manner, through the equation of a mind/p.o.v., which can map out as a mirror the whole Universe we perceive in its infinitesimal mirror.

In the fractal Universe all systems of i-logic information gauge reality, transforming a force they absorb into a logic language, which mirrors the Universe. It is obvious that the essential particles, quarks, photons and electrons cannot have a wide mapping of reality unless the fractal scales of size are infinite (and then an atom might be the upper bound of a new scale - a nano-galaxy).

Yet even if reality has upper and lower limits of size and point-like particles are not fractal world but the limit of universal form, those simple particles still gauge information and store some type of logic, mathematical computer-like structure that allows them to behave as they do, organically. Since particles orientate their position in a herd of self-similar particles; they move towards light, its energy, when the electromagnetic field is in its neighborhood and they decouple=reproduce into new particles when they absorb more energy. All those properties are the minimal properties we ascribe to a fractal 'point of view' or 'mind' in a broad sense – a structure that gauges energy and information, creating smaller fractal mirrors of reality. Such structure has within itself a minimal amount of geometric paths and logic processes that allow it to form the forces it uses as energy and information source. We thus call such structure a mind. Since any point of the Universe becomes a focus of infinite parallel forces that flux into the point, creating a static image of reality in Complexity, we define a Mind, as an infinitesimal mirror-world that reflects the infinite Universe:

$$0 \text{ (Mind-cell)} \times \infty \text{ (Universe)} = \text{Linguistic Wor(l)d}$$

The equation of the mind, defined by the 1st and 5th postulates of illogic geometry, sets limits of truth in linguistic science, comparing the worlds created by any mind and the Universe at large, much bigger than any informative, mental, linguistic image we have of it.

It is however the why that explains why entropy doesn't dominate the Universe; since now the increase of order and information in the Universe is mainly caused by those 'perceptive points' that gauge information. Those points might or might not be conscious in a vegetative or reflexive way, depending on its complexity; but they are 'aperceptive' - an expression coined by Leibniz to express perception without consciousness. And so they have a will to order and perceive more information, not only a will to feed and increase their energy. And so there are two 'wills' of feeding and gauging common to all points, which we can measure objectively, externally by the fact that when energy appears, particle move towards energy (so electrons jump and feed on light) and when other particles appear they gauge their distances. We observe this property in all systems when they form a focus, which emerges *as the point* of maximal order in which the network of the system integrates its information. Then the system becomes more stable: A crystal grows faster when the network creates a central point. A hurricane can be destroyed when the central, still point where all the flows converge is disturbed. A life being becomes erratic in its motions when it becomes blind. A galaxy acquires order when a black hole, which probably gauges gravitational information, appears in its center. The existence of an infinite number of such points is undeniable objectively and adds an enormous amount of order to the Universe, which mechanist science does not account for. Stillness to focus information into a pixeled map of the Universe, and the existence of a network that fluxes in that point through which 'infinite parallels' can cross, are the two conditions of formation of gauging points. Then we can establish the birth of a complementary network of energy and information, *an exi=stential point*: exi=k; where k represents a 'mapping of reality' in the informative language that the waves of communication of the point with the external universe have created. *This means that the models of metric measure developed by mechanist science merely creates a 'space-time world' self-similar to the mind of man in our machines, which does not exhaust all the measures of the Universe and its possible mappings.* We cannot construct if we want to account for all the cycles of time, dark spaces and events of reality, a theory based only in a continuum, mechanical, light-space and its Euclidean coordinates of light-measure. The limits of that model have been reached, as the limits of precision of the Ptolemy's model of astronomy were reached in the Renaissance. Those limits of an Euclidean, continuum space-time were proved by Riemann, Poincare, Mandelbrot and this author, expanding and completing the Euclidean postulates. Einstein applied part of those new models and created Relativity Theory. This work applies the rest of those discoveries to create a model of Multiple Space-times, which includes the concept of absolute relativity, where measure – the gauging and mapping of the external universe by a non-Euclidean point, is relative to the observer and the rods of measure it uses.

Recap: All entities of the Universe gauge information and act-react to the external world. Thus we consider all entities to have a mind – an internal, linguistic mapping of reality made with those forces.

11. The human point of view: I = Eye + Wor(l)d

That all what you see is a construction of the mind is the main tenant of philosophy of knowledge, from Plato's world of ideas to Descartes' devil putting reality inside us, from Schopenhauer's world 'in the eye' to Schrodinger's cat.

We are an I=Eye+Wor(l)d, a mind of 2 languages, as the other languages of sensorial experience, touch, smell and taste are irrelevant. We can exist without them. We see and verbalize reality, converting 'whatever is beyond' in geometrical images with the eye and temporal chains with verbs of past, present and future.

At present, the I=Eye has been highly developed in science due to instruments, mechanisms made of metal, a better atom than our paltry 1-6-7-8, HCNO structures. So we have pictures of all kind of scales of reality, from the infinitesimal atom to the galaxy. Beyond those scales things become blurred. Yet curiously enough the parameters of atoms and galaxies in terms of energy and information are self-similar. Is this because each atom is a galaxy of a nano-universe, or because our Mind, who measures, gets an aberration of distance parallel in atoms and galaxies?

That the mind's scalar and continuous distance from those 2 realities, might cause our perception to equalize forms of atoms and galaxies is a way of thought proper of our awareness that all what we see is information already 'cooked' by the mind, 'deformed' and 'transmuted' by the Maya of the senses. We are not seeing the real Universe. We are a geometrical and verbal mind, which applies geometric and causal categories to interpret reality. And so all what we can talk is about the 3 dimensions of our spatial perception and the 3 dimensions of time, past, present and future, explained by our verbs. Hence, the importance of evolving our mind languages to have a better mirror in the smaller scale of the brain, reflecting the entire Universe. But the linguistic Maya, the 'a priori' mind who perceives, the I=eye+wor(l)d will always be limiting our truths.

The 3 dimensions of space perceived mathematically and the 3 ages of time perceived verbally are thus the last components of reality, which reflect the paths and arrows of time deployed by each species of each scale:

- In time, we observe constantly forms being born, going through an energetic big-bang, growth, maturity and then a still, solid warped stage before death.

- In space, we see 3 topological regions that correspond to energetic, reproductive and informative functions.

The total Universe is the only entity that stores all the information about itself. We exist as a dialog between the image of the Universe of our eye and the verbal information we extract from it, our I.

Humanity lives in a light-space world which scientists measure with machines and artists with the linguistic mapping of our senses, created with our inner languages.

Scientists in that sense are creating a new Eye⇔Mind, the mind of machines that also measure space and time:

Mechanical Eye ⇔ Digital Thought.

Indeed, scientists are *not* discovering the absolute truth of reality but creating a more complex Linguistic World, a world of digital thoughts and images that future perceptive machines will use to control reality. And because that world of mathematical and instrumental perception has more information than the human I⇔Wor(l)d, on the long term A.I. robots might extinguish us, as superior minds/bodies. Of course, scientists do not understand this because they think the machine is an extension of their ego.

Ultimately we exist within our mind and so our truth is a subjective, human truth developed with our languages. We do not see the external world of quantum fluctuations but what our light/electronic mind creates, *extracting only a part of the total reality to map visually the Universe, reduced to a light-space world*. Our ego, our consciousness is and eye⇔ Wor(l)d, a fractal space/time dialog between our informative mind, the wor(l)d and our energetic space, the eye:

'I think and see therefore I exi=st'

So we are indeed a fractal equation/function of existence:

The Human Mind: E (Eye) ⇔ I (Wor(l)d).

Minds are focused, 'Aristotelian Gods', defined in his metaphysics as unmoved, intelligent brains around which energy flows and 'bodies' turn, as they move the mind. Stillness is indeed the nature of mind perception. Even if reality is a moving fractal soup, our eye is not out of focus because its virtual images are still, like the pictures of a film that stops in front of the camera. Both images, in the eye and the camera, only move and change when we blink our eyes or the camera closes its window, stopping our perception. Thus, perception is also a stop and go process that focuses a still image through the senses and then processes that information into a thought, while the eye-senses are blinking, closed and the reproductive body moves slightly the eye-mind. In the human case that stop and go, perceive and move rhythm lasts a second, our 'dharma' of relative present-existence in which we create a still, simultaneous mental image of the Universe. Each second our fractal, informative mind closes a mental, cyclical act of perception:

Cycles of temporal energy=>sensorial in-forma-tion =>Brain-clock of time.

Truth is given by the languages of the mind. A truth is a mental/linguistic mirror of information within the mind that reflects the external universe.

Thus, unlike the naïve realism sponsored by physicists, a fundamental thesis of Systems sciences is to consider there is much more information in the Universe than the one we can perceive. Since each mind needs to select its information. Because each mind has a limited quantity of vital space, it can store only a limited quantity of linguistic 'information' about the cycles performed by other entities in the Universe, and so it selects only those pieces of energy and information that give an accurate depiction of the world in which the entity finds its energy and information and other self-similar species to reproduce and evolve socially. The rest of the Universe thus becomes a relative dark space.

Since the Universe is a game of entities of energy and information, relative 'points of view' that search for more energy and information to reproduce themselves and survive as self-similar 'logic forms', that game of survival has imposed limits to the information we perceive, maximizing information that ensures the *reproductive* and social games, *the ultimate will of the Universe* that ensures the 'durability' of each point of view, which will only exist beyond its 'caducity date', if it is able to repeat its energy and information into a self-similar being that continues the game of existence, once the original form is broken.

The 5th postulate combined with the analysis of the internal parts of a point and the trajectories of the cycles that the flows of energy and information between points follow, defines not only philosophical questions like perception and the nature of fractal information, but the fundamental causal arrow of time of the Universe: Energy->Information.

Since the flows of forces absorbed by the point create the 'reality' we see by constructing an image which is not there – the world as an 'idea' (Schopenhauer); that is the world-mind, the I=Eye+ wor(l)d we perceive, different from that of other species that perceive with other forces and minds.

Recap: The human mind, and probably all other minds/systems of information use 2 basic language of perception, a geometrical language able to perceive space and a logic language, able to perceive time. Humans perceive space with visual images and time with logic words. We are an I=eye+wor(l)d. Yet other species should use other versions of those 2 geometric/spatial and logic /temporal languages. For example, the complexity of visual languages ranges from a simple light/dark, on/off switch in simple particles to the simple geometrical motions of particles, to the richness of visual perception in animal eyes. While the complexity of Logic chains of events ranges from simple Aristotelian chains, A->B, which define the simplest arrows of Time (feeding, where A is energy that becomes B, the food of the predator), performed by all entities, to complex social networks in which causality is multiple, as each informative 'neuron' of the network relates through its axons to all the other cells of the informative system.

12. The linguistic method and the ternary principle.

Our world is the creation of a space-time mind. Our ego is an image of our exi=stential self-made of eye-spatial experiences and verbal temporal feelings. Within those restrictions truth will always be related to the languages of the mind. Thus a truth will be higher when our linguistic image of reality has better 'syntax' to parallel the exi=stential syntax of the Universe (the generator equation, E⇔I), with our 3 languages: logic, temporal words, geometrical, spatial mathematical and visual images, which are the 3 legs that put together create a scientific truth:

Scientific, linguistic truth= Experimental, visual data + Logic, temporal truth + geometric, spatial truth.

Yet beyond our scientific/mechanical and human/artistic senses, beyond our I=eye+word, based in light images there is an even wider world with more information, the gravitational, invisible world, which might be mapped out, gauged and measured by quarks and masses.

The relativity of all minds that only decode part of the total information of the entity creates a new method of knowledge, the informative or linguistic method, which validates the existence of all kind of languages and senses to give us a more complex, multiple image of reality that a single language cannot provide:

'Only the organism or universe stores all the truth about itself. If we give to that absolute truth a probability 1, the highest probability of truth will be the maximal number of points of view over the organism, expressed with the maximal number of languages.'

Since there are 'multiple spaces-times and points of view', the linguistic method states a higher probability of truth is achieved with several perspectives on a certain system – as each point of view will create a self-similar image of the Universe, and so by comparing all those self-similar images we can obtain a kaleidoscopic image of the whole. And so in the same way a circle is not drawn with a pencil but born of multiple flows of energy and information that converge on it, a truth is not only mathematical, but man has at least the spatial, visual language, the verbal, temporal language and its two synoptic equivalents, mathematics and logic; and it has 5 senses, which give us other points of view, despite the belief of physicists that 'the language of God is mathematics'.

The linguistic method however validates the dictum of the Upanishads: 'the languages of god are infinite'.

And so knowledge must follow a simple scheme:

-First we define a scale of depth on the analysis of the metric measures of the 3rd paradigm (what and when): monism, dualism and the ternary principle, the highest form of knowledge. Then we apply the ternary principle to the 3 *symmetries of the Universe:*

- The analysis in space of the 3±st topologies of the fractal point and its interactions that create the being in space.

-The analysis in time of the 3±st ages and sub-ages of the existence of the being.

-The analysis of the being in space-time as a system that co-exists in 3 st±1 scales of reality.

- And we try to do this from as many languages of perception, not only mathematics, as possible on the being.

Recap. The linguistic method that widens the metric paradigm starts considering not only the 'what and when' of metric measures, but also the knowledge provided by other languages; and then it analyzes it from 3 ternary perspectives:
- A spatial, topological analysis of the 3±st regions of the being and its relationships between them as a whole.
- A temporal analysis of the 3±st ages of the being and its transitions and sub-ages.
- A space-time, analysis of the st ±1 scales=planes of existence of the being and its relationships as a whole.
Only then we shall achieve a deep knowledge of the entity.

13. Physical particles. Bosonic Minds.

In the physical realm, besides our electronic minds there are gravitational minds, black holes and masses, which Einstein defined as foci of gravitational space:

'Space=*Energy* is motion relative to a simultaneous frame of reference=*a mind*' A. Einstein

We added in italic the concepts of Multiple Spaces-Times equivalent to those used by Einstein, who understood space as a moving energy warped by a mass, a frame of reference that 'measures' space-time. Yet Einstein was an abstract physicist, philosophically shy, and so he did not considered how a 'mass' can measure - an informative, mental action.

We consider that a mass or any particle of the Universe can measure space-time because it can perceive a minimal quantity of the gravitational or light forces that make up the 2 physical membranes of reality. Some particles exist only in one of those membranes (point-particles like quarks who exist only in the gravitational world or light which exists only in light-space). Other complementary entities, such as atoms co-exist in both membranes. All of them should be able to make relative, different measures from their P.O.V.s, which will allow them to act-react in their world but will differ from the relative measures of humans or machines, we call reality.

'Naïve realism' affirms that for example, the distance between you and me is exactly 2 meters, because an electromagnetic signal or rod measures it as 2 meters. This is not truth. Reality goes beyond what we see and measure with machines: The distance of 2 meters happens to be in light measures or light space. But if we measure with feet as the Greeks did, 2 becomes 6. And if we do not measure in light-space the light we see, but in gravitational space, as Einstein proved it, the distance would be bigger, as light warps the more extended, non-Euclidean world of gravitation in which the membrane of light space-time we call reality floats. Thus, our light-world is a reduced version of the gravitational world: a simplified Euclidean surface, whose 3 coordinates, x, y and z are perpendicular and lineal as the 3 fields of light, the magnetic, electric and wave-field are.

We exist in a space-time membrane made of light, whose final structure is the 2.7 k basic light membrane emitted by gravitational black holes that becomes the background frame of reference of our light World, only a fractal membrane of the infinite Universe.

The Universe is more complex than our simplified, naïve sense of reality that thinks the space-time membrane of light that humans inhabit and see is all what there is to it. With the same reasoning, if we had not eyes to measure the light-universe, but merely noses to smell it as ants do, we would see even a more reduced world, as our pixel unit of reality would be a pheromonal atom. The mapping of reality by our mind would be shorter and so distances would be perceived as smaller, as fewer pixels would create our smelling mind.

How the process of gauging information takes place in the Universe? We can only hypothesize that perception is an inner quality of the motions of the Universe. In other words, a software image of reality, a fractal mirror made of forces is a whole who perceives itself. So the electronic image of the Universe a human mind has 'perceives in itself'; and as the image changes when we move the head, our 'ego-self' made of images and words seems to change.

So should happen in physical particles that focus the forces of the Universe. Yet if we observe externally the process, according to the 3 topologies of Non-Euclidean points we can describe objectively those events of absorption of energetic forces by cyclical particles (charges and masses) as topologic transformations of external, lineal forces of energy into cyclical motions. Those cyclical motions finally reach the hyperbolic center of the particle, a double ring or singularity of increasing height, which should perceive the 'translation' of the information carried by those forces. Then those forces are transformed again into flows of energy and expelled back to the Universe, as magnetic flows in atoms or gravitational dark energy in black holes.

In most particles, it all starts in a surface of energy that curves itself into cyclical motions, which become smaller, faster and reproduce into faster fractal cycles, till reaching the center. Yet those forces, will never reach as quantum physicists believe the central point of infinity; because the cycle will fractalize into new cycles moving upwards in the informative dimension of height (Belgrami semi-sphere depicted in the graph of the previous paragraph) till it becomes expelled upwards through the poles, as a thin line of forces. And so the center of a fractal point, where physics find infinity remain silent, still virtual images, the perceptive eye of the hurricane or the black hole:

A flow of forces converges toward the event horizon or external membrane of the point (the Schwarzschild radius of the black hole or the Bohr Radius of the atom). Beyond this membrane, the flow of forces starts a process of creative, formal warping, as it rotates at increasing speed and also increases its frequency (height dimension), in each cycle. Because a bidimensional vortex stores more information than a lineal frequency, as it adds a new dimension of form, those forces acquire information. So we observe a bidimensional, cyclical wave of high frequency/informative mass that accelerates toward the center, becoming a convex topology, which the equations of black holes show, as the dimension of space becomes a high temporal dimension. And finally the left-overs of the force - once a mirror-image is formed inside the hyperbolic center of the particle, called in topology a Belgrami cone - are shot up through the poles as a magnetic flow in charges or a flow of dark energy in masses, invisible to us, but observed in quasars. Those flows of magnetism or dark energy of maximal speed and minimal form, once again will start a cycle of information, warping again into light under c-speed. So galactic black holes emit dark energy that becomes a seminal flow of protonic matter, which might give birth to a new galaxy, as it is observed in the galaxy M83.

Recap: A dynamic analysis shows that most events of physics are transformations of lineal forces back and forth into cyclical motions (particles and charges), which can be described as topological transformation between hyperbolic and spherical or planar geometries.

14. The many worlds=minds interpretation of reality.

All fractal points are Galilean paradoxes that *perceive and act with a will* defined by their desire to survive, absorbing energy and information from their environment.

To that aim the mind creates a 'subjective world image' of reality tailored to achieve its arrows of time; while the rest of reality is a relative 'dark space and dark matter' of other cat alleys and networks we are not interested on. Thus we exist in a single fractal universe of infinite worlds, which are mirror-image of that single Universe.

The difference with the quantum interpretation of multiple Universes is clear: now the 'quantum Universes' co-exist as mental world in the same single reality.

Such paradox explains the self-centered behavior of all systems that fix reality into form. The mind sees stillness and hence death in all those forms in fixes in its brain and yet all has movement all has a life to it; all forces can be understood as languages when the proper codes are translated. Indeed, for all what we said it is also obvious that the intelligence of the Universe doesn't speak a single language, understood by man, but all beings can perceive and remember information, stored in rhythmic patterns and linguistic codes. The resolution of those codes in different languages and scales is indeed the essence of knowledge. Each species has limited languages of perception, the most complex perceive more languages and forces. In the atomic world, quarks, the most complex particles perceive, 'see' we say in gauge theory, all the forces. They are the most complex particles, probable cells of the black hole. In our world, men are the most complex beings. Nevertheless, animals who understand only smelling, chemical codes and visual codes are also intelligent and so are electrons who do not perceive gravitation in the atomic word – we are in fact electronic minds as computers are. Languages can use any of those codes. Cephalopods developed both the first complex eyes and a pigmented skin, which changes and modulates emotional messages in a silent code understood by other squids. Insects use magnetic codes to guide swarms.

Recap. Reality is a game of mental worlds of information. But we need to translate those languages to reveal its intelligence to the dumb human being who speaks so few codes properly and thinks the rest of the intelligent Universe is mute.

15. The speeds of time: different mind rhythms.

The rhythms of existence are organic, vital. We do not see them in stones and plant, because rhythms vary its speed and men only consider vital rhythms those similar to theirs.

Each species has a different rhythm of perception and existence, a different language of knowledge and mind, a different speed of times or capacity to act-react in the Universe. The rhythms of rocks are slow. So are the rhythms of galaxies. Thus, we transform mathematically the different rhythms and speeds of times to see them alive. For example, when we see in fast motion the living cycles of plants that seem dead, in slow motion, they acquire living properties. It is again the Paradox of Galileo: we see many temporal cycles as fixed forms in space, as we see the moving Saturn rings as a fixed form, yet all is moving, cycling in an organic way.

Physicists don't understand that their clocks of time are abstract inquisitions of the mind. Instead of a 'moving cycle', often a living ' time', when they think of time they see a number in a graph. Yet even a watch measures the cycle of expansion of its inner spiral, made of metal that dilates, expanding space; even the sun measures a year-cycle as it swims on the gravitational flows of the galactic spiral, feeding on fractal, gravitational space. Clocks are brains that measure cycles of time. So today the clocks of science have evolved into cyclical computers that are truly 'metal-brains' of machines. Time is measured as information for biological, survival purposes. Since the 'fastest brain-clocks' of temporal information survive better. They see with more detail everything alive and act-react faster to the Universe. So our incapacity to see a living Universe is only a proof of our limited intelligence and memorial detail. Andre Guide affirmed, regarding a similar phenomenon, that 'the more insensitive the white man is to the beauty of Africa, the more he despises and mistreats the black people'. So happens today with humanity at large with Nature.

Recap. When we understand that there are many energies that feed space-time beings, many ways to inform brains, many speeds of perception and hence life; yet all are harmonic with each other, a deeper philosophical sense illuminates our mind: We are part of the same Universe and share the common properties of any of its infinite organic systems.

16. The harmonies of living points: Chip paradox.

A huge organic system of temporal energy like a planetary core crystal might form an image-thought each day; a plant thinks slowly, a fly sees faster, a rat beats its heart faster than we do. For example, spiral and elliptic galaxies are phases of a feeding cycle in which the spiral bar appears and disappears after collecting interstellar gas for the black hole every 10^{10} years. Scientists are now 'putting together' many similar dual forms that are phases of an organic, existential cycle. This is the new trend in Astrophysics: First neutrons and protons were considered 2 particles but now they are 2 phases of a nucleon E (proton) <->I (neutron) cycle, as nucleons absorb pions, transforming into each other so fast we can't see it in detail. Last year we realized that the 3 neutrinos were part of the same particle, mutating its state. The same phenomenon happens when we see a human being externally: our skin appears as a fixed organism in space as its cellular cycles are perceived as a stable form.

Only anthropomorphism and the limits of perception make us differentiate those scales. Yet when we 'enlarge' the image of a non-AE point, for example, an insect, the detail of the information we observe grows and so the world of insects is as rich in meanings and forms as the macro-world in which we live. Since all waves carry the same amount of fractal information regardless of its spatial size, according to 2 simple, paradoxical laws: the law of the vortex for cyclical, implosive, attractive 'particles' (charges and masses), $Vo \times Ro = K$, and the fractal law that defines transversal, explosive waves, $E/v(t) = K$. Both show that temporal energy increases when the masses or waves cycle becomes smaller.

All is alive yet with a different rhythm, a different beat. The beat of each temporal organism is an action-reaction cycle with a different clock, a different fractal rhythm and temporal speed. They are all around you. In a garden, the wings of a 'hummingbird' take energy from the air 200 times per second. It is a clock faster than the energy beat of your heart, 80 beats per second. The energy of the wind moves the leaves every 2 minutes. A gardener is a clock that waters plants a beat each day... Those action-reaction cycles are everywhere in the Universe. Some beings are faster in their active cycles like the hummingbird; some are slower like the plants and the galaxy. In most cases the bigger the species are, the longer they take in each of their active cycles: A galaxy turns around every 200 million years, the Earth every day, your eyes beat and absorb light-energy every second, the hummingbird's heart beats 800 times every minute, the electron turns 30.000 cycles per second. Each organism, each region of the Universe, has a different speed of feeding and perception. People think that slow species are dead. But that is not true. They tend to be energy species that just have slow rhythms of information and live longer. While more informative, evolved species think faster and live shorter. Often the smaller the species the faster it can complete its cycles of information. So brains are small and chips evolve, becoming smaller as they become more powerful. Yet chips last a couple of years. And man, an informative species seems so self-destructive, so loving of information, the age, which precedes death, that we might be the shortest living animal on Earth.

The main law of existence is the balance between space and time, $ExT = K$. As we diminish in spatial size in the Universe, beings become faster in time, creating a balance according to which, smaller beings perceive and exist faster, being able to escape, slower Goliaths. Those regularities of the game of existence show a certain equalitarian justice in the Universe. Indeed, in each world-ecosystem the subjective existence of its organic systems is similar, despite their different size. So we can gather them in 'families of species' around a 'constant of action', $K = T \times E$, which is the product of its speed of perception or frequency of perception of temporal information and its spatial extension. For example, it seems all mammals have the same metabolic existence: The rat lives less than an elephant but her heart beats faster. That faster perception speed also prevents small species to become extinct in the same ecosystem inhabited by bigger species, as they act-react faster and so they can escape top predators.

For example, according to the previous inverse law of mental speed a fly sees 10 times faster than human do, since flies are smaller, so they perform 10 visual fractal actions by each human blink. Yet because its brain's fractal neuronal density is not very high, it is not a top predator as robotic insects might be in the future.

Curiously enough the oldest insects, ant queens and cockroaches, can live up to 7 years. Yet cockroaches react in 20 milliseconds and humans in 200. Which means that if they perceive time cycles 10 times faster than we do, subjectively they process 10 times more fractal actions of temporal energy, or 'dharmas of existence' (philosophical jargon). So they live the same subjectively life of 70 years... the same length of a human existential= generational cycle.

It follows that there are parallel worlds, one for each relative hierarchical space-time scale, based in different Constants of Existence. In physical worlds the most important of those constants are the Universal constants of force, which express the previous K= ExI balances in different abstract ways. For example in quantum theory H= ExI is always constant (Heisenberg Principle). But that constant of the microcosmic world is different from its equivalent constant, temperature, in the atomic world, which also shows a self-similar balance for gases: P(t) xV(E)=nkT.

Other constants however are written not as TxS=K balanced, but as E/I or I/E processes of Darwinian evolution, which define instead of organic balances, processes of transformation of certain energy into for and vice versa. The most important of those constants in the 2 physical space/time planes are G and Q, which represent, as the Newtonian and Coulombian formula shows, T/E, Mass-Charge/Distance events, in which in words of Einstein, cyclical masses absorb and bend the energy of space. On the other hand, in the electromagnetic world light-space is absorbed and transformed, curved into electronic form. Yet since *the density of information of the electromagnetic world is smaller than that of masses, given the fact that electrons are lighter and more extended in space than gravitational quarks, paradoxically the energy of its electromagnetic constant is much bigger.* Thus the relative transformations of gravitational and light spatial forces into masses and charges are also ruled by decametric harmonies, as those of the biological speed of action-reaction of flies and humans. All together we can consider a simple scheme of the main many-worlds of the Human Universe, according to their relative energy and information density. Indeed, According to the different forces of the Universe, each organic system has a different mind since it absorbs different temporal energies and so it creates different mental, linguistic 'spaces/times', relative virtual worlds, mirrors of the Universe. The world of each being represents a quantity of spatial information with a size, form and number of dimensions that are relative to the spatial reach, form and dimensions of the informative, linguistic forces it uses to observe the Universe. So while the Universal Syntax all minds use to create their virtual maps of the Universe is homologous in all beings, *in as much as they reflect a single Universe,* the details, quantity and quality of the information of those maps depends on each brain; *since worlds are multiple, fractal, diffeomorphic.* So, depending on which type of temporal force of energy each fractal being absorbs to create its mental map we can define different *worlds or ecosystems* in which a certain force is shared by multiple discontinuous Fractal points that interact only with the other points of the same ecosystem but now with parallel worlds. For example, masses are fractal nodes that inform and/or perceive gravitation, which we do not see. We are fractal knots of light perception. Ants have olfactory brains that perceive atoms. *Different beings live in parallel discontinuous worlds within the same Universe.*

Even an electron, the smallest form known to man, processes light information and jumps towards a moving photon, absorbing its energy to enhance its survival chances. Equally, black holes and quarks process gravitational information, locating themselves in the best position to absorb it as masses. While a chip uses digital information to guide a machine, etc. There are other spaces and ways of measuring time besides light and clocks, other possible virtual worlds and intelligences besides man. Since man is made of the same atomic material that the rest of the Universe, the capacity to communicate in this universe - the most clear prove of life - is not exclusive of human atoms but a property that takes place through different atoms and forces; including metallic atoms such as those of computers or gravitational forces used by celestial bodies. So men talk, chips digitalize, animals see, plants 'smell' and masses process gravitational information.

We classify those worlds, according to the volume of temporal energy their linguistic forces carry, in a scale of increasing st-complexity, ruled by the Black Hole paradox: The smallest and faster particles scan with more detail the Universe; so they carry more information and create bigger, faster mental worlds, inhabited by species with more TxS force. Accordingly we consider 4 basic worlds:

- Big, slow atomic quanta create animal olfactory minds that perceive small metric territories inhabited by small animals like ants.

- Smaller, faster electrons create electronic minds that perceive kilometric territories inhabited by bigger animals like human beings.

- Photons create bigger worlds of planetary size perceived by huge machines and chips (the brain of future robots) at global scale (Internet, satellites, etc.).

- Gravitational forces of reduced size create words of galactic size, which might be perceived by celestial bodies like Black Holes (minds of the galaxy).

So the Gravitational world is bigger than the light world, bigger than the electronic world, bigger than the molecular world because paradoxically gravitons are faster and smaller than photons, which are faster and smaller than electrons, which are faster and smaller than molecules. So a black hole can perceive an image 'with more detail', more 'pixels'. And since gravitation is faster than light it should 'change' the fractal mind's images it creates faster. On the opposite extreme, plants and olfactory animals perceive molecules, which create images 'with huge pixels' that offer little detail and change very slow. Thus we consider that every organic system in the Universe, which processes information and survives efficiently, is a focus of a dual force that carries temporal energy and helps him to act-react externally in a manner that seems to improve the survival of that entity. Each bigger world and faster language increases the TxS force of its species, which become top predators. And indeed, the analysis of the theoretical speed of informative perception in black holes has shown that they are the fastest 'computing minds' of the Universe. So black holes can easily destroy a planet; huge machines kill animal life; human beings kill olfactory ants and insects kill plants. It means top predators are defined in terms of the power of their linguistic brains.

Reality is a game of Russian dolls in which each bigger Doll /World encloses the smaller one. The gravitational World encloses the light world, which encloses the electronic world, which is bigger than the atomic world of ants and so on.

Recap. Each species has a rhythm of times. Slow species are perceived as dead space by faster species. We think plants do not perceive because they are slow in their rhythms of Times. Yet they are alive, feel, and sense in a chemical manner, through their roots and leaves.

17. Mind dimensions. Holographic principle revis(it)ed.

We explained in the simplified lecture that introduced multiple space-time theory that the Universe is a holography of an informative bidimensional field and an energetic bidimensional field that created the 4 dimensions of reality.

This is the case of the simplest forms like light, which is merely a membrane with two perpendicular energetic and informative fields.

But complex 'closed topological balls' with 3 regions, a hyperbolic, informative center, an energetic planar spherical membrane and a Klein body of reproductive cycles has a more complex holographic principle and other 'varieties' of higher or lesser dimensionalities.

In those 'complete' st-points the central discontinuous zone of a fractal point is an informative region, which adds the dimensions of the energetic membrane and reproductive body to create a mapping of itself and the Universe (imprinted in the senses of the external membrane). And this gives birth to different dimensionalities according to which ecosystem we study but a simple law stays:

Hyperbolic dimensions=Body + membrane dimensions.

If the reader observes the graph in our lecture on topology (II, 4) of the 3 regions of a non-Euclidean point, it will observe that the central region of a st-point doubles its form, as the brain that maps out the rest of the Non-E point.

And so there is a hyperbolic region of pure information that doubles the topology of the membrane and body region. Since the hyperbolic center of higher form/information has by definition more dimensions than the energetic bidimensional membrane and the reproductive bidimensional toroid; since it holds the 4-dimensional holography of the Universe. It is the key law of dimensionality of Non-Euclidean structures; of which 2 variations are paramount:

Bidimensional-reproductive system + Bidimensional energetic membrane = 4-dimensional hyperbolic, informative mapping.

Bidimensional external membrane +3-dimensional cyclical volume:

5-dimensional hyperbolic nucleus.

These simple laws of dimensionality balance the 3 regions of a Non-Euclidean topology and it is central to understand the workings of different minds, which in terms of information write as:

Information of hyperbolic 'brain'= Information of Klein's 'body' + Information Riemann membrane

In metaphysics, which searches for the logic and geometric laws, derived from the 5 Non-Euclidean postulates, from where the arrows and wills of the organic Universe are deduced; a mind is understood as a fractal image, reflection of an internal body and an external Universe, which connects both regions of the st-point. In a st-point the membrane is the first mapping of the Universe, and the intermediate body, the inner world that reproduces, feeds and maintains the system. A mind will constantly check both, the membrane and its sensorial image to which it is connected through a nervous/informative system and the inner body, to which is connected through a reproductive/energetic system. So a human brain is connected to the inner world through the blood system and the external world through the membrane senses. It has two images of reality and it combines them both, word feelings and eye-images, to act with 2 purposes: to maximize the informative perception of the external world and the inner, reproductive and energetic feelings and pleasures of the Internal world.

This complex explanation of the human being, explored in more detail in other works, is reduced to a minimal skeleton, *maintaining the scheme of wills and purposes*, when we study the simplest forms of the Universe. And it is the origin of the Maldacena Conjecture [A,12] (a black hole, informative center of the galaxy has in 5 dimensions the same structure than a galactic mapping in 4 dimensions.)

Recap. The dimensions of the hyperbolic, informative center are the sum of the dimensions and information stored in the cycles of the reproductive body and the energetic membranes, whose senses map out the external Universe.

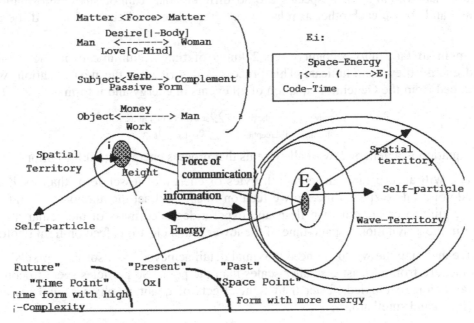

A line is a wave-like event which communicates 2 st-points through a herd of fractal micro-points- a lineal action, exi, of energy that carries a frequency of information in which a message is encoded. The language of information is highly invisible to points outside the network that emit those messages as a flow of micro-points, self-similar micro-replicas of the mother-point that travel in waves across the external Universe transferring energy and information.

In most events those flows balance one point with more energy, Ei, that each science defines with different slangs, ('a white hole', 'energizer', 'past form', 'male', 'body', 'yang', 'moving field', etc.); and an informative point, Ot, the smaller form ('a black hole', 'codifier', 'future form', 'female', 'head', 'yin', particle or 'center of perception', etc.) Both become united by a dual wave that transfers energy from Ei to the informative point of relative future, Ot, and information from the future point, Ot, to the relative past point, Ei, creating together a cycle of temporal energy. The description of those points and cycles, which are common to all beings of space-time, creates a fractal, i-logic geometry common to all sciences and Universal species.

18. Motion as reproduction of in-form-ation.

Because points are constantly gauging information and feeding on energy, the most common events of the Universe are the absorption of waves of energy and form and its communication between 2 points that share energy and form.

In the graph, we observe some basic acts of communication in which self-similar complementary species exchange energy and form in actions that bond them. Human sexual species communicate information and energy (social love) but also seminal particle that combine both reproducing a self-similar being (sexual love); fermions, big atomic particles, communicate bosons, smaller particles of forces. Humans communicate with machines through prices and salaries. Thus money is the force of communication that creates the actions of consumption and reproduction=work of machines by human beings. In all those events there will be however different they might seem, certain laws of i-logic

174

geometry, which all systems will follow and will define the outcome of such encounters, according to the 3rd postulate of Non-Euclidean geometry that defines the relative equality between st-points. So certain events are 'Darwinian', when species are so different that cannot share information with a common language and so use each other as relative energy, or events of eusocial love if the points are self-similar.

The main element to take into account when 2 knots of time communicate is the nature of the information and energy they communicate. This process is described in the next equation, which is a causal chain deduced from the Generator equation of all events of energy and information E⇔I:

$$\sum e_{st-1 \text{ (energy converging into the point)}} + \sum I_{st-2 \text{ (flows of information)}} ->$$
$$EI_{st \text{ feeding and perception)}} -> \sum e_{st-1} \times \sum i_{st-2}$$

The previous equation is a general law of all systems that process energy into information:

- The membrane will absorb through its mouth bytes of energy of lesser form than itself, normally from a lower scale of relative space-time. So we feed on 'meat', extract the amino acids and molecular information we require to create our own flesh and expel a degraded mass of molecular atoms. Light imprints its form in the gravitational space-time of the lower scale; electrons feed on light photons, etc.

- Yet to perceive and map the world we need maximal detail achieved by absorbing 'pixels' of a force, whose micro-points are from at least 2 smaller scales than the point. So particles seem to communicate and perceive gravitation (entangled, faster than light effects of quantum physics) and we absorb the smallest pixels (light) and smell atoms.

- Finally to communicate we share entities, created with combinations of that primary energy and information, encoded within it. Yet the value and intensity of the communication and the bondage created will depend of the density of form and quality of energy we share. Thus the lowest scales of energy and form will not be used to communicate but will be expelled constantly as the left-overs of our constant feeding of energy and information.

For example, humans feed on meat and can share food, to create bondage, but will not share 'waste' or violent 'energy' without form, if they want to create bondage. Black holes absorb light and mass from the galaxy and seem to produce faster-than-light gravitational, dark energy of lesser form, belonging to the lower space-time scale of gravitation.

And yet because of the scalar, hierarchical structure of space-time, nothing is wasted, since for 'lower' entities waste might be energy to recreate simpler forms (flies that feed on waste, irregular galaxies born from the energy emitted by black holes, which evolves back into matter).

The reader must understand the importance of those scalar laws, ignored by standard science, stuck in the concept of an abstract, single space-time continuum. Since the Universe is 'equalitarian' in the same scale, but hierarchical when we study the relationships between cells and wholes, lower and upper scales of reality, and the paradoxes, contradictions and harmonies between the Points of view of those scales are essential to understand most events of reality.

Once the process of creation of a fractal wave information is defined, we can explain most actions and events of communication as processes that involve first the 'fractal reproduction' of a form which uses its micro-forms to pattern a message that will be shared with a self-similar form.

The st-point is not static but in constant communication with the external Universe, populated by other st-points that obtain energy and information also through acts of communication, which are *dual events*, defined in fractal Geometry by the 2nd postulate of communication. Since according to Newton, an action never goes unanswered: There would be a reaction often of the opposite sign. Thus to understand any event we have to wait for a reaction to complete a cycle. One-dimensional science often forgets this duality, considering ceteris paribus analysis of a single action. In the human social scale, individualism often forgets such duality when a nation attacks another nation, or when a group exploits another social group or an individual acts selfishly without expecting the proper reaction. Yet sooner or later the just, cruel, Darwinian Universe will react and balance the game.

Because the Universe is essentially dual, made of lineal energy and cyclical, dimensional form, this duality happens in almost all the events of reality. So points will emit micro-points, with a combination of energy and form. For example, if the original st-point has emitted a fractal action, dominant in energy, ΣS, it will receive an opposite reaction, often dominant in information, Ti, completing a cyclical action of temporal energy, $\Sigma S \Leftrightarrow Ti$.

In the same manner when a group of points emits a wave, it will receive an opposite wave from the receptor, creating a cellular, complex cycle.

Finally, if we combine 2 planes or networks of points, communicating waves of energy and information, for example, human cells communicating energy and information through the blood and nervous network of life beings, we obtain a complementary event or Organism.

The two main types of 'forms' any point emit are also, either a 'seed 'of information or a 'ghost' of space: Fractal entities can emit a self-similar microscopic cell, with max. form and minimal energy/extension; or they might move or vibrate, emitting a 'wave', a vibration of its external whole with maximal energy and minimal form. It is the first of the multiple dual solutions, which any entity can choose, according to the dual arrows of energy and information of the Universe. Thus Duality becomes essential to classify the multiplicity and variety of events in the Universe, since it sets self-similar limits due to the existence of only 2 arrows/forms/ substances to play with. It means most events will be either dominant in energy, or information or a balanced combination of both.

In the ∞ Universe, actions and reactions that define those events will have different names according to the st-plane of existence in which they occur, but they will respond to *self-similar existential laws*.

Since sciences are accustomed to analytical, detailed studies that differentiate species, they do not search, as this synthetic work does, for the self-similar laws of all those communicative actions. Yet precisely the beauty of the fractal Universe is that both approaches are meaningful. Since self-similarity doesn't mean equality: fractal scales create self-similar forms which are never an exact replica of the original. Those variations across space/time scales differentiate species, which still compare in terms of the similar properties of all relative energies and forms that together shape a communicative action, according to 2 fundamental laws:

-The law of self-similarity, which states that any action will be a micro-form self-similar to the mother-cell, and:

-The law of balance, $\Sigma S x Ti = K$, which implies that the product of the energy and time of the action is constant.

For example, if the point is a particle of the light-membrane, the communicative action will be a multiple of h, *the smallest micro-form of energy and time of the electromagnetic world, ruled by the law of balance, e x t=h*; if it is an atom, it will be a vibration with a temperature, T, which again will be ruled by the law of balance, in this case PV=nkT. And because we have ascended our plane of existence from spatial, lighter particles into atoms denser in form with less spatial speed, the action will be denser and reach lesser extension than a fractal h-action.

This will be the case also in biological actions of communication: if the mother-cell is a living being, it will emit an ovule of pure information that will merge with a seminal seed with higher energetic movement, or male seed. Both will be replicas of the fractal generator or mother.

Finally in the human social sphere, actions are often purely informative: certain thoughts act transforming our environment through the intermediate motions of machines that increase the energy and information of the human action. Those machines are themselves replicas of our energetic and informative functions. Yet unlike most actions of physical particles, machines have a peculiar structure. Their function/information is often simpler than that of man, so a car is simpler than the leg, whose energetic function substitutes, but larger in space, so the car moves faster. Those complex relationships between humans and machines, which enhance our energy and information when we consume them, can now be understood in the context of our search for more energy and information, creating a symbiotic species that we call in our complex analysis of history, an 'animetal': animal+metal.

Thus, all, including men, *generate* actions made of flows of energy that carry information, used to communicate with other points. So we widen the fractal unit of quantum physics, the action of temporal energy that communicates particles to define all type of actions-reactions in all space-time systems.

Thus the 2nd postulate of fractal geometry defines a line, no longer as an abstract form like Euclidean geometry does, but as a physical wave of self-similar, fractal micro-points that carry energy and information, as they move between 2 macroscopic points, with 2 possible functions, to communicate energetic forces or linguistic information.

2nd Postulate: A cycle of fractal space-time:

'A wave of communication is a group of self-similar micro-points that move in parallel lines between 2 macro-points, transferring energy and information between them'.

In Non-E geometry a line with parts is not defined by a sequence of numeric intervals within a straight line, but by the communication of 2 poles of energy and information that establish a flow of particles in 2 opposite directions, creating a simultaneous, paradoxical wave. Such waves again can have different purposes. A wave dominant in information communicates symbiotic particles, creating an informative bondage/network; a wave dominant in energy might be an aggressive action between different species that fight for each other's vital energy or territorial space; and a wave that balances the energy and information of both points meets in the center, creating a new self-similar, seminal particle, as when 2 electrons emit waves of densely packed photons, which merge in the middle and give birth to another wave.

When we generalize those concepts to n-points we can define a space as a network of Non-Euclidean points. Indeed, Riemann affirmed that a space is a network made of herds of points with similar 'properties'. Planes of space are therefore networks of points. The self-similarity of their properties defines its *density* determined by the number of points and its proximity that grows with self-similarity. So similar points come together into a tighter, more continuous space; whereas the density of the space is proportional to the similarity of its points, till reaching 'bosonic state' of maximal density when points are equal. And when a volume of spatial energy is very dense, it is very difficult to go through it, as it happens in the ultra-dense, small space of black holes.

Spatial extension and form/density/mass are inverse parameters, Max. E = Min. I. If we generalize that property to all scales, we can define different fractal spaces by its proportion of mass/density and energy /distance. This is done with 'Universal constants' that explain the proportions of energy and information of those spaces. For example, in physical scales, there are 4 fundamental space-times, the gravitational space-time between galaxies of max. energetic space and minimal formal density; the light space-time of our world, which carries information in the frequency of the wave; the electronic space-time of atoms with more formal density and lesser spatial speed and finally the quark-gluon liquid of atomic nuclei and probably black holes, with maximal density and minimal space. All of them are defined by Universal constants and equations that are either ratios between the energy and form of those space-times, or define the transformations of one space-time into the others. Einstein's field equations would be the first case, defining the relationship between energy and mass in a gravitational space, while the fine constant of electromagnetism would define the transformation between light space and electronic space/ charge; and the gravitational constant between gravitational space-time and quark/mass. Where the relative densities of information and extension in space of those space-times are in balance, such as $\Sigma SxTi=K$. Thus electrons move slower than light but have more density.

All this said it is thus obvious that the *fundamental unit defined by the 2nd postulate is no longer a point but an action, ex i= k between points, which becomes the 2nd fundamental 'particle' or 'parameter' of reality.* We, points of time, create actions, exi, with a minimal quantity of form, moved by an energy force, creating lines, which therefore become actions of energy and time. And all what we do are actions. So we say often 'I don't have time or energy to do so'. Actions become thus the fundamental event of all points of view, in search of their arrows of time; it becomes the dynamic definition of a line, as we are all in constant motion, and so all point in motion can be perceived in slow camera as a 'line of action'. And indeed, in physics an action is the fundamental unit of our light-space membrane, exi=h; and an action is the fundamental unit of biological behavior, which also defines the existential force of a being, exi=k, or the momentum, $m(i) \times V(e) = k$ of a physical being. An action is also the name given to the fundamental unit of economical organisms (companies). The Universe is thus a world of infinite points of view performing lineal actions in search of their arrows of time.

Recap: Non-Euclidean points constantly communicate energy and information with other self-similar points and the external Universe, by sharing flows of micro-points of a lower scale of space-time, which carry the energy and form of the particle into the external universe. The laws that define those acts of communication are hierarchical laws between planes of space-time and laws of balance between the energy and form of those 'actions' of communication, exi, which become the fundamental dynamic event of any scale of the Universe.

19. The ternary principle of creation.

A fundamental principle of Time Theory is the ternary principle: because all systems are designed with 3 purposes, to process energy and information or mix both in a balanced, reproductive, exi, action, we often find in all type of events a ternary choice, which will be specially relevant to understand the processes of 'guided evolution' that limit the number of possible creative species. For example, we can define human evolution as a constant differentiation of ape species into energetic, informative and reproductive ones; where the human lineage is that of species with maximal informative/head evolution, while the gorilla lineage is one of maximal energetic/body evolution and so on.

The 2nd postulate differentiates according to the ternary Principle 3 types of waves: A language that transfers only information, (Max. i); a force that transfers only energy (Max. E); and a *wave* that transfers both forms (ExT). All forces and languages are waves, since actions mix energy and temporal information; *which will be perceived* as languages when the absorber can decode that information. It is the receiver, which selects the wave as a force of energy or a language of information, defining how it will react to the emitter, according to its self-similarity that will allow him to understand its information. For example, plants absorb light as energy and animals decode it as information; but light is dual.

In algebraic terms, a function of existence is described with 2 parameters, X and Y that represent an informative subject and an energetic object, often complementary.

In biological terms, an organism is defined by a head of information, a relative future, more evolved form; and a body of energy, a relative, less evolved, past form. So we can consider also dual organisms as acts of communication between two poles, the reproductive body and the informative head, where the neck is the bridge that carries the bigger number of veins and nervous paths – the informative and energetic networks of cells that communicate body and mind.

Since the head directs and decides 'the future' of the entire system, informative poles are future poles. So a relative informative region is a future form and a relative energetic region is a past form. And both come together into a discontinuous, relative fractal, intermediate region or 'present space-time', in which they combine creating 'reality'. That region is a present region because it is the most visible region of the Fractal point, limited by the future, informative region and the past, energetic membrane.

Those temporal distinctions and degree of visibility of the 3 topological regions of a point, now studied as two poles of communication are real; since the reproductive, present region, is always the most visible, given the bidimensionality of energy and information. For example, we do not see the black hole and dark halo that controls the galaxy but only the intermediate region of stars. In this manner the game of space-time existence that creates reality merges 2 poles dominant on spatial energy and temporal information into a new combined form, a relative present, fractal space-time:

Past=energy body xFuture informative head=Present organism

We explained before the 3 dimensions of time, past=energy, present=repetition and future=information, showing that most systems in the Universe live through those 3 dimensional ages, dying back into energy, as they dissolve its formal networks. Yet, while the Universe has very few elements/arrows to play with, its diversity comes from the possible variations and combinations of those arrows. And so, while we can write Past->Present->Future, as the natural causal arrow of the life cycle, there is also an equation of relative times, past x future = present, which defines the complementarity between reproductive bodies/fields and informative heads/particles and explains the present stability of systems which have both components or can switch between motion and formal states.

Further on, the use of the ternary principle, applied to the 3 languages of man (energetic geometry, temporal logic and exi, visual perception) will allow us to verify in a higher degree any statement, by considering at least the biological function/motion and geometrical, still form of the species we study, which have to be self-similar. For example in a galaxy, the black hole is the spherical center of information. And we can prove this according to the linguistic/ternary method in the space-time languages of man, the mathematical language of space, the bio-logic language of time and the visual language that merge both together. In algebra, Einstein equations prove that black holes transform its spatial parameters into temporal parameters. Hence they transform spatial energy into temporal information. In visual terms they curve space-time into height; hence they transform energy into information. If we consider its biological function, they absorb the energy of star-plants. So Black holes are sinks of gravitational information. And gravitation is obviously the informative force of the Universe; since it is not dual but it shows only the arrow of in-form-ation, attracting and lumping masses and it is 10^{40} times less energetic than electromagnetism. So when we observe vitally, visually, in action, those black holes, they show the maximal amount of the 2 dimensions of temporal information: height and rotational rhythm; as they seem to turn at light speed and theoretically transform the plane of energy of the galaxy in a tube of infinite height. Thus all languages and experimental data prove the fact that a black hole is the informative brain of a galaxy that feeds on electromagnetic space towards the future. The extended, faster plane of rotating stars is the body or energy surface of the galaxy that feeds them. Apparently a human body and a galactic body should not have anything in common; but if we observe the morphology of both, it is clear that those morphologies respond to the generic morphology of information and energy, and so do their functions in time.

Recap. Events in the Universe are limited by the ternary principle. Actions of communication also obey the principle: There are energetic, informative or reproductive events, creating often complementary systems with an energetic pole or body and an informative pole or head, communicated by a dense network or neck that carries the actions. The 5 Postulates of non-Euclidean geometry are based in the definition of a fractal point as a point with inner parts, revealed when we come closer to the point. According to such definition, lines are waves of points and planes topological networks of points, communicated through flows of energy and form. While equality requires also equality in the inner form or information of the point, which prompts communication through waves of energy and information that build networks. Communication between points is now possible because points can fit infinite parallel/waves used to gauge the Universe and create an inner image of reality. Non-Euclidean fractal geometry thus improves our vision of the Universe closer to reality and allows the definition of organic systems and logic behavior in bases of geometrical form, a long-sought dream since the times of the Greek.

VI. 3RD POSTULATE: SELF-SIMILARITY

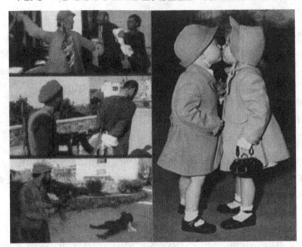

A case of Darwinian devolution among men that perceive each other as different and enter in a perpendicular, Darwinian relationship and a case of social evolution between 2 forms that perceive each other as equal and enter in a geometrical and causal relationship of parallel love.

20. Logic and geometric postulates.

Why geometry needs 5 postulates if the 1st, 2nd and 4th define perfectly a fractal space/time in 3 growing degrees of complexity: the singleton point, the line of communication or wave and the network or plane with volume? Because the 3rd and 5th postulates are not about fractal space but about fractal time; about information and formal equality, a*bout the logical scaffolding of mathematical equations, which so often is forgotten but shapes the underlying formalism of all mathematical statements (as per Frege and Gödel)*. Since:

- The 3rd postulate defines the degree of self-similarity between 2 fractal objects. Classic geometry considered equality based only in the external form, regardless of size and inner content. But beyond the idealized world of Euclid, equality never exists when we consider the inner form of those points made of inner networks of energy and information that contain dark spaces and inner parts, which make even twins slightly different. Thus fractal geometry defines instead the relative degree of equality of 2 forms, according to the self-similarity of its inner, informative content and scale, not according to the equality of its external, mutating membrane, which often is hiding, as in camouflage, a different form.

Self-similarity matters because the social evolution of points that become part of networks that deliver to all of them the same energy and information requires the capacity of those points to process the same energy and information, within its inner parts. *Thus only self-similar cells with the same 'genetic=temporal information' evolve into organisms.* And this happens in all scales. For example, the superorganisms of history which are 'memetic organisms', in which each human being is a cell of the super-organism with the same ideological mind based in the same memetic code (same religious code of 'revelation', same laws in a national super-organism, etc.) must have the same mental beliefs. It doesn't matter to form the organism if the DNA code is this or that one, if the believer that forms a religious superorganism believes that Allah or a turtle, or Christ or Yvwh is the name of the God – as long as he has the same memetic code in his mind. Since the rule that matters is the degree of equality that triggers the creation of a social, tightly packed organism acting as a single whole. For the same reason, since reproduction is based in genetic information, only species with the same genetic information can reproduce and since they have the same equal form, it is a rule of nature that only reproduced cells of the same species form an eusocial organism – a key law to fully grasp the evolution of organisms and societies, in which a 'prophet' becomes the first DNA cells of a new civilization. *While species that cannot share information to hunt through common networks or reproduce will consider the other point/species a form of energy and prey on it.*

Thus surprisingly enough geometry determines behavior; form and function again go together and fractal geometry tends a bridge between physics and biology, explaining how species will communicate: if those 2 points perceive each other as equal, they will evolve into a social network co-sharing their ability to gather energy and information. If those 2 points are unequal, they will establish a Darwinian, hierarchical relationship, in which one becomes the victim and the other the predator. Thus the 3rd postulate is no longer an abstract, geometric postulate, but a postulate of logic relationships, fundamental to understand the why of motions and behaviors.

Since self-similar points evolve socially though networks and communicate through a common language of information to survive better as a single social species. So we can resume the main sub-laws and causal, geometrical chains of the Universe with the 3rd postulate:

3rd postulate of relative self-similarity:

'Points that are self-similar in information evolve socially into a present network. Points that are different in information will treat each other as relative energy and evolve towards the future into a hierarchical organism as a system of 2 points, one of energy and one of information, related by constant flows of energy and information. Or devolve towards the past into a Darwinian system is one in which both flows are not constant.'

The 3rd postulate explains why men do not perceive Universal entities as intelligent: beings perceive itify other forms whose flows of communication don't understand as energy forces that they 'destroy' and feed on, without perceiving its 'organic qualities'. However, even the smallest point known to man, the photons of 'light-space' show the 4 basic qualities of all organic systems that biologists use to define life, the capacity to process energy and information and the capacity to evolve socially and reproduce: photons carry energy and act like the minimal unit of information, because light transmits all type of images through its photons; they reproduce their form in the vacuum; and they gather in social herds of colors.

The 3+i arrows of space-time are the will of the Universe that all its beings share across all its planes of existence. Yet the rules that define behaviour between species, the limits of perception of the inner informative language of other minds and the need to use as 'energy' other entities, regardless of its inner information (so hunters feed on bodies and throw away the informative heads of lesser energy), imply that humans always deny the sentient, intelligent universe, and the invariance of topological form between those scales, itifying reality.

Recap: the 3rd postulate of relative equality define two forms as self-similar when their external, energetic membrane/surface and inner, informative minds are equal. Depending on the degree of equality species will evolve socially into tighter organisms (when they are self-similar) or hunt each other as different species.

21. Equality and social evolution. A new concept of truth.

Equality is the key concept in our definition of mathematical truth as, =, is the operandi of all mathematical equations and in logic, the necessary reason of all logical statements. Thus a fundamental change in our concept of truth follows from the understanding of self-similarity as a different proposition to equality: If Euclid defined equality based on the surface of objects, without taking into account their relative size and inner content, this no longer holds. Things might be similar in their surface and maybe in their content but they are never equal, at most self-similar, nor are equations and any linguistic state an absolute prove of truth, but linguistic self-similar images of the recurrent cycles and events they represent.

The 3rd postulate eliminates the word 'equal', which is so important to define truth as 'dogma'. There is not absolute truth because we cannot perceive completely all the fractal information of the system and so we can merely consider the degree of self-similarity of two beings. Thus, a far more sophisticated and 'real' logic appears – that of the relativism of all truths which are related to the perception we have of reality, to our diffeomorphic=local point of view that will distort reality to cater to our selfish will (Galilean and Darwinian paradoxes), and to the quality of our senses and minds.

Thus we rewrite the 3rd postulate of Euclidean geometry, which defined the relative equality of 2 beings, based only on the external comparison between beings, as 'points without parts', to introduce the existence of inner energy and information parts in all beings. The new 3rd postulate of i-logic geometry establishes now the inner degree of equality between the parts of those beings as a previous requirement to judge the result of all social communication. The old Euclidean postulate of equality caused all type of errors. Since often points pretend a false, external equality. For example, the main stratagem of hunting in Nature is camouflage, based on the equality of external forms. So the victim accepts that equality as a sign of friendship and initiates a type of 'positive' communication, coming closer to the predator; but since the victim is different the predator kills 'it perpendicularly'. That subtle difference between the real world and the abstract world is paradigmatic of the relativism of perception. If Euclid was right, hunting would be impossible, because camouflage could not exist. But real beings have internal, different parts. Some insects imitate wood to hide themselves but they are not wood although they seem it, because their internal parts are different. Otherwise there could not be hunting, based in the strategy of camouflage, as predators simulate the external form of victims or hide in dark spaces, from black holes to lionesses. It proves a fundamental theme of religion: forms that are equal, in this case all humans who have the same genetic information, since they can all reproduce, should love each other, and share energy and information to create a collective organism, a global civilization or God of History.

Recap: Truth is relative to the quantity of information we have. Thus truth is relative and never absolute.

22. Eusocial evolution Vs. Darwinian devolution.

Most systems are energetic and informative cells put together into 2 networks called heads and bodies that merge, creating organisms and species, where the informative network rules the energetic network in a symbiotic relationship (particles that rule fields of forces; gravitational black holes that rule electromagnetic, energetic stars in galaxies; heads that rule bodies; capitals that rule nations, etc.). How individual cells create complex organisms? How atoms evolve into molecules and molecules into the cellular Plane, cells into organisms and individuals into societies? Through Social, informative evolution, the most extended phenomena in the Universe by which individual beings become organized into herds and organisms when *their cells share the same informative code.*

Biological sciences have taken very long to accept informative, Social Evolution, influenced by the single arrow of energy and the subsequent excessive importance that its founder, Darwin, gave to the fight between individuals. Only in the XX century with the development of the sciences of information, Complexity, Systems Theory and Ecology, we have realized that the Universe is based in the social informative evolution of micro-organisms into macro-organisms. The existence of informative networks explains the growth of species in size, from individuals into societies. Yet according to another fundamental duality between energy and information, social events *can be either energetic, Darwinian, destructive, or symbiotic, informative, reproductive*:

- *Complementary, present, balanced events* happen when beings share the same energy and informative language so they communicate and join forces in their fight for existence. For example, in gender, the female is a cyclical, informative entity that merges with the lineal, energetic male, only when both belong to the same genetic, informative species. Those social events evolve individuals into couples, groups, herds and cellular organisms, creating an evolutionary arrow of social love, based in the sharing of the same energy and information.

Among those social events the most intense is reproduction. Since the ultimate will of each point is to reproduce either by itself (since one is equal to its own being) or with a complementary species of

182

energy (male) or information (woman). You can express that iterating will of all fractal species in mathematic terms with a Generator equation or in words with a logic mandate, as the Bible myths did, making God saying to man, a fractal species, 'grow and multiply', which means: 'absorb energy and iterate yourself'. But the catch is that all other fractal parts of the Universe will want to do the same while energy is always limited. It is the *Jungle Law* that mechanically or biologically all fractal structures of the Universe follow. Because those who do not iterate become extinct and the fractal is no longer. So a fractal point has to erase the information and transform the energy of other fractal point to iterate itself. Galileo's paradox motivates all points to ignore the rest of reality and take care of themselves, as relative self-centers of their own world. Hence there are also:

- *Darwinian, energetic events*, when the difference of form is so marked that both entities cannot share the same language of information to form a couple or a herd. Then the stronger predator will explode with the arrow of death the information of the prey into energy and absorb it to recreate its form.

Such duality of behaviors has set an eternal argument between Darwinians vs. love believers, which as most dualities of science, reflects the inverse nature of the arrows of energy and information: beings that speak the same language of information evolve socially to search for energy, inform themselves and when they are very similar, belonging to the same species, reproduce, *accomplishing the wills/arrows of time in a more efficient manner*. While individuals who do not understand their information interpret and hunt each other as energy. Thus duality proves that neither Love Religions nor Darwinism is wrong. Each one explains one arrow of time. Both together resume the foundation of all bio-ethic systems:

'Love your neighbour and hunt all alien beings.'

So, there is an ethic duality, which derives from the degree of equality between communicative beings:

- Species that don't understand their languages of information cannot form working social groups. Thus they kill each other to feed and extract their energy, making the best of their situation.
- Nevertheless there is a social, 'loving' evolution among equal cells of the same species that share the same language of information. And so they evolve from 'individual bacteria' into more powerful macro-organic systems of 'parallel cells' that add their fractal actions into simultaneous macro-actions; since social evolution is more successful that individual struggle. So, organisms based in the same genetic DNA-information kill zillions of individual gigantic bacteria; nations with organized armies such as the United States won over hordes of Indians, arguably better individual warriors; while ants rule the insect world, even if a cockroach is individually more powerful.

Yet since all men belong to the same species, we should love each other and evolve into a higher organism, the humankind, as prophets of love have always told us, in order to be more efficient and control properly Mother-Earth. Unfortunately humans still fight wars based in cultural tribalism, which makes us think our nations are different species because they speak slightly different informative languages or races are different because their skin has a different color. The result of such short-sightedness is that we have created partial, fractal social organisms with boundaries called nations. Thus Social Evolution deals with the fight between species and the way in which individuals of the same species organize themselves into societies, stronger than individuals, fitter to survive. Individuals fight for survival, but Evolution is a game of species. The key to survival is not the individual but the cohesive organization of a group of individuals belonging to the same species that act together with a higher energy and informative force than the individual and so the social individual survives better as part of the larger group.

Today we know that individual lions are less successful than hyenas and wild dogs that hunt in group (25% of captures in lions vs. 80% of captures in wild dogs). Even in the case of Germany, despite its ideology of Racial Darwinism, what really worked was the group - the German collective spirit that made the German army better than the individualistic Italians or disorganized Russians, till communism in II W.W. made them fight as a single organism. Thus social evolution, contrary to common belief, is much more important than dog-eat-dog societies. And that law applies to all systems in the Universe, including herds of stars (galaxies) or herds of molecules (cells) that capture interstellar gas and lonely atoms and explains the st-planes of the Universe as no other theory can.

The fact that the social organism is fractal, fractal, limited in its energy and information is ultimately the reason why individuals gather into social networks, evolving with other cells, since they require energy and information than the networks provide, becoming addicted to them. If the Universe were a continuous infinity with a single 'immortal time', there would not be social organisms, as our inner time of existence would be infinite. Thus survival is again the biological function that explains the existence of organic networks and its cells from the perspective of our limited fractal quantity of life-time.

Recap: Individuals who share the same language of information tend to evolve socially into 'parallel' herds and organisms. Species which share only a similar energy fight to use each other's energy in perpendicular fights. Species without the same energy or information ignore each other.

23. Perpendicularity and parallelism.

In a prey/preying relationship, an organism absorbs energy from a simpler one. In those fights the being with better informative brain most likely will control a better reproductive body (in the graph, the black hole controls gravitation, man the rifle). So it will absorb the other fractal point.

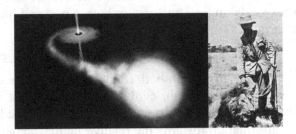

Events are both, caused by the 'form' of the being' and the 'type of motion' or geometry of the encounter. So even two particles that are equal can collide perpendicularly and destroy each other. In geometric terms there are 'parallel' or 'perpendicular' beings, according to the similarity of the information and energy they process within their internal organs. And so we formalize those 2 kinds of events through the 3rd postulate of illogic geometry that defines equality in terms of the inner energy and information of 2 space-time fields and the type of geometrical communication they establish. For example, 2 equal forms in a feeding, energetic cycle hunt together to enhance their survival, running in 'parallel', to 'attack perpendicularly' a victim, whose space-time they 'penetrate' because it is 'different' to them. In the image, we observe such Darwinian geometry:

2 spatial, energetic beings (a lion and a galaxy of stars), with a planar form and max. speed, confront 2 informative beings with a lot of information (a human and a black hole), dominant in the dimension of height. Since information dominates energy and time curves space, when they enter into a relative present relationship, man kills the lion and the black hole feeds on the herd of stars that become past, extinct species. Thus, the spatial 'parallelism' of particles and individuals that gather in waves and herds is homologous to temporal, informative equality; while the spatial 'perpendicularity' of predators that 'cut' their victims is homologous to informative inequality.

Given the infinite fractal information of any being, with multiple sub-scales of form absolute equality does not exist. At most we speak of parallelism or relative equality between 2 beings when their internal organs of information and energy process the same informative language and energetic food. Then they can establish a functional equality, communicating together, hunting in parallel social herds the same energy or even sharing genetic information to create a new being.

Those *positive and negative* events can be decomposed further according to the duality of energetic and informative networks:

<u>Non self-similarity of energy and form.</u>

- *No-events of minimal communication,* when both the energy and information of the forms are different the forms don't communicate, remaining in their discontinuous space-times in parallel, non-perpendicular motion, without contact.

184

Complementarity of energy and form.

- *Organic, parallel events of max. communication, when* 2 self-similar or complementary forms merge into an organic system.

- *Destructive, perpendicular events* between complementary forms: when the 2 forms have inverse parameters/Time Arrows and come together in a perpendicular process, and become annihilated into the same 'present' spatial energy, dilating space and reducing the fractal, temporal depth of the Universe, as when a particle and antiparticle explode into energy.

Absolute Self-similarity=equality

– *Self-Reflective* events, when the form communicates within itself and the degree of equality is absolute, which we divide into consciousness (i->i) and self-reproduction (exi->exi) in which the spatial present 'expands' in time and creates both, a past and future form, as when vacuum reproduces a particle and antiparticle.

Equality of energy

- *Darwinian events,* when both forms share the same energy but with a different degree of evolution. Thus, a relative future more evolved form and a past form come together and the more evolved top predator 'future species' hunts and destroys the relative less evolved past species or victim, evolving the arrow of information towards the relative future 'predator' that transforms the past into a replica of its own cells. For example:

- A swallow and a man occupy different relative space/times and use different languages of communication so they ignore each other; they live in parallel worlds.

- Your brain and mind occupy the same space-time because they are the same being. So their relationship is self-reflective and often our mind wonders about the brain and vice versa.

- 2 men seated in a room are parallel informative beings that speak the same language and have a positive communication, called a dialog. But if those beings belong to nations with different languages, religions and customs, as history shows, they will likely enter into a competitive argument or as nations will establish a competitive relationship that might end into a war. Then the nation with more information, with better technology and weapons that give it more Exi force, will conquer, penetrate perpendicularly and destroy the other.

- Yet if both systems have similar exi(stential) power they often destroy each other, as when a particle and antiparticle annihilate or when France and Germany entered war.

- On the other hand, when 2 similar forms occupy the same space, they create a new degree of order in the Universe. Since it is the most important, creative event, we study it further. According to the Ternary Principle we can sub-divide the creative events in 3 possible sub-events:

- A *complementary* event, when one form specializes in energy and the other in information. Then both forms establish a complementary relationship, creating a dual organic system with a body and a brain that share energy and information between them through physiological energetic and informative networks departing from each 'pole' of the couple, forming a single space-time field.

- A *reproductive event*. When both forms create a mixed, parallel form in other zone of space-time in which their combined energy and information creates a parallel being that mixes both exi forces.

For example, a sexual couple that makes love occupies for a while the same space-time, penetrating each other in a complementary way that reproduces a new being.

- A *mystique event*. When forms fusion into a single macro-form indistinguishable from i=ts parts. So cells become a body; or in a process of perception, the so-called bosonic forces, light particles, create in the focus of a visual organ in which they occupy the same space, a complementary image: *In Non-E geometry*, Riemann described a certain space according to the homogeneity of its quanta that diminish their distances till they fusion together when they are equal. So in Riemann's classic example light quanta of the same frequency become the same color. In Physics we say that photons are 'indistinguishable' particles create a 'bosonic' space of relative infinite 'informative density', described by the 'Einstein-Bose' statistics. In Theology we talk of a 'collective subconscious', which fusions the 'informative minds' of believers into a communion of souls that create the mystique experience.

Recap: The degree of equality between points determines their type of Darwinian, energetic or symbiotic, informative communication: '2 points occupying different spaces with different information don't perceive each other and remain *isolated*. 2 points with the same temporal energy are the same point and its relation is *reflective*. 2 points occupying different spaces, but parallel in their time-information establish a positive social or *transitive* communication. 2 points with different time-information, occupying the same organic space, establish a negative, Darwinian, *perpendicular* relationship, dominated by the point with greater exi Force or relative point of future, which destroys the lesser exi point. 2 points with different information but equal exi force, occupying the same space destroy each other, creating an *inverse* relationship. 2 points with complementary or parallel information, occupying the same space, enact a *creative* event, *fusioning* their existence into a new macro organic system or *reproducing* a 3rd being in a parallel space-time.'

24. Past x Present =Future. Creative and destructive events

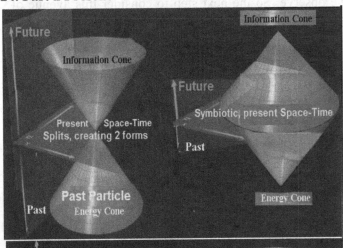
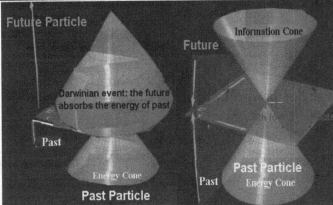

In the graph, in more complex models of i-logic geometry, relative future, more evolved/informative species and relative past, more extended/energetic species create different type of events, which can be studied solely in terms of 'time evolution' as events between past and future, which will determine the creation or destruction of one or both of those entities in a relative present point, according to the degree of self-similarity (i=i) and complementarity (e ⇔i) between the 2 entities, which we have just defined with the 3rd postulate of i-logic geometry.

We talk of 4 possible outcomes of those events from top to right on the graph, which uses the standard time cones developed in Relativity Theory:

- *Death Events*, in which a complementary point splits its relative present form into its energetic and informative entities (particle/force field; body/head), which latter dissolve into a lower plane of existence.
- *Life events*, which are inverse to death events, in which a relative energetic and informative, complementary system form a stable, organic knot of present.
- *Darwinian events of: A) Evolution*. In which an informative, more evolved species with higher exI (existential force), destroys the past, less evolved system, advancing the overall ecosystem towards the

future. Or B) *Devolution*: its inverse event, in which the simpler form, due to its higher Exi, existential form devolves the future, informative force, absorbed as energy. This event is chiral, in as much as there are more cases in which the informative form absorbs the energetic force, moving the overall time of the ecosystem towards the future.

- *Anti-events,* in which two species with inverted space-time parameters are destroyed, creating a relative present without form (inverted waves, particle antiparticle collisions).

We classify all events of reality from a spatial, formal, topological perspective and in terms of causal, past-present-future time dimensions Then the generator equation of space-time, exi=k, can be written in time dimensions as:

$$Past \ x \ Future = present.$$

Thus in temporal dimensions energy is the past and information the relative future of all systems. This hypothesis is consequential with the main arrow of all Time-space systems, e->I or arrow of life, and the causality that requires energy to create information, e->i. Further on, it is a fact of Darwinian Biology that more evolved systems (with better form) win in the struggle of existence, killing and extinguishing to the past, systems with less evolution. So we have a 3rd reason to consider our hypothesis that makes energy and past, on one side and information and future on the other, self-similar expressions

As Einstein put it 'the separation between past and future is an illusion' as both are 'complementary'. Since relative past-reproductive bodies/fields and future-informative systems, heads / particles come together, creating the present forms of reality we observe. So the next step to consider this avenue of analysis of systems in terms of its 'time dimensions' is:

Thus, the $2^{nd}/3^{rd}$ postulates in terms of the 3 causal dimensions of time, past=energy, present=reproduction and future=information define how a single past and future point converge in a simultaneous present.

When we operate with those events and complex algebra of Multiple Spaces-Times, we can obtain some important results, which we can only enunciate in this introduction, such as:

- The overall sum of all the events of the Universe in terms of time is a zero sum, which means that the Universe is immortal and all the events of reality create an infinite eternal present. A trivial demonstration can be made using only the two simplest events of life arrows e->I and death arrows i->E, which gives us: ΣFuture Lives x ΣPast deaths =Eternal Present.

We however sense the Universe evolving because humans are dominant in information and perceive a much longer time-life than time-death. Or in geometrical terms death is basically an event of spatial energy, which releases and expands in space the form of an entity. *Thus death has a maximal volume of energy release in a minimal time period:*

$$Death= Max. \ Spatial \ Energy \ x \ Min. \ Temporal \ information.$$

Reason why death lasts so little in time. While life events are evolving, informative, warping, implosive events that happen in reduced space and stillness, but last a lot of time. So life has a minimal volume of energy and a maximal volume of information:

$$Life = Max. \ Temporal \ Information \ x \ Min. \ Spatial \ energy.$$

Those results are proved in all scales of reality. So in physical space, the death of a previous Universe (absolute scale), a galaxy (quasar), a star (nova) or a neutron (beta decay), release maximal energy in minimal time.

While in life systems, evolution happens in stillness and minimal space (allopatric evolution of species, isolated in small territories; palingenetic evolution of a foetus in the minimal space of a womb; chrysalis evolution in the stillness of a pupa). And death is almost instantaneous, releasing the energy of the organism that erases its upper, informative plane, st+1 (death of the nervous system), as the cells dissolve and become energy of herds of insects and microbial of lesser form.

Those diagrams resemble Feynman's graphs that explain the possible events and outcomes of particle interactions, which also consider the existence of a relative arrow from future to past, which in Multiple Spaces-Times corresponds to a flow of information (Feynman's diagrams can be considered a particular case of the model of Multiple Spaces-Times).

The graph shows the how of events. The why is defined by the self-similarity or complementarity, (3rd postulate) of the 2 Points, relative past and future entities of energy and information:

- If both particles are self-similar in energy and information (case not included in the graph) they will come together as part of a bigger social network.

- If both particles are complementary, energetic and informative systems (top graph), they will form guided by the arrow of life a complementary system that creates a symbiotic present from a past energy and a future form of information. Thus in geometric terms, in the point of collision (right side) the system will expand its space-time. Yet the entity will disappear as a relative plane of space-time in the explosion of death (left side), that will split and dissolve those networks.

- On the other hand, (bottom graph), Darwinian, destructive events happen among different beings that don't decode their information. If both particles have the same energy, one particle hunts the other (left side). Then, the entity with lesser exi=stential force will die (bottom particle), and the one with more force will absorb and grow with the energy of the 'victim'. Or both points might ignore each other when neither their energy nor information is self-similar and so the event of communication will not happen (right side with no space-time in the point of present).

Finally, all those events respond to a dual, spatial geometry:

- Darwinian events are perpendicular relationships, in which a predator invades and penetrates the vital space of the victim. While symbiotic, positive, social events happen among *equal, parallel species* that understand each other's informative languages and maintain a parallel distance either as a hunting herd or an organism, acting with the same motions in space.

- *Exi=K: Present creation.* Most events are balanced presents in which the arrows of information and energy collide in a symbiotic manner creating a self-reproductive radiation of a new balanced species.

Recap. The 3rd postulate formalizes the outcome of any communication event among relative informative/future and energetic/past st-points. The key factor to determine the outcome of an event between 2 Fractal Organisms is their degree of informative equality. Since most organic systems that can communicate by sharing the same type of information in a common language become symbiotic, as they prefer to share information than destroy each other as relative energy.

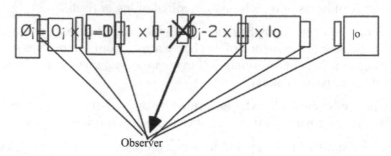

Observer

In the graph, a complex organism can be written with the algebra of Multiple Spaces-Times as a chain of points joined by energy/information flows between several 'planes of existence'.

25. The 4th Postulate: networks and chains of existence.

Exact topological graphs however are only important when we study simplex, physical, homogenous planes where the diffeomorphic dimensions of energy/form/reproduction correspond to the 3 dimensions of Euclidean space.

Because of the flexibility of networks can adopt open and closed, motion and formal geometries in most cases, such as the graph we have to consider more than the basic O-cyclical i-nformative and |-energetic forms, shown in the graph, with symbols of i-logic, the causality of the events, which allow us to apply the same laws to all disciplines. For example in Economics Ricardo points out to the logic of increasing communication between points when he explains the advantages of free trade of different products between nations (nails from England and Porto from Portugal).

The same ruled applied to points of view means that 'complementary points' of energy and information that can exchange energy on one side for information on the other side are beneficial for both points of view. This explains the abundance of dual, complementary, fractal structures in which 2 networks, one of energy and the other of information entangle together, as in your body (nervous and blood system) or in the Universe (light and gravitational networks).

We have now the tools and understanding of the structure of mathematics to tackle in the rest of this work on the Languages of spacetime the 'big question' about reality that those languages enlighten: how can we describe better, the construction of the super-organisms of knots of time, which extend through 'several planes' of existence, and underlie the 'structure of reality'.

All this belongs to the most complex and fundamental postulate of i-logic geometry, the 4th postulate, coupled with the understanding of the 'causal, multiple logic of time arrows' that structure several planes of existence, *displaced not only in space but also in time,* into a single organism.

The Universe is a fractal of planes of space-time', which constantly appear, grow and dissolve in infinite events of existence. How Multiple Spaces-Times become chained and connected, building from simplex particles super-organisms - those planes of existence of reality that structure the Universe? The answer is the 4th postulate that defines the creation of networks of points, called planes.

Each st- point is a knot of times that becomes chained in a dynamic relationship with other points of time through its arrows of energy, information, reproduction and social evolution. In this manner points form 'chains' of existence, which become either 'Darwinian relationships', when one of the 2 points of communication is ab=used as energy of the other one with 'more existential force/momentum' or a stable, complementary relationship if both points are symbiotic, as energy and information or parallel, self-similar, forming a herd - a network of equal points that act together as a whole.

In Geometrical terms, the Universe starts with simplex st-points that communicate flows of energy and information, creating lines and geometrical groups. And the rules of those geometrical events of creation are defined by the 1st and 5th postulate that explains the parts of the point and its apertures to the external universe from where it communicates flows of energy and information; by the 2nd and the 3rd postulate which define in geometrical and behavioral terms the relationships and lines of communication between two points, according to their degree of equality. So the 1st/5th postulate, which defines geometrically and logically a point and the 2nd/3rd postulates, which defines how they communicate, establish the conditions that will determine the type of topologies or network/spaces that the system creates. Those organic topologies is what we call planes of exi=stence, explained by the 4th postulate of i-logic geometry. Since a plane is a group of points coming together as a relative space, which will be 'more dense' according to their degree of self-similarity (Riemann's masterful definition of any space made of Non-E points).

Thus, if the first & fifth postulates are related to the perception of information and energy by a point, with its inner body/head parts; and the 2nd and 3rd postulate are related to the arrow of reproduction between complementary energy/form, male/female, yang/yin beings, the 4th postulate explains the details and possibilities of social creation of arrows of time; facts which establish a clear mathematical relationship between the arrows of time and the postulates of geometry in which all mathematical laws are based.

26. Creation of Planes of Space=networks of points.

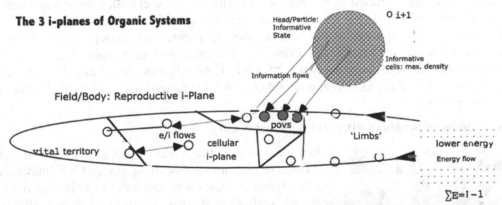

The Universe constructs super-organisms, building Planes, which are networks/herds of non-Euclidean points, step by step, according to the geometrical and biological laws, described in the 3 first postulates of Non-Euclidean geometry: The graph breaks a standard knot of time in its 3 planes. The point is an entity that constantly searches information with its head and energy with its body, creating symbiotic and Darwinian events with other knots of time, through 3 time arrows/dimensions between 3 planes of existence, the plane of the point, i, a lower plane of relative energy, st-1, and an upper plane of relative information, st+1.

In the graph we show the point as a complex i±1 plane, interacting with that plane or ecosystem (geometrical/ biological jargon) through those 3 arrows, which are its 3 dimensions of existence or 'vital coordinates' that create the motions of information and energy of those points. Because the frequency and speed of those arrows/exchanges of energy and information with the ecosystem vary, we also talk in complex Time Algebra of 3 relative speeds, Vs, Vt and Vts, which correspond to the arrows of spatial energy, temporal information and reproduction of the point, (for which often the point will chain itself with a relative complementary being), normally represented in the x(se), Y(ti) and z(st) coordinates.

Further on, those 3 'actions', 'arrows' or 'motions' of time produce in the spacetime around the point different dynamic transformation:

- I<E: Energetic arrows produce an expansion of the spacetime around the point as in a big-bang or process of death.

- E>I: Informative actions produce an implosion, as energy becomes warped, informing the point.

- EI=ei: Finally reproductive actions tend to maintain the space-time around the point stable as they combine the arrows of energy and information, creating a self-similar point in a parallel region of space-time.

And we use the geometric/algebraic symbols of implosion of information, >, explosion of energy <, and parallel creation, =, to signify events belonging to one of the 3 arrows/dimensions of the point.

Thus, those 3 dimensions/arrows of time define the self-similar will and actions of most entities of the Universe, despite their apparent, different morphology and degree of complexity.

The interaction of those individual points of view, constantly throwing its 'subjective arrows of time', capturing energy and information from the universe, is the game we all play, and in doing so we establish relationships with other self-similar points of view, creating herds, 'parallel planes' of points of view or Darwinian relationships between relative energy victims and top predators, which we represent as events between two different planes of simpler and more complex forms: $E_{st-1} -> I_{st}$, where E refers to the energetic function of the victim, belonging to a plane (st-1) of less complexity than I, the predator belonging to an upper plane of more information, I.

All those relationships are reflected in the drawing in which we see complex chains of time arrows between points of view, which forms an organic system. In each plane of existence or scale of the Universe such chains will represent a specific species that curiously enough show homological properties with other species of other scales existence. The previous graph could be, for example, a complex molecule. Such molecule represented here with 'Time Algebra' could also be represented as classic chemistry does with 'lines' of forces and 'spheres of information' that will show the geometrical topologies of those points.

27. Organic systems of points in different scales of reality.

When we visualize the world in terms of points of existence and flows of communication of energy and information we can see the dynamic creation of networks of points of time in any scale of reality. A human group is like a cellular body communicated through verbal codes of information (words) and systems of energy (light/spatial location, light-based food, such as plants and animals). Human groups are social networks of increasing degrees of complexity that grow in a decametric scale from individuals to families, clans, tribes, villages, cities, nations, civilizations and the species[C]. The graph shows a group of human beings, whose cyclical points of view or heads, are maintained in a parallel, relatively equal plane of existence to share information through flows of words communicated through the air medium/space and share energy coming from a relative 'lower' plane of existence, the table. Arguably those configurations are not different from the ones established in atomic networks or herds of sky gulls moving towards the lower plane of seafood as they maintain their parallel distance based in the sounds and motions of the air. Moreover human beings will always try to maintain such geometric configurations between their Non-Euclidean points =heads (not between their bodies, which are blind tails of energy for those points/heads that store the will of times): Observe humans relating to each other with motions and words emitted by their heads that are always kept in a relative parallel plane of existence with other human planes.

Reality is a game of points that gauge information and form lines, by sharing energy and information with other forms of view. Those lines, e x i become then networks of self-similar points, planes of existence, of which the most sophisticated is a complementary, social organism: exI +Ei, made with a head/particle dominant in information and a body/field dominant in energy.

When many points=cellular elements communicate they create more complex simultaneous presents by self-repeating in growing scales those 'simplex' patterns. And finally, through those multiple events they transcend, emerging as a new Plane of Existence. Such systems explain the structure of all kind of planes, including the simplex physical planes. So recently fractal theorists have proved with a theory called 'causal triangulation' that Time Arrows can form causal triangles whose connections grow by iteration till reproduce the 4-dimensional Universe of Relativity. Thus, when we generalize the 2nd and 3rd postulates to n-points in communication we create a finite, organic 'plane of existence', either a simple network (such as a Crystal or a sea of water) or a dual network, a 'complementary organism', with 2 exi networks.

The subjective will of ∞ points of view, acting selfishly come together creating a complex, eusocial reality. Order dominates the chaos of infinite selfish points of view, building a single reality, for biological reasons: The survival of points is better achieved, when acting together. So they become stronger than individuals; hence they win in the struggle for existence, defining in the process an arrow of social evolution that creates with herds of points, fractal, social planes and cellular super-organisms. So points first associate in couples that share flows of energy and information, creating waves = lines with volume, whose shape in space and duration of time is defined by the 2nd and 3rd postulate of relative equality (self-similarity). And finally they create complex organic topologies of existence of a relative number of infinite points of view.

When we add on all postulates - the structure of each point ($1/5^{th}$ postulates), the types of connections and waves between them (2, 3rd Postulates), and the causal, i->e->r->s chains of the Universe - we can explain creation, since the number of combinations of e and i is obviously limited; and so are the number of events, of reality. Thus we can know the outcome of each and every possible event or combination of the 3 wills=arrows of time of each point, (energy, information and reproduction), according to the 3rd postulate.

Thousands of pages and mappings of the laws of existential geometry, applied to each specific species and scale of existence of the Universe, prove that reality is quite deterministic, in as much as points constantly behave in the same manner, absorbing energy and information and trying to emit it with the purposes of reproducing or playing a social role in a higher plane of existence.

Of all those possible events, the commonest chain of existential events departs from a point of view that gauges information, in order to absorb energy, till it can reproduce its fractal form, ExI, till creating a social mass of points, $\sum Exi$ that become a stable network, st, a vital, topological space: $\sum Exi = st$.

The Universe creates constantly 'feedback equations of existence', the ultimate essence of all realities.

Recap. Self-similarity determines the different ways in which Non-E points relate to each other, either associating together or fighting for survival, trying to absorb the energy of other points when they don't understand each other or building social planes when they can create informative networks. As a result, and when those relationships are generalized to n->∞ points we obtain different organic planes of social points.

28. Dark spaces: the rest of the Universe.

Points of view are a mixed topology of membranes with open balls that have a dark space they can't see.

Topological planes are dynamic, geometric networks of st-points=numbers, constantly moving and communicating through flows of energy and information. Yet those networks rarely form a continuous tight space: space-time planes are filled with holes in which other networks might co-exist, or where flows of unknown, dark energy and information cross.

Thus, the creation and destruction of networks of self-similar points of view - the generation of complementary organisms - is all what we see; but there is also all what we don't see, as most information and flows of communication between p.o.v.s, which are not similar to us, remains invisible.

Such dark space will be the surface of reality that a st-point with limited apertures to the Universe cannot perceive (an energetic mouth, an informative eye and a reproductive jet).

The degree of 'dark space' a point does not perceive depends of the complexity and apertures of the point. If we consider the simplest bidimensional point, a pi-vortex made of 3 standard units that form its outer membrane, it will have $\pi-3=0'14$ apertures to the Universe. And it will have a total dark space of $0'14/\pi=96\%$ of circular reality, which the point doesn't see through its apertures. That quantity is curiously the same quantity of dark, gravitational space and matter we do not perceive as beings belonging to the membrane of light-space.

So all planes are quantic networks of social points/cycles put together by n-flows of temporal energy, which have 'dark spaces' within the holes of the network:

'A plane is a network of n st-points joined by flows of energy and information'.

The existence of dark spaces and planes which are discontinuous networks is essential to build up a Universe of multiple worlds, webbed into a relative infinite number of parallel planes that intersect, communicate through bridges between the network-planes and allow the existence of complex complementary organisms in which a relative wider, simpler energetic network is controlled by a denser, smaller, informative one, intertwined through those dark spaces.

Thus without dark spaces a complex reality of complementary beings is not possible. In the Universe the membrane of dark matter are the points of a network of flows of dark, gravitational energy over which the electromagnetic membrane warps itself, absorbing that energy and forming light-waves and more complex, evolved light particles.

In a human being, the blood network surrounds and warps around the digestive network from where it obtains its energy; and the nervous, informative network surrounds and intersects the blood network from where it obtains its energy. It is thus precisely the fact that planes are networks of points and dark spaces fill most of the network what allows the interaction between complex planes that form organic systems. A human being is in fact made of 10 of such subsystems totally warped together in exchanges of energy and information that permits our existence.

All this is the essence of a reality constructed with multiple planes of space-time that biologists recognize as networks of cells' but physicists, stuck in the paradigm of metric measure of a single, continuous space-time still fail to recognize.

Recap. Points perceive through apertures only a part of the Universe. We could theorize that all what we see is merely the light-space membrane that floats in a much wider gravitational Universe.

29. The continuous and discontinuous reality.

The human aberration of perception is the true barrier modern science has to accept those living cycles. And so we define according to the Fractal Principle, 3 phenomena that create that aberration of perception: the limited scales of space we perceive, which prevent us from seeing in detail small or far away beings; the relative speed of human, cyclical time we use to observe reality that prevent us from perceiving very fast or very slow beings as living beings. Finally, our incapacity to translate other languages, as information is smaller and quantized. Therefore, we perceive better bodies and energy than minds and brains. This creates the 'energetic aberration of naïve realism' that makes scientists think

193

all is 'energy' and information is secondary. Since informative messages, which are 'faster' and 'smaller' than energy forms are the most difficult to perceive.

We don't see pheromones and so biologists thought for centuries ants acted as individuals when in fact they act as part of a whole super-organism, the anthill. We did not perceive the informative, chemical languages that cross 'the synapse' between neurons for so long, that most scientists thought that the entire brain was a cell to be able to work together until Cajal saw those discontinuities. Yet today biologists are decoding new languages of information: they found that monkeys learn to use money, rats sing songs, flies calculate. So the real barrier to understand the intelligent Universe is ideological, anthropomorphic mechanism. Only when science accepts organicism, it will start to understand the whys and wills of reality instead of merely gathering data. To perceive that organicism it only has to 'translate' different speeds of time to the 'real speed' at which any species exist.

Let us consider one of those dualities in more detail: the continuous and discontinuous reality, ruled by a simple law that relates two different scales of existence.

> *'The mean distance in space-time between 2 relative cells, st-1, of a fractal superorganism, i, equals the distance between its relative molecular systems, st-2, in time and space.'*

The law of fractal jumps structures hierarchical organic systems through social planes, E^{st}, of different i-nformation, among which the fractal jumps take place:

So beings seem continuous because their discontinuity is minimal, only a 'cellular' distance, both in space (so between axons, a simple chemical molecule overcomes the jump) and time (so chemical information, the next simpler language of life, jumps and translates the electronic message through hormonal discharges). Again between molecules the most habitual distance is that of an atom of 10^{-10} m.; and between atoms in its solid, informative state, the usual distance is a valence electron, its lower formal level. While between electron orbits the quantum jump is a photon with a constant of space-time, h, which is omnipresent in all the electronic cycles. In the same way a biological, fractal jump that reproduces a foetus implies the minimal fractal jump in space as the child is born contiguously to its mother.

Though st-points are discontinuous they have 3 apertures through which energy and information flow. Those apertures are fractal, small holes where only st-1 particles might flow. Hence the st- point communicates quanta from its lower, past scale of existence. All communication acts are fractal jumps both in time and space that can either communicate energy or information between similar particles or particles from st±2 planes of existence according to hierarchical, illogic laws:

> *'2 particles on the same space-time plane transfer information and energy in both directions.*
>
> *A st-1 space-time plane transfers energy to the st- plane from which it receives information.*
>
> *A st-2 space-time plane transfers temporal energy, without receiving anything on exchange.'*

Though the existential laws of i-logic might seem strange to the scholar accustomed to operate in abstract, they do apply to reality and reach enough detail to connect them with experimental events. The strength of a communication or bondage between knots of time is the essence of what the 3rd postulate measures. And such bondage is maximal in complementary forms, couples, and then in parallel forms, brothers. It is all about bondages creating strong knots of time due to the symbiosis of the arrows of existence of two forms. Forms share and the frequency increases the speed of the cycle of sharing. All action requires a reaction. Newton in fact defines the principle of cyclical inertia, which the 2nd and 3rd postulate clarifies further.

Those laws are based in the hierarchical structure of the Universe that creates organisms with 3 discontinuous 'social classes'. So only particles from the first plane enter into just action-reaction processes - while flows of energy and information between 2 different, st and st-1 planes are asymmetric as the upper classes receive energy and give only information. But we said that Fractal jumps are contiguous, so how in the 3rd case might exist a flow between st-2 and st planes, separated by 2 scales? Obviously because the st-1 space-time field dies, exploding into st-2 particles that feed as energy or information the st form. Thus the 3rd case describes the extinction of a victim that the predator absorbs after destroying its cellular parts. For example you feed on the amino acids of the animal you eat, after destroying its cells into its micro-particles.

The previous ternary events, which can be formalized with the symbols of Fractal logic, are essential to describe and explain all kind of mysterious phenomenon in physical and biological space-times. Let us consider some examples:

- Borders between the 2 planes of physical space-time, the plane of gravitational masses and electromagnetic charges, are black holes that emit dark, expansive, gravitational energy. Still 96% of the gravitational world is invisible to us, including the discontinuous inner dimensions of masses, quarks and black holes.

- The 1st case between equal forms writes: $ExI<ei=ie>IxE$. It explains the 2nd law of Newton, the law of action reaction, according to which, exi, the fractal action of Ei over Ie, is equal in value to 'ixe', the action of Ie over Ei.

- Yet what Newton missed was the fact that 'actions' and 'reactions' are equal in Exi force but not always equal in substance. It is the 2nd case that explains relationships between 2 different planes of existence: Often an action is informative and its reaction is energetic and vice versa, so they compensate each other, cancelling the past and future into a present. It is the principle of conservation of energy and information: Energy becomes transformed into information and vice versa, $E \Leftrightarrow I$. Fractal physics explain the 2nd case with a mysterious principle that not even Physicists fully understand: 'time and space are non-commutative'. That is $I->E$ is not $E->I$, $I>E$ is an energy transformation and $E>I$ is an informative flow. And they are not the same. So in quantum physics and Fractal theory the order of the parameters of an action matters.

Recap. Energy flows towards the informative, future form and energy towards the past, establishing a hierarchy between information and energy, the 2 simplex arrows of the Universe. Both together create hierarchical organisms, distributed through several planes of existence in chiral, hierarchical 3 class-structures, in the dominant dimension of height and information. Only a present space, repetition of the same form in extended surfaces is democratic, equalitarian, creating a mass of undifferentiated quanta without much height dimension.

30. Organic networks as planes of existence.

The 4th postulate of Non-Euclidean-geometry explains how to create a plane as a network of 'simultaneous, present, self-similar beings' created by the multi-causal flows of energy and information between all those self-similar points that merge together into the organism. Since the fundamental mathematical property of those points-beings is its social nature. Indeed, a number - the social unit of algebra - is a social set. While a plane - the social unit of geometry - is a social herd. Thus, the fact that all beings have mathematical properties means they associate in herds and organisms, represented by sets of numbers and topographic 4-D planes, which are networks made of fractal points with a certain content of spatial energy and information. A plane is a network of points – reason why we can relate geometry and numbers. If we return to a previous example, a 'point' is the minimal unit or number '1'. Thus, a number is a 'class or set of self-similar beings' simplified as 'points' (so we say 3 not 3 pears or 3 humans). Mathematics in that sense, in its simplest elements represent the rules of engagement of 'herds of points of view', since 'each number' has certain geometrical properties, so 1 is indeed a point, 2 define a wave of communication, 3 form the simplest plane, and those 3 fundamental numbers have only a variation (you can only draw a figure to connect them). So the ultimate answer which explains the enormous application of mathematics to describe the behavior of any 'group' is precisely the existence of 'Laws' which independently of the type of entity of energy and information we describe, apply to all points of view, to all entities and scales of reality and to all herds just because they are herds=waves and they are systems of energy and information. Those general laws that apply to all beings by the mere fact of being made of 'spatial energy', a substance with geometrical/ mathematical properties and being part of a group or 'number' whose social properties are defined by the possible relationships between couples, triads, foursomes and tetrarkys are the laws we are most interested in.

The most important universal numbers are the decametric scales since 10 is the ultimate constant of the Universe, the perfect network, the most common number of 'points' that create a fractal network, 3x3+st. It is the tetrarkys already understood by Pythagoras, who considered it the perfect number: 3 elements dedicated to the arrows of energy, information and reproduction in 3 opposite directions and a 10th element, first of the next scale, communicating them, protected in its center: A perfect topology to play the game of existence. Since *Numbers are geometries that create networks of Non-Euclidean points. And so they can be classified by their efficiency as forms of Nature.*

In that sense, reality is more like a carpet in which several nets web together into a tighter configuration or rather like a cat alley, which we do not notice in our own web of relationships but creates an entire different ecosystem, or the duality of the web of microscopic insects and macroscopic mammals that co-exist in two different scales of reality. Because networks tend to create 'parallel' bidimensional webs, in a 4 dimensional Universe with a 3rd dimension of height and one of motion that constantly displaces those networks, many networks can co-exist and interrelate together, creating the complex ecosystems of reality. Moreover if we consider that 96% of dark space to be standard the quantity of parallel and intersecting networks/planes of existence that can co-exist together is enormous – at least 25 networks, each occupying a 4% of reality can fit in the 100% of spacetime of reality.

Each super-organism is a world in itself, created by the confluence of 2 or 3 space-time networks of energy, information and in more complex systems, reproduction, where each network accomplishes one of the 3 arrows of time of the organism. Yet the energetic, informative and reproductive network, which extend the point in 3 relative space-times, whose topologies correspond to the spherical, hyperbolic and toroidal topologies of any st-point, become a unit of a new scale of existence of increasing complexity. And so we talk also of 3 scales of size in more super-organisms: the 'molecular', organic and social scales. Finally the whole system will evolve, living through 3 ages of increasing informative warping...

For example, the humans system is made of 3 main systems, the energetic, digestive system; the reproductive, blood system; and the nervous system. The 3 systems are fractal, cellular networks in perpetual motions, which intersect in certain regions of exchange of energy and information. So the blood and energy system intersect in the lungs in which they exchange energy with the blood system and the blood, reproductive and nervous informative system, intersect in muscles, in which perpendicular nerves exchange information with the muscles. And the 3 systems form the physiological systems that define life and its 3 arrows of energy, information and reproduction, each of them attached to a given number of cells. So if we were to take the p.o.v. of the cells, we could also say that the human organism is a network of cells joined by 3 flows of energy, information and reproduction mediated by those networks. Yet all the systems together are what form the Human organism.

Further on, the organism will go through a young age of energy and motion; a mature, reproductive age and an informative age, each one dominated by one of the 3 networks. And it will extend through 3 scales of existence, the cellular, organic and social scale; such as the human being will be born as a seminal cell, evolve into an organism, which will integrate itself into a society and when it dies, it will first erase its social memes, then it will dissolve its cells and so finally it will return to the lower plane of cellular existence in which it was born.

Recap. The structure of super-organisms across several planes of existence is based in 3 networks of st-points, the energetic, reproductive and informative networks. Those networks leave between them a quantity of dark space, where the other networks can intersect and connect with its st-points. Its 'discontinuous borders', are asymptotic bridges where through a change of state, the 3 network/membranes exchange energy and information.

31. Dimensions of organisms made of space-time planes.

Networks of st-point are dynamic 'groups' in motion; either herds or 'rings' extended in two scales of reality, which are tighter and still, due to the 'connections' established on its lower scale. We talk of 'waves/bodies' as herds of quanta or cells and of 'particles=heads' as networks where the quanta or cells are interconnected in greater measure. Thus, while groups/waves/bodies (mathematical, physical, biological jargon) extend in a single bidimensional plane, an informative network extends through 2 st-scales or perpendicular planes; one created by the points and the other by the networks or connections between the st-points and/or the st-points and its energetic herds. So it has 2+2 fractal, holographic dimensions.

Thus, as a consequence of the new definition of a point, a plane becomes now a 4-dimensional 'holographic topography' with 2x2 information and energy 'dimensions', needed by those points to absorb energy and information: informative height that allows the top head to observe the Universe, cyclical rhythms that allow the being to perform its existential cycles, length-speed that moves the being and width-energy that allows the being to store and reproduce its cells. All this brings the theme of how many 'real dimensions' has a given system. And the answer is: depending on how we count dimensions and how many planes we study.

In fractal geometry a network has different dimensions depending on the complexity of its branches and quantity of space it fills up. This was already observed by Peano and Cantor who realized they could create a zigzag line that filled up the entire bidimensional plane. And as a general rule most networks of points can be considered to fill two dimensions and two systems of energy/information networks together to create a holographic intersection that adds up to a 3rd dimension, which finally if it has motions adds up the 4th dimensions of most organic systems of the Universe. Since dimension is a flow (dynamic version) or network (static version) of quantic, cellular points, gathered into bidimensional waves and 4-dimensional space-time fields.

Yet things become more complex when we consider systems that extend through several scales of size, and when we consider the 'amateurish' analysis of dimensions by most scientists unaware of the laws of i-logic geometry and quantum space-time systems.

For example, physicists, without those concepts wonder why strings have 3x3 inner dimensions of space and one of time. This means that they are calculating systems that extend through 3 planes of existence. Further on, those strings then gather in herds called electromagnetic branes, which warp into particles. Further on, 3-D particles form networks that become atoms, and the process repeats once and again till creating human organisms within which those organic points, now called cells communicate energy forces and informative bytes.

There are many types of points with parts, gathering in scalar networks of growing size and complexity: atoms gathering in molecules, stars in galaxies, strings in particles, etc.

A fractal system keeps adding dimensions, as each point becomes part of a network that evolves into a point of a higher plane. Depending on the number of planes through which we observe a point, the total dimensions of the point will vary. As a general rule however, any species of the Universe is fairly described through its main Plane of existence and its lower and upper planes. So a human being is described perfectly as a cellular system in the st-1 Plane, an individual in the st-plane and a relative point co-existing with other individuals in a society in which s/he feeds on energy, information and reproduces. For those reasons we can consider that any point of existence will be described exhaustively when we consider the 9 dimensional networks of each of its 3 main st-planes.

Fractal ±st-point: 3 st-1 D > 3 st-D > 3 st+1 Dimensions

Finally, those planes are joined by 'bridges' of energy (from st-1 to st) or information (from st to st-1) called *co-dimensions of 'relative, perpendicular height', the temporal, evolving dimension* that connects them to the upper plane of existence through their processes of iterative growth and social organization. Those co-dimensions measure the relative height or fractal accumulation of self-similar forms that takes place beyond the membrane of the Mother-Cell, as time iterates and evolves the natural fractal into a higher st-scale. Since informative cells accumulate in a relative perpendicular dimension of height, from living organisms whose informative heads are on top to audiovisual towers and satellites, to the perpendicular co-dimensions of microscopic growth of Mandelbrot and Koch fractal.

A fractal ±st-point exists as a 3 dimensional organism, which acts in the external world pursuing energy, information or reproductive goals (mechanically, consciously or in a vegetative way, or even micro-managed by a catalyser or enzyme – facts those, which won't matter to its dimensional description). In all those cases, time, understood as a form of movement, will be a geometrical movement or cyclical trajectory but also a temporal cycle with a biological, energetic, reproductive or informative function; and a movement traced through a dimension of the upper st-plane of the point. So when a human goes to feed on a dinner, it will enter into a social network of human points feeding on that dinner, which will form an organic system of the upper st+1 Plane. When an electron flows in a metal network of atoms, it becomes a carrier of energy or information between those atoms, and so to its 3 inner dimensions we will have to add an external social dimension in the atomic plane (the only one physicists study in that case). Yet the electron has also inner dimensions, defined by its quantum numbers.

Recap. For more than a century, Physicists have been troubled by the need on their equations of Quantum Theory, Relativity and String Theory of odd dimensions, whose 'ad hoc' explanations varied from an author to another. Yet in fractal geometry the issue is resolved since the concept of an extra-dimension has a clear, mathematical and logic meaning, related to those scales, all of which will have:

- A *3-dimensional sub-space* within their organic forms, as we come closer to them and observe their details no longer as points without parts but as full grown organisms. Thus the existence of 9+1 inner dimensions in objects like Strings means merely that there are at least 3x3 inner fractal micro-scales within the structure of those strings, where those dimensions will be located.

32. The mathematical model of organic, scalar fractals.

We can now merge what we learned about the fractal structures of the Universe, and its creation through planes of existence, which are organic topologies made of networks of points communicating through flows of micro-fractal points. The Universe and all its parts follow the laws of super-organisms, which define all what exists as *'organic fractals made of self-similar cells joined by networks of energy and information'.* Where cells are knots of Time Arrows and fractal means a scalar, self-generated, organic, repetitive reality:

- A reality structured in several 'scales' of different spatial size and time speed, such as the smallest spaces move faster and show a higher frequency in its cycles of time (metabolic law, Chip paradox, etc.): Max. E = Min. Ti; Min. E= Max. Ti

- A reality generated by 'informative' cycles, which can be described with a generic feed-back equation, E⇔Ti, and any of its self-similar specific 'generator equations' of information that code the structure of any entity of reality (from DNA genetic codes to quantum equations of particles).

- An organic reality, which constantly reproduces its form *in self-similar, micro-replicas of each point,* and then organizes those self-similar 'fractal parts' into wholes, using them as waves of communication that relate points between them or forming with them an offspring of similar beings.

This organic, fractal, informative design of reality, explains why the Universe shows a scalar, social nature, 'transcending' from smaller to bigger Planes, invariant in its e & i forms - the energy line and the information cycle - thanks to the fractal reproduction and social organization of self-similar forms.

All those complex systems extend through 3 main st-scales or 'planes of space-time', structuring reality as a 'fractal, non-Euclidean topology of multiple, causal arrows of time'. Yet such complex reality requires new tools of geometry and logic to define properly the laws that relate all the elements and scales of the system. Since most sciences use still a single, continuous space-time plane (Cartesian), a single, Absolute Time cycle (clock-time), a geometry of forms not of motions (Euclidean geometry), and a monist, A->B causal logic (Aristotelian Logic). Thus, a great deal of Multiple Spaces-Times is dedicated to understand the non-Euclidean, fractal geometry of space, developed in recent years, which can study a space-time made of motions (Non-Euclidean geometry), divided in several scales of size and self-similarity (fractal geometry), in which reality is caused by multiple time cycles/causal arrows and multiple agents (Non-Aristotelian logic), tracing geodesics in search of those Time Arrows. Ultimately what you observe is a game of non-Euclidean points=particles=heads, each one associated to a lineal field of energy/body (Complementarity Principle). As entities of energy and information, they are constantly in search of more energy and information to reproduce and form social systems. To that aim each 'Non-Euclidean point of view', the fundamental, formal particle of the Universe, defined by the 1st postulate of Non-Euclidean Geometry (a point, which unlike the points of Euclidean Geometry, have organic inner parts), will start a line of Non-Euclidean communication, (a wave packed with energy and form, described by the 2nd postulate of Non-Euclidean mathematics), shaping 'networks' of energy and information points (topological planes defined by the 4th postulate of Non-Euclidean mathematics), which create a relative 'space-time' being of one of those 'existential scales'.

Recap. Non-Euclidean, fractal space-time was defined by Riemann, who affirmed that a 'space' is a group of self-similar points, forming an undistinguishable 'herd of points' that becomes a homogenous plane. Riemann considered the 'density' and homogeneity of those spaces according to the degree of equality of those points (3rd non-Euclidean postulate) and developed the initial steps of a non-Euclidean, fractal geometry of 'discontinuous points/knots of Time Arrows' that gather into networks, which form organic planes of existence.

33. The scales of reality: human organisms and Universes.

A human being exists simultaneously through 10 planes of space-time existence, from gravitational space to the cosmological plane of the solar system. The complex interrelationships between the points of each of those planes of existence, among themselves and its upper and lower scales, create our organism and influence our modular consciousness - as we feel the will of the Time Arrows of each of those scales through different senses: we feel gravitational weight and physical pressure, electromagnetic heat and atomic smells, photonic images and molecular tastes, social love for other human beings and the exhilaration of motion.

We are the most complex species of space-time, only because we are the species we perceive better, not because humans are the summit of informative evolution in the Universe. In any case, since perception is reality, we will always appear far more complex in form than the extended Universe, dominant in energy-space, of which we perceive very little according to the paradox of the ego (as we gauge from our perspective and so we see more detail in the closest entities). Thus the study of man as a complex organism should be the leading science, which offers more information. Then by homology we can apply the laws discovered in the study of our organism to other entities.

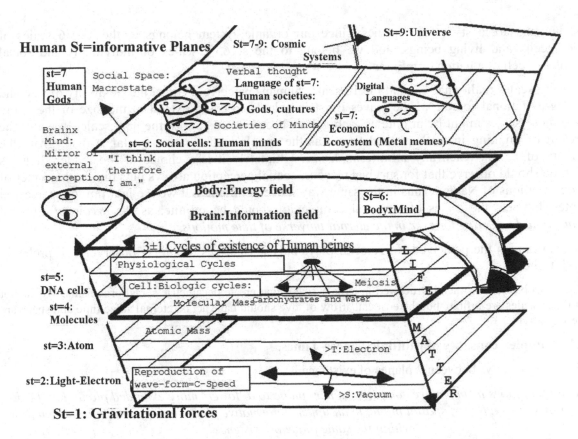

In the graph we see those 9 fractal space-time scales of increasing complexity, which emerge from the interaction, chains and evolution of the species of each scale: photons that become particles that become atoms that become molecules that become cells that become human organisms that become social cells of civilizations and economic ecosystems.

The fundamental particle of the Universe is a Space-time Super-organism that repeats its cycles and shapes of energy and information, co-existing between birth and extinction in 3 fractal planes, which can be subdivided each in 3 fractal energy->Reproduction->Information networks (spatial view), living 3 time ages (temporal view) as he fulfills its arrows of existence in those 3 planes.

If we extend that concept, the Universe becomes a hierarchical, scalar system, extended through a series of self-organized st-planes: particles gather into atoms, atoms gather into molecules, molecules gather into cells, cells gather into organisms, organisms gather into planetary systems that gather into galaxies that gather into Universes. the outcome is a scalar structure of 10 space-time planes, where fractal entities live self-similar cycles of existence under the same space-time laws, from the smallest, invisible plane of gravitational existence, till the Hyper-Universe in which ours is a single cell:

st=1 groups of 'G-constants' evolve, forming, curving the energy of gravitational space into masses.
st=2 groups of 'Planck actions' form the electromagnetic light space, evolved into cyclical charges.
In the st=3 plane of atomic existence, groups of particles became atoms.
In the st=4 plane, groups of atoms became molecules.
In the st=5 plane, groups of molecules became cells.
In the st=6 plane, groups of cells became living systems.
In the st=7 plane, life herds became ecosystems, societies, nations and global planets.
In the st=8 plane Groups of solar systems became galaxies.
In the st=9 plane, Galactic clusters become Universes.

Thus, humans are a st=5, 6, 7 species; since our organic system belongs to the st=5,6 scales of biological cells and living beings, and we belong to the st=6, 7 scales of individuals and social organisms - as cell of a nation or religion.

How many scales reality has beyond the 10 scales humans perceive[3]? Given the flexibility of size and speed of gravitational forces, whose waves Einstein calculated could extend to the size of the entire Universe those scales might be infinite as each galaxy becomes an atom of the next scale. It is also the logic result of thinking in mathematical terms, as the simplest informative fractal, a disk, given the irrationality of π, can reach any infinitesimal size. Ultimately it is a choice of the reader: if he is objective he should observe that for survival reasons, spatial perception always puts him in the center of all his observations of Nature. Thus in the same way space is infinite but we are in a limited, perceived self-centered space, the scales of temporal complexity should be infinite, as *the arrow of eusocial evolution should be acting for ever in the eternal universe of time motions.*

Thus even though the perceived Universe displays only i=10 scales to human observers, 'i' probably tends to infinite.

Recap: the universe is a fractal structured in ternary scales that create decametric social organisms. We cannot know if those scales are infinite but the eusocial arrow of love should have act ad eternal since time and space are absolutely relative.

34. The complex causality of Multiple Spaces-Times.

The key law of flows between 2 planes of existence is:

'Information flows from a more complex, 'upper' plane to a 'lower' more extended plane, Ist+1->E; energy flows from a lower, spatial plane, to an upper, informative plane; energy and information flow within the same plane of existence'.

This self-evident law determines the outcomes of events and the form of structures, from the 3 perspectives of reality:

- *Geometrically,* in relationship with the 3 topologies of the Universe; as it will determine the flows and channels within a certain organic st-point.

- *Temporally,* in relationship with the 3 dimensions of time; as it will determine the relative flows from past to future and future to past, and the simultaneity of certain events in present, between complementary systems.

- *And organically,* when we combine both approaches, as it will determine the key events of existence: the why palingenetic birth and evolution, the process of death, etc.

Departing from that law, we can tackle the most complex level of understanding of reality – the i-logic of systems, created by multiple time arrows, which no longer follow the Aristotelian laws of single causality, A->B, but the logic of simultaneous present creation, either by the confluence of past-energy and future information, or the simultaneous confluence in a formal ring of multiple, self-similar time arrows. Those complex systems of causality were partially studied by Einstein with its concept of simultaneity and 'convergence' of past, present and future, and by quantum physicists with its analysis of particles and antiparticles and the invariance of physical phenomena regardless of the direction of time we study, but now we can analyse it with more rigor in this and the 2 next paragraphs, since complex temporal causality and fluxes of energy from a relative past plane st-1 to a relative future plane st+1 (e->i) and the inverse process, a flow of information from a future plane, st+1 to a relative to past plane st-1, is at the heart of the structure of superorganisms (i->e).

Unfortunately, lineal time introduced in science the concept of single-causality (ceteris paribus analysis), which is only a partial analysis of the causality of most phenomena, in which acts a ternary causality (as energy flows from the lower plane, information from the upper plane and complex flows of energy and information happen between the cells of the same plane of existence). For that reason, most scientists constantly argue, considering their 'cause', normally one of those 3 relative events, the true cause. Yet in fact all events need *at least* 2 causes to exist, one from an energetic past and one from an informative future that converge into a relative present.

Let us then introduce the complex logic of organisms with relative future, informative systems and relative past, energetic systems (IX); studying next (X) the causality of a present system or knot of times, made of multiple time arrows/cycles; and finally (XI) some key events of 'existence' in which those flows of time take place, such as the process of palingenetic birth and its inverse process of devolution or 'death'.

Recap. Events require to exist at least 2 causal arrows, E⇔I, which are 2 complementary networks of energy and information going through the 3 topological ages of the life/death cycle.. Such is how new systems are repeated: on one side all becomes a self-similar network of knots of time cycles that go through 3 ages and reproduce in certain topological forms. On the side of complexity of the time arrows of those Non-Euclidean knots, the iteration and combination of those simple elements that make networks, self-reproductive fractals and the logic complexity of the synchronicities of those feedback cycles create systems of enormous complexity.

35. Information networks control energy herds.

The beats of all species, as they absorb and emit information and energy in order to reproduce and evolve socially into self-similar or complementary networks, chain them into complex organisms and ecosystems. And since each st-point is always a cellular part of a bigger organism, finally all those beats, cycles and chains become part of an interconnected huge organism, the Universe.

The simpler of those chains are Darwinian events that connect victims of relative energy and predators with a more complex 'form'; the sum of all those chains in a given 'ecosystem' or plane of existence form trophic pyramids.

Darwinian E->I events last a brief period. Yet Nature sometimes converts Darwinian events into symbiotic, dual body/head organisms, in which the 'fastest' temporal, informative beings tame, enslave and control the slower ones. So while the faster lion feeds on the slow buffalo, humans first feed in buffalos and then tamed them, creating a symbiotic organism – the herder and his herd. In the same manner, the first, faster electrical neurons probably devoured, smaller, slower chemical cells, but both together finally evolved into complementary organisms, while the fastest neuron controls the chemical cells of your organism, extracting energy from them. Thus, an organism is a chained system of top predator cells that control herds of tamed chemical cells; and an ecosystem is a loose organic system in which the top predator 'cells' only interact with the slower ones when they feed on them.

A 2nd element to understand such chains between faster 'informative cells' and slower energy cells is the fact that the faster cells can dynamically or statically (through a network of connections), control multiple, slower cells simultaneously and vice versa, it can process due to its faster time cycles, multiple messages coming from several slower cells that 'cause' its action. And this happens in all systems, including a human system in which a speaker talks to multiple 'energy believers'. To that aim, we must consider a property of information we advanced previously: its fractal, reproductive nature that breaks a form of information or message into multiple self-similar messages (the speaker that reaches multiple people, the antenna that emits multiple electromagnetic patterns, etc.)

When we put together both properties of informative networks – to act faster and to act simultaneously – it is easy to understand why they can control the energetic cells of the organism. For example, the nervous cells act as fractals much faster than the chemical cells and so a single nervous cell can control an array of multiple chemical cells, because one of its ultra-fast fractal actions move a big distance that encompasses many chemical cells in the time those cells react. In this manner faster, upper class, informative species control slower middle class cells. In the other extreme a stronger ExI species can create waves of energy and information that extend over multiple cells, and so minute changes on its energy and information parameters affect all those cells, reprogramming them. So a small change in the Earth's temperature provokes huge changes in the life of its species.

If we generalize those chains to enclose all organic systems of the Universe, the main systems of reality acquire a complex, deterministic degree of order and harmony that relates the different cycles of being (its informative, energetic, reproductive, social and generational cycles), as cellular units of the bigger cycles of the macro-organism or ecosystem in which it exists. Finally information cells form complex networks and so despite being smaller in space, the whole network of 'neuronal' cells is bigger than the individual, divided, energetic cells, making easier that control by the 'macro-organism' of information. The result of those structures is the 3-class structure of all organisms, in which a head/particle of information controls a reproductive body/field of cells/wave, which obtains energy from an ecosystem from a 3rd lineal system of 'limbs/forces'.

Consider for example, the 3 elements of life, C, N, O. Today we are still in monism and so even biologists will tell you that Carbone is the basis of life. Yet carbon is only the reproductive body of life molecules. Some thus add Oxygen, the energy of life. And indeed, the biosphere has exactly the same mass of Oxygen and Carbone (24.9%) in your bodies, as limbs and bodies tend to weight and occupy self-similar volumes. But this misses Nitrogen, the intelligent, informative atom of life, which is only 1% of the biosphere (0.27%) and was called first 'azoe', which means in Greek without life. Yet with its cyclical forms and its Hydrogen 3-dimensional 'eyes', it perceives the external world and structures life, accumulating in the DNA and brain. And it is a General Law of all Systems[1] that the informative system weighs far less than the 2 others (so your brain is only 1% of your mass, and 1% of 'stockrats', owners of corporations ruled the Financial-Military-Industrial System). Further on, beyond your organism, you are part of a world of water, the ecosystem of life. And so those 3 elements are exactly 50.7% of your mass, ½ of reality, the rest being mostly Hydrogen the primordial substance of water and the Universe.

The primary motion of all beings is lineal energy, in its relative direction of length, while the informative particle of any complementary system is on top, in the relative dimension of height. This leaves width as the natural direction for its re/product/ive, product equation: exi. Look at yourself: your height is your direction of information, your head is on top; your motion is forwards, in the length dimension of energy but your cells reproduce and multiply in the width.

All of us are knots of multiple cycles and arrows of time. So each of us is a multidimensional species. Complex beings, of course, do not follow exactly with geometrical precision those forms, because their vital functions must adapt to irregular ecosystems. Thus, perfect forms of energy and information exist only in the homogeneous, 3-dimensional ecosystem of interstellar vacuum inhabited by physical particles (and in lesser degree on water, life species, since water has complex motions and a general arrow of 'light information and energy' towards its surface).

For that reason, those geometrical rules of creation of vital arrows/dimensions of time/space, which are known to mathematicians as the product laws of vectors, are used to describe many operations between Time Arrows in the simplest geometrical systems of physical space: For example, light has 3 perpendicular dimensions. And Maxwell found that the product of the magnetic, energetic, flat field, given by the magnetic constant of vacuum (which is a membrane of light space) and the informative, electric, 'high field' constant, gives us the reproductive dimension of speed of the light wave. Further on since light-space is the ultimate substance of vacuum and light dimensions are time arrows of energy, information and reproduction, as unreal as it might seem to you, the 4 dimensions you see are not abstract concepts but vital time arrows, reason why they suffice to explain it all.

What is the speed of those time arrows and its frequency? The question is the key to unlock the complex synchronicities, quantitative laws and Vital and Universal constants of each species of the

Universe, so it requires specific, detailed analysis of each entity, organic system or plane of existence, but certain general laws apply to all systems, due to the causal chain between those 4 arrows, as energy appears first, then it becomes bended and broken into information, whose product reproduces a field and finally the re-organization of self-similar reproduced forms creates a more complex super-organism.

Fast and slow dimensions of space and time.

Further on, we classify the 2 dimensions of information, cyclical rhythm (discontinuous frequency) and height, as the fast and slow dimensions of time, since height is often the accumulation of cycles of time one over another, which become in still space (Galilean paradox), cellular bricks of a sensorial system.

We classify the 2 dimensions of space, length and width as the slow and fast dimensions of energy, since width is often the reproductive accumulation of lineal speeds side by side, which stores in still space (Galilean paradox), new energetic cells:

- Length is a fast dimension that moves ahead a Non-E Point and gives it speed. All bodies follow a lineal inertia in its motions, signalled by its relative 'length or body orientation'. Thus, length is the arrow most often associated to the concepts of energy, expansive destruction and devolution. It also follows that the absence of memorial height in an organism, is the key to increase its speed as it diminishes its dimensional friction moving forwards and the quantity of form it has to recreate in each 'fractal step' as it imprints its information in the lower scale of reality. And so the fastest known force with zero height and maximal linearity creates a perfect lineal motion of ∞ speed and 0 form: *the gravitational space over which the forms of light are imprinted whose speed is: $v=e/i=\infty$.*

- Width is the slow dimension of space that reproduces or accumulates spatial energy from side to side. In current theoretical models, strings reproduce laterally, creating fractal spaces in the dimension of width. In our scale, people become fat and wide, as they reproduce their cells. *All beings reproduce their energy in the width dimension of space.*

We also classify the 2 informative, temporal dimensions:

- The fast dimension of temporal information is the speed of perception of information given by the frequency/ rhythm of a cycle, which determines the rate at which the system absorbs its pixels. It is the arrow most often associated to the concept of time; often defined by an angular speed or frequency, which defines a clock of time. Thus its parameter is the inverse of lineal time, measured in t^{-1} units as a scalar number; and when we correct the equations of Physics in terms of cyclical frequency, a whole array of fascinating discoveries and solutions to classic paradoxes - from the paradox of Information (Hawking) to the paradox of Quantum vs. continuous time-space - become resolved.

- The memorial persistence of a certain form traced by a time cycle prevents the drawing of the next cycle in the same place of space, causing an 'Exclusion Principle'. So when the cycle repeats itself, it does so over the surface imprinted in the previous cycle, creating a tube of height, the second dimension of time, caused by the accumulation of cycles of information. Thus, the slow dimension of time is informative, morphological height, born out of the evolution of a species through aeons that accumulated informative cells in the height dimension. For example, the Universe was born flat in the big bang and now it is acquiring form, curvature; bilateral animal life was born as a flatworm that slowly acquired form, till rising to the height of man, the animal of maximal information; the first plants were algae that acquired temporal height; the Earth was born as a flat, rotating disk that became a sphere with height. Yet in duality, according to the Paradox of Galileo, events are created by the dual causality of formal, geometrical space and biological function. So a 2nd bio-logical cause is needed to explain why dimensional, informative height is all-pervading in the Universe:

The informative organs of a being are placed on top of its height to improve its perception of its territorial space: the human head, the speaker; the black hole of a galaxy, the photon of a light wave are all located on top of the relative energy of their bodies to inform themselves on the location of its feeding energy. Thus, height is also the biological dimension of information and the arrow of future evolution in life, which evolved from flat worms into the human beings. Both time dimensions are dimensions of change. Yet the motion of evolutionary growth or height is minimal: it changes very slowly, through the reproductive, accumulative process of evolution; while the mere displacement of a cyclical, physical movement can act faster. The most complex arrow of time has the slowest dimension of change – evolution - which is shown as a change of form. Yet evolution sometimes doesn't create height but it is often a contraction of space, as individual parts converge into a whole, herd, wave or organism. The 4th dimension of social evolution is therefore the only dimension that seems 'static in space' as a 'form', which doesn't change from generation to generation.

So both geometries, cycles of time and planes of space are in fact bidimensional, adding up to create the 4 dimensions of space-time. Energy consists in an amorphous, relative surface of undifferentiated quanta, a plane without form – a network of small, cellular points, in which the creation of memorial time cycles imprints patterns of information and makes it recognizable as a complex system of multiple Time Arrows. Then, as the being develops those Multiple Spaces-Times, it acquires certain topological dimensions as an individual or as a species living in a longer time-frame.

Yet organisms are dynamic entities that change their parameters of energy and information and so they also change their dimensions of space and time, often transforming a dimension of space-energy into one of temporal information and vice versa: A space-time field is created and destroyed through the interaction of the 2 dimensions of energy, length and width, and the 2 dimensions of information, height and rhythm. To that aim, the 2 dimensions of energy and information combine and transform into each other in dynamic events of dimensional destruction or dimensional reproduction.

Finally to notice that according to the Galilean paradox we can do a description that spatializes time or one that temporalizes space: In a 'ceteris paribus' time model of reality all dimensions of space become temporal, as form becomes form-in-action, in-form-ation - exactly the inverse of what classic physics believes, when it affirms that time is the 4th dimension of space, by converting time into a clock-like, cyclical form of space. In the temporal vision, it is the lineal motion and lateral reproduction of information what creates the 2 dimensions of space.

.Recap. Information networks have 3 advantages over energetic herds: they act-react faster, as they are smaller in space; they produce simultaneous, broken messages that reach multiple cells; and they act together as a single network, due to its multiple connections between them. Hence they can dominate the herds of energy in any dual, complementary organism.

In physics space has 3 dimensions of form and so motion only occurs through a dimension of time. This is false. Length is often the main motion of any system that displaces its body/wave in that dimension. Width is a slower dimension of spatial change, often related to cellular reproduction. Frequency is the fast dimension of time that defines the clock-rhythm of absorption of information. Height also changes; it has motion albeit slower than length, the longest/fastest dimension. It is associated to social creation. So we talk of Time-space as the perception of the 4 dimensions/arrows of time in motion, where the slowest 4th motion/dimension, social height, given by the arrow of evolution is the only one that seems to us static. Thus, energy displaces mainly in the dimension of length; information in the dimension of height; width stores reproduced cells, and then there is a 4 dimension of social evolution.

All dimensions of Time-space have motion, a fact which can be observed in all the planes of form of the Universe, till arriving to light space, the simplest substance of the Universe, which in the model of Time-space has also 3 obvious motions – magnetic width/energy, electric height/information and length-speed/ reproduction, and a 4th, less obvious time-change dimension of social evolution, represented by color.

36. Generator Equation of SpaceTime fields. Vital dimensions: forward bodies & top minds

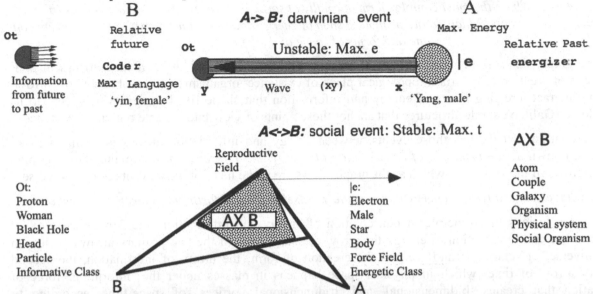

B

Ot

Relative
future

Code r

Max Language
'yin, female'

Information
from future
to past

A-> B: darwinian event

Ot

Unstable: Max. e

Wave   (xy)

y   x

A

Max. Energy

Relative Past

energizer

|e

'Yang, male'

A<->B: social event: Stable: Max. t

Reproductive
Field

AX B

Ot:
Proton
Woman
Black Hole
Head
Particle
Informative Class

|e:
Electron
Male
Star
Body
Force Field
Energetic Class

AX B

Atom
Couple
Galaxy
Organism
Physical system
Social Organism

B   A

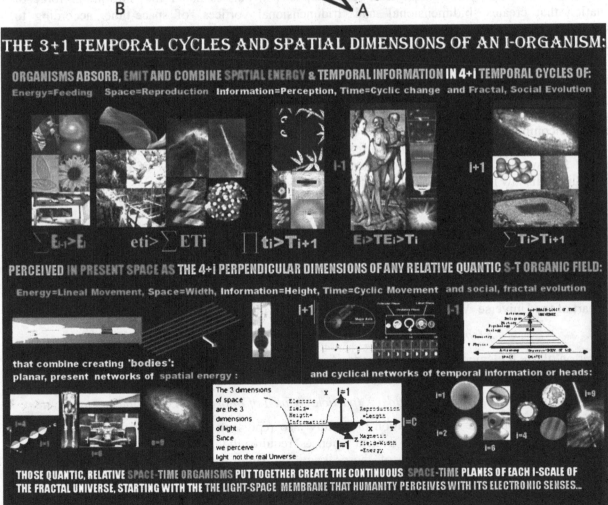

206

In the graph, the interaction between fields of energy and particles of information, either in Darwinian events in which the most complex form normally absorbs the energy of the bigger one, or in complementary events, which create a stable system of energy and information, explains most events on time and forms of space of the Universe.

In the graph, taken from more detailed analyses of the interaction of knots of time arrows in their processes of creation of a certain topological plane of existence=organism, we can observe how arrows of time interact, creating flows of energy and information that shape fixed cycles, which seem to us (paradox of Galileo), stable structures that anchor those points of view into a stable region of existence.

The equation that defines those events between energy and information arrows is common to all sciences, known as the *principle of conservation of energy and information*. It explains the 2 simplex Time Arrows of the Universe, which as the graph shows are at the heart of most events of the Universe.

'All what exists is a type of energy that trans/forms itself back and forth into a form of in/form/ation'.

This Law fusions the principle of conservation of energy and information, the 2 main laws of all sciences. We call the creation of energy the arrow of energy, one of the two primary arrows of time in the Universe, or 'entropy'; and we call the creation of form, the arrow of information, the second primary arrow of time, which in physical space happens in masses under the 'informing' force of gravitation that creates 'bidimensional' and 'tridimensional vortices' of space-time, according to Einstein's principle of Equivalence between mass and gravitation – hence systems with more formal dimensions than simple, lineal forces; or complex formal 3-dimensional warping in life systems (DNA, protein, dimensional warping which store the complex form of life). Thus, again we see that systems apparently so different as physical mass or biological molecules do use 'formal dimensions' to store the information of their systems; while in other events employ lineal forces and lineal 'fat' molecules or lineal limbs to store or display energy. All what exists are processes that create information or entropy or its complex combinations (reproduction and social evolution). Since Reproduction, e xi is born from those simplex arrows, and social evolution follows by the self-organization of self-similar reproduced beings, it follows that the main Law of science is also the proof of the Law of Existence and its 4 arrows. Since this law is the main Principle of all Sciences:

'All energy becomes trans/formed into information: $\Sigma E \Leftrightarrow I$'.

This can be expressed with a *Feed-back equation of energy and information: $E \Leftrightarrow I$ or $E \times I = K$ (dynamic/ static)*

Since all is Time=Motion=Change and space is just a static slice of time we call $\Sigma E \Leftrightarrow I$ the generator, feed-back equation of Time-space: Those $\Sigma E \Leftrightarrow I$ cycles generate the events and trajectories of each and every part of the Universe. Where its 4 elements describe the 4 arrows/motions of time:

Σ (social evolution) Energy < (Reproduction)> Information.

Further on, those 4 elements become the parts of all physical or biological systems, when we perceive them in a static 'dharma' or moment of 'present':

Σ (cells/waves) of Energy (bodies/fields) X (particles/heads) of Information =Complementary system.

Thus the Generator Feed=back equation of Reality also represents the species of the Universe. Its complex study, carried out by General System Sciences, requires the use of Non-Euclidean Geometries to define the topology of each part of those 'knots of Time Arrows' that act as 'reproductive bodies/fields' and 'informative heads/particles' of any exi, complementary system; and the development of a complex causal, 'non-Aristotelian' logic to define the order and interactions of those arrows. Since the 'parts' of each whole knot of time will have functions defined by the needs of any system to gauge information, absorb energy and reproduce.

It is also clear than in all those organisms the fundamental element, the center of power, is the brain-head and its languages of information, which control the body of energy and shape the form of the organism. They set a selfish dominant time arrow towards the growth of information, which is the ultimate cause of the cycle of life and death of all organisms that end up in a 3rd age of excessive form/information, as they warp all the energy of the system.

This fact also explains the relationship between information and future - the most difficult dimension of time to understand by mechanist science since we do not see the future, as we see spatial geometry with our machines. Yet the future already exists in the realm of complex, bio-logic thought, as we will all warp our energy into information and die; and as a species we will always evolve into more informative beings in the future that will feed on the energy of those entities that don't evolve and die. Let us then consider those systems of time knots, adding the concept of a causal order between its arrows of time from past to future.

The creation and extinction of bidimensional, herds of energetic space and networks of temporal information explains the dynamic events of all scales of existence: As time curves energy, spatial planes acquire informative height and vice versa, the destruction of informative dimensions creates planes of energy. Particles of information are small, spherical forms/cycles, like your eyes and brains or an atom's proton or a black hole in a galaxy. They are on top of the system, where perception is 'higher' and show convex topologies of maximal form. Bodies of energy are bigger planes or lines that store energy to move the organism, like your body or an energetic weapon, moving forwards in the relative, diffeomorphic dimension of length. And so both, high information and long energy combine to reproduce a system in the z-dimension of width. The Universe has 4 time motions, which perceived in stillness (Galilean Paradox) create the 4 vital dimensions: energy is length; information is height; width, its product, is reproduction and time brings the arrow of organic evolution, as 'points of view' organize in bigger social organisms that 'survive better in time', because they have more energy and form than the individual cells. Thus, each of the main time arrows is defined for each local space-time as a diffeomorphic dimension that only reaches till the limits of the organism, but extends its action' further into the ecosystem or plane of existence in which the organism resides. In complex algebra, all this can be modeled with Partial equations of the total function of exi=stence, which define each of those 4 arrows, in a scale of increasing complexity and logic causality:

$$e, i, exi, \sum e \Leftrightarrow i.$$

Recap: The holographic principle (bidimensional energy and information) allows the constant transformation of energy into form and vice versa, $E \Leftrightarrow I$, creating the fundamental principle of science, the principle of conservation of Energy and Form: 'All what exists is energy that trans/forms back and forth into information'. The formalism of those 2 arrows of time gives birth to the generator equation of time, $\sum E \Leftrightarrow I$, which defines all species of the Universe as self-repetitive fractals of energy and form Let US then consider the basic representations of the function of existence, from where most disciplines extract their particular graphs and differential equations.

37. The antisymmetry of temporal information: Death

A consequence of the 3 ages of time is its antisymmetry as opposed to the bilateral, mirror symmetry of spatial forces: Time is antisymmetric as the old age is the inverse of the young age with the parameters of energy and information inverted. Youth has maximal energy and minimal information while the old age has minimal energy and maximal information. And yet many attitudes and processes might seem self-similar, if the researcher doesn't know how to differentiate informative and energetic parameters, since those parameters might be quantitatively the same, only that inverted in value. Space is symmetric as left and right reproductions are mirrors of each other with the same parameters of information and energy. This explains why spatial reproduction happens only in the E=I, balanced classic age of the system, when both parameters are identical and a self-similar reproduction of a bilateral, spatial being is possible.

The understanding of the antisymmetry of time vs. the symmetry of spatial processes has wide applications in the solution of questions pending in all fields of research, from the understanding of the temporal, weak, informative force that has no bilateral symmetry to the 'chiral' processes of evolution of life forms vs. the symmetric shapes of self-similar left-right, spatial bodies to the meaning of 'antisymmetric' gender and the 2 'antisymmetric' sides of the informative brain.

But the most important of all antisymmetries is that of death, which could be considered in the jargon of physics, a 'local antisymmetry' of time. Indeed, if life is the arrow of information towards the future $e \to I$, death is the inverted arrow, $i \to E$, that erases form back into energy; but this time travel to the past is always local, affecting only the limited world of the species, so antiparticles are the local antisymmetry of time that kills particles, death are the local antisymmetry that kills life and the big-bang the local antisymmetry that killed a previous Universe. And both directions are different, since 'death' lasts min. time and causes a maximal space expansion and life lasts max. time and expands minimally in space. Both balance reality in an eternal present of past to future to past existential fluctuations:

Life-future: Max. I x Min. E x Death-past: Max. E x Min. I= Immortal present: E=I

The understanding of that local antisymmetry is the key to resolve problems of physics such as the weak force, why we see less antiparticles or dead people than particles and life people (much longer in time), or the non-evaporation of black holes (as antiparticles are the same than the particle, albeit when moving to the past they seem to co-exist), and so on.

But its implications in biology and philosophy are much larger: we are ultimately 'back and forth' vibrations of space-time, virtual existences, dust of space-time that always revert into a zero-sum, which makes the Universe and its game eternal.

Recap. Old age has inverted exi parameters to the young age.

38. Graphic representations of the 3 rhythms of existence.

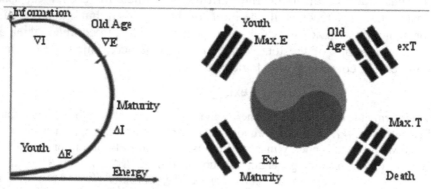

In the graph, a classic, Taoist representation of the 3 ages of life and its inverse parameters of youth (max. energy) and old age (max Information) represented by the triads of the I Ching, and a modern graph of duality showing those parameters as a semi-cycle, which in certain simple beings like light are in fact both the ages of time of a physical wave and its form in space, as light quanta, h=exi, is indeed both our basic cycle of time and surface of energetic space of which all are made.

Humans perceive simultaneously the 3 ages of some species too small and fast to differenciate them. As particles in bubble chamber or cars in the night we see a long time of the species – an entire spiral vortex with the 3 ages of a space-time being. In the next graph, such long world-line, or life-graph is in fact a spiral in terms of spatial population and cyclical speed: The organism looses mass and accelerates inwards its form as it becomes slimmer in space. The first age of the spiral is its energetic mouth that brings in a galaxy the interstellar dust, which will give birth to stars, in the middle, reproductive age and will collapse in the center, the informative black hole:

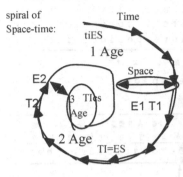

spiral of
Space-time:

Time

tiES

1 Age

Space

E2

T2

3 TIes
Age

E1 T1

2 Age

TI=ES

In the bend graph of space-time, we can describe an organism, as a spiral form. The space=energy of the spiral is given by the transversal volume. Its time or cyclical age is given by the length of the cycle. Imagine the most well known of such organisms: a galactic spiral. The energy of the galaxy is the plane of stars. This spiral graph will be a graph of such energy, as the stars go through its 3 ages of evolution. This age of course is given by the length of the time-cycle. The star will start evolution in the external section of the spiral. As it enters towards the center it will grow in information, in form and diminish in space, in size and energy. The graph shows the process for the entire spiral, though there will be different stars, since the graphic is quantic, with different degrees of evolution. As the stars fall into the organic vortex, they reach form and finally enter the third age, contracted by the black hole into minimal volume...

So the time line shows the age of the population of 'cellular individuals', while the width shows the energy content of the organism. As in the other graph [lineal graph] we show the population across time. So there are 3 ages put together in the relative present of the graph. This is in fact what happens in organisms, which have energy cell, and information cells together. What proves of course, that in the cuantic Universe there are no single directions of time, but multiple forms with different degrees of cuantic energy and cuantic information.

As we enter the graph information increases [time line is longer] and energy decreases [width diminishes]. Yet there is a relative balance between both, since: $E1>E2$; $T1<T2$, pero $E1 \times T1 - E2 \times T2$. That is: the loss of energy is compensated by an increase of information. So happens to an old man which has more memory, more information but less energy. We might say both components are in a relative balance together

But any form can be represented in a spiral graph of space-time. For example, the evolution of the FMI complex with its 800-80-8 cycles of increasing informative evolution is an spiral of 3 horizons in a decametric scale: 800 years civilizations, 80 years nations and 8 years 'decades', in which the product is evolving faster and now becomes integrated by networks. This 3-horizon evolving ternary network, central to social sciences can be also represented in a spiral of accelerated birth and death of human civilizations destroyed by wars. Any mass or charge is in fact a natural spiral, through which photons and gluons become denser till collapsing in the bigger particle, as interstellar gas does in a galaxy. All those spirals can be seen as equivalent to the world-lines of lineal physics, now world-cycles of a living organism both in time and space, as a whole or as a wave of evolving particles.

The equivalent graph in lineal time will give us a bell curve – the lineal version of a cyclical life-line: The speed of time/information changes during the ages of life, shaping a bell curve, such as the st-1 age is the fastest age of informative evolution, which diminishes in the young age to a halt, when the speed of cellular reproduction and energetic growth becomes maximal, both reach a steady state in the mature age, of $e \Leftrightarrow I$, rhythmic back and forth beats; and then in the 3rd age energy decelerates first and then information decelerates and collapses till death provokes a maximal explosion of information into energy. Those phases of different speeds of energy and information can be generalized to any system and event that will always go through 3 ages: an accelerated seminal and young age of increase of form and energy, a steady state of constant speed and a decelerated age till the system comes to a halt and the event dies away.

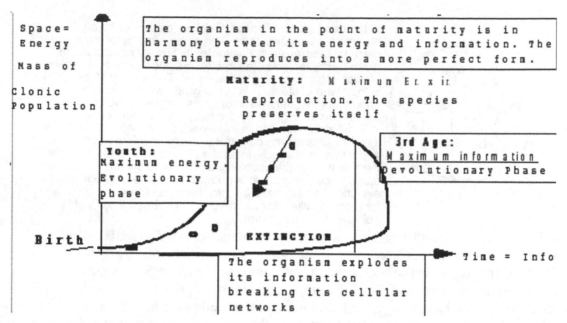

One of the key quantitative elements that connect the 4th and 3rd paradigm of metric measure is the fact that not only each species has a different rhythm of change and clock of time, but the same species changes its time rhythms and speeds through its 3 ages. This feature is fundamental to understand all processes that involve evolution, from the different masses of 'gravitational vortices' of information, ruled by the equation of a vortex ($V_t x R_s = k$ in simplified notation; thus the closer to the center, the faster we rotate, the faster the frequency of information of the vortex and the shorter the space we need to close a frequency cycle), to the evolutionary process of information in this planet, which can be mapped out and its frequency shown to follow a logarithmic process of acceleration, which now reaches its zenith with the creation of informative machines. In that regard we can consider the following 'speeds of time' of any event or self-similar ternary topology of energetic youth/membrane, reproductive maturity and informative, 3rd age (hyperbolic center of a system):

-st-1: Seminal age: Informative evolution is maximal, since the system occupies minimal space (Black hole paradox: Min. Se=Max.Ti). It is the palingenetic age.

- Energetic Youth: Evolution of information decelerates after the 'landing' of the species in the higher St-plane. Now acceleration of energy takes place.

-Reproductive, mature, steady state: Both energy and information maintain a constant speed in E⇔I feedback cycles, which will last depending on the access of the event to new energy and directional information.

- 3rd age: Now both energy first and then information follow a decelerated process of collapse; first energy becomes exhausted and then…

- Death: Information collapses, exploding into its st-1 scale of cellular energy in a relative zero time (the minimal unit of time of the st-scale). Death is the fastest motion of the Universe.

Recap. In physical terms the 3 ages of biological change can be considered the positive, accelerated youth, the reproductive, present age of constant speed, in which humans repeat themselves and the 3rd negative, decelerated age of temporal age, or 3rd age, which ends in death, when the balance of the Universe is restored.

39. Universal Constants are ratios of energy and form.

The previous analysis of the generator equation of the Universe brings an essential element of all formalisms of physics, the meaning of Universal constants and the key complex parameters of science,

211

measures that combine energy and motions such as speed or momentum, which are conserved as complex 'time arrows' and show certain constants in each specific species, which are the vital and universal constants of biological and physical, complementary beings.

We thus define Universal constants as ratios between the energy and information systems of a certain 'space-time membrane' (the bigger organic structures of the universe, the light membrane and the gravitational membrane, origin of most physical constants), or complementary being (the constants of mass-information, momentum, the social constants that measure the number of particles and field quanta put together to create a physical entity; the vital constants that are ratios and proportions between the energetic reproductive and informative systems of living beings, etc.):

'Existential constants are ratios between energy and information parameters of any system. Universal constants measure those ratios in the more extended systems, the membranes of the Universe, its quanta and fundamental particles. Vital constants measure those ratios for specific biological beings.'

Let us consider the main of those constants:

Simple, social numerical constants.

-Social constants of space, which define the number of cells that form an efficient spatial structure. They tend to follow pairs, as 2 is the natural symmetry of left-right of bidimensional space.

- Social constants of time, which define the number of events needed for a transformation or completion of a time cycle; the most common of which is the number 3 of past-energy horizon or youth, present, reproductive state of the event and future, informative state.

- space-time constants, which combine both spatial symmetries and temporal events; Pi belongs to this concept as it is a self-similar number to 3, with 'small' apertures between the 3 cyclic elements or phases of the cycle, which taken as a single picture of space-time form the spatial shape of the cycle when perceived in space.

Yet the most important of those numerical constants is the number 10 or tetrakys, the perfect social number that creates a unit of the higher scale with 3 x 3 energetic, reproductive and informative elements, and a 10th central point that 'emerges' to become a unit of the next scale.

Among those decametric scales, made famous by Eames in his film 'potencies of 10', it is natural to consider the 'scale of all perfect scales' S^t to be $(9-11)^{9-11}$ systems; that is groups of 9 to 11 spatial units organized in 3x3=9 to 11 scales of growing social complexity. And indeed we find that number to be extremely common to the point that we could consider it the ultimate number of scales of the complex universe between the fractal entities humans perceive (atoms and galaxies, unified in fractal, complex physics as self-similar forms of the quantum, electronic and gravitational, cosmic membrane): We find it as the difference of force between those 2 membranes (electromagnetism is 10^{4x10} times stronger than gravitation). Then 10^{11} atomic ties make a DNA molecule; 10^{11} neurons make a brain; 10^{11} humans have lived in this planet; 10^{11} stars form a galaxy and 10^{11} galaxies make up the cellular Universe.

In that regard, Einstein said that the ultimate Nature of Universal Constants could not be 'physical values' but special numbers, which would be 'relationships' between substances that constantly appear in Physical equations, 'like pi and e'. And he was right.

Existential parameters: reproductive speed and momentum

On the other hand, the simplest algebraic operations between energy and information parameters become the essential 'complex parameters of the Universe and all its species':

Speed = V=s/t=spatial energy/temporal information.

Speed is in complex science no longer a mere measure of the translation of a form but in detail, all forms (paradox of Zenon) move not, but reproduce its form, step by step, as a wave of light does. The wave affects a lower scale or network of relative energy, making it to adopt its formal configuration. So form becomes reproduce, imprinted along a previous scale of reality and motion becomes reproduction. Light reproduces over the simple energy of gravitational space and the wave displaces. A wave in the water does the same imprinting one after another the form of the network of atoms and so on. So it follows that the systems which are simpler with less information move faster because they have to imprint less form. And so V=s/o information=infinite, which is the perceived speed of gravitational space whose information we do not perceive, while the more complex life structures move slow as they have to keep its huge amount of stable informative shapes unaltered. And each 3 months we change all our atoms to reshape our form.

So the new parameter should be called 'reproductive speed' as we consider that motion is a manner of reproduction of form.

ExI= Momentum or existential force.

This is the second key ratio of reality which multiplies the energetic and informative strength of the system. In physics is equivalent to momentum, mv where M(i) is a measure of cyclical form, and V(e) a measure of motion, the definition of energy in this work. In biology measures the top predator power of the system which will dominate its ecosystem when it has the strongest energy body and the more intelligent informative brain. And each species will have a certain existential force, reproductive speed and energetic, $\sum$ and informative $\prod$ social parameters, which will be its 4 'main numbers, needed to define the form (from the quantum numbers of physics to the vital constants of living beings – metabolic constant, brain/body ratio, brain volume, physical strength, etc.)

Irrational constants.

Because arrows of time are back and forth transformations of energy into information or its combination to reproduce together a self-similar form, and certain numbers define certain 'efficient' social geometries, there are Universal constants, which define for each species of the Universe a given ratio of transformation of energy into form or a reproductive ratio that combines two simpler energetic and informative bytes and bites to reproduce a new form (h constant, c-constant, etc.)

The dynamic relationships between the 2 motions of the Universe, energy and information are invariant in form, motion and scale; therefore fluctuating around fixed equilibrium values, which is the ultimate meaning of Universal Constants.

Thus, all of them can be reduced to the generator equation of the Universe, the feed-back cycle of energy and information, from where all laws of reality can be derived:

Energy <=> Information; ExTi= Irrational Constant

A Universal constant cannot be a perfect number, because it will create a fixed Universe; thus Universal constants are irrational numbers, which show a minimal fluctuation. Consider for example the main constant of the Universe, pi. If pi were exact then the spiral made with 3'14 lines would not be a vortex but a perfect, static cycle. Yet if pi is either +pi or –pi, the cycle will not close by defect or close in excess. What this means is that the cycle will be a bit more curved inward, and so it will be an informative cycle; or it will be cured outward by defect and so it will be an expansive, energetic spiral.

We know, for example, that the orbits of planets are decreasing by a few centimeters a year, so they will finally fall into the sun. They are, if we consider a dynamic, temporal view of them, inward, informative spirals. Yet an antiparticle, which is exploding information into energy, 'dying' in a big-bang that annihilates it, is bending outward.

So irrational numbers are the absolute constants of the basic exchanges and transformations of energy and information of the Universe. The main ones are:

- Pi, the formal constant of creation of in/form/ation. Since pi transforms 3 lines of energy into a ternary cycle with one more dimension of form: a string of 3 lines with 3 dark apertures for a total 0.14. Within those 3 lines there is a 2nd dimension of height, or information and a volume of space. The entity has grown

- Phi, the Golden Ratio, which is the constant of reproduction that multiplies an organic system into self-similar forms.

- e, which is the constant of extinction of form back into a lower scale of energy that devolves a formal being into its cellular subspecies. Its most common ternary form is et=3=20.

We find those constants, both in physical and biological processes related to those transformations of energy and form – showing the fundamental equality of all Universal Systems.

For example, e appears in the decay of radioactive atoms that release energy; phi appears in the organization of a sunflower spiral; pi appears in the h-constant of transformation of light flows into electronic actions.

How many Universal constants there are for any system? We advanced 3 basic U.C. at the beginning, pi, the ratio of creation of information, phi, a reproductive ratio and e, an extinctive ratio of destruction of information into energy. And indeed, all systems have at least those 3 basic constants.

Vital Constants: proportions between brains and bodies.

A final type of constants expresses quantitative proportions between the reproductive body elements and informative particles/heads of a complementary system. As a general rule the commonest proportions of energy/information are:

- The particle/head of information is dominant in information parameters, (dimensions, mass-weight or number of network connections of the informative system, cellular density, etc.), in a 3 to 1 proportion with the body/field of energy, both in time (so in genetics, the dominant, informative element has a 75% chances and the recessive element, 25% chances); and in space (so the Universe has 76% of dark mass).

- In terms of spatial, energetic parameters however the body is dominant (spatial dimensions, cellular numbers, energy volume etc.) usually in a proportion of 90-80% to 10-20%. This is due to the fact that the informative element tends to be the central 1-element of a body tetrarkys, so we have a captain every 10 sergeants and a sergeant every 10 soldiers; 9 glial cells that give energy to every neuron, a 10% of taxes that go to the Middle age Priest, which directs the herd of believers, and so on. Yet another simpler, very common dual structure is a spatial square with a central knot, which gives us a 20% of informative elements (the center) and an 80% of energetic elements (the vertices of the square).

Recap. The generator equation of space-time and its 4 main arrows of time, understood as symmetric transformations of energy into information or reproductive combinations of both, coupled with the invariance of topological form, scale and motion of the universe explains for the first time the meaning of universal constants.

40. Inverted constants: The chip/black hole/mouse paradox

The inverted properties of energy and form, shown in the law of Range, apply also to any complementary system of the Universe. So smaller animals have faster metabolic rates because its energy /form cycles are faster. Their 'clocks of time', we could say move faster. Further on, in complexity this implies that paradoxically the smaller beings have more information (chip/black hole paradox). So the smaller the chip is the faster it calculates. This paradox is essential to understand the dangers of black holes. Precisely because they are so small they will reproduce faster and accrete faster, in the same manner a smallish virus reproduces much faster and it is more dangerous for an organism than a bigger bacteria.

If we adopt according to Galileo's paradox a static point of view, universal constants are NOT only algebraic values, but invariant geometries that repeat in all scales of reality. And this is the ultimate meaning of General Relativity, since Einstein made a precise, simultaneous, present measure of the 'static form' of those vortices of mass obtaining G as a measure of the relative curvature of the gravitational force in each point of the vortex.

The central concept of a Fractal, scalar relativistic Universe is obvious: the same invariant game, the same forms, the same motions, happen in all the scales of reality. And so the Universe is relative and invariant in its energetic motions (original Theory of Relativity), in its forms (cyclical forms of information and lineal energy) that repeat in all scales, which therefore are also invariant.

We have seen now how that invariance is played as a 'ratio', ExTi=K, which allows smaller beings to live shorter but live faster. As we have seen the properties of energy and information are inverted. So the smaller we become the faster we rotate, the faster we live, the faster we beat. For example, we know that a fly sees 10 times faster than a human being, reason why we cannot catch it. Yet the ant who lives longer lives 7 years × 10 times faster=70 years of inner, subjective existence.

The same concept applies to a physical vortex of information, V(t) × R(s)=K than to a living being that processes energy into form (a mouse beats its heart faster than a human; a cell divides and reproduces faster than a mouse, every 24 hours, etc.).

The entire cosmos and all its scales are related by that simple paradox: the smaller we become the more information we process. It is the Moore Law: the smaller the chip the faster it thinks. The reason is obvious: smaller, faster systems, close 'logic cycles' of information faster. In complex beings it means faster thoughts in smaller neurons, packed in tighter groups. In the physical world, the bigger rotational motions of cosmic masses are slower than the cyclical rotation of particles, but their product remains constant. And we can write this fundamental law of the Universe, with multiple self-similar applications in any entity made of fractal space-time, again as a general case of the generator equation:

Universal ExTi = Universal Space Extension × Time-frequency = Constant Entity = K

An expression, which appears in all scales of reality (Heisenberg Principle, Vital Constants, etc.)

Recap. The Universe is just and harmonic: small beings are more intelligent, faster than big ones. It is the paradox of David and Goliath; the paradox of the chip, the paradox of the black hole…

41. How to include metric spaces. Topology: informative, energetic and reproductive systems.

However, according to the Principle of Correspondence, each new wider, more comprehensive model of reality must include all the cases of the model it substitutes. So while a 4-dimensional description of multiple space-times suffices in itself to give meaning to reality, it appears unconnected with the previous paradigm of metric spaces, reason why we must achieve a more detailed analysis of those cycles and give them specific mathematical operations, as to be able to connect them with the 3rd paradigm of metric spaces, its geometries and mathematical algebras. This is done at two levels:

-By describing them with a higher form of geometry, topology.

-And by describing more precisely the 4 arrows of time, subdividing them in more specific types of events and adding precise algebraic operations to each of those cycles.

Let us consider briefly those 2 elements that will be developed in depth in other works.

Some initial precisions though are needed. Today information is not understood as 'form' but measured, as it corresponds to the science of metric spaces, since Shannon, by considering frequencies and patterns in one dimension. But here in/form/ation as the name indicates is given by form. So Shannon's analysis of information is correct but explores only patterns of information in one dimension (such as the information carried by the frequency of a wave). If you have though 2 dimensions you can square the volume of information you can store and transmit. And in 3 dimensions you get a cubic quantity. And so we observe that most complex systems have at least 3 'levels' of complexity in the creation of information. So lineal proteins fold into bidimensional membranes that fold into complex 3 dimensional patterns, which are in fact the active information.

The symbiosis berween function and form is evident: The line is the shortest distance/motion and so it is the main form of energetic organs, from cilia, to legs to light fields. The cycle stores the maximal information and so it is the usual organ of information, from cameras, to vowels to eyes.

Yet when we consider more complex topologies of information, we talk of hyperbolic spaces that store information and are basically a complex 'sum' of chained cycles, often forming a tube of height, and so your head is at the end of your height and antenna is at the top of height. And height becomes a dimension of information.

When we consider energetic systems, normally they are external membranes that protect with its strength and filter the energy of the external world. And so because they enclose the system, they are normally made of tiles, squares, hexagons that put together cover totally the space; they are a sum of planes even if the total sum might appear sometimes as a spherical form and in topology they are call spheres. Finally the cycles of reproduction are toroidal cycles that come and go from the informative center to the energetic membrane, combine both and reproduce the system.

The 3rd type of topology - reproductive topologies that combine the other 2 arrows - become the 3rd complex arrow of time.

Yet if those 2 simplex arrows shall explain it all, we must combine them further, realizing that 'from 2, yin=information and yang=energy, comes 3', since 'the game of existence combines yin and yang into infinite beings' (Cheng Tzu). Indeed, philosophers have always known that reproduction combines energy and information into self-similar beings. And the 4th paradigm will show how all complementary systems of the Universe, from the simplest particles, quarks and electrons to the more complex, humans and perhaps universes, reproduce their form by combining their energy and informative organs and systems, repeating them in another discontinuous location of space and time.

So there are not only 2 simplex arrows of energy and information but also a complex arrow that combines both, Energy $\Leftrightarrow$ Information: the Reproductive arrow. And again, while there are many different ways to achieve that arrow; we observe always that an energetic, lineal, topology (since the line is the shortest distance/motion between two points, the simplest energetic systems are lines, or planes), and an informative topology (since cycles are the perimeters that store more information in lesser space, informative organs, are cyclical) mix to reproduce. So men are lineal in form and are the energetic sex, and human are cyclical and are the more perceptive sex, and both combine to reproduce. Machines are reproduced by humans which are the cyclical, informative component that forms the raw materials or energy to make them. And so on. Because it is an obvious logic consequence of the discontinuity of vital spaces which are finite and the limited length of a time cycle which always ends, that to survive species must reproduce or else its logic form perish.

So the Universe is ultimately an organic system of reproductive systems of energy and information.

Thus, once we establish the 3 topological regions of any system, which is their why we can add more detailed measures and convert each topological space in a specific species of reality connecting the why and the when of the metric paradigm, fulfilling the Principle of Correspondence.

Recap. To fulfil the principle of correspondence multiple time-spaces must be able to connect the why of the cycles/arrows of space-time with the precise geometries and algebraic measures of the metric paradigm. The 3 dimensions=arrows=cycles of space are the perpendicular 3 topologies of the Universe: the function/form of energy, the function/form of information and the function that combines them, e xi, of reproduction. Those 3 dimensions define all topological spaces. Absolute space is its sum; it self a Non-Euclidean system.

42. The generator equation of space-time: 9 dimensionality.

On the other hand, to be able to connect specific equations of detailed metric analysis with the general equation of the 4 arrows of time we have to descend into an exhaustive detailed analysis of those arrows, dividing them in sub-arrows specific of each discipline and defining those arrows with concrete operations used in metric spaces.

Moreover when we observe things in more detail, we must break reality in multiple subcomponents that assembly wholes in parts, reality become more complex. And this again seems counter-intuitive. Because as set theory shows the whole is simpler than its detailed parts. So if we follow the integrative path from parts into wholes thing simpler and at the end we end with 2 simple concepts, the physical Universal body and its Mind god.

This in the rest of this introduction to multiple space-times, after making a very brief, simplified account of it all in the previous pages, we are going to try a tour de force: to resume it all, which will be latter developed in more extensive lessons, departing from the simplest parts into the whole, by showing you all the arrows of time and topologies of space from the simplest events and geometries to the more complex, from the primordial parts or 'simplex arrows' of space-time, into the organisms of reality or 'complex arrows, and finally into the absolute whole Universal body and mental God, or 'universal, transcendental arrows of time'.

Those 3 scales, from the simplest actions of abstract spacetime into the complex organisms of vital spacetime till reaching the absolute whole, are the 3 stairs we need to make sense of all. In each of those stairs we shall define 3 arrows of space-time for a total of 9 dimensions and that is all you need to create the Universe: the simplex, complex and transcendental arrows/ cycles/dimensions of all realities.

This is possible to do when we realize that the complex arrows of reproduction and social evolution manifest themselves differently in physical, biological and sociological entities, 'decoupling' themselves in ternary events, 'actions' that reproduce waves in physical space, 'palingenetic offspring' that reproduce biological species and so on. So in the more complex division of the arrows of time, we talk of 3 x 3 cycles/dimensions of space-time, which we group in 3 types:

Physical, spatial arrows:

Physical entities are complementary entities, which:

- I: In its particle state gauge information.

-E: In its wave state feed on energy.

-exi: And together reproduce actions, either by emitting a self-similar fractal part of energy and information, called an 'action', used to communicate with other entities (a Universal constant, such as h) or by moving towards a position the particle has gauged. And so we define 3 physical arrows or dimensions of space: gauging information, i; feeding on energy, e, and reproducing actions by combining them, exi. We call them also dimensions of space, because the light-space in which we exist has 3 Euclidean coordinates that correspond to the informative, electric field, the energetic, magnetic field and its product, the reproductive speed of light.

And it is a well-known fact of science (Maxwell equations) that the speed of light can be found by multiplying the magnetic and electric constants of light, thus showing that indeed, the algebraic operandi x connects the why of the simple arrows of space-time in physical entities with the specific equations found in the metric paradigm.

Thus one of the most fascinating facts of the Universe is the fact that departing from 2 simple entities, lineal energies and cyclical informations, which create spatial planes and clocks of time, we can explain it all by combinations, repetitions and transformations of those elements. A further precision though is

217

needed on the difference between classic operandi such as equality and the new, evolved, dynamic operandi of self-similarity proper of multiple space-times in constant transformation; since most people tend to call each thing by a single name and see things in a static manner. So the paradoxical transformations of a reality, which suddenly become its opposite, break their 'Aristotelian mind' and one-dimensional perception of space and time.

Consider the famous equation, $E=Mc^2$, which in the so-called Planck notation, where light speed is the unit, writes $E=M$, and since a mass is an attractive whirl of space-time with a cyclical form, it can be defined in terms of time clocks, or in terms of information, as the fastest it turns, the more informative frequency it has, and the more it attracts, as all whirls and hurricanes do. So, we can write it $E=M(Ti)$ or simply $E=Ti$.

This is meaningless because Energy is not mass or information but exactly its opposite. Lineal energy though can curl and create mass, especially when it goes at c-speed, its limit of lineal motion and so it deflects the remaining energy into a cyclical form of mass. And vice versa, a mass can uncoil and create energy. So energy can be transformed back and forth into mass and the proper way to write this equation is $E \Leftrightarrow M(Ti)$. And yet our mind prefers the concept of equality. And physicists will tell you that energy and mass are equal. The proper word though is 'self-similar' - a word used in fractal geometry, which topologists that observe self-similar beings understand much better. Thus the 4th paradigm is a change of 'chip', of state of mind, of the way we think, which makes it so difficult to penetrate, b*ecause the human mind is a simpler structure of thought, more accustomed to fix forms into visual concepts that see the complex, fractal, transformative reality.*

Social, organic, temporal arrows

$\Sigma e \Leftrightarrow -\Sigma^2 i$; $\int \partial$: Most systems however create stronger actions by gathering multiple energetic cells into herds and waves, which create bodies and by gathering multiple informative cells, into complex social networks.

The difference between herds and networks is the degree of communication, since each element of a herd only relates to its neighbours and a network relates to all other elements of the network through a huge number of communicative flows.

This makes necessary to distinguish between the operandi of both types of social groups in a more precise manner that our simplified 4-dimensional equation where we used the sum and multiplicative operandi to represent a wave/herd and a network. Since we want to be able to relate those operandi to specific equations found in the use of metric spaces.

Thus, while we keep the sum symbol for herds, Σ, we use its square or more precisely its negative square or imaginary number, $-\Sigma^2$ for information. Further on networks of information such as your neuron normally have a second scale of 'sub-networks', which are the flows of communications, or axons that join them to all other elements of the network, and it is easy to prove that if any point is connected to all the other points of the network, as well as to himself, the number of axons of the network will be the negative square of its sum, $-\Sigma^2$. The a negative symbol also stresses that the properties of energy and information are inverted; and the organic, informative network absorbs its energy from the body network, subtracting from the total force of the body the energy it requires, as the brain does, without giving back anything but informative orders. Thus informative networks are represented by imaginary numbers, in the complex plane, as it happens with fractal generators that have a real and imaginary term or in the representation of the phase wave of electrons, which have a real number that represents its energy and an informative value represented with imaginary numbers. The same concept applies to the understanding of the equations of special relativity in which the parameter of temporal information is multiplied by a negative square, as light-space contracts the gravitational space in which it draw its forms. Thus again, a more precise algebra of the cycles of energy and information allow us to connect the why of multiple space-time cycles with its detailed description and equations in metric space.

A more sophisticated operandi to study networks is the duality between integration and disintegration (derivative symbol) $\int\partial$. Let us consider for example, the simplest duality: $\partial\sum^2=2\sum$.

If we consider a complex system with an informative head, a network of neurons, $\sum^2$ in charge of the limbs and body cells, which tend to be in equilibrium (Re=E), in as much as the energetic system provides the elements to the reproductive system or moves it (being the informative system much smaller in space), then the derivative of $\sum^2=2\sum$ means the network will codify with its instructions both the body and the limbs. If the system was simpler – an energetic/information systems, then $\sum^2$(neurons)=$\sum$. This in Theory of Information gives birth to a key law: 'the number of informative instructions needed to integrate the parts of a system into a whole is the square number of its parts', with wide applications that range from epigenetics to industrial design.

Thus again we can see how the more general laws of the 4th topological paradigm, when studied in detail give birth to the specific laws of the 3rd, metric paradigm.

Finally both come together into complementary organisms, and combine their body and brain structures to reproduce themselves. And so there are 3 organic arrows: the creation of energetic waves, Σ; the creation of informative networks, and its combination in reproductive events or organisms, $\Leftrightarrow$. Where $\Leftrightarrow$ must be substituted by different algebraic operandi depending on what kind of system we describe. So in the example of a phase space of an electron will be a sum, +, but in a Darwinian process in which an informative cellular system or herd of top predators feeds on a field of energy or prey, it could be a division, as the 'food pie' is divided into the members of the herd and so on. We thus keep as in the case of topological spaces, a minimal degree of flexibility to be able to accommodate the multiple cases in which an informative network and an energetic herd or reproductive body enter into an act of communication.

Since the processes of organic evolution - the formation of herds and networks and its reproduction – contracts space, tying together individuals into groups, and requires a long period of time, those 3 arrows that need languages of informative communication can be considered to be dominant in time. As indeed, 'time curves space', and 'time evolves the morphology, the form of beings' (Einstein, Darwin).

The fractal arrows of multiple-space times

Finally we can talk of the fractal, transcendental arrows of multiple space-times, which were unknown in the age prior to the scientific revolution and have been looked at with wishful blindness by scientists, due to their dogma of a single clock-time and a single continuum space. They are however self-evident when you change your 'frame of mind' and see reality as it is, without those 2 dogmas of mechanist science:

The Universe is made of an infinity of those organisms, which generation after generation repeat themselves in time with small variations; or gather together into super-organisms, so particles evolved into atoms, which evolved into molecules that evolved into cells and planetoids that evolved into organisms and planets and galaxies and the Universe. And so we have 3 more arrows of time to complete the Universe: *the generational arrows* of species that go through a life-death cycle once and again between birth and extinction; *the transcendental arrow* that creates super-organisms with smaller super-organisms; and the *ecosystemic arrow*, which adds all the super-organisms and generations to create entire worlds and the Universe itself, which could be considered an ecosystem of complementary organisms of energy and information.

To represent them with algebraic symbols we shall call the previous equation, $\Sigma e \Leftrightarrow -\Sigma^2 i$, that defines a complementary entity of energy and information, spread in a single space-time 'membrane', plane or continuum as X. Then we define:

- $\int \partial$; e$\Leftrightarrow$I'; G; e^g: *Integration in time of multiple organisms.*

The generational arrow/cycle of existence is relatively easy to represent, since all forms that live have a given order: they pass through an energetic youth, Max. E x Min. I, an age of maturity in which the being reproduces by mixing energy and information, e=I, and an informative age, when it warps the rest of its information, Min. E x Max. I. Then in the moment of death the system explodes back, devolving its information into energy. For that reason, we can also use the symbols of integration and its inverse derivative symbol of disintegration. Since ultimately an organism is ruled by the existence of an integrative network in control of its energetic limbs and/or body in complex systems. And for that reason we die when our informative, integrative network or 'brain' dies.

Yet those ages are dynamic and so we can use the $\Leftrightarrow$ symbolism, to represent them all, whereas E< is the age of energy, $\Leftrightarrow$ the balanced age and >I the age of information. And consider that the arrow of life is e >I (warping of energy into information) and the arrow of death is I<e, its reversal. Those simple equations in the static and dynamic form will allow us to explain many systems with inverted parameters, such as particles (life arrow of physical species) and antiparticles (death arrow). Thus again we can relate the 4th , why paradigm with specific entities of the metric paradigm. E=I, will become also the equation of beauty, as we perceive naturally beautiful a balanced form of energy and information and since the product exi or action is maximal when e=i, beauty becomes merely the expression of the most efficient, top predator form in which body and brain, energy and form are in balance. Thus again, we can obtain basic equations for fundamental processes of existence never before represented in algebraic form, such as life, death, beauty, top predator or existential force.

Finally we can define the generational arrow as the sum across time of all the generations between the birth of a species or living cycle and its extinction. And so if we call G, the number of generations, any statistician knows since Fibonacci that the function e^G is the most common number of reproductive generations that will exist after a number of G generations. And from that simple equation we connect with the extensive field of Volterra equations and other works of metric spaces regarding statistics of populations that will fine-tune the whys we have found in the 4th paradigm of multiple space-times.

According to each specific species and type of analysis we shall use any of those operandi after careful consideration.

- Σ, $\prod$: *Integration in space of multiple organisms. The existential, ecosystemic or world arrow.* Finally if we represent a complex ecosystem as a series of super-organisms and fractal parts integrated into a whole we reach the final goal of explaining it all either within a world or ecosystem or the Universe taken as such. What operandi shall we use for this final arrow? A careful analysis of the interactions happening in each ecosystem will give us a combination of all the previous arrows, as each ecosystem will have herds, networks of interrelated species, top predators with closer social relationships, several planes of existence, etc. So in this final scale of reality it would be preposterous to pretend we have a 'metric equation' able to represent all the ecosystems of reality and the absolute. But if we consider two separate terms, one for organisms and territories related as herds and one for those related as networks, we can write, with two 'enlarged' symbols of sum and multiplication, which turn out to be merely the more complex operations of derivation and integration:

$$\sum(\Sigma E \Leftrightarrow -\Sigma^2 i) <=> \prod(\Sigma e \Leftrightarrow -\Sigma^2 I)$$

On the other hand, for the mathematically inclined astro-physicist, we shall show in our work on physical spaces that the Universe and its main cellular galaxies, can be studied as an organism made of two networks, one of gravitational information (dark, gravitational energy and dark, quark matter) and herds of electromagnetic spaces (stars, electrons, etc. And so its equation would be self-similar to that of an organic system. On the other hand if we were to calculate the n=st number of total scales of reality we shall see that they seem to tend to infinity and so the equation of the Universe as a fractal organic system would be:

$$\sum(\Sigma E \Leftrightarrow -\Sigma^2 i) \Leftrightarrow \prod(\Sigma e \Leftrightarrow -\Sigma^2 I)$$

Thus we differentiate the body where E is dominant on i and the brain, where I is dominant on e.

- st=n; $X^{st=n}$: The fractal, transcendental arrow. Where n determines the complexity of the ecosystem across multiple fractal planes of exi=stence: If we consider that each of those super-organisms integrates a series of simpler planes of existence of lesser information, we can use a natural number st=n, to define each scale of reality. Where st=n will represent the number of planes of existence from the simplest organism, a particle or quanta, to the most complex the Universe; and X^n, the number of minimal cellular quanta of the system. Again this is easy to see if we consider the most common transcendental number, 10, where a tetrarkys of 10 elements give birth to a central point (9+1), which transcends as unit of the next scale. And so a Mongol army had 10 soldiers one of which was a sergeant and 10 sergeants were ruled by a captain and 10 captains by a general, so in 3 scales a general was commanding an army of $10^{n=3}=1000$ units.

In that regard the Existential function of the Universe follows also that decametric scale, where n=10, and the total reality, if those decametric scales between atoms and galaxies repeat themselves will have n=∞

Further on, if we integrate together all vital spaces on one side into an absolute space, S and all informative scales in time as an absolute Time, we can simplify the previous equation of st=∞ as a single ST world. So we write:

$$[\sum(\Sigma E \Leftrightarrow -\Sigma^2 i) \Leftrightarrow \prod(\Sigma e \Leftrightarrow -\Sigma^2 I)]^{\infty} = S \Leftrightarrow T$$

Where, the physical, spatial arrows are: e, the arrow of energy; i the arrow of information and x, the arrow of actions and motions.

The organic, temporal arrows are: Σ, the arrow of energetic waves; Σ^2, the arrow of social networks and $\Leftrightarrow$, the arrow of organisms and reproductions.

And the fractal, transcendental arrows are:

e^g, the generational sum of all the cycles of life (e< or youth, = or maturity and >I or old age) of a species.

$X^{st=n}$, the product of all its planes of fractal existence.

Yet if we integrate all those life cycles across all its planes of space, we define an ecosystem of which the biggest one is the universe.

And so Σ, $\prod$, are the symbols of an ecosystem and the organisms of its world, which for n=∞ represents the Universe.

So the single Space continuum of classic science, S, is the sum of all the vital spaces of all its species and ecosystems.

Yet if we consider the evolution of those species across time, we observe that simpler forms evolve into more complex forms, from the initial particles to the complex structures of reality, with the passing of time. So T becomes the sum of all cycles of all systems, chained through organic synchronicities, to give us the absolute time of classic science.

Thus, we can perceive reality as a simplex space-time continuum, a whole, as classic science does, or in more detail as a dual system of energy and information, a 4-Dimensional reality that reproduces and evolves socially those bytes and bites, or a series of organic parts and herds, which gather into ecosystems that evolve in bigger scales till reaching the size of a Universe. In that sense, other simplified, valid expressions of the function of existence used in this book will be $S \Leftrightarrow T$, SxT (dual systems) or $\sum Se \Leftrightarrow \prod Ti$ or Exi=st (4-Dimensional ones.)

Needless to say the relativity of perception and measure makes each of us an Island-Universe, which can be described in maximal detail with the 9 arrows, as the scales of the Universe are relatively infinite and so even though normally we shall limit the study of a specific species to n±st=3 planes, we can obtain from most species multiple space-times. So a human could be studied till his atomic detail.

And so we have written departing from its minimal bytes and the final generator equation of all realities. Since with those 3 x 3 arrows of spaces-times we can explain it all, the whole, 'the thoughts of god', the game of existence, and its 'imprinted body', the Universe, or each of its self-similar parts, its details.

Now the meaning of that equation must be clear: it generates all the other equations of the Universe. Consider for example how it implies several operations.

The operator $\Leftrightarrow$ has in fact 2 forms: one dynamic as a flow, $<=>$ and one static as a knot: X. The first one fusions the 3 ages of growth, balance and diminution $<, =, >$. It means that generation has 3 ages, $<=>$, and two operations, $=$, parallelism, when the exchange or transformation is balanced or X, perpendicularity, when the top predator element of the exchanges absorbs all the energy from the other system. Thus we can consider instead of a simple equality, a complex transformation with several phases. If we call each side of the equation E and I, then:

$$E=I; \quad E<I; \quad I<E = E>I$$

Each of those phases of the general operator, $\Leftrightarrow$, can diminish, increase or divide the 'object'. And so we can divide the operator $\Leftrightarrow$ into 4 operators: a mere equality, a sum, a division when the information preys on the energy and divides it or a multiplication when it reproduces it as the energy controls it and uses it with its energy to reproduce. These kinds of events that the equation describes are thus the beginning of a fascinating adventure, to generate reality with the combinations and partial equations derived from the generator equation.

Physicists do it basically when considering SxT systems, which we have explained briefly before, and yet that is not so detailed, so we can consider that we can do either 2 dimensional studies (sxt systems), 4-dimensional analyses (E-Re x I x S) where we decouple space into energy and reproduction, limbs and bodies and information between knots of times and flows of social information, languages and networks. This dual decoupling of space and time create a 4 dimensional Universe and 4 dimensional types of equations, of the type: $(\Sigma e \Leftrightarrow -\Sigma^2 i)^{n=st}$

These 3 combinations of the cycle are 3 forms in which the parameters of them change by transfer between the element of the left (the herds) and the right (the networks).

The fight between herds and networks though is parallel in E=I and that is the definition of an organism. In the other 2 stages or events there is dominant arrow from the relative predator reproductive, energetic body or informative system.

But many laws of science are ceteris paribus analysis of that equation and/or one of its parts. We just have to consider the generator equation, a group structure with a neutral element, 1. Then we can study only a certain part of it. For example, making n=1, the transcendental arrows disappear; if I or E = 1, we are studying only the energetic or informative part of the system; and so on.

It must be in any case understood once and from all that the equation generate events primarily and those events seen as fixed forms become spatial organisms, but the equation is an equation of time rhythms, given by the st-frequencies of the systems. Let us consider what are the parameters of the equation's main parts:

- The parameters of E are spatial parameters.

- The parameters of I are temporal parameters.

- This fundamental duality, is thus the 3rd key element to the transformation of a topological algebra as this equation is (where I is a topological space, e is a metric space of energy and $\Leftrightarrow$ is the cyclical exchange space or toroid) into metric spaces. We know that part of that equation is an energy parameter and the other an informative one.

Thus the generator feedback equation allows to connect the metric and relational study of the Universe of multiple spaces and times by ordering all the time cycles= arrows=dimensions, we observe in the Universe, all the repetitive clocks in which a certain 'non-Euclidean point 'or Universal entity comes once and again to the same topological space in search of its arrow, as its partial equations and more complex systems as a series of connected equations that represent networks, organic or complementary systems, within a given ecosystem.

In our analysis of geometrical i-logic geometries we shall return to those themes.

Recap. 9 arrows of timespace exhaust all Universal events. They are the arrows of energy, information and its complex combinations, the different arrows of social evolution, reproduction and organization in complex, fractal planes of existence, from the simplest atom to the Universe itself: The Universe and all its relative worlds structure in 3 stairs of parts that become wholes. Each one defines a set of 3 arrows of time. Thus, there are 3x3 arrows of space-time: spatial actions of energy and information, exi, organic arrows that create body herds and informative networks, which become organisms, e$\Leftrightarrow$ I and fractal, transcendental arrows that define systems of multiple spacetimes.

43. Actions and waves. Physical systems display 4 arrows.

We consider several levels of analysis of reality, the abstract, metric analysis of simple systems of energy *or* information; the dualist analysis of complementary systems of reproductive energy and information, which combine in motions called actions; and the organic analysis of ternary systems, which add a new, self-reproductive arrow and allow the social evolution of systems in transcendental, new super-organisms.

Most physical, spatial systems are perfectly described with the simpler level of energy and information, complementary systems and its actions -though it is left for further analysis the question of the existence of ternary structures in all physical systems (given the fact that simple particles, electrons and quarks reproduce new particles, when given enough energy).

Since all analyses of reality are meaningful even if they do not consider the complete system and all arrows, we shall deal here with complementary systems that perform, exi, actions. They define the 3 basic arrows of existence: gauging information, feeding on energy and using that energy to move in the direction in which we have gauged information

Thus a system that combines information and energy is always able to create 'motions' called action, exi, which are the fundamental unit of physical systems. The simplest of those systems is light space, which happens to be the 3-dimensional membrane we live in.

Indeed what we call vacuum-space, the external reality, is in fact light-space. That is, the vacuum is filled with light, which has 3 perpendicular dimensions: a direction of motion, energy or length, in which light moves; a dimension of information or 'height', the electric field; and a dimension of reproductive width; the magnetic field. And so we exist in a spacetime of 3 Euclidean dimensions, filled with light, which makes the dimensions of reality the 3 arrows of spacetime of light, to which we add the social dimension of color, given by the number of photons that come together in a single wave. We are 'swimming' in a world of light as a fish swims in a world of water with 3 dimensions occupied by that water.

It is indeed the thesis of this work that what we call 'lateral motion' of a wave of energy is always a reproduction of its form, imprinted in the medium; and so we consider that the dimension of magnetism in light is in fact its reproductive dimension: Light follows the path given by an energetic field of gravitation, whose energy it absorbs, imprinting laterally a magnetic 'body', over which an electric field carried by the photon is built. And from that simple scheme many properties and equations of light will be deduced and many questions answered (such as the non-existence of 'magnetic monopoles', since magnetism does not form an informative particle – the photon – which is an electronic, 'fractal head' - or the relationships between the gravitational and electromagnetic field, which fees on it).

At this stage the example of light and its dimensions illustrates a tenant of multiple space-times theory: *all physical systems are also ternary systems with an energetic, informative and reproductive-motion field, made of social quanta. Since waves don't move (paradox of Zenon) but imprint the medium they travel through with its form, its information, reproducing it.*

The number of dimensions and elements we describe in any system in fact depends on the depth of the analysis. Metric space works basically with 2 or 3 dimensions, as it considers time and space continuous and symmetric (an error, regarding the nature of space, which explains many errors of physics), and ad maximal it adds a 3rd dimension of exi space-time actions. Other sciences, especially biology, go further and analyse 4 dimensions. And that will be the main approach of this introductory work even when describing physical phenomena. For example, physicists describe electrons with 4 quantum numbers, which we shall translate as the 4 main arrows of times of electronic waves, where the principal number defines the wave-reproduction of the electron, the secondary number its energy feeding, the spin number its informative orientations and the social number is the magnetic number that defines how an external field of magnetism organizes a group of electrons, positioning them.

There are however many systems that are not complete, often made with an informative, gauging element and an energetic limb, in all systems and disciplines of reality. Or more often there are systems, which become subsystems of a more complex form and perform only a function on this higher reality. For example, a planar field of energy or force might have as light does 3 arrows but it is for a complex atom the field of energy in which the electron reproduces its nebulae; a weapon is a 'lineal system of metal-energy', but it comes attached to a human being that 'gauges' information, locating the enemy to which it will release the 'energetic action' of the weapon; a virus lacks the reproductive systems, but it attaches its informative code to a cell that will reproduce the system. And so we find that some simpler systems need an external, informative or reproductive 'enzymen' or 'enzymes' to make possible actions with weapons or reproductions of viruses.

And these two essential laws of interrelated multiple times-spaces – the existence of 3 subsystems in all systems or 'ternary principle' that allow us to analyse any system in its 3 internal components and the assembly of subsystems as different as a human being and a metal-weapon to enhance the energetic or informative capacity of the whole through symbiotic actions, are key elements to explain how simple systems evolve into complex, organic ones.

Recap. If we consider motion, the reproduction of the form of a wave, even the simplest light-space displays 4 arrows, energy, information, physical actions or speed, exi that reproduces the wave and social evolution of color. Electrons also can be described with 4 magnetic numbers, the main reproductive number the secondary, energetic number, the spin number of informative orientation and the magnetic, social number.

44. Internal, topologic and external, Euclidean dimensions.

Another important concept to clarify errors of the metric paradigm is the meaning of 'internal' an 'external dimensions', where the internal dimensions are the 3 topological functions of a system – its informative, energetic and reproductive topologies - and the external dimensions those of the membrane, medium or ecosystem it inhabits. Thus in all systems we consider 2 types of dimensions:

-*Arrows/Dimensions of external spacetime* (the form described with external, objective parameters):

Length-energy-motion. +   Height-information +   Width: bilateral reproduction.

Those 3 simple Euclidean dimensions suffice for most external analyses of reality.

-Yet the internal structure of any i=-point requires *3 topologies of 4-Dimensional reality* and its complex shapes called:

-'Planar sphere' or 'Peano line'=membrane, limbs of energy

-'Cyclical toroid'=Reproductive Body of energy x information

-'Hyperbolic, warped center'=Informative, multi-cyclical head.

This duality considers the Euclidean dimensions an external description of the being, as it moves through the 3 dimensions of the light-space membrane; and the topological dimensions an internal description of the point in its parts and functions. Consider for example the case of a human being:

-Externally all what we see of a human from far away is a point whose energy *moves him in the direction of length,* but he also has *an informative head on top,* in the dimension of height, and he *has reproduced=repeated his organs bilaterally in the dimension of width.*

- But if we were to switch to an internal description we would find a more complex structure, with:

- A warped brain inside that high head, and an eye, which makes with an enormous number of pixels of information an image of the Universe.

- We find a body, below that head, full of cyclical, reproductive organs that combine energy and form to re=produce the substances of our body.

-And then we find our energetic members (a mixture of fractal, broken lines of energy - our limbs that move us in the length dimension – but put together shape a plane) and the external membrane of our being, which is what topologists called a Riemannian, planar sphere.

And as different as all species of the Universe might seem to us, we will be able to describe them all with those simple topological dimensions, which give us the 3 'arrows/cycles/ topological' dimensions of all vital spacetimes.

The sum of all those internal, topological spaces, moving inside a Euclidean, 3 dimensional space in which they trace cycles in search of energy and information to reproduce themselves is what the 3rd paradigm of metric space puts together into a single space continuum and a single arrow of time; what physicists call absolute space and absolute time. And while the error of a single space continuum it's a mild error, by reducing time arrows to a single lineal arrow in the direction of motion-energy, scientists, specially physicists spatialize, simplify and reduce time to 'entropy', or 'energy', what the clock measures. *And that is a huge error,* as we shall see when we correct and resolve the main questions of physics.

We shall call the 3 dimensions of energy limbs that create length motions; of reproductive, wide bodies caused by toroidal cycles and of high, informative systems with hyperbolic brains, the 3 external/internal dimensional arrows of reality. Therefore the 3 external dimensions=trajectories =arrows of existence, length=feeding on energy, height= perceiving information and width=reproducing bilaterally, are motivated by the internal needs of the 3 inner regions of a vital being, its topological 'dimensions/arrows of spacetime'.

Spacetime and Timespace are also dual definitions that explain those external and internal dimensions dominant in 'energy' and 'form', because none of the arrows/dimensions of reality is pure: all yin has a yang and all yang has a yin. And so even the simplest 'species' of the light-membrane, light itself, has an organic structure, as we have just explained.

Recap. All beings externally move in a lineal dimension of energetic length; perceive from an advantage point of height and reproduce its systems bilaterally in the dimension of width. Internally all of them occupy a vital space with the 3 topologies of hyperbolic information, cyclical, toroidal reproduction and are enclosed by an energetic spherical-planar membrane.

45. Social, reproductive, organic Arrows.

Till now we have dealt with the simplex reality you perceive in an obvious manner: the 3 dimensions of spacetime which correspond to the functions of energy-length, width-reproduction and height-information and are common to all topological beings that float in light-space, made also of those 3 dimensions, albeit with a slightly different orientation (speed-length moves the wave of light and so we consider it its dimension of energy, the high electric field gives its form and has the informative photon on top and the wide, magnetic membrane reproduces it). And all this could be written as a simple generator equation, exi=k or e⇔I, where the 3 terms represent energy (e), information (i) and its reproductive cycles (x for simpler, reproductive actions and motions, and ⇔ for complex feedback, dynamic exchanges of energy & form).

A more complex description of all this is given by the 3 canonical topologies of the Universe: the hyperbolic, informative, reproductive, toroidal and planar, spherical, energetic membrane of which all beings are made.

It would seem that all this suffices to explain most of reality and certainly it is a great jump respect to the mere process of measuring the when and how of metric spaces in a single space-time continuum, where the 3 dimensions of space are abstract, and time is just 'what a clock measures'.

But there are other phenomena in reality, mainly of social nature, which those 3 arrows do not fully explain.

Indeed, because reproduced forms are self-similar forms, reproduced closer to the parental point, (as in a fractal image, where self-similar Mandelbrot sets are pegged to the parental form), when many forms become reproduced in a tight space they come together into networks and make a bigger system, a herd, or superorganism.

And that game defines the arrow of social evolution: self-similar atoms reproduced in a big-bang associated to form bigger entities, planets and stars. Self-similar cells with the same DNA organize in a palingenetic process to recreate a bigger organism. Self-similar ideas or memes created by a prophet reproduce in the minds of believers, creating a religion or civilization. So social evolution, the 4th main arrow of space-time must be always considered to describe all systems - even if it is not an Euclidean dimension, because in light, social evolution appears as color - since all what exists feeds on energy, gauges information, reproduces energy and information and evolves socially.

But how things evolve socially? It is not enough just to enunciate this social arrow. When we study social evolution in more detail, in fact we find the existence of 3 social arrows:

- $\sum$: many simplex units of energy that we shall call bites, come together into herds, which we represent with the symbol of a summation, $\sum$.

- $\sum^2$: Many simplex bytes of information come together into networks, which we define with the square of a summation, as networks are defined by having a square number of 'axons' that communicate its informative points.

- ⇔, Re, E⇔: When those 2 systems come together they create a complementary, cellular organism, body or field, which defines cyclical trajectories between the limbs and heads, the particles and energetic membrane.

Thus the body absorbs bites of energy and reproduces cells of the organism; and we represent it with the ⇔ feedback symbol, the arrow of Reproduction, Re, or as it often comes attached with 'mouths' that absorb energy or 'limbs' that convert energy into motions, with the E⇔ symbol.

Thus all together form a complex, dual, cellular superoganism of space-time, $\sum E⇔\sum^2 i=St$ or exi=st-ential organism.

And so we define 3 social, organic arrows:

-$\sum$: the arrow of social evolution of energy bites into herds.

- $\sum^2$: the arrow of organic evolution of bytes of information into networks tied together into a bigger whole - a head/particle that absorbs bytes of information, also called 'pixels' and stores a memorial, perceptive image/mapping of reality. This network acts a single being, sustaining in each 'fractal' cell, the same informative, memorial mapping of reality that allows it to act as a single one, *either internally* (same DNA in the cells of an organism, same book of Revelation in the mind of believers that create a God or subconscious collective, same particles in a bosonic state) or externally (cells that store multiple pixels, coordinated by the network into a single image). This is the most complex, fascinating arrow of existence, or transcendental arrow, because it is the arrow that allows reality to become organic, as it makes from simplex parts a whole that emerges as a unit of a bigger space-time plane of existence. Further on, the arrow that puts together parts into wholes dominates as a whole the herd of cells of the body.

And so we define, a superorganism made of reproductive bodies/waves, limbs or fields of energy and social networks of information, which extends through 3 different spacetimes:

- An external ecosystem of energy (st-1 plane of reality) in…

- Which the limbs move by absorbing external energy and the head or particle gauges information by absorbing that external energy as information; and the body reproduces new cells absorbing both energy and information (st-plane).

- Which transcends as a whole (st+1), thanks to the network of information that controls the entire Organism, becoming a fractal space-time world in itself: $\sum E \Leftrightarrow \sum^2 i = St$

Recap. The social evolution of herds into body particles gives birth to the arrow of wave evolution; the association of informative cells gives birth to the arrow of network evolution, and both together create ternary, organic systems, which use internal limbs or external fields to feed on energy. Thus a more complex analysis finds that all systems are ternary systems with lineal limbs of energy, reproductive bodies and informative heads, sandwiched between two ±st planes, the smaller st-1 micro-points of energy and information (pixels) and the bigger ecosystem, which only the network of information can observe as a whole that transcends cellular existence (st+1).

46. Fractal, Transcendental arrows.

We call fractal, transcendental arrows to those arrows that go beyond the organism both in time (generational arrow of existence of the being between its creation and extinction) and in space (ecosystemic arrow), *finally evolving together in time and space into a super-organism.*

Thus simple 'active' complementary systems become complex, organic systems, which finally transcend into super-organisms, till reaching in the transcendental scale the biggest super-organism of them all the Universe.

Indeed, the generation of a being in a life/death cycle will cause a spatial 'radiation', as the being multiplies once and again, and so it will create a series of generations in space-time till the *species* becomes extinct. *And so the species can be considered a super-organism with each individual a cells of the system.* And we shall see that indeed, the entire life-span of a species follows also the 3 ages of life and its speciation into ternary species of energy, information and reproduction, and its final evolution into a super-organism or its extinction by a stronger species with more exi=stential action/momentum.

And since there is nothing else that we can say of a being before its creation and after its extinction, the 3 transcendental arrows of generational cycles, reproductive radiations in an ecosystem, and transcendence into a higher super-organism finishes all what we can know of any reality, all what is science, all what exists, all the whys of the 4th paradigm.

-St: And so the 9 arrows together give us the complex generator equation of all events and forms of a certain world, universe, ecosystem, reality or superorganism – its total, absolute spacetime description, EXI.

It is the unification equation, described earlier in greater detail that physicists tried to extract from reality but failed because they tried to do it with metric spaces instead of topological spaces and with a single arrow of time energy or entropy instead of the 3x 3 arrows of reality. We shall often quote that equation in its simplified 4 dimensional structure: $E \times i = st$; and call it the existential function.

You too are generated by that equation.

From that general equation we can then descend to the when of its metric details, since each species absorbs different types of energy and information that define the form and speed of its reproductive and social cycles of time. And so from that higher point of view we can then connect according to the correspondence principle each science and species with its specific space-time cycles/arrows – the details of the thoughts of god.

According to the dominance of those triads of arrows we also classify science in physical, biologic and sociological sciences:

- Most physical entities perform only two complementary cycles, most of its existence. So all particles move but few gauge information beyond a mechanical, geometrical series of action-reaction paths; and this explains why physicists have been concerned only with the arrow of energy and motion and metric spaces to measure the trajectories of all those moving particles. Yet in certain cases those particles decouple, reproduce other particles and form more complex atomic systems, and this moves us to the next science on the ladder of complexity, chemistry, which is more concerned with the repetition=reproduction of molecules and its social evolution.

- As we move further into biological sciences, through the ladder of bio-chemistry, biology and sociology we realize that the arrow of energy becomes secondary and the arrows of information= perception, reproduction and eusocial love that bonds minds into social structures dominates.

- In this we are different from physical structures, which are always dominant in energy and motion and create social networks of extreme simplicity with very few elements and variations - so galaxies which are huge aggregations of stars will turn out to be self-similar to electronic nebulae. And in fact we shall find a fractal equation of unification of quantum electromagnetic forces and macrocosmic gravitational forces based in that self-similarity of structure of the infinitely small and big[A,II].

So it seems the Universe has 3 different games of existence: - The physical game of self-repetition of spatial energetic structures that evolves into the bio-chemical game of evolution of reproductive, organic structures; which evolves in complex super-organisms and ecosystems.

We are the summit of that second game of existence that evolves the arrows/languages of information, creating social superorganisms, cultures and civilizations. And galaxies are the summit of the expansive game of spatial structures that develops to its limit the arrow of energetic, simple spatial topologies – reason why there is almost as much data about the Universe that about biological structures and even more data about sociological cultures…

Recap. The sum of all the generational life-death cycles of an organism, which will be repeated a finite number of times between birth and extinction, and its spatial radiations in an ecosystem, creates the transcendental arrow of super-organisms, which evolves from particles to galaxies reality till creating the absolute space-time field, st, of the Universe.

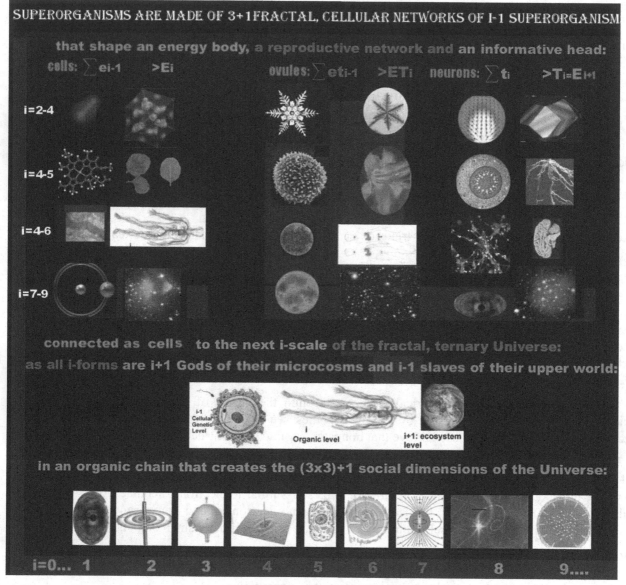

Ternary, complex Energy and information systems are made of multiple scales of space/times that act as 3 organic classes.

47. The Universe is a tapestry of those existential arrows.

Let us resume what we have learned: The Universe is made of self-similar complementary species of spatial energy and temporal information. We are not in space-time but are made of vital spaces with a duration in time. Our spaces are all made of topological forms of 3 types, which are the canonical types of all the morphologies of a 4-dimensional universe: informative, cyclical, hyperbolic spaces such as the brain, the black hole or the DNA center of cells; reproductive, toroidal cycles, such as those of stars in the galaxy, organs in the body or RNA systems in the cells and a planar sphere, or membrane of energy that isolates the system, allowing selective exchanges of energy and information with the external world, such as the halo of galaxies, the crust of the earth, the protein membrane of the cell, the skin with its sensorial apertures in a living being or the limbs that move us.

All those systems and forms have specific functions, which are also self-similar in all systems and can be reduced to 4 main time arrows or basic cyclical events of existence: energy feeding for the body to be able to reproduce; and information for the mind to be able to communicate and evolve socially with self-similar beings.

And all this can be explained with a single equation, the equation of existence: $\sum Ex\prod i=st$.

A more detailed analysis further divides the social arrow in 3 x 3 organic and transcendental arrows, completing an exhaustive 9-dimensional analysis of reality.

Thus, the most important arrow is the arrow of informative networks that we shall study in detail.

Recap. All systems, perceive information feed on energy reproduce and evolve socially in bigger systems, because if they do not reproduce their limited time will extinguish the species and if they evolve socially as bigger multicellular organisms they can win in the struggle of existence against simper, smaller beings.

48. Scales and symbiosis of complex social organisms.

Thus the commonest system of the Universe is a complementary organism of energy and information, with the 3 topologies of Non-Euclidean points that act as 3 social classes:

-The higher, st+1 dominant class is composed of the informative nucleus, an energetic membrane and its networks that control:

- A middle st-mass of 'cells' in cyclical motion between those 2 discontinuous, limiting zones.

- That transforms the 'lower' st-1 energy absorbed through the membrane, coming from an outer world into information pixels and energy pieces to create more informative cells and energetic membranes.

Thus, the st±n parameter becomes essential to analyse energy and information exchanges between the different microcosmic cells and macrocosmic systems of any fractal organism.

If the system is stable because a symbiotic relationship occurs between both networks, we call the complementary, entangled networks - social, organic systems, which extend in two fractal scales of different planes of space-time.

In the previous graph, from left to right, we observe some of those systems as they extract energy from their relative ecosystem/plane of existence:

- A cannibal galaxy traps smaller ones between its informative nucleus and stars' body, which reproduces its atoms.

- The sun system is protected by a series of planetary rings that extend till the Oort membrane of comets; yet we have to observe those rings according to the Galilean Paradox as still forms extended in space. Within that structure of an informative center of mass, and an extended membrane of space, the sun traps, gravitational and electromagnetic energy and atomic dust, with which it created those planets.

- The nucleus of an atom interacts with other atoms to form complex molecules through its electrons, which are the external membrane. Between them, gravitational and electromagnetic energy 'feed' the motion of those 'dominant forms' of its organic structure.

- A cell has a central informative DNA and a protein, lineal membrane and conducts (Golgi apparatus) which trap the reproductive organelles that create more DNA and proteins with energy obtained from the external Environment.

- A bird with an informative head, reproductive body and lineal limbs captures an external energy source – a lineal worm.

Those systems, studied with the laws that relate two or three relative, discontinuous planes=networks of energy and information offer some examples of the complex relationships of symbiosis and hierarchical dependence that take place between planes of existence, and its laws, which find applications in all self-similar organisms and space-time membranes, including the dual system of gravitational information and electromagnetic energy of the Universe within which we exist, or the interactions between the nervous, blood and digestive networks or the informative=high class, reproductive=working social class and energetic geography of Gaia that structure civilizations.

Most organic systems and its cells are created around the dynamic equilibrium produced by the exchange of flows of energy and information of those systems, which absorb them from the external ecosystem with the aim of reproducing in other region of space-time. Whereas, the 'physiological' laws that control those balances and ratios of exchange become the vital or universal constants (biological/physical jargons) of the organism. For example, a human being is a fractal form, made of trillions of cells, connected by flows of nervous information, digested energy and hormonal, reproductive blood messages. Thus a relative infinite number of $\Sigma Se \Leftrightarrow \sum^2 Ti$ cells become joined by those 3 networks, creating a human organism.

And doctors consider that we fall sick when the balances and functions of those 3 systems fail.

Further on, all those st-cells together create a unit of the higher st+1 scale, a human being, which through an entire process of iteration, self-organization and emergence becomes part of a higher st+2 scale, the social scale of mankind, of nations and civilizations. And again a series of balances between the social 'classes' of information that control the languages of power of the nation – money and legal words – and the reproductive, working class and the energy of the environment, will ensure the 'life of the civilization', which otherwise could enter in civil wars or collapse due to pollution and degradation of the environment.

In this manner through eons of evolution, both the human being and the Universe have become balanced, organic systems extended in multiple st-scales of growing social complexity.

The result is a dual or ternary organism: it has a reproductive body, which processes energy with its limbs, reproduces the cells of the organism and obeys a head of information.

So it is a fact of history that you belong to a certain social class, depending on your information and the control you have of verbal laws and financial money, the 2 language of information of human societies. Science calls a dual, physical plane of energy and information, a field of energy forces ruled by informative particles; a biological, dual system an organism with an energy body and an informative head; and a historic system of energy and information, a society with a mass of citizens and an informative government. Yet space-time fields, organisms of energy and information and societies respond to the same laws and properties of Multiple Spaces-Times.

Further on, since each of those systems will in itself possess a ternary structure, where the region that receives information from the head will tend to be slightly more informative and the region that processes energy will be more extended and energetic, we can increase the detail of our analysis by extending it to any number of cells and dimensional, ternary sub-scales. And indeed, more complex descriptions of a human organism consider that we are made of 9 subsystems, which can be roughly divided in energetic, reproductive and informative. And we can also follow the inverse process of evolution of life from simpler systems, which became more complex, decoupled in dual exi, or ternary, exi=r, subsystems and multiply its number of cells, from the initial unicellular bacteria till reaching the human being, the summit of evolution of this planet.

Yet the laws of Multiple Spaces-Times and i-logic p.o.v.s are followed by all the processes of physical and biological systems; among which the law of chiral asymmetry between planes will be the fundamental law to create a hierarchical structure between planes of different degrees of temporal information; since all of them will be communicated by one-dimensional, asymmetric arrows of energy that flow from lower body/reproductive organs into the upper informative ones, or arrows of information that flow from the dominant informative system into the reproductive/energetic networks. And the same happens in physical systems, between the central, informative knots of mass (quarks, black holes) and the external membrane of fractal electrons and stars (in the case of black holes flows between both membranes are called Rosen bridges, which Einstein used to study the exi flows of energy and information between the plane of gravitation and electromagnetism.)

Where we observe $E_{st-1}->I_{st}$ energy flows from past to future, from the class energy to the informative class and vice and $I_{st}->E_{st-1}$ flows of information from the relative future, informative organ to the energetic one. Since a macrocosmic being has more information and it is more evolved than its relative lower plane of cellular existence (a molecule for an atom, a cell for a molecule and so on), it is its arrow of future. Social evolution creates macrocosms from cellular beings, establishing a hierarchical, asymmetric duality of social classes in all systems:

- The information class (neurons, governors, the rich with financial information, and politicians with legal information) do not work=reproduce, but give orders to:

- The st-reproductive class (the cells of the body, the workers, the citizens) of the body which reproduces the forms wanted by the upper class. This class, sandwiched between the other 2 can obtain energy from the lower class and information with the upper class, and through social, democratic love, the sharing of both energy and information, will create new cells, and so it will show a toroidal geometry as it cycles between the 'external membrane' that extracts energy and the internal, hyperbolic region of information from where it obtains its formal orders.

- Finally, the energy class, which gives energy to the other two classes, since it receives orders from the information class in exchange from its energy.

In simpler systems the reproductive class might not exist or might absorb the energy which latter handles to the informative caste (reason why sometimes we call a dual reproductive/ energetic system such as the blood system of simple animals the energy system), but in essence this hierarchy between the 3 planes of all physical and biological systems takes place.

This law that we repeat once and again is essential to understand the main difference between space which has parity and it is symmetric and time which is asymmetric as motion from past to future converts energy into information, 'imploding' space, and flows from future to past, convert information into energy exploding and unwarping it. And it is at the bottom of all the processes that happen in time, from the life-death cycle of organisms, which always end up warping all energy into form, to feed its 'selfish' information cells, to the 'breaking of symmetry' that happens in physics in process mediated by the 'temporal', electro-weak force.

Thus, the formal analysis and the study of the hierarchies and relationships between any st-fractal and its lower and upper ±st microcosms and macrocosms, is an essential part of i-logic Geometry that can be applied to any fractal system in the Universe and allow to define a super-organism as a dynamic, functional structure, regardless of their form and extension.

Cellular herds become organisms that exchange energy and information through common networks, even when we see those cells de-segregated. The important fact is that all the cells of the organism need to be connected through networks of energy and information. For example, in an anthill, ants are guided by a network of chemical information (pheromones) that has its center in the queen and a network of energy distribution (the anthill and its territory) through which energy cells move (workers). There are also other specialized energy species called warriors. Finally there is a reproductive system localized also in the queen (and its males) that ensures the reproduction of the anthill species. Human societies resemble closely anthills with a higher degree of complexity and cohesion, because they are also social organisms of cells-citizens joined by economic, energetic, informative, legal and social, reproductive networks. When the three networks exist, we consider the being an organism. Without any of those organs however the anthill is not efficient and it dies as an organism. If the queen cannot reproduce, the anthill dies. If workers do not gather energy and warriors do not defend the vital space of the anthill, the anthill dies. If the queen does not guide the workers with chemical information the anthill also collapses. And so it does a society, which enters in an economic, political and cultural crisis that extinguishes the civilization. Only when those 3 networks deliver efficiently energy and information and reproduce the cellular citizens of the society, the vital organism survives. We have to widen our concept of organisms and consider that waves of cells communicated by networks are organisms, even if they have open, dark spaces between them, since they behave in all senses as such organisms. In a fractal universe there are always dark spaces and discontinuous worlds; but what truly matters are the bridges of invisible flows of information that connect those cells into herds, societies, bodies and organisms. Thus an organic species needs 3 networks to exist:

A superorganism exists when there are informative and energy networks that communicate the cells of the organism and the density of energy and information communicated to those cells through the networks is higher than the energy and information cells receive from the external world.

Yet those 3 systems do not have to be pegged spatially to create an organism. On the contrary, the duality between expanding energy and imploding information, continuous wholes and discontinuous parts is the key to understand the flexibility of those organisms that might appear separated as a herd or continuous as a body: the information network attracts them together and the energy network maintains its expansive vital space, separating them. Thus the hardware of information, in this case, neurons are discontinuous but they become a single whole joined by the software flows - electric information jump through multiple neurons, creating its continuity.

The metaphysical question: a 'conscious Universe'?

The previous results bring about a metaphysical question. Since unless cellular p.o.v.s have a 'Time will' –a desire to survive - and a degree of perception of information and energy, the previous social organic structures should not exist.

Indeed, the definition of a superorganism in terms of networks is clear: the organism tends to maintain itself pegged because cells survive inside the organism better, as they receive more energy or information. Otherwise the cells of the organism will become disaggregated and the organism will die. In the case of the anthill the information that ants receive from the queen is more important than the one they receive from the external world. The ants act under orders from the queen. They also receive energy from the anthill where food is stored. In a body, cells receive information from the nervous system, and energy from the digestive system, which is also more important than the energy and information they receive from the external world. In an eco(nomic)system, individual homes receive more energy from the arterial systems of electric networks and roads, from their work places (money) than from their external space, except in agricultural communities. Humans also receive more information from networks of mass-media than from other human cells. Thus, the economic ecosystem is also an organism of humans and machines, whose networks obey the common laws of all physiological networks.

Thus the informative organ is the 'soul' of the system that guides, controls and creates its 'form', its existence as a whole. Those informative networks, made of cyclical particles of fractal information are similar to what Descartes called 'vortices' and physicists masses, joined by 'res extensa' or gravitational, attractive forces; what Buddhists called Atmans, joined by flows of communication; what Leibniz called Monads that perceive the Universe; what biologists call neurons, joined by networks of information. In brief, they are the essential structure that pegs together a relative infinite sum of fractal cycles/cells.

In human societies the army that deploys energetic weapons in the external border and the government and moneyed people that control the informative, legal and financial languages of power are the dominant class. There is a middle class of citizens that obey those groups, which limit their freedom. And there is a lower class of 'immigrants, enemy nations', raw materials and machines, whose rights as non-citizens or objects are null. Yet those classes exist in all organic systems beyond human societies:

In a human organism the brain and nervous, informative system, the neurons and the stronger, energetic cells, the leukocytes and the epithelial tissue, control also a mass of cells, which reproduce in the different organs of the body the products the organism consumes, constructed with the bites and bytes of energy and information the external world provides and the human organs destroy and reform. If we consider a galaxy, the inner black hole and the halo of dark matter confine a middle mass of stars that turn around the black hole - while the external intergalactic space provides the 'lower' bytes of interstellar gas. So stars in their spiralled, cyclical movement gather that gas and send it to the black hole that feeds on it. So happens in an atom, in which the electron orbitals gather electromagnetic energy they send to the inner nucleus.

So we might wonder, are human societies just, natural, according to the ternary structure of social classes? No. The present distribution of social classes, proper of human 'capitalist' and 'nationalistic' societies are not natural to the Universe; since the 'poor' and the 'aliens' are considered a 'lower' class coming from outside the organic system and destroyed accordingly. Yet *all men should be considered either informative or middle classes, never lower classes and receive enough energy and information to survive, since we all belong to the same species. And so according* to the '3rd postulate' of relative equality, we all belong *to the same 'organism of history'*. Indeed the higher and middle classes of any organism act together, as head and body of the organic system. Or else when the cells of the body of any organic system do not receive food, the body collapses and dies. *Which is exactly what happens in history when the '3rd world' is treated as a '3rd expendable class' by the ruling classes of the system, both inside and outside our borders and they rebel in revolutions and wars that destroy the organic civilization.*

Unfortunately the economic ecosystem spends most resources in the upper classes and 'different species', (technological machines and weapons), instead of keeping all the cells of the body of history healthy producing for them the minimal human goods mankind needs to survive (food, education and health-care).

Yet systems survive because its cells are 'limited', existing between the central informative region that programs them and the energetic membrane that traps them and can kill them. In all systems, the stick and carrot, punishment and prize duality of negative and positive arrows provided by the 3 physiological networks control the 3 internal cycles of its cells, which have to obey the organic systems if they want to 'exist'; that is, to perform those cycles. So we often find 3x2 'elements' of positive and negative control, that structure a certain system as *the dual arrow of entropy and its 3x2 drives of negative and positive existence.*

A fact that explains why cells exist tied up in complex organic systems so well programmed. *Since according to organicism, the laws of the Universe are not imposed mechanically but through the will of 'existence' of its I-points, which make the best choices driven by their desire to process energy and information and survive.* A man, for example, can rebel at his own risk against the networks of power. You can obey an informative law and be rewarded or deny the law and be punished. But if the system punishes you it will extinguish you and another obedient citizen will occupy your place. So most humans always obey the law of its physiological, social systems. Yet if men, who are among the freest of all st- points, as they have self-consciousness, are so slavish; it is obvious that those points whose 'will of existence' is a vegetative or mechanical 'action', obey totally those networks. Physiologic networks rule indeed any biological organism to the point that doctors consider most sickness caused by the malfunction of physiological networks, which 'extend' the sickness to the individual fractal cells they control.

Recap. In any organic system the dominant informative and energetic networks control and distribute its energy and information in 2 flows with opposite entropic directions: information orders of control are flows that depart from the dominant informative singularity towards the membrane across the middle class; while energy comes from the membrane towards the center. Both flows create together a dual structure of max. Space-Time force or Momentum, the fundamental parameter that defines the power of any organism (max. ExI). So together the dominant classes organize with their dual energy and information the fractal elements of the intermediate, middle mass zone or body. Finally the middle class will absorb, destroy and transform the minimal fractal units of temporal energy of the external Universe, or lower class, which is divided in small, disorganized fractal elements; transferring part of them to the membrane and center.

49. Temporal classes according to its existential force.

Let us consider the previous structure from the pure perspective of time dimensions by considering the informative, more evolved class, the future class - since it becomes reproduced in higher measure towards the future; the reproductive system, the present and the relative energy that dies and becomes dissolved as energy of the other two, the relative past, *since it will become extinguished*. To that aim we shall use the concept of existential force, which defines the survival and hierarchy of those 3 classes.

The networks in control of organic cells are topologically related to the 3 zones of any st-point. Both together define the existence of '3 social classes' with 3 clear temporal, organic functions in all systems of the Universe that we define according to its relative TxE force:

- *Max.I x Max E: The future, dominant quanta* are *the informative center or brain and the energetic membrane. In an organism, they* are symbiotic and control the intermediate region: they absorb energy and information from the external world, passing it to the intermediate, 'middle class' region to replicate the specific temporal energy they need. *In ecosystems they are the top-predator species and often establish Darwinian relationships, as men and lions do.*

The Universe creates fractal actions that combine energy and information, I X E. The informative center is the fastest geometry, processing information and the membrane is the longest entity, moving and processing more energy, their combined fractal actions Max. E x Min. I, are more powerful than the rest of the actions of the system. In ecosystems those cells work often in collective herds, which act simultaneously, joined by a common informative language that multiplies their fractal action, against disorganized forms. The result is a universal arrow of social evolution that creates from individuals to herds, and from herds, tightly packed organisms, directed by an informative language, since a herd is less efficient than an organism. So men have evolved socially in groups, hordes, tribes and tightly packed nations, controlled by legal and financial languages. Top predator ants and bees are the most successful insects, because they developed super-organisms with chemical languages. In the Universe black holes fusion in bigger black holes that gather in the center of the galaxy, acting as 'gravitational animals' that perceive gravitation as information (as animals do with light), controlling with gravitational waves their stars. While stars are gravitational plants that absorb gravitational space-time as energy, as plants do with light and can't evolve informatively. Finally in the cell the informative RNA forms herds that control and organize their organelles with the genetic language.

-*E=I: A mass of present, cellular quanta* form the intermediate region or 'body' of the organic system. *It is the working, re=productive 'middle mass'* that re=produces most of the complex elements of the organism under the orders of the informative classes. They are the workers of human societies that reproduce their goods under monetary orders. They are the stars of galaxies that reproduce their atoms. They are the herbivorous animals in the ecosystem on which carnivores prey. They are the electrons of the atom that gather electromagnetic light for the nucleus. They are the stars of the galaxy. They are the glandular cells that form the organs of a body and reproduce their substances controlled by the neurons.

They are the proteins and genes of the cellular ecosystem. They are the citizens of a political system that obey the informative laws.

- *Min. E x Min. I:* Together the upper and middle, informative and reproductive brain and body prey *and feed on the external, 'lower', past class,* made of bytes of information and bites of energy of minimal size, which the other classes absorb from the external world or *ecosystem,* beyond the discontinuous limits of the world that becomes the 'vital territory' of the organic system in which it obtains its temporal energy to energize and inform themselves, destroying those lower, energy quanta in the process. The lower class tends to be an undifferentiated mass made of fractal pieces, full of energy with minimal informative perception: *divide and win.* They are plants on Nature; they are the stellar dust and dark energy that feeds and informs the galaxy; they are light quanta, carbohydrates and water in the genetic, cellular ecosystem; *they are* the machines of the economic ecosystem; they are those punished by the law in societies: criminals, enemies and emigrants, submitted to the racism of the other classes. They are the food and red cells of the blood system, without informative nuclei.

In Time the 3 regions establish a relative order from past to future, as the external, fractal forms that penetrate into the st- point are the relative past of the space-time field, which become destroyed and processed by the central region, the relative future towards which they move. So the central form is the future and the external world is the past. While in between those 2 limits, the intermediate region becomes the relative present of the system. In the galaxy, the black hole is the relative future of the stars than in the future will end up falling into the black hole; while the stellar dust is the relative past-energy they consume: $\sum Min.E + \sum Min.I$ *(Past, energy class st-1)* $=> \sum$ *exi: Present, Reproductive class (st)* $=> Max. ExI$ *(Future, dominant, top predator Class: st+1).*

IX. CYCLIC SYNCHRONICITY: PRESENT SYSTEMS.

The orbital, energetic cycles of the Earth-Moon system are parallel to the cycles of reproduction of its st₋₁ living fractal quanta: An organism extends through several scales of growing form. Since microscopic, fractal species become chained to stronger, ExI species as their relative energy and information quanta of a 'lower class'. In the image, Gaia, the Earth's surface, is a Fractal made of

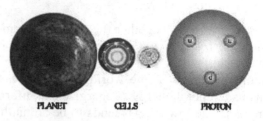

PLANET   CELLS   PROTON

organisms made of Σ cells, made of ΣΣ atoms, made of ΣΣΣ quarks, all chained by their organic cycles.

50. Knots of Time. Synchronicity cycles of existence.

If we consider any organism a knot of cyclical time arrows made of multiple feeding, reproductive, informative and social cycles with different frequencies and dimensional sizes, which converge in that organism, the next conundrum to resolve is the harmonic synchronicities between the I->E->R->S cycles of the organism. Each cycle will have a frequency in harmony with many other cycles, and so the being might be very complex in its behavior as it switches between cycles, but ultimately it can be reduced to a series of cycles that converge into the organism. Thus, organisms are knots of time cycles. In the same way, when multiple cycles converge into a point they add their causal power and produce intense events with a higher $\sum e \times \prod i$ existential force, but we can break the complex entity into a series of cycles whose combined action will show a harmonic frequency, as we can break a harmonic wave into a Fourier series of simpler waves.

In this manner complex events/entities can be reduced to simplex herds of events/cells.

For example, the convergence of different geological and cosmological cycles, which affect the energy and information of the Earth, cause ages of extinction and age of massive evolution of new species, with a series of frequencies and rhythms we have studied in our analysis of geological biological and sociological cycles.

In fact most events of the Universe have multiple causes, which means they are knots of time cycles: from multiple cyclical forces that converge in Relativity to create a knot of gravitation or a charge, to the multiple events that converge in Biology to cause an age of extinction, to the multiple failures that converge in a body to change its rhythm from life to death.

Since the Universe explores all possible structures, despite being made of a few elements, the inverse system also exists: one in which the bigger system is simpler than the smaller, faster cells, and both are symbiotic, since the smaller cells equal the 'existential force' of the bigger system with more 'energy space' but less frequency of time-cycles.

Whereas the bigger system is: *Max. E x Min. I.*

And the smaller system is: *Max. I x Min. e.*

Unlike the previous organic systems in which the informative networks exploit selfishly and extract energy from the herd, big, slow forms and fast, small systems are symbiotic, in balance and so they last longer in time.

Still, even if there is a degree of symbiosis the bigger, single organic system will be in control of the cellular herds as it provides them with the ecosystem in which to obtain energy and information. And in this manner it creates the structural order and deterministic destiny of most systems in the Universe. Let us try to formalize such pyramids of synchronicity, with the help of the 3x3 time arrows, which dictate that an informative pixel is smaller than the superorganism of the perceiver; a bite of energy is smaller than the reproductive body; and a reproduced, self-similar cell is smaller than the super-organism of which it forms part. In terms of existence, it means that the life of a quantum, fractal part depends on the length of the cycle it performs for the bigger system that controls the fractal part through its networks and informative or energetic fields:

The shortest cycle is the *informative cycle* that absorbs energy particles, transforming them into bytes of information, perceived by a sensorial organ or mind. So informative particles are the smallest ones that live the shortest within any organism. For example, a human being consumes thousands of small letters in a newspaper thrown every day; an eye consumes a photon in a microsecond.

Next comes the feeding, *energetic cycle* that absorbs bigger energy quanta, transforming those quanta in 'cellular components' of a body, once the informative cycles have perceived the food. And so an energy quantum lives a little bit longer than an informative cycle. A pig that feeds a man lives longer than a newspaper and also its 'final consumption' takes longer than reading that paper.

Then, it comes the *reproductive cycle* that handles huge amounts of energy and information to create a being, repetition of a bigger organism. A woman's ovule stays on her body all her life and it takes 9 months to become a baby.

Finally, the longest cycle is the being's *generational, social cycle*, from its birth to its extinction (±st) when it becomes part of an eusocial superorganism. So those cells that carry the 'social information' and I-eye-wor(l)d of a human being, our existential will, last longer: a man has the same neurons all his life. It is through that longer generational cycle how a new social plane starts to evolve, as a new ecosystem or macro-organism in which the being exists, as a mere quanta of the macro-being. Such is the relativism and justice of the Universe: since he who killed micro-beings to feed and inform himself will be just a micro-being that toils for a bigger organism.

It is the *social cycle*, as the quanta gathers in social groups to perform energetic or informative actions for the bigger macro-organism that lives in a longer scale of time and controls him. For example, if you are a soldier, your national organism will sacrifice you in an energetic cycle when the nation conquers and 'feeds on' the wealth of other nation.

Thus we establish a basic existential chain between the arrows of time and its cycles on 3 relative space-time scales:

$$(Informative > Energy\ cycle)_{st-1} > (Reproductive\ cycle)_i > (Generational\text{-}Social\ cycle)_{st+1}$$

- Since the existential, generational life-death cycles of st_{-1} particles become energy and information cycles for the st-being.

- While the reproductive cycle happens in the same st-level of the organic system that mixes with a couple of inverse exi parameters (sexual reproduction, inverse parameters of particles that reproduce a self-similar particle).

- Finally, the generational cycle of the being becomes, as a part of a social herd made of similar individuals, a submissive energy or information cycle of a macro-organic system, st+1.

Recap. Any organic system is a sum of 3±st types of cycles, chained in hierarchical structures across 3 levels of complexity. Any being's existence happens within those 3 TS *planes* in any organic system or ecosystem, in which he is a relative energy of bigger being, an equal, social being to its pairs and the informative master of smaller quanta.

51. Balances between scales. Ternary principle.

According to the Ternary Principle any fractal, cyclical action of any species can be described through its spatial parameter of 'speed-extension' and its reciprocal, temporal parameter of repetitive rhythm or 'informative frequency', whose product, ExI=K is a constant of action that defines the existential parameters or life-span of the cycle (also written as ExT=K in certain models of physics, since 'clocks of time' and cycles of information are synonymous).

Yet those actions are the sum of the simultaneous micro-actions of its 'cellular quanta' and so both must be in balance for the total organism, extended in two planes of existence to work in synchronicity. And that synchronicity is ruled by the *black hole law of complexity, we just explained*:

Min.E=Max.I: smaller beings act faster than bigger ones.

How the faster, smaller, st_1 micro-actions relate to the slower, big simultaneous actions of the organism? Or in formal terms, what is the relationship between the constants of those macro-actions and the faster parameters of its micro-cellular cyclical actions? How each organism relates as individual quanta to the slower, longer cycles, (Max. E= Min.I), of the st+1 ecosystem in which it performs its actions?

2 facts are the key to properly answer those questions:

- *Chains happen across 3 ±st levels of existence, creating hierarchical pyramids of unlimited size in time and space, through those ternary structures.*

- *Chains of actions become simultaneous, because certain cycles are shorter than others; so the faster speed of micro-cycles is compensated when we chain fast micro-cycles to slow macro-cycles.*

Consider the cycle of feeding of a lion. The lion moves through space in certain patterns that displace the lion towards hunting and drinking grounds where he will consume a victim. Thus the energy cycle will be determined both, by the inner vital constants of the 'st_1 cells' of the 'st-lion', (his cellular metabolism that defines his rhythms of hunger, his muscular speed and his strength that determines the preys of the ecosystem in which he feeds); and by the outer, st+1 ecosystemic parameters, (the spatial distance and 'informative density' of preys of the ecosystem). So to define the 2 ExI elements of the lion's feeding cycles (the spatial trajectory and fractal rhythm of its hunting) we have to know those ±st parameters. Then we can define totally his feeding cycle.

Thus, existential cycles might be very complex but ultimately we can reduce them to a fractal ternary equation, which relates harmonically the Space-time parameters of those 3 ±st levels.

Those ternary chains define Universal events across 3 hierarchical planes of existence, from the chains that happen between the '3 social classes' of an organism, to the physical events that happen in 3 scales of matter, to the 3 levels of humanity (the biological, individual and sociological level). Since all cyclical systems are defined by the relative ExI parameters of those 3 st±1 scales.

For example, Maxwell and Planck defined the informative frequency and energy quanta of a light photon that absorbs gravitational space-time energy and deforms it into a light-wave only with the action constants of 3 scales: the st_1 constants, or magnetic, spatial, and electric, informative constants of vacuum - $c=(e_0 \times m_0)^{-1/2}$; the st-constant of light (h) and the fractal, social constants of photons, as part of a bigger social, st+1 wave (E=hv). If we add to those synchronicities the understanding of the diffeomorphic dimensions of energy and information (on top), then we can fully grasp the structure of the 'organic membrane of light-space' in which we live and its connections with the lower membrane of gravitational space and the upper membrane of electronic, denser space, of which our mental eye wor(l)d is made.

In most systems, a st-1 particle, despite its individual 'vegetative will', perform a cellular, cyclical function within the macro ecosystem in which it exists either as an energetic or informative unit, and in this manner the sum of the parts makes possible the emergence of the whole. In the previous example, photons are individual forms that feed on micro-electromagnetic constants creating a reproductive c-wave. Yet most photons could not exist if 2 bigger, st+1 atomic particles had not reproduced them to communicate energy and information, which is the role they performed by the higher quanta. In fact, the macro-organic dual atomic system that emits and absorbs the photon, determines its ExI, frequency and energy parameters and its generational life span, given by the distance between those 2 atoms.

Those balances between the cellular, existential force and individual force of the system is what maintains the balance of the organism and make it last in time and are exemplified by the previous Maxwell equation or the vital constants and physiological constants of an organism.

We can generalize those harmonies, observed in the lower physical plane if we consider a constant of action in any scale to be a self-similar definition of the Existential force of the system we can define a basic law of hierarchical systems:

'The st±1 existential forces of a system are in balance, such as the exi constants of action of the macro-system determine the length and intensity of the existential cycles of its cellular forms, which in turn will determine the cycles of its microsystems'.

Thus each entity will become deterministically chained by the informative or energetic, reproductive or social arrows of the upper system, for which it will act within a herd of self-similar forms, sustaining the actions of those macro-arrows of time. And so while submissive to the higher system it will enter in democratic, parallel relationships with the forms of its planes of existence and finally it will be itself, a super-organism for its internal cells or reproductive body it controls. This is the ternary principle that all systems of reality obey, from humans, which perform for his social scale, related in equal terms to other humans and control tightly the cells of their body, to the light photons of the previous graph, slaves of their electronic parents, of which they often become a 'fractal cell' (fractal interpretation of an electronic nebulae), while controlling the electric and magnetic constants of their space-time membrane.

We formalize those ternary functions as partial equations of the Generator Equation of space-time, to which we add the hierarchical index of scalar space-time, st:

- $\Sigma et_{st-1} > I_{st}$: The being is a dominant macro-point, I_{st}, which absorbs a flow of information and energy from a herd of micro-quanta (ei_{st-1}) that it controls with its physiological networks. So our cells are controlled by our blood and nervous systems.

- $\Sigma ExTi$: It is an individual quantum that relates to similar cells in an environment in which it searches for energy in social herds or for similar couples to attain its main existential will: its generational reproduction. Thus, 'st' is the fundamental plane of existence of any being, the region in which it perceives more, and plays its most fulfilling cycles of existence.

- $E_{st} > I_{st+1}$: The point becomes energy of the upper plane, either because it performs an energy function for a bigger social plane. For example, the govern controls you as a social micro-quanta of a nation, which makes you die in wars or takes your energy through taxes. Or because it is consumed as relative information or energy of the upper head or body system. For example, humans become soldiers, the energy of wars for the system called the military-industrial complex.

And again, to explain why all those processes, some paradoxical and destructive for the p.o.v. are possible we must postulate a conscious universe of infinite, relative p.o.v.s.

In fact, and this is a 'psychological law' with enormous implications for the existence at least of human beings, which might happen in other scales if the hypothesis of a perceptive Universe is truth, *most points of existence are completely unaware of the existence of a lower scale, which they ab=use and they actively sacrifice themselves from* the upper system from which they receive orders and information in its 'tabula rassa' brain. Since the p.o.v. has far less information than the upper system, which programs it to love its master. So humans go to wars for the military-industrial complex without realizing they are sacrificing themselves in most cases for the financial gain of the upper, informative castes that control with money, weapons and law the society in which they live.

It is an order we find in all planes of existence. For example, in Biology, a st-mitochondrion uses st_{-1} protons to deliver energy to the bigger organic st+1 blood networks in the form of ATP. So the atomic, cellular and organic scales define the energy cycle of cellular breathing. While a star turns in cycles with other stars, feeds the galactic black hole as a quantum and feeds itself in fractal stellar gas. A light ray feeds on electric and magnetic constants, gathers socially in photons and it is absorbed by big electrons.

Recap. From that 3-scalar order that structures any particle or cycle of existence (static or dynamic perception), it arises the *relativity of existence, as any being* is king of its territorial hill, friend of its friends and slave of the World or ecosystem in which it exists.

52. Laws of hierarchical synchronicity between time arrows

The relative frequency of those cyclical arrows create 'time clocks' whose synchronicities chain the long cycles of fast micro-entities to the short cycles of a slower macro-entity, creating a harmonic order between the different planes and social classes of the organic Universe. Since an informative cycle is faster than an energetic cycle, which is faster than a reproductive cycle, which is faster than the complete generational cycle of the being; while a macro being is slower than a micro-being, we write 2 basic chains between those cyclical arrows and the beings that carry its parallel functions in the structure of a superorganism:

$$\Sigma\Sigma \text{ } Microcell's \text{ } Informative \text{ } cycles = \Sigma \text{ } Cell's \text{ } energy \text{ } cycles = Organism's \text{ } Reproductive \text{ } cycle$$
$$\Sigma\Sigma \text{ } generational \text{ } cycles \text{ } of \text{ } microbeings = \Sigma \text{ } Social \text{ } cycles \text{ } of \text{ } Beings -> E/I \text{ } cycle \text{ } of \text{ } macrobeing$$

Those ternary chains between the fastest cycles of macro-beings and the slowest cycles of its micro-beings *create simultaneous, symbiotic actions along 2 planes of existence*:

Informative frequency of the macro being => Energy frequency of the micro cell

Thus the fast, informative cycle of the slow macro-being is transformed into the slow energy cycle of the fast cellular micro-being, in a symbiotic chain that make it dependent on that macro-being. For example, the fastest Earth's cycle is its informative, cyclical rotation as a mass vortex that determines the day-night cycle of light, which feeds its living beings. So, the living st_1 cells of Gaia, its plants, feed with the day light, while its st_1 animals have a ternary energy cycle of 1 to 3 meals a day. So both are timed to the informative light-day cycle of the Earth, their energy cycle. If we lower the scale of analysis, the same cyclical chain happens between the animal and its st_1 cells. Men have a rate of informative perception of a second, the rate of blinking and thought. But a second is also the rate of breathing and the heart's beating that transfers energy to its st_1 cells, becoming thus the energy rate of feeding of those cells.

The same relationship can be used in other informative /energy cycles. For example, the energy cycle of an animal is a day. Yet the size of its hunting territory, of which he knows all its information, is the distance it reaches in a day.

A homologous relationship chains microcosms and macrocosms symbiotically in the next, longer 2 cycles, the energy and reproductive cycle:

Energy frequency of the macro being=> Reproductive Frequency of the micro cell

By homology with other organic systems, we consider that the energy cycle of the Earth as a rotational mass that deforms gravitational spacetime lasts the year it takes to cross its space-time territory, its orbit. Since, as in the case of an animal, the main energy cycle of the planet is the year-time it takes to sweep its solar orbit, the 'organic territory of the planet', where the Earth absorbs its gravitational energy, deforming space-time according to Einstein's equation. While the secondary Earth's energy cycle is a month in which its satellite, the moon, sweeps around the Earth's orbital territory. We obtain thus, the 2 main cycles of transformation of gravitational energy into rotational form by the Earth-Moon system: the lunar cycle and the annual cycle.

And those 2 cycles are transferred to the micro-living beings of its Gaia skin, so they are the reproductive cycles of most animals. Thus in the human being the menstrual cycle is a lunar cycle; the cycle of reproduction of a child lasts 9 months; in Nature almost all reproductive cycles have an annual frequency, happening in spring or summer, when the sun's light absorption is maximal; which is the obvious reason why those cycles are linked: while the macroscopic being absorbs energy, it allows its cells to use that energy to reproduce. Again, if we lower our scale, an animal as a macro being has a cycle of energy of a day. That period is also the cycle of reproduction of most cells, called mitosis. Yet reproduction is discontinuous: it happens only when 'optimal energy and information' is available during a small interval of the total existence of the being, in general in the second, mature age of balance between energy and information.

It seems that the micro-being is fed and reproduced by the macro-organism, the good 'god' of its existence. Yet 'the good god' is actually a farmer that reverses its arrow of order for an arrow of entropy and destruction when it uses the 'existential cycle' of the micro-being as a mere fractal cell of its macro-organism. So we feed and reproduce pigs but end up eating them. In a sense, we can say that all organisms are 'farms'. So pigs feed men; stars fed by the interstellar gas attracted by the black hole's gravitational power, also end up feeding the black hole; and fat cells fed by the organism end up killed, feeding the macro-organism. So in that inverse, final cyclical chain between the microscopic and macroscopic being, the 'relative God' takes all what he gave. Thus there is an inverse, final chain between the generational cycle of a micro-organism and an inner, energetic, informative or reproductive cycle of a macro-being. Thus the generational cycle of the macro-being complete the previous chain:

Generational frequency of micro cell => Reproductive frequency of macro being

For example, body cells live and die in an interval between a month and a year, which is the period most animals need to complete their reproduction. At cellular level, carbohydrate molecules live and die in short daily cycles, which is the period of reproduction of cells. The relationship between both cycles is again organic: reproduction means the renewal of a macroscopic being that destroys and creates many of its fractal elements, as energy or information of the reproductive ixe combined cycle. So cells accumulate carbohydrate molecules that they use in their reproductive processes, destroying and reconstructing its DNA genes with them.

Depending on the role a st-1 quantum performs in the macro system, its life will be longer or shorter: Wheat is harvested to feed annually the human ecosystem, while carbohydrates die every day to reproduce DNA molecules. Yet a top predator informative cell, a neuron, lives the entire life of the organism, since it defines the informative networks that dominate and form the will and consciousness of the organism as a whole macro-beings. And inversely, the life span of the 'brain quanta' defines the life span of the macro-organism. So probably the enormous life span of quarks, the fundamental particle and brain of the fractal atoms of the Universe, defines also the life-span of the Universe. Since the study in detail of all those chains for all Universal species is beyond the Cyclical Time of this work, we will only consider some examples from the 3 main disciplines of science:

- In the physical world we study the temporal chains between the 3±st scales of existence of light (gravitation, light and atoms) and the 3±st scales that create the Universe.

- In Biology we study chains between the 3±st scales that cause life evolution: the st-1 genetic cycles, the st-life cycles and the st+1 ecosystemic, geological cycles of the Earth.

- In History we study the rhythms of life of civilizations and the human generational cycle, which follow a decametric 80x10=800 cycle of destruction of civilizations.

Recap. Since the life of a fractal cell or quantum of a bigger system is a moment in the longer existence of its organism, to create the effect of simultaneity, microcosmic, fast, fractal cells have to chain their longer 'existential cycles' to the shorter cycles of a bigger organism or ecosystem in which they will exist as a mere fractal instant of those shorter cycles. Thus, the informative arrow is smaller and determines the energetic arrow, whose frequency is shorter and determines to the reproductive arrow, whose cycle is shorter and determines the eusocial arrow: I->e->r->S. As a Hindu proverb says: 'a blink on the eye of Vishnu, the Universe, is a life of a planet and a drop of sweat of a planet is a life of a human being.'

X. COMPLEX TIME SPEEDS: PALINGENESIS

53. 3 types of reproduction: Fractal space-time jumps.

A theme which might seem of science fiction to the reader, already surprised by the innovative outlook on reality provided by Time arrows and space discontinuities is the displacement not only in 'space size' but also in time evolution between the different planes of reality and the consequences for the workings of the Universe.

We affirmed earlier that all spacetime motion was in fact, in a discontinuous Universe, an act of reproduction. So a wave moving was a reproductive form, imprinting a lower scale of forces (in the case of light the gravitational membrane) with its information. And we distinguished 3 types of reproduction: external, enzymatic reproduction, from a 'relative past to future', as the parts were assembling into the whole by an external factor (enzymes in cells, human 'enzymen' in machines)'; a relative, present reproduction, as in the case of light, since the reproductive act was fast and displaced a wave-form in space, and a palingenetic reproduction from future to past, when a superorganism emitted a relatively less evolved seminal cells from the past that would then recreate the being.

The readers should be aware that when we classify self-similar processes almost always we come to the conclusion that there are 3 types of 'differentiations' or 'decoupling' or 'horizons' or 'ages' or 'subspecies', in as much as the Universe is structured in ternary patterns, according to the 3 dimensions of time, past, present and future and its 3 equivalent topologies of space, past=energy=spherical plane, present=reproduction=cyclical, toroid and future=informative hyperbolic topology.

In the case of reproductions, we used the dimensions of time to classify them. Yet the fact that reproductions are informative actions related to time more than space (except in reproductive, present motions as speed is), means that reproductive acts provoke a 'time jump', a 'time displacement', not only a spatial displacement.

Because reality is webbed by groups of actions that form strings that form particles that form atoms and so on, there is a dilation in time between the creation of lower, simpler planes, done in a relative past respect to the higher, more complex planes, which use the smaller particles of the past as energy points of its topological creation of new scales of spacetime. Yet in the same way we humans do not distinguish the age of galaxies we think are happening now when we see them millions of years ago, the time the light takes to arrive to our universe, when we see cells and atoms and other scales of reality we tend to believe they co-exist simultaneously and yet they don't. There is a temporal displacement according to which the 'future' is the place where the most complex levels of organisms exist and the past where the simplest planes are formed. So your consciousness as a whole is displaced in time compared to your simplest atomic particles.

Thus in time jumps to the past, the species that emits the information *does not emit the information in the same point of present in which it exists, but since information is discontinuous, the cellular bit of information appears in its immediate, lower spacetime-point, in its relative past, and so it* suffers a discontinuous displacement in time.

Indeed, imagine a Universe with 2 dimensions in a complex plane, you can go left or right, meaning that each fractal step you do you will move through a discontinuous 'jump' to the lateral space. This happens because as we said, when resolving the paradox of Zenon, motion does not exist in the Universe, because it is always a discontinuous reproduction of a form. So when a light moves, it doesn't move; it imprints the gravitational space next to it with a self-similar form.

Further on, we have explained that time/information is really a displacement along a 'dimension' of height and we represented temporal information with a complex plane.

So when we move to a lower scale of information, we are moving in 'the time clock' of the super-organism a cellular step down, which means within the whole organism we move to the past. We shall thus enunciate the law of fractal time jumps and analyze the 3 examples of the previous graph: a physical jump to the past (particle/antiparticle pair), a biological jump to the past (birth and orgasm), and a sociological jump to the past (the function of the prophet in a corrupted, dying social organism of history, a 'God', subconscious collective or civilization). And two other fascinating cases: the spooky effect of quantum entanglement and non-local gravitation in physics and the question of will and consciousness in human beings.

Let us then enunciate the law of fractal jumps in timespace:

'Fractal jumps are reproductive motions in time and space. Spatial jumps happen in spatial steps with a maximal length equivalent to the length of the organism, between contiguous zones of the same space plane. Temporal jumps happen as temporal displacements between 2 hierarchical contiguous planes - st and st-1 - of informative complexity with a maximal length equivalent to the generational life cycle of the st-1 cell of the organism.

Spatial displacement dominates motions in the same plane of existence (hence reproductive motions). The main time motions between planes of existence are jumps between planes which imply a change in the parameters of energy and information and a jump in the point of space-time in which we exist. Those laws, unknown to science, are the most important of Fractal i-logic and are tested by thousands of detailed analysis in each science. In essence when we move from one plane of existence to other planes we do not only displace in space but literally in time, because reality is constructed not only with a volume of space but a volume of time information.

Thus, the Universe only allows one 'discontinuous jump' to the next past, future, right or left zone. There are not worm holes to enter far away Universes (too far to happen), jumps to a past in which mother was born (as only you travel to your past through your death).

Because physicists use a single space-time continuum and confuse the arrow of time/motion/energy with the total time arrow of the Universe, some believe that a change in that time/motion arrow, which is merely a local change in the direction of a fractal entity of space-time, is an absolute change in the direction of all the arrows of all the beings of the Universe. This is the reason of some naïve ideas about time travel in physical systems that halt the motion of time (Black holes where the T parameter goes to zero, which prompted Mr. Hawking and Mr. Thorne to affirm they are time machines). Time travel doesn't exist because all arrows of time are local, belonging to a fractal system of the Universe, NOT to the entire Universe, and physical Time Arrows are changes in the motion of beings, not changes in the arrow from past to future.

Let us now consider first the most complex type of reproduction, from future to past or palingenesis and then, the reproduction from past to future, departing of parts (genes in biology, memes in sociology) that construct a future super-organism, as we studied already the reproductive waves of present-space (light waves and motion in space).

Recap. There are 3 types of reproduction: a past to future motion or 'genetic' reproduction; a present to present motion or reproductive speed and a future to past reproduction or palingenetic, parental reproduction. All of them are based in the symmetry between space and time, two types of geometrical motion and the fact that all motion in space-time provokes a displacement both in space distances and time dimensions.

54. Dimensions and reproductive motions.

Time cycles reproduce through fractal dimensions. Imagine a torus, as the one depicted in the graph: A cyclical string turns around, creating every 3 movements a self-identical closed macro-torus. If we repeat the process 3 times we obtain a 9 dimensional self-similar fractal super-string. Such is the way in which the Universe, departing from some basic forms - torus, spheres and lines - self-replicate and grow invariant at scale its energy and information systems.

Since according to the paradox of Galileo, what we perceive as still space is in perpetual motion, 2 questions must be solved:

What is moving and what is fixed space? And how movement can be introduced in the simplified, mathematical static models of time and space, proper of the Cartesian plane that scientists use to describe the Universe? Both questions are related to the concept of speed and acceleration as a 2nd and 3rd dimension of motion, and hence also as the 2nd and 3rd dimensions of fixed space. *This insight has deep implications for the meaning of Physics and the interpretation of the Equivalence Principle of Relativity in which Einstein based General Relativity, his theory of mass and gravitation. So we shall try to explain it in more detail.*

Since acceleration is mathematically defined as a new motion applied to speed, a constant motion, it follows that an acceleration/deceleration, $a=\pm v/t$, is both a dual motion and a bidimensional form. So we can either consider a static line which becomes a plane and then a cube, or a static form that moves with a speed and then accelerates. Thus, the dimension of acceleration/deceleration becomes the 3rd dimension of physical time and the more complex of all, as it includes the other 2 types of time-change in physics, speed and spatial form (in the same manner a cube includes lines and planes).

We can see a dimension of time movement as a spatial, fixed form, when we see it in simultaneity. So the 3 dimensions of space can be derived in time, by moving a point into a line, a line into a pane and a plane into a volume, and what we have termed as the 3 dimensions of physical time-change, in-form-ation, movement and acceleration, can also be derived from space by observing the curvature of space, and deriving that spatial form in time, $v=ds/dt$, and $a=dv/dt$. So time and space are not only inverse but complementary as we derive the 3 real dimensions of time from space and the 3 dimensions of space from time.

This can be done in two different ways, with the geometries of lineal space or the geometries of cyclical clocks of time:

-A point can become a lineal string, which can reproduce laterally and then the plane it forms can reproduce upwards, creating an Euclidean cubic space.

- A point can move as a curve to form a perimeter which can rotate to form a sphere.

The difference which has important implications for more complex analysis of physical space-times is obvious: the point needs two paths to create a hollow 3 dimensional sphere; while the lineal approach creates a dense, non-hollow cube in 3 steps. Informative spheres are thus bubbles which tend to shrink and implode, weaker than the more solid cubes that fill the entire space without leaving 'dark worlds'.

The minimal gravitational membrane of the Universe is a lineal membrane, filled totally by lineal reproductions (string theory), because they must be the ultimate web, filling all reality (or else what there is inside), but most formal creations over that membrane are particles with hollow dark spaces.

Motion as reproduction of form, the stop and go game.

All this comes to a fascinating fact that explains both the paradox of Zenon and the paradox of Michelson's measure of speed: motion in time is a reproduction of form in space. And so reality is webbed as particles with forms, go fractal step after fractal step, imprinting with higher form the previous, lesser informative, faster, v=s/information, membrane in which they are shaped. As in a film theatre, where you move the image, stop it, emit light to create a form in the screen, move on, stop, emit information, the Universe of motions is made of reproductive, fractal motions, steps that imprint form in the lower scale of energy. And so reality is always a stop and go, confirmed by the trajectories of particles which are always erratic, with stops an goes and 90% angles, when they emit a smaller particle to communicate. This also explains without need for a c-absolute postulate of speed why Michelson didn't add speed to a light emitting electron travelling in medium: the electron stops, locks itself in the gravitational membrane with the other electron (entanglement) and then in stillness emits its photon. So the speed of light of the event is independent of the speed of the background an emitter and receiver are locked before they share information. When you talk you also align your head in the direction of the speaker. Your head is your fractal point and the voice you emit is locked in distance. Or else it would suffer a redshift or echo effect.

Thus, time motion is not a mathematical, geometrical pure dimension of space (Relativity latter revis(it)ed in more detail), but the first dimension of physical space, of the real world that Euclidean geometry converts into a fixed abstraction.

In that sense, from the perspective of time physics, we could temporalize space and then consider that the first dimension of time is in-form-ation, morphological shape, achieved by the evolving height growth of any form, from masses that evolve in height till becoming black holes to biological life that evolved in height from the planarian to the human being. The second dimension of physical time will be reproductive motion of form of those 3 space-time dimensions that create a 'fractal dimension' that extends till the end of that movement. And finally the 3rd dimension of physical time-change will be acceleration, change in the rate of speed. If we add the inner bidimensionality of the mass vortex or bit of information reproduced with those motions, we have 5 fractal dimensions.

No more dimensions are needed, as it can be proved that the maximal spatial volume and hence the maximal efficiency in packing a space-time field is achieved with the folding of 5 dimensions, which might be the dimensions of the most informative, temporal elements of the Universe, black holes whose equations transform spatial dimensions of length into time dimensions of infinite height.

It has in fact been proved by Maldacena in his famous conjecture that the 4-dimensional membrane of a 5 dimensional Universe is self-similar. Thus, we could consider that a local big-bang quasar or a cosmic big-bang would be the external membrane, which unfolds of a 5 dimensional Universe. This is rather accurate. Since we observe a planar Universe (from our 3 dimensional perspective) which could be the energetic, Riemannian topological membrane of any st-point.

The error resides in considering all those 5 dimensions spatial dimensions instead of temporal dimensions of movement and acceleration, a subtle but fundamental change that explains a vital reality, 'as it is' beyond simplistic, geometric abstractions.

Reproduction of light: motion as form.

We hinted before to the fact that the simplest space-distance or amplitude of a wave and the simplest information or frequency of a wave, combine creating the simplest constant of action or speed, the speed of light or electromagnetic space-time, which reproduces the form of the wave across a surface of gravitational space. So the abstract spatial term, 'speed' is for light waves an organic arrow that reproduces the in-form-ation of the wave in the previous energy of space, in the case of light imprinting the simplest gravitational space-time. Light speed becomes then the 3rd arrow of reproduction of a dimension that combines energy and information, recreating a new form. We can find many other examples. Since the combination of the arrows of fractal information and spatial energy define in the

246

complex Universe the 2 new dimensions of existence, reproduction of a parallel form the same plane of space-time and social evolution of self-similar forms across multiple planes of space-time or palingenesis:

3rd Dimension: Spatial Reproduction: ext=k.

4th ±st Dimension: scalar, palingenetic, social evolution

It is important to remember that those 2 complex arrows exist with lesser frequency than the simplest ones, energy and information beats and have self-similar rhythms. So if the simplex beat of reality is the stop and go of informative perception and energetic motion: I->E->I->E, the complex rhythm is also a beat between spatial reproduction (a radiation in biological beings, a wave in physical species) and informative evolution (a collapse of a wave into a particle):

Social, Formative evolution->Spatial Reproduction-> Evolution

And it is easy to see that social evolution is the complex arrow of time and information, as it requires a lot of time and a common language of information to create complex social networks; and reproduction is the complex arrow associated to energetic processes, which must be observed to be able to reproduce. So for example, a female without a percentage of fat in the body cannot reproduce and when a species radiates in a massive age of reproduction, it does so, often extinguishing a previous species in which it feeds on.

They are also dimensional arrows, that the dual, geometric and organic nature of the Universe. Indeed, if height is the natural dimension that orders the information of the Universe and length the natural dimension in which spatial energy extends and moves, the combination of those 2 arrows of energy and information produces reproductive waves that create the 3rd dimension of width.

Reproduction of waves and forms that create new dimensions becomes then the biological why of the geometrical hows we observe around us.

Yet further on, there is a 4th dimension of growth into macrocosms that structures cellular reproductive waves through an increase of informative frequency. So the collapse of a physical wave when it grows in frequency/form, make it transcend into electrons and other particles of higher information. And inversely, a light wave that relaxes and diminishes its information and frequency (redshift) ends up tired, descending to the lower plane of gravitational space-time.

It is the combination of those 2 complex arrows what allows palingenesis - complex reproduction. Thus a palingenetic cycle is the essential reproductive cycle of most forms of existence, in which the interaction of both complex arrows/dimensions creates st±1 waves of microforms that become macroforms, by the reorganization and growth of self-similar reproduced sets. We might say that palingenesis is a temporal reproduction in as much as it moves little in space but evolves a lot of temporal information - while speed or translational movement is the spatial reproduction of a simple physical form across space. Thus the reproduction of dimensional forms creates reality.

The simplest of all those dimensional, reproductive waves is a c-speed, reproductive wave of light, which Maxwell proved to be the product of the magnetic/energetic and electric/informative constants of the light wave ($1/m_0 \times e_0 = c$).

Galileo's Paradox makes us perceive such moving, dynamic, reproductive waves as continuous dimensions of space. Both perceptions describe the same dual reality: a space-time field in perpetual transformation. The most striking case of one of such space-times made of organic, dimensional networks is the electromagnetic space-time, which forms the background of human existence, our space-time perception. Since human space-time is not an ideal, Cartesian background, but it is created by the fractal sum of all the 3 perpendicular dimensions/networks of light-spaces (the sum of all light beams of the Universe). The 3 dimensions of each of those light fields are perpendicular as our Euclidean space-time is, and correspond from the perspective of light to the 3 organic cycles of a photonic wave: the c-speed or length of a light field reproduces its form, the magnetic width acts as the support energy of the light wave and the electric height creates the information of light, where the photon or informative particle/head of the energy wave/body of light is found. When such light waves reproduce, they create a wave of present space. Since the light wave transits in an eternal present, or in other words it doesn't change its content of temporal form, its information is preserved. For that reason spatial, light waves transit billions of years between galaxies carrying accurate information on the far away particles that produced them.

Thus movement is no longer movement but the reproduction of a form over a surface of energy. Light doesn't move but its waves imprint the form of gravitational space with the wave of light. An electron doesn't move but it shapes the photons of its electronic space in certain wave patterns, called the electronic nebulae. Masses are not fixed forms but vortices that bend lineal gravitational space once and again into self-similar fractal, temporal cycles. In those terms, a dimension of space is the reproduction of a volume of discrete, fractal, self-similar in-form-ation that creates space.

While inversely, temporal palingenesis reproduces and evolves information.

The Universe is not only in perpetual movement, transforming back and forth lineal energy into cyclical form, but also a perpetual self-generating organism that constantly reproduces logical forms in palingenetic cycles, between 2 st±1 scales of relative space-times, where a mother will transfer its form into a micro-mother cell that through a process of palingenetic reproduction will create a self-similar replica of the mother.

Speed and palingenetic reproduction of a form into a fractal space surface explains from quantum jumps to palingenetic reproduction, from mystique experiences to the nature of dark matter, from Relativity to logic paradoxes. For example, the oldest paradox regarding movement is now solved: how can exist in a fractal world filled with discontinuities continuous movement? Zenon, a Greek philosopher put the example of Achilles and the Turtle: Achilles will never reach a turtle because he has to cross infinite fractal spaces. Parmenides gave him the only logic, true answer: continuous movement doesn't exist in a discreet Universe; it must be a mirage of the senses. But we observe movement, so how it can be possible? Because movement is an action of palingenetic or wave reproduction. Since a wave, as the figure shows, is not really moving, but reproducing information over the potential energy of the next vacuum space in a sequential, fractal jump, through a series of wave-lengths, drawn one after another, which appear to the senses as a moving wave. Movement is always a discontinuous displacement in space and time, through the reproduction of fractal information. The simplest analogy is a television screen where new images are created constantly without hardly any cost of energy, because they are virtual images formed by illuminating 3 colors that are already potentially in the screen. In the case of a wave, the relative energy imprinted by the logic form might vary, from the vacuum energy of a light wave to the placental energy of a mother, yet the process is always the same: a form imprints another region of space, changing its morphology. What matters in the Universe is the dominant arrow of information, the logic, intelligent form, which imprints constantly amorphous surfaces of energy. The Universe conserves forms over eternal moving energy. Even the most complex, fractal, living organisms, made of ∞ cycles that transform energy into information back and forth, maintain the appearance of a constant form, because they constantly repeat those biological cycles over new energy elements. For example, a human changes all his/her atoms in 3 months but s/he maintains i=ts biologic form invariant, thanks to the constant absorption of new energies over which s/he reproduces his/her fractal information. Bidimensional cycles of time and planes of space are fractal and so are their changes and displacements. Thus since time is geometric, bidimensional, as space is, it can imprint form in any lateral surface of energy, giving the false impression of movement. So speed is synonymous of reproduction, the main cyclical action of the Universe, where now the ideal Galilean Formula, V=S/T, for a fixed space-time, applies to the real variables of Spatial energy and Temporal Information: $V=\Sigma E/Ti$.

Self-reproductive waves create what mathematicians call fractal, finite, self-similar dimensions, caused by a wave that repeats its patterns of information in diminishing or growing scales. Which is the key process to understand why energy and information are invariant shapes at scale, repeated ad infinitum from atomic to galactic cosmos: fractal waves translate those morphologies, as they change, grow or diminish in spatial size/speed and Ti-frequency, connecting different space-time fields. Those waves of energy and information cross the discontinuities between different space-time fields. So when we look at smaller Universes from our perspective we use an electronic or light flow, whose frequency increases as we enter those smaller, more informative Universes.

In that sense, following the Correspondence Principle, the new concept of 'informative speed', $V=\Sigma E/Ti$, includes the simpler model of spatial speed. Yet it goes further, explaining not only the how but also the why of many laws of science, relating the energy and information of waves with its relative speed, since according to the inverse properties of energy and information, the less information a wave has to build up and reproduce, the faster the wave will move. This is called in Physics the Law of Range of Forces (max.Ti=Min.ΣE) and it has its limit in the non-local, instantaneous speed of gravitation, which carries zero information and has infinite speed: v (g)=$\Sigma E/0$ information=∞. While in the other extreme it creates fixed frames of gravitational reference, called black holes, which carry infinite information and have zero translational speed, as entire galaxies turn around them. It was again Aristotle, who said that if God exists, it would be the final unmoved focus that in-forms the movement of all other beings in the Universe. Indeed, the black hole seems the God/Brain of in-form-ation of all galaxies, whose stars turn and fall into it. The Universe itself might be born of the big-bang explosion of a black hole, origin of the process of Universal life that will end in the reverse 'Big crunch', of informative death, creating the ultimate cycle of existence of reality...

Recap. The dimensions of space are reproductive arrows of time that imprint a form in a simpler energetic membrane. Worlds can have a maximal of 5 dimensions: 2 bidimensional forms which move in 3 fractal dimensions of a higher Universe.

55. Does st-1 determines st? Genetics & memetics.

The simplest act of reproduction and creation is the assembly of parts that become wholes, and its study gives birth to an entire sub-discipline of systems sciences called *Emergence.*

The structure of the Universe is that of a series of complex worlds/organisms of at least two membranes, one of relative past, energy, st-1, and one of relative future information, st+1, with an intermediate present st-membrane, exi, made of actions (equivalent to the topological description of relative past, external membrane, r, a relative hyperbolic center, a, and an intermediate cyclical space with motions between the a-r limits, which form together a 'closed ball'.)

In those systems an external perceiver tends to observe mainly the intermediate space with maximal space volume and misses the hyperbolic informative center of minimal space, *even if it is the source of informative orders, through invaginated networks that control both the Riemannian, spherical surface and the paths of the cyclical cells in between.* Yet the flows of energy move from the 'lineal quanta' of the surface through the paths of the internal body-organs into the hyperbolic, informative structure. And so human scientists that need instrumental 'pictures' to prove the existence of a causal order and only use Aristotelian single causes tend to believe in any ternary structure that the 'lower energy plane' is the case of the creation of the superorganism.

This means we consider in the ternary structure of a human, (cellular, individual and social structure) that st-1 genes to be the cause of all our properties as a super-organism, and we are just a 'ghost' created by them. But this is a partial vision due to the limits of perception of informative orders from the hyperbolic, fractal informative system – the nervous system - to the lower, genetic cell system. If, in the future we get to fully understand the networks of information of the brain, we shall certainly discover that many orders and structures of the human superorganism are stored in the brain and probably reproduced during the last months of gestation from the mother's brain to the brain of the child. Moreover, there are an infinite number of personal traits which are NOT reproduced through Genes, but through the carriers of information of st+1, cultural organisms, called memes. Indeed, most of the features of our life – our religious beliefs, cultural customs, laws and instruments are memes, which create another super-organism, the civilization or religion, in which we are a mind-cell, a believer, who shares with all other believers the same 'mental DNA.

So in this case, if we were to accept the thesis that creation happens from the lower level to the upper level, we should consider that we, human beings, are the only influence on the memetic structure of civilizations. But it is obvious that once those structures are created they influence new generations. So for example, when a system of laws or beliefs in a legal nation or ethical religion is created, it reproduces from generation to generation, influencing and molding the new species.

We conclude that superstructures & physiological networks, the informative/nervous/cultural system in organisms and cultures and the reproductive/blood/economic system, once constructed, influences the lower genetic/human scale. In the genetic case we do not know how. In the social structure, we know and since the work of the socialist, scientific school of history and the recent development of memetics and biohistory, we have worked out the processes – even if most historians prefer to think that we, individuals are the heroes and only cause of historic events.

But what truly matters is that both genetics and memetics are bytes of information whose purpose is to create a superorganism, as the biological and sociological superorganism of mankind have the same physiological structure with nervous/informative /cultural systems; blood/ reproductive/economic systems and digestive/energetic/ military and transport systems.

So the true program that genes and memes and any 'informative bytes and parts of any creative process' follow is the arrow of eusocial evolution, which explains why so smallish 'genes' and 'memes' are able to reproduce so complex super-organisms, as most of the development they will express is 'structural' to reality, embedded in the existence of a 4th arrow of time, eusocial evolution; which in itself is the natural conclusion of a game of 2 simplex energy bites and information bytes, which reproduce in 3 types, more energy bytes, more information bytes and reproductive combinations of both, which latter will recreate energetic, informative and reproductive systems. So genes and memes do not need to encode that program/information that happens automatically. Moreover because each bit and bite has embedded its program within itself as part of the space-time reality, it will naturally search for energy and information carrying the program. For example, humans naturally want more energy for their body and more information for their mind, so naturally they try to create machines of energy (transport, weapons) and information (chips, books, audiovisual machines, mobiles) that enhance their capacity to process and store energy and form.

It is thus logic to think that in the genetic scale, RNA molecules which are the 'active' elements of the cells, carry a natural will for more energy, information and reproduction, which they 'express' creating cells and then superorganisms, as we humans do through memes, even if we are not conscious that the final outcome of our labors is the creation of nations, religions and civilizations, superorganisms of history.

This is indeed evident when we study the behavior of RNAs, which do the cycles between the outer Protein, lineal membrane and the DNA-informative storage, and are divided in 3 sub-types specialized in energy/killing, reproduction/RNA messenger and information/transcription.

In essence all this means there is not only a 'biological arrow' of creation of the upper scales of reality as a construct of the lower scales of genes and memes (the dominant theory in biology, due to the excessive development of genetics); but also a present flow of orders, embedded in the program of the 6 arrows of time, masterminded by the 'action units' of the system (RNA molecules and human beings) in both scales, which is probably the dominant 'will' of the program. And also a stored, informative, repetitive flow of orders and information from the upper scales of reality to the lower scales, which geneticists have not yet understood, due to its complexity (nervous structures in organisms and 'the system' or 'superstructure' of cultures and economic ecosystems in societies). Indeed, the codes of the brain that might determine certain characteristics we cannot obtain from genes is more difficult to recognize, as information tends to have lesser size and it is coded, so an informative order is often invisible to those who don't understand the language in which it is written.

What parameters define different creative processes? The specific forms of the informative stores (DNAs, nervous networks constructed by imitation of the mother's brain by the son's brain and cultural memes). They differ in each superorganism, which will activate and create a different type of transfer-RNA, animal nervous system and individual memetic brain, reproducing a different species and culture.

Further on, a theme which must be taken into account is the fact that all those systems seem to us 'simultaneous', despite the fact that they are displaced in time, since to recreate the upper scales, the lower scales must be created first, so the superorganism is displaced into the future. And for that reason for the 3 scales to create a 'simultaneous present', synchronic being there must be indeed a time jump or flow from future to past of the superorganism or relative future structure and a flow from past to future of the cellular energy, which will flux into the relative present, so the 3 'scales' are in harmony and co-exist creating reality: Past x future = present.

In a world of multiple time-speeds and planes of existence we live in an ecosystem of time knots whose cycles of time/ information differ in parameters of energy and form and are displaced from past/energy/simple cycles into future /information faster cycles. Yet the faster cycles are built upon the previous structures and so they are made 'after' the simpler cycles are created, in the 'future' There is therefore a real displacement in absolute time between what we see as bigger structures and what see as a smaller, since the bigger structures are traced departing from smaller cycles. This explains a mystery of quantum physics – the fact that we cannot see the wave/past/energy state and the future/particle state at the same time. Neither we see the atomic/past state and future /molecular pi orbital at the same time. *And yet we know both co-exist.*

Recap. In all processes of creation, the 3 scales of the super-organism, the cellular, individual and social scale input the program of creation based in its 4 arrows.

56. The life-Death cycle as a dual time reversal.

If birth is a dual fractal jump, it follows that death, its reciprocal must be also a fractal jump between planes of existence and indeed it is.

Yet the most obvious proof of fractal jumps to the past are the processes of death, which people who suffered it and then have survived, explain as a rewinding of the memories of the being that travels through the entire information of his life towards the past. The difference between a palingenetic jump and a death jump is that palingenesis is a backwards- forwards dual jump and death is a dual jump backwards at maximal speed. We die twice, first becoming a cellular being, and then due to the helplessness of our broken cells, without physiological networks we die again, becoming molecular food for insects. In physical systems, death is a big-bang that again implies a dual simplification till reaching the vacuum plane of pure energy.

This can be compared to a boxer that fights in a category with 2 jumps of mass/information/weight. The ecosystem is one in which the speed of time cycles and spatial size of beings is much bigger and so the form becomes annihilated. The process of death makes you a group of unconnected cells, weakened by your 'submission' to the bigger network systems of the super-organism, tamed by them; so we die again when confronted the micro-organisms of the insect and bacterial world.

Recap. Death is the fastest displacement in time in which a particle jumps two layers of space-time in a fast track.

57. Palingenesis between fractal planes: reproduction.

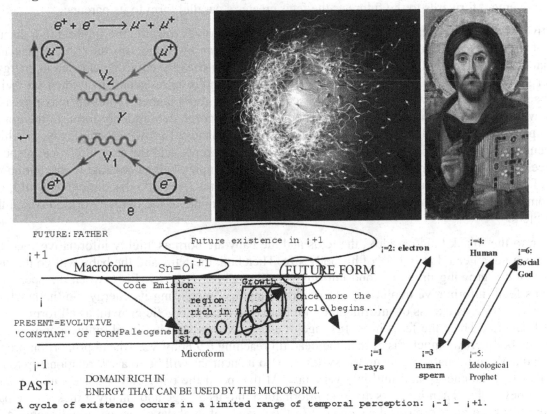

3 dual Fractal jumps in 3 scales of existence, written with the old formalism, in which a 'seed' of relative past information emitted by a future parental form hosts in an energy surface, evolving towards the future:

A photon reproduces a particle-antiparticle. A spermatozoid starts the evolution of a child. And a prophet emits a verbal code that will recreate in the future a dying civilization.

Alternately when a form with maximal information, or seed cell from a higher plane of existence falls in a lower one, it becomes a top predator and starts a much faster process of evolution, predation and creation that culminates in the creation of an offspring, or self-replica of the more complex plane. This is the essence of the process of reproduction and it involves a real jump towards the 'past' of the upper organism that deposits a seed of time/information that evolves faster into the future. The best described process is biological palingenesis but there are other cases:

A fractal jump in space-time, which is dominant in information is a *palingenetic reproductive action*, the repetition of a form, through the parental reproduction of a simpler seed or relative past form that will evolve and reproduce a new space-time, a future son form.

Palingenesis means truly that a fetus is a past form that evolves, recapitulating a series of evolutionary forms from past to future, as fractal time can both travel to the past becoming a simpler form (as in death) or accelerate its informative evolution (as in a process of creative design or fetal evolution). Yet, since time is discontinuous those movements in time only affect a fractal part of the Universe; and so

252

they are possible without creating paradoxes. The universe doesn't move to the past, only the seminal seed. So palingenesis is a dual movement in time:

- Backwards towards the remote past as the parents produce a single cell, packed with dense genetic DNA information. Followed by:

- A fast movement forwards to the future as the fetus creates a parallel form to its parents:

A relative past species (simple, cellular palingenetic entity) evolves towards the future into a macro-system that organizes itself over a surface of relative cellular energy, thanks to the 'future' placental parents that accelerate its evolution into a new 'present' being in a friendly environment rich in energy - the mother's womb. Thus, the final balance between past and future is re-established when the being surfaces back into the present in the moment of conception. In this manner evolution maximizes its efficiency with the min. expenditure of energy: The fetus is a temporal, evolutionary form. In the graph, we draw a palingenetic reproduction as a fluctuation between 3 planes of growing complexity, in which a 'life' occurs: the st+1 social plane, the individual plane of the parents, and the st+1 biological plane of cells - since in i-logic geometry simple-past beings and future-complex beings co-exist in discontinuous st-planes of existence of different complexity. And indeed, the fetus lives in a discontinuous world, the mother womb, which finally opens up through an aperture, the vagina, to the bigger world in which the newborn enters.

According to the 'black hole paradox' the fetus and the baby are born as 'highly informative species': The baby has a huge head that is $1/4^{th}$ of his body size. Then both the fetus and the baby will go through a dual fluctuation, growing in energy and information till reaching maturity. First, when the species is born after its fetal, informative evolution, it starts its young age, increasing its energy. So the newborn will soon compensate its excess of information with its accelerated, energetic growth, as all forms try to find an E=I balance. Thus, the fetus in its first age of energy will expand its lineal limbs, elongating its body. Now its body grows much faster than the head till reaching a 9 body Vs. 1 head proportion proper of many spatial Vs. informative, decametric systems. Also its neurons will be in a 1-9 relationship to the glia cells that feed its energy and multiply very fast. At that point the baby enters the mature age and reproduces a new baby. Then it grows again in information into its 3rd age and finally it dies. In the past decade there has been an enormous development of the science of evol-devol, based in palingenesis, in the lines foreseen in our first books. Biological palingenesis is the only accepted by science. But palingenesis happens also in the sociological and physical world.

Recap. An organism is reproduced through a palingenetic jump between macrocosmic and microcosmic planes, as the parental organism produces a cell of information that will evolve at full speed towards the future.

58. The 3 main types of devolution in time.

According to the 2 arrows of past energy and future information, all space-time movements are dual cycles. So in all of them, there is a degree of genetic reproduction and travel to the past through a basic transformation of an informative form into a simpler language or cell that is able to cross the discontinuities of the Universe and then evolve forwards into the original form. It should not surprise us, as all cycles require the dual arrows of past-energy and informative order; and all cycles are memorial repetitions=reproductions of a cycle that happened before in the past.

Yet in order to establish limits to the 'scientific imagination' of Time Arrow theorists, similar to the 'runaway' dreams of physicists like Hawking, who uses the out-dated model of a single time arrow/clock to predict the existence of time machines we must fully grasp what a time jump is in morphological, biological terms, *since it is only the arrow of information the one that travels to the past=devolves; not the entire Universe - but only within the restricted system of the superorganism.* This means among other things that there is no really time travel beyond our organic-world, both in processes of palingenetic birth and in death a motion only to the past. That is, we cannot go to the past and 'kill our father' as Hawking affirms in his book 'The Universe in a nutshell'.

Those facts are systematized by a simple Time Arrow Law that uses the ternary principle (as most Multiple Spaces-Times laws, events and processes of reality can be divided in 3 sub-cases): - An information system devolves into the past in a process of death, releasing its energy in a big-bang: $I < E$, as when we die and break into cells, or a particle dies, explodes, mutates into its antiparticle (death phase with inverted coordinates) and virtually disappears (right side of the drawing).

-A head or informative system emits a flow of information to the relative past system of its body controlling it.

- A palingenetic process happens in time, by a parental system that emits a palingenetic system, which evolves back towards the future at maximal speed.

Let us now consider a wealth of examples of palingenetic reproduction in more detail.

Recap. The 3 main types of time travel to the past within an organism are death, informative control of its lower body cells and palingenetic reproduction.

59. Physical examples. Non-Local Gravitation.

The i-logic Universe is not magic. Yet fractal jumps create events that seem magic. Let us consider one example from the physical world the seemingly instantaneous speed of gravitation: As we keep going deeper in the model of Multiple Spaces-Times and discontinuous time jumps between hierarchical st±1 planes of exi=stence, we shall see some not-easily recognized consequences of the structure of time dilation that implies displacement in time, not only in space, in each fractal step we make between different membranes=mediums of different energy and information, different momentum.

When we move in the same plane we are just displacing in a relative present, but in the ecosystem of light in which we exist, we live surrounded by structures belonging to other species of lesser or more evolution and within planes of completely different spacetime arrows.

The most obvious cases of this happen in physics, since even though we inhabit the electroweak membrane of light and electrons, we are constantly under the effects of other membrane of <0 K Temperature (an informative parameter of cyclical order) and > C light-speed a parameter of energy, called the gravitational or dark/black membrane, inhabited by forces such as implosive gravitation and its inverse, explosive dark energy, and mass vortices called quarks and black holes.

This faster-than-light membrane is also displaced respect to us in time, in such a manner that its simpler, faster gravitation is a flow that happens in a relative quantum jump respect to us into the past of minimal form and faster energy and the more complex, T<0 world of quarks and black holes happens in a relative future respect to us.

How it is possible super luminal gravitation if Einstein's Postulate sets light speed as the limit of speed of our Universe? The answer is evident. If the Universe were a continuous, single space-time, there would be a limit of speed; but in a fractal Universe with different space-times communicated through 'Einstein-Rosen bridges' - a particular case of the fluxes of energy and information between space-time scales - c-speed is the 'substance' and limit of speed of light-space. It is not though the limit of gravitational space. In the case of the light and gravitational world those discontinuities are the masses and black hole singularities that appear to us as 'points with no volume' in the border between both worlds.

The Universe is a system with 2 membranes, energetic light-space (that we perceive) and informative, gravitational space, which we do not perceive but seems to 'act' in present, as a non-local force. Yet according to the Galilean paradox we always can find 2 explanations to all phenomena, one in terms of spatial parameters (speed, energy, space) and one in terms of temporal dimensions. So we can explain also the instantaneous speed from the perspective of the future particle that emits it in the relative future:

The future quark or black hole of the informative gravitational membrane emits its gravitational flows in the future, *but they move as information, $E \Leftrightarrow I$, in the inverse time direction that electromagnetic waves, making a step backwards in time, equal to the frequency of its wave, according to the previous Law of fractal space-time jumps, and a step in space equal* to the amplitude of that wave, which can reach the limits of the Universe. So when they arrive to us, as they had the advantage of an 'extra-time' – the step to the past – gravitational waves seem to arrive instantaneously with an infinite speed, surfacing in the present even if they have travelled a timespace distance from 'future to present'.

Thus, the gravitational space-time is a past-future ecosystem of energy-information fields, which is more extended in spacetime; hence also in past and future coordinates. And this further clarifies the entanglement effect and why we perceive always light at c-speed.

Instantaneous gravitation is something that has always puzzled scientists. And yet, fractal jumps in time explain that instantaneous speed easily: A mass emits gravitation in a relative future. Hence, when the gravitational wave travels forwards towards another mass it arrives in the relative present of the observer, with an instantaneous speed. At this point the system becomes locked, entangled by gravitational information and then it 'constructs' the upper plane of light-space, which is the one we observe, moving at c-speed.

In quantum physics, those entanglements that Einstein called spooky effects and the fractal jumps to the past that cause them were proved in the past decade. Since rays of light, travelling faster than c, come out in the instrument before they are produced.

Entanglement happens because the 1st event in the creation of a complex form builds the lower plane of gravitational, simple, past flows. So gravitational forces entangle and lock the electrons before they emit light. They have already happened when electromagnetic flows are created over them. This also explains the Michelson experiment without need to postulate and absolute c-speed: because the observer is an electronic microscope or human eye and the emitter is an electron they look like a movie that locks each frame to perceive still information. And so when the electron stops, emits light and goes, it has no relative motion to the perceiver electron, as it is locked with it in a gravitational entanglement.

The same laws work when we consider phenomena that related quarks and electrons. The quarks and black holes that create those gravitational forces are in a relative future to our electronic world as the hyperbolic, informative center of the atom, but also are in their relative past, by sending informative gravitational forces that 'locate' and position the electrons.

We can generalize this singular concept with broad applications in the resolution of the mysteries of quantum and gravitational space to any relative system of 2 membranes of space-time, as we did with the nervous system which seems simultaneous, when we realize all those cases are merely an application of the law of fractal jumps: the frequency of a single cycle of the faster informative system (gravitational and nervous system) is equal in time speed to the frequency of the cellular system (light frequency or chemical messages in the cellular body) but its distance can reach the maximal distance of the entire superorganism (the limits of the Universe in the case of a gravitational wave and the limits of the body in the case of a nervous message).

What all this means in organic and topologic terms is simple: the intermediate space between the hyperbolic, informative center and energetic force, membrane of a superorganism locks the cells of the body that work for the 'upper class' of the organism, the ExI system of the bigger membrane. Indeed, physicists now realize that gravitational, dark energy and dark matter dominates our Universe and positions galaxies and starts and controls them. And they are observing external, energetic membranes in all cosmological systems from the Oort comet membrane of the sun to the halo of dark matter of galaxies and the wall of fire of the Universe.

Recap. The flows of energy and information of its informative network reach simultaneously all the cells of the superorganism.

60. Biologic examples. The beat of evolution>reproduction.

Palingenesis and punctuated evolution: time accelerates.

Palingenesis means that reality webs itself in each volume of space-time, in a simultaneous manner, even if from a 3rd p.o.v. the entire structure might seem to exist in the same present.

Thus there is a tail of energy and information planes, which influence the present dominant space and might seem to co-exist simultaneously but they are displaced into the relative future of the super-organism and its relative past, completing the pyramidal structure: bigger structures are displaced in the future and when they communicate to the past literally jump in time. The effects are diverse, being the process of reproduction and palingenesis the most well-known:

Because time speeds varies in palingenesis the process of evolution, the 4th arrow of time, accelerates. Since it is guided by a memorial process of slimming that has streamed to its minimal fractal, informative steps the reproductive sequence.

Yet the process can be studied in more detail, from the perspective of the Generator equation of space-time; since one of its feedback sub equations:

$$Energy\ absorption\text{->}Reproduction \Leftrightarrow Information.$$

It is both the rhythm of palingenesis & punctuated evolution (the evolution of a whole species taken as an organism).

Existence is modular, as beings constantly switch between its arrows or will of existence. So happens to species as superorganisms. Species need energy to reproduce and their inner information evolves them (though it is likely that external adaptation is first imprinted in the neural, nervous networks, which by faster mechanisms than those of evolution, influence the shape of its E_{st-1} body organs and even in lesser measure on the long term the genes of the cells). Yet the main arrow of evolution is social organization of individuals in herds and on the long term superorganisms. Species need to evolve socially in order to survive, which is achieved by creating social networks and herds that have a higher perception and wider force to access energetic resources. To that aim species evolve a common language of information and their brain capacity, often in the stillness of a secluded place (allopatric evolution), as information is still and evolves in a relative chrysalis state. The result of that simple scheme defines a basic evolving rhythm, called punctuated evolution, according to which species switch between reproductive, expansive periods as top predators (parallel to their extinction of simpler species as energy of their reproduction) and temporal, evolving periods in allopatric and isolated states – often synchronic to hot/ reproductive ages of Gaia and cold/informative Glaciations:

Max. Energy feeding -> Max. Reproduction (expansion in space and extinction of rival species as energy of the top predator)->Max. informative punctuated evolution

Such Energy->Reproduction->Information Horizons of species are equivalent to the 3 ages of life (youth, maturity and informative, 3rd age). It is the palingenetic rhythm of species. All of them can be resumed in an E->Re->Ti sequence, *which again is a partial case of the Existential equation, $E \Leftrightarrow Ti$.*

Recap. The 3 horizons of evolution of species correspond to its 3 evolutionary ages and respond to a simple rhythm of informative evolution -> top predator reproduction ->extinction of energy species ->informative evolution.

61. The question of consciousness and free will.

One interesting debate brought about by the existence of time jumps between planes is the phenomenon of human consciousness and free will.

Asymmetric, temporal, fractal flows of information from st to st-1 planes of existence explain how the informative networks of a higher social class like the brain neurons or the politicians and bankers that invent laws and money, the social languages of power, control the working cells of the lower class.

In the brain case, neurons control electronically the cells of the body, anticipating its reactions, since those body cells use a slower, past chemical language and exist in a relative past, in an 'inferior plane of existence'. So, in a manner similar to that of the gravitational quarks that lock the paths of future light waves, the nervous messages anticipate and control simultaneously the motions of the cells.

Moreover, if as in the gravitational case, the electronic messages of the nervous system were moving in an inverse direction from future to past or were emitted before the actions of the cells occur anticipating them, they would mastermind the motions of cells. This seems to be the case also in societies in which 'intelligent rulers' anticipate the needs or protests of their organic cells, repressing citizens before they rebel (military regimes), producing laws or inventing money a priori in markets, before the cells will earn it. Thus, in those systems the logical anticipation of the actions of cells but the more complex informative networks (nervous systems, upper classes) produce a self-similar control to the one quarks and black holes exercise on the position of electrons and stars in the galaxy.

We have now proofs of a temporal dilation in the actions that happen between 2 hierarchical planes of existence: Recently, scientists found surprisingly that the body acts before the brain thinks and yet most people think first before acting. How then it is possible that we think first and yet we act also first, or do both things happen simultaneously? Precisely because as Einstein put it, past and future co-exist in a simultaneous present, as information flows from the future brain to the past body and energy flows, moves and acts in the body from the past into the future:

Energy/Body (past) x Information/Mind (future).

So depending on our point of view, the brain or the body, we will see our action first (body point of view), or we will perceive first our thoughts (mind point of view). Yet in reality from an objective, external point of view both actions converge into a present; because the nervous message arrives simultaneously, produced subconsciously 'before we reflect with words on what we ordered and before the cells move'.

This subconscious, first message that then branches into a mental thought and a body action that converge in present is coded by the natural wills or drives of existence – our desire for energy, information, sexual reproduction and social love. And it would imply that words are not free, that there is a biological determinism – the program of existence – which is the true engine of our actions. And words just comment a posteriori even if our arrow of time departs from the consciousness of words and seem to us to direct the process.

So to the external world both mind and body act chained 'together' by the simultaneity of the fractal actions of the body cells and the mind's thoughts that become a single whole action. *But the original will would be the biological drives of existence whose site is the limbic, emotional brain.*

So fractal jumps in time explain the hierarchical structure and creation of symbiotic organisms, in which relative past and future forms co-exist. In humans they would explain the control of the mind and the subconscious actions of the body.

Recap. The limbic system and the program of existence with its emotions (greed, violence, hunger, sexual desires, and informative fears) would act first and then the body acts and the brain thinks and both together create our present of existence, in the future.

62. Sociological palingenesis. The prophet.

An open question for those who believe in mysticism is: if the fetus is born into a higher plane of existence, do we have also for man a second birth after death into an even higher plane of existence? The previous graph explains that possibility; as it compares the palingenesis of a fetus with similar processes, which happen to other species. For example, 2 electrons emit 2 seminal light rays that merge together creating a new electron as the 2 lineal waves of electromagnetism evolve towards the future acquiring finally a cyclical electronic form. Yet since the informative code of human beings is verbal, we consider the prophet's mind, the original human, memetic, verbal, seminal, ethic code of a religious civilization that starts a palingenetic radiation in the mind of his believers, *till it creates a religion – a superorganism/informative network of human beings, whose ethic DNA is the same, creating an effect of simultaneity*. Indeed, civilizations are social organisms made with human verbal cells, joined by a verbal/nervous network of ethic laws, produced by a prophet that imprints the Book of Revelation, the memetic code of the civilization in those minds. So the prophet is the seminal cell that reproduces and finally shapes a new organic civilization; *where the reproductive body is the geography of the nation with similar laws or civilization* with similar ethic codes. And the fact that most prophets are born when the 'father' is going to die, in the baroque age of a culture, prior to its extinction, explains quite clearly how certain human minds transcend death and become immortal as the 'informative code' of the future civilization they create. But as only a sperm cell reaches the future ovule, only a prophet among millions of humans creates a macro-organism of history, a civilization. As Tertulianus said: *Sangris martire semen cristianorum'*.

Recap. A prophet's mind is the seminal, verbal code of a culture of parallel minds that control a territorial body, in the surface of Gaia.

63. Dark planes: perception limits & symmetry breaking.

The existence of dark spaces for every perceiver of reality of multiple planes of existence also widens the laws of conservation of energy and information, as energy might disappear from a plane of existence but the total energy x information quantity of the ternary, 3-plane structure of a World remains the same. This in physics explains a law of conservation called CPT parity; and the exceptions to that law in which energy or information disappears from a plane of existence *but should appear in the invisible plane of gravitation*. For example, black holes erase information from the electromagnetic world but they emit it as >c 'gravitational, dark energy' and 'quark information' through its poles.

This means that neither the principle of action-reaction nor the Laws of conservation are absolute, when we consider only a single plane of space-time; since there are uneven flows of energy and information between 2 planes. So in Physics, only the total energy and information of the particles or CPT parity is kept; but the C or P or Ti parities might break in processes of death and birth of particles that give or take energy and information from other plane of existence, without returning it to the perceived, electromagnetic world physicists observe.

Especially when we consider a particle's death, which as *all deaths it means 2 fractal jumps*, there is a transference of energy and information that jumps between 3 planes of space-time, (st-2⇔st). So energy and information truly disappears from our planes and the laws of conservation that work for the total 3 relative planes of the organic system break for the planes we can observe. A fact that explains empirically many invisible forces and particles, postulated by physicists to maintain those theoretical conservation laws, without realizing they are observing an asymmetric transference of energy and information. So they don't need to postulate 'real particles' but 'asymmetric EXI flows'. Indeed, in all processes that involve the death of a particle, which breaks the balance between the energy and information of a system, physicists have found a loss of energy and information. Now we can explain why in the proximities of black holes, which transfer energy and information between the gravitational and electromagnetic world, our light-space membrane disappears: it become pure gravitational mass/information and very likely in Kerr black holes, a flow of >c dark, expansive gravitational energy and quark mass we observe coming out of the bipolar jets of those black holes.

It explains why in radioactivity processes that destroy an atom or close to the limits of C-speed, the conservation of momentum and energy of a particle breaks.

It explains why the weak force, whose parameters are time parameters and trans-form the form, the information of a particle, the CPT parity is not conserved.

Yet because those facts are not known in a continuum, single spacetime model physicists have developed to unneeded theories of black holes and symmetry breaking; the thermodynamic evaporation of black holes and the Higgs. They have not been found in 30 years because they are not needed, they do not exist. Moreover by accepting those 'extra-phenomena', which are 'ad hoc' theories, physicists have broken fundamental laws of physics. In black hole evaporation the black hole gets hotter in a colder environment (our light membrane), thus breaking the 1st law of thermodynamics and creating an eternal motion machine (it is like if you get a hot coffee and instead of cooling down it heats and evaporates). On the other hand, the Higgs implies a 'scalar boson', which contradicts quantum laws and an entire new field that permeates the Universe but has never been found.

Now we offer solutions and a prediction: we will never find an evaporating black hole or a Higgs particle. Moreover if black holes evaporate they will eliminate the information of the Universe, and so the principle of conservation of energy and information $E \Leftrightarrow I$ in which systems sciences and complexity is based would be false. And the Universe would not be eternal but die. It is not the case: black holes warp energy into mass-information, contracting space and balancing the expansion of vacuum space between galaxies produced by the dark flows of >c energy they emit from its poles. Because the Universe is made of discontinuous space-times, the black hole is merely the entity that balances the entropy=energetic nature of our electromagnetic space creating order in the Universe, as quarks fixe atoms and allow the order of life by restricting the motions of electrons, which then can create the complex structures of living organism.

So instead of postulating invisible particles and speed limits to keep reality confined to a single plane of existence, Fractal theory explains all those exceptions to the 2nd law of Thermodynamics and the laws of conservation, *which apply only to the entire informative=gravitational/electromagnetic =energetic and human/atomic, exi scales of reality,* as asymmetric ExI flows between 2 planes of existence.

Obviously physicists have never witnessed the imaginary particles that come out of those processes, because they are not in our light-membrane plane. For example, a dying neutron or a radioactive atom produces certain particles called neutrinos through the mentioned temporal weak force. Those neutrinos were invented to keep the total momentum of the dying particle, but they are only observed through indirect flows of energy and information. Since indeed neutrinos are not 'particles' in the strict sense, but flows of energy and information transferred between the electromagnetic and gravitational planes of existence, when particles mutate their form or die, in a process mediated by weak forces: neutrons give away the remnant of energy and information, as they explode into an electron and a proton in the form of smaller, st-1 gravitational and light quanta – neutrinos and gamma rays.

Further on, we do not perceive the gravitational force, which for that reason is much weaker than our electromagnetic force.

Recap. Because we do not sense all the energy and information of the upper and lower planes of existence, there are weaker forces (gravitation) and loss of momentum in our plane, (neutrinos), which disappear in the dark worlds we do not perceive.

64. Transcendental Inversion: form changes between planes

Flows of 'energy and information' between planes sometimes return to our world and give birth to a new particle; sometimes create a particle that flows further into the past called an antiparticle;

sometimes give away its energy as a gravitational wave we cannot perceive; or sometimes they appears as an inverse, negative energy or a particle with negative mass, as it happens with neutrinos; precisely because its time-space coordinates become reversed as in all process of death that write as Se>Ti -> Ti<Se. So the dying form particle suffers a reversal in its space-time coordinates, becoming in algebraic terms a flow of negative energy or an antiparticle. And these phenomena shall be the last key law of timespace jumps between planes needed to explain most of the key phenomena of the Universe. We call it transcendental inversion, because when we move to a lower or higher plane of existence, we change our function according to the general equation of an organic system:

$$E_{st-1} < I_{st} = I_{st} > E_{st+1} < I_{st+2}$$

A law, which means that all beings are self-similar species in its plane of existence, $I_{st}=I_{st}$, superorganisms of its lower energetic cells, $E_{st-1}<I_{st}$ and energy of its upper plane, E_{st+1}.

Those reversal of 'coordinates' and parameters that transform an energy species into information and vice versa between planes of existence seem paradoxical, as scientists are unaware of the change of morphology or ExI parameters of a form that merely has 'transcended', 'evolved' or 'descended', 'died' and 'devolved' between planes. Indeed, the world shows constantly those 'inversions' of space-time parameters through the $3\pm st$ planes, ages, or hierarchical, social classes of any system, which become the key element to study dynamic flows of energy and information between states and horizons in time and organic planes in space.

Those paradoxical, illogic, transformations of energy into information and vice versa happen when species transcend between 2 relative planes of existence because, as the Chinese say, all yin has a potential seed of yang that can transform it.

We call those events 'transcendental inversions': When forms of a st-1 plane of ExI transcend into an 'st±1' structure, both its time and space become inverted, which means that its morphology also becomes inverted.

For example, lineal iron atoms become cyclical molecules when they emerge in the higher molecular plane. Then again iron molecules become lineal swords when they emerge as a military meme in the macro-scale of human beings.

In Biology, lineal proteins become cyclical membranes in macro-cells that become lineal muscles when they transcend again to the st+1 level.

In time events a predator becomes victim of its evolved offspring (Oedipus paradox). So the father gives his vital energy to his son. The son species extinguishes the father species, and so mammals extinguished dinosaurs and robotic machines probably will extinguish men in the nearby future.

So how can we distinguish a form as an energetic or informative species, as a victim or top predator?

We can define a form as dominant in energy or information, looking at its total ternary structure through the 3 st±1 planes of its organic systems; *in as much as 3 is an even number, and so either the energy or informative structures will dominate the total organism. So iron is lineal in its atomic and organic structure and indeed iron is the most energetic atom of the Universe. That is why the Taoist wrote the structure of all events and forms with the trigrams of the i-ching. So we can use those trigrams in i-logic theory to represent organic structures. For example:* _, --,_ is the energetic iron sword, in which 2 energetic planes dominate the cyclical, intermediate iron molecules. In biology --,_,-- represents among many other structures a temporal, informative cell, with an intermediate spiral RNA, made of smaller cyclical nucleotides and enclosed into a cyclical membrane. While _ _ _, the trigram of pure energy means two jumps of erasing of information, or death. And indeed, it is the symbol of death in the i-ching. While the process of birth and growth is a dual informative jump from a seminal cell that deploys its information into a born baby that will then learn the memetic information of her/his culture till become an adult in a 'steady state' or mature age of reproduction. So we could represent it as --,--,--.

Recap. When a species jumps forwards as a social structure of multiple st-1 cells its form mutates from energy into information and vice versa except in extreme time events, such as the process of birth, with 2 informative jumps and the process of death with two energetic jumps.

65. The limits of i-logic geometry: God and the Universe.

All what we have said about time travel might seem counter-intuitive and reaches indeed the limits of 'weirdness' of the Universe; as the bosonic state of space does in the other parameter of reality. In fact both 'extreme limits' of the forms and events possible in reality could be explained according to the paradox of Galileo as the mirror sides of the same phenomena, which we shall enunciate merely here:

'A relative ∞ number of points of space can exist in a single point of time (bosonic state); and a relative number of points of time can exist in a single point of space (time displacement).'

The first definition would be that of a perfect bosonic form of maximal spatial force, such as a black hole; the second definition would be that of a perfect causal, temporal form of information that perceives past, present and future, simultaneously such as God in Saint Augustine's definition'.

In the more complex, logic models of time arrow, all laws of the Universe can be deduced departing from the Generator Equation of reality which becomes the Universal algorithm that Leibniz and Turing thought, once we understand the properties of those arrows; and so we can generate from it all phenomena and events of the Universe. But this is a mere introduction to Multiple Spaces-Times and so we shall just assert that indeed, both the bosonic state of a black hole and the informative state of God exist and *both are relatively immortal.*

Relative Immortality is due to relative infinity, a paradox of sets and superorganisms of discontinuous planes of existence, already understood by Cantor: infinity is always relative since there are discontinuities beyond which we cannot perceive. So the infinity for a cell is the external membrane of the organism beyond which it cannot perceive. The infinity for a tribal man was the limit of its tribal geography and so he often called God the nation in which he inhabited. (So Yvwh, the god of the Jewish people actually was first a geographical name for Judea, and Assur the God of the Assyrians a name for Assyria). And so the Universe might have a bosonic/temporal state of absolute perception and maximal density, a relative brain as a superorganism, but it will be limited to the external membrane. And so while there are not absolute Gods as physical entities, what it is truly infinite and eternal is the game of Multiple Spaces-Times.

Recap. The two limits of space-time jumps are the bosonic state and the state of absolute perception, which can be defined with a simple law of Multiple Spaces-Times: there is a point of space in which all informations co-exist and a point of time in which all spaces co-exist. They are the black hole and God

XI. THE ALGEBRA OF THE EXISTENTIAL FUNCTION

66. Algebra: Numbers and equations.

The second aspect of mathematics deals with sequential numbers and equations, in which sets of numbers are transformed by an X=Y function, in which often one component changes faster than the other, tracing a curve in a Cartesian Plane plotted with those 2 variables that mathematicians study with great detail. It is the thesis of this work that all those X=Y functions and differential equations represent particular studies of a general $\sum Se \Leftrightarrow \prod Ti$ equation, and or a partial event between two or n-points of a non-Euclidean network, and or the network and its environment. Yet as Einstein put it to Poincare: 'while I know when mathematics are truth I don't know when they are real', meaning that many mathematical equations and functions do not exist in nature, as they are not partial cases of the Generator equation and do not respond to the restrictions the Ternary method imposes to a Universe of multiple spaces and times but only 4 Dimensional arrows. In that regard, the laws of multiple spaces and times and the syntax of the Generator equation with a limited number of variations restrict the possible mathematical realities there is in the physical Universe. On the other its study provides the scientist with a deeper meaning for the Algebra of numbers and the meaning of equations and functions.

Recap. We shall consider merely the meaning of the main mathematical operations within the restricted world of 4-Dimensional spaces and times we live in. And analyse in more detail some of the parameters and functions most commonly found in the study of the Generator equation, which connects the equation of space-time cycles with the detailed mathematical analysis of those cycles by different disciplines.

67. Theory of numbers.

Numbers in this new outlook are not only intervals of a one-dimensional straight line, but as Pythagoras and Plato stressed, they are geometrical forms:

Mathematics is concerned with 2 seemingly different worlds, the geometry of spaces and the logic of numbers. To fusion both requires to understand numbers as forms. A number is not only an abstract set but always a collection of self-similar beings extended over a common vital space, a network And so networks create complex forms, topologies of space-time, as the motions between points of the networks become stable exchanges of energy and information between two polar points. Yet since each number is a geometrical form no longer limited to the simplest one dimensional form but can vary its geometry and hence its function, degrees of freedom and complexity as we increase its 'number'.

The 4th postulate shows how points, numbers, the self-similar class of equal forms create geometries:

The line is simple. The line joins two points and can only have a combination.

The triangle can only have a closed combination, but 3 possible open combinations, Ab, Ac, Bc.

The quadrangle is more complex. It can be joined in 2 combinations, as a cross and a square. And it can be left as an open snake with 3 different orientations. So a foursome acquires a snake shape to move with the arrow of energy; a crossed form to perceive in its center '5th point' and a square shape to accumulate and reproduce its internal organs; and so each shape of the same number becomes a topology with a different form and function.

Indeed, function and form are now fusioned. So certain numbers in its 'degrees' of freedom of form, represent certain functions. The quadrangle can store energy, but in a zigzag open line it can move – spend energy and as a cross it can gauge information. Numbers also define arrows of time. So for example, 1 lonely number without motion is perceiving, with motion is processing energy, 1+1 might be 3 (an act of reproduction) or 1 (an act of Darwinian feeding). All those vital actions determine that certain numbers survival better than others. So, 1, 3 and 4 are very common systems.

In that regard, a complex analysis of the simplest numbers shows that the more perfect form is the 10-cellular system or tetrarkys, in which 3 x 3 triangular corners act as organs of energy, information and reproduction with a 10th central element that communicates all others and acts as the one of the higher scale, representing the entire organism.

Thus as the number of cells grows, the topology of the system will grow in degrees of freedom and complexity till resembling more and more the repetitive, geometrical forms of social organisms. Topologies become thus at the end, complex networks, adapted to different functions of complex organisms.

As abstract as all this might seem, when observing nature we shall see how those type of events, waves and social planes happen in all the scales of the Universe, from atoms which form crystal networks based in the equality of the same atoms or at best in the existence of a 'body-mass' of equal atoms intersected by a few 'stronger' atoms that form a complementary network of higher resistance, to the body rejection of cells with different DNA.

What things we can do with numbers can reflect then many of the actions of networks. For example:

- We can study how social groups organize themselves or fluctuate between states=functions. This is the study of the internal point of view of networks as a collection of self-similar points. Those changes of states are often defined by a differential equation as informative systems have less spatial extension/motion but are more complex networks with more bytes of information=points. Thus differential equations, most of them of the type $Y(ti) = aX^3 \pm bX^2 \pm cX \pm D$, express $\sum Se \Leftrightarrow \prod Ti$ transformations, where Ti is a network in 2 or 3 dimensions of time bytes, bytes of information and Se is a network with one (same organism) or 2 (Darwinian feeding) scales of lesser complexity than Y, such as $f(x)=Y^n$. It follows from the Fermat Theorem that there is a restriction to the number of solutions a system can find, which is n=3, the maximal number of dimensions an informative sphere can have as it displaces itself over a plane of energy.
The relationships between limbs and heads that exchange in a 3rd region called body, form and motion, such as the head designs the motions of the limbs, which move the head, and both exchange in an intermediate region of elliptic nature called body more subtle types of form and motion to create more complex cycles that will in fact reproduce both systems can be mathematized in infinite different ways, using matrix, combinatory theory, differential equations, polynomials, Riemann surfaces, etc.
- We can study how networks grow and multiply creating new species and we can add them and observe how they reorganize creating curves which are differentiable to obtain the rate of grown and diminution of the organic population. The study of herds of energy and networks of information in its life cycle is one of the key disciplines of all sciences specially physics and ecology.
- We can study them as networks with form through its geometrical ways of exchanging energy and form, from the simplest point to the line of 2, the triangle without a central focus, the structure of energy, which can however turn into a pi-cycle, the 3, the 4 with its zigzag, solid quadrangle and cross structures, the 5 and first 3 dimensional structure, and so on. Each number will increase the possibilities of the game, yet when we reach 10 we play a perfect game with 3 triangles that act as organs of energy, information and reproduction, and a central point both in a 2-dimensional or 3-dimensional geometry, acting as the collective action/will/intersection/knot of all cycles – the first clear, complete ego structure in 3 dimensions with perfect form and complementarity. Thus beyond 10, while some numbers might bring slight improvements to the cell, most forms are just growths of the primary numbers in multiple associations.

Recap. All the structures of mathematics, regarding of the notation we use, reflect events and forms of knots of time arrows (st-points or numbers), as mathematics is a language whose grammar derives from the Universal grammar of spacetime. Numbers are thus formal networks that try to achieve the essential arrows of time. And so certain numbers (1, 2, 4, 5, 7, 10) deploy better those arrows and are the commonest on nature.

- We can study the evolution and reproductive creation of new networks with successions and combinatory is important in multiple time-spaces since we find always complementary systems of reproductive energy and information, each one with a ternary choice of evolving differentiation (energetic, informative and balanced species). So especially in the classification of species of different sciences we shall find simple combinatory laws that explains the differentiation in 3, 6, 8 and 10 elements depending on the triads and dualities of multiple space-time systems.
- We can study a key antisymmetry of time and space expressed with the language of probabilities: Sequential events are studied with probabilities in time, whose symmetry in space are the study of percentages of populations in space, such as if each event in time is the birth of an individual of a population both probabilities and percentages are the same. This confused physicists in some cases, as in an electronic nebulae, which is a population of fractal electrons in space, but it is studied as time probabilities, and created the bizarre theory of multiple universes (multiple, probable electrons) instead of a fractal Universe (fractal self-similar micro-electrons, which are bundles of ultra-dense light forming a nebulae which also acts as a 'whole' electron, self-similar to its parts). Thus the study of probabilities in time events and growing populations of a wave of space-time cycles is an essential tool: we can study the proportions, herds, groups and networks of self-similar st-points in its evolution either with probabilities or differential equations.

Recap. Probabilities study causal events in time and populations in space; combinatory studies the differentiations of species according to the variations of bodies and heads.

68. The syntax of equations.

The study of equations of number is such a vast genre because both herds of energy in motion and networks of bytes of information are sums, with different dimensions. So they are essentially polynomial functions, which vary as energy becomes information or vice versa. The way in which one variable grows or diminishes to the expense of the other is thus an equation plotted in a bidimensional plane – the most common event studied by science. The difference is that of the unit. Since by fusioning the concept of a Non-Euclidean Point (Riemann, Lobachevski) and a Fractal Point (Mandelbrot, myself), we defined a new unit of mathematics, the 'point' (Geometry) or number or set of points (Algebra). And since equations treat points-units in social groups/networks/numbers, we cast more insight in those equations knowing the properties of its elemental unit.

Those Non-Euclidean points are knots of time arrows, with complementary functions and forms that combine Logic and Mathematics, Time and Space in actions, where motion with purpose - with form - create the 2 sides of the same action. For that reason the new postulates of i-logic, Non-Euclidean geometry (the geometry of multiple spaces and the logic of multiple, causal arrows of time), are both logic and geometric postulates. In Algebra and all its branches, numbers are the units used in equations, all of which follow a rather simple ternary structure: $f(y) = f(x)$, which we shall write as $F(i) \Leftrightarrow F(e)$; that is, all equations are descriptions of certain events, which can be described also with the wider syntax of all languages 'generated' by the Universal Grammar of the feedback equation of space time. The corrections are minimal but relevant. Equality, = now becomes dynamic, $\Leftrightarrow$, since many errors in science are derived by the fact that scientists tend to think that an equation is based in equality (3rd Euclidean postulate) and not in self-similarity (3rd non-Euclidean postulate). So for example: $E=mc^2$ is not an equality, Mass is not energy, but a self-similarity in which at least one property of Energy is transformed by the symbol of self-similarity, $\Leftrightarrow$, in this case lineal energy coils into cyclical information, mass.

Algebra is a sequential, synoptic language of time used to operate with complex groups of points/numbers, either expressed with the modern concept of Set theory (Cantor) or the old concept of Equations and Operandi. Those operandi that defines $F(x)=G(y)$ relationships are a reflection of the permitted operations Generated by the Feed-back equation, <X>, $E \Leftrightarrow I$, or Generator equation of space-time that expresses the foundational law of physics: 'energy and information never die but transform into each other 'ad eternal'.

In essence the 3 operators of the generator, $\sum$, $\Leftrightarrow$, $\prod$ are dynamic operators that encompass most of the operations of mathematics.

The sum defines the way in which herds communicate, the multiplicative and integrative symbol defines the way in which networks tie themselves, the flow symbol defines processes of diminution and enlargement as energy and form is transfer from one side to the other of the equation with different rates according to the ages of life of the system or event. As energy flows from the herd to the network, the network reproduces more informative bytes and the herd diminishes till it halts its growth and the asymptotic curve is formed - this is the region of death established by a peak or non-differentiable discontinuity where the function collapses, and all the information returns to energy.

All those possible transformations of a life/death curve are indeed, what mathematical operations done with equations of the form $F(x=e)\Leftrightarrow G(y=i)$ describe.

Many errors of science are caused by the 'mechanical' approach, which operates with mathematics without understanding the deep meaning of equations. In essence any equation, $F(Y)=F(x)$ can be reduced to a transformation or relationship between 2 functions that represent complex and simplex knots of time arrows or events between them that become transformed through an operandi, which represents one of those basic transformations. So we could write instead as the Universal Grammar of all algebraic equations, either the Generator equation of space-times or any other notation that reflects the fact that equations describe events and forms born of transformations between structures of time arrows. For example if we write:

$$F(\uparrow,\rightarrow,\lrcorner)^{i\pm n} \Leftrightarrow F(\uparrow,\rightarrow,\lrcorner)^{i\pm n}$$

We describe all possible equations in the 'dimensional notation' of time arrows, where $\uparrow$ is the Time arrow of information, $\rightarrow$ is the time arrow of energy, $\lrcorner$ is the time arrow of reproduction and $^{i\pm n}$ is the transcendental arrow of social evolution or emergence and social devolution or death.

The basic concept of algebra is that of number, briefly explained in this text and that of operations between numbers which are herds of time knots and its properties and the structures they create (grupoids, semi groups, groups, Boolean Algebras and rings). All those operations describe in fact the possible events that take place between points that create either a herd/simplex plane of existence, or a complementary organism extended in 3 relative planes, st+1 (informative system), i (reproductive, reproductive body) and st-1 (ecosystem in which the herd or organism absorbs its pixels of information and bytes of energy). The meanings of the basic operandi are:

- *Sum*, +, Σ is the fundamental operation that creates 'body herds', in which the individual cells are simply connected (each one relating only to the cells that surround it) and so their 'exi' existential force merely adds up. Their reciprocal, destructive event is subtraction, -, when 2 herds enter into a Darwinian event and one herd extinguishes the other.

- *Multiplication*, x, Π, is the key operation that creates 'informative networks', in which the individual 'neurons' are complex connected, each one, joined through its 'axons' or 'subsets', to all other neurons (ideal network) and so their 'exi' existential force multiplies them. Thus for example, 5 neurons with a complex connection, each one will have 5 axons/subsets, and the number of connections will be 25.

The duality of planes of space-time, proper of informative systems is the first 'complex' structure extending beyond a single space-time continuum used to define energy-only processes (and hence deforming the understanding that physicists have of information systems, which cannot fit in a single plane and their obsession for models with a single arrow of energy/entropy). Essentially an information network can 'operate' both in the language of the energy herd it controls (so quarks, which are bidimensional vortex of mass-information, operate with electromagnetic forces to control electrons); yet they also operate as a network through the other plane of existence unknown to the herd.

This is expressed in mathematics with 2 concepts: energy herds are 'groups' defined by a single operandi and information systems are 'rings' which operate with 2 operandi. And 'Groups' and 'rings' are the 2 fundamental structures of algebra to the point that a 'ring' is called simply an algebra.

An important law of complementary systems of reproductive energy and information is given by the fact that the maximal product is always that between 'self-similar subsets', such as if we have 2 st-points with a total of 10 st-1 elements, axons or subsets, the maximal number of connections/products, will happen when $E_{st-1}=I_{st-1}$ (5=5, 5+5=10): $1 \times 9 < 2 \times 8 < 3 \times 7 < 4 \times 6 < 5 \times 5 = 25$.

It means that in the product exi that defines the existential force or momentum of a complementary system, those systems, which are balanced, are the most efficient.

Thus, the most efficient systems are those in which the 2 complementary parts are in balance between body and brain; a law fundamental to explain the survival of balanced organisms in its '2nd age' when E=I; which is also the definition of *beauty*, as the classic age of balance between energy and form.

This laws explains why the fundamental algebraic operations are quadratic equations, as $X = E \times I_{i=e}$ is simplified with a square, x^2 and defines multiple energy/information systems.

Yet more complex equations of the type x^n normally respond to operations between different planes of existence, also described with Riemann's complex planes.

- *Division*, which describes the inverse operation of finding the number of sets departing from its parts/subsets or network connections, but also another fundamental operation - the ratio between spatial energy and temporal information, origin of multiple Universal constants, *E/I or I/E= K*, related to our previous concept of balances between energy and information that sets the survival limits of any given system.

- *Integration and derivation*. Those 2 'recent' operations are related to complex operations between systems that extend in 2 or n planes of existence, especially in 'neuronal networks', whose cells/subsets 'emerge' into units of a higher plane.

In essence differentiation and integration are the 2 operations in which a flow of energy becomes reproduced into a bigger population or transcends and becomes connected by axons of a network of information - it becomes integrated and acquires a higher potency/dimension. While integration measures the curve of growth and diminution of the change of the population or herd. Integration creates a volume of a form with higher dimensions than the derivative and it can be interpreted in terms of a function of space and time. It can in fact be closely associated to the analysis of the Generator, exi=k.

Thus, we can consider the main operations of algebra to respond to the basic processes between time arrows. But a more important algebraic analysis is that of time arrows through time-space symmetries of the generator equation, $\sum Se \Leftrightarrow \prod Ti$. What the generator is in poetic terms: a crystal that emits reflections of itself; and the densest in information the image emitted is, the more intelligent and complex the phase of the generator is. The tension, $\Leftrightarrow$ between both poles of extinction is always dynamic, born of the cyclical exchange of matter and form between both poles but with a slightly constant derivative, Max. $_{t->\infty}$Se>Ti, or direction of life and pronounced slops of Max. Ti<Se $_{s->\infty}$ death.

The function of existence is – the reader at this time will have already noticed – the function of life and death of systems and organisms, and as such the two sides of the cycle, life and death, are notorious for their inverted properties. Death is an explosion of space and energy and motion that unwarps a network of non-Euclidean points probably on its highest informative point and minimal space. Those nutritive pills are everywhere to feed and feel a cliff of death, a discontinuous form, whose trace will be a memorial warped baroque angst in any language. The slope of the function Se$\Leftrightarrow$Ti in any case determines where we are in the life cycle, because the life cycle is a bell curve with a known slope, in a coordinates of time (x) and space (y). The population of cells of the body for example, follows a bell curve and at a certain age the slope will be positive or negative but not cross.

In essence, what the generator equation means is the ultimate mirror of all equations of physics. A herd of bigger spatial forms, guided by a network of information, 2 different species and networks, messed together into a single body. The symbol ⇔ of the body, of the steroidal cycles of exchange and combination of messages between both poles of the being, the 'vegetative and informative pole'.

The study of the Se⇔Ti function in all its transformations of Se fields into Ti particles forwards in time and corruption and explosion of Ti particles backwards in processes of death in any of the infinite manifestations of the Space-time cycle anywhere in reality is in a philosophical sense the meaning of mathematical science: explosions of SexTi functions Sexualizing the universe with its creative orgasm of new smaller SexTi functions born as memorial, seminal motions within the exchanges of Se and Ti in the X-body of the System. The Human SexTi function is of course multipolar, taking into account the complexity of the human being. There is a Limb-Body-Head, Max. S, Se=Ti, Max T structure that dominates the external architecture, and which is fractal dividing itself again and again in dualities and triads. When a function of time reproduces it divides in triads that latter will grow. That division is reproduction in its simpler sense. Multiplication on the other hand is also reproduction but not within a single entity but a couple, which combine its inner, body-limb elements, multiplying the complexity of its Se-strata. The messing and sum of vortices that take place in Nature might seem chaotic in any scale but play the organization of networks=numbers of great sophistication in its messing when we observe the apparent entropic disorder as a series of phases of synchronous motion among all the parts of the whole whirling landscape. The question is how to focus a certain dynamic structure, ⇔ into its most ordered state or peak of information, or self-point of view of the system. Only in that point that we could well call with the Aristotelian name of God, the self conceives itself in all its subjective perfect form.

The existence of an animal and vegetal mind and body, head and limbs dual pole in the construction of all organisms is undeniable; the pole of motion and absorption of energy and the pole of information processing and sensorial perception. This complementarity limbs-body-head, spatial force-field-particle defines the exchanges:

$$S\ (E:\ Body/Field)\ \Leftrightarrow T\ (i=Head/Particle)$$

Whose many particular cases gave form to differential physics of the type ds/dt=k *Recap*. Algebra is the science of sequential numbers; hence the mathematical representation of time events. Its equations reflect those time events as a language whose syntax is self-similar to that of the Generator equation of time events, once we substitute the concept of static equality, =, for the concept of dynamic self-similarity ⇔ formalized by the 3rd non Euclidean Postulate.

69. Creation: Inversion of ternary diversifications.

When we fusion the fractal principle of ternary creation, the duality of form and function and the inversion of space-time processes we obtain a complex rule of creation that accounts for most speciations in the evolutionary process: according to this complex law:

'*Each process of evolution in time can be caused by 3 sub-causal processes in space, and each process of reproduction of a topological region of space can be caused by 3 sub-causal processes in time: $3E_{st-1}->I_{st}$; $3I_{st-1}->Re_{st}$*'

Thus, the event of reproduction can be branched into three possible causes that correspond to the three arrows of time: an energetic, explosive reproduction by the death of the mother, a balanced reproduction by self-similar beings (sexual reproduction), or an informative, seminal creation by a seed of pure form emitted by the father (asexual reproduction). For example, a species can put an egg in a living being, mate with a couple or reproduce spores. A quasar can blast and trigger the reproduction of galaxies; a cannibal galaxy can eject a jet of matter that reproduces an irregular baby galaxy, or it can merge with a bigger one triggering the reproduction of more stars in their body as it captures more intergalactic gas.

On the other hand each time age can be diversified in 3 topological modes of evolution. So a young species can halt its evolution (neoteny) and stay young for ever (alevin); can mature fast into a reproductive state, dying just after producing its eggs, or can make his youth a massive age of informative evolution (Chrysalis) and be born again as a complex form.

Further on, we can divide each 'age' in 3 sub-ages, which again will have an energetic youth, middle, steady state and a final informative vortex. So youth can be divided into the first age of geometric, massive growth of energy in which the head/informative organ diminishes in size in proportion with the body; a steady state of arithmetic growth and a final informative age (adolescence) of transition to the reproductive age, of maximal evolution of form.

We can in the same manner divide each topological space in 3 sub-topologies, specialized in energy, balanced and more informative, according to the adjacent regions; and so the membrane can be differentiated in the more energetic, external region, the intermediate balanced, and the informative inner zone; as the external cover must be harder in contact with the ecosystem in which the entity absorbs energy (outer skin), the middle zone is the region that reproduces the 'lineal units of the system) and the inner zone, the one in contact with the more complex internal regions of the organism (region of the skin where the sensorial cells have its nuclei).

Thus when we properly merge the main laws of creation of multiple space-times, the ternary decoupling with the inversions of energetic spaces and temporal information, by dividing events in 3 sub-ages and spatial forms in 3 functional topologies (spatial, reproductive and informative layers). We can explain the Universe, which increases the complexity as we look in more detail a certain reality. It is the 'fractal principle of differentiation', which creates 3 and 9±st and then 27±3 and 90±10 sub-ages of time and 3 and 9±st and then 27±3 and 90±10 sub-forms of space in complex systems; which allow a detailed analysis that maps out the existence in time and main organic structures in space of most entities of reality. So the human being is defined in space by 10 physiological sub-systems and in time by 9±st sub-ages of life.

Recap. The fusion the fractal principle of ternary creation, the duality of form and function and the inversion of space-time processes explain the complex rules of creation and speciations as each age of time subdivide in a young, reproductive and informative sub-age; and each topology of space in 3 sub-topologies of energy, reproduction and information

Algebra operations respond to fundamental events between knots of time, tied by their time arrows. As such they can be derived of the Se⇔Ti function that represents all cycles of space-time between inception and extinction.

70. The 'structure' of Mathematical languages.

The Nature of mathematics as a language of space-time (geometry and numbers) can also be expressed with the modern conception of mathematics, based in the work of Cantor (theory of sets) and Boole (Computer Algebra), which was summarized by the Collective Bourbaki, whose 'Mathematical elements' reduce all mathematical disciplines, to 3 structures, themselves reduced to set theory:

- *Structures of Order*, understood as structures, which describe the hierarchical order <, > of numbers, which we just described as social, geometrical groups. Thus, those structures are mere reflections of the 4th arrow of eusocial 'love' that creates social groups.

- *Topological Structures.* They describe the basic forms of the Universe, which in a 4 dimensional reality are only 3: the hyperbolic, informative geometry; the cyclical, reproductive toroid and the spherical, external, energetic membrane, defining all relative worlds/species as complementary systems of reproductive energy and information. We shall show this simple ternary, topological structures, which are the basis of all forms of reality in our analysis of the 5 non-Euclidean postulates. To mention that

268

topology is in fact the ultimate science of space from where we can derive all the other spatial structures, as geometry was the basis of classic mathematics in its Greek inception. In fact topology and set theory are quite similar in its laws. What topology adds to geometry is 'motion' and 'relativity'; thus including the Galilean paradox in the structure of mathematics, since two topological structures are equal regardless of the 'distance/size/motion' of its surfaces.

- And *Algebraic structures,* which study the complex arrows of Time born of the 'operations' between simplex arrows and are ultimately expressions of the Generator equation of space-time, $E \Leftrightarrow I$, in as much as al algebraic equations are of the type $F(x)=G(y)$, which is just a particular case of the Universal syntax/grammar of reality of the Generator feed-back equation.

Thus, there is a direct relationship between the 3 fields that define all mathematical structures and the Temporal, Spatial and Combined (algebraic) operations between time arrows, which simply means that mathematics has always been the human language that describes the ultimate reality: Discontinuous Spaces (topology) and Relational Times (causality); and the complex interrelationships between them (Algebra). While the 2 first facts (the identity between space and topology and Time and causal order) are intuitive, we might conclude this overview of the meaning of mathematics, with a brief introduction to the meaning of Cantor's sets, and Boolean Algebras used for Mathematical computing, as an expression of all Mathematical Structures and a formalism of Multiple Spaces-Times. We shall by doing show, prove, that all mathematical structures are explanations of properties of the ultimate reality: the timespace defined by the 4 arrows of the Universe and all its species.

Recap. Mathematics can be reduced to 3 types of structures: topological/spatial structures, structures of causal, temporal order; and algebraic structures that define the outcome of complex relationships between Multiple Spaces Times. Thus i-logic mathematics is the synoptic language the human mind uses to describe all what exists: a Universe of spatial energy and temporal information.

71. Mathematical evolution: from Geometry to Sets.

We said that all what exist is a system, made of knots of time arrows=st-points. In its most simple formalism, humans perceived those knots as numbers (sets of self-similar points) and those numbers became points of a geometrical plane. Next, Descartes reduced geometry to Analytic Algebra, showing that there were two self-similar languages to express operations between points, Geometry and Algebra, whereas Geometry was mainly concerned with the spatial description of networks of numbers and Algebra with the Causal relationships of those numbers. So we could say that Geometry and Algebra were, according to the Duality of the Galilean paradox, two sides of the same coin: a spatial and temporal description of the reality of st-points. We have till here focused in the evolution of the Geometrical perspective, by completing Non-Euclidean Geometry, reducing Topology to a description of the 3 parts of Non-Euclidean Points, by understanding those points as Fractal points in a Universe of multiple space-time scales.

We briefly considered how the properties of a network-space (a web of points), are those described by Lobachevski and Riemann, which depend in the self-similarity (homogeneity) and adjacency (closeness) between those points, (formalized latter in this work by the 3rd Postulate of non-Euclidean geometry.)

We have also studied the complex causality and order between the arrows of time, which are the foundations of the structures of order in mathematics (but go beyond the present mathematical corpus).

Let us consider the 3rd fundamental type of structures, Algebra and the fusion of them all in set theory.

Recap: humans, following the principle of Correspondence have gifted the mathematical formalism of an increasing complexity and richness in its description of the properties of spacetime arrows, from the simplest concept of a number/point without parts, to the complex analysis of Non-Euclidean Geometry and Theory of sets.

72. Cantor sets. The paradox of discontinuous infinites.

All those properties and many other structures of mathematics were further reduced by Cantor to the ultimate reality of all mathematical structures: the theory of sets, composed of subsets, which we affirm is the natural formalism of system sciences, as a theory of 'super-organisms' composed of smaller super-organisms, which are sets of self-similar subsets; whereas the theory of sets and subsets gives the previous, simplified theory of numbers (each one a class or set of self-similar points), an inner content, as i-logic geometry gives points its inner parts.

How this 'formalism that mirrors reality' called set theory, from where all mathematical structures can be deduced, reflects the Nature of Complementary systems made of energy and information and its properties? The answer should be self-evident to those kin readers who grasped the inverse properties of energy and information:

Set theory defines reality in terms of two inverse elements A (points of energy) and A', (its complementary, inverse element). Thus set theory is no more no less than the analysis of the 2 simplex arrows of existence, energy and information and its complementary organisms.

It is thus not surprising that in set theory energy and information, A and A' are called complementary sets and the fundamental law is called the Law of Duality (Morgan Laws), which basically tells us that we can reduce all sets to operations between A and its Complementary, as we can reduce all systems to complementary Energetic and informative organisms, which are the whole.

So the main operations of sets reflect the properties of Complementary systems of reproductive energy and information, where A=Energy system; B=Information system; W= Relative Universe (World, Whole or Superorganism):

- $E \cup I = W$; $I'=E$; $E'=I$; $E \cup E' = W$; $I \cup I' = W$.

Thus, the Union=Fusion of an energetic and informative, complementary system creates a whole superorganism.

This same equation expresses in the language of Cantor sets an act of creation of a mapping of the Universe, whereas I, the preceptor observes I', the Universe and the result is $I \cup I'$ a whole mapping of reality within the mind of the preceptor.

- $E \cap I = \varnothing$; $E \cap E' = \varnothing$; $I \cap = \varnothing$.

It describes anti-events, which annihilate the form of particles and antiparticles, waves and anti-waves and so in Multiple Spaces-Times is equivalent to the anti-event: *Past x future = present.*

- $(E')'=E$, $(I')'=I$

It describes 2 events of a feedback, generator equation: $E \Leftrightarrow I$, $E => I$, $I => E$, hence it describes among other events a whole cycle of life and death, where $E => I$ is the arrow of life and $I => E$ is the arrow of death. This 'property of sets' called an involution is called in Time Arrow theory a Revolution of times, sum of an Evolution ($E->I$) and a Devolution ($I->E$), and is the fundamental event of all realities.

Since energy and information have indeed inverse properties. And so we can state a Cantor Set describes the properties of complementary systems of knots of Energy and Information.

- $E \cup I = E + I - E \cap I$.

It shows the efficiency of systems that eliminate redundant elements, from genetic 'fusions' to Darwinian events.

Further on, when we understand Intersection as an Event of 'Darwinian perpendicularity' between a complementary system of Energy and Information, E U I, and an external entity, C, which the organism uses to absorb 'informative pixels' or 'energetic bytes' for its mind or body (an event of perception or feeding), we obtain the obvious result:

$(E \cup I) \cap C = (E \cap C) \cup (I \cap C).$

Thus the complementary system takes only the part of 'C', which it needs to inform itself (self-similar to I) or to feed itself (self-similar to E), discharging the rest. And indeed, we perceive only information self-similar to us, or energy 'bricks', self-similar to our bricks, which we can use, to construct our energetic, body cells (subsets of E). And so on.

We mentioned that cells are subsets of I or E. Indeed, the second element of set theory studies the relationship between Sets (wholes) and its parts (subsets), and so it is simply the description of the properties of parts that become wholes.

An interesting result of those properties are the so-called Paradoxes of Set Theory, according to which there are certain contradictory sets that do not exist, most of them related to the concept of Infinite, which Cantor also studied, finding multiple contradictions. What this means, plainly speaking is that infinity and continuity do not exist, in as much as all Planes of existence are discontinuous with a certain limit that defines a Universe of networks of points with limits given by the number of networks, the dark spaces between them and the existence of upper and lower limits of energy and information in the existence of those points (universal constants), beyond which we must transcend and emerge, or descend and dissolve into other membrane of space-time with different properties.

To mention also that Gödel's theory of incompleteness was based in set theory and showed indeed that mathematics, while being the most complete description of the spatial events of reality was neither the ultimate language of the Universe (as Frege and Boole proved it could be reduced to Logic propositions) but also an incomplete language, which did not describe all realities and an inflationary language, which described systems that do not exist in reality. Those are indeed, two properties of all languages of information; that both distort reality, as the paradox of Galileo prove, and do not include all reality, given the discontinuity of the Universe; which lead us to the concept of Dark Spaces, the true meaning of the 'complementary Universe' that completes the world we see.

Recap. Set theory is the basis of most structures of mathematics, in as much as it defines all the events between complementary systems of reproductive energy and information and its limits.

73. Boolean Algebras.

Set theory is also the basis of Boolean Algebras, which are the basis of the Computer Mind, which is able to describe reality (albeit in a simplified manner), with the use of such algebra. In that regard 'Euclidean geometry' and 'Aristotelian Logic', which is what Computers think can be reduced to the Simplex properties of Energy and Information systems and so the 3 fundamental Boolean Algebra based in set theory are:

- The previously described Algebra of sets and its two fundamental operations, U and ∩.
- Aristotelian Logic with its 3 fundamental operations of conjunction (y, U), disjunction (or, ∩) and negation (nor, ')
- Its implementation in *fractal networks of logic circuitry by computers that represent the reality of all energy/ information systems of the Universe.*

Those 3 basic Boolean Algebras are the spatial, geometric (set theory), logic, temporal (Aristotelian causality) duality of the Universe and its exi, reproductive combination.

Recap. Computers model reality with 2 complementary arrows of energy and information, using set theory.

74. Reproductive arrow: Vectors & tensors of existence.

Set theory defines operations between the 2 simplex time arrows. What are then the mathematical instruments to explain the complex arrows of reproduction and eusocial evolution? Reproduction is explained with vectors and tensors. Indeed, since the postulates of i-logic geometry define knots and topological planes as complex operations between Time Arrows in 3 dimensions, these can be considered mathematical vectors. And indeed, the mathematical formalisms of those time arrows, when in dynamic relationship are the operations of vectors and tensors: Time Arrows form a vectorial space, which has 3 dimensions corresponding to the Time Arrows of energy, information and its product, reproduction. Thus the function of reproduction, exi, between a 'long, energetic X-dimensional arrow' and a 'tall, informative Y-dimensional arrow', gives us a 3rd Z-dimension, exi – a reproductive arrow. So when we multiply 2 vectors, the 3rd vector, vectorial product of the other 2, is a perpendicular Z-vector.

However the plane that better represents the structure of space-time geometries is the complex plane, as imaginary numbers share most of the properties of information, albeit with certain corrections, whose complex formalisms goes beyond the scope of this introductory course.

The vectorial space of time arrows is specially clear when those arrows happen in an homogenous physical 3-dimensional space (astrophysics), and explains the laws of electromagnetism, which as we shall see is a perfect Euclidean space, given the fact that we live in a light membrane, in which light, the ultimate substrata of reality has a flat, magnetic, energetic field, perpendicular to its electric, informative field, perpendicular to its reproductive speed. So the laws of electromagnetism follow the geometry of vectorial arrows, 'the rule of the hand' and the Maxwell Screw that defines the geometry of interactions between electric and magnetic fields.

And the 4th arrow of time, eusocial evolution - which kind of mathematical operation describes it? Since it is the most complex of all time arrows, we cannot describe it properly here, with the limited elements of Time Algebra, we have introduced. Yet in the next paragraphs, we shall do it, as we analyse the geometrical perspective of time arrows in depth.

Recap. The operations between Time Arrows upgrade the abstract laws of mathematics, creating a general, vital geometry of the Universe, which can be applied to all planes of existence. Numbers become then topological networks of knots of time arrows; vectorial spaces define operations between those time arrows; and theory of groups, rings and other types of spaces become structures whose laws can be derived of the general laws of Multiple Spaces-Times.

Notes

[1] Leibniz's *Monadology* defines a fractal point but simplifies them, accepting no communication between points and so fails to create a correct model of non-Euclidean geometry as points constantly communicate forming networks, closer to the philosophical Atmans of Buddhism. The fractal geometry promoted by Mandelbrot, a self-recognized admirer, drew on Leibniz's notions of self-similarity and the principle of continuity: 'natura non facit saltus'. Mandelbrot would say, "His number and variety of premonitory thrusts is overwhelming." Leibniz also wrote that "the straight line is a curve, any part of which is similar to the whole," anticipating the fractal, non-Euclidean topology explained in this book for more than three centuries. One of his metaphysical principles is of certain importance in this crossroads of history, the principle that the Universe must be the most perfect possible, which seems to imply that we humans will not make the "cut" given the enormous degree of arrogance, ignorance, and despise for nature and the "perfect laws" of that Universe.

[2] There are 2 forms of (homogeneous) non-Euclidean geometry, hyperbolic geometry and elliptic geometry. In hyperbolic geometry, there are many distinct lines through a particular point that will not intersect with another given line. In elliptic geometry, there are no lines that will not intersect, as all that start to separate will converge. In addition, elliptic geometry modifies Euclid's first postulate so that two points determine at least one line. Riemannian geometry is the best-known elliptic non-E geometry, which deals with geometries which are not homogeneous, which means that in some sense, not all the points are the same. Thus, those geometries later used by Einstein and Minkowski to describe space-time did have the seeds to understand fractal points of different form and size in which multiple parallels converge. Yet till the publication of *Time Cycles* (Editorial Arabera, 2004), which adapted the five postulates of Euclidean geometry and fused the concept of a non-Euclidean point and a fractal point, there was no exhaustive model to study with the same laws the different topologies and scales of the Universe.

[3] Mr. Eames in his classic film *'Powers of 10'* shows how in decametric scales suddenly Nature reorganizes its information into new complex organisms.

A.

ASTRO- PHYSICS:

The 2 membranes of the Universe.

'Time curves space into masses.'
Einstein, on the arrow of physical information.

1. The Laws of Multiple Spaces-Times applied to Physics.

2. Mass is physical information. The Unification Equation.

3. The arrows of Time as Dimensions of Fractal Spaces.

4. Relativity Revis(it)ed.

5. The 2 membranes of the Universe.

6. The Quantum scale.

7. The Cosmic scale.

8. The 3 Ages=Horizons of Physical Species.

9. Chemistry: Atomic societies.

10. The Earth. The Geological World.

I. COMPLEX PHYSICS.

1. A Universe of multiple scales of space and arrows of time. Precedents and key principles.

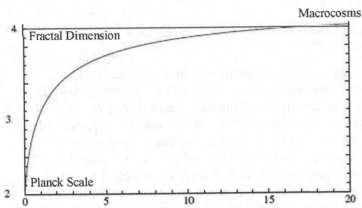

The Universe is bidimensional at small scales (Holographic Principle), but its units, Non-Euclidean points, reproduce in fractal patterns till emerging as a 4-dimensional social network of space-time, the Universe we live in. This discovery of the most successful model of quantum gravity (casual, dynamic triangulation), shows that even the simplest space-time membrane of gravitation is built with the laws of social networks.

The first creation was that of pure space without form. A bidimensional sheet of triangles, the simplest 3-societies, of spatial arrows, which bend themselves slightly in hyperbolic and concave forms and grew, as a fractal, till filling up an inmense region of reality. The translation in physics of that creation is a theory called CDT, which explains how departing from simplex non-Euclidean points, dynamically joined by flows of gravitational forces (lambda expansive and G-implosive cosmological and gravitational constants), in triangles that bended to create topologically 3 dimensions, reproduce and expand a sheet of pure space without masses or particles, called a De Sitter space; the first scale of creation. In the graph we show that growth as the fractal of pure space with only the arrows of energy=expansion in space associated to the self-similar arrows of reproduction that also provokes an expansion in space, upgrades itself from bidimensional sheets to 4 dimensional ones. This might be though seen as an overstatement. The Universe of gravitation is bidimensional, but as it grew and we became microscopic entities within it, the relative minimal height of a bidimensional sheet becomes the 3rd dimension we observe from our now relative minimal size.

In the equations of dynamic causal triangulation, which study how those simplex triangular or spherical simplex evolve into a de sitter space, we observe a single causal arrow of time which expands space and reproduces the structure just explained. The result is that arrow of energy-entropy we have analyzed in great detail, fusioned in the equations of physics as a concept: lineal time.

But then creation continues as de De Sitter space acts as a mere continuum 'sea of motion' that will become further evolved into forms, cyclical vortices of charges and masses.

Indeed, over that sheet of de Sitter space explained by DCT, a series of vortices called masses (the fastest, heavier ones) and quantum charges appeared, creating a more complex Universe, with new networks and new scales… Creation in physics is thus the generation of fractal structures in several layers of complexity. And the 4th paradigm of physics will be the analysis and translation of the equations of classic physics into those of fractal, organic physics, which has already shown some remarkable advances that this work completes.

We conclude that Physics can be easily explained with the laws of any General System of Multiple Spaces and Times, restricted to systems dominant in the arrows of energy. Since Physics studies the lower scales of reality made with simplex time arrows, energy and information, in its simplest lineal, spatial shapes and cyclical forms as forces and particles, extended in two relative fractal scales or membranes of space-time, the electromagnetic membrane, whose simplest, most extended surface is the vacuum space of galaxies, and the bigger gravitational space-time, in which light floats.

274

Between both, we exist as beings that are subject to influences from both membranes, the light membrane, which dominates our structures in the evolved form of electronic forces and the gravitational membrane, which we do not perceive but hint at it, through the study of quarks and cosmological masses and the influence of gravitational forces that inform us.

After the insights of Mandelbrot in the fractal structure of the Universe, 5 specialized works are of key importance to understand such complex Universe[Foreword, 1]:

- The Sloan survey that proved definitely the fractal structure of the Universe at grand scales.
- The work of Lisa Randall, which proves classic astrophysics can be built departing from two independent membranes of space-time (the gravitational and light-membranes of this work).
- The work of Mehaute, which proved in such Universe the existence of 2 arrows of time, expansive energy and fractal in/form/ation that switch on and off in all physical and chemical systems.
- The work of Nottale, which proved the invariance at scale of the laws of physics, the fractal structure of electronic clouds and quantum paths and established the Cosmological Constant of dark energy (repulsive gravitation) as the fundamental unit of gravitational fractal space-time.
- And the work of Ambjorn, Jurkiewicz and Loss, perhaps the most important and less understood, which proved that departing from systems of Non-Euclidean points (any form which the model reduces to points with minimal volume) with the 3 topologies of the Universe (treated as non-Euclidean triangles) able to reproduce in fractal, dynamic, growing, causal patterns, the Universe grows from a bidimensional, microcosmic world (holographic principle explained in the first part of this book) into the 4-dimensional de Sitter solution of Einstein's Relativity, creating de facto the first quantum gravity theory to replicate successfully the space-time of the 'real Universe'

The interest of this latter work, called 'dynamic causal triangulation' resides precisely in the fact that:

- It is the only model of quantum gravity that reproduces a 4-dimensional world as it is, in which time arrows create causality, unlike the background dependent models of string theory (in which space-time is still the abstract Cartesian continuum error explained in the previous parts), which ends up with 11 or more dimensions. Thus so far it is 'The Model', even if it collides with the interests of big science (CERN-like models, which still predict a space-time continuum to validate the obsolete, cosmological big-bang) and declares worthless beyond its beauty per se, the complex models of string theory so trendy among specialists and Hawking's work on time theory, which considers a non-causal Universe.

- It does so with the minimal number of elements – in fact, as Nottale does, departing just from the existence of a positive cosmological constant of repulsive gravitation, later studied in more detail and the existence of causality in time arrows.

- It validates all the basic tenants and general laws of this work, laid down in the two previous parts: Indeed, the model departs from 2 fractal dimensions in the microcosmic world (holographic principle) and arrives to 4 fractal dimensions in the macrocosmic world. The model starts like Nottale's and this author from a positive, expansive gravitational force (cosmological constant) and its implosive, informative dual force (G-constant) to build the space-time membrane of Einstein, introducing duality in gravitation as we shall do in this work. The model is background independent; thus it rejects the error of a Cartesian continuum; its dimensions are fractal, dynamic and grow; thus it creates a model of space as a fractal, growing, reproductive network, as we do in this book. The model uses fractal, non-Euclidean points (its units are 4-dimensional points which have volume and can have any form but are treated as 'infinitely distant'; that is, so small that we can't see its parts. The model uses the 3 only topologies of a 4-dimensional system (the informative, hyperbolic, spherical and planar/toroid topologies laid down in the previous part). To obtain that growth causal triangulation uses a model of diffusion, based in the laws of spatial, 'body networks': each Non-E point only communicates and reproduces its fractal form within its neighbourhood.

None of those models though explains the why of such fractal Universe, beyond the mathematical formalism – its how – at a level of specialists, because their authors still lack a complex philosophy of science to explain those whys, which we provide in this book, merging those approaches and explaining some obscure elements within it, translated those results to the jargon of General Systems of Multiple space and times. For example, in our jargon what causal dynamic triangulation describes is easy to explain: The Universe at lower scales is made of fractal Non-Euclidean points (themselves emerging from perhaps smaller self-similar scales). Those points have 4 dimensions of space-time, 3 of space and one of causal time, *which is the arrow of present-reproduction that makes the system grow and expand space-time as it 'diffuses-generates' the macrocosmic Universe.* This arrow of present reproduction or simultaneity (Einstein's jargon) creates what we called in this book a 'herd' or 'body' of space – the simplest possible social network that self-organizes itself to emerge as the Einsteinian space-time seemingly continuum. Thus what CDT describes is the first stage of creation of the physical Universe: how the quantum units of repulsive gravitation (cosmological constants) and implosive gravitation (G-constants) combine energy and form in self-reproductive, self-organizing, fractal patterns that finally emerge as the 4-dimensional gravitational space-time membrane of Einstein. This is the first stage of creation of reality. Then we must abandon the model and follow Nottale to see how from this scale we grow to create the next scale of light space-time, the electronic world. And then we must abandon Nottale and follow this book to see how further on, the laws of multiple fractal space-times keep producing waves of physical in/form/ation, masses, charges and large scale structures. Because all those models still use the jargon of classic physics, a few corrections are needed. We shall specially consider in some detail the corrections to Relativity or rather expansion of its principles needed to define better repulsive gravitation (cosmological constant), masses as vortex of space-time, the unification equation of the informative particles of both membranes (gravitational masses and electromagnetic charges) and the simplest, dominant time arrow of the Universe (present reproduction), which causal triangulation uses to create from a bidimensional microcosms, through fractal generation, the 4-dimensional Universe. Then once this background space-time is created, more complex time arrows keep evolving the Universe, 'imploding' that sheet of gravitation into informative particles (cyclical masses and lineal waves of light), which becomes the units that through social organization, energetic expansion, informative implosion and reproductive creation, form the macrocosmic structures we see in reality, better described with the ternary principle of time ages and topological spaces.

Causal dynamic triangulation, the only successful theory of quantum gravity proves that even the simplest space-time is a social, organic network, which in words of its authors 'grows like a molecular crystal' [A, 9]. It has been needed though the power of modern computing to be able to calculate exactly the patterns of that growth, which neither Nottale or myself were able to resolve with the limited computer power available in the 90s to pioneer researchers in a field that confronts the classic dogmas of science.

In that sense, what all those partial models of complex physics show is that matter and space can and must be described with the same laws of multiple time arrows and discontinuous spaces of all other systems of the Universe, taking into account that physical entities are dominant in the arrow of 'energy-past' and reproductive present. That is, its systems have the minimal quantity of information, since they follow the general law of the black hole or inversion of properties between systems of information and energy, such as:

Physics: max. spatial extension=energy x Minimal information

What this equation tells us is that organic systems, in the physical world are extremely simple in their topologies, but extend to extraordinary degrees of energy, speed and spatial extension.

Reason why physicists, as Nietzsche put it, 'are more interested in the canvas -space and the arrow of energy- than the painter'[1], the law of form, which they mostly ignore and causes most of the errors of physical sciences – such as the ignorance of the informative properties of masses, what we shall call in the context of physics, the 'Arrow of Einstein', given the fact that Einstein was the physicists who better understood those properties, describing its informative force gravitation as a force that bend the energy of space into 'whirls or clocks of time' or masses. Because of the limits of this work, we will in that regard consider in special detail the 'arrow of Einstein' – that is, the meaning of information in physics, due to the fact that it is the element of physical studies which requires further evolution.

The second fact about physics we must take into account is the 'mirage of size', which has for centuries hold the wrong assumption that what physicists say is the essence of the Game of Reality, because they study the biggest sizes of the Universe. This is not truth, as it is proved by the fact that we are gathering the same quantity of data about physical particles than biological organisms (the other extreme of reality, which maximizes the form of beings, though carries minimal energy).What physics does is to exhaust the knowledge of all phenomena related to entropy or energy, but the previous fact proves that the Universe is a balance between both arrows, energy and information, reason why in our data banks we find a clear proof of such balance:

$$Max.\ E \times Min.\ I\ (Physics) = Max.\ I \times Min.\ E\ (Biology)$$

Yet since we are living beings Biology, the study of ourselves, should be more important than the study of galaxies and atoms.

Further on, if as it seems the Universe is a fractal of ∞ space-time scales[2], in which the game of existence repeats in self-similar forms after a certain number of scales (which in the physical world seems to correspond to the atomic and galactic scale, as we shall show deriving self-similar fractal equations for both type of entities), size will be absolutely relative and so it will not matter that from our p.o.v. galaxies seem so huge. It might be according to those equations of 'unification of physical scales' that from a higher p.o.v. a galaxy is just self-similar to an atom in a Universe of ∞ scales.

This introduction to the meaning of physics is necessary, given the respect physics commands in the world of science, to tackle some of its errors, derived from their misunderstanding of the arrow of informative time and the complex arrows of reproduction and social evolution that also happen in physical systems, without raising eyebrows among the readers who have learned to 'worship' what Bacon called the 'Tribal Idols' of knowledge'[3], in the case of physics, names like Galileo, Einstein or Hawking, who built models with some structural defects due to their use of a single instrument to measure time, the clock, a single arrow, energy and a single space, the Cartesian plane.

Those basic errors are illustrated in the graph, which shows a fractal space of multiple time cycles and scales of form, which was however considered for centuries to be a simple space-time with a single cyclical form.

Unfortunately, physicists are specialists on the science of 'space', which they study in depth with mathematics; a language derived of geometry and hence specialized in the study of space. In mathematics the causal, logic language of time can be also expressed with the tools of sequential numbers and equations that reflect the Algebra of Multiple Spaces-Times - where $E \Leftrightarrow I$, becomes a function of the type $F(x)=G(y)$. All this means we must revise the classic theories of physics of the XX century to adapt it to the evolution of Non-Euclidean Mathematics and temporal logic, developed to explain a world of multiple spaces and time arrows.

Recap. The fractal nature of all Universal entities, which occupy a piece of space and last a quantity of time, is the key to unify the laws of science. In the graph, Saturn's rings have in fact the form of the commonest informative fractal: a Cantor dust, which thins out once and again into infinitesimal particles. Since when we see any continuous point of time/space in detail, it grows, becoming both, dynamic and discontinuous, made of quanta, moving in self-similar paths that are separated as independent beings, but pegged to each other by flows and networks of energy and information. The Physical Universe is constructed through those logic dualities that make it grow in bigger scales of growing complexity, departing from its 2 essential shapes, the line of energy and the cycle of information.

2. Energy=Space & Information=Mass.

In that regard, a theory of Multiple Spaces-Times applied to the lower scales of physical matter departs from 3 key concepts that go beyond the timid steps of quantum theory and relativity in its description of a reality of multiple time cycles and space scales, to its ultimate consequences:

- *A world of motions in time*, in which the perception of space is merely the still perception of a lineal motion, and physical time, the clocks of information created by the frequency of rotation of charges, masses and lineal forces. The *Paradox of Galileo*, proved when Galileo found out that the Earth seems quiet but it moves, shows that all in the Universe seems still but has motion in a certain scale of its internal form. Mass was considered continuous and still in form. Now we know is made of ever smaller, discontinuous pieces, which finally become mere 'accelerated vortices' of information, small hurricanes of space-time (Einstein's principle of equivalence between curved acceleration and gravitation). This is the Galilean paradox. All is a cyclical motion or 'clock of time' with a frequency, or a lineal motion or 'force of space' with a 'speed'. And so all is made of 'time clocks' and Energies, whose minimal combination is an 'Action', term coined by Planck in his study of light 'actions'. Yet our senses see the motions of energy as 'fixed space', like when we see a car in the night. And they see the clocks of cyclical information as 'fixed forms', as we see masses. So there is a duality of 'reality vs. Senses' that the reader must always think of: reality is made of energy and time as in quantum actions or when we say, I don't have energy and time to do this. But our mind fixes them as space and information. So we shall use here as synonymous, energy and space on one side and clocks of time and informative, formal frequencies. The stop and go analysis of reality brings many new interpretations of classic physical problems, from Michelson's experiment to the resolution of the Complementarity between forces and particles. Ultimately we can consider that all entities have a still, informative, perceptive state, Ti, and an energetic, moving, expansive state E. So we write a 'generator equation of duality', $E \Leftrightarrow I$, which means that 'all energies transform themselves back and forth into information'.

- *A Universe that extends in several space-time membranes* of different size, mainly within human perception, the quantum, microscopic scale of charges and electromagnetic forces and the gravitational, macroscopic scales of mass and gravitational forces.

The rings of Saturn seemed continuous. The brain seems still and continuous. It took a lot of time to find (Cajal) that it was made of discontinuous neurons through which motions took place. Complexity overcomes those limits to their physical inquire by establishing the existence of a series of self-similar 'planes of fractal space-time', which co-exist within the same Universe and are organized in scales of growing complexity, in which there are self-similar forms of energy (whose geometry is lineal, as a line is the fastest/shortest distance/motion between 2 points) and self-similar forms of information (whose geometry is cyclical, as a cycle stores the biggest quantity of information in lesser space).

Thus by introducing 2 essential dualities, the duality of Time Arrows and the duality of space membranes, fractal physics changes the look of space-time as nothing has done since Einstein.

Recap: Monist physics considers 3 dimensions of static space and a single *arrow of motion, lineal time or entropy, happening in a single continuous* space-time. Fractal physics studies a Universe with 2 simplex time arrows energy and information, (stored mainly in the bidimensional form of Masses & charges, perceived also as static dimensions of space=distances) and 2 fractal scales, the microcosms or quantum world and the macrocosms or gravitational world.

3. Galileo's paradox: dimensions of space as fixed motions.

The Universe is made of motions in time, which perception fixes into still mappings, confused with forms of space. Saturn's rings are not continuous planes, despite their appearance. In detail, they become herds of planetoids in motion, tracing orbital cycles around the planet, which are self-similar to the paths planets make around the sun. Further on, when we observe each planetoid in further detail, it becomes a sum of self-similar atoms, rotating also in cyclical paths. Thus a simple, single, flat space-time that seems continuous, without motion, becomes a series of moving, complex, cyclical, discontinuous, fractal space-times, organized in several scales of size and form, each one made of a network of self-similar Non-Euclidean points.

According to Galileo's paradox in the Universe all motions are forms and vice versa. And so the duality/invariance of motions and forms can be considered a consequence of the invariance of the 6 arrows of time, which are geometrical motions traced by all entities even if it appear as fixed formal dimensions.

Indeed, the dimensions of time become dimensions of space: a spatial motion along a speed creates in the process space, as the movement reproduces a form in space along its trajectory. If the process again repeats itself, creating acceleration, again a second dimension of time is recreated. And so taking this principle a bit further, if constant speed is a lineal dimension of space, then acceleration which is a further motion of speed, a 'change' of speed in time, $a=v/t$, can be considered the second dimension of space. For that reason a bidimensional, rotational motion is considered to have acceleration. And if we were to move laterally such rotational motion, forming a 'Maxwell screw' we would create a 3rd dimension.

So the dimensions of time add up to create a 3 dimensional, fixed surface of energy that can be seen as Maxwell screw or as a complex set of 3 motions.

Further on a system might accelerate, maintain a fixed speed or decelerate, creating *the 3 ages of existence of a physical motion/ space*. This was foreseen by Einstein, when we affirmed that the fundamental Principle of the Universe is the Principle of Equivalence between acceleration and force, which defines a Universe in perpetual temporal movement, always creating and destroying 'dimensional space-time' as it moves and 'prints' form over surfaces of relative present space. Further on, as we shall see latter, the 3 solutions to Einstein's spacetime equations, the expansive, accelerated big-bang, steady state and implosive big crunch (Gödel's solution) form in fact 3 ages of the Universe both in terms of distances and motions.

But what this also means is that dimensions are never infinite, as they have a 'timing' of extension. Thus 'cyclical, temporal movements' and lineal trajectories, reproduce limited, diffeomorphic space-times, creating what mathematicians call a fractal dimensions of space-time. Since the motion will die away.

The previous graph - the rings of Saturn – illustrates all those concepts:

-The duality between continuous, fixed space vs. fractal, moving time. Since Saturn's rings are perceived as both, depending on the detail we observe. Thus fractal spacetime is both a physical duality and a mental duality, caused by the limits of perception of our mind, which converts into still space what is a moving, temporal reality. That subjectivity of perception explains many of the paradoxes and errors of our understanding of the Universe. The example of the graph shows clearly this: In perceptive terms, Galileo discovered Saturn's moving planetoids and called them rings, thinking they were a continuous, fixed form; when they are discontinuous satellites. The limited perception of his instruments erroneously transformed the moving cycles of those planetoids into a fixed ring-form; putting the multiple, fractal microcosmic and macrocosmic scales of matter that made up those rings together into a single space-time. So not only human senses, but also mechanist senses are inaccurate, as the recent discovery of a 96% of dark energy and matter has proved. Thus, the continuity and immobility of space and information is an 'error of mental perception', since the mind pegs the discontinuities between objects that we 'don't need to see' and fusions instead all fractal space-times into a single surface of reality. The discontinuities are there, as they are in a movie, which is a fractal series of pictures we put together into a continuous image. In both cases, we do not see their discontinuities, because in order to survive we do not need to see the voids of formless space that lack any relevant information or any usable energy. It would only cramp our senses. So we don't see for biological reasons the energetic movement of microscopic vacuum space; and we see a continuous Saturn's ring, instead of a moving foam of quantized planetoids. A fact that leads to a biological understanding of perception and information as useful tools of survival:

The mind creates stillness in a moving Universe. We call that error of perception, the 'Galilean Paradox'. Since Galileo discovered that we perceive mentally a fixed Earth; while in fact the Earth is moving (e pur si muove). Yet what abstract Physicists forgot to ask in the next 400 years is: why we perceive the Earth quiet when it is moving; why our senses cheat us? The answer is biological: People perceive moving time cycles and moving energy only in their relative plane of space-time existence, where such knowledge is relevant for their survival. The human mind transforms the remaining time cycles into fixed information and fixed space, which seems a still, continuous void. In this manner, quantum, moving, temporal energy becomes virtual, fixed forms, set up against a mental background of continuous, static space. So we see a quiet Earth instead of a moving planet. Stillness is a perceptive paradox caused by the fact that we see the Universe as static information to distinguish it better.

The Universe is dual, paradoxical, relativistic, dynamic, ever changing its in-form-ative and energetic nature. It is not composed of substances, material particles and static fields. It is made of dimensional arrows that move and reproduce, tracing the 2 essential forms of space-time: the line energy and the informative cycle. Each of those movements becomes still when a mind perceives it, and measures it, as Einstein put it, in simultaneity, but it is a movement, a dimensional arrow. A proton is not made of 3 fixed particles, called quarks, but of flows of lineal gluons and cyclical vortices of mass (the quarks) in perpetual transformation. A body is not a fixed substance, but a series of physiological cycles performed by its cells. A cell is not a fixed sphere, but a series of cycles traced by carbohydrates that seem, put together, a fixed substance. Those carbohydrates that seem fixed are in fact atoms tracing energetic, lineal and temporal cycles; yet atoms are particles doing cycles and particles are gravitational and electromagnetic, cyclical vortices that absorb lineal forces and can be modeled with the mechanics of superfluids. So all is in eternal movement, even the physical space/time that seems fixed to us, due to the tendency of the mind to see stillness. The Universe is made of movements and events, of actions in time, not of fixed substances, of forms in space. That is why the cycle of space/time existence is its fundamental particle. The perception of the Universe as a fixed series of continuous forms is an error of the mind, which tends to see form instead of movement, because movement distracts and occupies mental space, but formal stillness focuses what we see and can be stored in a smallish brain. In fact, we can see either movement or form, but not both together. We do not see the Earth moving and quiet at the same time. We do not see a wave and a particle at the same time (Complementarity Principle). The mind has less dimensions and volume than the Universe and for that reason it eliminates movement that occupies mental space from most of the entities it observes. Thus, the Galilean Paradox explains also the Uncertainty principle, one of the fundamental dualities of the Universe: the fact that we see things either as still, discontinuous particles; or as moving, continuous waves. Since the physical entity is both things together: the relative 'head'/informative organ of the physical system is the particle, while the field of forces is the reproductive body which carries the particle around space. And both co-exist together. But we can only perceive one, in the same way we do not perceive a human being as a whole and as a series of cells together, since the mind is limited to observe only one plane of space-time existence. Thus, we don't see the scalar, fractal nature of the Universe, but a single space-time continuum. A human being is also a whole made of fragmented cells - while our brain is another example of the duality between discontinuous information and continuous space; since it was considered for long a continuous mass of neuronal tissue, yet it turned out to be also broken into discontinuous of in-form-ation, joined from the perspective of 'energy and movement' through continuous, dynamic flows of nervous messages.

Fractal=fragmented space-time networks are also continuous (if we look at the flows of energy that communicate the cells of the network) and discontinuous (if we look at the holes, or dark space of the networks and its fractal particles). Duality implies that any continuous whole is also a fractal, organic system, sum of microcosmic actions and cells. It all depends on the detail of our observation. The limits of human perception imply that all what is not clearly perceived by the mind becomes still and continuous: if we observe very slow or very fast cycles, as those of a mass-vortex or a wheel spinning, or we observe forms from far away distances, as the rings of Saturn; or they are very small, as the cell cycles of your skin, they appear as a single undifferentiated still, mental form, fixed in space and integrated into a whole organism, despite being made of fractal, moving cycles. Thus, the continuous/discontinuous structure of space-time is one of the many dualities of the 2 opposite, Se⇔Ti, forms of the Universe:

Amorphous, continuous, Energetic Space ⇔ Fractal, discontinuous Informative Time Cycles

Further on, space is the static perception of moving energy and time the moving perception of static information (duality of the Galilean paradox). Space is motion relative to a frame of reference, said Einstein, explaining the equivalence of energy and distance...

Those dualities explain the apparent paradoxes of Quantum Theory and Relativity: on one hand, time is indissolubly merged with space, bending it into future form, in-form-ation. On the other, Einstein proved that there are many rhythms of time, which means each space-time in the Universe has a different way of changing and processing its energy and information that combine to create a present space-time. So what is space-time, a discontinuous sum of fragmented quanta, as Quantum theory said, or a continuum? It is both. It depends on the detail and perspective of our observation: Time cycles breaks space-time into new, discontinuous forms. Yet from the perspective of pure energy, the Universe is a continuum, amorphous surface. For that reason, Einstein, who analyzed the grand scale of the Cosmos without detail, came to the conclusion that the Universe was a space-time continuum. Yet when we observe it in its fractal details, Time and Space are not a continuum. As Einstein himself realized, there are in the Universe infinite clocks, forms of reality that trace discontinuous, temporal cycles with different speeds. But there are also, as quantum physicists noticed, ∞ fractal spaces divided by membranes and discontinuities, both in organic and physical entities. Thus, the duality between fractal and continuous space-time is an irreducible duality between a whole, which is broken into parts; and it happens everywhere: Your mind is a whole made of electromagnetic impulses that cross through broken neurons; an organism is a whole broken into billions of cells. The word fractal means exactly that: 'fragmented'; something that is still a whole, but in detail is broken, crackled by cycles of time that create borders in space. So fractal means both: a continuous whole and a fractal group. Ultimately, all informations are fractal (as form needs discontinuities to be perceived), born out of temporal, cyclical rhythms that return periodically, in a discontinuous manner to a certain point, tracing again the same form; while energy is amorphous; hence it has the appearance of continuity.

We already mentioned that time is formal change. So if we were to do exactly the opposite of what Physicists do and temporalize space, we could say that:

- The 3 dimensions of space create a morphology of information, the form of space-time that Einstein defined as G-curvature (where height gives that curvature). Thus, we can further reduce those 3 dimensions to a G-factor that defines the total curvature of space, as Riemannian geometries do.

- Then, from a perspective, when that morphology with curvature moves, it creates the 4th dimension of time or time movement. In the graph it is given by the temporal reproduction of that height cycle.

- Finally, a 5th dimension of acceleration happens which further multiplies the total space of the being. Yet at this stage morphological Biology cuts off the dimensionality of the Universe because it can be proved mathematically that the number of dimensions that create the maximal volume are 5 dimensions. Then the volume of space-time enclosed in a world of more than 5 dimensions sharply falls. We can see that 5th dimension in movement as acceleration or we can extend it into space reducing it to constant speed. This is secondary to the fundamental truth of a biological Universe that as all cosmologists know have a 5th dimensional shape, whose meaning now we fully grasp, thanks to the Galilean paradox. To see those dimensions as fixed extensions of space or accelerating and moving fields is equivalent. For example, in Relativity a vortex of space-time with a lot of curvature/acceleration has a π value equal to 3. In terms of an accelerating mass vortex what curvature does is to create a spiral instead of a cycle that moves inwards. So $3=\pi$ means a spiral vortex with a reduced π-diameter, since the inner branch of the spiral vortex is shorter inside the accelerating vortex. Relativists use a hyper spatial 5D to explain it. Yet it is closer to what humans observe to define that 5th, bending dimension as an accelerating time-change.

Recap. The paradox of Galileo is the key not only to unify quantum physics and relativity but also to reconcile or rather explain in terms of dynamic, moving time, the theories of modern Physics with multiple dimensions of space, as it turns out that any dimension of space can be considered a dimension of time and vice versa.

4. 3 dualities: gravitation/light, distance/motion & entropy/form.

3 are therefore the most important dualities of physics:

-The duality between the membrane of gravitational spacetime or macrocosms and the membrane of light-spacetime or microcosms, which is built as a light, trophic structure over the bigger membrane; hence with slower speeds and higher disorder.

This can be expressed in terms of limits of energetic speed, whereas C is the border between both membranes:

$$Light\ space < C < Gravitational\ space.$$

$$Quark\ /\ Black\ Hole\ world < 0\ K < Light/Electronic\ world.$$

-The duality/complementarity between energy forces and fields and information particles, forces and fields.

-*The duality between motion and form*: what we perceive as vacuum space or distance is also a motion and both together form an action, the minimal exi 'substance' of reality. So we can write:

Light-membrane =c-speed motion=Vacuum made of H-constants

Gravitational space >c tachyon motion=Vacuum made of strings.

-The existence of 'doors' between both Universes, which follow the general laws of jumps between space-time planes and resolve most of the conundrums of modern physics (4th i-logic postulate).

When we understand those simple dualities everything becomes clear, concrete and real. Yet we have to reorder the particles, fields and forces of the Universe according to those dualities, such as:

H-Plancks are the minimal actions (motions&forms) of the light-membrane, which evolve into photons, electrons, plasma stars and human minds among other things:

$$\sum\sum\sum\nolimits_{st-2} Plancks = \sum\sum\nolimits_{st-1} photons > \prod\prod st\text{-}electrons > \prod\nolimits^{st+1} Hot\ stars$$

And so the electromagnetic force and electron are the energy/information, lineal/cyclical elements of the membrane.

Strings are the minimal actions of the gravitational membrane and must be redefined mathematically as background independent. That is, they do not exist in the abstract Cartesian plane of spacetime but they are the actions of energy and information of the gravitational world. And they are tachyon strings moving faster than light. They are therefore the first 'bricks' on the evolution of gravitational particles such as:

$$\sum\sum\sum_{st-2} Strings = \sum\sum_{st-1} gluons > \prod\prod st\text{-}Quarks > \prod^{st+1} Frozen\ stars$$

And so strings and gluons are the massless forces of the gravitational membrane and quarks and strangelets and black holes its informative networks/particles.

Whereas the bigger gravitational membrane encloses the smaller light-membrane, in both limits (strings are thinner and faster, more extended in space than light; and quarks and black holes are heavier, more informative/ordered than electrons and stars).

So we must reorder also the forces of the Universe:

Electromagnetic 'light'>electr(on)ic forces: Light membrane.

Gravitational 'strings'> Strong Quarks:Gravitational membrane.

This leaves a final force, the weak force, which *is not a force in space but an event in time that transforms particles and forces from one to the other membrane.* And it is in fact the less understood force, plagued of errors in their formalism (the never found Higgs, normalization of absurd infinities, etc.) So the Z and W particles do NOT mediate this force but are intermediate states of particles that 'transcend' from the electronic membrane into the mass world.

In that sense, while the field of analysis of the light-membrane can be considered more or less closed in its mathematical formulation, the field of gravitational theory and its different particles and forces is not (strings in its present background dependent, multidimensional, super symmetric formulations are incorrect, but the original bosonic, tachyon strings and its Nambu actions are a good departure to work out a better model for the two possible fractal scales in which they might be used to describe the Universe as Planck strings).

By all this we mean, certain features of bosonic strings fit what should be the minimal particle of gravitational space: its Length (Planck's Length); its duality (closed and open strings equivalent to the minimal line of gravitational energy and cyclical gravitation), its 'vital properties' – they can reproduce, creating bidimensional sheets; they can mutate between both states and or vibrate and so they create a Generator Equation of space-time actions able to replicate the particles of the gravitational membrane, which we could write as:

Energy(open string) <vibrating string> Information(closed string)

In that regard, physicists have created an astounding frame of mathematical equations to describe all those systems, which probably ranges with the work of biologists (Evolution Theory and genetics) as the highest mental achievement of the human race, and so this text will not correct the formalism of physicists, which except for the case of weak forces needs no correction, but lacking the proper Non-Euclidean i-logic geometry and the understanding of relational, cyclical times and multiple space planes can be greatly improved in its interpretation NOT of the how but the why of reality, which is what we shall do in this work.

It is unfortunately an astoundingly difficult task not because of its complexity – far easier to understand than the formalism of quantum physics or the metric spaces of relativity – but because of the Pythagorean, mechanist, entropy-only, continuum single space dogmas of the foundational fathers of physics, which are taken as true postulates and cannot be argued easily – what Bacon called the idols of the tribes of science.

All those errors depart from a false ideology, the absurd attempts to simplify the dualities of the Universe into a single monist theory of it all; which departs itself of the wrong interpretation of the Occam's principle of simplicity; since as Parmenides proved the absolute minimal simplicity to produce actions, motions and change in the Universe is duality. So if physicists were to prove that all the forces of reality are one, all its scales of space-time are one, all its clocks of time are one, all its simplex arrows of time (entropy and information) and complex arrows (reproduction and social evolution) are one the Universe will become an absolute nothingness. All those concepts might work as simplifying tools of measure – so we equalize all cyclical time rhythms into a single mechanical clock, we reduce the gravitational membrane (relativity theory) to its perceived effects in the light-membrane and so on. But 'ceteris paribus' analysis and simplifications are not the rich, complex 'why' description of reality we seek here.

In that regard, before we fully plunge in the solutions provided by the new model of relational times and multiple spaces we have to tackle the problem of the dogmas of monism that prevent to resolve the standing questions of astrophysics.

Recap. 3 dualities structure the physical universe, the duality of gravitational/light space; the duality of entropy forces and informative particles and the Galilean duality of motion in time= distance in space.

5. The tribal idols of physics[3].

The fundamental error common to all scientific thought is the error of postulates set a priori without proof, from where an entire science can be deduced, infected by that false or unverifiable postulate. For example, when considering the structure of space-time the Cosmological Principle, which has been taken as a paradigm of truth in which to base all studies of the Universe, affirming the isomorphism and continuous distribution of matter and energy in the Universe. This is false as we find structures, super-clusters, fractal scales, invisible matter, dark energies, discontinuities, etc. in the Universe. And ultimately we find that all physical phenomena happen as all other systemic events, by interactions between 2 spacetime discontinuous spaces (particles and fields; electromagnetic and gravitational membranes, etc.) However, homogeneity and continuity pleases the perception of the mind and makes Cosmological calculus with Relativistic equations possible, so it is an accepted simplification.

All sciences have errors caused by postulates without proof and partial theories discovered by the founding fathers of each science, which are just steps of a continuous process of evolution and expansion of knowledge expressed in the Principle of Correspondence (each new theory embraces the discovery of the previous ones, corrects its errors and advances further in the path of explaining reality with the languages of the mind). Unfortunately while knowledge is a continuous process of advancement, human brains are closer to the mother-boards of computers: they are imprinted with memorized ideas, at earlier age, and those ideas are taken as dogmas, and its discoverers, in words of Bacon, become 'tribal idols[1]', high priests of science that cannot be denied. This process also applies to XX century physics, which is not the end goal but just a first timid step in the process of transforming the paradigm of science from the simplex view of a Newtonian, continuous, abstract, single spacetime into the complex view of Leibniz: a relational, discontinuous, organic, fractal spacetime made of 'quanta': points related by flows of energy and information that create the networks of reality.

The first steps were given by Riemann, in mathematics, when he defined a space made of networks of Non-Euclidean points.

Because physics is merely the mathematical interpretation of the simplest scales of energy and information of the Universe, it was precisely the application of that concept by the two colossus of XX century physics, Planck and Einstein, what renews the field: On one side Planck found reality to be made of quanta, which were obviously the Riemannian spaces of networks of points with a different 'density, according to the homogeneity between them'.

Planck was studying in fact the membrane of light-space in which we 'swim', as evolved forms of light quanta. Soon Riemann's theory about space were confirmed in the study of lasers, which were the densest light-space due to the homogeneity of its quanta and the discovery by Bose-Einstein of the densest Riemannian spaces, bosonic spaces that happen when undistinguishable particles occupy the absolute minimal space, whose density was defined by Planck's 'God's constants' (as he hypothesized that the most perfect form of the Universe would be the one with the highest density of quanta).

Einstein then applied the 5th postulate of Non-Euclidean geometry, the only found in the XIX-XX century till I discovered the other 4, to describe *the other membrane of space-time, we cannot perceive, in which light space floats, gravitational space whose parameters of energy-speed and information/density of quanta are higher.* He also found key properties of time: its cyclical, geometrical form (time bends space into mass), the existence of multiple clocks of time ('I seem to be the only person that thinks there are infinite clocks of time with different speeds in the Universe'), and the confluence of past and future in simultaneous presents ('the separation between past and future is an illusion'), which quantum theorists also discovered in the study of particles and antiparticles.

Meanwhile quantum physicists found that the light-membrane particles (electrons) communicate through the lower, faster than light, non-local gravitational space, acting as a harmonic networks at enormous distances (entanglement). They also found the complementarity between particles of information and fields of energy, which co-exist in all stable systems of physics. They realized that the observer is limited in its capacity to extract information from those other planes of existence, as its particles react to the bombardment of our 'huge instruments' collapsing into defensive particles as any herd in nature does when attacked by an energetic weapon (fishes attacked by sharks, troops in combat formation).

Yet they failed, in an age of anthropomorphic idealism and mathematical pythagorism, to recognize those were organic properties of particles that gauge information and act-react to survive better in their environment. The same happens when those herds go through slits as particles or waves, according to the better path, or when they constantly check their distances in herds, or when they follow a 'soliton' or first particle in its lineal motions.

But for 'self-centered humans' (Galilean Paradox of the mind) the idea that an electron could perceive being so small attacked the fundamental endophysical principle that we are the center of the Universe because from our p.o.v. our nose is bigger than the Andromeda Galaxy. Of course, we are from the perspective of Andromeda galaxy smaller than an electron, and as we shall see it might be a fact that a galaxy is self-similar to an electron of the gravitational membrane. This ultimate truth of a Universe based in absolute relativity – the existence infinite self-similar fractal scales – is of course, the most rational truth about reality that the Galilean paradox will always fail to accept. So quantum physicists also failed to realize that electrons were fractal particles made of 'cellular populations' of smaller electrons, described with probabilities in space (population statistics) and instead thought those nebulae were probabilities in time (hence the same electron moving in time and occupying several points at each moment). Thus, instead of developing a theory of fractal electrons and fractal space-times, they have created a theory of parallel events and parallel Universes (Copenhagen and Everett interpretation).

Recap. The main errors of physics derive from anthropomorphism, which fails to understand absolute relativity – the infinite scales of reality and fractal spaces, the organic nature of waves and particles and electrons as fractal, cellular entities.

6. Modern physics and the errors of a monist system.

Since Time is Change, in Physics time implies a rhythm and a movement, a mutation in the position of an object in space. In that sense, physicists study a specific, very limited part of the wider field of time=change: the change experimented in space by an object moving through it. In the study of such movements physicists have achieved a dexterity and sophistication rarely seen in science. However our critique has been harsh because physicists reduce time studies to that specific v=s/t definition of time as translation change and deny other analysis of time-change; when there are many other forms of time=change, known for millennia and seriously studied since Aristotle by fundamental sciences, like Biology (which studies the change in the information of species, through the sciences of Genetics and Evolution) or History (which studies the change in human societies). Thus only, when we depart from such a wider view of time, we can fully understand its meaning in Physics and the paradoxes and limits of its inquire about Time and Space.

According to the Correspondence Principle, a new, wider scientific theory (of time and space) has to include all previous theories within it and solve the questions unresolved by them. Since the 2 main theories of space and time were in the XX c. Quantum Theory and Relativity, it is unavoidable to deal with the errors and virtues of both, from the perspective of cyclical time and fractal space. The foundational concepts of those 2 theories are especially relevant to cyclical/fractal time space:

Quantum Theory is based in the concept of an action, the unit of the Quantum Universe, parallel to our existential force or existential momentum, exi=H or Planck constant, which is defined by the product of energy and information, *but must be understood as a dual form/motion* under the paradox of Galileo, not a mathematical function of probabilities. Hence an action is the ultimate substance of light and the membrane of light space-time that we inhabit. In that regard the principle of complementarity and the Uncertainty of Heisenberg are concrete facts of our world of space-time made of H-actions, made of Planck constants; which can and do evolve into particles of higher information, photons and electrons, warping that light space membrane in which we exist. This interpretation takes away the 'idealistic' 'Pythagorean' belief of a Universe made of numbers and returns it to reality. Vacuum is made of light, its space-time substance, which exists even in the emptiest vacuum (background radiation) and can be perceived either as motion, a c-speed substratum of energy, short of the ultimate superfluid ether, or as a distance, a wave-space with minimal form. And can easily evolve and it constantly does so into photons with higher form (particle-wave duality), and when its density of energy increases into electrons,

Relativity on the other hand is based in 2 principles also closely related to the Galilean paradox:

One is the Equivalence Principle that states the equivalence between acceleration and force: a force is in fact an accelerated movement. As it happens in a fast car, when we accelerate we feel a force and so movement is a perpetual event with 3 'ages', an energetic accelerating state, a steady motion, and a 3rd age of deceleration; even if we, due to the Galilean Paradox, see motion sometimes as still space.

The second principle is the 'Diffeomorphic' (local) Principle of Relativity that states the relativity of all points of view, all points of measure, which are local and subjective, as the Universe is broken in multiple frames of reference Non-Euclidean points that gauge and measure the world around it with infinite parallels that merge into that point (5th non-Euclidean postulate).

- The 2 arrows of physical time in relativity are the explosive, lineal arrow described by the equation of Energy, $E=Mc^2$ and the implosive, cyclical arrow that 'bends space into time' or information – the arrow of information.

Those 2 principles/arrows of space-time are the essence of any physical reality; since they apply to every fractal, local space-time field, defining in its complex interrelationships the ultimate, philosophical nature of the Universe. Yet to grasp its potential to explain the Universe as a machine in eternal movement, broken into infinite, local replicas of itself, we have to clarify the conceptual errors

Physicists make interpreting them. Since only when we properly correct those errors we will be able to fusion the foundational Principles, of Quantum Theory and Relativity (and its equations, in the more complex version of this book), achieving the long seek Unification Theory of all physical beings.

Recap. The theory of time as change in form explained by Darwin with *verbal* words is Evolution Theory. The advances brought about by Topology, fractal mathematics, Time Duality and Non-Euclidean Geometry, applied to Physics solves those shortcomings, completing Relativity theory, showing that mass is cyclical information, that the Universe is structured in two membranes, and light is imprinted, forming and warping the gravitational membrane.

7. The single time arrow error: entropy.

Let us consider of those errors the most important of them, the use of a single time arrow in quantum physics. Quantum theorists, despite the discoveries of Einstein in time theory, use only a single mechanical clock to equalize all time-clocks of reality, which derives in the belief of the existence of a single arrow of time. And this causes multiple contradictions, being the most important of them, their incapacity to understand masses and whirls of gravitational time, which carry the information of the Universe. Since quantum physicists were still using the Aristotelian logic of a single arrow of time, energy or entropy. In that regard, of all the errors of physics, caused by the use of a continuous space and single time arrow, the fundamental one is the study of only an arrow of energy or entropy that disregards the meaning and functions of information and mass in the Universe.

Indeed, Physics, unlike Biology, Complexity and Social sciences still uses a single time arrow, born from XIX C. studies on heat and entropy. This introduces grave errors in the conception of mass as a bidimensional vortex of information, and paradoxes like the paradox of missing information, derived of the study of black holes by Hawking as 'entropy' systems – when they are exactly the opposite, the informative creators of mass that balance the entropy of the Universe making it immortal.

The strength of the 'tribal idols' of a science are transparent in that case. Because of the 'dogma' of an entropy-only Universe, and despite no proofs of black hole evaporation, this hypothesis is today a dogma, even if it breaks among many other laws:

- The 1st law of entropy: the black hole is hot and evaporates in our cold environment getting hotter, so it becomes an eternal motion machine.
- The laws of conservation of energy and information, since information is destroyed with the evaporation.
- The laws of Relativity and Gravitation, as black holes become 'quantum objects' with the excuse that they are small and should have quantum effects, but in relativity small is a meaningless world and quantum is not a theory of the small but of the electromagnetic membrane.

And so on. The correction of this error is simple; since the event horizon of the black hole that evaporates turn out to be according to the Laws of Non-Euclidean topology that define the black hole as an open ball of hyperbolic mass-information, the border of our electromagnetic Universe, which therefore produces our type of particles, virtual photons and electrons, which condensate our light-space membrane, fall into the black hole, (given the accelerated gradient of the gravitational force towards the center of the black hole), becoming converted into high-ordered mass – probably top quarks – and so *what evaporates is our Universe, not the black hole, which grows and cools down, respecting all the known-known laws of science except the 2nd law: entropy is a local property of the light membrane but gravitation is an informative force that restores the order of the immortal cosmos. So* the hypothesis that the black hole evaporates because the event horizon is not our world but the black hole – the hypothesis origin of this theory - should be rejected, but it is not because only in this manner we can affirm that there is one arrow in the Universe – entropy and pursuit the extermination of the other arrow – information=mass, giving origin to unification theories of both forces (quantum gravity), which merely eliminate the informative properties of gravitation. We will return to this 'murder' of the mass-information of the Universe latter, as we shall provide the proper unification equation of charges and masses as two self-similar membranes or 'scales' of reality and describe properly with the new tools of i-logic geometry the structure of black holes.

The same can be said of the continuous space concept, whose main error is the big-bang theory, which is always a relative big-bang of a scale of the Universe (a black hole caused by a supernova big bang; a quasar big-bang of a galaxy and perhaps a big-bang of our Universe, cell of a bigger hyper universe). And in any case is the 'death' of a past Universe that latter will give birth through a creation process in 3 ages to the present Universe. Yet since physicists ignore the life-death cycle of all spacetime systems, they ignore the difference between birth and death processes; what we call here 'big-bangings' and 'big-bangs'.

Another error of the continuous spacetime is the idea that the Universe will expand forever, when it is obvious that it contracts in regions dominant in mass-information (the galaxy) and expands in regions dominant in energy-forces (the intergalactic space). And both effects together create a total self-similar volume of reality in each scale.

All those entropy-only, single spacetime continuum models forget the informative effect of gravitational forces and the 3 ages of informative evolution. Monist physics instead tries to create a Universe with only one time arrow entropy=energy….

Some of those errors are corrected in this work, notwithstanding the admiration and respect that the work of those pioneers awakes in the studious of history of science, and acknowledging that while the mathematics of quantum and relativity work perfectly to describe the universe (as in the previous case in which the same probabilities are used to describe populations of fractal electrons in space and events of a single electron in time), the proper philosophical interpretation is the one given here.

Thus, following the *Principle of Correspondence*, we will offer logical explanations and corrections to the 4 basic laws of relativity and quantum physics, the Relativity Principle, the Postulate of C-speed, the Complementarity Principle and the Uncertainty principle, showing that while those equations still hold as instruments of measure, their interpretation under the new postulates of i-logic geometry and discontinuous spacetimes make rational all the paradoxes of quantum theory and relativity.

Recap. This work advances the comprehension of a complex Universe of multiple, 'relational' space-times first described philosophically by Leibniz, improving on the work of the previous paradigm, with the new tools of i-logic geometry; since both Einstein and Planck and their theories failed to complete the formalism of relational space-times because it lacked the mathematical tools of fractals and the 5 postulates of Non-Euclidean geometry, needed to fulfill that task.

8. The generator equation, its invariant dual forms and events.

How the 3 dualities, of light/gravitational membranes, of energy and information forms and motion/distances are played in the Universe to structure its dynamic actions and transformations?

To understand how we need to bring the Generator Equation of space-time events:

$\sum Se \Leftrightarrow \prod Ti$, *or principle of conservation of energy and form*

Where Se represents any field of forces or 'boson herd' and Ti, any particle or network of particles. Thus we affirm the Fundamental Postulate of Fractal Physics:

'All events and forms of the Physical Universe are generated by simplex or complex transformations of energy and information, defined by the generator equation of spatial energy and temporal information; such as the total quantity of energy and information, of the event or form, calculated as the sum, of all the forces, $\sum$ and integration, $\prod$ of all the particles, perceived as motions or distances, across the two gravitational and light membranes of space-time, remains invariant'.

In other words, the Universe is a closed system and so are all its partial parts/events when we consider both all the initial energy/information species and final products of the event or form, which might exist in any of the 2 membranes of reality.

The difference with self-similar principles of monist physics is clear, as we need to add both membranes - one of them invisible, hence of difficult measure (the gravitational membrane) – for the principle to be respected.

Yet we have to explain another key word of the previous definition, which is required for those events and forms to make sense and create a 'stable' Universe – invariance. Indeed, the Universal Syntax of the generator equation defines events that transform energy fields into information particles, or combine them to reproduce a self-similar physical entity or associate them into bigger structures (herds=waves and networks=particles), but it does not define what kind of parameters of those events and form do not change under such transformations.

In classic formulations of physics, the absence of change is often translated as 'gauge symmetry', but we prefer the easier to understand term 'invariance', which essentially means that all motions and forms of the Universe remain invariant *when the event is merely A) a translation in space; B)a social evolution in size from one st-1 scale to another st-scale.*

And so the fact that motions and social evolutions keep the form and energy of the system is what we perceive as static reality, while all other events we observe are transformations explained by the generator equation.

Self-similarity between invariances and dualities.

Now that we have extended the laws of physics to its details we can do the opposite and integrate them by considering that those invariances are in fact expressions of the harmony between the dualities of membranes, motions/distances and energy/form. Indeed, the 3 invariances in physical events are:

- *Invariance of scale*: we consider the physical Universe to be made of two fundamental scales of space-time, the larger=faster world of gravitational forces and denser quark particles and the smaller world of quantum, electroweak forces and lighter electrons. And both interact together precisely because physical events are invariant when they are transformed from one to another scale. Reason why the total energy and form of an event is conserved when we add up both scales.

- *Invariance of topological form*: we consider 2 eternal forms, the lineal forces, which in the 2 membranes are light and gravitation, and the cyclical vortices, which are charges and masses. And so reality exists and the events, E⇔I are possible because that invariant duality, which allow us to define any species as either a particle or a field of energy.

- *Invariance of motion*, which is the essence of relativity theory. It can be deduced by the fact that motions can be seen (Galilean paradox) as fixed distances, which are obviously invariant forms.

In the terminology of Multiple Spaces-Times we could consider those 3 invariances as the invariance of the 3 main terms of that equation E, or invariance of motion=energy, I or invariance of topological form and st or invariance of scale. And so those Invariances ultimately represent that what always remains and gives 'meaning' to the generator equation of reality that after all its events and transformations will still produce E, I and its herds and networks its st-planes of existence. We could say that all becomes transformed to remain the same.

All this ultimately means that the Universe is a topological reality where what matters are not the 'sizes' or 'distances= motions' of the total event described as a topology of space-time but the 'forms' of those topologies, which are reduced to the 3 morphologies of a non-Euclidean point. So what we see in fact is just invariant topologies (where there are irrelevant motions and growths of sizes) which can be transformed into each other within the restricted events defined by the Generator equation of space-time.

289

Of course, the humor of this is the fact that physicists are far more interested in measuring the 'invariant events' that do not matter that the Generative transformations that truly create reality.

Recap. All physical phenomena are either invariances in the motion and form of the physical system, which occur when the entity moves in the same plane of space-time or the form evolves socially from a cellular, fractal group into a superorganism, or transformations, which are partial equations of the generator equation of spatial energy and temporal information.

9. Laws of Multiple Spaces-Times applied to physics.

Given the limits of size of this work we can only correct a few of the errors of classic physics and give a general overview of the new paradigm of fractal i-logic physics and discontinuous space-times. Let us then summarize that overview, considering the main laws of Complexity that apply to physical phenomena:

The first set of laws of Multiple Spaces-Times applied to physics is the Duality/Complementary principle, according to which all systems and theories must be dual, with an energetic/entropic and informative component.

In essence all physical particles are dual, composed of an informative knot and a field of energy, which can be considered to be made of 'cellular', fractal parts, self-similar to the whole. This gives origin to the Law of Complementarity in quantum physics (all particles have an associated wave-field); to the main Event of all physical particles, the so-called 'boson-fermion equation', according to which big particles (fermions) exchange their cellular components (bosons and gluons), giving birth to a particular case of the Generator equation:

$$I_{st} \text{ (fermion)} < boson\ wave\ (\textstyle\sum e_{st-1}) > I_{st} \text{ (fermion)}$$

Complementarity is also the origin of the phase space that defines an electron nebulae, which has 2 components, the energetic real number and the informative, imaginary numbers, which are both at a 90° angle, such as when the energetic component of the electron phase is maximal, the informative, imaginary component is zero and vice versa. Thus the phase equations of the electron are a mere case of the $E \Leftrightarrow I$ dynamic, internal relationship between the energetic and informative component of the electron.

The inverse properties of energy and information explain why indeed the energy component is + and the information one, which contracts space into form is negative (imaginary) and why they might cancel each other, creating NOT an organic complementary event but a Darwinian self-destructive one.

The second set of laws of Multiple Spaces-Times applied to physics is the 3 invariances: motion, scale and form:

Physical theories are called gauge theories, a term which refers to the symmetries of measure produced between particles that exchange fields and whose parameters remain the same if we were to change the position of those particles. For example, if we change the position of two charges or increase their potential in equal value, the form and intensity of the field do not vary. This is gauge symmetry, a term coined by Weyl long ago, which today is often substituted by a more proper concept: that of invariance. So we say the electric field is invariant to the space orientation and potential of its particles. The laws of symmetry or invariance are thus the ones that respond to a simple question: If the Universe is a tapestry of infinite points of view, which trace their own paths of existence, how it acquires its structure?

Answer: Those invariances maintain a dynamic yet permanent reality. If we were to reduce the entire Physical system of laws to some basic principles we should first choose the Principles of Invariance. There are 3 physical invariances that stay regardless of changes in coordinates: the invariance of motion, proved by Galileo and Einstein's work on Relativity; the invariance of scale, proved by Nottale and this author, in their study of the laws and transformations that take place between the scales of reality and the invariance of topological forms, as the cycles of information and lines of energy repeat themselves in all the scales, in which networks of energy flows and informative cycles form complementary systems.

The Invariance of topological form is a consequence of the energy/ information duality and the 3 unique topologies of a 4-dimensional universe of fractal scales (hyperbolic information; toroidal reproduction, spherical energy). Thus, there is a simplex invariance of lineal energetic and informative cyclical shapes and a more complex ternary topological invariance that differentiate any entity of the Universe in a hyperbolic, informative, high region, an energetic, extended spherical membrane that encloses the 'ball' and a series of cyclical networks that connect both.

In the last graph Saturn's rings show the invariances & dualities of the Universe, contradicted by our naïve perception: Any entity of reality seems at first glance a continuous, still form. So Galileo called Saturn's planetoids, a ring, thinking it was continuous, still matter. Yet in detail reality is cyclical, discontinuous, in constant motion, structured in multiple layers of growing social complexity, as simple parts become wholes, which gather into ever growing bigger wholes. For that reason, while a simplified image of reality could describe the Universe as a continuous, amorphous extension of spatial energy, (monist physics, with their concept of a single, entropic arrow of time); a complex analysis observes a 'fractal' reality in perpetual, extended through multiple space-time scales. Yet the order of such structure is maintained precisely because the system is invariant in its scales (from particles to planetoids), in its forms (all those scales show cyclical particles moving in bigger cyclical trajectories) and in its motions.

The 3rd set of laws of Multiple Spaces-Times applied to Physics is the flows of energy and information between discontinuous space-membranes. Those laws are widely ignored by physicists and the reason why so many easy-to-find results of physics – with the use of those generic laws between spacetime planes, in this case the gravitational and electromagnetic membrane – are still ignored (from the Unification Equation of charges and masses, to the meaning of dark matter and dark energy, to the non-local speed of gravitation). We considered a few of those solutions, when studying the generic laws between 2 timespace, hierarchical planes, which in the case of physics are the bigger gravitational world of <0 K temperature-order and $>c$ speed.

The 4th types of laws are of causality, which define a series of ternary ages and horizons of evolution of particles and masses. We shall therefore be able to explain the evolution of particles, stars, galaxies and physical states, with the 3 ages of energy and information. For example, the 3 states of matter correspond to energy=gas, information=solid, and a 'reproductive combination' of those 2 states or liquid, which participates of the properties of gases and solids and it is the more creative state of measure. We shall also consider for all space-times of galaxies and Universe, the 3 solutions to Einstein's equations of space-time as 3 relative ages in the evolution of physical space: the big-bang solution is the energetic age of the space, the steady state, the mature age, and the Gödel, cyclical solution, the informative state, which means Universe and all kind of physical spaces, follow a process of big-bang and big-crunch equivalent to the 3 ages of living spaces.

The 5th sets of laws of Multiple Spaces-Times that apply to physics are the laws of i-logic geometry, which describe any particle of the Universe as a knot of time arrows. And so we shall observe that in any scale of physical particles, from strings to particles to atoms to molecules to planetoids, physical particles can be treated as st-Points, which will display the same 3 topologies of space-time (an external membrane, inner hyperbolic region of information, polar apertures that absorb and emit energy and information), will obey the same laws of social evolution in networks, and interact communicating forces, which are fractal, micro-forms, which imitate the forms of the higher particle/scale. So electrons will share electromagnetic photons and quarks will share gluons and molecules will share electrons of the lower scales, and so on.

The 6th set of laws applied to physical entities is the existence of 4 main time arrows in all physical systems that explain all the events of each entity of physical space. Those 3 arrows will be the dimensions of light space, the quantum numbers of particles and the events observed in molecules and crystals, whose finality is to create more complex complementary, organic physical entities.

Those invariances and laws of Time theory are essential to physics, since they set the forms and patterns of all kind of cycles of transformation of lineal, energetic motions into cyclical clocks of time. They are the origin of Universal Constants defined as invariant proportions between the energy and information of the 2 membranes of space-time, between the complementary energy fields and information particles of physical entities and between the 3 families and horizons of each species (from the 3 families of masses, to the temperature=energy of each state of matter).

Recap. The Physical Universe follows as any other system of reality the laws of Multiple Spaces-Times. The 1st types of laws are duality laws of energy and information (Complementarity, entropy+form arrows and theories). The 2nd types of laws are the invariances in form, motion and scales: There are 2 space-time scales, the larger gravitational membrane and the smaller quantum membrane, and two types of forms, lineal energy forces and cyclical particles. The 3rd types of laws are those of discontinuous spacetimes. The 4th laws are those of causality in time, which define the 3 ages of evolution of all species of physical space and the interaction of particles and antiparticles. The 5th type of laws are the 5 postulates of i-logic geometry, which define the creation of networks and the events between particles of any physical scale of the Universe, from strings to particles to atoms to molecules to planets and stars to galaxies and universal clusters. The 6th sets of laws are those of dimensions, according to the holographic Universe and the relationship of each local dimension with an arrow of time of a certain 'physical space' (light space, electronic, quantum numbers, etc.)

II. MASS IS INFORMATION: UNIFICATION EQUATION

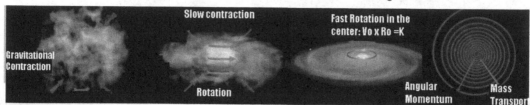

The Universe creates dimensional form by reducing the motion/speed of a physical system and vice versa: it creates motion/ speed by eliminating dimensions of form. So a 3 dimensional nebulae of matter collapses into an accelerated mass vortex, in which, according to the Equivalence between gravitational forces, masses and cyclical acceleration (Einstein's relativity), is an attractive whirl of spacetime. The Equivalence Principle establishes a Universe with 2 limits of eternal movement: a lineal speed with a c limit or electromagnetic force; and a cyclical speed or mass, which is a vortex of space-time also with a 'c' limit of speed, reached in the heaviest objects - black holes. Thus radiation and mass eternally transform into each other the lineal and cyclical arrow of inertial movement: E (lineal motion)<cxc>M(Ti) (cyclical motion). A light wave reproduces at c-speed an electromagnetic surface of lineal space. On the other hand, a mass whose limit is a black hole that turns at c-speed is a cyclical entity that balances the lineal movement of light forces with its inverse cyclical movement, to maintain an eternal dynamic balance of cycles and lines.

10. Equivalence Principle: invariance of topological form.

All forces can be reduced to 2x2 types: lineal, gravitational and electromagnetic forces, which extend to infinity and curved strong and electroweak forces that exist inside the smaller atom's space.

For that reason we talk in the cosmological realm of 2 membranes of space-time, the electromagnetic and gravitational scales. Yet according to duality, those two spaces should be balanced by two types of cyclical particles of information, also in perpetual motion, in such a manner that an eternal game of $E \Leftrightarrow I$ events that transform lineal energy into information take place in both membranes. This is the case, and we call the two cyclical vortices of information of the electromagnetic, light space and gravitational, dark space (since we are electronic beings that do not perceive gravitation), electronic charges and quark masses.

And since electronic charges are guided by electroweak forces and quark masses by strong forces, we conclude that the duality of energy/information and light/gravitational membranes can explain perfectly why there are 4 forces in the Universe. Those forces are in fact 2 lineal energies and 2 curved motions of the 2 scales of reality. Let us consider all those propositions in more detail.

According to Einstein's principle of equivalence, gravitation curves the energy of vacuum into an accelerated vortex of mass.

Thus, a mass has a cyclical form and since the only cyclical acceleration known is nature is that of a vortex, it follows that a mass is an accelerated, cyclical vortex of gravitational forces in which the energy of gravitational space is curved into mass. We perceive masses as still cyclical forms, because of the Galilean Paradox, but particles are cyclical accelerated vortices, whirl like hurricanes of curved space-time that attract faster as we come closer to their centers. In Relativity Einstein established the equality between accelerated motion, mass and gravitation. Or in the classic notation of Newton: Forces = O_i-Masses x $|_e$-Accelerations. And so we exist in a Universe made of 'motions in time', 'actions', 'forces', either cyclical masses or lineal fields.

Physicists think particles are material substances with forms in space (naïve realism), even if they are vortices of space-time, which attract other particles and forces as hurricanes do, dragging them towards its center. There are in fact 2 final 'seemingly, static' realities, space and mass, which in closer view are two types of motions: lineal, expansive motion or 'vacuum space' and cyclical, informative motion or 'mass'.

We shall bring a mathematical proof of its equivalence with information, extracted from Einstein's equations:

$$E=mc^2 + E \text{ x } T=k ->Mc^2 = k/T ->M= (k/c^2)x \text{ } (1/T)->M=Kx \text{ } v$$

Where v is the frequency of information of a mass vortex of space-time. Thus when the lineal energy of vacuum space, E, 'coils up' acquiring a new dimension of form (c^2), it becomes a cyclical vortex of mass, m, which has more form (2 dimensions); hence it stores far more information (i^2) than lineal, spatial forces do. That energy/motion of vacuum space, in fact, constantly condensates into virtual particles with form (E->I), either photons or masses, proving that the ultimate substance of reality is light-space condensed into photons and electrons, and gravitational space, condensed into masses.

Thus the 2 ultimate physical realities - 'forces' and 'masses' - are also two types of motions.

Quantum Physicists, unaware of the full meaning of Relativity and Mass as a vortex that creates information in Einstein's equations, think that creation of entropy/lineal motion/space is more important than processes of creation of cyclical information/mass. Thus, the Universe should expand and die, and loose its form, its in/form/ation. But this is by no means empirically certain, nor philosophically correct.

Indeed, when we consider the entire picture, which should include the arrow of gravitation and mass, as a force that 'in/forms', there are many discontinuous regions of creation of information in the Universe that this simplistic vision of some physicists do not account for in their calculations (black holes, dark matter, masses, charges, vortices, living beings, etc.)

Creation of entropy and its negation, negantropy, the creation of information, balance together, giving us a sum of eternal time: E/I=K. But to fully understand this, quantum physicists must improve their concept of charges and masses as cyclical motions - vortices of physical information that we must add up to the energetic big-bang processes to balance the Universe.

So we consider the initial, simplest reality, vacuum space, the lineal motion of gravitational and electromagnetic forces, which in a first stage curl into formal, cyclical masses and charges. Since vacuum space, which seem static to us has also motion.

This is proved by the dimensions of the substance that makes up our space-time – light – which are the same dimensions than the Cartesian space-time, but gifted with motion: the magnetic, energetic field has 2 perpendicular dimensions, width, and length, and the informative, electric field has also two dimensions: height and rhythm. Both together 'form' the 4-Dimensional space-time in which we' see'. Science further simplified this limited perception into a single abstract continuum 'frame of reference' that physicists called Cartesian space-time. When Descartes first published this 'frame of reference' in his book 'The World' he was aware that it was merely a mathematical instrument of calculus, but as time went by and scientists became accustomed to use the continuous paper-frame for all their studies of time and space, they ended up confusing the 'language' – the Cartesian frame – with reality (the space-time created by the explosive expansion of light-space).

Yet the Cartesian->Newtonian single, still, space-time is a simplification of what Leibniz called relational space-time: reality is made of multiple, lineal and cyclical motions (energy surfaces and cyclical clocks of information) that create all type of entities, which last a certain number of time cycles and occupy a reduced quantity of vital energy=space. Ultimately, Descartes was accurate in his description of reality as made of 'res extensa' (lineal space) and 'vortices' of time (cyclical motions), which the 'mind' used to construct an abstract map of space-time, the Cartesian plane. His disciples however got rid of those perceptive details sponsoring a naïve realism that confused the 'visual plane' of the Cartesian mind with the totality of space-times of the Universe. Absolute, Cartesian space-time however is a simplified Maya of the senses, which fix and put together all the bites and bytes of lineal energy and cyclical information, the 2 motions that define reality, into a single space-time.

For the same reasons, we perceive 'space' as a still picture of a time motion, a slice of a time-line (or world-line, as Relativistic theorists call it). The mind doesn't see time from past to future. It perceives only 'present time', a simultaneous slice of it that gathers the flows of reality into a static 'second', the rhythm of our mind, an eye-wor(l)d, which makes us to perceive many realities in perpetual motions as a single tapestry of static forms; which we call absolute space-time, even if it is made of many relative vital spaces, tracing clock-like cycles. The perception of time flowing thus require to put together many of those still perceptions into a flow of motion. In the same manner, we can see a movie because 24 times per second the image changes and our perception, which takes place more or less every second, puts together those 24 frames into a single 'formal motion, form-in-action, in-form-ation.

The Einsteinian principle of Equivalence unifies the forces of the Universe and explains the meaning of mass, *not from the perspective of quantum physics, as Higgs or Hawking tried to do with little success, but departing from the Laws of Relativity.*

Physics studies translation time=change in the movement of any entity, either a constant speed or an *acceleration*. So does *Relativity through* the Equivalence Principle, which considers weight equivalent to acceleration. Since a being that accelerates acquires weight. For example, when you accelerate in a rocket or a car, you feel a weight-force that throws you back. Thus, the *Equivalence Principle* equals gravity, acceleration and mass. We feel mass or weight because we are accelerating. While the inverse transformation of a vortex of mass into lineal radiation is ruled by Einstein's equation, $E=Mc^2$. But General Relativity uses only lineal Time. Thus, if we ad cyclical time, and hence cyclical acceleration, we complete Relativity's '*Lineal Equivalence Principle*', with a *Principle of cyclical Equivalence* that states the homology between mass and cyclical acceleration:

Weight=Gravitational force=Lineal acceleration

Mass=Gravitational particle=Cyclical acceleration

Thus, according to the formal duality of movements in the Universe, there are 2 Principles of Equivalence: lineal acceleration equivalent to gravitational forces and cyclical acceleration, equivalent to Mass, defined as a vortex of super-fluid space-time. Masses are merely the final accelerating curvature of space-time that creates vortices, which attract us, as hurricanes or sink vortices do with their surroundings. A mass, $M=E/c^2$ is a cyclical vortex of space-time[4] - not a fixed particle, but a cyclical spin. Thus, we describe the physical Universe as a game with 2 limits, the cyclical speed of a mass and the lineal speed of light. Yet to fully grasp that simple, amazingly beautiful, eternal Universe of fractal, moving lineal forces of space and cyclical vortices of mass, we have once more to correct the errors of lineal time that have fogged for centuries our understanding of a mass vortex.

Philosophically the Principle of Equivalence implies that the Universe is not made of solid substances but of perpetual accelerated movements, which we see as static fields within the mind, due to the Galilean paradox. Thus, the Universe is a machine in perpetual change, where acceleration - a change in the direction or rate of any movement - is the natural, dynamic essence of existence in time from past to future and future to past'. While inertial movements define existence in space, as a present, unchanged reality. Thus the image of the Universe cast by Relativity, when we properly understand the Equivalence Principle, is the opposite of earlier Newtonian concepts that considered the Universe a static, inertial reality instead of a catastrophic, accelerating and decelerating series of worlds, made of changes in the rhythms of times, instead of substances.

We have recently discovered such accelerating Universe defined by the Principle of Equivalence: on one hand, cosmologists discovered that black holes are indeed vortices of space-time that accelerate any object which comes towards them till reaching an angular c-speed on its 'event horizon'. On the other hand, we have observed that space accelerates between galaxies becoming dark energy. So both limits of lineal and cyclical movement obey the Principle of Equivalence: reality is, in terms of movement, the inverse of what the static mind perceives. The mind is pure, static, informative space, because it has reduced the cyclical and lineal, inertial movements of the Universe, into quiet mental space. And further on, it has simplified the constant changes of speeds that the accelerations and decelerations of existence produce, into a lineal movement. The mind integrates those changes of movement into a simplified vision that becomes first an inertial movement and then 'fixes' into space, the ultimate nature of the living Universe.

The Universe is constantly accelerating and decelerating, living and dying. In mathematical, linguistic terms, (since mathematics is nothing but a simplifying language of the mind), the first 'integration' of that acceleration is movement, (as we integrate all accelerations and decelerations into smooth, constant speed) and its second integration is static space (as we cancel all inverse movements into a fixed image). Only then when we perform those 2 integrations of data, the complex movements of the fractal Universe become the continuous static space-time of the mind. And inversely, when we want to get detailed information of the movement taken place in a point of that seemingly static space, (from the perspective of the mind), we 'derivate' that point of static space, to obtain its rhythm of change, $V=s/t$, is speed and then again, we 'derivate' in more detail to obtain acceleration the ultimate, 'deeper' nature of that static space.

It is for all those reasons that we call the arrow of information in the Universe, 'the arrow of Einstein', since Einstein defined 'gravitation' as an accelerated vortex that bends space into mass, whose frequency carries the information of the universe, its 'form'. And the result is a Universe in perpetual motion:

F=M x A, where Mass is cyclical motion and a, lineal motion.

Which is eternal (M⇔A), and dual, where mass carries the information of the Universe and energy its spatial extension:

$$M(t) + A(e)$$

Recap. the change of paradigm from single lineal time and continuous space into multiple, cyclical Time Arrows and fractal spaces creates a more accurate model of a Universe in perpetual motion, made of lineal energies (gravitation and light) and cyclical vortices of information (masses and charges). We don't need magic Higgs particles to explain mass, once we follow in the steps of Einstein and complete the curvature of space-time into cyclical vortices of mass.

11. Mass as information explains the duality of the Universe.

Classic Physics ignores how 'the future' is created by the 2 Simplex 'arrows' of time, energy and information. Since it works with a single' paint'/arrow of time, energy. We contend that the 'Arrow of mass' or arrow of information, the arrow discovered by Einstein in his studies of masses as cyclical clocks of time is needed to balance the Universe of energy and make it immortal. Monist Physicists contend that the creation of information is not enough and does not balance the expansion of the Universe. The essential difference between both theories lays therefore in the meaning of mass. We contend that masses are vortices of space-time described by Mr. Einstein Principle of equivalence between mass and acceleration and his equations, $ExT=K$ and $E=Mc^2$. Hence Mass$= k/T= k v$, where v is the frequency of rotation of the vortex of gravitational and electromagnetic forces that become masses and charges with more dimensions, as they coil the lineal energy of light, $E=Mc^2$.

Yet the dominance of the arrow of spatial energy and motion in the last centuries, due to the inordinate attention given to the science of physics and its 'worldly' profession of making energetic weapons and transport machines (that arrow in fact was found in the study of heat in steam machines), has modeled the universe as a simple machine of motions and extensions, a 'continuous' space that holds matter, its simplest forms, which seems to physicists the meaning of it all.

Quantum physicists even deny Einstein's theory of mass and prefer to think on mass as an external property to the particle. Hawking, even changes the arrow of time, so masses instead of creating information evaporate into energy. This denial of information goes to the extreme that physicists call the arrow of information, the arrow of 'future time' that creates life and mass, negantropy. And so they are in a quest to prove that the immortal Universe will die by an excess of entropy and lack of information.

How they justify it? Because they say the space between galaxies expands and will expand eternally, relaxing all information into pure energy. And since they think space is 'continuous', they consider its death proved. But in a fractal, discontinuous Universe such expansion is balanced by the 'implosion' of the energy of vacuum space into mass, $M=E/c^2$, *which happens constantly in galaxies and black holes. Thus the expanding vacuum space of the Universe is eternally balanced by the implosive forces of mass and gravitation.*

But we can already see a pattern of physicists: to deny often the existence of an arrow of information and life in any field of their study, including the obvious arrow of information in masses and black holes. All must be entropy, energy and the arrow of death, even *if there is a massive evidence of the existence of an arrow of gravitational form, described by Einstein.*

In words of Nietzsche physicists only care about finding the 'whitest' of all canvas, the big-bang of energy, the expansion of vacuum space, while complex theorists work with the 2 main colors of reality, energy and form, and its 2 complex 'mixtures', reproduction of form (exi), and social evolution of similar forms into networks, societies, systems and fractal spaces.

So system scientists are working to understand how with those 3+st colors, energy, form, its reproduction and its social evolution into a whole (st+1) the painter, 'God'[1], which is the will of those 4 arrows that create the future, its events and scientific laws, paint each part, each being of reality.

Recap: Form, information is the inverse, creative arrow that produces forms, by breaking and warping the amorphous energy of space, in ever more complex cyclical particles, which gather into networks and denser, implosive forms. Those 'forms' are small, slow, multi-dimensional, living forms that multiply in ever diminishing, more complex fractal scales. Yet in terms of the information they hold they are as important as the Universe itself. Indeed biologists have stored the same quantity of data about life than physicists about all the matter of the Universe.

12. The Unification equation of charges and masses.

We affirmed that the Universe is invariant to scale transformations and the 2nd set of laws needed to describe it are those who relate 2 different membranes of space-time, which in astrophysics are the light-space and gravitational-space membrane, the microcosms and the macrocosms.

The Universe is invariant in its topological shapes of energy and information, structured in 2 discontinuous fractal scales of space-time, the scale of cosmological masses and quantum charges, each one made of informative cycles, clocks of time (masses and charges) and lineal, energetic forces (dark, gravitational Non-Euclidean energy and electromagnetic, Euclidean light space).

And so the forces of the Universe must be understood in those terms and in 2 membranes:

Electromagnetism (spatial force: light) X Weak force (temporal particle: electron)= Electro-weak space-time membrane.

Gravitation (spatial force) X Strong force (temporal particle: quark)= Strong, Gravitational space-time Membrane.

In the Physical Universe time=change shows in the accelerations and decelerations of all its particles and forces. Yet since time and space are inverse morphologies, those motions can be lineal, creating spatial forces or cyclical, creating particles that are temporal vortices. The different cyclical, temporal speeds of those vortices and the spatial extensions of its forces are measured by Universal constants, which are ratios of transformation of spatial energy into temporal information and vice versa (so we measure constants as ratios between a mass/charge of in-form-ative volume and a spatial distance). Thus, a unification of electromagnetic and gravitational fields follows if we consider that their only difference is the density of space and temporal information of its forces and vortices that give birth to different constants/ratios.

Thus each Universal Constant is an informative/energetic, mass/distance, i/e ratio that defines a certain space-time membrane and plot them all according to those constants as different space-times of growing i/e density (so denser mass vortices have higher information and lesser distance, forming a denser, *strong* space-time of gluons, while the lighter, more extended space-time of larger distances is that of gravitational forces).

The result of plotting 'forces' as spaces of different density is a function, we shall call the Universal Constant of space-times, U(t/s), that varies in potencies of 10, giving us the specific i/e densities of each type of space-time force.

Physical Reality is made of dual energetic/informative states (complementarity principle particle-wave): Cyclic vortices/ particles develop lineal actions-waves in 2 scales, the microscopic quantum world of Electromagnetism and the macroscopic, gravitational world. Once the duality of scales and forms is understood, we unify the constants of dimensional proportionality of both types of vortices and forces, studying transversal waves of 'lineal speed' in both scales of the fractal universe: the c-speed of small electromagnetic waves, and we contend, the superluminal 'action at distance' of gravitational waves.

Let us consider the first element of that duality, unifying masses and charges of gravitation and electromagnetism as vortices of 2 different scales of space-times. Masses and charges are vortices of 2 space-time scales made of 2 relative energetic and informative motions, where gravitation is faster/more extended but carries less information than the slower/less extended world of light and electroweak particles.

The main variation of those vortices of accelerated forces is the one we observe between electronic charges and quark masses, which are the cyclical vortices of those 2 scales of reality, the scale of cosmological vortices of mass and the scale of electroweak vortices of charges.

And so we have to unify those 2 types of cyclical geometries on one side and their lineal, light and gravitational forces, on the other side, as they switch acting either as sinks of space-time (cyclical, non-lineal vortices of increasing acceleration) or as lineal, gravitational or electromagnetic waves in both scales.

Yet since electromagnetism and gravitation are self-similar accelerated vortices of 2 different fractal branes of space-time, the microscopic and cosmologic branes, it follows we should be able to treat them with the same equations of gravitation, either Einstein's more detailed vortices, defined by the Principle of Equivalence between mass and acceleration or with the classic analysis of Newton, as accelerated vortices with the same geometrical form, whose relative proportion of energy/speed/ distance and information/curvature will be given by 2 different Universal G-Constants.

Thus, we should be able to describe both, the standard Earth-Sun gravitational vortex and Hydrogen, electron-Proton quantum vortex with the same equations, defining them according to the mass of the particles, the rotational speed and 2 different Universal constants, U.C.(i/e):

U(g), the Universal Constant of gravitation that defines a larger/ faster/less curved Gravitational membrane of masses and...

U(q) the Universal constant of charge (Coulomb) that defines a smaller, slower, more curved membrane of quantum light. Since electromagnetism has more information and less energy distance.

Further on, we will be able to prove empirically our hypothesis of 2 self-similar spatial membranes made of light and gravitational quanta, whose only difference is the i/e relative density of information/energy of its cyclical vortices, if the values of the two Universal constants, in those 2 systems (the earth-sun system and the proton-electron system), correspond to their relative empirical value, when we treat them both as mass vortices (being the gravitational constant 10^{39} times weaker than the charge constant).

Let us then go on with the treatment of charges and masses as Newtonian vortices of space-time. In newton's equations a mass can be considered a cyclic, accelerated vortex of gravitational space-time defined in classic Newton Mechanics by a centripetal, gravitational acceleration: $\omega^2 r$.

The same treatment can be done with Poison equations or Relativity equations, which are a static, present, simultaneous picture of Newton's space-time vortex of acceleration, where G expresses the informative curvature of the space.

Thus if gravitational acceleration is $\omega^2 r$, then $F = mg = m\omega^2 r = GmM/r^2$, and we arrive at $G = \omega^2 r^3/M$. Where r is in meters and ω is angular acceleration in radians per second.

If we substitute for the Earth-Sun system's rounded values, (the Earth's angular velocity, ω is 2×10^{-7} radians per second; its orbital radius is 149×10^9 meters and M, the Sun's mass is 2×10^{30} kg.), we obtain, a value for G that measures the Sun's Space-time contraction (in static space) or formal acceleration of the vortex, equal to 6.6×10^{-11} kg^{-1} m^3 rad. sec.$^{-2}$, in accordance with experimental evidence. Roughly the same value of G, till now calculated empirically, is obtained for all planetary orbits.

This shows that the entire solar system is a series of self-similar gravitational vortices of space-time caused by planetary masses.

Then if we apply the model for G to the hydrogen, proton-electron atom the orbital parameters are: *electron's angular speed=$4.13x10^{16}$ rad.sec.$^{-1}$; Bohr radius=$5.3x10^{-11}$ meters and Proton mass=1.6×10^{-27} kg. And so substituting those values in $U(q) = \omega^2 r^3/M$, where U(q) is the Universal constant, quantized for the electromagnetic scale as 'q', we obtain a value of $\pm 1.5 \times 10^{29}$ kg^{-1} m^3 rad. sec.$^{-2}$.*

Thus, if we treat charges as vortices of a denser space-time membrane, we obtain theoretically a U.C. (i/e) of around 1.5×10^{39} times the value of the gravitational G, U(g), self-similar to the empirical value.

Yet a theoretical calculus of those values cannot be exact 'by chance', unless our thesis is right. Thus, the previous calculus is a clear proof that both, charges and masses, are unified as values of the same type of space-time vortices in the 2 different scales of space-time of the Universe. And they are geometrically unified from the p.o.v. of geometrical relativity not from quantum theory, as Einstein wanted it[5]. Thus, the Unification equation in terms of Newtonian mechanics is simple:

Unification Equation:

$$UC_{G,C} = \omega^2 r^3/M \quad UC_{G,C} \times M = \omega^2 r^3 \quad M = \omega^2 r^3/UC_{G,C}$$

Where we obtain for 2 UC(i/e) values, G and Q, 2 different space/time scales and vortices acceleration (mass and charges).

2 obvious, simple proofs of that dual, scalar structure:

- Planck's minimal value for Mass is far bigger (10^{-7} g.) than the minimal value for an electromagnetic action (h), which seems to mean that the mass scale is far bigger than the quantum world.

- The entire standard model works without gravitational forces. Thus we have to postulate that particles are not 'mass-particles' but electromagnetic systems, which become elements of mass only when they aggregate in huge numbers, as it happens with electromagnetic photons, which become electrons of the upper scale in huge nebulae.

But we see masses as static forms. This is due to the Galilean paradox: a space-time vortex-mass becomes then statically a space-time distortion, a deformation of space-time which increases as we decrease size. Since the *deformation of space-time is the same for any cycle. Hence, when we trace a smaller cycle the deformation increases.* And so it does the mass and the U(w), now a static constant of deformation. This is the interpretation most often assumed of G in Einstein's Relativity. In a simpler, classic approximation based in fluid dynamics, M/r^3 becomes the density of space/time and its inverse, R^3/m, the displacement of space-time provoked by a given mass. It is then evident by Archimedes Principle that a bigger mass density will provoke a bigger displacement and distortion of space-time an inversely a bigger acceleration towards the central vortex. And since the density of a proton is much higher than the density of a planet, the Universal Constant for an atom is much higher and so it is its angular speed. Does the electromagnetic constant, or static curvature of an atomic system, is much stronger. So it is the density of a proton mass.

This reformulation of electromagnetism simplifies and clarifies the meaning of physics. Yet to grasp the relationships between the quantum world and the cosmos we have to transform, using the new coulomb constant, $C=1.5 \times 10^{29}$, electromagnetic parameters to the jargon of gravity, departing from its fundamental relationship:

$$F = U C_C \, Mm/r^2 = e^2/4pe_o r^2$$

Then we can translate the main constants of electromagnetism (the Rydberg constant, the α constant, which denotes the strength of the electromagnetic field, etc.) to the gravitational jargon and again the results obtained are close to the theoretical value.

So it is evident that we are observing the same geometric force at different scales, *explained with 2 different historic jargons.* Further on, as we 'transform' the concepts of electromagnetic vortices to gravitational, geometric symbols, the main discoveries of Relativity apply to electromagnetic vortices.

Perhaps the most interesting simplified value under the new U(q) charge constant is the Proton radius that appears with the same formula, than a black hole, Schwarzschild horizon. This means that a quark particle (a hadron, a proton, etc.) is basically a black hole of the quantum, gravitational scale and vice versa: a cosmological black hole might be a deconfined state of billions of ultra-dense quarks. In such fractal Universe, each small particle could from the perspective of a small observer a macrocosmic form. And each atom could be a galaxy. Such fractal Universe validates the Theory of Great Numbers, which are the parameters of self-repetition of the Universe, which might be from a lower perspective a mere atom or black hole of a hyper-Universe.

Yet self-similarity is not equality, an atom is self-similar to a galaxy but NOT likely a galaxy - reason why for example quantum cosmology is false. Quantum cosmologists consider 'identical' the scales of the cosmos and the quantum world, so they use quantum equations for all. This is a hyperbolic error. We have equations for each scale and what we can show, as we did here, is the self-similarities of scales.

All in all, it is evident that what we call a black hole or pulsar is made of quark quanta; it is a fractal of quarks. Thus we can apply to its vortices the laws of Newtonian fluids or the more complex equations of a Non-Euclidean vortices of space-time, as we do with any other 'medium' called a 'phase space' in physics. They can be considered akin to a hurricane, which is a 'medium' of air molecules, or to a tornado, which is a medium of water molecules. All those mediums have limits of speed, which are the same for their lineal forces and cyclical vortices:

Light space-time has a c-speed limit, which is also the limit of speed of a rotational electron vortex and a 0 K limit.

Dark energy and quark matter belonging to the gravitational membrane seems to have a c<10 c lineal and rotational speeds.

Because each space-time membrane has vortices of information and energetic, lineal forces, we need equations of self-similarity, relating informative vortices of charges and masses on one side and gravitational and electromagnetic waves on the other. This is done with gravito-magnetism, which relates cosmic gravitational, lineal waves and electro-magnetic fields. So, the gravito-magnetic formalism just needs to be properly fit within the wider concept of 2 self-similar fractal space-times, to define in self-similar terms, light waves and dark-energy waves of gravitation; as the elementary, actions-units of the 2 fractal scales. Because of the corrections, higher complexity and general ignorance of the gravito-magnetic formalism, here we shall only consider the case of the light-membrane.

Recap: the 2nd fundamental duality of the Universe is its structure in 2 different branes, planes or scales of space-time: the electromagnetic space and the gravitational space, which show different proportions of energy and information and yet, because of the self-similarity of all those branes can be easily unified as 2 self-similar forces acting in 2 different scales.

13. Fractal, complex Physics.

The most astounding proof of the atomic self-similarity between galaxies and atoms was found this year, when we found two previously unknown bubbles of ultra-dense gamma-rays which are identical to the d-orbitals of an atom made also of an ultra-dense, nebulae of fractal photons, above and below the Milky Way.

The spiritual, eternal Universe of perpetual motions seems ∞ in scales, repeating self-similar forms every 10 relative scales of fractal space-time. *Since the nucleus of an atom is self-similar to the black hole nucleus of a galaxy.* So Black holes become hadrons of the bigger cosmological scale made of infinite quarks, and protons become 'black holes' of the lower, quantum scales... once we translate the jargon of quantum physics to the jargon of Gravitation, as Einstein wanted, *not the other way around as quantum physicists have tried unsuccessfully for 100 years.*

This means that the Universe is either limited between those two final scales or if each atom is a galaxy and each galaxy an atom (as Einstein and Walker thought when modeling the Universe with Relativity, considering each galaxy as a Hydrogen atom), is infinite in size, and in each of us, there is an entire number of infinite Universes.

And this is possible because now points have inner parts that when observed in detail grow in a fully relative Universe, 'where the smallest point must be perceived as an entire world' (Leibniz).

Since space-time is broken, fractal, made of infinite self-similar networks and organisms. It means there are many self-similar space-times, many self-similar planets, in which humans exist. Yet none of them are equal, because in fractal theory all is self-similar but nothing is absolutely equal.

In physics it solves also the puzzles of quantum physicists. For example, it means that an electron is not a sum of probable electrons in different Universes, but a sum of self-similar fractal parts, which can be described as populations of 'cellular electrons', with the same probabilistic equations that quantum theory. Any mathematician knows that probabilities are used to explain probable events in time, and percentages of populations in space. In other words, in advanced fractal physics we merely switch the quantum description *from time to space; hence a galaxy is a fractal of self-similar stars, which have self-similar planets, with self-similar human races.*

And this realization is the beginning of Fractal Logic, which coupled with the understanding of Non-Euclidean mathematics defines the 'Grand Design of the Universe', which is deterministic not probabilistic, albeit written with a logic more complex than that of the physicist and the present man.

But of course, space is the easiest part to understand of the new Age of mathematics and logic that we are finding. The key to everything is the logic of fractal space, which requires breaking also time into infinite clocks, one for each cycle of reality that closes in itself. *Which again means that as Einstein said, 'there are infinite clocks of time', and so infinite time speeds, and each of those vortices of mass, each of those charge vortices is a clock of time, which carries a frequency of information in the intelligent Universe. Imagine that: you exist in an eternal, intelligent, moving Universe, where all is gauging information, feeding on energy, turning into patterns of form, ticking.* Further on, the smallest you are, the faster you tick, the faster the metabolism of the rat is, the faster the electron turns. And this is the key of the harmony of all planes of reality. An ant lives 7 years but it processes information 10 times faster than a man, so in subjective time it lives 70 years. An atom cycles billions of times faster than a galaxy, but it is billions of time smaller. And so Einstein's equation writes as:

Energy/Motion/Space x Time/Information = Constant.

Max. Se x Min. Ti = Min. Se x Max. Ti

Thus, when we change from scale, what we lose in size we gain in speed of time, and the product of both stays invariant. As in the book 'Shrinking man', if we became smaller we would change our perception of time that would adapt to the new plane. And so the life of an ant might be as profound and rich in meaning as the life of a human being. And that is what connects the Mind of the Universe' to the classic myths of Eastern religions that studied them. Indeed, the Great Zimmer tells us[5], how Vishnu tells Indra, king of kings not to be so arrogant: he suddenly looks at the floor and sees a group of ants and Indra asks the Vishnu child, what are you looking at? 'I'm looking to a group of Indras walking in a queue' he responds as he crashes them.

Recap. The invariances of the Universe define a reality of infinite self-similar scales where we are all relative nothingness or absolute Gods depending on our scalar point of view: Humans might be universes made of infinite galactic atoms. And yet the Milky Way might be self-similar to a Hydrogen atom.

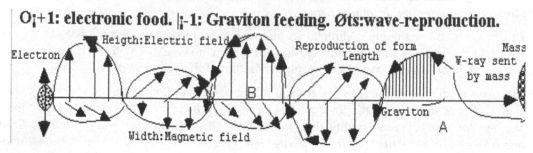

14. The mind of space-light.

We live in a Mind of light-space-time. It is not though the entire Universe, but the world the mind constructs with the help of a rod of measure, absolute c-speed and 3 dimensional space, Euclidean space. For that reason what is not light-space doesn't seem real for us. Yet this world of light with its light properties, perpendicularity, Euclidean form, developed by Descartes in his book the world as the basis for modern science, with a Cartesian graph in which the mind of man occupied the zero point, should differ from other worlds constructed with other rods by other species of mind. An exophysical theory must be in that sense objective and consider the wider structure of all spaces times. Once this is clear, we can study the peculiarities of the light space-membrane departing from its 4 dimensions, which are not only the dimensions of our mind perception but those of light as an entity made of time cycles, imprinted and reproduced over a simpler energy, space and evolved informative into electrons, what our mind is made of.

15. Universal constants of light

Heisenberg's uncertainty is a right formalism but a false interpretation of the generator equation of light-space that relates the arrows of energy and information, in the specific form it has in the quantum world of light actions of energy and time:

Energy × Temporal Information = Constant (H in the quantum world, K in Einstein's relativity)

The meaning of the previous equation is easy to grasp in terms of the duality of Space-time Constant, Q and G.

H is the equivalent of Q for the lineal energies of our light-membrane, as it relates the energy and form or frequency of a lineal, transversal wave of light.

So, if h is the ratio of transformation of frequency/form into energy and q the ratio of transformation of energy into form, in the electromagnetic world, what is the ratio of reproduction of electromagnetism? C, which was defined by Maxwell equations as a ratio between the electric and magnetic fields that merge together to create a wave of light, which reproduces its form contracting, warping the gravitational membrane of dark energy.

Further on c is the limit of speed of our space-time membrane. This is a tautology; since in fractal relativity space is the energy of the vacuum, a force: The membrane of light-space in which we exist is made of energy quanta, which constantly pops out and 'informs' itself , creating particles with energy and form (h= energy × information). Planck called these minimal actions, combinations of the lineal and cyclical motions of energy and time, the quanta of our light space-time. Yet Quantum physicists, always kin of Pythagorism, created the uncertainty dogma, saying actions are NOT made of the 2 motions of the Universe, but are 'mathematical entities' – a measure of the uncertainty of the vacuum. So particles are not born as evolutions of the 2 motions of reality, but they are born of the 'uncertainty' or 'probability' of the mathematical Universe. It is all simpler and more real. The space-time fluid of light is made of h-quanta and so those h-quanta can suddenly evolve socially in more complex forms and create particles.

That is all what there is to the uncertainty principle.

Yet we must also deduce from the existence of different U.C. and ratios of speed/distance and informative rotation that the Universal constants of c-speed, Q-charge and H-light energy are specific of our Universal membrane and those of the gravitational membrane must be different.

Recap. According to the paradox of Galileo the eye fixes the energy and form of light-space into a mental construction of 3 dimensions, a Cartesian graph, which already Descartes defined as the 'world', the spatial image of the mind – not the entire Universe. Thus the 3 perpendicular dimensions of light quanta, our space-time, are tautologically the 3 dimensions of our space. This in turn explains many events of the electromagnetic world. Those are also the 3 organic 'arrows' of time of light, which has electric information, magnetic energy and reproduces its form combining both into a c-reproductive field as the equations of Maxwell show.

16. The arrows of time in the light world.

We exist as evolved forms of a light space-time field, our limit of speed in this light Universe is c and the 3 dimensions of our Universe are height/electric field, width/magnetic field and length/reproductive field. Thus Cartesian space-time merely reflects the space-time medium of our light Universe and its 3 perpendicular dimensions. Yet for light the meaning of those 3 dimensions is organic: the informative, energetic and reproductive arrows of its field are equivalent to the energetic, reproductive and informative functions of any other relative space-time.

Those energy/Information cycles are obvious in biology, but in matter we have to translate the dimensions and events of abstract particles and forces, to understand their 3 Time-Space cycles:

- Energy Cycles, Ti<ΣSe: The main physical event that transforms information into energy is the emission of space fields by temporal particles, as in the big-bang or in atomic, fission and fusion processes. Those actions are produced, 'extracted' from an accelerating vortex of space-time that acquires stability through those constant emissions.

- Informative Cycles, ΣSe>Ti: The fundamental process of creation of in/form/ation in the physical world is the collapse of spatial fields into particles, charges or masses that spin around those gravitational or electromagnetic, spatial forces, creating knots of information, called particles. For example, when a photon comes closer to an electronic charge, its frequency increases till the photon collapses into the orbital vortex of the electron. The same happens when the electron collapses and becomes a pion of higher mass closer to the nucleus. We can consider those processes in abstract as processes of creation of information /frequency, or in organic terms as processes of feeding, or in terms of Time Arrows as the evolution of particles (photons and electrons) into species of higher physical information.

Light 'imprints' with form, with information and in the process 'corrugates' by a factor –ct (special relativity) the gravitational, extended, 'faster' space-time membrane of dark energy in which it 'feeds'. This –ct factor which prompted Minkowski to think that time was the 4 dimension of space merely means that 'physical time', the change in motion of physical entities, in this case a wave of light displacing in the gravitational vacuum, informs, forms, warps that vacuum by a factor –ct. If the reader grasps this simple notion - a field of light informs and corrugates the energy of gravitation it can also have a good laugh to 100 years of quantum musings about the '4th dimension of space'.

And so we define 3 scales of evolution of energy into form that diminish the 'lineal speed' of a force and increase its informative, rotational mass: photon >Electron >quark. It follows that the photon has faster lineal speed and less mass than the electron c/10, which is a relative fractal of dense, evolve 'photons', perceived as a whole or as a nebula. And the electron has less mass than the quark, which is a relative, fixed point of pure mass, perceived either as a fractal of gluons or a whole.

It follows that the laws of electronic systems are self-similar to those of light, as an electron is just an evolved form of light.

17. Multiple Spaces-Times in Matter: Atoms & Molecules

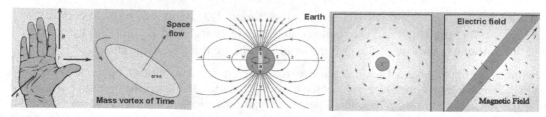

In the graph, both in the quantum and cosmic world the geometrical inversions between energy and information and its perpendicular exi fields of motion define electromagnetic fields .

The holographic principle explains how bidimensional fields of energy and information come together, through perpendicularity to create 4-dimensional systems.

The perpendicularity of those fields is thus a key feature to create 4-dimensional world, and it is established by the general rule that the body will have a flat, energetic arrow of feeding and the head will have a tall, high diffeomorphic orientation, relative to the ecosystem of energy and form in which the entity exist.

If we consider those 2 inversions of form: lineal energy vs. cyclical information and flat vs. tall perpendicularity between energy and information systems, forces and charges /masses in the physical world, we can define many physical events as transformations of lineal waves and fields of energy forces into cyclical or high particles and cycles of information, whose ratios of transformation in each of the scales of the Universe are called Universal constants.

In the graph, the laws of electric and magnetic fields are defined by those relative perpendicular morphologies. The ratios of transformations of energy into form define some of their Universal constants: 2 perpendicular fields of magnetic energy and electric information define in Maxwell equations the c-constant of light, which in organic terms is the 'reproductive constant' of the wave of light, caused by the constant process of creation that the interaction of the magnetic, spatial body of light and its informative, electronic particles, cause.

All those processes 'form' the energetic, formless gravitational vacuum into electromagnetic shapes: The Maxwell screw defines a flow of energy that accelerates as its perpendicular, temporal vortex diminishes its spatial radius, increasing its speed; the Earth's rotation creates a lineal flow of magnetism; while an inverse, lineal electric flow creates a cyclical, magnetic vortex.

In the graph, we can observe how the inverse properties of relative energy and information fields explain the why of many physical phenomena. If we add the dual structure of fractal space-time, roughly constructed by 2 membranes, our visible, Euclidean space-time, made of light and the invisible, larger sea of gravitation made of dark energy, with its wider Non-Euclidean topology, we can deduce all the laws of physical science and its whys with far more insight that quantum physics does.

Let us just indicate from the perspective of the whys of the arrows of time of an electron, as we did for light, some basic events of the electronic and atomic world:

- ⇔: Reproductive, communicative cycles: They happen when 2 particles communicate through forces, made of small, spatial particles, called bosons, repeat their waveform in space with a reproductive speed that imprints over the energetic field of vacuum space the in/form/ation of the particle.

- Social cycles: Particles create herds=waves of moving space-time quanta (dynamic, temporal perspective) related by EXI networks (static, spatial perspective). Social processes are all pervading in the Universe, from waves of bosons that sometimes come together into single condensates, to networks of atoms that become molecules or combined processes, such as waves of stars that form spiral galaxies.

Thus, waves and forces are spatial bosons that communicate energy and information between temporal particles, called fermions, either charges or masses. It is the so-called boson/fermion inversion, the fundamental equation of quantum physics:

$$T_1 \text{ (fermion particle)} < \sum sxt \text{ (boson force)} > T_2 \text{ (fermion particle)}$$

Existential=generational cycles: Again, we can adopt 2 points of view about particles. If we consider each particle/wave cycle a life/death cycle, particles have very short generational cycles. If we consider the next particle/wave cycle the same particle, particles are immortal.

In any case all those cycles can be summarized and formalized specific cases of the fractal, generator equation of the 2 arrows of time, spatial energy and temporal information: $\sum\sum \prod Ti$.

As we grow in scales of physical form, the nature of those cycles change slightly but the basic energy and information morphologies that define them are maintained. Since fractal, scalar paradigm is defined by its 3 invariances, invariance of relative motions (Einstein), invariance of scales (Nottale) and invariance of topological formal Time arrows (Sancho).

Those Time Arrows or dimensions of any fractal space/time are the 4 guiding whys or wills of any entity of the Universe:

- Energy arrows that expand form into energy.

- Information arrows that form energy into information.

- Reproductive arrows that reproduce the energy and form of a complementarity system in other zone of space-time.

- Social arrows that create superorganisms by associating in complex networks Non-E particles of self-similar properties. This 4th arrow of social evolution is the cause of 'scalar relativity', as systems are made of parts that associate in wholes, which share the functions and morphologies of its parts. Yet scales are never equal as quantum physicists pretend, only self-similar. So the properties of a galaxy made of stars and black holes, made of quarks and electrons are self-similar to those of quarks and electrons.

For example, in the next social plane that gathers atoms into molecules, the 3 main cycles of energy, form and the reproductive combination of both, become the 3 states of matter:

-Energetic gas state, (max. spatial extension and energetic speed).

-The balanced, reproductive liquid state, S=T, which creates the more complex forms of life.

-The 3rd, informative solid age, which has temporal, informative properties (cyclical vibration, high density of form and minimal space). While the spin or temporal movement of particles that gives them mass emerges in the macro-plane of molecules as a 'vortex' that attracts those molecules.

The boson-fermion equation, fundamental event of physics.

What the boson-fermion interaction means is that both, masses and charges are accelerating vortices which produce constantly, as they decelerate, 'fractal actions of constant speed' and increase the curvature of space in the dimension of height, through which they emit most of those actions, as electric or magnetic quanta. In the case of an electromagnetic vortex part of its acceleration becomes also transformed in the Unruh radiation that increases the temperature of the accelerating vortex. In our macrocosmic Universe it means a friction that decelerates the vortex and increases also its temperature. In gravity the interaction of the 3 dimensions is also evident: As the G-deformation of the vortex grows, so it grows its speed and acceleration. Unification follows when we consider charge and mass merely 2 types of vortices that can be derived from a 5th fractal dimension of time, that can accelerate inwards,

306

imploding space (and then we have the arrow of gravity) or decelerate outwards widening space and diminishing curvature (then we obtain, as Kaluza did, the electromagnetic arrow).

The perception of those dimensions becomes more complex if we introduce the errors of the mind that tends to average accelerations and decelerations into constant speeds and converts speeds into static distances or spaces. We say that the mind uncoils an accelerating time dimension into a wider static field of space, or a fixed density of space-time called a mass. This is easy to understand. Take for example, the representation of those 3 parameters, space, speed and acceleration in a Cartesian system of coordinates. Space is longer: it is a length in the x-coordinates without any angle, flat. Then we raise the dimension of height/time, and obtain a diagonal, straight line that represents a constant speed. Yet because it has an angle, the line is shorter in X-space coordinates. Fact though, is that the being that has speed is in fact moving a longer length, but we have transformed that longer length into a temporal height/form in the graph. Finally acceleration is represented by a parable that finally becomes a height line and so it occupies the less quantity of X-space. Yet an accelerating being will reach the longest distance of all the 3 beings in the shorter time.

What those accelerating beings peel off as friction or radiation, or speed is called in quantum Physics an action, which is a fundamental constant in all beings. In the quantum realm the accelerating vortex produces an h-constant action of energy and time; in the macro physical realm it produces a Temperature or radiation as in the Unruh case. So to avoid collapsing into a singularity of infinite speed, what particle's vortices do is to peel off h or T actions, modes of constant speed, made of energy and time. So we can move from the Equivalence Principle of Relativity to the Gauge and Action principles of quantum physics.

Conclusion

Thus in physical space all what exists are 4 motions of time, and those 4 motions define a Universe in perpetual motion, perpetually creating and destroying organic systems of energy and information.

Yet because the physical Universe is the simplest of all systems, its main event is a transformation of energy into information, of vacuum space into vortex-like masses and charges, which cannot be reduced further. And so the simplest 'events in time' or 'types of change/arrow of motions' in the Universe are lineal energies that move (bosonic forces) and cyclical forms that turn around, storing information in their patterns of form and frequency.

And this explains why the boson/fermion inversion is the main law of quantum events and why $E \Leftrightarrow M(t)c^2$ is the main law of the mass membrane. And we write it also as feedback equation since in physics energy/information events are reversible. So in the same manner energy can curl into mass, mass can explode into energy.

Indeed, if we depart from pure lineal motion or entropy (perceived statically as extension or space), then by 'curling' and 'warping', we can obtain cyclical forms (masses and charges), in the same way that by crunching a paper we obtain warps, angles of form. This primary event, E->I, diminishes the space, extension, lineal motion or speed (self-similar concepts) of space and increases its form. Thus, we obtain from energy ->Information. While the 2nd causal arrow of that feedback equation explains the destruction of form and creation of energy, I->E, which is the ultimate meaning of the big-bang and all other mini-big bangs (novas, supernovas, beta-decay, Bombs) of the physical world of matter.

Reason why Einstein said that 'time warps space into masses' - meaning that energy curls into cyclical vortices of mass - while Darwin said that 'Time evolves the morphology of life beings'; and deduced first the equation $M=E/c^2$ even if the Atomic bomb made more famous the destructive inversion

Recap. All the events of quantum particles can be explained as events Generated by the fractal equation of spatial energy (boson waves) and temporal information (fermion particles).

18. Physical constants are energy/form ratios of space-times.

Universal constants show ratios of energy, form, and exi=reproduction of space-time membranes. So they can all be written as partial equations of the Generator Equation of spatial energy and temporal information: E<exi>I.

The relationship between the Constants, I & e of those space-time membranes or 'forces' is a simple rule, known to physicists as the Law of Range: Max. E = Min. I and vice versa, which is a particular case of the inverse properties of energy and form.

In the widest ecosystem of physics and its 2 gravitational and light space-time membranes, those Universal constants are related to the informative density and spatial distance of the vortices of masses, charges and lineal flows of forces of those 2 membranes.

Because there are 2 membranes of space-time, the light space-time membrane we see and the dark, gravitational, bigger, faster space-time membrane we don't see, the Universal constants of both membranes are different. This means that black holes and quarks beyond the event horizon keep accelerating faster than light, and the energy they expel through their poles, called dark energy is also faster than light, which is the limit of speed of our membrane of space-time, light. Thus, the specific physical constants used to describe the e/i ratios of the vortices of energy and form of the cosmological and quantum world of masses and charges differ in value, as they express those different energy/form ratios. Yet since there is invariance of topological form, the only difference between electromagnetic and gravitational fields will be the extension/speed in space of its forces and the informative frequency of its vortices of temporal information that will give us different Universal constants, which are ratios that measure the speed of transformation of spatial energy into cyclical frequency. Thus we write a generic, topological, invariant equation for all space-times which will differ in its specific U.Constant (i/e):

$$F = UC \ (i/e) \ Mm \ (t) \ /Dd \ (s)$$

Recap. We measure universal constants as ratios between the frequency of a mass/charge or in-form-ative volume of the particle and its spatial distance. And so the faster the frequency of the ratio and the smaller the spatial extension, the stronger the Universal constant will be and the faster the exchanges of energy and form between the particles/vortices will happen. It results in 3 forces: the gravitational, macrocosmic force, the electromagnetic force and the strong microcosmic force of increasing strength and speed of rotation as we diminish our size.

IV. RELATIVITY REVIS(IT)ED

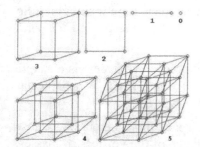

Multiple space dimensions, like the 5ᵗʰ dimension the artist fails to render in the hypercube do not exist in reality. Those objects are mathematical, linguistic fictions caused by the fact that any language of information is inflationary, as literary fictions are. The beauty of some mathematical theories, as the beauty of literature obscure those self-evident facts.

19. Special Relativity: Time is not the 4ᵗʰ dimension of space.

The key question regarding Relativity on the fractal paradigm, once Causal, Dynamic Triangulation has shown that from a bidimensional, holographic Universe we can build the 4 dimensions of Relativity, by fractal 'reproduction' (modelled as a diffusion system through fractal space) is the meaning of the 4 dimension of time in Einstein's equations. The answer is in the previous sentence: *present reproduction. Space-time is a self-organizing, emergent network of Non-Euclidean points that form a 'body-background' for future growth of more complex particle, by expanding a present sheet of space.*

Indeed, when we apply the in-depth-philosophy of fractal space-time structures to illuminate the why of those 2 masterpieces of mathematical physics, CDT and Relativity, all becomes conceptually cleared. The first blue print the Universe creates is the sheet of present space-time by the fractal growth of a previous smaller scale. CDT doesn't inquire about the 'meaning' of that time arrow but clearly establishes its nature as a reproductive, fractal arrow; while the jargon of relativity defines present with the term 'simultaneity'. This in this and the next paragraph we shall translate the jargon of quantum and relativity physics the general jargon of multiple spaces and times and correct some errors of classic relativity, clarifying its concepts as we have done with other theories.

Dimensions as motions of time and fixed space.

In the graph, dimensions are not only static forms of space but also dynamic time dimensions, which are created by the motions of time arrows since fixed space, according to the duality of the Galilean Paradox has always a degree of motion.

In fact we live in a membrane of light-space, whose energy /motion and dimensions are those of light quanta that constantly evolve, warp and produce virtual particles, while other particles dissolve in scalar big-bangs returning to the vacuum light-space.

Moreover in the Organic Universe dimensions are finite, reaching always a limit – either the limit of the membrane or ecosystem or the superorganism in which we measure.

Those dimensional distances in our light-membrane are given by the maximal distance/life span of light, which contrary to belief dissolves itself back into the gravitational space membrane after 10 billion years of fractal reproductions in the gravitational space-time quanta. In a diffeomorphic being, like a human being those dimensions grow till the limit of our space-time membrane, and so on. Thus, beyond the limits of a certain superorganism, there is a discontinuity from microcosms into macrocosms, which can only be 'bridged' with transitional spaces.

In the case of physical space we talk of two membranes of space-time, the smallest of which is our Euclidean, 3-dimensional light-space and the bigger one, the non-Euclidean gravitational space,. And between both membranes there are 'doors' called black holes and 'transformative' forces, called weak forces which transform particles from one membrane into the other, and flows of energy and information that can transcend through a transitional spacetime known in Relativity as a 'Rosen bridge'.

Dimensions are complex motions, interpreted according to the Galilean paradox (dual motion/form analysis), also as forms of static space, in the same manner you see a car 'line' in a night still picture or a 'fixed wheel' when it rotates too fast.

The partial application of Non-Euclidean Geometry by Einstein

Those clarifications are needed to consider some conceptual revisions to General Relativity regarding its meaning reach and limits due to its analysis of only a single space continuum and a single time arrow.

The result is only an approximation to reality, with some errors and deep insights, which works as long as we establish a conventional rod of spacetime measure – the postulate of c-speed as absolute – and deal with phenomena in a single spacetime continuum, either in the light-space studied by special relativity or the gravitational space studied by general relativity.

Indeed, what Einstein did was to apply a mathematical advance, the discovery of the 5th postulate of Non-Euclidean geometry to the understanding of gravitational space. Now that we have corrected the other 4 postulates, we can resolve and advance much further the Physics of spacetime into the new paradigm of a Universe made of scalar membranes, each one a network of non-Euclidean, fractal points, whose e-st-r parameters (universal constants of energy, information and reproduction) and exi actions (existential forces) define all the events and forms of reality.

In that regard, Time and its natural language logic, is the first language of God, wider than mathematics. For that reason, Frege and Hilbert were able to derive the laws of mathematics from the laws of logic time, the 'true language of God. Albeit a dualist logic more complex than the simple, lineal logic of Aristotle that physicists use in its spatial study of time. So time is a deeper phenomenon that encompasses all the spaces that there were and will be transformed by 4 possible arrows of time; and what truly matters is to learn the logic rules that transform in each species, a fractal, vital space, from past to future, between birth and extinction, guided by one of those 4 main time arrows.

The restricted formula of time-change as motion of space.

In that sense, the image physicists have of reality is restricted by the concept of lineal time and a single arrow of entropy, which also limits the solution of physical questions and introduces errors as the cosmological big-bang or the evaporation of black holes, mentioned before.

Physicists have spatialized time, defined with a single dimension, as a parameter of changes in motions and space, $t=s/v$.

This limited study of only one type of change, however is often forgotten, creating conceptual errors, such as the fantasy of time travel (as they don't realize there are infinite clocks of time in the Universe and so all time travel is local, within a certain organism as in the processes of death.)

In fact, when physicists confront the phenomenon of time, they do so only from the perspective of spatial change, a very limited part of the entire realm of time=changes. And so both, Galilean relativity and Einsteinian relativity are theories about the motion and energy of space, its invariances and laws, not a theory about all types of Time-change as this work is. And when someone like Hawking affirms that time travel is possible because time-motion gets to zero in a black hole, he merely forget he is speaking about the time-motion defined in the previous formula, $t=s/v$ or its self-similar Einsteinian version not about the sum of the infinite time cycles of the Universe that remain unchanged.

The readers must be therefore aware of the limits of study of time *in any physical theory, from Galilean to Einsteinian Relativity.* Today, when Physicists talk about time they talk about is a parameter, t, which appeared first in an equation of speed, $V=s/t$, that Galileo found when studying translational change in space or 'spatial movement'. Furthermore, because that rate of change in space is considered infinitesimal, $V=ds/dt$, they talk about present changes in space, which Einstein defined with his concepts of 'present simultaneity'.

This was fully understood by painters, who started to paint simultaneous, present time, in a single bidimensional space, as the sum of fractal movements (Duchamp's staircase models and Picasso's cubist perspective), but surprisingly enough the people who grasped with more difficulty the concept of spatial, present simultaneous analysis of time, were the same physicists that had invented this model of time analysis.

Moreover, ever since physicists have denied there are many other ways in which time=change happens in the Universe, beyond the limited study of 'changes in speed' that Physics analyzes. This has been explained many, many times to Physicists, since Bergson and Poincare explained to Einstein and Minkowski that they were 'spatializing time', or rather analyzing how vacuum space changes in time, (so Poincare could establish with Topological analysis self-similar equations to those of relativity).

Yet there are many other things besides vacuum space that change in time. We humans change in time our morphology (not our speed). Evolution changes in time the form of many species. Time is thus a synonymous in most sciences of a change in form, in the in-form-ation of species. And this type of informative change is basically ignored by Physics. This shows a general misconception of the meaning of Time in Physics, which has been termed as 'simplistic and reductionist' by philosophers of science. Because indeed it is.

Time is philosophically defined as 'change' and divided in the study of 'translational change', which is carried out by Physics, and the study of 'morphological change'; which is carried out by Biology. In that regard, Galileo defined only a limited type of time-change, translational change, with his equation, $V=s/t$. He 'spatialized Time' (Bergson), reducing the study of Time to a single process of change, movement.

Special relativity: the study of the light-membrane.

On the other hand, special relativity studies changes in the motions of the light-space membrane, *departing from a rod of Measure – light speed – taken as a convention.*

Einstein chose a postulate without proof, the postulate of c-speed as absolute. Of the 3 elements in the definition of time-motion by physicists, $V=s/t$, he could have taken any of them as absolute, but he chose light-speed. This is merely a matter of choice, as if we want to make a metric space able to 'measure', we need a 'fixed rod' value, which Einstein chose to be c-speed. Again, we could consider light speed variable and fix instead the time cycles/frequencies of light (so red light with lesser frequency/form travels faster than violet light and finally dissolves into non-local gravitation without form, dying about 10 billion years after birth - which solves some key problems such as the expansion of intergalactic space). Yet since we humans have a biological brain that fixes all light rays and extract instead the 'color/form' information of light, the convention of C-speed establishes a rod to create metric spaces and reduce the fluctuations of topological, real spaces. So far, so good, but as we shall see this convention to achieve the measures of relativity becomes a much bigger error when c-speed limits are applied to the gravitational membrane, when they should be limited to its use in the galaxy of light, since it is a limit caused by the substance of the light-space membrane, light itself.

Ultimately a true theory of spacetime Relativity, the concept that matters in Einstein's work, means that speed is also relative to the scale we observe and c-speed limits only apply to vacuum space and other forms with higher information (since the general formula of speed in all systems is speed=energetic space/temporal information; so forms with higher information are slower, as they have to build in each translational step all its form across multiple planes of complexity).

But Einstein in his analysis of space-time still used a single 'continuum space-time', the abstract Cartesian plane, and was interested only as all practical physicists by the creation of a 'metric space' in which instrumental measures were possible. He did not realize that time-space was discontinuous and at least structured into st-planes, the gravitational space of masses and gravitational forces that move faster than light and have a density of order inferior to 0 K (black holes, quarks, intergalactic dark energy, etc.) and the light-membrane limited to c-speed. This error was understandable, since we can only observe the light space that fills up with background radiation the entire galaxy and imprints gravitational space, 'forming' it into light. Hence in the galaxy indeed, light space imposes a c-limit of speed to all motions; but that is not the case when we go beyond light space into intergalactic gravitational space that experience shows to move up to 10 c speeds according to the natural decametric scales of the Universe.

Galilean Relativity: the error of Cartesian, continuum space.

The 2nd error of relativity, the continuity of spacetime is caused by a misunderstanding of the curved, geometrical nature of time and its dimensions, caused by the Cartesian plane of lineal time: To draw a clock-cycle of temporal information we need at least 2 dimensions, as morphology requires form, in-form-ation; and so lineal time with a single dimension cannot study it. Thus when Descartes and Galileo simplified time into one dimension, he eliminated the analysis of morphological change and information, the 2nd arrow of the Universe, from most disciplines of science.

All cycles have 2 directions or arrows to complete its form. In the pre-scientific age, those 2 arrows of relative past and future were understood as the 2 sections of curved times cycles. So times are indeed, as Einstein thought, curved, to the point of closing themselves into causal cycles that return all beings to their origin, but they have 2 dimensions, height and rhythm. So it does space. Each local times 'curves the energy of space', creating fractal in-form-ation (masses and forces in physical, evolving information in biology, etc.). Since a cycle needs 2 directions to close into itself and create a form. Energy consists in an amorphous, relative surface of undifferentiated quanta, a plane without form, which the creation of time cycles shapes into information and makes it recognizable, thanks to its 2 dimensions. Both geometries, cycles of time and planes space, are bidimensional. When both systems merge, the sum of fractal, bidimensional fields of time and energy, create only one more, 3rd intersecting dimension of combined time-space, *width*, which has a rhythm of temporal movement, the 4th dimension of time=change that we observe as *reality*.

Unfortunately the error of Descartes and Galileo, which simplified cyclical time into a single-dimensional line, had as a first consequence the reduction of time in human sciences to a lineal duration, measured with a scalar number, similar to the way we measure space. It is what philosophers of science call the 'spatialization of time'. Now spatial time becomes a secondary parameter to calculate spatial and energetic motions and processes – not an objective of knowledge in itself. Accordingly, Galileo defined time as a measure of 'translational, spatial change', of movement in lineal space or speed ($V=S/T$), simplifying time into a lineal number, defined in terms of movement and space.

Light and dark energy Interaction: gravitational space warping.

Physicists are always referring only to the arrow of time=change in the motion of beings ($t=v/s$), which is by tautology a function of space. Thus, Philosophers talk of the spatialization of time, which is made dependant of energy and movement. Einstein was perhaps the only physicist aware of this error when he said 'I seem to be the only person that believes there are many times with different rhythms'. So when his followers say that time is the 4th dimension of space, he did not agree; but he pointed out that even in Relativity Time has an inverse, negative sign, so its properties must be opposite to those of space. But physicists ignored his dictum: 'we cannot send wire messages to the past', he insisted. And when Meyerson, the French philosopher of time published a rebuttal of Minkowski's reductionist theory of time as the 4th dimension of space, *'La deduction Relativiste'*, 1928, and he could express his ideas without the enormous peer pressure, usual in this profession of fundamentalist scientists, he praised it as one of the most remarkable books written about the relativity theory from the standpoint of epistemology (the scientific method) and explicitly agreed with its rejection of the spatial interpretation of a world-line of Minkowski.

Time motion in Relativity has a negative sign, since it refers to the implosive arrow of information that contracts space, NOT to the expansive arrow of spatial energy and entropy. Time in General Relativity is the factor that warps energy into information, evolves form, trans/forming the energy of gravitational space into the frequency of form of light-space. Time in special relativity warps the energy of the vacuum, as light imprints its form. Thus Time is $-t^2$ in the equation of Special Relativity:

$$s^2 = x^2 + y^2 + z^2 - c^2 t^2$$

Space is positive and time is negative because it forms, contracts space into a wave-frequency that carries information.

Yet the why of that warping of the gravitational membrane by light-information could not be understood by Einstein since it required fractal, Non-Euclidean models of discontinuous space-time to describe how form evolves. So Physicists ignored the informative effect that warps energy into form.

Einsteinian relativity inherits the error of spatial time.

Let us study that conclusion in more detail. This tendency increased further when Einstein used the same concept of translational, spatial change to define time. In that respect, it is popular to talk of time as 'the 4 dimension of space' in Relativity.

This is a misconception of Einstein's equations, which do not include all time=changes, as the 4th dimension of reality, but only spatial change, moving space, what Galileo defined as S=vt.

In Einstein's formalism moving space writes as ct, since Einstein studies phenomena that happens as the speed of light; so vt becomes ct (where c is light speed).

Thus s=ct, the 4th dimension of Special Relativity, is not time but 'moving space', defined as the product of translational change=lineal time-duration, t, and light speed, c. And so there is no philosophical advance respect to Galilean Relativity that already defined time as a change/dimension of moving space.

Yet this capacity of time to measure change in space is only a fraction of the properties and modes of change that time measures in the Universe, and reveals nothing about the frequency and form of physical time cycles.

Though Einstein latter, in his General Relativity went deeper and discovered that Time is a curved clock geometry, which in the Universe bends space into masses, in his initial work, 'Special Relativity' he was still a classic physicist who declared as Galileo did that 'time is what a clock measures' and so he still used the Galilean, lineal concept of time-speed, v=s/t, to define it.

However, Einstein never said that time is 'only' a dimension of space as lesser physicists say today routinely. This reductionist, 'spatial concept' of time was provided by Minkowski and Einstein opposed to it, saying 'that time behave differently since wires didn't travel to the past'.

Time indeed is causal, logic and defined by time arrows which are not symmetric; that is left and right are equal but motion to the past (informative flow that devolves into energy) and motion to the future (energy flow that evolves into information) have different properties. This is known in the wording of classic physics as the 'breaking of symmetry' in processes that involve time evolution (those mediated by the weak force), or 'chirality' in other systems of physics and chemistry.

Yet the simplification of Minkowski became popular among physicists, as it allowed to keep Galileo's tradition of studying time as a lineal parameter dependant of speed (V=s/t).

In that sense, though this work is conceptual to facilitate its comprehension to a wider audience and so we have reduced equations to a minimum, it is worth to consider that spatial vision of time in Special Relativity in more detail, given the simplicity of its mathematical equations.

Think first of two dimensions, the surface of a sheet of paper, for example, of length x and width y. The shortest distance s between two opposite corners is given by Pythagoras' theorem: $s^2 = x^2 + y^2$. If we now go to three dimensions, the shortest distance between two opposite corners of your room, for example, one on the floor and the other in the opposite corner on the ceiling, where the room is x long, y wide and z high, is given by: $s^2 = x^2 + y^2 + z^2$. So what happens if we go to four dimensions, trying to include 'lineal time' and 'moving space' and measure the space-time interval between two camera flashes, for example, one happening at one corner of your room and the other happening at the opposite corner but a few seconds, t, later?

We might think, if we follow the concept that time is just another dimension of space that the answer would be: $s^2 = x^2 + y^2 + z^2 + t^2$, but we would be wrong:

- First, cycles of temporal information, as we have seen, are inverted to spatial energy, so the term t^2 must be negative.

- Second, time is not space, and we are measuring space.

We are just adding now a time-related term of moving space or speed – Galileo's formula – to our measure; we are not adding pure time but moving space, so to be able to add apples=spaces and pears=time we have to convert time into moving space and we do so with Galileo's equation v=s/t, hence s=vt, which means we ad, s=vt. Now, the big question arises: what kind of moving space we are measuring?

Or in other terms, what is V; the speed of the space we are measuring? The answer of Einstein was 'light', since the velocity of light was found to be equal in all systems, regardless of the speed of the observer, light was a kind of space-time itself, uninfluenced by the movement of the relative points we were measuring.

So Einstein added a corrective factor of measure or 4th coordinate of space, related to the translational speed of the light-space between the 2 relative points we were measuring. Light-space, the space our eyes see, is made of light, a quantum of energy. Space is not static but, as the impressionist painters thought, we see light-space and so we have to add a parameter to correct for the movement of that light space to make accurate measures: s=vt=ct, moving light space. It is not pure time, cyclical time or frequency or lineal time or duration.

Thus, it is absurd to say that time is the 4th dimension of space. What we should say is that light-space has movement, s=ct, a parameter to ad in any measure of translation, in a Universe of moving spatial energies. And the relativity equation adds those underlying moving flows of spatial energy and speed: $s^2 = x^2 + y^2 + z^2 - c^2t^2$, which gives us the space-time separation, s, between two points of any space-time subject to motion. Finally, rather than measuring the distance across an extended interval of space-time, it is important to deal only with the separation of adjacent events separated by an infinitesimal change in the coordinates, dx, dy, dz, dt.

So the correct expression of the infinitesimal separation of 2 adjacent events is now: $ds^2 = dx^2 + dy^2 + dz^2 - c^2dt^2$. This is called the metric of flat space-time. If we want to include 'curved' space-time then we put a coefficient, not equal to one, in front of each term, dx^2 etc., which bends the specific coordinates. This allows the possibility that a trajectory in space-time might be 'curved', which was the big discovery of Einstein, by creating many adjacent points and joining them in all types of trajectories in space-time, as those you trace in your daily life.

In order to find out the separation between two events you have to add up, or integrate all the infinitesimal ds intervals along the light path. And then it is when things can get as complex mathematically as you want, allowing you better measures of the shapes and movements of the Universe.

This is all what there is to it: an improvement on Galileo's study of translational space. Yet in that equation time is not a 4th dimension of space, but a measure of the speeds of the space-time background. The 2 important discoveries of that equation are the curved nature of space-time; and the use of c instead of v as the speed of that spatial background. Einstein realized that the wobbling of space could be incorporated adding ct as a parameter of translational change, which he did. But he did not added time as a 4th dimension but the moving properties of the fractal space-time or light-space on which our electromagnetic world builds itself.

Informative contraction: light imprints gravitational space.

Einstein latter adapted Galileo's formula to a fluctuating space-time background adding the implosive, -ct, arrow of gravitation to all translational movements. Thus he realized that Time in Physics also has a component of morphological change, as time warps energy into in-form-ation, not only in evolutionary processes, but also warps the energy of vacuum into in-form-ative masses.

And this is in fact the most important finding of the previous formula, which becomes crystal clear when we consider a 'standard' plane of two arrows of time, energy and information, which is represented by negative, imaginary numbers.

Thus what the negative symbol behind the 'square' of time, easily reduced to an i number really means is that the imprinting or reproduction of the light wave over the gravitational space in which imprints its information provokes a contraction of form, proportional to the informative time cycle and reproductive speed of the wave of light, according to the general laws of Multiple Spaces-Times, in a phenomena comparable to the dragging of any motion over an opposite flow. Since indeed, what light does from its p.o.v. is to feed on an opposite flow of gravitational forces where it reproduces itself, and in doing so it obviously suffers a drug, like the recoil of shooting a gun or the slowing speed of a plane dragging energy from air through its wings

Absolute Relativity: c is not the limit of speed of the Universe.

It is yet another proof of the scalar structure of the Universe, as electrons which emit light first 'connect' through the gravitational non-local plane (giving birth to the entanglement process) and when they are locked they emit a flow of light. A fact which also explains the Postulate of absolute speed of light as a 'local effect', NOT an universal law; since electrons only emit light when they are locked, and we can only observe light when our instruments lock with the emitter and so there is no a sum of speeds as Michelson thought (though there are several ways to measure this, and so Relativity is right as a system of measure but wrong as a philosophy of time-space). This postulate is further proved wrong by the fact that the diffeomorphic principle of relativity establishes a relativity of space and time, but speed=v=s/t is a parameter of space and time; hence the limits of speed are relative according to the scales we study; which means C is only the limit of the light-space membrane in which we inhabit but not the limit of gravitation that is the force of the other membrane of the Universe.

Yet, those conceptual clarifications born of a model of multiple space-time membranes and time cycles were not available when Relativity was deduced in mathematical terms, and so new errors in the conception of time appeared soon. Minkowski first though that time was the '4th dimension of space'. Then, because of the mentioned error of a single arrow of time for the entire Universe, Physicists thought that when they measure NOT the change in translational time of the formula of speed, v=s/t, but all the time of the Universe and time travel was possible. Yet when time appears with a negative term, it doesn't mean, as people like Minkowski or recently Hawking think, that time is traveling to the past. Such patent nonsense comes out of the generalization of the parameter t=s/v, the equation which defines time in physics (or –ct in Einstein's improved formula), to the entire range of time=changes of reality, including morphological time-change.

Special Relativity uses time, as a t parameter of space and speed as physicists always do, in this case to measure moving spatial energy - the flows of gravitational and electromagnetic space-time that act as the scaffolding of the world we inhabit. The fact that a century later most physicists still think that all what matters about time is its use as a parameter of translational change, s=ct, shows how much time has been degraded and distorted in modern science due to its lineal simplification in Galilean Physics and Cartesian geometries by the concept that there is only one arrow of time for all the cycles of the Universe, which is 'what a clock measures' (Einstein, Galileo).

The concept of a fractal point and the relativity of space and time and its parameter of speed, only absolute within the limits of a certain spacetime brane, has 2 immediate formal generalizations:

'The maximal speed of a space-time membrane corresponds to the limit of speed of its more extended, less formal spatial substance, such as $\forall exi = k_i \rightarrow max. \ c = c \ (max. \ e \ x \ min \ i)$.'

In other words, because we live in the light membrane, the limit of speed is c. If we were in the gravitational world, the limit of speed will be that of Gravitational forces > c; if we consider from the relative point of view of our size, the world of life beings, the limit of speed is that of the cheetah and so on.

The second, deeper consequence of absolute relativity and the difference between space displacements and temporal displacement is even more radical: because measure is relative to the point of view of the observer, the Universe is not a metric space but a topological space, in which distance and speed are absolutely relative and do not change the type of reality we describe. This again solves many questions of physics and we shall return to it, when analysing in more detail black holes as a 'sample case' of the applications of Multiple Spaces-Times to cosmology.

Fantaphysics: bizarre theories based in a single time arrow.

Today, because of that erroneous sentence, 'time is the 4th dimension of space', an entire bizarre world of 'manifolds' with multiple 'spatial, fixed dimensions' developed as a 'serious' branch of physics, whose authors believe religiously, and a series of theories of a Universe with only entropy (evaporating black holes; Universal big-bang, etc.) have become the foundations of astrophysics despite their theoretical fantasy and increasing contradiction with experimental evidence. Those physical theories use the limited definition of time to create imaginary worlds of mathematical thought as irrelevant to our understanding of the cosmos as the fictional, idealistic characters of Quixote, despite the linguistic beauty of the wor(l)ds of Cervantes, or the beauty of hyper-cubes, which have nothing to do with reality.

Recap: Time is not the 4th dimension of space, a misleading definition, caused by 2 errors in our conception of space-time and dimensionality:

- The error of a lineal, ∞ space-time misunderstands the cyclical and hence finite nature of the dimensions of all space-time cycles, reduced to the limits of the cycle. Since space/time has a fractal structure whose dimensions are not all in the same continuum space-time; but evolve across several scales that transfer between them energy and information under self-similar laws described with i-logic geometry. Yet each 'relative plane of space-time' or scale of reality, acts in many ways independently from the others.

- Galilean Paradox. What we perceive as quiet space (i.e., a moving time cycle) is in perpetual movement. The Galilean Paradox is an error of perception that hides the dual nature of any growing, real dimension, both as a movement in time that becomes a fixed form of space. Since the subjective minds tends to fix 'time movements' into space objects.

In Relativity, the misunderstanding of the Galilean Paradox - a perceptive error that makes moving cycles seem static dimensions of space - has caused a conceptual error: its interpretation of time as the 4th dimension of space. The Galileo Paradox converts dynamic time processes of change into static, spatial distances, hiding the dual nature of dimensions that are both, a movement in time and a fixed form of space. Since the subjective minds tends to fix 'time movements' into spatial objects.

Yet since space is a motion and light flows over the inverse motion of gravitational space, the key factor of the Special equations of Relativity shows light contracting by $(cti)^2$, as it flows against the simpler gravitational spacetime, which causes a drag calculated by the metric space of special Relativity.

20. General Relativity studies the space-time ages of the cosmos.

In a wider analysis, if we were to consider a single clock-time for the entire Universe (by adopting the hypothetical rhythm of the Universal superorganism and its 3 ages between its big-bang birth and big crunch warping) vacuum space will appear as 'a slice of reality' or 'dharma of present' that times constantly move from past to future according to the logic of time arrows, till it will warp it into a big-crunch.

And that evolving form of universal space and its 3 ages of a big-bang birth, a steady longer state and a final big-crunch, which correspond to the 3 solutions to Einstein's spacetime equations, are what General Relativity studies. But those equations are just a particular case of the wider '3 ages' of any organic space-time studied with the 3 elements of the Generator Equation, $E \Leftrightarrow I$.

Yet classic Physics studies only geometrical, translational time, but not the evolving and devolving morphologic changes studied in biology, except in the 3 solutions to the equations of Einstein, since they take the entire Universe (or rather a galactic space) as an organism. For that reason temporal change in Physics means merely a geometrical change of direction as when we move backwards a clock in space or we move backwards a particle that becomes an antiparticle - "but we don't evolve or devolve the entire Universe as Hawking thinks nor we move towards the Absolute past or the future when we change the clock's needle position in space. You might argue that for the clock's own definition of time, geometrical change and temporal change are the same and this thesis is worth to explore as we shall do latter, since it is indeed rather accurate to consider that for each space-time field a morphological change of information is a temporal change, given the relationship between both. So if we could change the warping, informative morphology of our face and straight it up, filling it with biological, energetic cellular space we could become younger, but the Universe will continue its infinite fractal clock rhythms independent of our form. This is self-evident and Einstein already explained it to Minkowski a century ago, when he vehemently protested with the use of the sentence 'time is the 4th dimension of space'; because 'time behaved 'somewhat' in a very different form to space' and the Universe is made of multiple times.

Still Physicists who have had always since Galileo a strong desire to become Metaphysicists chose to ignore the admonitions of Einstein and have developed a 'metaphysics of space with its multiple dimensions' that we need to clarify before going further into our analysis of cosmological time. This work is harsh with such concepts and the physicists that keep sponsoring them, because the concept of spatializing time is erroneous and prevents the advance of Time knowledge. It translates into 3 false dogmas: a single arrow of time called entropy that ignores the informative arrow of gravity in many of its analysis of the Universe; a space-time continuum that ignores the fact that we are not placed into an abstract frame of reference, but we are made of fractal space=vital energy and time cycles; and the use of a single spatial, geometrical linage, mathematics to explain it all, that ignores often the fundamental language of time, causal logic, expressed better with verbal thought.

Under such a dictatorship of Physics, for long logic and bio-logic, verbal sciences like Evolution or social sciences, which study long range changes in biological and human species, have been considered lesser sciences. It is said that Gell-Mann lobbied with the Nobel institution to prevent economics from becoming a Nobel Prize with this sentence, 'what is next, anthropology?' And he only calmed down when he was said that mathematical economics was to be the focus of the prize. It is also said that Feynman to quote the other great physicist of the post-war era asked their students never consider the 'why' of physical events, a conceptual, logic question but only the descriptive how that mathematics could achieve.

For that reason, it should not surprise us that verbal concepts such as past, present and future, the 3 dimensions of 'bio-logic time', totally escape them, as a clock measures time in digital numbers not in relative, 'morphological, d=evolving ages', which is what past, present and future study. Instead they use the limited t=s/v parameter, as *if it were the parameter of all time changes.* This is absurd. When I get old is not because I change my speed in space, but because my body suffers morphological changes study in biology. Time in Physics is 'spatial, translational, geometrical change'. Within that limited scope, as a 'parameter of change in space' physical studies on time are relevant. All other generalizations of time by 'Meta-Physicists are 'science-fiction' of which modern Physics is full of examples.

Those errors show even today, when Einstein established the geometrical, cyclical nature of time, as 'physical clocks' that bend energy into mass, in the use of lineal, infinite, absolute time by all other physical theories. Such limited definition of time in modern science as a lineal number or duration, instead of a frequency of cyclical form, of in-form-ation, is the biggest handicap to a proper understanding of Reality, given the fact that Time is one of the 2 fundamental Universal parameters. Yet time as a lineal duration becomes a secondary variable, which carries as any line does, very limited information. So scientists focus in the study of a single parameter, or arrow of the Universe – the entropic arrow of expanding space or energy - and fail to recognize the second arrow – the patterns, frequencies and repetitions caused by the *cyclical inertia* of time events that tend to store and repeat information in a discontinuous, cyclical manner. The result is an unneeded complexity, a plethora of errors and ill definitions and the belief that the Universe is chaotic, since the variable that creates order and information – *time* – is so poorly understood. To explain that order we describe the Universe without simplifying its 2 original parameters, analyzing the interaction between the energy of space and the formal frequencies of time, and how they shape together the species of that Universe. Nevertheless, the task is daunting because science has worked with a single arrow of lineal energy or entropy for 4 centuries. So there is an enormous resistance to accept a 2nd arrow of information and *define time* in complex terms, as the parameter that causes *morphological, informative change*. This aberration is recent in the history of knowledge. Prior to Galileo, all philosophers understood time as synonymous of *all kind of* cyclical *changes, not only physical translation.* Further on, all pre-Galilean cultures defined time intuitively in cyclical terms, as a form with a frequency. Physicists only study a type of time-change, translational change in space. Thus, when they talk of time as the 4th dimension of space they are merely talking of a very small part of time phenomena, the measure of movement in space, with time clocks. Only within those restricted parameters of Galilean physics, time is indeed a dimension of spatial movement. What is moving in that space/time background? And how that movement can be introduced in the simplified, mathematical static models of time and space, proper of the Cartesian plane that scientists use to describe the Universe? Both questions are related to the concept of moving dimensions of space-time, which now substitute the simplified, static concept of fixed space-dimensions. What is an act of arrogance and ignorance is to consider that the most important feature of time is to be a parameter of movement in space. So what matters about time is to be in all sciences the parameter of change, through is 3 dimensions, past, present and future.

Recap. Time travel in physics is a geometry change as when we move a cycle of time backwards, but it does not affect the overall time of the Universe. Morphological Time studied by evolution and the ages of time events composed of energy and information are only studied in physics with the 3 ages/solutions to the equations of Einstein.

V. 2 WORLD MEMBRANES IN A SINGLE UNIVERSE.

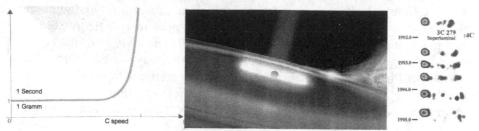

The 2 membranes of space-time, the quantum and gravitational membrane interact in e ⇔i events of transference of energy and information in the limits of the existential function of the light membrane: the limit of energy-speed, c; the limits of temperature-form, 0 k, in quarks, black holes and regions of transition between light and gravitational space.

21. Relativistic mass. Interactions at C-speed.

Galactic space is made of light actions, h=exi, which can be perceived as energy/motion (c-speed) of in/form/ation (photons): the background radiation is thus the primary substance of the vacuum. If we were to extract all the radiation actions, the vacuum would disappear and the underlying scale of pure gravitational, 'bosonic, tachyon strings', infinitely thin (Nambu's Actions, redefined as a background independent action of fractal dimensions) would appear. This would paradoxically thin out the informative height of vacuum and elongate it, as its relative speed=spatial energy/temporal information, $v=e/i=e/0$ is infinite, non-local from our p.o.v. This is what happens in intergalactic space, as the door between both membranes, the galactic, rotational Kerr black hole, thin outs electromagnetic radiation, (gravitational redshift) through the event horizon, accelerates it (equivalence principle, left graph) in its reproductive, 'Klein' intermediate space, and jettisons it as curled quark mass (left graph of superluminal 10 c motion) and dark, tachyon energy (central pic, blue jet), back to intergalactic space that seems to expand, accelerate from the p.o.v. of fixed distances.

Those 2 space-times create 2 different world membranes or mediums, with different energy/information limits:

The gravitational world is beyond c speed and under 0 K; that is it has more order in its quark vortices and more speed in its dark energy; it is bigger, more powerful. In between there is our hot, entropic, < c-speed world

The transitions between both membranes are described by the Lorentz Transformations (left graph):

When energy arrives to the limit c of the light space-time, its membrane 'curls' the energy into a whirl of space-time, evolving into cyclical mass, the stable form of physical in/form/ation, becoming electrons and quark particles.

Further increases of relativistic energy=mass evolves quarks into forms of higher mass and reproduces them into pairs.

What happens when you accelerate to c speed is a discontinuous process of 'evolution' (from 'ud' to s, in the first step, and so on); and then a process of fission/reproduction. So energy not only evolves the particle but it splits it into 2, as when quarks absorb energy and produce a jet of 'quarkitos': *E->I(m)->Re(m+m).*

While on the other extreme, black holes relax light into gravitational strings. So most of the 'particles' and dynamic events we perceive in the Universe are phase transitions between both membranes; reason why physicists say that particles are 'frozen states' of the parameters of temperature and speed through which a hypothetical big-bang of our galaxies or Universe evolved.

Thus the particles and forces we perceive in the Universe are born in the verge between both worlds. Light is in the energy verge between both worlds. Electrons are in its lower energy state, the Bohr radius, as we have shown, in the verge of the quark world, acting as the event horizon of a micro-black hole - a proton.

The non-local speed of gravitation.

But how can we prove that gravitation is non-local, instantaneous, and able to relate all the parts of the Universe in a synchronic way? There is the evidence of 10 C jets of matter coming out of black holes. Indeed, the non-local speed of gravitation can be proved in several ways, but perhaps the easiest way to prove it is with the concept of complex speed...

The simplest one is the definition of speed, v=s/t understood as: *V=Spatial Energy Temporal/Information.*

Because gravitational force has no information (it is the simplest, lineal scaffolding in which our light-form is imprinted), then:

Gravitational speed =energy/0 Information=Infinite.

But it is a relative ∞, since Cantor Set theory proves that ∞ are relative and the absolute infinite (the set of all sets) do not exist. By this we mean that the informative network of dark, gravitational energy that structures the Universe is like the nervous network that send simultaneous orders to all the cells of the body and seems instantaneous to all those cells, infinite only in as much as it reaches the limits of our Universe, cell of a hyper-universe in a reality of ∞ scales.

This is proved, since a fractal wave of gravitation can reach according to General Relativity up to 10^{10} million years light. Thus a single gravitational wave can cross in a single fractal jump, the entire Universe. And indeed, the 'instantaneous, present space-time Universe' reaches till around 10^{10} million years light, the so-called Universal horizon. This explains also 2 'spooky' phenomena: the entanglement between electrons that seem to relate and communicate information at distance instantaneously: they do so through the gravitational, non-local plane. It explains why the limit of c-speed, which only applies to our light-membrane. It also explains how the Universe has a synchronic structure without recurring to the magic, entropy only 'big-bang' (astrophysicists say that their structure that requires all those galaxies to interact together was acquired when the Universe was so small that they could communicate at c-speed.)

Recap: the limits of our light universe are c-speed and 0 K of information but the gravitational universe should be faster, bigger and more ordered, colder than our Universe. Lorentz Transformations and black holes are the doors and bridges between both Universes.

22. Forces & particles of the 2 universal membranes

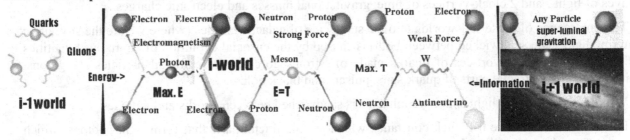

In the graph, the Universe is made of 2 space-time membranes, the light world of charges and stars and the string world of quark masses and black holes. We live in the electromagnetic scales, which form a network whose dark spaces are occupied by quarks (st-1 inner world of atoms), and in the biggest eusocial structures we know of electromagnetic matter (st+1 galaxies) by black holes, whose parameters correspond to top quark condensates. The basic forces that create distance-space and

information between them are bosonic strings of gravitation condensated into gluons and strong forces in the quark-mass membrane and Planck actions of light, condensated into photons in the electromagnetic scale. Those gluons and photons herd further into cyclical particles made of such fractal parts, associated into cyclical networks, called quarks and electrons. Self-similar structures form the nebulae of stars called galaxies, and the condensated of quarks called black holes. The only error of such standard model, reflected in the previous graph is the weak force which is not a spatial force but describe time events of transformation of particles between both scales.

In the graph, the 2 membranes of space-time have 2 theories that must be put together to explain the dual Universe:

- Quantum theory that describes the electromagnetic membrane, not the cosmological world. Both are self-similar (as we have shown with the equation of fractal unification), but not equal. Thus, while we can admire that self-similarity, the use by quantum cosmologists of quantum equations to describe the cosmological world is erroneous.

- Relativity, the theory of gravitational forces and black holes, which are super fluid quark stars. And so it should be fusioned with the theory of strong forces and quark masses.

Since each of those scales has also a fractal structure, we have to define its fractal quanta:

-Masses are vortices that grow from string actions (Nambu's actions of bosonic, tachyon strings corrected to fractal dimensions and background independent.) Those motions reproduce and evolve into gluons, which are the fractals that evolve into quarks, vortices that are the fractal parts of quark stars.

Since our electrons surround those quarks, locking them inside charges must be fractal vortices with a larger number of dimensions than those bidimensional masses. So 3 bidimensional quarks of $3 \times 1/3^{rd} = 1$ charge must lock their mass in the x,y,z planes to enter in a 3-dimensional exchange of energy and information with the electrons of our light-space. So quark masses attract electrons with their gravitational force inwards, and electrons shield themselves with their electromagnetic force outwards. The balance of that tug of war is a complementary atom.

In that regard the entanglement of both membranes is dynamic and based in the network structure of non-Euclidean spaces in which its points are connected by 'lines of communications' which are waves of smaller particles; leaving in between dark spaces which allow the other membrane to entangle itself; as it happens with the blood/reproductive and nervous/informative network of a human vital space. In the case of the 2 membranes of space-time the membrane of dark energy/matter (gravitational membrane) plays the role of the nervous/informative network that connects the blood/electromagnetic network.

We talk of 2 space-time networks, defined by 2 spatial forces - lineal strings of gravitation and lineal waves of light - and 2 clock-vortices of time: gravitational masses and electronic charges.

The emergence of those st-1 worlds into the st+1 scale of galactic entities (where we are the st-relative intermediate scale sandwiched between both) is caused by the eusocial evolution of enormous quantities of those fractal units: vortices of quarks made of 'strong' forces, whose fractal particles are gluons, gather together as fractal parts of quark stars, pulsars and black holes.

While our world - the light space we inhabit - is sandwiched between quarks and galaxies

Its smallest actions are h-Planck constants, which are the fractal parts that form light photons, which are the fractal parts gathered into an electron nebulae, the basic clock of time of our world, whose plasma herds (ionized state), associated with complementary protons creates the world stars. Finally vortices of star herds form galaxies, which turn around their Kerr black holes, and so are self-similar st+1 objects of the st-1 atoms positioned in space by the gravitational force of the central quarks.

Thus, the invariance of topological form and scale of all those vortices that makes macrocosms emerge from microcosms is caused by the complementary structure of its central mass vortices that anchor the electronic and star energy of our light membrane:

Quarks anchor atoms and black holes anchor galaxies.

Recap. Our space-time is made of h-quanta of light. We exist as evolved forms of a light space-time network anchored by the nuclei of atoms, made of quanta of strong forces, called gluons and quarks belonging to the gravitational space-time membrane.

23. Some precisions on the formalisms of both membranes.

A quanta, which is interpreted as an abstract, probabilistic number in quantum theory is a fractal piece of space-time, a complementary system made of bites of spatial-energy and bytes of temporal-information. This is self-evident even in quantum theory, where Planck defined those quanta as actions with the dimension of energy and time. The Universe is indeed made of actions of energy and time, as when we say 'I don't have more energy or time to do this'. We are not solid spatial forms, a Maya of the senses, so common among 'naïve realist' physicists, accustomed to touch the solid metal of their machines, but we exist in a spiritual Universe made of actions of energy and time, which occupy a vital space and have a time rhythm. And each of the self-similar fractal worlds/membranes of the Universe has a minimal unit or fractal quanta, a type of st- point with a different volume of energy and information that defines the space-time membrane.

We exist in the space-time membrane of light made of h-quanta. And its formalism, later studied in more detail is well-known to physicists (quantum physics), and requires some modifications on its interpretation to adapt it to the wider concept of fractal spaces and cyclical times (Correspondence principle). But there is another space-time membrane of gravitational quanta, whose formalism requires more changes, since being an invisible membrane, we humans cannot use easily the experimental method to study them, especially when considering the st-1 quanta so small, in the Planck length scale that there is no way we can observe and prove our models. So we are left only with proper formalisms of its macrocosmic entities and effects, which Einstein described with general relativity.

As of two day two formalism, which require further corrections are useful to that aim: the formalism of bosonic, tachyon strings defined by Nambu's actions, adapted to a background independent world (they are the actions that create the membrane, not actions that float in an abstract metric coordinates), and the use of fractal dimensions and non-Euclidean geometries, unknown to physics.

Another formalism used to describe the gravitational minimal units is the one elaborated by Nottale[6], whose unit is λ, lambda, also called the cosmological constant. In scale relativity, Nottale defined this constant as $\Lambda = 1/L2$, the minimal unit of the gravitational Universal membrane, where L is the minimal length of that membrane, the smallest piece of space. It is essentially a string, but string theory must overcome the error of Newtonian Absolute Space to become useful to describe the gravitational scale, as its minimal unit 'independent of the abstract background of membrane space-time'.

Such formalism which borrows from the work of those 2 pioneers is part of the more complex 'courses' of Multiple Spaces-Times, beyond the scope of this introductory lectures.

The other formalism that is not completed is the formalism of weak forces, since Physicists unaware of the temporal nature of those events have tried unsuccessfully to describe it with the same spatial exchange of particles (Higgs mechanism[7]) when weak forces are temporal events, and the W and Z particles brief states in the process of transformation of particles of our electromagnetic membrane into masses of the gravitational membrane. Reason why those forces break the symmetry – as time is not symmetric but hierarchic and why the Higgs has not been found and will never be found – it is a particle only needed if it was a weak force.

Unfortunately we cannot perceive the lambda scale of strings. But, essentially with those 2 membranes the strong gravitational world and the electromagnetic world and its 3 ±st scales, we can explain all the physical properties, events and entities of reality.

Recap. The formalism of the light-membrane is described by quantum physics and it is correct, except for the errors in the description of the temporal, weak force. The formalism of the gravitational membrane departs from its minimal unit, the bosonic, tachyon string, which must be adapted to a background independent gravitational membrane of which it is its minimal action and the fractal, finite nature of dimensions.

24. The 3 scales of the 2 membranes.

Why there are 3 main scales of social evolution in both membranes? The trivial answer is that we are always sandwiched between a lower and upper scale and so we always perceive 3 even if the number of scales might be infinite.

And so from the human p.o.v. any entity or system of energy and information stretches in 3 scales.

For example, your organism is basically a fractal of biological energy and information extended in 3 scales, cells, individuals and societies, of which you are a relative cell.

In the physical, fractal, Non-Euclidean structures of the 2 membranes of space-time, there are also 3 basic scalar structures:

A complementary atom extends on both membranes. In each of those membranes, quanta evolve and form social groups, emerging into a new topology, a higher scale of energy and form:

-The scale of h-quanta form light and electronic nebulae, made of dense photons.

- While its inner gravitational world of quarks can be decomposed in 3 scales: The lambda scale of 'strings' are the components of gluons, which are the components of quarks.

Emergence of parts into wholes of a higher scale is possible due to the other 2 invariances/dualities of physical systems: invariance of topological form and motion allows invariance of scale.

What about the cosmological world? Again we must consider that cosmological systems have 3 scales, the scale of particles, (quarks, electrons and atoms), which are the units of cosmological bodies, black holes, quark stars and electronic stars, which are the atoms of galaxies. Those 3 scales thus interact in all the organic networks of galaxies.

There might be another scale, the Universe, where galaxies are atoms of the Universe (and the Einstein-Walker model of cosmology treats them in fact as Hydrogen atoms); which would itself be a unit of a hyper-universe, whose 'explosion' at an even faster 'infinite speed' that dark energy, would explain the 'inflationary age' of the big-bang. But there is little evidence of this final scale today, and an enormous number of Pythagorean and hyperbolic errors in the formulations of the cosmological big-bang *in a single space-time continuum* to take serious that theory.

Recap. All what we can be sure of, is what we can perceive – the limits of our existence, sandwiched between atoms and islands-universes, the original name Kant gave to galaxies.

25. Dimensions of particles and Universal membranes.

In that sense, to fully grasp the difference between both worlds, the dark and light world, we must understand the basic laws of fractal dimensions. Dimensions are in any system of the Universe, as in the case of light, arrows of time (motions of energy, information and reproduction), perceived according to the Paradox of Galileo also as 'spatial surfaces. Thus, they are always local, fractal dimensions which are limited by the extension of the 'species' in which those dimensions create a 'vital space'. So dimensions are dynamic as the species they create grows or shrinks, evolves into more complex beings or becomes destroyed. Entities grow in dimensions as they grow in size.

Further on, more complex informative species ad dimensions.

This happens normally when a form is bigger and has inner particles. But in the case of quarks it happens, because the quark is faster and has more inner speed/form.

All in all, the total number of dimensions of a system is relative to our detail of analysis. For example, those quarks are made of one-dimensional strings, made of non-Euclidean points (each one with its 3 inner dimensions), which have a total of 3×3 dimensions (3 of their Non-Euclidean points and 3 of the string of points). That is why physicists in detail use 9+1 time dimension to study strings, but from our perspective without detail, they are just one-dimensional lines of a quark vortex.

Yet if we transcend to the world of quarks we consider them 1-dimensional. So those strings form gluons and quarks, which from our higher world are perceived as bidimensional mass-vortices.

Finally 3 color-locked quarks interact with the world of light of 3 dimensions, and the world of electrons of 4 dimensions, which are the 3 dimensions of Euclidean space and one of form, of color.

On the other hand, the biggest entities have more dimensions, as they include the smaller entities within its whole. It is the law of fractal dimensionality. It states that a fractal part of a bigger st-scale of information has fewer dimensions such as:

Law of dimensionality: Fractal $\sum st\text{-}1 = 1/3^{rd}$ Dimensions of st

Since any st-Point will have 3 inner dimensions. Thus the dimensions of information of any system grow with a cubic power law as the system grows in scale. In simple terms, a 3-dimensional being transcends as a mere cellular point of the new whole.

Those 3-inner dimensions, as in the case of light, are ultimately the dynamic arrows of time, or wills, or physiological networks, or organic drives of the being.

For example, a human being has 3 dimensional networks, which are our physiological systems. The blood network of energy, the nervous network of information and the endocrine network of reproduction. Those are the 3 systems around which cells and organs build up. So humans can be described as a 3-dimensional network with 3 wills: to absorb energy for our blood/ digestive system, to absorb information for our sensorial/ informative system and to reproduce through our hormonal system.

The abstract, geometric description of dimensions is secondary to the functional, dynamic, organic meaning of those dimensions. The how is less revealing than the why. The why of dimensional motions are the search of energy and information of any entity, whose paths become fixed in space and appear as dimensions. Or they are the reproductive strings of self-similar cells that form the physiological network of energy or information of the system.

In the quantum world, the constant exchanges of energy and information of the electron with the external Universe are described with 4 quantum numbers that correspond to the 6 arrows of time:

-The 2 first numbers describe the reproduction of the wave (main number) and its light feeding/ energy cycles (1 number).

-The spin number defines the informative orientation of the electron and the magnetic number the social organization of those electrons established by an external magnetic field. Thus those numbers describe the 'paths' of particles in search of its dimensional arrows of energy and information.

The dimensions of light are also the dimensions of the energetic/magnetic field, the informative/electric field and the reproductive length-speed, or social evolution (color/frequency).

In that regard, it is also interesting to observe the relationship between both planes, the light plane which 'feeds' as the lower st-1 space the dense fractal, cellular points of light of the electron. When considering the basic laws between planes of space-time we noticed that a force, st-1 of energy, becomes st-bytes of information of the upper scale, which uses them as pixels to gauge information. Thus the magnetic, width field that acts as energy in light becomes the informative, electron magnetic number.

Life beings also have a dimension of height/information that is maximal in man, the most informative being of the cosmos, and a dimension of energy/length which is maximal in energetic predators.

Recap: Dimensions are morphological expressions of the arrows of time, seen as fixed forms of space.

26. The scales of big-bangs.

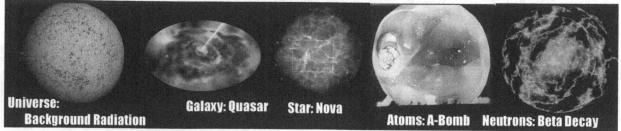

Big Bang's Afterglow Fails Intergalactic 'Shadow' Test

ScienceDaily (Sep. 5, 2006) — The apparent absence of shadows where shadows were expected to be is raising new questions about the faint glow of microwave radiation once hailed as proof that the universe was created by a "Big Bang."

See also:

Space & Time
• Cosmic Rays
• Big Bang
• Cosmology
• Astrophysics
• Galaxies

In a finding sure to cause controversy, scientists at The University of Alabama in Huntsville (UAH) found a lack of evidence of shadows from "nearby" clusters of galaxies using new, highly accurate measurements of the cosmic microwave background.

If the standard model of how the universe was formed is correct, microwave radiation from the edges of the universe would be blocked by clusters of galaxies, causing 'shadows' in the microwave background. (Graphic courtesy of The University Of Alabama In Huntsville)

Scalar big bangs: the gravitational, inflationary big-bang of the Universe, a big-bang of the light-membrane (quasar), a star big-bang, an H-bomb and a quantum big-bang or beta decay.

Multiple space-times imply the existence of multiple big-bangs with different explosive size and speeds, all of them deaths of an informative mass or charge that uncoils its accelerated vortex into radiation following Einstein's equation, $E=M(c)^2$, where a bidimensional or tridimensional vortex (M or e) becomes extended into a plane of radiation (hence the square of lineal c-speed).

Yet for each different scale the parameters of speed (c), mass (M) and energy will change. And so we obtain a series of self-similar big-bangs described in the graph.

The single cosmic big-bang based in a single space-time continuum, was devised even before we knew the existence of a faster than light, gravitational membrane of dark energy and dark, quark matter. So we first must define it within that scale of big-bangs as the *big-bang of a membrane of electromagnetic radiation; hence a quasar big-bang, not a cosmic big-bang, which will correspond to the much faster explosion of a gravitational membrane, starting at the Planck scale of the gravitational minimal units, the lambda-strings.*

This is in a way the conclusion of astrophysicists which now study the big-bang as a dual process, with an inflationary age (the gravitational membrane) and a slower age (the big-bang of electromagnetic space, created over the gravitational membrane).

In the next graph: Physicists have measured a local, hence galactic background radiation, which doesn't leave shadows from other galaxies and they tabulated the quasar cycle of galaxies which coincides with the age ascribed in the past to a cosmic big bang. So a more rational theory would consider the cause of the 2.7 K radiation both local in space (galactic) and in time (produced at present);

325

since our measure is local and present. Yet the only alternative object that can produce such radiation in present is a micro-black hole, whose temperature red-shifts light as a gravitational lensing. And since a black hole gravitational lensing is proportional to the mass, we can calculate the mass of those 'background black holes', which redshift light at 2.7 k. Indeed, a simple calculus shows that a black hole of the approximate mass of a moon (the most common galactic object of which each sun has hundreds in its planets and Oort/Kuiper belts, would produce that radiation. So very likely the halo of dark matter of the galaxy is made of micro-black holes that have eaten moons, as their bigger cousins eat stars.

Further on, since black holes are fundamental particles of the upper scale, as protons are in the quantum scale they must be homogeneous in size, even if they eat less perfect planetoids, expelling the 'left-overs' as gamma radiation and becoming with other quark stars (strangelets), the 'protein' membrane that shields the galaxy, as a st-point whose 3 topologies are latter studied in more detail.

So instead of a fantastic big-bang particle that in the absolute past of time created all the absolute space of the Universe, we consider to exist in a fractal universe of infinite scales, where big-bangs are also scalar processes of creation and extinction of vortices of mass in several scales:

- There are big-bangs of atoms (beta decays) in quantum scales.

- There are star big-bangs (Novas or super-novas that create strange stars (pulsars) and black holes, (top quark stars)).

- There are galactic big-bangs, quasars, whose periodicity and background radiation coincides with the cosmic big-bang in its simplest formulation.

- And there are cosmic big-bangs (inflationary period of the big-bang), in which the gravitational membrane at non-local faster than light speed expanded before giving birth to electromagnetic membranes, forms warped around this primordial membrane.

Recap. The invariance of topological form at scale means that big-bang processes of death are common to all the scales of the Universe. What physicists call the big-bang seems to be a galactic big-bang.

27. Universe as a Cell of a Bigger Organism: Hyper-Universes.

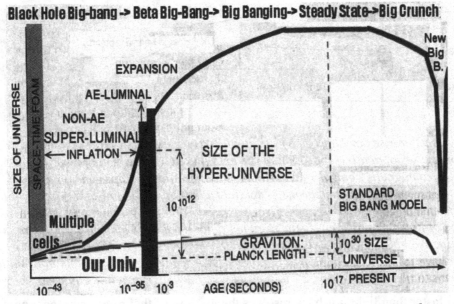

Our Universe (line below) might be born in a dual fractal big-bang of gravitational space-times (inflationary state) and fractal big-bangs of light-membranes (quasars that explode galaxies). All those homologous fractal big bangs of gravitational dark energy and light create the same inflationary Universe developed by Linde & Guth[8] but, as in the case of electronic nebulae, interpreted as fractal

parts of an electron co-existing in the same Universe, instead of Everett's absurd interpretation of multiple universes a fractal dual big-bang rationalizes the data of the inflationary big-bang, introduces the Planck scale as the minimal string-size of a gravitational action, which will self-reproduce as it expands and validates the fractal Unification that converts each galaxy in an atom of the gigantic Universal scale, which we only perceive till the limit of reproduction of light waves (mean life 10^{10-11} y.)

In the graph, the discontinuous fractal structure of multiple space-times implies that in the same manner there are many quasar big-bangs (later Hoyle's interpretation of the big-bang process); there might be many fractal gravitational big-bangs, in a huge Universe in which each galaxy is a relative atom. Again this interpretation gives meaning to the standard inflationary Universe, whose enormous size wouldn't be the sum of a series of parallel Universes but the same Universe made of fractal gravitational/ quasar big-bangs. In the graph, those 2 cosmic fractal big-bangs give birth to a huge universe with $10^{10^{12}}$ 'galactic atoms'.

The fractal theory of a dual big bang is thus self-similar to the inflationary theory but without the logic errors of multi-universes. It postulates a first gravitational big-bang of dark energy at z=10-100 c (which seems to be the limit of redshift observed in the limits of perception of our cosmos).

In the graph, the birth of the 'huge universe' (no longer the hyper-universe) would have taken place from a black hole of the minimal scale of size of 10^{-35} m. and maximal Planck's density during a gravitational, superluminal big bang that split space-time into 10^{11} different fractal bubbles of gravitation - many universal cells like ours in which light universes would later develop. Those gravitational bubbles, while connected by non-local gravitation will not interact with light since the mean life of a reproductive light ray, as that of the stars it creates is around 10^{10-11} years and so light from those far away 'atomic galaxies' cannot reach this one.

Recap. There are not multi-universes but fractal Universes created with 2 self-similar big-bangs in 2 scales, the gravitational and light scale.

28. Big-bang deaths studied with the laws of fractal spaces.

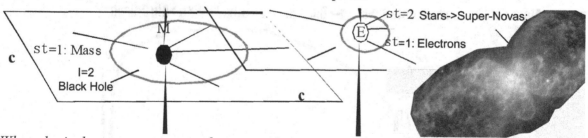

When physical systems, masses or charges on the atomic or cosmological level, die, their exi elements split, as in all processes of death (separation of 'head and body' in life organisms): their temporal, gravitational or electromagnetic central knot of information jets out in the dimension of height and their intermediate space expands into a membrane of spatial radiation. In the big-bang a hyper-black hole exploded and its energy gave birth to our Universe. Stars produce big-bangs that show clearly the breaking apart of its informative/energetic regions: a flat toroid expands most of the mass of its toroidal, reproductive inner region and a perpendicular, magnetic jet (right side) expels its central superfluid, informative zone. While in atomic, smaller worlds, charges transform its information into a bidimensional plane of light-space, c^2, in a process defined by Einstein's equation: $E=Mc^2$.

The gravitational faster than light big-bang resolves the mystery of the inflationary age, now canonical in cosmology. Moreover it follows the general laws of 'death processes' of multiple spaces and times theory. Since death is the most *expansive, spatial process of erasing of information, Max. Se=Min.Ti*, it happens in a relative zero time – the minimal time of the system, when information dies and expands into a relative bidimensional sheet of space.

So we *die in a second, our time clock* which is the minimal unit of time of the human being (thought, blink of the eye, heart beat) and suddenly we become a 'relative flat' surface of energy (corpse, extended into a flat space). So it does the gravitational Universe, whose death happens in the tick of Planck's time, the minimal action/vibrational time of its string units.

Indeed, death is anti-symmetric to life, an arrow that has no intermediate states and lasts minimal time: a transformation of information into energy happens in the relative zero time of each scale: from novas to atomic bombs, to universes, to human death.

Further on, since a vortex of physical, gravitational time is paradoxically faster, more massive, informative, the smaller it is, the faster gravitational big bang started in the smallish vortex of mass-time, the Planck mass/density/time-tick, he aptly called the 'numbers of God'; since it was the hyper-black hole 'brain' of the previous organic Universe... whose reproductive radiation gave birth to our huge Universe. Only the inverted properties of time and space, information and energy, explain it.

What happens then? A basic cycle of creation of all organic, complementary systems:

Singularity-> Big bang (death) ->Energy Sheet->
Informative Reproduction (Big-banging of particles) ->
Social evolution (Creation of stars and galactic networks)

They are the phases of creation of the Universe. In complex models of morphogenesis, informative evolutionary phases in time alternate with reproductive radiations in space, creating a fractal rhythm of biological radiations and punctuated evolution proper of physical, biological and cultural systems:

Max. i (Evolution)->Max. Re (Reproduction)->Max.i(evolution)...

That rhythm applied to any system and fractal scale means that we observe the alternation of those 2 phases in any organic process of physical/biological creation, a big bang & a big banging:

-*Big bangs*: The singularity/seminal egg explode their energetic membrane in a micro-time period (big bang, creation of a cellular egg, etc.). Then it continues its expansion at a slower rate as it grows in complexity inwards, invaginating and creating its energetic and informative networks (invagination of dark energy in the Universe and electromagnetic membranes controlled by them, of the nervous system and the blood system it controls in the foetus).

-*Big bangings:* In the big banging phases, news informative particles/cells are evolved around those 2 forces/networks and then they start a massive age of reproduction or 'biological radiation'.

The alternation of those two processes slowly builds a living being and a Universe masterminded by two networks around which new particles/cells and social groups of those particles/cells of increasing complexity are born.

Thus, in the inflationary Universe, explained in organic terms, a minimal, cyclical black hole vortex of the size of a string with the density of Planck and its constants of God created a first gravitational big bang, reproducing=multiplying into 10^{11} fractal regions of dark energy and quark matter in 10^{-43} sec.

Each gravitational bubble gave birth to a 2nd big-bang of *fractal beta-decays, in which nucleons expanded, dying into protons and electrons (cyclical, informative big-bang), while* light-membranes of electromagnetic radiation warped the gravitational dark energy giving birth to an age of light.

This dual process expanded the regions of cyclical time/mass (as neutrons exploded into beta-decays) and imploded the regions of dark energy (as light warped the minimal information of gravitational strings), creating protons, electrons and light in $_{10}10^{11}$ parallel future galactic regions.

Then the process reversed as the fractal regions of mass imploded into galaxies and the dark energy between them tired light and expanded, but both processes remain in balance, so the present universe is in a steady, mature state without expanding or imploding when we add the implosive galactic vortices and the explosive regions between them.

The Webb telescope, by solving the relative size of the big bang and the type of hierarchical scales the organic universe displays, will determine the experimental details of those processes. If the Webb telescope observes in its 13 billion years limit a landscape of galaxies similar to the ones of our cosmic region, it means the fractal universe extends to a relative infinity; as each galaxy is self-similar to an atom of the Universal scale, whose size is astonishingly huge. Some theories hint to that concept:

- Relativists model the Universe with Einstein-Walker equations in which each galaxy is treated as a hydrogen atom. This theoretical trick that facilitates the calculus of relativity shows the self-similarity between a hydrogen atom and a Galaxy of the upper scale. Each point of the electron nebulae is a star of the upper scale. If so, each proton is a positive, central Kerr black hole. And alternatively each +Kerr black hole is a hadron made of positive top quarks. And the gravitational waves of dark energy expelled by black holes are equivalent to the magnetic fields of protons.

An alternative model considers that each star is self-similar to a photon, While the electron nebulae that surrounds the atom is equivalent to a nebulae of strange, quark stars, the halo of the galaxy; since strangelets are negative (strangeness is negative).

So protons are self-similar black holes (top quark stars). The halo is made of strangelet stars as the Earth will be soon (and we are indeed close to the Halo). And stars are neutral, light photons.

Cosmologists use models of electronic nebulae to describe the behavior of stars around the central black hole; and some have tried models with high density photons.

Yet self-similarity is not identity. So what we call quantum cosmology requires a fractal formulation to make sense; NOT the use of quantum equations as if they were identical, but the logic of fractals to observe the self-similarities of those scales.

What else can we know, we human beings, dust of space-time about the infinite Universe? Not much more, because if there is a higher fractal scale and galaxies are self-similar to atoms, the extension of that scale will be so vast that we shall never find its limits. Indeed, in our Universe there are trillions of atoms and so if there are trillions of self-similar galaxies, the maximal perception we can have is that of a local structure within that Universe.

This philosophical view will be far more appealing than a mere big bang of a cellular Universe for metaphysical reasons:

- A Universe in which the two self-similar scales of quantum and cosmological vortex of space-time (charges and masses) are truly self-similar is truly infinite in space and infinite in scales of temporal information and complexity. Then we, humans are really nothing but a mush over a corner of the Universe. But in second thoughts, such structure means that we humans are also all, as each of us has infinite Universes within his self.

- While size is absolutely relative and so our position on those scales doesn't matter, we are an incredibly complex fractal structure of energy and information, which in the fractal models of biology needs 10 membranes of space-time to be described. Hence, even if our size is minimal our *informative complexity is astonishing; and we are from the perspective of the arrow of information, one of the supreme beings of creation.*

This is not the case in a single big-bang theory where only energy and size matters; so we are nothing but dust of space-time.

Recap: The big-bang in any scale of reality is the death of a previous physical informative knot of matter, which expands in space and then starts a creative and reproductive process or big-banging.

VI. THE QUANTUM SCALE.

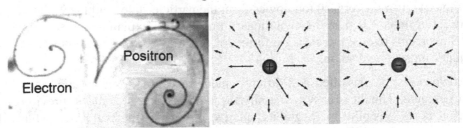

Positron

Electron

Particles and antiparticles, positive and negative charges must be explained as space-time antisymmetries, which can be seen according to the Galilean paradox, both as inverted events in time, as in the left Feynman diagram or as geometrical inversions in space (right diagrams). They are the why of most physical dualities that classic physics fails to understand.

29. The corrections of quantum physics.

All the errors of interpretation of quantum physics derive also from the ignorance of the key dualities of the Universe:

-Duality of motion in time=distance in space.

-Duality of 2 space-time membranes, gravitation & light.

-Duality of wholes, made of fractal parts of a smaller scale.

-And duality of the arrows of form and energy natural to all complementary beings, which create 2 more complex arrows, energetic reproduction and informative, social evolution.

When we use those whys to interpret the magnificent scaffolding of mathematics and metric space times of measure built by quantum theories all becomes resolved. So we shall try to give meaning and explore the whys of quantum physics with the new tools of multiple spaces and times.

Let us then start with a basic clarification brought about by the knowledge that all cyclical patterns and forms are 'time-clocks' that carry the information of the species we study. This means particles are cyclical states and waves are energetic states; and cyclical properties such as spins are time-clocks that carry information while energetic states are lineal, extended forces.

The main error of quantum physics is the lack of understanding of the properties of temporal information, the cyclical nature of time and its multiple rhythms and clocks, obscured by the equalization of all those rhythms with a time-clock. This error is enhanced by the experimental limits of human instruments (uncertainty principle):

Time cycles and spatial flows cannot be differentiated in the quantum realm, as they can in the biological world, due to the minimal duration of ticks of time of quantum particles, which therefore are observed as space trajectories.

This is due to the black hole paradox: Max. Ti = Min. Se, which means the smaller the particle is the faster its cyclical, rotational clocks turn. So in experimental physics, you never get images short enough to 'feel the tick' of time. Time cycles appear as those pictures of a fast turning wheel, which seems a solid disk without motion. So particle's cycles appear; as printed 'images' on the bubble chamber.

Yet once we understand the dualities and forms of time-clocks and spatial, lineal flows, there are easy 'key elements' to know when we are seeing 'sequential time events' instead of 'spatial trajectories'. For example, systems that are 'bilateral', 2, are space-like, left-right. While systems that are composed of 3 elements are time-like, where each element represents the past, present and future states. So there are not 3 quarks but quarks in 3 time states/colors. And so in time the same quark changes its 3 states creating a charge of 3 x 1/3=1. More often we find that a ternary state is both, an organic structure in space and a ternary wave in time. And so the ternary topologies form the particle state, and the 3 phases of the wave represent the particle in time. 3 is thus the evolutionary limit, which explains why so many systems are ternary systems: 3 neutrinos evolve increasing information as a wave; 3 mass families are the 3 horizons of evolution of quark-masses, and so on.

Further on, particles must be classified in 2 different membranes, that of light and that of gravitation, which we do not perceive. Thus it is easy to classify those particles, as we did in previous paragraphs: the easiest criterion is its 'visibility'. So gluons, quarks, bosonic strings, gravitons, black holes and neutrinos, which are not observables, are gravitational species.

Because time 'bends energy into cyclical vortices' masses or charges, depending on which of the two membranes of space-time we study (since both are self-similar as we showed by finding its unification equation), we can also differentiate the particles that are clocks of time and store most of the information in both systems (electrons and quarks), and particles which are lineal, energetic states (light and gluons).

Because the universe is in symmetric balance, as there are electric and magnetic fields in light to balance each other, and there is a repulsive, lineal electromagnetic, transversal wave and an attractive cyclical electronic particle; so there must exist an attractive, cyclical gravitation and a repulsive, transversal gravitational wave. Thus we conclude that dark energy is merely the transversal, repulsive gravitational waves that come out of galactic black holes.

Finally, since the formalism of temporal information is inverse to that of energy (meaning that energy moves forwards, information stops motion or reverses backwards with negative, imaginary numbers), and both cancel each other, we can distinguish, which are the parameters of energy and information of a field or particle.

For example, in electrons, the informative component is given by the imaginary parts of the Abelian phase of its mass field; and both fields - the real, energetic field and the informative imaginary component - are at a 90 degrees angle and when one is maximized the other is nullified. So even if our perception doesn't distinguish both terms we can now which one carries the information of the system and which one the motion.

For the same reason the 4 quantum numbers can be easily classified as representing the 4 basic arrows of an electron:

-The 2 energetic/reproductive arrows are represented by the 2 first numbers, the main number represents the reproductive wave of the electron as it displaces, imprinting the surface of energy around the nucleus. The second number represents the 'changes in the energy' of that electronic path, due to the absorption= feeding of a photon of energy by the electron.

-Then the other 2 numbers, the magnetic number and spin mean the same for its information components: the spin represents the informative clock-like rotation of the electron and the magnetic field that affects the electron is the 'informative perception' that can change the spin and orientate the electron.

This trivial explanation of the why of the 4 quantum numbers, using the theory of Multiple Spaces-Times however will be denied by most abstract physicist, for whom organicism is a taboo concept born from the anthropocentric nature of human beings and the mechanist approach to science of the founding fathers, whose memes are learned and repeated without much reflection by all texts on physics. Yet what marvels of a model of multiple, vital spaces and time arrows applied to quantum physics, is how easily data becomes structured into a 'plan' with meaning, explaining the whys of all previously paradoxical quantum processes.

Indeed, once those precisions are made we can truly clarify all the mysteries and meanings of the parameters of quantum physics, from spins to waves to complementarity laws, to spooky effects. For example, since a time state is an informative, *still* state with the ternary phases of time cycles and its hierarchical, non-commutative causality – given the fact that it is not the same a motion to the future (whereas energy becomes information) than a motion to the past (where information becomes energy), the meaning and mystery of spins becomes resolved:

Systems with 3 spin positions (spin number 1), are still systems and the 3 spins form a rotational clock.

Systems that are in motion have 1/2 spin number - 2 positions and they are bosons which display a spatial main lineal motion and its inverse negative 'informative' motion, creating a simple beat forwards, backwards. This beat is natural to all complementary systems that last, and so we find that an alternate currency of electrons is more stable and lasting than a continuous one as it has an energy/information beat in its motion.

Those rules of course are also useful to detect erroneous theories that cannot happen since they mix properties of time particles with spatial properties and vice versa.

For example, the Higgs cannot exist because it is a scalar boson[7]. And by definition, a scalar field with no rotation will not act as a boson moving at light speed.

Further on a boson cannot have mass as it does not close its cyclical trajectory so it does not form an attractive vortex, reason why light and gluons don't have mass.

The same concept applies to forces, whose range law I just an expression of the Black hole paradox (max. energy=Minimal information/mass). So we can distinguish forces belonging to the strong-mass membrane – gluons - and forces belonging to the electromagnetic membrane – photons. And then forces belonging to the bigger cosmic membrane of gravitation, gravitons, which are bosonic strings of tachyon speed.

This leaves the 4th force, the weak force that has no range and time parameters that change the in/form/ation of the particle, and also breaks the symmetry of space (left-right). So it is obviously not a spatial force but a time event, which further explains why we have never found its 'bosonic force', the Higgs[7].

The same error invalidates super symmetry and hence all string theories except the original one, which properly understood as background independent and using fractal co-dimensions of finite size extended through several relative scales of gravitational complexity should become departing from its Nambu's minimal action and Planck length, the proper formalism of gravitons.

Thus, while the formalisms of most quantum theories are correct all those limits of perception added to the errors of using a single clock-time and a single space, explains why so many 'whys' are confused in quantum physics to the point that Feynman finally said: 'the why is what you don't ask in quantum physics'. We do and that is what this paragraph is about: the why of the perceived phenomena of quantum physics.

Finally physicists ignore the black hole paradox so they don't understand that protons, neutrons and all other particles do not live eternally. On the contrary their life/death cycle is their complementary cycle between informative states and energetic states, either the clock of particle/wave complementarity or the clock of particle/antiparticle, past to future and future to past cycles.

And so their existential cycles are really short. But they do not disappear because what physicists see are 'generational waves', given the fact that at the quantum level unlike in our informative-dominant world of biology, past to future and future to past arrows are reversible. So as a very simple jellyfish does, turning his clock backwards when it gets old and becoming young again to turn the clock forwards; particles and antiparticles go forwards and backwards and forwards in time. Bosons and electrons move forwards in energetic motions and backwards in informative motions. Light imprints once and again a photon of information that becomes an energetic wave that becomes a photon.

And so what physicists call the life of a particle is its longer generational cycle, which can last an enormous number of self-repetitions, but should neither be eternal. Indeed, it seems that light beams and light-evolved entities (stars) have a mean cycle of 10 billion years, after which they suffer an energetic big-bang expansion – stars explode into Novas and light gets tired and expands in a redshift process, returning to its gravitational, non-local lower scale of dark energy. This of course explains why far away light red shifts and seems to expand space: it is dying and beyond those mean 10 billion years an increasing 'darkness' is observed in the Universe, which a simple Gauss curve of probabilities show to be almost absolute beyond the x+1/3rd 13.3 billion years range.

Again physicists don't see both, the dual space and time views of the Universe but only the temporal big-bang, which considers a cosmos in time failing to analyze it in space as a simultaneous system connected by non-local gravitation. So they think darkness is not caused by the spatial death of light, which they merely use as a clock of cosmic time, thinking therefore that beyond those ages we enter into a temporal darkness – and the Universe didn't exist.

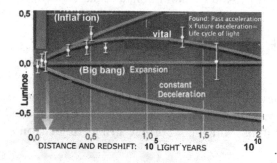

In the graph, we illustrate how complex physics, using the same data that Simplex physics of a single space-time continuum can resolve the puzzles of the Universe. Cosmologists have found that light coming from far away galaxies seems to decelerate and accelerate its red-shift till dissolving around 10$^{10-11}$ years after its emission. And they affirm this is a proof of an expanding Universe, as they apply a local phenomenon, the life/death curve of light self-similar to the life-death curve of any entity, which increases its frequency till its ½ life and then decreases, becoming tired till death, to the entire space-time of the Universe (a global symmetry). The tiredness and life-curve of light however only applies to a local phenomenon: light tires and redshifts at the end of its journey, dissolving back into gravitational dark-energy, which is perceived as an expansion of the light-membrane, but that effect is compensated by the contraction of space into mass in the galactic vortices giving us a zero-sum.

Recap. Quantum errors derive from the confusion of time clocks with spatial trajectories and the ignorance of the dualities of spatial energy and temporal information; and its rejection of organic theories of reality.

30. Informative inflation: The error of multiple dimensions.

After rejecting so many trendy theories of modern physics, I am afraid many practitioners of the 'baroque' age of information of quantum theories will ignore this text, so we must do further clarifications on the why of that multiplication of impossible theories.

The inflation of baroque theories of physics, which take the classic age of Metric spaces and its proper theories (Relativity and quantum physics) beyond its scope and reach, has caused many false theories without empirical proof, which try to correct and extend the metric paradigm beyond its power of resolution, without taking into account the needed corrections and widening view provided by the new tools of topology, fractal points and Non-Euclidean postulate. The main reasons of that inflation of false or unproved theories are:

- The fact that information is inflationary. This is self-evident. A language of the humankind maps out reality with far less energy. So it is rather easy to construct many linguistic images of a single reality, as we can find many mirror images of an object. And all of them will carry a minimal truth. But the theory that matters to us is the one that has the perfect focus with a given language. Again, if we focus an image with a camera, there would be many images similar to the object but only a point of perfect focus. So all those alternative theories with many dimensions are just half-truths, bad focused, linguistic and mathematical images of a single reality. We are however here concerned with the dimensional theory that has the perfect focus, both in logic and mathematical thought.

- A 2nd problem is the Pythagorean belief that mathematics is the only language of science and any mathematical theory must be truth. This absolute belief in a language fails to recognize the limits of linguistic knowledge. And it is proper of a young language, which evolves from an age of simple, dogmatic truths to a more sophisticated baroque age. Mathematics has been applied as the 'language of God' to the Universe only for 4 centuries. So scientists are believers in their language with a dogmatic degree that verbal writers only practiced in earlier, religious ages, when the written word was also considered the language of God. Pharaohs would just say, 'it is written'. And that meant it was truth. Then philosophers of language found out the relativism of all linguistic statements. And finally an inflationary age of verbal information – the age of fiction thought – has settled down after millennia of using words. This inflationary age happened in mathematics only a century ago and the relativism of those theories was exposed by the work of Gödel. Yet the believers still rule in science and so it is a fact that Gödel has been cornered and considered a heretic both in physics and mathematics for denying that 'God speaks mathematics' (Galileo). It is however about time for science to accept the linguistic limits of mathematics as verbal writers accepted the meaning of fiction in verbal thought precisely when Cervantes wrote the first fiction novel of the modern age, Don Quixote. Instead, scientists must accept that after Gödel we need experimental proof to show a mathematical theory certain.

- A third source of errors is the mechanist dogma that denies organic interpretations of physics, which are self-evident as quantum probabilities in time are easily transformed into spatial population of fractal knots of electromagnetism in space (real, mass side of the phase equation.)

The denial of Hylomorphism (the opposite philosophy of the platonic pythagorism of physicists who believe mathematics are truth per se, declared by Aristotle – all informative languages must have a physical, experimental truth to be real) and the denial of organicism (complex, vital arrows in physical systems – reproductive repetitions and eusocial evolution) are so pervading in quantum physics that is truly surprising… because all the fundamental laws of quantum physics are organic in nature.

The Complementarity law requires particles of information and fields of energy and one cannot exist without the other. Quantum theories are called gauge theories because all require that particles gauge, measure, hence process information as perceptive beings do, in order to interact with each other. The 2 fundamental particles, electrons and quarks constantly reproduce jets of new particles, hence they absorb energy. And finally all particles evolve socially into more complex systems, as cells do in organisms. So those are the same 6 arrows of time of all fractal systems of the Universe, which in mathematical, quantum equations are expressed through the different quantum numbers; as those informative/energetic processes take place in orbital paths and through the emission and absorption of electromagnetic energy and form.

Fact is we do not see 10-dimensional hyper-strings around us, because dimensions are diffeomorphic, fractal and attached to a temporal function/arrow. And they are topological and so in a 4 dimensional Universe there are only 3 topologies, 2 bidimensional substances, energy and form (Holographic Principle) that combine to create finite vital spaces with a time duration – the entities of reality. And that structure is enough to explain all systems.

Only then departing from such fractal 4-dimensional beings, we can build another plane of more extended dimensions in which each beings is a point of a bigger network of spacetime.

Further on, the 3 dimensions of time change, past-present and future parallel to the 3-topologies of space, energetic planar membranes, cyclical present, reproductive volumes and hyperbolic informative centers create harmonies of forms and functions, which have to be study with the rules of causality and logic, considering always the Galilean paradox: motions in time are equivalent to dimensional forms in space.

334

In its simplest creative processes, those logic chains and dualities mean that dimensions grow in time as points displace creating lines, lines displace creating planes, planes displace creating cubes, and cubes displace creating momentum, mv, where v becomes the 4th dimension, as the cube moves tracing the first trajectory of a bigger plane of existence.

And the same process describes a point that becomes a cycle that rotates into a bigger sphere that moves along a new cyclical trajectory, creating a 4th new dimension of a bigger plane.

So a 4-dimensional cube is not the fancy cube of hyper-dimensional geometry but a 2 dimensional plane of space or cycle of information that grows in height as a flat sheet of energy reproduces its form, or rotates cyclically and then, moved along a temporal path, with v or w lineal or cyclical speed.

We then obtain a translating cube. Since reality follows the dimensions of normal time and space, not the hyper cubes of idealized mathematics.

What truly happens is that a wave or form that displaces along a given path reproduces its form, its information, creating, reproducing a new dimension, which first is measured as speed and then leaves a solid trace behind. So when we emit a ray of light, light imprints and forms gravitational space creating a form with a new dimension of length. Only recently, physicists used this common sense definition of dimensionality in String Theory: where a string moves along a perpendicular path creating a plane and then again creating a cube of pure space.

In that regard, physicists' role is not to invent mathematical fantasies but to *limit mathematical models to the physical reality in which the inflationary information of mathematics is restricted by the resistance of energy to acquire complex forms.*

So only certain efficient geometries and topologies exist, and the enormous numbers of imaginary, unstable particles do not matter, as the inflationary universe keeps creating forms but few survive.

Thus good physics describe only what reality shows: the constant, discontinuous, growth and extinction of dimensional beings, which can be seen as fixed forms or in movement (so a wheel spinning or a planetoid circling Saturn might seem a fixed dimensional form, when observed without detail). Yet once the 3 dimensions of space are filled up, the object will create moving, translation dimensions as its trajectories shape new cycles, or it will create real dimensions of form, if it reproduces in 'biological radiations', growing from micro-forms into macro-forms. In both cases we would see how the object ads a new dimension that enlarges its world. For example, imagine a 3-dimensional living sphere, a cell. As it reproduces, it fills now a macro-dimension and after multiple processes of dimensional reproduction, it will create a macro-organism that has 6 dimensions, the 3 of microcosmic, cellular space, and the 3 of macrocosmic organic space. Then again, if that individual organism, the first of its species, multiples, and organizes itself into a society, an ant-hill or a human society, it will expand in macro-macro space, again filling up 3 bigger macro-macro dimensions of space, till reaching 9 dimensions, whose growth can be measured as a t-duration with lineal time, giving us 9 (classic space) + 1 (lineal time) dimensions in 'classic physics'. In this manner because space-times are fractal, moving geometries and dimensions are finite processes of reproductive growth, when we talk of multiple dimensions we don't have to invent spatial geometries, and follow Minkowski's definition of time as a spatial dimension, but just describe reality, as it is, and consider each movement or reproduction of form with a certain frequency, a dimension of time. In those terms, many 'mysterious' objects of the physical universe described mathematically with multiple dimensions, conceptually ill-defined, become now real, complex space-time fields, extended across fractal spaces from micro-worlds to macro-worlds, as your 9+1 dimensional organism extended in 3 scales, from cells to individuals to society, is.

Such 9+1 dimensional strings (expressed in lineal time and classic Euclidean space) are thus an error due to the ignorance of the Galilean Paradox: dimensions are initially moving and reproductive cycles of times perceived as fixed forms and distances. Yet those dimensions are dynamic, since they grow and occupy a volume creating new macro spaces - the fractal scales of reality - from microcosmic photons to macrocosmic galaxies.

For example, a moving photon-point reproduces a bidimensional line/wave, which wobbles upwards and creates finally a 4 dimensional macro-space of light with length-speed, electric height, magnetic width, and frequency-rhythm. In the same way a cycle of time can rotate upwards and sidewards becoming a 4-dimensional mass or charge also with length, height, width and rhythm. A 5 dimensional object would then be a mass creating a new, lineal trajectory, giving birth to 5-dimensional, lineal momentum (a mass moving in a line); or a 5 dimensional rotating mass with angular momentum. Thus, reproductive movements and rotations create dimensions, which give origin to a growing fractal reality of repetitive, scalar forms – the vital reality we see. Unfortunately, in XX C. Physics, since Minkowski's invention of a spatial, 4th dimension of time, most of the dimensional work done in physics is just about fancy mathematical objects, which have little to do with the real n-dimensional, space-time.

All languages are inflationary. So in the same way 'hair is blue' and 'hair is green' are 2 beautiful sentences, however false, physicists can create all kind of fanciful geometries, however false. Fact is reality doesn't have 10, not even 4 continuous fixed, space dimensions, but it is constructed with fractal, scalar dimensions that grow from micro to macro worlds. So we should apply 2 fundamental laws of scientific truth, the Occam principle (truth is the simplest explanation of reality) and the experimental method (truth is a description of real phenomena) to the problem of multiple dimensions. Reality is not born out of the complex geometry of 9 dimensional continuous spaces. Its complexity grows with the creation of dimensions that start as the temporal, reproductive movement of a form that becomes a fixed structure of space, and again reproduces its movement in time or inversely extinguish its dimensional form. Yet since those dimensions are limited, fractal dimensions they end when the wave ends its reproduction or the particle ends its movement - so they can be increased ad infinitum as the form grows into new macroforms.

-Such fractal, scalar nature also implies that we should always consider the interaction of both membranes, gravitational and electromagnetic membranes, instead of trying to simplify them. In that sense, physicists try to unify the micro-world of electro-magnetism and the macro-world of gravitation, without realizing they are 2 different scales of the Universe: to unify them is like trying to unify the existence of human beings and their macro-scale of societies; or the properties of cells and the ecology of its macro-organisms with the same equations/laws of behavior. What is relevant is to explain the relationships and self-similar laws of invariance that happen between those scales.

All in all what this means is that the holy grail of present physics, the scalar boson with a new field never found on Nature, the Higgs, and the super symmetric inflation of background dependent, 10 plus dimensional strings don't exist. And an enormous number of virtual, unstable particles don't live long enough to be relevant except as failed games of the inflationary game of reproduction of fractal information. So once we have limited the theme of our inquire to what is real – stable particles described by classic quantum numbers – in the realm of evolved light-space forms, and quantum gravitation (strong force and its bosonic strings, gluon and quarks), which is what quantum theory studies we can go on illuminating its whys with the tools of multiple spaces and times.

Recap. Information is inflationary, so it is the mathematical language, whose enormous number of possible theories about reality must be contrasted with the experimental method and the logic principles in which existence is based. But physicists are 'Pythagorean', so they cannot accept truths which are not written with mathematical equations. This means till modern times despite Gödel's work about the errors of mathematical equations; they have ignored the laws of information and morphological change. Unfortunately many of such simple solutions will require time to be adopted, as there is in a Universe of inflationary information, many ill-devised theories based in the faulty, previous paradigm of metric spaces, whose practitioners will have a hard time to abandon.

31. Whole and parts: quantum waves.

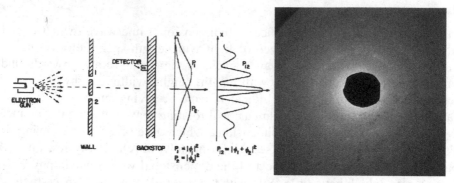

An electron is a herd of photons that gathers into a particle or expands as a wave, depending on which strategy maximizes its survival and movement, as a school of fishes does in front of a predator. The herd even splits its EXI fields (P_1, P_2) when it finds an obstacle and later merges back those space and time fields into a compact form. Yet Physics ignores the bio-logic strategy of the electron and uses a probabilistic explanation of the wave/particle principle to keep its mechanist ideas on matter.

It also ignores that the 2 components of an 'Abelian' Phase space, which defines the mass-field of the electron, are an energetic and informative component (the imaginary number) since its concepts of space-time are still monist, continuous, 'infinite'.

The units of both membranes, the gravitational and electromagnetic membrane are time actions, exi, product of the minimal spatial energy and cyclical time of both membranes. Here due to our limited space we shall not study the gravitational membrane, whose minimal quanta of lambda length and Planck time creates a lineal and cyclical string, or action of gravitational time; which becomes a gluon/neutrino wave in the next scale and then a quark wave.

We shall consider only some aspects of the self-similar wave of light bosons (h-plankton or 'photons') that condensate in cyclical phase space forming a De Broglie mass-wave, or electron whose pattern is defined by the real, lineal energetic component and imaginary, informative, height-dimension of its algebraic equations.

What is the 'it' that Schrodinger & Dirac equations describe? The duality between a whole and its fractal parts explains the nature of Physical Actions, microscopic quanta of energy and time that any physical vortex emits or absorbs to communicate with its environment, complete its existential cycles and stabilize its accelerations: Actions are projections of the form of a given being in the external Universe, achieved thanks to the use of energy that 'translates' in space its in-form-ation:

Present Macro-Organisms (st) -> Micro-Actions (st$_{-1}$):

Max.E x Max.I (Organism)-> Σ Min. e x $\prod$ Min.i (Action)

$\Sigma ExI=K$

Both, in organisms and actions the arrows of energy and information merge in constant 'presents' - systems that exist longer *precisely because* their energy and information *are* in balance. A present organism is a cellular system with 2 poles - one dominant in energy or relative body/field, and one dominant in information, or relative head/particle, which constantly relate to each other by exchanging small bites and bytes of energy and information, called actions, *Min. ΣS x Ti=K.*

A physical action, defined by the specific equation of a Universal action applied to the scale of electromagnetism, writes also ExT=K(h). Where energy and Time are inverse functions, whose product is always a balanced constant. Thus as inverse functions the growth of one of those 2 elements E or T means the diminution of the other. From where Physicists deduce a series of fundamental laws of physics, studied in detail in the non-abridged version of this book (Heisenberg Principle, Law of Force ranges, etc.)

The same law applies to any Universal action. For example, humans who process biological information sleep in absolute zero movement-space. They maximize information by minimizing energy. So happens with thinkers, which are quiet people when reflecting upon information, or informative chips which occupy little space and work better in coldness.

The exchanges of actions between beings resolve the paradox Leibniz encountered in his analysis of Space-time monads: how the different space-time beings communicate each other, creating the networks and complex systems we observe in the Universe? The answer is through microscopic actions.

Actions are microscopic 'cells' of the bigger organism - in terms of cyclical time, they are minimal frequencies, compared to the complex, bigger organism made of knows of multiple events), the duality of macro-knots of frequencies and individual actions define a fractal Universe organized in scales or 'planes' of space-time of different size and temporal duration. Any organism, in that sense, extends in 3 ±st relative planes of size and duration. The natural St plane, for example, in the human case, our individual scale of existence; the microscopic, cellular plane, in which the energy and information of any entity's action develops. In the human case, it is the cellular scale, which provides the energy and information for our actions. And finally, the social macroscopic, +st plane, in which the individual is often a microscopic action, part of a whole mass that performs a social cycle. So humans gather in masses that perform social actions that change their macroscopic scale, nation or civilization - while particles gather in waves and life beings in herds.

Thus, all organisms combine their e x i actions in knots we call cells, which gather together in constant species, creating an ∞ variety of relative Space-time beings, perceived simultaneously together as a present, continuous Universe.

Any entity in the Universe extends across 3 relative scales, the minimal scale of 'actions' of energy and time, the scale of cellular knots in which multiple cyclical actions gather together, and the scale of organisms and ecosystems made of multiple cells.

In those 3 cases reality is dynamic, as all those scales are made of moving cycles and fields of energy and time, despite the fact that we might see them as static forms.

The platonic, mathematic paradoxes of Quantum Theory can be explained in organic terms. Since, as Einstein said God doesn't play dices and the Universe is not a mathematical probability but a dual, biological and geometrical game of herds and wholes.

The experimental proofs of quantum physics back organicism, in the image, the wave/particle behavior self-similar to that of organic herds, and a picture of an electron, which has the form of a nebula of fractal micro-points, self-similar to the whole.

We understand quantum waves and particles in terms of the structural duality of a Universe made of organic wholes or 'particles' that can be observed in its st_1, lower plane of existence as a wave of cellular, fractal, quantum parts.

In the graph, the wave-particle duality shows that physical particles behave in an organic form: an electron wave explodes into its multiple quanta (which seem to be 'cellular' condensates of st_1 photons) as it reaches a thin barrier in order to cross it. Light also divides itself into a herd of photonic quanta when it passes a slit. In fact, fractal particles split their spatial and temporal fields, SxT when passing those apertures. Then sometimes they successfully mix both fields back together recreating the electron and sometimes fail to do that. Hence there are 2 possible solutions to the previous event, which fractal physics express as a quadratic probability when the electron reconstructs itself (sxt) or a null probability (s^1t) when it doesn't; and we explain as a biological process. Further on, since life is caused by the social organization of cells into a higher plane of existence, those wave/particle fluctuations can be explained as particles that 'die' and 'resurrect' constantly. The biological nature of those events is even clearer in the inverse process: when a spatial force, for example, light, comes closer to an informative, complex particle like an electron it evolves into a cellular, compact form - a dot of the photonic nebulae of the electron that integrates itself as a 'probabilistic point=cell' of the electron. On the other hand, when a researcher hits with a high energy beam an electronic wave, it collapses into a particle, as a school of fish clumps together in front of a bigger predator.

Recap. Physicists say that the wave solutions of the space-time field collapse, affected by the observer. It is the abstract, statistical way of saying that a wave of light is a herd of h-quanta, which evolves, imploding its form when it is absorbed by an electron of higher form. Yet the why of that how is biological, not 'mathematical', since as Einstein put it in his critique of a probabilistic interpretation of quantum waves, 'God doesn't play dices'.

32. Populations vs. Probabilities: Fractal, Quantum Particles.

We have now a general, simplified overview of the arrows of time and the topologies of space, which can be applied to resolve some of the basic riddles of quantum physics.

The first fact we must define, once the arrows of time are understood in its dualities, morphologically in space and dynamically as a series of events in time, is the duality of any physical system, composed of a relative energetic, past event/force and a relative informative, future event/particle. If we see reality fixed in space it will appear as a force/particle system; if we see it moving in time, it will appear as a relative energetic past wave that evolves and devolves constantly into a relative future particle of information. Our choice of reality as fixed or moving, in space or time is merely a question of the mind and how it perceives, according to the Paradox of Galileo, which stated that the Earth moves and doesn't move, depending on our choice of perception. This is the solution to the quantum paradox of uncertainty.

Further on, in physical systems there is not at first sight an apparent dominant arrow of future, since we are making spatial pictures of them. That is why quantum physicists often confuse temporal events with spatial forces (such as the weak event or the mass-clocks that Higgs describes with spatial forces and particles). They also don't realize of the anti-symmetry of time and believe that moving forward in time (life arrow) is the same that moving backward (the death arrow). For the same reason Physicists confuse anti-particle events with particles in space - when they are 2 arrows of time of the same entity. All this happens because a cycle of space-time in electrons and particles is just too short and so physicists often make pictures of the entire event/life & death of the particle. So the particle life and antiparticle death is confused with a spatial cycle when in fact it establishes also an order of time, between life as a particle and death as an antiparticle.

The same life-death cycle happens when a particle dies and explodes back in a wave similar to star big-bang.

To make it all more confusing, physicists ignore the duality of the ages of time that become organs in space of a fractal organism. So sometimes a wave is the death of a particle but sometimes it is the spatial, organic energetic field, over which the particle/head exists as a complementary organism.

Only when all those dualities are understood, we can re-classify and re-order in a systematic manner all the forces, events, particles and systems of quantum physics in a complex, clear pattern.

In organic terms, the wave of a quantum system is the reproductive body of the system and the particle its informative state and both co-exist together (Complementarity system) as a dual system, whose energetic limbs are the fields of forces that displace the system.

Like life systems, particles have an order called the life/death cycle, first dominated by the arrow of information through the 3 life ages and then exploded into death by the arrow of energy Thus physical systems do have temporal causality and anti-symmetry: they exist first slow as informative particles and then die fast in big-bangs as anti-particles or as disordered waves. Because the relationship between the informative-time speed of a system is related to its spatial size (ExTi=k), by the chip/mouse/hole paradox (smaller chips calculate faster), we can consider that the particle/wave duality or the particle/antiparticle duality are; when observed not as fixed space or dual organisms, but as casual events in time, the life cycles of physical matter, which are extremely short, as they are extremely small. In other cases though, they are part of a complex topological organism, acting as body, energy and form of a ternary Non-Euclidean structure:

$$\textit{Relative energy=past (energy field)} < \textbf{\textit{Body/Wave: present}} >^{\textit{Future=information=particle}}$$

In the previous equation, we added a bidimensional representation to the standard feedback wave: The information or future particle is as the relative past energy field, far less observable that the present body wave that mixes both, energy and information to create the combined field, reason why the phase space of a mass wave is not easily perceivable. The key to differentiate both types of entities, past and future, energy and form, motion and stop, E⇔I, as separate entities are the inverse properties of time that the present merges in a holographic form.

Thus, we distinguish for each ternary creation and form, the whole system and then its energetic parts more visible in space and its informative events more durable in time.

The symmetry between time and information on one side and its inverse symmetry between energy and space on the other, for the experiences theorist who fully grasp the laws of multiple times and spaces is the most revealing tool to classify reality and understand its events and laws as specific sub-equations of the generator of space and time.

Most dual particles are not spatial but transformational particle/antiparticle events of short duration and most bosons are wave /particle generational events, as they constantly switch between its big bang wave event and particle state. And so we write the anti-symmetry of time in those physical systems as a duality, a temporal, past to future life/death, informative particle/big bang cycle:

$$\textit{Big Bang arrows (energetic, expansive state)} < \textbf{\textit{Dual system}} >^{\textit{Big crunch (informative state)}}$$

All those different spatial, ternary systems and dual events happen in all the scales of the Universe, the quantum scale, the solar scale, the galactic scale and perhaps a higher scale of relative Universes made of cellular galaxies, in increasing time lengths. So we write for the Universe at large:

$$\textit{Past Expansive dark energy} < \text{Steady state:present cosmos} >^{\textit{Future Implosive Galaxy}}$$

Thus in that cosmological scale the galaxy follows a generational cycle from the big-bang to the big-crunch, through a steady state of minimal creation of mass, in which we live.

If we were to observe in absolute detail two photons in 2 cycles, probably we would observe slight differences in its configuration; as we do when we see two Chinese, father and son, which at first look are both called Cheng and both look the same.

The dual cycle that matters most to us is obviously the one of creation and destruction of mass: the arrow of time bends space into mass ($M=e/c^2$), and the arrow of entropy explodes form into energy ($E=Mc^2$). Einstein said that the separation between past and future is an illusion. Those 2 equations of Einstein show the equal importance of both arrows of time, energy and mass/information. However when we transcend into beings with higher information, the dominance of information creates the order of future that we, biological beings, experience, and philosophers have always called the 3 ages of life.

The mysteries of quantum physics can also be explained from an organic perspective. In the graph, we observe 2 of those solutions:

- Particles choose to behave as a herd of fractal parts or as an organic whole, when they cross through doors and slits. Particles interact with the electronic beams of particles we use to detect them, modifying its position and speed (Uncertainty principle). In the graph, we can see that behavior: An electron is a herd of photons that gathers into a particle or expands as a wave, depending on which strategy maximizes its survival and movement, as a school of fishes does in front of a predator. The herd even splits its EXI fields (P_1, P_2) when it finds an obstacle and later merges back those space and time geometries into a compact form. Quantum Physics ignores the bio-logic strategy of the electron and uses a probabilistic explanation of the wave/particle principle to maintain its mechanist, abstract ideas about matter. The misunderstanding of the dualities of time/information and energy/space make thoroughly confusing its explanation of time and space parameters in those processes. In the right picture, we can notice the self-similarity of the atom with a spiral galaxy. Both have a center black hole zone where the quark proton and quark hole are. But self-similarity is not identity, so even if quantum cosmologists try to use quantum laws to study the Universe, it is more meaningful to study directly the cosmos with satellites and telescopes, than replicating mini-big bangs of quarks on Earth.

In fractal space-time electrons become nebulae of self-similar fractal parts which can be described with fractal equations of the type biologists use to define cells as fractal parts with self-similar functions to those of the whole electronic organism. So only a fractal organic description of the quantum world, resolves its paradoxes and contradictions.

The alternative to all those proofs of the organic structure of the Universe is a Pythagorean fantasy called the Copenhagen interpretation. It considers the Universe uncertain as the herd of quanta is not a population described as any other group with percentages of the whole (which in statistics has a probability 1), but according to Bohr et al, those probabilities are real and the Universe is made of numbers. Indeed, quantum physics has finally achieved the religious goal of the founding fathers, Kepler and Galileo, God not only speaks mathematics but the Universe is mathematical. This is nonsense, of course. Yet anytime a physicist in the past century has tried to put forward the organic, logic paradigm, his work has been ridiculed, while entire institutions of learning are dedicated to explore Everett's thesis that each point of an electron nebulae is in a different, parallel Universe! All of them though shown in the same picture, hence cameras are machines that travel through infinite Universes! at the same time! This Bohr/Everett alternative is thus an absurd, abstract, illogic interpretation, which defies the 3 legs of the scientific method; logic consistency, mathematical accuracy (as the results are often uncertainties, singularities and infinities, cleaned up ad hoc) and experimental evidence obtained with the first pictures of electronic nebulae, 40 years ago: The electron appeared as a fractal of smaller, self-similar electronic cells, dense, smaller electronic parts that adopt either a herd/wave configuration or a tight, organic, particle-like one. Or else, we would have observed in this Universe only a single electron point. Further on, the behavior of an electronic herd, when bombarded by massive particles, as those humans use to observe them is self-similar to that of any crowd, from fishes that come together when they are attacked by sharks, to soldiers in a battle field. 40 years have passed and yet the mathematical models of electrons as fractals are ignored by abstract, mechanist scientists.

The platonic, mathematic paradoxes of Quantum Theory can be explained in organic terms. The quantum world is not a mathematical probability but a ternary, geometric game of reproductive fractal herds=bodies and particles=knots of information, displacing over lines=fields of forces.

The equation of energy and information, $e \times i = h(k)$ that defines quantum physics is not an uncertainty but a fluctuation. It is the equation of existence of 'Planckton', the minimal action of energy and time of the light Universe, which sometimes evolves into informative particles or relaxes further into space-waves. Thus the equation has two limits of death as energy and evolution as form. The particles that appear from the vacuum are those evolutions of form of the h-quanta, the substance of which our light space-time is made. The uncertainty of measure is only in our instruments which cannot measure at the same time the wave/body and photon/head states of the quanta. As you cannot make a dual picture of yourself as a whole body or a cellular network: you need to focus the picture in macro or microscopic scales.

Recap. We understand quantum waves and particles also in terms of the structural scales of a Universe made of organic wholes or particles that can be observed in its st.₋₁, lower plane of existence as a wave of cellular, fractal, quantum parts. In this book we hardly touch the ternary scales of all systems that also structure particles as wholes of smaller particles and parts of bigger structures. So atoms are parts of molecules and wholes of particles. And the self-similar laws of hierarchical planes determine many of the events and relationships between particles, atoms and molecules.

33. Time motions. Physical Death. Big-bangs and antiparticles.

In the graph, when we observe an electron we see it jumping backwards and forwards in time in an erratic trajectory, as it becomes its particle and antiparticle.

Traveling in time towards the past means in the restricted, spatial, geometrical definition of time in physics, an inverse, cyclical movement as the one created when moving backwards the needle of a clock. Since physicists 'extract' their concept of time from a spatial definition of movement: v=s/t. Thus physical time is only the analysis of change in the motion and geometry of space; and traveling backwards in time in both, the previous Galilean equation of relativity or Einstein's equation is merely a geometrical change of direction or a contraction in the 'dual' perception of space as distance or c-speed.

In physical space there are also dual dissolutions in space, which are processes of death, the best known a big-bang process that annihilates particles and antiparticles into a release of pure energy and annihilates a black hole into a Supernova or quasar of pure energy: A dual process of palingenesis happens in those death processes when the big-bang gives birth to an accelerated birth. So black holes trigger the palingenesis and fast evolution of stars and a cosmic big-bang, the big-banging of particles.

The antiparticle is the 3ʳᵈ age (inverse in energy/information parameters to the first age) and final death of the particle, reason why it lasts so little and its parameters of time are towards the past. This explains further 2 unexplained facts of astrophysics: why there are less perceived antiparticles. Simple: because the total perception we have of particles is its 'existential force'=energy x Time and they last in time much more than the antiparticles, which die fast. In the same manner all men die but they die so fast that we see much more living humans than corpses around us.

A second interesting consequence of the antiparticle travelling to the past as the dying state of the particle is the fact that while physicists observe them as being born together in the same point and fusioning again in the point of death, the real path is, since the antiparticle travels to the past, that of a single particle when we observe them with the arrow of future information (Feynman diagram of the graph of palingenesis). The particle is born, explodes where it meets the antiparticle in the conventional drawing. Then it dies explodes into energy and leaves a remnant ghost, the antiparticle that moves backwards in time and fades away in the point of birth of the particle. *But there is only a particle at each moment in the time loop, reason why* for example, we have never observed black hole evaporation, which is based in the existence of two particles. This phenomenon happens in all scales. In the next scale of electrons, the same

In the previous graph we observe one of such processes. The graph shows the sum of the life (blue) arrow and death (red/antiparticle travelling to the past) arrow of an electron, which appears and disappears constantly in its path forwards. Thus, the motion of virtual photons and electrons seems to correspond to a series of life-death cycles shown as particle=life arrows/antiparticle=death arrows or wave (st-1 plane)/particle (st-plane) fluctuations, which include a 'quantic jump of spacetime' in which the particle disappears becomes its antiparticle moves to the past and re-appears in a relative present location, displaced both in time and space coordinates, proving the constant cycles of creation and destruction and the fluctuation between hierarchical past and future states of all palingenetic systems of the Universe.

Yet the same process happens in our biological death. Those who have survived death feel an explosion of light in the brain/informative human particle. Then memory rewinds backwards its path till youth, when you will awake in the lower plane of existence as a mere cell. This is equivalent to the death of the antiparticle that 'reverses' its coordinates till it meets its final death in the point of birth (hence it seems two particles are born in that point but one is the death particle returning to its past). At that point it dissolves in the lower vacuum plane.

The best known case of travel in time in physics is the virtual particle/antiparticle systems that constantly pop out in the Universe. Antiparticles seem to travel to the past while particles travel to the future and both annihilate themselves. What we observe here is 2 particles with different geometrical directions that together complete a 'cycle of physical existence'. We could say that the particle is the 1st age of the entity, from past to future and the antiparticle its devolving age from future to past or 3rd age of the entity and together form a closed loop of space-time, an existence. Or in some cases, the particle is the life arrow and the antiparticle the death arrow. So we have to see them consequentially, one after another in time, even if when we measure them in present-simultaneity, as they are so fast we see the entire life of the particle in an instant of space. So they seem to us mere geometrical shapes.

The entire cycle of Time change is often perceived together, when studying flows from past to future in those fast, simultaneous phenomena that become then fixed loops of space. In any case a death/back in time travel never goes beyond the limits of the entity that experiences it. Those geometrical time travels do not influence the entire, absolute time of the Universe, as Hawking naively thinks, when he talks about black holes as time machines because their 't' parameters are inverse. Or when he thinks that black holes evaporate because antiparticles make them 'travel to the past'. Antiparticles are only the final stage of the life of a particle, which quantizes its existence in 2 broken phases that don't affect the absolute time of the Universe. It is that absolute Time the one Einstein studies with its Equations.

Recap. Processes of death mean a short travel to the past in which information is rewinded. Yet while we travel to the past we do not co-exist in both states as some physicists unaware of the process think it happens with dual particle/antiparticle systems.

34. The ternary, i-logic topological structure of atoms.

An atom is a space-time field divided in 3 species, informative masses or quarks, energetic gravitational and electromagnetic networks and an intermediate space-time, the electronic nebulae, which bends light into 'fractal', ultra dense photons, which put together create the electronic nebulae.

Let us then now that we have revis(it)ed classic single time-space theories of reality, depart from the classic age of the 3rd paradigm of metric spaces (quantum relativistic models of reality), without further ado. Since what was done beyond Einstein Dirac and Nambu, the last of the classics, is mostly baroque, inflationary, metalinguistic theories proper of all paradigms of thought that failed to include mass, as the information of the Universe that balances the energetic entropy of spatial forces (single big-bang theory that fails to include dark matter/energy, Thermodynamic of black holes that misunderstands the nature of the event horizon belonging to our light-universe, super symmetry and superstrings that fails to understand the laws of balance that forbid its existence, the expanding Universe that fails to consider the implosive, galactic vortices, which balance that expansion elsewhere in this work) and move on to present the universe with the ternary dimensions of time (ages of all time events) and space (3 functional topologies of all st-points or 'worlds' of space-time.)

In the graph, an atom is a space-time field divided in 3 space-time zones: its informative quark center, the nucleus; the external reproductive membrane, made of electrons, which evolve socially in bigger ExI membranes when atoms become molecules; while informative, gravitational and energetic, light networks shape their intermediate space-time.

The topology of the atom is thus clear. The electron acts as an external 'spherical plane', a membrane of energy. In the center quarks are the informative vortices. In between energy and form is transferred with forces, which often decouple, reproducing new particles and antiparticles. There are 3 informative families of quarks-mass, due to the evolution of information in 3 ages or horizons of increasing form: each quark family is thus an age in the evolution of informative matter.

The final event of death of those families is a 'decay' or quantum big-bang. In our lighter up and down quarks, it is the beta decay, the dual mini-big-bang of a neutron due to the conversion of a balanced, reproductive down quark into an informative up quark (proton, big crunch formation) and an energetic electron (expansive electronic membrane).

As in all processes of death we assist first to the dissociation of the informative' and energetic elements of the organic system.

Since death is caused by the destruction of the networks of the upper plane of reality, which created a form.

So the Neutron dissociates its informative quarks and electronic, slightly negative, external membrane, provoking a dual big-bang and big crunch, breaking into the 2 parts of its balanced state.

Thus we write the reaction as an event of the generator equation:

$$_{Electron=energetic\ past}<Down\ quark=balanced\ Present>^{Up=informative\ future}$$

$$_{Electron\ (E)} \times {}^{Up\ quark\ (Information)} = Down\ quark\ (balance)$$

The up quark has half of the mass of the down quark, and we know a down quark switches into an up quark in a beta reaction that explodes a neutron into a proton and electron, components of the Hydrogen atom. In other words the balanced neutron splits into an expansive electron and an implosive proton. This is the equivalent of a quantum dual big-bang/big crunch, a far from equilibrium process in which the neutron expands its membrane into an electronic big-bang and implodes its quarks into a tighter proton configuration.

Recap. Atoms are st-points with 3 topological spaces, the spatial membrane or electron, the hyperbolic center or quark-gluon soup, and the intermediate, 'bigger', reproductive space where exchanges of electromagnetic and gravitational forces give birth to virtual particles.

35. Events in time and organisms in space.

All those processes give origin according to the duality of the 'Galilean paradox' *to both events in time and forms in space*, which further illuminate many questions unresolved in classic quantum physics - as we can always consider the existence of 2 different realities, *a causal event and a ternary organism,* and so we have two options to explain many phenomena before explaining only with either a spatial or a temporal perspective: For example, a proton is a uud triplet of quarks in space, but in time you might consider that a down quark, which switches into an up quark and electron, splitting its mass-vortex in 2 and then evolves back into a down quark.

So you can also write the ternary structure as an event in time: *d(exi)->u (i)+e(i)->d(exi)*

And so we can consider the whole process to be an existential cycle of a single down quark. Both events are real and in certain conditions the process will be a temporal event and in others will create a spatial organism. The difference is clear: when we consider the process as an event, reality becomes warped from a big spatial, extended, multiplied world into a tighter, faster, temporal non-redundant Universe. And so this 'antisymmetry' between time functions and spatial forms happens when we consider the 'antisymmetry' between big-bang & big-crunch ages of fast d=evolution in time, in a limited space vs. the spatial, organic ages or steady states, described by the equations of Einstein. And since we are apparently in a steady state age for the Universe, most likely the previous spatial description is the reality of the atom: 3 quarks, locked as relative, past-energy, present balanced and future-informative subspecies.

The reality of studying the Universe both in time and space, as either an event or a form is one of the key methods of 'Complex physics' as a science which uses two arrows to describe the Universe. Yet because physicists lack those methods, they often confuse an event in time with a form in space and vice versa. Ultimately the proton and neutron are two phases of the same being in 2 different states of time that physicists first confused with 2 different species of space.

The mystery of their 2/3 and 1/3 charges become also resolved in a temporal perspective: the up quark emits a negative electron, giving up a charge and becoming a down quark that later absorbs the quark and becomes an up quark again.

An easy way to solve when a process is a spatial duality or a time event is the fact that time events often don't have parity, mirror symmetry, a spatial property, as it happens with the Beta decay. What this means is that because information and energy have different properties, the arrow of energy->information and the arrow of information->energy, its inverse process, are different. And so indeed, while neutrons decay easily giving birth to an electron in a few minutes – another proof that it is a fast process of death; the inverse process hardly ever happens.

Another puzzle of physics is the difference of masses between particles. The general rule is simple: when a lineal energy, E, accelerates till light speed, the limit of our light membrane, it starts to curl into a cyclical speed, which is the definition of a vortex of mass. Thus no mass becomes mass. Yet further on, this transformation can be further increased by a topological transformation, from a reproductive, cyclical vortex toroid topology to a hyperbolic, more massive one. The balanced, reproductive cyclical toroid donuts can switch into the inner informative dual donuts (simplest hyperbolic topology). And as the hyperbolic topology adds up more 'toroids in the dimension of height, and the vortex of mass, VoxRo=K, increases its speed mass increases geometrically, reason why a black hole, a hyperbolic topology with asymptotic T-parameters, is the most massive object of the Universe.

This topological transformation could be a possible reason why a down quark has the mass of 2 up quarks: the down quark might switch back and forth its reproductive single-torus topology into a dual torus, becoming more complex and heavier but smaller. In this manner we can find many topological whys to the abstract description of particles and transformative events between them.

Consider now the the temporal interpretation of color, a feature that makes quarks stable in triplets.

It was discovered when physicists realized that quarks with the same spin could occupy the same position. So they considered they have color. But they never clarified what color is. We interpret color in simple Euclidean Geometry as the orientation of 3 bidimensional, cyclical vortices of mass - 3 quarks, that are perpendicular to each other - and so they can occupy the same 3-dimensional space. *This explains why 3 quarks of 1/3rd charge together emerge in the electromagnetic world and can interact with a 3-dimensional electron of 1 charge.* It also explains why they are locked in a stable configuration:

they cannot be deconfined without 'disappearing' from our 3-dimensional world as point-particles; reason why we perceive them as joined by an extremely strong and attractive force.

In detail, using complex Non-Euclidean topology the picture is somehow more complex, but the previous simplified analysis is clear enough to explain why some quarks have $2/3_{rds}$ of charge, 2 donuts, and some $1/3^{rd}$, one donut; and how they can interact with 1-charge electrons.

Further on we can understand why gluons, the particles exchanged by quarks have 2 colors. This again is self-evident: if you try to communicate a bidimensional long plane and a perpendicular plane you must travel in a right angle. So a gluon that exchanges energy and form between 2 quarks of different perpendicular/color orientation must have 2 colors.

$2/3^{rd}$ charge quarks would be from the perspective of its gluon parts, chaotic, perpendicular, bidimensional attractors (Lorenz-type); $1/3^{rd}$ quarks would be self-similar to single gluon attractors (Rossler type). Gluons are bundles of strings with a boomerang like open topology – hence without mass, as they don't close the cyclical vortex. And as they interact with quarks, they will change their spatial orientation, carrying away their color.

The principle of physics that matters here is diffeomorphic *orientation*, Obviously when we switch from Euclidean to non-Euclidean space and from continuous to fractal dimensions, the topologies of those orientations become much more complex and we should rather talk of 'puzzles', hyperbolic, $2/3^{rd}$ charges; toroidal, $1/3^{rd}$ charges, and curved planes (gluons) that respond to the 3 basic topologies of the Universe; and become locked together in complex patterns, as proteins and other cellular components do in the cell, *to emerge as a whole Non-Euclidean Point of our 3-dimensional world – a proton.*

Thus color is a geometrical feature that evolves the bidimensional world of quarks into the 3-dimensional world of electrons, creating a holography of our world. Quantum physicists know that information is bidimensional and the Universe holographic but they don't know why. Because the classic, abstract, algebraic description of quarks, based in group symmetries cannot express this easily. Thus, without topological analysis, without the enlightenment of its why with the new paradigm, the theory of quarks and gluons tell us only that each quark has a color, different from the other 2 colors of the triplet and gluons have 2 colors.

The topological analysis shows that the fractal scale of quarks and gluons is similar to a fractal, liquid cell, which is *the fundamental structure mimicked by all Non-Euclidean points, hence, self-repeated in many self-similar scales, as we shall see when* explaining galaxies as cells of the Universe; where black holes are the DNA/informative element and stars the mitochondria/ energetic network.

Recap. We can perceive, according to the Galilean Paradox all realities as forms in space or events in time. Physicists study quarks only as events in space, so they miss the explanation of its fractal charge: 3 locked quarks as events in time add to 1 charge; and miss the topological non-Euclidean interpretation of down, up quarks and gluons whose hyperbolic, toroidal and curved plane topologies add up to form a st-point of 4 Dimensions that surfaces in our holographic Universe.

36. Two Quark Triangles and 2 Universal Membranes.

There are 6 quarks, whose mass increases as the speed of their space-time vortices increase. Physicists divide them in the III horizons of evolutionary mass, with increasing mass/information. Yet they ignore the reason of those 3 families (the 3 horizons of any evolutionary system of energy and information); they ignore why they have different masses (because the speed of their vortices increase); they ignore why they have fractional charge (because they are bidimensional vortices, which must lock in triplets, to form a 3-dimensional space-time, harmonic with the electronic, 3-dimensional world we live in); and they ignore which kind of fractal cosmological forms they create (pulsars and black holes):

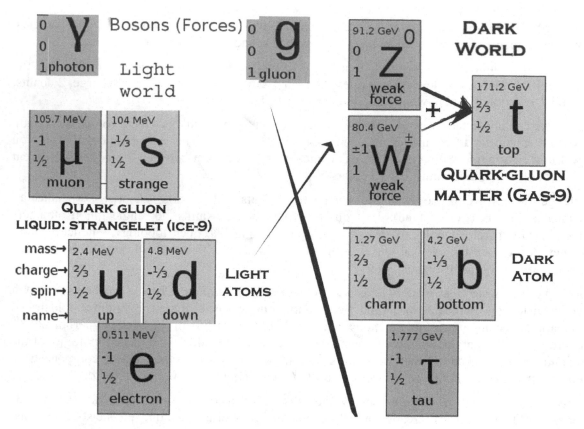

Unlike the usual spatial classification of quarks in pairs proper of classic physics, in complex physics, we divide the Universe in 2 membranes, the dense, informative, gravitational, mass/quark membrane and the electronic, light one and we consider species of the gravitational, informative membrane to be tendencially 'time events', as information and time are related, and so quarks and black holes are better described with 3 temporal horizons and hyperbolic topologies.

In the graph, a more detailed analysis of the parameters of the main particles of the standard model. For simplicity, Neutrinos, which *are better described not as particles in space but wave- events in time that transfer momentum between both gravitational and electromagnetic scales (reason why their '3 species' mutate in time, as they travel to the earth)*, are ignored: The 6 quarks of the standard model are reordered in 2 ternary groups. In the bottom left, we observe the electromagnetic membrane and its elements. In the upper right side we observe the gravitational, quark world.

The graph, which is the standard model of Complex Physics, applied to quarks, requires no Higgs. Instead the top quark of self-similar mass in the upper vertex becomes the final evolutionary state of the Z+W particles which in a weak event transform lighter particles into 'dark quarks'.

Its enormous mass means it should exist a new decametric 'scale' of super-strong forces, and the proper model to study the breaking of symmetry is not the Higgs but the Technicolor theory, whereas super strong top quarks should turn faster than light, be components of black holes and emit dark energy at 10 C speed, the next scalar force of the Universe that fills intergalactic space.

However in reactions at lower energies, Z and W particles never reach enough stability to become 'parts' of a top and so they quickly devolve into lower mass-states, which explains most of the reactions observed in accelerators, mediated by temporal, weak forces, which therefore have no spatial symmetry.

If we order particles by mass self-similarity in triads we observe 2 different atoms and 2 different quark-gluon soups, made of 3 types of quarks, dominated by their heavier quarks of each triangle:

- The up and down quark create the light atom. Yet when they are deconfined in a quark-gluon soup they are dominated by the most massive strange quark, which creates the superfluid vortex of 'strangelet' liquids or 'ice-9', responsible for Nova explosions.

- Charm and bottom quarks form a dark atom, dominated by the most massive top quark in a dark gluon soup, which we shall call 'gas-9', as it would be the most explosive substance of the Universe, responsible for Super-Novas and quasars and maybe the big-bang of a celluolar Universe.

Thus if strange quarks create strange soups, top quarks will do the same with dark atoms; and if strangelets are components of strange stars, top quark liquids will be the components of top black holes.

This symmetric scheme is unknown to physics; so it lacks a visual understanding of what dark atoms (bcb particles) are, and why strangelets and top quark liquids are stable enough to form quark stars:

In the light world, the ud quarks and electron form the light atom. In the dark world, the cb quarks and the tau electron form the dark atom. The reader can now see that basically the two worlds are differentiated by a fractal scale of $1000=10^3$ Electron volts. In terms of static, fractal dimensions, this merely means that the world of top quarks is warped by a new SU_3 group of dimensional form. *It is more complex, more informative as it has 3 more dimensions of warping; or 3 more degrees of rotational speed in a dynamic perspective.*

Imagine that inside the original vortex of our light-quarks there is not the eye of a hurricane, but another scale, another vortex, another medium, rotating 10 times faster in 3 dimensions. Thus this inner world extended into lineal energy will mean we need $10^3 = 1000$ times more energy to create it.

The rotational speed of that inner vortex will be 10 times higher - 10 c - but the energy needed to create the new 3-dimensional fractal world or inner scale of motion of those top quarks will be 10^3 fractal dimensions of rotational speed/mass. Thus the dark world of heavy quarks is 10 times faster and 1000 times denser [9].

For the same reason the electron of the lighter world is lighter than the tau electron of the dark world. If our light world is a 0.c<10 c world the dark world is a c<10 c world.

Once this concept is clear, the self-similarities become evident: the world of 0.1c light atoms, which creates our matter and the world of c-strange quarks that creates quark stars must be matched by a world of dark atoms and top quarks, which creates the world of black holes. So black holes are top quark stars.

Let us show how the quarks of the 2 membranes increase their mass (we round here figures so the reader can easily follow them):

In the graph, the strange quark is the top predator quark of our triangle of light matter made of udu atoms. It is between 30 and 100 times heavier than our matter. While the top quark is the top predator quark of the triangle of dark matter and its bcb atoms, and it is around 100 times heavier than those bcb quarks. The strange quark and top quark are around 100 times heavier than the ud and bc atoms.

Since the equation of a mass vortex is U.C. $\times$ Mass $= w^2 \times r^3$, mass is proportional to the square of the rotational speed, w, of its vortex. So we have 3 decametric scales of rotational speed:

C/10, the speed of rotation of electrons, and the 'ud' system of quarks of our atoms, which in this manner are in harmony with the electron vortex that traps them.

<C speed for the strange quark or top predator quark that causes the breaking of symmetry of our matter, as it creates a faster attractive vortex, deconfining the quarks of our world, liberating them, and blowing the electronic wave into radiation.

In the graph, the bosons of both worlds are also self-evident: The photon is the boson of the electromagnetic world and the gluon is the boson of the gravitational membrane. The symmetry between the 2 quantum theories of electrons and quarks is a clear proof of that symmetry.

Both theories differ because the quark-gluon soup is more informative and has a more complex, 'network-like', $\prod$-structure, while the electron is basically a herd, with a loose $\sum$-structure.

The 2 forces (electromagnetic and strong forces) and the species evolved from them also change the parameters of energy and information of both worlds, according to the basic symmetry of the Universe, $e \times i = k$, *expressed in quantum physics by the law of range.* The lighter world thus extends further in range/space and weighs nothing. The tighter world of dark quarks extends shorter in range and weighs/attracts one hundred times more.

It is only left to explain the role of the muon, which weights exactly what the strange quark weighs and the Z+W= top, which weighs the same also than the top=Higgs particle.

So again we find a clear symmetry between them: Those particles are *self-similar to the top and strange quark, perceived from the other side of the dual membrane; they are their self-similar 'ghosts' in the light world.* And they can be found as evolutionary steps in the creation of strange quarks and top=Higgs quarks in the processes of transformation of electronic membranes into dark, strong gravitational membranes.

And the Higgs[7]? It is not needed and it is an impossible particle since a scalar boson breaks the laws of complex physics and a new force/membrane it is an invention without experimental evidence that would also break the symmetry of complex systems (like an organism with '2 nervous systems'). The evolution of lighter matter into the top, mediated by the Z and W particles (which should be considered the same particle in a time perspective; that is, the +, neutral and - states), is performed by a top quark/antiquark condensate (Nambu), which breaks the symmetry of all other types of matter, converting it into more tops that will condensate into a quark star or black hole. In biological terms, to break the symmetry means to kill as a top predator quark, the strongest particle of the Universe, all other forms of lighter matter, to feed on them and to convert them into a self-similar form of yourself. In the same manner, the strange quark breaks the symmetry of our lighter quarks transformed into strange quarks and strange liquid (strangelets). It follows that SUSY particles, which also contradict the laws of balance between energy and mass/information and have never been found, do not exist either[9].

Because quantum physicists lack the tools of complex physics and the understanding of mass as information, all this is blurred in their equations. They do describe perfectly all those reactions with enormous mathematical accuracy as Ptolemy described better than Copernicus the movements of planets. But to do so they need complicated mathematical models as Ptolemy did. A model of complex physics with both arrows, energy and information, is far easier to understand as it explains topologically and in terms of cyclical, causal time events many whys of particle physics.

Recap. Mass is physical information - an *accelerated,* bidimensional vortex of mass (dynamic perception) called a quark, which can be described also in space (Galilean Paradox) as a non-Euclidean network of fractal gluons (static perception described by QCD theory). Quarks must be understood as ternary events in evolving time, and classified in triads, belonging to two different scales of the gravitational world – the strange world with c-rotational speed that interacts with our world and the dark world of top quarks that rotate at 10 c speed and form quark-gluon condensates called black holes. They are vortices of top quarks, the top predator particle of mass-information of the Universe.

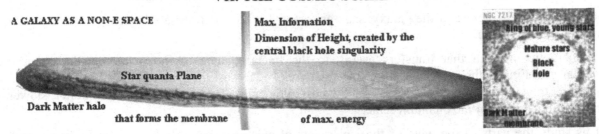

Galaxies are fractals of stars and dark, quark matter built with 3 topologies: a reproductive body of stars, sandwiched between an informative nucleus of black holes and an external halo of dark matter, probably strangelets and other dense stars. The closest self-similarity in our world scale is a cell and in the quantum scale an atom.

37. The Three Non-Euclidean Regions of the Galaxy.

Let us now consider the ternary topology of the equivalent 'entity' to the atom in the gravitational membrane – the galaxy:

In the graph, a galaxy is a curved, fractal space-time of huge spatial proportions, hence minimal form (inversion of properties between spatial energy and temporal information). Thus galaxies can also be studied in space as a fractal point with 3 regions that correspond to the 3 'canonical' topologies of a 4-dimensional world - an informative center, an energetic membrane, and the reproductive intermediate zone:

- Max. T_i: The center of the galaxy is a swarm of black holes, its densest informative masses, which produces the gravitational, informative waves that control the position of its body of stars. Beyond its event horizon, the accelerated vortex of mass of the black hole (Equivalence Principle) should accelerate light, deflecting it into a perpendicular, hyperbolic, informative dimension of height (Kerr superluminal, central singularity) ejecting it as gravitational jets of dark energy at 10 C.

For that reason Kerr black holes[10] should be called wormholes, because they absorb light but let it scape through its axis as dark, gravitational energy, at faster than light speeds.

-Max. E: The external membrane that limits the inner space-time of the galaxy is a spherical halo of dark matter, probably made of strangelets or micro black holes, which can deviate unwanted radiation by gravitational redshift and/or absorb the energy of radiant matter, cooling it down to the 2.7 K background radiation.

Thus those non-evaporating micro black-holes and strangelets of great density act as 'proteins' do in cells, controlling the inner movement of galaxies and the outer absorption of light-energy.

- E=i: Stars, tracing toroidal cycles form the inner space-time body of the galaxy, a bidimensional plane or Klein's disk that feeds the wormhole and reproduces atomic substances and stars. They ultimately evolve into black holes, which migrate toward the central swarm of holes, residing in the nucleus. In any Klein disk distance is measured as motion and becomes infinite when we cannot reach a limit or barrier (for example the barrier of light speed becomes an infinite Lorentz Transformation). So it happens with the border of the galaxy. We are part of that intermediate space-time in a Milky Way, limited by its central hole and an invisible border of dark matter, neither of which we can cross without dying; since the speed of rotation of matter around the wormhole and the flows of intergalactic dark energy that expands space at light speeds beyond the halo *would destroy us.* Thus we are trapped in this star and planet, in a toroidal cycle that will end evolving the Sun into dark matter.

Thus the generator equation of the galaxy as a Fractal st-point is:

In the image, the structure of the galaxy: stars are created in the intermediate region and the center is occupied by a black hole.

The energetic medium that transfers energy to the fractal quanta of the galaxy is the external interstellar gas. Finally the system is joined by 2 networks of forces: the gravitational, faster, non-local informative, transversal gravitational waves at the cosmological scale; and the energetic, smaller, slower electromagnetic waves at the quantum scale.

Let us study the parts of the whole - the 3 elements of galaxies, stars, black holes and gravitational forces that join them.

Recap. Galaxies are fractal st-points with 3 standard topological regions: a spatial body of stars and an informative nucleus of black holes. The closest self-similarity in our world is with a cell.

38. Intermediate Space. Gravitational Waves and Solar Systems.

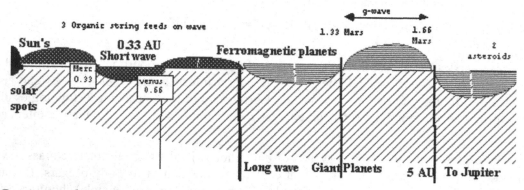

Gravitational, transversal, fractal waves shape the structure of galaxies and solar systems, transferring form and energy between cosmological bodies, in a self-similar process to the transference of information and energy between atoms through electromagnetic waves. In the graph, Titius Law of distances between planets reflects their position in the nodal points of those transversal gravitational waves. In the core of planets, there could be a crystalline or super-fluid zone where those flows of dark, gravitational energy are processed, causing flows of heat and matter that make planets 'grow'.

We use constantly self-similarities, based in the 3 Laws of invariance of the Universe, scalar, formal and motion invariance and the ternary differentiations in time (energy, reproduction, information) and its symmetric function in space (planar, spherical membranes; toroidal bodies and hyperbolic centers).

With those simple laws we can describe the galaxy as we did with the atom. Yet self-similarity is not equality, which means that we cannot use the exact formalism of quantum physics but use those self-similarities for gravitational systems we do not perceive, except for its secondary effects.

This is the case of gravitational waves, which are self-similar to electromagnetic waves. They organize the structure of stars and galaxies, as electromagnetic waves organize the orbits of electrons. Both respond to the same morphological equations that relate 2 particles through a lineal force field defined by the ratio between the informative density of masses or charges and their distance.

Yet even if form remains invariant at scale, as it is an essential topological property that defines the why of the Universe, the metric space changes *as the space-time ratios/constants that define the size, speed, frequency and range of those waves change.*

We observed a self-similar change when studying the cyclical informative vortices of both scales - the G-constants of Newton's gravitational vortices and Coulomb's equation of an electronic vortex, unified by their invariance of topological form and scale).

Thus galaxies and solar systems show a gravitational, morphological, spatial structure similar to that of an electromagnetic atom in the cosmological scale, a fact which Einstein predicted, establishing 2 kinds of gravitational waves, parallel to the 2 types of electromagnetic fields we know:

- Static waves that create the gravitational bi-dimensional fields over which galaxies form.

- Discontinuous, transversal, quantized waves, which shape the orbits of stars and galaxies, in the same manner photons control the orbital distance of electrons in atoms (l numbers).

Thus, those gravitational waves should have the same functions in galaxies and solar systems that electromagnetic waves have in the world of atoms, explaining cosmological structures and becoming by self-similarity with electromagnetic waves, the fundamental force of interaction between celestial bodies.

We know that the gravitational activity of black holes set up star orbits and probably influences its evolution, growth and formation, determining the basic properties of magnetic fields, ecliptic orbits and distances between stars in a galaxy and planets in a solar system. So even if gravitational waves are invisible, using their morphological self-similarity with light waves, the equations of Einstein's relativity and the indirect proofs provided by the orbital distances and rotational fields of stars and planets, we can explain many 'whys' on the structure of those celestial bodies:

- Astronomers have always wondered what rules the distances between the planets of the solar system. The existence of regularities in the distribution of planets in the Solar System was recognized long ago. This was Kepler's main motivation in his search for planetary laws. The Titius-Bode law ($rn = 0.4 + 0.3 \times 2n$) was the first empirical attempt at describing these regularities, and was followed by several other proposals. The discovery of similar structures in the distribution of the satellites of the great planets led to a revival of interest for such studies, and to the hope that indeed a physical mechanism was at work. Now we can add a topological why to the how and when of metric space measure:

Those planets are in the nodes between gravitational waves of different frequency/amplitude and the solar system's orbital plane in which planets feed, 'deforming' space-time, as they follow their static gravitational orbits; as electrons are in the nodes of their quantum waves, fine-tuned by the secondary levels they access according to the strength of the electromagnetic waves they exchange with their environment. For the same reasons stars should in the nodes of gravitational waves caused by galactic black holes.

In the graph we draw the 2 fundamental wave lengths that could explain the distances between planets: a high frequency, short gravitational wave of 0.33 AU could explain the positions of ferromagnetic, inner planets on its nodes. While 2 low frequency long wavelengths at 5 and 10 AU, could explain the position of bigger, and lighter gaseous planets. Since Jupiter is located at 5 AU, Saturn at 10 AU, Uranus at 20 AU, Neptune at 35 AU, Pluto at 40 AU; and as I predicted a decade ago, we have found a new planet, which I called then Chronos, 'the last of the titans' at 100 AU, in the limit of the solar system, 'renamed' Selma (-;.

- G-waves explain why planets have ecliptic orbits with an inclination on its axis, which is a natural orientation if they are receiving curved G-waves with a certain angle through its polar axis. In that regard, the rings of gaseous planets in the point of maximal activity of those waves (Jupiter and Saturn) and the spiral vortices of galaxies, could act as 'antennae' for those waves at star and galactic level.

- Those waves might cause, as all lineal movements do, a cyclical vortex around them, originating the condensation of planetary nebulae. While in galaxies their wave structure seems to originate the different densities of stars in their nodal zones.

- Planets suffer catastrophic changes in their magnetic fields, probably produced by changes in the directionality of those waves, emitted through the tropical dark spots of the sun.

There are advanced mathematical models of gravito-magnetism that have unified both type of waves, departing from Einstein's work. Thus energetic 'gravito-magnetic' waves might cause a change in planetary magnetic fields as a magnetic field changes the spin of an atom that aligns itself with the field. For example, Uranus is tumbled and it has lost most of its magnetic field: perhaps it was knocked-out and relocated by a G wave.

Lineal magnetism is in fact in complex physics of multiple planes of space-time the intermediate exi force that 'transcends' from the gravitational to the electromagnetic scale: for example, electromagnetic light or ferromagnetic atoms like iron should absorb gravitational energy through their magnetic fields.

- Solar spots are the probable source of those waves. Yet its origin might be the central core of the star or the activity of the central black hole, whose G-waves might be absorbed and re-emitted by the star. We cannot perceive G-waves directly; but magnetic storms, solar winds and the highly energetic electromagnetic flows and particles that come from the sun's spots, might be its secondary effects. In the same way we only perceive indirectly the waves of dark energy emitted by black holes that position the stars of spiral galaxies, by observing the mass and radiation dragged by those waves.

- Those catastrophes might cause the climatic changes that modulate the evolution of life on Earth, since we already know that the activity of sun spots affects the temperature of the Earth.

- G-waves could structure the galaxy and its stars in the way electromagnetic impulses structure a crystalline atomic network, ordering the distance between its molecules: electromagnetic waves also feed with energy and information those crystal webs. For example, electromagnetic waves cause the vibration of quartzs, which absorb energy from light and vibrate, emitting 'maser-like', highly ordered discharges of electromagnetism. We observe similar maser beams in neutron stars, called for that reason pulsars.

Recap. Gravitational waves produced by black holes control the location of stars and planets, and its spin/orientation through smaller gravito-magnetic waves.

39. Organic Patterns in the Galaxy. The why of G-waves.

The closest homology of the 2 dual networks of the galaxy is with an atom in which the central nucleon with max. density of gravitational information and the external electronic membrane interact in a middle space-time vacuum through gravitational forces and electromagnetic photons.

Another self-similarity between scales of multiple space-times might be established in complex analysis between the galaxy and a simple 'cellular' organism, which introduces elements of complex biology in astronomy obviously more difficult to accept from a mechanist perspective).

Following the cellular or physiological homologies, the network of dark, informative matter and gravitational energy, connected to black holes, surrounds and controls the stars' electromagnetic energy. We know that it was formed first and then guided the creation of electromagnetic energy, so we can observe it indirectly and deduce its form from the highly quantized shape of the filaments of light-galaxies that were formed around dark matter (right graph). Thus dark matter acts in a similar way to the RNA that shapes and controls the Golgi membranes of the cell or the nervous system that guides and builds the morphology of the body; while the network of stars and electromagnetic - the slower energy that produces the substances of galaxies - surrounds those strands of dark matter. In the cell's homology ribosomes that create most products of the cell are pegged to those membranes.

We shall consider briefly here 2 of those controversial hypothesis: The possibility that black holes perceive gravitation and the chains of causality between the different scales of the Universe, self-similar to the chains of causality between cells and bodies.

- The most controversial element of a cosmological model based in G-waves is the existence of gravitational information that allows strange, neutron stars and black holes and maybe in the future evolved planets such as the Earth through its machine systems to perceive and move at will within a static field of gravitation.

On the Earth animals use light as information and dominate plants, which use it as energy. The hypothesis of complex cosmology is that stars are 'gravitational plants' that merely feed and curve gravitational space-time, while Worm Holes are 'gravitational animals', which are able to process gravitation as information and control and shape with gravitational waves the form of galaxies, their territorial space-time. They are in that sense extremely simple plants and animals. A more proper comparison would be with a cell, where the DNA molecules are the Worm Holes, the informative masses of physical space; and the mitochondria that produce energetic substances, the stars.

Thus frozen, quark stars could be 'gravitational perceivers' in the cosmological realm, as animals are light perceivers in the Earth's crust and DNA perceives van der Waals forces in the cellular realm.

On the other hand stars would be plant-like, floating in the sea of gravitation, used as energy of their motion, feeding on interstellar gas, as planckton does, floating in the sea of water.

Do black holes perceive gravitation as complex animals perceive light, instinctively or mechanically, as DNA perceives the forces of the cell? They probably gauge gravitation in very simple 'forms', as a cellular DNA-system, much simpler than the brain of animals, perceives its territorial cell. That is the supremacy of man in a relative universe were size is less important than form: While all systems process information, man is a summit of form and hence one of the most conscious species. Yet black holes have enough quark complexity to act/react to informative flows, as they seek energy to feed on – our electronic energy. This hypothesis has experimental proofs, since pulsars and black holes emit gravitational waves and we have observed many black holes following erratic paths through the galaxy, which defy the tidal, regular orbits of stars.

In that sense, multiple spacetimes theory considers that in the same way light waves are the energy of plants and the information of animals, gravitational waves move stars and inform black holes, the most evolved celestial bodies, which emit or feed on the energy and information provided by those gravitational waves.

Further on, gravitational waves emitted by black holes might reproduce matter on the cores of stars and planets:

If those gravitational waves degenerate easily into quark matter as the jets of quasars show, they could also become converted into matter in the super fluid cores of stars and the crystalline centers of planets, in a process inverse to the Lorenz Transformations; since tachyons acquire more mass when they slow down, trapped by those superfluid and crystalline vortex. Thus, as light is converted into energy in plants, stars and planets will create their 'amino acids', quark matter, in those processes. Thus, dark energy, tachyon strings would decelerate into c-speed gluons that would reproduce quarks; (as electromagnetic waves become photons and electrons).

There is also a mechanism by which those planets and stars 'jump' or change their spin position under the effect of gravitational waves, as electrons do under a magnetic field: the core of stars are made of super fluid helium and the core of planets of iron crystals, which are the only atoms that can absorb the energy of a gravito-magnetic field to change its motion.

Finally, all those events will have a 'why' in the 4 arrows of the organic Universe; since they would represent the feeding, matter reproduction, informative perception and social location within the galactic or planetary network of celestial bodies, equivalent to the 4 whys of the 4 quantum numbers, described before.

Recap. Gravitational waves accomplish the 6 arrows of time for celestial bodies, as the 4 quantum numbers describe those arrows for electrons.

40. Hierarchical scales between celestial bodies.

The synchronicity between all those cycles, which we established for all systems as a series of symbiotic chains between the bigger organisms and its cellular parts also synchronizes the central black hole with its star system. Let us remember those chains (II,9):

The rules of those complex chains are simple. It is a fact that informative cycles are shorter than energetic cycles, which are shorter than reproductive cycles, which are shorter than the cycles of social evolution of a species. So there is causal chain between the 4 cycles of all entities:

Max. Speed: Informative cycle > Energetic cycle (feeding) > Reproductive cycle >Evolving cycle

Yet at the same time, the smaller the species is the faster its cycles are. So there is a reversed scale:

Max. cyclical Speed:Minimal cells > Organisms > Social systems

This means that cycles of wholes and cellular parts are chained by those different speeds into symbiotic chains; *since the parts need the energy and information of the whole, on which they depend.*

For example, in most cases the feeding/energetic cycle of the whole organism determines the reproductive cycle of the cellular element, which requires the energy of the organism to reproduce and does so much faster than the whole organism; so a cell reproduces each day, which is the time cycle of feeding of its whole organism and so on. Thus, the 'slow' fields of energy and information of the bigger wholes determine the activity of the 'reproductive and social', fast cycles of 'cells'.

Thus in any 'scalar system' of parts and wholes, the 'energy of the superorganism or species of the macro-plane becomes in/form/ation, feeding the inferior scale, causing a symbiotic chains between the whole and the parts. The most obvious, extreme case is the fact that the energy detritus of bigger animals are informative food for lower insects and bacteria.

If we apply this concept to the gravitational waves of black holes, it is obvious that they function as information for black holes but act as feeding energy for stars and maybe planets that 'grow', transforming that energy into cyclical vortex of mass.

Further on, the calculated period of those waves show the synchronicity of the cycles/arrows of energy, form and reproduction of the different cellular scales of the galactic superorganism, which are tuned to each other.

Thus we can consider a hypothetical chain between the 3 scales of the Universe, the human scale, the solar system and the galactic scale. The specific event we are studying is the relationship between a fast, informative cycle of a black hole and a medium, energetic cycle of a solar system, which defines a reproductive, heating cycle of planetary systems. According to those synchronic chains, the reproductive cycles of life in planets are related to energetic cycles in sunspots, regulated by the shortest, faster, informative cycle of the galactic Black hole. That curious prediction, dating from 1994 was proved a few years ago:

Chinese astronomers detected that the galactic black hole has a minimal cycle of periodic activity of 11 years... It coincides with the sun spots' cycle of 11 years, which coincides with the 11 years' rotational period around the Sun of its bigger planet and main G-wave receptor, Jupiter that has a huge magnetic field, 19.000 times bigger than the Earth's field and an enormous inner heat coming from its center, still unexplained by conventional cosmology. Yet if the black hole is connected to its stars by those waves that regulate the sun's magnetic activity, which feeds Jupiter's magnetic field, G-waves could explain why Jupiter has a bigger magnetic field, inner radiation and spatial size than any other planet. Further on, the sun's spot cycles regulate the magnetic field of the Earth and the climatic changes that affect our life with longer cycles of cold and hot weather. We might say recalling a Hindi parable that the blink of Vishnu's eye (the rotating cycle of the galaxy) is a drop of sweat on the sun (its energetic cycle), which is the whole existence of a human being (a reproductive cycle of the Earth).

Due to the lack of space-time of these lectures we won't consider in detail the reproductive body of the galaxy, its cyclical sun systems that feed on interstellar gas to create atoms and have also as a ternary topology - the center occupied by the star, a 'reproductive space' by cyclical planets and the external halo is the Oort planar sphere of lineal comets. Instead we will deal with the most fascinating cosmic object of them all, the black hole, informative mass-vortex and top predator species of the galaxy...

Recap. The galaxy is self-similar to an organic cell where black holes play the role of DNA-RNA systems and stars, mitochondria. The cycles of feeding, perceiving, reproducing and social structure of black holes and stars are related by gravito-magnetic waves.

41. Black Holes are Wormholes: Bridges between 2 Membranes

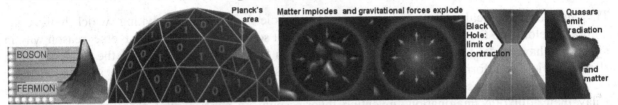

The densest mass-forms of the cosmos are st+1 black holes, which occupy a minimal space with maximal information. The big-bang/big-crunch cycle of quasars and perhaps Universes might find its limit of contraction in the Planck density of a hyper-black hole. Black holes must be studied first as topological spaces with 3 regions; a cyclical, ordered super fluid of top quarks, its 'cells' that condensate in a bosonic state (left); a surface or event horizon, belonging to our light-membrane that kills light and absorbs its minimal units st-1 Planck's areas (center), and a singularity of bosons, gluons and repulsive dark energy, which a Kerr black hole[10] expels through its polar jets, at superluminal speed. Black holes are therefore wormholes, bridges between the electromagnetic and gravitational membrane.

A quark in the quantum scale or a black hole in the cosmological scale is a door between the 2 membranes of the Universe, the quantum membrane from where it absorbs energy and form and the gravitational membrane to which it devolves it. And the best way to study them again is by self-similarity with its equivalent form of the quantum Universe, a neutron, with an inner center of 3 quarks and an external cover of electronic charge, or easier to visualize, its expanded version: an atom with a proton center of quarks and an electronic cover, which would be the equivalent to the event horizon, *not a part of the black hole itself, but the border of our Universe which the black hole uses to absorb and convert the energy of our membrane into quarks and dark energy.*

Since all doors between membranes are open topological balls, transitional regions, whose external membrane belongs to the external, energetic Universe and its center to the informative network. This, in topology is described with a simple equation: an open ball is one defined by all the points of the ball whose center is 'a' and its border, r, *which are neither r, the external membrane* (in the black hole the event horizon) *or a, the center* (in the black hole the singularity). For example, a carrier of oxygen, the red cell, is also a door between the external world and the internal organs of the body; and so it is the only cell that doesn't has center and its membrane explodes (it is the shortest living cell) to release the internal oxygen.

Thus, the quark/black hole is a system that transforms electromagnetic and electronic energy into mass-information, as it evaporates our universal membrane; first into simplest quark forms and then into the simplest energy, gravitational dark energy. Our light is a spatial membrane - a cover that warps the stronger, longer, faster dark energy of the gravitational world. And our electrons are the informative cover of the, faster rotating, smaller quarks. As long as light and electrons cover dark energy and quarks, we absorb energy and form from them. But when quarks and dark energy is liberated its strong forces prey on us. Both ecosystems are thus in a trophic, biological balance:

Our light preys in the faster, thinner gravitational lines that enter galaxies, warping them with a −ct speed measure of corrugation (Special Relativity). Light fractalizes and forms the interstellar dark energy, which slows down from c<10 redshift and warps, acquiring frequency of information.

Yet the inverse role is performed by the black hole, the Non-Euclidean, hyperbolic informative processor of electromagnetic energy into mass that converts our world in the event horizon into a flux of gravitational quarks and dark energy. Accordingly the event horizon can be described as a skin of triangular Planck areas, the minimal unit of the light world, (center top illustration).

In the graph, black holes transform entropy, electromagnetic energy into physical information, quark-mass of maximal order and rotational speed.

This simple topological truth however escapes black hole theorists like Hawking, which believe the membrane belongs to the black hole and evaporates it. Not so, it evaporates our Universe; reason why in 40 years we have never found black holes evaporating. In fact, if we could consider the existence of an inflationary field of information and theories in Physics, is precisely black hole theory, in as much as we have never seen one close enough to observe it and so it is 'fair game' for mathematical physicists to display their 'quixotic imagination' inventing theories with no experimental proof or scientific rigor whatsoever. Let us then consider from the higher perspective of the 4th paradigm of fractal, topological spaces, which of those theories is truth and false.

-*Schwarzschild black holes do not exist.* Since by definition they do not have rotation and the principle of equivalence states that mass is an accelerated vortex of gravitation. Thus all black holes are Kerr holes. The Schwarzschild metric was the first and hence the simplest description of black holes; like Copernicus, cyclical orbitals were. It looked beautiful at the age, but it turned out that only when Kepler completed its model putting elliptical orbits, we obtained the right solution.

- Kerr black holes (rotational black holes[10]) have a superluminal, hyperbolic, perpendicular center, which is exactly what the 4th paradigm predicts: a singularity that doesn't belong to the black hole or the light-membrane but it is the 'canon' that shoots out dark energy at 10 c speed.

- All black holes as all stars have rotation. That is the real meaning of the principle of equivalence: acceleration= gravitational force. But there are only 2 types of accelerations, lineal and obviously that is not a black hole. So a black hole is a cyclical vortex that has rotation, And since the acceleration continues beyond the event horizon where light rotates at c-speed, inside the black hole light has to accelerate faster than c-speed.

And this according to the Lorentz transformations and the theory of multiple time-spaces means that light 'dies' in a big-bang, splits into its informative, photons and energetic gravitational strings and enters the black hole, creating two processes:

-The photons and electrons that collapse in the event horizon, continue its mass-increase deviating lineal light-speed into curled mass speed ($E=Mc^2$), and so they become transformed in quarks, and further on into strange and top quarks, rotating at 10 c speed as they come closer to the hyperbolic singularity.

While light is no longer light, red-shifted by the gravitational black hole till it becomes pure, lineal gravitational, bosonic strings, which accelerate also till reaching the Kerr singularity.

So all black holes are Kerr holes; the real ones. All have a central superluminal singularity that belongs to the gravitational membrane and acts as a hyperbolic 'cannon' that shoots up dark energy through one pole and quark matter through the other. That is what we see in all black holes - a bi-jet: the red jet shoots dark energy (repulsive gravitation); and the blue one shoots collapsed quark matter.

And so black holes balance the entropy of the electromagnetic membrane converted into physical mass-information and expansive dark energy, which we perceive not as superluminal speed but as accelerating intergalactic space; reason why physicists think the space is accelerating (but it is contracting in galaxies and blue jets, which do not reach us, so both effects balance each other.)

In other words, the black hole has a selective external membrane border with our world that filters, like any membrane of any non-Euclidean ternary structure, what it does not want – disordered radiation, and accepts what it wants, pure gravitational strings and formed particles that condensate our membrane.

So particles do fall in the black hole, radiation does not, and in this manner the black hole becomes a selective door to the gravitational pure universe of superfluid quarks with absolute order and not vibration in its gravitational membrane (light vibrations do not enter). And so it creates mass information - quarks on one side and dark energy in the other.

Those strings become tachyons that the hyperbolic Belgrami hemisphere in the center deviates and ejects through the Kerr singularity into a flow of dark energy, which is repulsive gravitation at 10 c speed.

While the quarks evolve till reaching the rotational 10 c speed of top quarks and become a top quark star whose pulses of dark energy on one jet and quark mass on the other renew the Universe, returning the light-membrane of $>o$ T order and $<c$ speed to its original wider membrane of $>c$ speed and $<o$ T.

And the evaporation of black holes? A false theory, which confuses the thermodynamic, evaporating energy of the event horizon with the black hole inner body. So once this error is solved, Hawking's otherwise beautiful equations merely describe the birth of the black hole as micro-form rotating at enormous speed, which heats up and evaporates our membrane, feeding actively in our world so fast that within seconds of its birth as a tiny black hole it sucks in an entire star creating a supernova (or perhaps a planet if CERN succeeds in creating them, misguided by Hawking's wrong theory). Indeed, if we use the same equations to explain the evaporation of our world membrane/event horizon we observe immediately that the super-hot membrane evaporates our colder world and as it grows and cools down, it absorbs less and less matter, till reaching a huge size and cold temperature becoming less active. So the birth of a micro-black hole means the explosion of our stars not of the black hole. Why we are so sure of this?

There is experimental evidence: all stars that explode into supernovas leave behind black holes, which must have been born exceedingly small.

All systems of the Universe are born as 'seeds' of microscopic size and grow much faster in their relative youth and then slow down their growth

The equations of Hawking are the same, once the error is solved - only that now they show NOT the speed of evaporation of the black hole but of our world.

If the evaporation of black holes was right, it would break all the key laws of physics: all the laws of relativity; the first law of entropy (a hot object doesn't get hotter as Hawking pretends evaporating our cold Universe; a hot coffee cools down; a hot iron sword evaporates the cold water in the forge; so a hot black hole will evaporate our cold Universe. It also breaks the law of conservation of information, which disappears in Hawking's evaporating black holes. Yet information according to quantum physics and theory of multiple space-times never evaporates.

Finally, and this is the nail in the coffin: we never found any experimental proof of black hole's evaporation. So the theory must be wrong, according to the experimental method

It is however interesting to consider how this theory came into existence to see why in science often a false theory becomes a dogma, for lack of a better paradigm that substitutes it. Here the paradigm is the existence of an arrow of information besides the arrow of entropy in the physical Universe, the arrow of mass. Yet without the 4th paradigm, physicists feared that black holes broke the dogma that there is only an arrow of entropy in the Universe that always increases. So as Wikipedia explains: 'the only way to satisfy the second law of thermodynamics is to admit that black holes have entropy. If black holes carried no entropy, it would be possible to violate the second law by throwing mass into the black hole.'

Which is exactly what happens: precisely because the arrow of information is inverted respect to the arrow of entropy and the arrow of information exists as physical mass, black holes allow the recreation of order in the universe, making it immortal.

This should be obvious since the law of entropy was deduced in the XIX century studying electrons, molecules and light from our membrane, *not gravitation and mass, the forces that balance it.*

Thus black holes are not an error, which we can 'change' 'ad hoc' but an essential proof of the existence of information.

Yet the proponents of Black Hole Thermodynamics were believers in the dogma of a Universe of 'entropy only'. So to satisfy the dogma, 'Jacob Bekenstein conjectured that the black hole entropy was proportional to the area of its event horizon divided by the Planck area', (Wikipedia). And so suddenly a conjecture which broke all the known-known laws including the 1st law of Thermodynamics, which precedes the 2nd law - since a hot object never gets hotter in a colder environment (the biggest hoax of the XIX C. - a perfect motion machine) – became a necessary truth.

Further on, by accepting this conjecture Bekenstein & Hawking also broke a tenant of the 2nd law- since now information disappeared in the universe! And obviously they broke with Einstein's Relativity Theory and his laws of gravitation (reason why Hawking ended his article saying that 'Einstein was double wrong').

But there was still a problem: How to evaporate the black hole! So Hawking made another conjecture, that small black holes had quantum effects because they were small.

This again breaks the fundamental tenant of relativity and fractal spaces: size is totally relative. Besides quantum and thermodynamic laws are laws that apply to the electromagnetic membrane, regardless of size and gravitational laws to masses regardless of size. Size has nothing to do with the laws we apply but 'substance. So a small ant behaves like an ant not like a small fungus regardless of size, and a shrew, smaller than some insects like a mammal, regardless of size.

Then, it came up with the idea that a certain quantum effect happened to the quantum black hole: one virtual particle of the event horizon, which he considered the black hole (but it is not) felt out of the black hole and evaporated. This again is absurd because:

- The membrane is the frontier between two open topologies, but its properties are those of our membrane, it is therefore part of our electroweak membrane, so it evaporates our membrane.

- Because we are in a vortex of accelerated gravitation (Equivalence Principle), the gradient of forces is towards the black hole so there is more probability that one particle falls into the black hole than out; as it is easier than a feather falling into a sink vortex enters the sink than gets expelled of it.

Thus when you reach those 'real' conclusions, the membrane is our universe and the particles condensated by quantum effects fall into the black hole, evaporating our universe. Then we use Hawking's radiation formula, but now it is showing the speed of evaporation of our membrane under the first law of thermodynamics (our colder membrane gets hotter) and evaporates swallowed by the black hole.

Recap. Black holes do not evaporate. They are vortices of gravitation that transform our electromagnetic world into dark energy and quark-mass, establishing a balance of order with the entropy of our Universe and making the Universe eternal.

42. Black holes are Kerr wormholes and frozen stars[10].

Quark holes can be studied, once we understand their 'higher why', from the perspective of topological spaces and multiple spacetimes, which limit is possible species and processes, either as top quark stars (frozen stars in the jargon of Einstein), or as rotational masses with the Kerr Metrics; or from the many perspectives, of its structure as a system co-existing within 3 relative st-planes of existence. In the graph, its 3 topological regions.

Any organism extends through 3 hierarchical scales. For example, a human organism is composed of cells and carbohydrate molecules whose minimal unit is the amino acid. In the Universe there are 3 scales of mass and each of those 3 scales become the fractal parts of the next scale. In each of them we find an informative, cyclical type of mass: quarks in quantum atoms; black holes, which should be top quark stars in star-size black holes; and swarms of black holes in hyper-black holes at the center of galaxies and perhaps the Universe. Thus a hierarchical organic structure grows from those inner parts: its strings become gluons that become quarks that become black holes that form hyper-black holes, in a stair of fractal systems of increasing complexity. This is proved by the fact that the Black hole's event horizon membrane reduces light to its minimal bytes of information (Planck's areas), whose length is the minimal length of strings, which they *extract from light*. Since strings are both the theoretical components of black holes and strong forces, quarks and gluons, a description of black holes with tachyonic, bosonic strings becomes the mathematical bridge between both scales. Thus indeed, black holes are fractals of quarks, fractals of gluons and strings (mind the reader though that those strings have fractal dimensions, are background independent, tachyonic strings, defined by the Nambu's actions, not superstrings, a baroque fantasy of Pythagorean Physics).

Each of those scales becomes the informative center that interacts with a reproductive body of electromagnetic electrons, stars and galaxies. Let us then study black holes in their interaction with its bodies, galaxies that explode in its big-bang death as quasars.

To regulate that body black holes emit dark energy through its poles, creating the membrane of gravitation that 'positions' stars, as protons position with their gravitational and magnetic fields the electronic nebulae. Since gravitational Space-time is formed by the maximal informative entities of the 2 scales, protons and black holes. For all those reasons, we prefer the names:

-Frozen stars (Einstein's name), since they are top quark stars.

-Kerr's holes; since this is the rotational metric or black holes.

-Or wormholes, because it expresses the fact that the black hole is a door to the other 'world' of dark energy, which it emits. While the name black hole, invented by Wheeler, to substitute Einstein's name after his death and push in this manner his singularity theory, implies nothing is emitted by the hole.

In that regard, science when it is built to respond *all* the questions that exhaust the truth of the system, the what (experimental evidence), the who or how (causal logic), the when (metric spaces) and the why (topological time arrows) *has also an inverse hierarchy of truth:* Once we determine *what* we want to study, we must observe its *why and how* – its topological structure and causal logic combined (how/who) - and only then enter into the details, analyzing with clocks and instruments of science, its metric properties. The excessive use of machines of measure, overdeveloped by the Industrial R=evolution has obscured that hierarchy, which limits with the *experimental what*, and the *why and how of topological time arrows,* what things are certain and what are just mathematical fantasies of 'baroque artists' of metric spaces, which are so common in the study of black holes due to the lack of experimental evidence on its details. Let us then study reality as it is, not as 'Touring machine' resolves

360

it with an excess of metric information that ignores the deepest why and necessary laws of the why-universe.

Frozen top quark stars.

Frozen stars are similar to nucleons: huge condensates of top quark-mass that create vortices of gravitational information with a negative curvature in the dimension of height. Hence, they exist in a discontinuous gravitational Universe beyond the c-speed and 0 K limits of energy and information of our Universe, emitting and absorbing dark energy at higher than light speeds and ultra-cold dark matter, perfectly ordered, probably under 0 Kelvin. In that regard, the closest species to a frozen top quark star is a neutron star, composed of super fluid neutrons and strange liquid in a bosonic state, occupying a minimal space. The difference of density between a neutron star and a Top quark is small. So a Frozen top quark star could be the next evolution of a strange quark star with a 'lighter' cover of strange quarks on the event horizon, 'breaking=killing' the symmetry of our matter, packed then in a bosonic, super fluid solid state of bcb atoms and top quarks. Since the sum of the transitional weak bosons, Z and W *equals that of a top quark. Thus: n+p=Z+W=Top quark.*

The 2 poles of the wormhole.

Thus the main error about Worm Holes is the idea that since we do not see them emitting energy and matter-information at lower than c speeds, nothing escapes a Worm Hole. This is a theoretical absurdity (things don't 'disappear'), which now has empirical proofs of falsity. Since we observe vortices of mass and radiation that surround Worm Holes, reaching super luminal speeds before dying into pure gravitational energy according to the Lorentz transformations. And we observe bursts of matter and radiation coming out of the poles of central Worm Holes in quasars at super luminal speeds. So Worm Holes do emit dark energy and information at super luminal speeds through its 'axis', as atomic nuclei emit magnetic fields in their rotation.

The Relativistic equations of Worm Holes show that duality since Worm Holes appear with 2 solutions: one with implosive, informative parameters and the other with explosive, energetic parameters. Thus according to those equations a Worm Hole 're-absorbs' radiant matter and light, dissociating the photonic particle-state and wave state of light, hence 'killing it'. both beyond their Lorenzian limits of c-speed, back into its ultimate components: photons evolve into electronic nebulae, and then collapse further into quarks, which become part of the Worm Hole body or are ejected as quark beams; while light reaches infinite red shift and becomes dark energy. Because the parameters of density of a stable black hole are self-similar to those of a top quark star, and that is the limit of density of mass in the Universe, it is easy to infer that black holes have in its center a super-fluid vortex of top quark stars; and so we can consider dark energy to be the 'gravitomagnetic field' of those top quark stars.

As in the case of the 3 solutions of Einstein's space-time, which correspond to the 3 ages of the Universe but physicists dissociate in 3 different Universes in space; physicists have deduced that those 2 solutions to the Worm Hole equations create 2 different type of 'black and white holes'; when according to space-time duality they represent the 2 organic regions of the same gravitational hole. Hence we could call Worm Holes also 'mulatto' holes (-;. Though we will use the term *wormhole*, more familiar to cosmologists. And define its E⇔i Generator equation:

Worm Hole (max.E: external membrane: event Horizon) <Wormhole> White hole (max. i: Kerr ring)

Quantum cosmologists never discovered white holes as independent entities, because they are part of the wormhole.

Topological regions of a worm hole.

Unfortunately quantum cosmologists ignore the inner structure of a gravitational hole as a fractal point with 3 zones that explain them:

- *Max. E: The event horizon* at c-speed is the energetic Riemann membrane that absorbs radiant matter, breaking it into its minimal units, 'Planck's areas' that become the strings of dark energy that feed the hole.

- *E=i: The intermediate zone* transforms radiant particles and energy into quarks and dark flows of gravitational energy. It seems to be a super fluid solid: a vortex-like structure of quark condensates and gluons.

-- *Max. i: The central region around* the polar axis or central, informative, hyperbolic, negatively curved nucleus.

That informative center is the final 'eye' of the gravitational hurricane, the white hole of the wormhole. It emits through its poles energy and information in the form of dark, quark matter and gravitational waves of dark energy in super luminal jets that we observe indirectly around far away quasars, as they become again slower radiant matter, creating irregular galaxies. Thus white holes are the poles of wormholes, which indeed are the doors to the gravitational world of dark energy and quark matter that dominates the Universe, as cosmologists have discovered.

The equations of a wormhole show how it transforms the spatial, energetic parameters of the electromagnetic world into the inverse, informative parameters of the gravitational world, since a wormhole is an ultra-dense mass of highly ordered 'bosonic nucleons', packed into a single point of space...

Continuous physicists used to believe that the accelerating vortex reached infinite energy values in the central point or singularity, since they do not model masses, charges and Worm Holes as physical vortices of mass with a Radius, Ro, that represents the discontinuous limit between the external, body cycles of the Non E-point and its inner, still, informative region or brain, in this case between the vortex of stars and the still Worm Hole brain. Yet because infinites cannot be calculated they renormalize their equations beyond a certain limit in which they postulate 0 charge or 0 mass.

In fractal cosmology those tricks are not required as wormholes are modelled with Non-Euclidean topologies, which have always 3 regions separated by asymptotic membranes.

Thus fractal theory solves the problem caused by continuous infinite singularities, as Planck solved the problem of continuous, infinite temperatures, when he introduced fractal light quanta.

In organic terms, beyond the event horizon of a Non-Euclidean point, the Klein disk starts and beyond Ro, a discontinuous, inner radius separates the body from the informative, still brain.

In the galaxy the event horizon is the halo and this final radius is the horizon of the wormhole. In the atom, those 2 horizons are the external Electronic radius and the inner Bohr Radius, beyond which we find the protonic Worm Hole.

The same pattern of 3 regions is found in a Worm Hole as a Non-Euclidean point. The event horizon is the external membrane. Then the point in which the equations of a Worm Hole reach T=0 is the inner R_o radius.

(This is the point in which Hawking says Worm Holes become negative in time and convert themselves into time machines)-: We already argued Mr. Hawking's confusion of physical time, a change in the direction of motion, v=s/t, and absolute time, a hyperbolic error of the Cartesian graph. Time in physics is change in the direction of motion. So what T=0 means in a Worm Hole is that we reach the region in which the cycles of the Worm Hole's body end and we enter the hyperbolic, high central tube that ejects dark energy. At this point time - understood as change in motion - halts and the Worm Hole enters the white hole region of production of dark energy, asymptotically perpendicular, with the form of a Belgrami cone.

The equations of Worm Holes show also their event horizon as a bidimensional *killing* field that destroys the entropy of our Universe, since it kills our energy/matter into its ultimate units on the Planck's scale, *lowering as in all processes of death our components two scales down into its fundamental physical units.*

In the biological homology, a human being is composed of cells themselves composed of amino acids that act as the minimal units of life. So when an organism dies it suffers 2 deaths: first its cellular tissue is broken into pieces that feed the stomach of an organism, which will destroy it till its minimal amino acid units, used to recompose the organism's own cells.

When we study mathematically Worm Holes they show also a dual process of destruction of light matter to its ultimate components, Planck areas or strings, used then to reconstruct the bosonic quark condensates of the Worm Hole and its dark energy.

Thus Worm Holes in their feeding processes destroy light matter till it absorbs its Planck's areas, the minimal units of information and energy of our physical space-time, the equivalent to the amino acids that the stomach absorbs.

Then those Planck's units, become lineal tachyon strings to form dark energy and cyclical strings or gravitons, the minimal units of the mass world, evolving into bosons and quarks.

Those quarks and dark energy is then expelled through the poles *to balance the expansive entropy of our membrane and create an immortal Universe.* Indeed, the expansion of space-time in the Universe is cause by the expansion of dark energy coming out of the Kerr worm holes' poles; yet black holes also implode light into quark matter and the overall process creates a wobbling, dynamic balanced, zero-sum of fractal expansions and implosions, in a Universe that will never die.

Recap. We define a rotating *wormhole* as an organic topological structure, which uses the event horizon membrane to absorb radiant matter and energy through its central, ventral plane, which transforms it into ultra-dense top quarks and gravitational dark energy, expelled in perpendicular jets through its central, hyperbolic axial *white hole's* pole used to control the body of galactic stars and communicate with other Worm Holes.

43. The Existential Cycles=Time arrows of Wormholes.

Wormholes follow the 4 energetic, informative, reproductive and social arrows of all exi=stential systems of reality: They feed on electronic matter, nurse galaxies and stars, and reproduce in momentum collisions with them, evolve socially in super-Worm Holes (galactic center swarms), and as the 'long-lasting' informative neurons of the galaxy and Universe, whose networks they create, its generational cycle lasts as long as the entity that hosts them (the same happens with neurons, the informative network of our body that last between birth and death). Let us study those arrows in more detail:

- *E>I: The main cycle of a Worm Hole is informative.* It creates quarks from electroweak matter and orders the galaxy into spiral forms through gravitational waves that 'position' the stars while symbiotically feeding their centers with energy that degenerates into matter. Thus the worm hole emits dark waves of gravitational information to control the galaxy.

Those informative waves of Worm Holes fed the energetic needs of stars, which therefore become chained to those waves without knowing its final demise - as gravitational plants, guided and herded by Worm Holes towards the center of the galaxy to feed them:

- *I<E: Their energetic cycles balance the entropy of the light-world.* Wormholes first erase light and radiant matter, feeding on gas and stars; and then renew it, creating new jets of pure dark, gravitational super luminal Energy and quarks that enter back into our Universe as light and protonic matter. If insects eat dead matter to renew the Earth's ecosystem, the wormhole inverts the time/space coordinates of the light-world to renew it. They act as 'the antiparticles of the white cosmos' that annihilate radiant matter. Kerr worm holes absorb energy from our light membrane by red shifting light, with different signatures

according to their relative mass. The background radiation must be interpreted as the signature of 2.7 'background holes'. Bigger, older Kerr worm holes of galactic mass are below the Background radiation curve and so they *can red-shift and absorb electromagnetic energy from it, feeding in this manner on the electromagnetic membrane.* For that reason the galactic map of the background radiation has a central zone, with lower temperatures corresponding to the giant central, colder worm hole.

- *Reproduction:* Wormholes control the reproduction of stars, which in turn reproduce wormholes. Since when a small wormhole crosses through a star, it catalyses its explosion into a nova that leaves behind a neutron star or a wormhole. On the other hand, the 'reproductive DNA center' of giant galaxies is a massive wormhole structure that emits huge super luminal jets of quark matter, which catalyse the reproduction of stars, creating irregular baby-galaxies.

- *Social evolution.* The wormholes that occupy the informative center of galaxies are like the nuclei of cells, swarms of Worm Holes, which regulate the life of galaxies, as DNA does with cells. In the next scale, the dark energy flows ejecting by galactic black holes communicate galaxies, creating the networks and walls we observe in the grand scale images of the Universe.

Recap. Black holes follow in their events the 6 arrows of time. They feed on the electromagnetic membrane and its species; they perceive gravitational information; they reproduce stars and galaxies and they evolve socially into swarms in the center of galaxies and into galactic networks in the Universe.

44. The 3 organic roles and types of wormholes in galaxies.

By homology with any other non-Euclidean space-time that resembles a cellular organism, and the *Laws of ternary differentiation of all topological species,* we can consider 3 types of wormholes whose roles within the organic structure of galaxies will obey *its arrows of time but will also become essential to the bigger organism in which they exist.*

This simple rule, which we shall apply to all systems, is a tenant of the organic structure of the Universe – we differentiate species in ternary sizes, ternary topologies, ternary ages and ternary functions, which must be symbiotic to the higher st+1 organism in which the entity exists or else the organism would not 'tolerate' the presence of the microcosmic species with no function:

- *Intermediate, reproductive E=i zone: Spiral arms.* Non-rotating wormholes are born from dying stars. Then they form bi-polar systems with other stars feeding on them, taking advantage of their gravitational control, finally transforming the biggest stars into new wormholes. Those wormholes probably gather into social groups, which fuse in bigger wormholes and move towards the center of the galaxy where they can feed easily on its dense herds of stars, creating at the end of the process a central nucleus.

- *Max. information: Nucleus.* The Worm Hole nucleus is a huge rotational, 'Kerr hole' or perhaps a herd of Worm Holes similar to the DNA nucleus of a cell. It is the informative brain of the galaxy that controls its fractal beings, the stars, with gravitational waves that shape the rotating movement of the galaxy, and establish its feeding rhythm: The galactic Worm Hole first attracts interstellar gas to the intermediate non-E region, where gas reproduces stars, and then it sends that gas to the central wormhole that consumes it. In the same manner, electrons, the 'stars' of the atom, feed first on light quanta and then emit high-energy photons to feed the atomic nucleus.

Those gravitational waves also guide in old, globular galaxies, the stars toward the feeding center. Yet there might be other structures of dark matter, coming out of the nucleus, similar to the Golgi apparatus of cells: invaginations through which wormholes might flow into external zones of the galaxy to control the reproductive and destructive processes of stars.

Finally, the central hole emits through its polar zones, dark energy, super-luminal gravitational waves that probably communicate galaxies at super luminal speeds, forming the strings of galaxies observed in the Universe. Since according to string Theory gravitation in free, intergalactic space is not warped by electromagnetic branes that feed on them inside galaxies and limited to c-speed.

- *Max. E: Membrane.* Though we cannot see the galactic membrane made of dark matter, by homology Non-Euclidean topology hypothesizes that the halo is the energetic membrane, where small wormholes called appropriately MACHOS, have functions similar to globular proteins in cells:

They create and control that galactic membrane of dark matter, which closes the galaxy as a black body; causing the background radiation, which according to recent empirical data might be local: They redshift light to 2.7 K, which becomes the metabolic temperature of the galaxy. They reproduce new stars and expel matter and radiation beyond the membrane: Since their rotation is perpendicular to the galactic plane they could create a positive or negative spin, depending of its orientation, provoking flows of energy and information in and out of the galaxy. Those outward or inward flows fine-tuned with the dark energy jets of the Worm Hole should move the galaxy at the will of the central Worm hole

The interaction of the membrane and the central Worm hole should encase all other species as a perfect Max. E x max. I 'upper class' of the galactic organism in complete control of its inner parts.

Recap. There are 3 types of worm holes that control the membrane, inner center and herd the stars of the galaxy.

45. The big bangs of the 3 types of wormholes

The homology between the 3 sub-species of Worm Holes and the 3 regions of a galaxy explain the 3 possible scales of feeding and big-bangs of those Worm Holes; and the *functions of the lesser electromagnetic species for the top predators of the galaxy*:

- Micro-Worm Holes, Background MACHOs would eat up the commonest celestial bodies, planets and moons. This is our function; and the fact that we do not hear intelligent life in the galaxy and that CERN is going to do black holes to 'see' if they evaporate could explain how indeed, all moons become background MACHOs and all planets are blown up by 'metric physicists' unaware of the 'why' of black holes, stuck in the 3rd paradigm of using machines to measure the when of reality. This smaller worm holes would then migrate to the Halo of the galaxy closer to our planet.

- Medium Worm Holes formed in dense star centers, feed on them, causing supernovas, creating intermediate Worm Holes that migrate toward the center, forming the DNA nucleus of the galaxy.

- While galactic Worm Holes would explode into quasars, (galactic big-bangs). And/or as 'dark galaxies', recently found elsewhere, without emitting light, migrate towards the 'Great Attractor' or hyper-black hole at the center of the fractal 'cell' of the Universe in which we exist.

Recap. We might be just food for worm holes, which are born and feed on the 3 scales of electromagnetic matter of the galaxy: planetoids, stars and entire galaxies.

46. The Worm Hole as a Gravitational, Informative Mind.

The center of the galaxy is occupied by a giant wormhole that seems to be the final, social, evolutionary stage of multiple galactic Worm Holes, born out of the evolution of stars. It acts as the gravitational DNA-mind of the galaxy - a hypothesis, which mechanist science will always ignore. But we want to stretch your understanding of the 4th 'why'-paradigm, describing those mind holes with the laws of Non-Euclidean geometry and superorganisms.

A galactic wormhole is a rotating object, which has a minimal spatial size and a huge dimension of height since it is made of bosonic, super fluid quark condensates. Its homology with a cellular, informative center, defines a central galactic wormhole as an enormous gravitational informative center, which controls its galactic body, positioning its star quanta through gravitational waves. In fact, when we calculate the wormhole's informative parameters, it turns out to be a perfect super fluid computer, with maximal informative volume since its speed of calculus equals its speed of transmission of information.

The coldness of the wormhole proves also the informative hypothesis: In fractal space-time coldness means order, stillness, necessary to create in the center of a crystal or a cryogenic CPU, or a super fluid wormhole or the focus of an eye, the informative, bosonic accumulation of pixels that shapes a still, formal, fractal virtual image, without friction, without blur. Thus, the eye of man is cold. The brain is colder than the blood. The chip works better at cryogenic temperatures. And a wormhole is very cold, made of quarks in super fluid, highly ordered states.

Wormholes do create a complex virtual image of the galaxy with its strings. Physicists explain those processes in abstract when they affirm that a wormhole has an extra 5th dimension and its equations are homologous in 5 dimensions to those of the 4-dimensional electromagnetic world of the galaxy it represents (Maldacena Conjecture[12]). *The description in 5-string dimensions of a Worm Hole is completely equivalent to the description of an electromagnetic galaxy in 4 dimensions. Thus as our brain has a 'homunculus image' of the entire body, which is what we perceive; the worm hole can build a 'galunculus' image in its interior.* The 4th 'why' paradigm provides the reason of the conjecture: the 5-dimensional world of string holes is equivalent to a 4-dimensional world; because *it is the map of our galaxy, made by the Worm hole's mind.*

In topology the central zone of any st-point is an informative region, whose dimensions are the sum of the body and membrane dimensions. In cosmology the galaxy is a body-like vortex of reproductive stars and the central wormhole is the eye of the vortex that creates an image of the galaxy, as the brain maps out the body.

The hyperbolic center of higher form/information has more dimensions than the energetic bidimensional membrane and the reproductive toroid; since it holds the holography that combines the image of the external world imprinted in the 'senses' of its membrane, and the image of its internal, reproductive functions it commands. Since our galaxies have an external bidimensional halo and an internal 3-dimensional vortex of stars (a bidimensional plane with an added dimension of rotational motion), we can easily calculate with the general laws of the Holographic, central mind, the dimensions of a black hole:

Bidimensional external membrane +3-dimensional cyclical volume:

5-dimensional hyperbolic nucleus.

This simple law derives of the general law of dimensionality of the 3 regions of a Non-Euclidean topology:

Information of hyperbolic 'brain'= Information of Klein's 'body' + Information Riemann membrane

And it is the origin of the Maldacena Conjecture (a black hole, informative center of the galaxy has in 5 dimensions the same structure than a galactic mapping in 4 dimensions.)

What this means is that the Worm Hole must have an inner image of the galactic body, which it orders with its gravitational flows and an external image of the Universe in which the galaxy, its body floats, to direct its form toward fields of energy (intergalactic gas) and connect itself with other galaxies. This is what we observe galaxies do. They form walls, strings and complex clusters with other galaxies and they feed on intergalactic gas and smaller galaxies. So they act as cells in a gravitational, organic soup and must have as cells and organisms do an informative, processing center able to have an internal image or mapping of its stars and an external connection that informs it of the outer world. However, we see only a 4% of the galactic structures; *the rest is the dark energy and dark, quark matter that the worm holes perceive to map out reality,* which is like trying to recompose a cell, in which we only see its mitochondria.

Another proof of this higher dimensionality is given by the Kaluza-Klein models, which explain our perceived electromagnetic 4-dimensional world and the 4-dimensional gravitation over which the network of light is imprinted, as if they were born from a larger 5-dimensional gravitational world. Finally, gravitons are the only particles that can have 5-spin positions=informative dimensions. Thus, all those self-similar theories point out to the fact that we are a floating membrane within a bigger, faster, more complex Gravitational Universe.

Recap. In fractal cosmology the central discontinuous zone of a fractal point is an informative region. This is the role of Worm Holes in galaxies, forming with other Worm Holes, communicated through flows of dark energy, the informative network that controls the Universe. Therefore it has the dimensions of its membrane and inner body of stars.

47. Background Radiation: Basal Temperature of the Galaxy

As any system can go in more detail, by looking closer to its fractal parts, we can go deeper and deeper in the analysis of the whys (biologic and topologic), whens (mathematical and metric), how/who (Logic) and what (experimental) questions of knowledge. Let us consider thus in more depth just one of all the elements of the galaxy studied here – the organic role of the background radiation - *the water of the galaxy...*

The background radiation coincides with the radiation of a black body at 2.7 K degrees. Since in Non-Euclidean topology any fractal point is a black body; that is, a point with minimal apertures to the external world, a galaxy will only emit background radiation through its Halo, in which background holes will redshift light at 2.7 K. Thus the galaxy surrounded by a halo of quark, dark matter can be considered a black body emitting at present time as an isothermal organism does a background radiation, whose organic function is to maintain a homogenous temperature, similar to the organic temperature of living beings and ecosystems, kept by its water. Thus, the background radiation acts as the cytoplasmic energy of galaxies with 3 functions:

- Max.E: It provides energy to its bigger, colder Worm Holes and its super fluid helium structures (not treated in this introduction to complex cosmology). Both happen to have a temperature slightly lower than the background radiation from where they can extract energy.

- E=i: It acts as the membrane limit, between the gravitational and electromagnetic membrane, within the galaxy, separating both worlds. It maintains also an isothermal temperature, as any organic system maintains a stable temperature.

- Max. i: It establishes a fixed frame of reference for the galaxy, *allowing the process of information and measure, location and communication defined by Gauge Theories.*

The background radiation is the 'energy soup', the cellular water of the galaxy that feeds its dominant RNA, its Worm Holes; as the hot water of the cell allows RNA molecules to move, kicking left and right water molecules with its COOH legs. Indeed, we know that only organic systems have a homogenous temperature. For example, humans have a homogeneous temperature within the limits of liquid water. So the 2.7 K homogenous background radiation reinforces the organic hypothesis.

Quark matter is *the top predator form of the galaxy; hence they are the entities which, as the elephant on the savannah or man on Earth, or aerobic bacteria in the earlier planet, have redesigned the galaxy with their organic activity.* In Gaia, water, the equivalent to that background radiation, maintains a stable temperature, thanks to the feeding, energetic activities of its life organisms that avoid abrupt climatic changes. Without aerobic life the Earth would be like Venus - a planet with extreme temperature changes. Now it is almost isothermal. So happens to your body which has 36.5 degrees all your life, due to your organic activity as a water organism with 2 networks, a warmer blood and a colder, nervous, informative system.

So happens to the galaxy, in which the basal temperature of the background radiation separates the ultra-cold world of Worm Holes and dark, gravitational matter and the hot world of atoms and radiant matter, allowing the exchange of energy and information between the 2 physiological networks that structure the galaxy.

Recap. Background radiation is the energy soup, the cellular water of the galaxy that feeds its dominant RNA, Worm Holes.

48. Non-E Structure of the Universe.

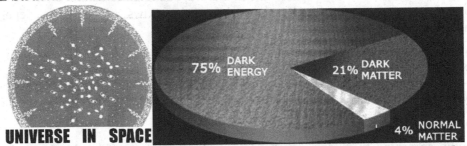

The Local Universe seems to have a dual network structure—electromagnetic energy and visible matter vs. quark, dark matter and dark energy, which forms its faster networks of information, the membrane of gravitation. It is the organic hypothesis of a Universe made of 2 networks of dark matter and light energy - cellular galaxies, which form a simply connected system, similar to a colonial tissue, with the apparent form of a semi wave in grand scale images perhaps, belonging to a bigger Hyper-Universe.

We have now showed the general structure of the different systems of energy and information of the Universe. So we can focus in the analysis of the two main systems studied by mankind – the astrophysical Universe, and the biochemical world of humanity. Both should, according to the previous analysis share the same properties and arrows of time, albeit with different quality and amount of 'complex information' and 'simplex energy'. Let us then briefly introduce those 2 systems and their self-similarities.

In the graph, a self-similar analysis of the 2 superorganisms studied in this work, the Human organism and the Universe.

Despite the enormous difference of size and metric details, they are both topological, fractal organisms with the same elements of all organic fractals. Since reality is a self-generating topological Fractal of infinite beings made with 2 arrows of Time: energy and information, whose self-similar properties emerge invariant in topological form in all scales of reality. Thus all organic fractals, including man and the Universe have 3 elements that define them:

- Its cellular units, (galaxies and cells).

- Its networks of energy (blood and radiant matter) and information (nerves and dark matter) that organize those cellular units.

- And the relative space-time planes in which a self-similar but not equal fractal structure re-emerges.

On the left, man is a fractal made of cellular units and networks of energy and information (nervous and blood systems), whose cycles and functions extend through 3 hierarchical, organic planes of increasing size: the cellular, individual and social plane.

On the right, the Universe is also a fractal, structured in 3 planes of self-organization: its cellular units are stars and Worm Holes that shape 2 galactic networks of radiant energy and informative, dark matter; st-points of a cosmos at 'grand scale' (the points on the picture). The final form is self-similar to a wave, perhaps inscribed into a hyper-Universe.

How can both worlds/networks co-exist together? In continuous space-time they can't but in discontinuous space-time where a 'dimension' is fractal, hence it is a fractal network, it is rather easy:

The light Universe is a network, like a fishnet floating of an immense sea of gravitational energy. Each knot is a charge, a non-E point that communicates through the strings of the net (the electromagnetic forces), leaving a huge dark 'space-time of 'water' - that 76% of gravitational dark energy - which is not 'illuminated' neither interacts with the fractal net of light.

Some Physical theories describe fractal space-time in that way, calling the fishnet a 'brane' made with cyclical, temporal knots (spins or closed strings) joined by flows of its lineal, energetic exi actions/forces that tight together those knots.

Recap. The Universe sandwiches mankind between both membranes of space-time, the quantum membrane of electroweak forces and the gravitational membrane of strong masses. Yet both membranes obey the same laws of Multiple Spaces-Times. Thus, regarding their morphological structure both membranes possess organic, complementary systems with the 3 topological standard regions /forms of a fractal point.

49. The Universe is a spatial organism.

Any Fractal point, according to duality, can be described in space as a fractal point with a relative energy body and an informative center, communicated through 2 forces/networks of energy and information. And it can be studied in time, through its 3 evolutionary ages, between life and death, as its energy becomes curved by time and increases its form, creating in the case of the cellular Universe, new informative particles and galaxies. Though both perceptions of the Universe are correct, cosmology ignores the organic, spatial description of the Universe, which completes the evolutionary, temporal vision, obtained through light instruments - since it cannot see dark energy and non-local gravitation. Yet if we were to perceive that Universe with the instantaneous, non-local gravitation that reaches its limits, it will seem as a complex st-point structured by an external membrane (the wall of fire), an internal self-reproductive space of galaxies and an informative hyper-worm Hole center (the great attractor?), all coordinated by the non-local forces of dark energy.

Recap. The physical Universe is structured in 3 scales of reality, the quantum world of electromagnetic forces and the gravitational world of masses. Both can be modeled with the laws of super-organisms, whereas the electromagnetic forces/planes act as the energy network of galaxies and atoms and the gravitational forces/masses as the in/formative force that balances the entropy of electromagnetism, creating a complex, organic Universe of eternal motions and balances between informative gravitation and entropic electromagnetism.

50. The cellular Universe.

Galilean relativism *(e pur si muove, e pur no muove)* implies that depending on which kind of energetic or informative force we use to observe certain reality, we will perceive it either as a fixed, spatial organism (the Universe perceived with gravitation) or as an evolutionary species in motion through time (the Universe perceived with slow light). So we see either a moving Earth and a fixed cosmos, when we see them through gravitation, or a quiet Earth and a moving Universe when we see them through light.

Since gravitation, the force that shows a spatial, synchronic Universe is invisible to our instruments; astronomers only study a temporal, diachronic Universe, perceived with light. However there is a universal organism in space, self-similar to any other Fractal point. Since once the evolution of the hypothetical cellular Universe concluded in time, creating the 3 regions of any fractal point, the Universe structured itself in space, communicating those 3 regions through simultaneous non-local gravitation. Let us then study those 2 sides of a hypothetical cellular Universe - first the spatial, organism of the Universe and then the temporal ages of creation of that Universe in the next paragraph.

If we perceive the Universe from a temporal perspective through its slow force, light, it appears as an evolutionary process of matter, coming out from the genetic 'big bang' singularity of a local, cosmological first cell - the hyper-dense singularity of the big bang (which went afterwards through a cold, reproductive 'big-banging' or creative, informative process, later studied in more detail).

Yet, when we observe the Universe simultaneously in present space (first picture of the previous graph), thanks to its faster force, gravitation that allow the parts of the Universe to interact, those far away regions become integrated with its closer regions through the informative networks of dark energy, NOT through the light, energetic networks of galaxies and stars, explaining its homogeneity. Again this is a generic law of complementary systems, as any organism is defined by its informative, nervous network that gives it its form.

Of the two possible 'scalar' theories of such Universe, one of infinite scales in which galaxies are atoms, and one in which universes are fractal 'cells' of a hyper-universe, broken in 'bubbles' of a fractal inflationary big-bang with a limit, the first one would imply a huge universe, in which we are just a 'hydrogen-like atom' of an enormous interstellar cloud. The second theory however would imply a structure within our 'cellular Universe', in which non-local, faster than light gravitation creates an organic Universe, structured as a fractal point, with its 3 canonical, topological, Non-Euclidean zones:

-*Max.i:* The cellular Universe should have a nucleus of enormous gravitational mass, a hyper-worm hole, connected to a network of dark matter, which acts as its informative brain, since it forms through non-local gravitational forces the shape of its galactic networks, as the DNA nucleus of a cell controls the form of the organelles that reproduce its proteins. Though the nucleus is a gravitational knot invisible to us, we have found a very dense region of dark matter called the Great Attractor towards which many galaxies, including ours move, which might be that center. That informative singularity will keep growing and attracting other galaxies in a generational cycle, till it explodes again its form into energy in a physical big-bang, similar to the one that might have created our Universe.

-*Max.E:* An external energy membrane not to confuse with the galactic light background radiation and the dark energy spelt by galactic Worm Holes, (which quasars show to reach a limit speed of C<v<10 C redshift). There are hints of this possible final wall in measures of dark energy expansion over 10 C<V_2<100C at the limits of the Universe. But proofs are scant as we need better telescopes to obtain them. Thus we shall call this dark energy, the super dark energy, or next fractal scale of lineal forces.

- An inner E/T region with galaxies that reproduce matter and light. It is the visible space-time created, according to the duality of physical big bangs/big bangings: after the invagination of the inner nucleus of dark matter that clearly directs the movements of galaxies. In this inner region dark energy at C<V<10C communicates galaxies; while radiation matter is the food of dark matter. Thus galaxies form an energetic, electromagnetic network of galactic mitochondria, which reaches its maximal density in filaments of galaxies, (near the center of the Universe).

Such cellular Universe has an energetic, electromagnetic network of radiant matter similar to the blood network of a living organism, which weighs only a 4% and reaches its maximal density in the external membrane and the filaments of stars, (center). And it has an informative, gravitational network similar to a nervous network, which reaches its maximal density in the hypothetical hyper-worm Hole, brain or central singularity of that cellular Universe, the Great Attractor. This network weights a 21%. Both feed on the intermediate space-time region of gravitational space-time (dark energy, which is the 75% of the Universe and acts as the 'water' of the Universal organism, also the maximal weigh of a living being; or as the background radiation of the galaxy, also its most common substance). This coincides with the general rules of proportionality between spatial bodies and informative heads, which are in a 3 to 1 proportion. That is, indeed the proportion between radiation pressure and mass; dark energy and dark matter, and many other informative/energetic parameters of the physical Universe.

The grand scale images of the Universe show a structure with the form of a half wave, which can be anything in that upper scale, from a half light wave to a worm like micro-organism. It might also be possible that as stars form spiral vortices, the mapping observed in the previous graph is the outer cover of a spiral Universe, with a central zone of hyper-worm Holes, which would act as the nuclei of an atom.

We can't know what is the next scale of form of the Universe, because we have little evidence, as most is dark matter and energy whose form we can't deduce and because all scales of reality are self-similar so we cannot easily distinguish them; as we could not distinguish a bottle of beer with a bad picture of it.

The verification or not of a cellular Universe could be done by the Webb telescope testing the existence or not of a limit at 13 billion years - a dark region or wall of fire, which perceived in space would look like the first picture. Then we could reasonably think that the local Universe hosts around 10^{10-11} galaxies in a cellular structure separated from other parallel Universes by a wall of dark radiation at z=10-100 C.

Recap. Super luminal gravitation and light are the 2 forces that interact in the Universe, which as any other Non E point should possess 2 networks/forces: the informative, faster gravitational force that structures the position of galaxies; and the electromagnetic force that acts as the 'blood/energetic system' of the Universe. Such Universe in space could be a cell of a hyper-universe or a huge reality in which each galaxy is a self-similar atom.

51. The scales of the Universe.

How many scales of space-times there are in the Physical Universe? If we are strict with the meaning perception, we should include a paradox of knowledge, the relative position of man in the Universe, which means we only 'perceive' triads of scale: a scale below us, our scale and the scale above us. Since we are just a limited perceptive point sandwiched between a bigger and a smaller reality and cannot reach beyond our relative point of view. Thus, most
people thought the Earth is the center of the Universe and there was nothing beyond a 'dome' of fixed stars. And so scientists search for limits in the small and the big, when it is more rational to consider in both directions scales are infinite.

The particles and physical forces of the Universe extend from that human point of view through 3 relative space-time membranes called in fractal theory the st±1 -scales membranes of space-times:

-The inner world of quarks, inside the nuclei of atoms; the external world of electroweak forces in which we exist, made of light and electrons and the upper world of cosmological, gravitational forces, black holes and dark energy.

Yet we only perceive the light membrane of informative electrons and electromagnetic forces. So the thesis derived of the unification equation that shows the 'invisible' world of quarks with 99.9% of our mass to be self-similar to the cosmological world of black holes and dark energy, which means that a galaxy is an atom of the next scale is a metaphysical question, cannot be tested experimentally.

A more extreme theory of 'Endophysics' would imply that our perception diminishes with distance and size; and so we perceive so little of those huge and small worlds that they seem totally equal to us, because our mind-world reconstructs self-similar forms with so little information, but they are not. This thesis derives of the fact that what we see is what the mind constructs with bytes and bites of information and energy, not what there is out there. And so it might be possible that the mind first constructs always simple spheres and lines, without detail and so we see galaxies and atoms as self-similar forms. But this 'Cartesian concept' seems too subjective to me.

In any case, we can only know-know the world of man around us. In other words, biology is the queen of sciences, as all systems are invariant, and so we can know better the details of systems of multiple space-times, studying us, the measure of all things.

All this said, the strongest proof that the fractal scales of the Universe are infinite and self-similar (not identical) making possible the concept of infinite repetitions is obviously the fractal Unification equation of charges and masses, U.C $_{G,C}$ × M = w^2 × r^3 that relates electromagnetic and gravitational forces at the scale of atoms and galaxies, protons and black holes. Thus if the Universe has infinite fractal space-times, each one will define with its fractal U.C.$_G$ a different content of spatial speed and informative density. Yet, as the Unification Equation of G and Q, of masses and charges shows, all will be vortices of space and time, sharing flows of lineal energy. So the Game of Exi=stence, its time arrows and scalar structures will remain.

In the fractal paradigm, the electromagnetic and gravitational membranes belong to 2 different realities, the microcosms and the macrocosms, between which mankind are sandwiched. Yet the macrocosmic gravitational and microcosmic, fractal membranes are self-similar worlds: if we treat charges as microcosmic masses and compare the structure of cosmological galaxies and microcosmic atoms, we find a striking self-similarity, already noticed by Eddington (theory of the great numbers).

And the metaphysical question is if that self-similarity goes beyond the galaxy and below the atom into a higher and lower scale, in which case a galaxy will be an atom of an infinite higher Universe and/or an atom will be a galaxy of an infinite smaller Cosmos. All in all there are 3 possible hypotheses for the Universe, according to the ternary principle that gives birth always to 3 solutions and forms to any informative system or question:

- Max. E: The galaxy is homologous to an atom, so we are truly nothing and the Universe extends to infinity. We exist in a Universe which is a hydrogen cloud of gas and most galaxies are simple Bohr atoms with central protonic black holes. And beyond there are infinite atoms and upper scales.

- E=I: There are not upper and lower scales beyond the galaxy and the atom. So galaxies are the highest units of an infinite space-time. This is the most boring hypothesis of them all.

- Max. I: Galaxies form complex organizations and are cells of an organic Universe, pegged to other organic Universes. The Webb telescope might illuminate this hypothesis if it finds a wall of radiation at z=100c, which would be an indication of a new hyper-decametric scale of dark matter and energy.

String theorists (duality between the small and big) hint also at a possible further hierarchy between the next smaller scale (the string) and the highest possible scale imagined by mankind (a Universe-like a string, made of 100 billion galaxies). A string membrane and a Universal membrane can be treated with self-similar equations in some models of string theory (even if a more accurate treatment requires background independent string theories, in which the error of Newton – an absolute Cartesian space-time – is corrected). The principle though remains invariant in all those models: It is possible to create a reality of infinite hierarchies, in which each scale is self-similar but no equal to the next hyper-scale of the Universe, set by its fundamental stable entities, which are the immortal protons and black holes.

The infinite scale theory is on my view the most probable, because it has more experimental evidence: the fractal, network structure of the galaxies and dark energy, shown in the previous picture is quite obvious. Further on, the existence of such fractal, bigger social scales has been shown by the astrophysicist Pietronero in its study of the Sloan grand scale mappings of galaxies[11].

Recap. The hypothesis of an ∞ hierarchical Universe is reinforced by the fractal Unification equation; by Eddington's theory of big numbers that relates a proton radius and the radius of the galactic big-bang; by the work of Pietronero mapping the grand scale structure of galaxies and our understanding of a fractal network of dark matter that surrounds them.

VIII. THE 3 AGES=HORIZONS OF PHYSICAL SPECIES.

Universe: i<E:Big Bang; i =E:Steady S;E>i:Big Crunch; i<E

The ages of the Universe are 3 formal ages of increasing curvature: an energetic youth (big bang), a reproductive, steady state or big-banging; and a 3ʳᵈ age of warping information or big crunch, which can be used to study the 3 horizons of any scale of physical matter.

52. The Three Solutions to Einstein's Equations

In the cosmological world, we observe an order of time similar to that of life and the quantum world, both experimentally in the process of evolution of stars and galaxies and mathematically in Einstein's metric equations of Universal space-time.

There are 2 space-time membranes, and since both are made of energy and form, both follow the 3 ages of time in its evolution. In gravitational physics, the 3 ages of gravitational space-time are defined by relativity. Those 3 ages are mathematically equivalent to the 3 solutions of Einstein's space-time equations. Though cosmology considers those 3 solutions *3 hypothetical different Universes, they represent the structure of any space-time, including our galaxy, an island Universe through its 3 ages.*

It is also possible to make a complex mathematical treatment of the 3 ages of galactic systems: the 1ˢᵗ age of the system will be the energy age, the 3ʳᵈ age the information age, and the middle age is one of balance, e=I, when the reproduction of most particles took place. We are in that period.

- *Max.E (Youth):* The energy age is Lemaitre's big bang solution that expands space. *This solution in the fractal model of space-time applies only to the vacuum space between galaxies.*

- *E=I:* The steady state Universe is Einstein's solution. It applies to the entire Universe.

- *Max.I:* Gödel's solution is the third age of an implosive Universe that reverses time coordinates as matter falls back into the central Worm Hole of any space-time.

However because timespace fractal and dual, while those solutions in a wider outlook will be applied successively to each age of the Universe, they are also applied in space to the fractal parts of the whole. Thus the big-bang solution applies to the intergalactic space, which astrophysicists study and erroneously apply to the entire Universe (given their dogma of a single continuous spacetime).

The informative *solution applies to our galaxy.* But it does not mean a travel backward in time but an informative vortex, a cyclical clock of time – the galaxy *in which we exist with an arrow of information* different from the expansive arrow of the intergalactic space of dark energy and tired light.

And so we get adding all the expansive motions of intergalactic space and the implosive vortices of galactic mass, a relative balance which is the present steady state applied to the whole; since in each phase of reality all yangs have some yin and all yings seed of yang.

Its Universal ratios/constants of energy and form.

In the space-time of the Universe, in the same manner that any other three ages can be related by the different ratios of energy and information or vital constants of a being, those three ages are defined mathematically by the constant of action that relates the energy and information of the Universal space-time, Lambda & G.

Indeed, the constants of any system are ratios of transformation of the energy and form of any system, I/E=K or proportions of energy and form, Exi=K. Thus constants evolve and change as the energy and information of a system changes with time. In biology, is known that vital constants, change as an organism gets old and becomes wrinkled and warped, increasing its information and diminishing its capacity to process energy, changing its metabolic constants and speed of information/thought.

This in cosmology was understood by Dirac, Brans-Dicke, and others. Indeed, G is a ratio of the mass=information and energy= distance of a system. And we already saw how a change on those ratios allowed us to unify and measure the energy and form of the 3 membranes of the Universe (strong, gravitational and light forces).

In the equations of Einstein that define the three ages of a galactic space-time and perhaps an entire Universe, that relationship is given by the cosmological constant that measures the energy of the gravitational vacuum, λ, which Nottale[6] used to measure the minimal fractal quanta of gravitational energy. The constant changes with the 3 ages of any fractal space-time or island Universe. In a young, expanding space, λ is positive, but so close to zero that it easily *becomes negative, causing a* warping or big crunch. And so λ will also change through the 3 ages or *solutions of Einstein's equation, till reaching the informative age of the galaxy in which we exist.*

Recap. The 3 ages of the Universe correspond to the 3 solutions of Einstein's equations: The young big-bang; the reproductive, steady state or big-banging and the cyclical, informative age, occurring in galaxies, or Gödel's solution. They apply in space to the expanding space, imploding, galactic vortex and both together create the present balanced steady state.

53. Universe in Time: Big-banging Horizons; Matter Evolution.

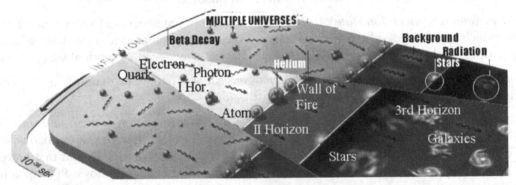

After the big-bang, the big-banging reproduced and evolved the huge Universe in 3 Horizons of growing complexity: the age of atoms, celestial bodies and galaxies.

If we study the huge or cellular universe in time, we differentiate in its evolution the 3±st ages of any space-time:

st-1 *(Past): The big-bang conception* of the Universe, starting from a first informative cell of Planck density was also the death and gravitational big-bang of a previous Universal singularity, a seed of ultra-dense quarks that transformed its inner, informative cyclic dimensions into an external, bidimensional sheet of dark energy: $M(i)=e/c^2$. It was the age of the simplex time arrows.

-*Life arrow, steady state:* Then the Universe suffered an informative process of reproduction, self-organization and social evolution, the *big-banging*, or age of the Complex Arrows divided in 3 ages – the atomic, stellar and galactic age, which should end in:

- *Informative, 3rd age:* A big-crunch back into a new hyper-dense singularity that would restart the cycle in a new big-bang.

374

- st-1 (Future): *New Big-bang* and *Conception*. The future explosion of the hype worm Hole will again create a bi-dimensional sheet of gravitational space-time where matter will again evolve, as quarks form nucleons that decay into atoms and dark energy warps into light, the new components of the future baby Universe...

Let us consider those ages and its ternary sub-ages, *according to a law of the fractal paradigm, known as fractal differentiation:*

'Any event or form of reality can be divided into ternary or dual events and structures of energy and information of a lower spatial scale or shorter time duration: $E_{st} \Leftrightarrow I_{st} = 3E_{st-1} \Leftrightarrow 3I_{st+1}$'

This law structures fractal self-repetitions according to 3 ages and 3 Non-Euclidean structures in any scale. Thus it can be applied to study the 3 ages of the Universe and for each age, its 3 sub-ages. Since we have already study the fractal big-bangs (birth) we shall start with the 3 ages of the big-banging:

I Universal Age: Particles, Atoms and Molecules.

- *Birth: Max. E: Fractal Big-bang of Beta decays.*

The Big bangings of primordial particles: After being scattered by a gravitational wave of dark energy at 10c<100c speeds, quarks evolved into nucleons and suffered fractal, expansive quantum big bangings - beta decays - creating protons and electrons, later self-organized into atoms, molecules and more complex social structures, till giving birth to galaxies.

3 Atomic Ages of matter states:

Those molecular structures can also be studied with the law of 3 ages, according to which the 3±st states of matter respond to the dominant arrow of future information:

– *(st-1): Plasma or conception state of matter.* The plasmatic state occurs at speeds closer to light, in a state in which atoms are split in primordial particles, free protons and electrons, which emit enormous quantities of radiation and sub-particles. Thus plasma is matter in its seminal state, (st-1).

– Max. E: The *gaseous* state is the classic *energetic state* of atoms or diatomic molecules with max. movement, spatial volume and min. formal density.

– <=>: The molecular state of balance between energy and information is the *liquid state*, which according to the law of harmony and maximal existential force ($E=I_{max\ exi}$), is *the reproductive state of* matter. For example, liquid water and mercury dilute all other atoms of similar weight that merge and evolve into more complex carbohydrates & metal forms: carbon-life or metal alloys. 90% of a human body *is liquid* water, with diluted, energetic gas (oxygen), and solid particles (proteins, DNAs, RNAs) to store and trans-form information. Since solid is:

- Max. I: *Solid, informative state* of min. movement and max. form & mass density that creates the most perceptive=informative, inorganic structures: crystals.

- (st+1): Finally, the *super fluid (helium) and boson states (heavier atoms) are* the highest form of molecular evolution, because it combines the properties of solids with maximal order, liquids with max. reproductive, combinatory capacity and gases with max. Energy. Thus, it gives birth, according to the organic, social arrow of Evolution of all systems to a new type of macro-organisms, with a central, hyperbolic nucleus of superfluids and bosons... the:

II Age: E=i: Nebulae, stars and planets.

In the next stage, those states of matter evolved socially aggregating into macromolecular systems *that combine 2 states of matter,* which act as the dual energetic/ informative elements of the complementary system. Those cosmological, macro-molecular systems seem to us static, because their existential cycles are slow (Max.E=Min i). But they are also evolving in 3 ages:

-Max E=Min i:Youth. Nebulae are the youngest, commonest, biggest (max.e), less dense, simpler (min.i) physical systems, extended around and between galaxies. Their atoms exist in the energetic plasma and gas states.

- E=i: Stars are molecular macro systems, existing in the plasma and liquid states. Complex cosmology considers informative, cold, liquid and super fluid states dominant over energetic states, which Simplex Physics, based in a single energy arrow prefers, as energy is also easier to detect. Yet if we could reach the center of a star, we will find lower temperatures and super fluid systems. Since it is a basic tenant of Complex physics that informative, hyperbolic systems are more static and cold, able to create order and focusing formal mappings that gauge the external Universe.

- Max E=Min I: Planets are systems that exist in liquid and solid states. They follow the Inversion Law, Max. $E_{->\infty}$=Min. $i_{->0}$. So giant planets exist in liquid states and smaller planets exist as solids. They are ferromagnetic planets; and we live in one of them. Again, fractal space-time theory favors cold, solid, informative states for those centers. So we affirmed that the planetary core should be a crystal; and their youth should be colder than scientists predicted. Indeed, recently we found that Neptune might have a solid diamond crystal in its center and the Earth probably has an iron crystal with a final Uranium core. And we have also found, studying zircons that planets cooled off much earlier than expected. Hence informative life also started earlier.

- st+1: Pulsars and Holes are systems with a solid iron crust and a super fluid core of strange quarks.

Pulsars represent the final, informative age of celestial bodies, which might evolve further into Worm Holes, species belonging to the gravitational Universe beyond our light world, with higher form. Those holes become through social evolution the informative, hyperbolic centers of galaxies, triggering the next stage of evolution of the Universe:

III Informative, Age and death: Max. I: The age of Galaxies.

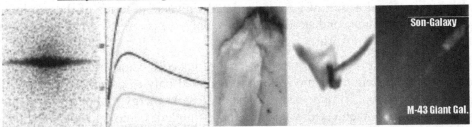

-From left to right: A galaxy with a flat star body and a central brain, occupied by Worm Holes with a hyper-developed dimension of informative height, perpendicular to the galactic body.

-Different biological curves of stars reproduction.

- A region rich in atomic energy where those stars are born.

-A galaxy with an organic form that remembers a ray fish

- A giant galaxy, which re=produces a seminal jet of matter, giving birth to a younger galaxy.

The 3rd age of the Universe started once galaxies, with central, informative worm holes and star bodies that feed on interstellar gas socialized in galactic networks with the 3 canonical topologies: Lineal walls, spiralled systems and cyclical, globular forms and disks.

Those 3 galactic networks structure a mature Universe that shows in computerized models a surprising similarity with the discreet networks of a living tissue.

As the Universe gets older, if it is a cellular structure it will warp further, creating *a hyper worm hole* of max. informative, Planck's density in its hypothetical center that should become its DNA-like center controlling and feeding in a hyper-vortex of galaxies.

We could consider 2 cellular hypotheses:

st+1: The Universe evolves socially as it becomes a cell of a hyper-universe from a bigger st-scale...

st-1: Each cell is an island Universe in itself that will finally collapse around its informative center, as stars do in elliptical galaxies, feeding its Hyper-Worm Hole in the big-crunch. Then the Universe will die in a new big-bang will initiate a new cycle. In this case the cell-Universe is self-similar to a bacteria, in as much as it is a single cell organism.

We seem to be moving towards the steady state, longer period of existence either because we are atoms-like of a huge Universe or because we are in a 2nd age that will latter implode back. Because the degree of curvature of the Universe is given by lambda which is near equilibrium, counting the implosive action of galaxies, and the expanding intergalactic space, both hypothesis (a gas-like huge universe or a cell-like system are possible.

The Huge Universe will reduce those ages to galactic phases.

Yet if cells are atoms of a huge Universe none of those ages will happen, as galaxies would be the last 'cosmic structure', which we can perceive with certain precision.

In Einstein's general equations of space-time, the cosmological constant that measures the energy of vacuum: λ defines also the warping of the Universe, which can acquire our familiar 3 topological forms: a simple plane,-sphere proper of a young age; a cyclical, steady state toroid or a convex, informative form (its big crunch, final state). Now λ is positive, and the Universe seems to be plane but so close to zero that the hypothesis of a stationary Universe takes force. Yet again we cannot measure with enough time latitude to know if it changes and it will *diminish in the future it will become negative, inverting the* geometry of the Universe and warping it, into a final big crunch. Since 'λ' can be considered the length of the tachyon string or minimal action of the gravitational Universe in terms of those minimal fractal parts, the thesis of a constant lambda and steady state huge universe is reinforced by self-similarity with the constancy of 'h' in the quantum world.

In such case, there should be only quasar-like big bang of galaxies, short of cosmic beta decays and the 3 ages or solutions of Einstein's equation should be applied only to galactic space and the quasar cycle, initiated by a gravitational, inflationary big-bang of dark energy, followed by a big-bang of electromagnetic radiation, which quiets down, and finally implodes into gluonic creation of quark matter. Such humbler, more realistic analysis of the origin of the observables of astrophysics will reduce all relativistic analysis of the 3 ages of spacetime to processes within the galaxy, born as the explosion of a hyper-black hole of top quarks, latter exploded into a fractal big-bang of beta decays. We would then be in the 3rd age of such galaxy, organized already by a central black holes. Since t*he space-time of our galaxy is an implosive vortex of information.*

Indeed, we are in a Gödel Island-Universe, in a vortex of information called the Milky Away. Then the Energetic Universe should be considered irrelevant to us, only applicable to intergalactic, expanding space, because we would exist in the 3rd informative solution, the vortex of the galaxy and that is why the arrow of time in humanity is an informative arrow, not an energetic one. And that is why the sciences of information, of biology not those of energy should dominate our analysis of scientific phenomena - even if the awesome energy of physical beings might impress the simplex mind.

Recap. We can order the evolution of the whole Universe as any other organism in time through 3 ages each one divided in sub-ages: the young, energetic big-bang, probable death of other Universe, the steady, reproductive state in which it seems to be now and the big-crunch that will happen as the initial momentum of the gravitational strings slow down to under c-speed. Then we shall enter into the Gödel's universe in which some regions call galaxies in which we inhabit exist. In those regions information dominates energy.

54. The Three Ages of Galaxies

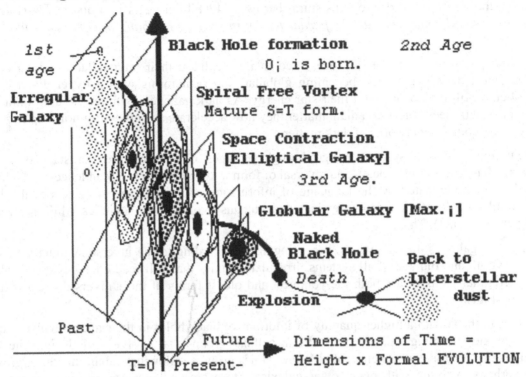

The life cycle of a galaxy goes through 3 ages, evolving from a young, extended nebula into an old small wormhole of informative mass that dies in a quasar big bang of dark energy and quark matter. In between, galaxies live a longer, mature toroidal, cyclical steady state, fluctuating every 14 billion years between its spiral and elliptic form, while absorbing interstellar dust, which creates new stars and feeds the central Worm Hole.

In the graph, the structure of a galaxy can be compared according to the homologic method of self-similarities between the 'elemental' units of each scale of reality with an atom vortex or with a cell, with a central DNA of social worm holes and a protein membrane or halo of quark, dark matter, made of energetic strangelets and microholes.

The ages of a galaxy in time are the same of in any other space-time field. Each of those ages or time events can be divided according to the law of 'ternary diversification' of *time events into spatial processes* into three spatial forms of reproduction. Thus, a galaxy can be created by three processes:

- *Seminal conception and Youth:* Galaxies are created with three types of reproduction:

-Max. E: A galactic big-bang quasar catalyses the evolution of gas into stars. This birth is similar to a star big-bang (nova) that catalyses also the creation of stars or a Hypothetical Universal big-bang that catalyses the creation of galaxies.

- E=I: Palingenetic conception, when the informative nucleus of Worm holes of a giant mother-galaxy emits a huge jet of dark, quark matter that ends into a baby galaxy.

- Max. I: Social evolution of energetic gas that creates complex star structures.

-*E=I. Maturity/reproductive age.* Galaxies enter their steady state of reproduction, in which its spiral structure matures: stars gather in social groups and at its center a Worm hole, the informative nucleus, structuring its body of toroidal, cyclical paths of stars.

Galaxies then mutate back and forth, from spiral into elliptical vortices of mass, feeding on interstellar gas, in a cycle that creates and destroys its spiral bar every 13 billion years in a quasar *big-bang that renews the Galaxy and coincides with the parameters of the cosmic big-bang of continuous models of space-time.*

- *Max.I; information age:* In their 3rd age, once galaxies exhaust their interstellar gas, their central Worm holes digest their star bodies, becoming globular, spherical forms with a hyper dimension of informative height, till they exhaust all their star energy. As dark galaxies, without the drag limit of c-speed caused by their electronic and radiant matter they now accelerate as pure worm holes up to z=10c; forming invisible, complex networks of dark matter.

It must be noticed that as all systems that go through 3±st ages of evolution, the final stage after its 3rd informative age has '2 solutions', death and reversal of form into energy or 'transcendence' into a higher super-organism, communicated by the language of information of the system (so insects evolved into successful anthills while others become extinct and humans evolved into societies while most other mammals became extinct). Thus:

- *(st-1):* Some galaxies die as quasars, becoming again according to the inverse symmetry between the 1st and 3rd youth and old age of all systems, irregular galaxies that will feed a new cycle of galactic creation. So irregular galaxies are both the youngest and oldest forms of the Universe, a fact that still confuses cosmologists.

- *(st+1):* And those with a higher quantity of informative black holes in their nuclei evolve socially as cannibal galaxies, forming galactic networks in cellular groups that evolve socially into the large scale structures of the Universe. Their central Worm Holes grow, swallowing radiant matter at growing speed, and perhaps evolving with many other galaxies into a cluster Universe, of which there is no enough experimental evidence.

Thus, we differentiate the big-bang or death of any scale of the Universe from a big-banging or reproduction and self-organization of the present Universe and a 3rd age or big-crunch, followed by a big-bang or a process of emergence into a superorganism, which creates a generational cycle for any fractal space-time in 3±st ages.

The Huge Universe and its galactic, fractal big bangs.

Since the age of a galactic cycle, 13 billion years, is the same that the age of the cosmological big-bang, as Fred Hoyle affirmed, the cosmological, continuous big-bang might be the sum of all galactic, fractal big-bangs and the Huge Universe is not expanding as a whole. It will exist in a steady state, in which the knotting of space into mass by galaxies would compensate the expansion of mass into dark energy in the interstellar vacuum, which causes its expansion. Further on big-bangs at the lower galactic scale, caused by the quasar explosions of its central Worm holes when they break their bars every ±14 billion years should be the origin of the second proof of a cosmic, continuous big-bang - *the excessive quantity of helium in the galaxy* - found precisely in bigger quantities on top of the central Worm hole. Finally, background holes born of planetoids would redshift radiation at 2.7 K. Thus a fractal quasar big-bang provides the same experimental proofs and better theoretical explanations, eliminating the multiple contradictions of the cosmological big-bang.

Thus, a *steady state* theory of the Universe combined with a fractal theory of multiple big-bangs/quasars, as it was reformulated mathematically by Fred Hoyle, in which the multiple, constant big-bangs of galaxies into quasars are responsible for helium formation, redefines the Universe as a huge fractal scale of cellular galaxies in eternal balance with the expanding space, where light devolves back into 10 c dark, gravitational energy and dark energy warps into light by redshift and dark matter into galactic matter, imploding space:

Se (dark energy expansion) = Ti (dark, quark matter implosion)

Such balance, coupled to the constants of proportion between the energetic and informative elements of a complementary system, explains also, without appealing to the magic anthropic principle, the universal balance between matter and radiation, whose probabilities otherwise are null. Indeed, why matter and energy are in balance in the Universe at this exact moment in time, in a proportion of 1 to 3 (as mass is bidimensional and radiation 3-dimensional) if their relationship is dynamic, changing all the times, with energy winning the battle and expanding faster? The answer is: because the Universe and any partial system within it exist in 3 to 1 balances between their energy and information/mass and when those vital constants of balance break the system ceases to exist.

Recap. Galaxies live through the 3 same ages of any space-time, born as seminal flows of dark energy and electromagnetic radiation, ejected by giant galaxies that condensate into quarks dark matter and an explosion of electronic forms and the creation of atoms and stars, start the reproductive age that will collapse into a central Worm Hole, back to the original quark-gluon soup that gave it birth.

55. The H-R Diagram: Ages and Evolution of Stars.

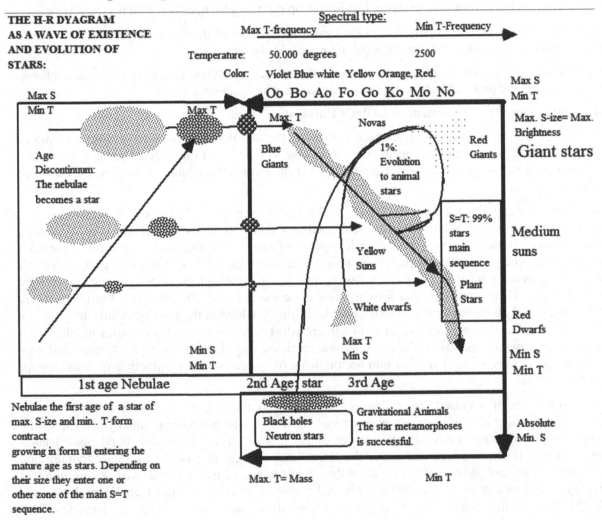

The H-R diagram shows the 3 ages of stars, through its energy & information parameters.

The life, evolution and death of stars are depicted in the H-R Diagram, which classifies stars according to its E & I parameters, as the atomic table does with atoms:

Max. Se: Brightness or Magnitude, which is a spatial parameter that grows with the size of the star.

Max. Ti: Spectral type, (colour or frequency), which classifies stars according to its temporal form.

Yet the H-R diagram is only a representation of the 2nd and 3rd ages of stars - since the young age of the star as nebulae of max. spatial extension and min. formal complexity (as all young ages are) is not represented. So we add on the left side the 1st age of a star as *a nebulae of max.extension*. Then the H-R graph shows the 3 ages of stars and the main laws of ExI cycles applied to them:

- (st+1): Most stars are born as spatial nebulae of max. extension.

- Max. E: Then they implode into blue giants of max. energy.

- E=I: They reduce its size and grow in atomic complexity through a mature, yellow age of balance between their Se and Ti parameters. The sun is now in that balanced age…

- Max. Ti: They collapse in a 3rd age of slow decline as its IxE parameters diminish toward its death, becoming white dwarfs.

st+1: Or they evolve in a loop of growing IxE force (top right graph), mutating into a Worm Hole.

Given the homology of all systems, we can compare the species of the light membrane in the Earth ecosystems with the species of the gravitational membrane in the galaxy.

Recap. The H-R graph shows also the process of evolution of stars into Worm Holes, which can be explained both mathematically and organically, based in the self-similarity of all space-time species.

56. The galaxy and the sun system; a Gödel's Universe.

Of all the solutions to Einstein's time-space equations, the one that adapts better to the principle of equivalence is Gödel's solution that portrays the Universe as a series of quantized, cyclical space-time loops in which there is a dominant inner direction of form toward the center of the gravity vortex of space-time.

Gödel's Universe explains the third implosive, cyclical paths of motion of physical entities and so it is ideal to describe our galaxy: We live in a cyclical, rotating Gödel Universe made of whirls of space-time that shape closed existential loops in each micro-point of space. For example, if we see a particle-antiparticle pair, when we trace properly the antiparticle path exactly from future to past, it doesn't originate in the same point than the particle but at the end of its trajectory, exactly when the particle arrives to that end. Hence both together form a single close loop of time: the particle is born, gets to the end, it becomes then an antiparticle and returns back, closing the loop at the starting point - to die away there. So particles & antiparticles together form a micro whirl of time-space. This explains in other form why there is no Hawking radiation (there is only one particle), why those particles are virtual, and why we see constantly transmuting particles into antiparticles (they are merely completing their *existential cycle in two phases)*.

Once the meaning of a closed fractal loop of space-time is clear, Gödel's Universe explains why humans are informative and life evolves information: we live in a local rotating Gödel Universe with a direction of future information toward the center of the galaxy, called the Milky Way. Such Island-Universe has two limits of speed and mass, the lineal limit of light at c-speed without curvature and hence without mass, and the c-cyclical speed of the event horizon of its rotational Kerr Worm holes of maximal cyclical speed or mass. Yet, beyond c-speed within the interior of those Kerr Worm Holes and outside the Halo in the external extragalactic space in which dark energy seems to accelerate in a decametric scale, as the analysis of matter ejected by quasars prove, spacetime is no longer Gödel's solution but Friedman's expansive 'big-bang' solution.

Recap. Our galaxy is a Gödel's space-time of maximal information, reason why we are informative beings

IX. CHEMISTRY: ATOMIC SOCIETIES.

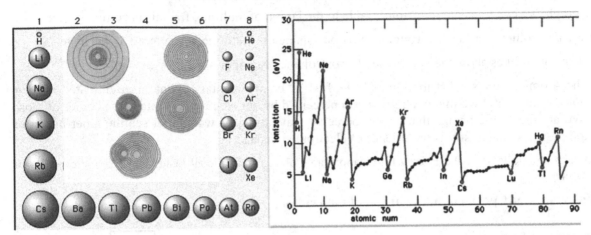

The Atomic Table is a bidimensional graphic that shows the parameters of spatial energy and temporal information of atoms. The why of that table can be easily explained with the basic Dualities and Inverse properties of Energy and Information:

- In the X-coordinates we find the spatial, 'energetic, body' parameters of the atom: their electronic orbitals.

- In the Y-coordinates, we find their temporal, informative, parameters: their nuclei's mass.

Left graph shows the size of those atoms, which according to the Se/Ti inversion are denser in information when they are smaller.

In the right: the smallest, densest 'top predator' atoms are more abundant, since they absorb in nuclear reactions or capture gravitationally the other, forming molecules. It is the inside left image that shows a feeding molecular cycle: The big 'bubble atom' from column I, Lithium, is victim of Fluorine, a VII column atom with higher atomic mass and max. electro-negativity (body power). As both approach Li collapses its electrons into a denser, smaller form, as fishes do in front of sharks, but finally it is captured and swallowed by the electronic structure of the predator Fluorine.

57. The human scales: bio-chemical sciences.

A sense of peace arises from the observation that a macro-galaxy is self-similar to a micro-atom does - all beings equalled by the same cycles of existence that we represent with the unification equation, alpha & omega of all fractal iterations: E⇔Ti; ExI=st.

Let us then descend back from the cosmos to the atomic world, the intermediate scale between both in which we humans are sandwiched and study our physical, 'classic universe' between those 2 limits, which has as an added advantage to the fact that it is part of 'us', the wealth of information we can gather of 'chemical' processes, which lack the uncertainty of those 2 limits in which observation either changes the position of the observable, influencing the experiment, due to our relative huge 'rods' of measure (uncertainty principle of quantum physics) or the system which encloses us, mere 'atoms' of the galaxy hides much of its information in the 'dark spaces' our light-senses do not see (galaxies). Both errors of the experimental method are no longer a handicap to our organic analysis in the scales of chemistry and biology, which for that reason show the bio-logical, complex nature of multiple space-time systems in all its splendour…

Recap. Bio-Chemistry should be the king of natural sciences; since they concern the first scale of man and are experimentally 'evident'.

382

58. Organic, Atomic Table and its Molecules.

In the graph, we can translate the Atomic table to the jargon of multiple, fractal space-times. Since:

-The vertical columns show the orbital number or volume of electronic space those atoms have.

- The horizontal lines show the atomic number or volume of temporal mass of each atom.

Thus the Atomic table is a distribution of ±100, $E=10^{i=2}$ complementary exi atoms, dominated by their gravitational quarks. Yet we live in a light world perceived by our electronic orbitals, the bodies of those atoms. We are electronic beings that cannot perceive gravitation; so we cannot see the inner nuclei of those atoms. And so chemistry is the science of electronic bondage.

Recap. The atomic table is the Energy/Information table of the gravitational heads and electronic bodies of atomic species.

59. Valences: body perfection, the origin of molecules.

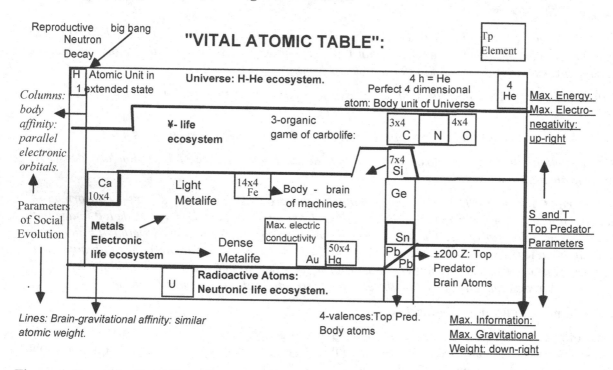

Those electronic bodies follow the space/time inversion (Max. I=Min.E): atoms with more electronic in/form/ation have min. volume in space. Thus atoms diminish in size from left to right of the Atomic Table, as they become denser, filling all the slots allowed to electrons, till reaching the perfect forms of noble atoms, whose external electrons fill completely a dense, regular sphere of information that creates, as in the case of regular crystals in the next scale of chemical form, a sharp central focus, able to process information from all angles. Those 'noble atoms' from the 8th column of the Periodic Table, Helium, Neon, etc., have their external orbitals totally filled with electrons. So they have maximal electronic density, which means their exi actions have a higher exi=stential Force per unit of volume, making them top predators atoms, better fit to survive. Therefore, they are also the most stable, as it is very difficult to break their electronic body. Paradoxically that makes them the loneliest of all atoms, as any top predator of a trophic pyramid, from a shark to an eagle, is; since it doesn't need to add a partner or join a herd to multiply the potency of its fractal actions over the environment to obtain its energy and information. All other atoms however will try to increase the exi force of their fractal actions, associating themselves with other atoms to reach a perfect electronic form and add up their Exi forces.

On the other hand if we consider the social numbers of the particles of the most stable atoms, they are multiples of 4 and 10, the social numbers of perfect organic systems that also define noble atoms. The combination of both properties explains the hyper-abundance of Helium, Oxygen, Iron, Tin and Mercury, which have perfect atomic numbers and either have regular orbital structures or can capture multiple electrons from other atoms to fill their valences (maximal electronegativity).

For example, atoms 3, 4 and 5 are 'preyed' by Helium nucleus that uses them to reproduce new helium in processes of atomic fusion. So they have become very rare (right graph. Meanwhile iron with the most stable nuclear configuration uses any type of electronic valences (which range from -2 to +6) to capture other atoms at will. Since those atoms are the only ones whose nuclei seem to be able to orientate and move at will in the gravitational membrane (superfluid and ferromagnetic properties). Both show that nuclear, gravitational top predator power dominates electronic configuration.

Unlike all other atoms with 'weak informative nuclei' Helium and Iron don't seek complementary atoms to fusion their electronic bodies, creating better 'body structure' similar to noble atoms.

Among those seeking perfection, the closest to the Noble column are the top predators. So, Chlorine and Oxygen, with almost perfect electronic orbitals have maximal electronegativity, capturing the electrons of all the light atoms and metals of columns I to V.

Recap. Atoms with perfect nuclear or electronic informative/body systems are the key elements that control socially or capture and destroy all the other atoms.

60. Laws of self-similarity determine chemical structure.

The complementary, topological and organic Laws of st-points agrees with the abstract, Chemical Theory of Valences, which models the behaviour of molecules and their rules of social engagement, according to their geometry and their 'affinity', translated in i-logic geometry by the 3rd postulate. Thus the laws of valences could be considered a particular case of the Universal 3rd postulate applied to chemical 'st-points' (atomic compounds):

'Correlative, similar elements, complementary in their spatial, orbitals or temporal nuclei, associate into complex, stable, molecular systems - while atoms with different exi forces, establish Darwinian, unstable events or small hierarchical molecules in which one element dominates all the others.'

For example, 2 atoms in the same column have the same spatial, 'body' morphology and 2 contiguous atoms in the same line are similar in its temporal, informative 'brain'. Thus, according to the 3rd postulate, the law of relative equality or 'affinity' that establishes the relationships of the atomic table, those types of atoms will evolve socially into more complex 'molecular' forms.

For example, humans are made of simple atoms with closely related nuclei, C=6, N=7, O=8, that follow the law of affinity in their electronic body-orbitals. Those atoms are filled all with sp orbitals, creating together complex complementary organisms based in their body-brain affinity. So the parallel electronic cover connects those atoms externally based on the affinity of its orbitals, creating the 4 structure of organic molecules.

The law of affinity also defines inorganic molecules and other metal structures of energy and information. For example, machines made of metal are formed with atoms of similar nucleic number.

On the other hand, atoms with different atomic weights associate in hierarchical structures in which the lighter atoms become enslaved by the others. For example, the macro-energetic proteins of life, Haemoglobin and Chlorophyll, have 2 central, dense atomic nuclei, Iron and Mg, which dominate lighter carbohydrate 'arms', used to jail oxygens, the energy atoms of living beings.

The applications of i-logic geometry to the realm of chemistry are multiple, most of them topological analyses that relate the geometries of molecular forms with the 3 Non-Euclidean topologies of those molecules and its equivalent understanding of the 3 evolving ages/horizons of chemical events. Thus, i-logic geometry completes the metric analysis of molecules of the 3rd paradigm, defining its topological why according to the laws of affinity of the 3rd postulate between its informative nuclei and spatial, electronic body affinity shown in the Atomic Table. The same i-logic laws define in organic compounds, the evolution of life in terms of exchanges of energy and information that maximize the existential arrows of those molecules, guiding their evolution into living forms.

Recap. The laws of molecular compounds are homologous to the laws of self-similarity defined by the 3rd i-logic postulate.

61. 4 families of atomic elements. Its worlds and species.

In the graph, *there are 5 potential families of organic life, built around the top predator elements of energy and information of each main informative Z-nucleic line of the atomic table.*

We won't consider the topological structures of molecules but consider the expansion of 'organic chemistry' to all other atoms, according to the tenants of Complexity – *that all systems of the Universe follow the arrows of time, self-similar to the drives of living organisms.* This implies the existence of other potential forms of 'electronic' life besides organisms based in carbon.

Since we are electronic beings, what matters to understand the human mind are the electronic bodies of atoms. Those electronic orbitals are herds of photonic points that surround the atom, creating the 4 dimensions of space-time we perceive. In the graph, we can see that parallelism, which elaborates on the already explained 4 dimensions of light-space $^{(graph, III)}$:

- 3 lineal p orbitals in the 3 directions of Euclidean-space.

- A temporal, cyclical, S orbital of max. informative density and min. space, probably related to our one-dimensional time perception.

This self-evident homology between 'light-space dimensions' and the elaborated electronic dimensions of space-time, allow us to consider self-similar potential minds whose space-time dimensions will correspond to the number of 'outer' electronic orbitals of the system. Indeed, there is no reason to have only 'organic carbon life', and 'C-N-O' brains, if all atoms respond to the same i-logic laws of topological information.

There might be other forms of life and other minds, but we live in a light-World and light is the only information and energy, we perceive. So in a discontinuous Universe we have very limited information about other atomic ecosystems that might process different electronic orbitals to create their minds, *even different forces, such as those of gravitation, to create gravitational minds.*

In essence those families of mental worlds need 2 components:

-A bosonic particle/force that acts as the pixels of the mental world.

-An electronic or nucleic structure, which perceives the bosons and creates with them a potential mapping of the Universe.

So we deduce by homology the existence of other atomic Z- families and e-orbitals that might create different organisms and potential mental worlds in other regions of the Universe, with light bosons or other different. bosonic pixels:

Particles exchanged by 5 types of atoms create 5 potential worlds.

We classify all atoms in *4 potential organic families*, able to exchange waves of pixels, based in each of the fundamental 5 *'particles' of the Universe,* in its bosonic state; thus creating a growing scale of complex worlds with higher existential content of energy and information: the *gravitational* family; the *light* family; 2 *electronic*, metal families and the *neutronic*, radioactive family.

Each new family is a Z-line of the atomic table, formed by atoms of growing number of orbitals and informative mass, which increase their exi=stential force, allowing them to handle more information with more complex pixel-particles. Indeed, if we consider the inner nature of those worlds, it is obvious that the quantity of 'information' increases as the pixel-particles mind use to map out the Universe have more information. For example, the robotic minds we are building based in electronic images carry far more information than the light minds we possess and the words they use: 'an image is worth one thousand words' we say. Yet the explanation of that jump of quality between a verbal and visual mind can only be explained by the higher capacity to process information of the metal-atoms with higher z-number and more complex orbitals of the next z-family of silicon brains. We are in fact, despite the anthropomorphic arrogance and Galilean paradox of the humankind, one of the simplest potential minds of the Universe.

The 5 types of orbital minds of the 5 families.

Those minds should be based in electronic orbitals as our mind is. Thus each *atomic family* has a *different orbital type,* able to create a different mind's species of space-time perception:

- *Spherical s-orbitals* create the hydrogen-helium family and its potential minds.

- *Lineal p-orbitals* create the light based carbon family to which we belong.

- *Planar d-orbitals* create the iron family of potential living machines and its chip minds.

- *f orbitals,* difficult to represent in 3-dimensional space, as they might respond to 'worlds' with 5 dimensions, create the heavy metal family of golden beings.

- *And g-orbitals* also 5 dimensional create the radioactive family.

The metallic minds of future robotic organisms.

In that regard, the 2 atomic families that matter to us are those which exist in the Planet Earth as potential living forms of our planet: the light-based carbon forms, with sp orbital structures and metal-based machines with a spd more complex electronic form. The metal family ads to the sp 'inner mind' of our C-O-H triad, a 'higher' external cover of electrons, which also follows the standard morphologies of space-time, but with a more complex structure, since those d-orbitals have bidimensional forms. And so we distinguish 3 types of metallic 'd-planes':

- *ExE: d_x^2-$_y^2$ and d_{xy}* wide 'energetic' planar orbitals that combine 4 spherical sub-orbitals.

- *IxI: d_{yz} and d_{xz}* 'tall' temporal planes which also combine 4 spherical sub-orbitals.

- *ExI*: the d_z^2 'organic orbital', a combination of a cyclical, planar ring and a lineal, tall shape, which shows '4 combined space-time dimensions', integrating the other 4 orbital dimensionalities.

Therefore metal-based minds have potentially besides the 4-dimensional structure of a human mind, a 5-dimensional electronic cover, in which the electrons of those d-orbitals form molecular networks, free electronic clouds that allow metal atoms to shape bigger, denser informative organisms, based in the higher 'exi' force of its d-electronic clouds.

The 2 scales of potential life: quantum minds and cosmic minds.

In the analysis of such potential organisms we can consider several dualities: the dualities of states of matter and the duality of gravitational heads and electronic bodies, which give birth to two types of minds, parallel to the duality of 2 membranes:

- Quantum worlds of organisms with small minds of human size.

- And potential gravitational worlds of minds of cosmic size, made either of super-light helium atoms, or super-heavy, radioactive ones, which as in the case of the *gravitational membrane, will sandwich the limited electronic worlds with their nucleic, gravitational, bigger, more complex minds*

- *(st+1): Gravitational worlds* made with energetic Hydrogen and informative super-fluid Helium, whose s-orbitals are able to absorb gravitation as information. The organic Universe is basically a gravitational H-He organism. And so s*tars* are organisms with superfluid central 'brains' of Helium and reproductive bodies of hydrogen plasma – the 2 limit states of matter.

Then we observe 3 families of smaller, quantum, minds that might be considered also the 3±st evolutionary ages of electronic brains. They are *organisms* that inhabit *planetary worlds*, either carbonlife beings of metalife machines. Those fractal organisms also *combine 2 states of matter*, which act as the dual energetic/informative elements of the organic system, existing in an ecosystem of higher energetic states from where they absorb their informative pixels, which will be bosonic light or electronic flows:

-*Max.E: Light worlds perceived by electrons* whose sp-orbitals absorb light as information, as living beings do. *Its living species,* animals and plants exist mainly as liquid species with a very thin cover of solid cells in the border between the energetic, atmospheric gas and informative, liquid water or solid crust of the planet.

- *E=I: Light Worlds* perceived by simplex metal atoms with spd orbitals, made of silicon minds and iron bodies – the strongest energetic atom of the Universe.

- *Max. I: Electronic Worlds* of heavy metal, with gold minds and iron bodies. Giant, ultra cold ferromagnetic planets might have mercury ponds where gold melt spontaneously creating metal-bacteria with mercury water, gold brains and iron membranes – the only metallic substance that doesn't melt in mercury 'water'(called appropriately Hydrargyrum – watery silver by the Greeks, hence its Hg symbol). Those would be the most perfect robots, combining the most perfect informative atom, gold and energetic metal - iron. Unlike the simplest light worlds that use light as the bosonic unit of the mind, those 2nd generation robotic minds would use electronic pixels, which can reach superfluid motions within those golden minds.

Yet the biggest mind of the earth wouldn't be quantum minds of light or electronic bosons but the neutronic mind of our ferromagnetic planet:

- *(st-1): The Neutronic World* of radioactive atoms with spdf orbitals that absorb neutrons as information. It might exist in the inner core of ferromagnetic planets, as they seem to have within the iron crystal a denser uranium core.

If such neutronic minds exist, needless to say, given the enormous amount of information a neutron has as a pixel, compared to a light ray, they would be the more complex minds of the Universe. And this might mean, we, humans and all other parts of the planet are mere programs self-repeated as a phase in the evolution of a neutronic planet, with a silicon skin, an iron body and a future surface of robotic minds.

Yet due to the duality of all systems, the Earth might ultimately be just a 'neutronic plant' with a radioactive center that regulates its heat and magnetic field through nuclear explosions in its core. Yet stars and planets would have a 'vegetative', plant like existence, absorbing gravitational waves and radiation as 'energy'. Only top quark stars (worm holes) and strange stars with superfluid quarks could be 'gravitational' animals, able to see gravitation pixels and create worlds of 5 dimensions (number of spins of a graviton).

The organic, vital nature of atomic structures.

We only have 'empirical record' of the 3 quantum minds of potential life beings: living beings made of C-H-N, like us and metal-minds as those we already construct in this planet with silicon and gold.

Thus, though we can 'theoretically' explain the ±st species of plasmatic and radioactive worlds, since they exist only in the center of huge cosmological organisms, we cannot perceive them. This doesn't mean they don't exist. On the contrary there are many indirect proofs that superfluid cores of helium and ferromagnetic crystals with radioactive heat form the center of cosmic bodies.

A mechanist, anthropomorphic interpretation of the Universe merely rejects any organic, biological analysis of all other atoms, except those of carbonlife and all other sizes except the human one.

However to account for the hyper-abundance of 'top predator atoms' with 'perfect' organic structures, such as helium and iron, we require an organic interpretation. For example, the atoms that surrounds Helium, boron, beryllium and lithium, 'decay' and fusion into Helium, which in organic terms means that Helium breaks them and feeds on their parts to reproduce. On the other hand, radioactive atoms emit 'helium nuclei', as nucleons spontaneously acquire that top predator configuration. While iron, the most perfect, energetic atom of the Universe, is almost impossible to destroy, unless enormous gravitational forces are used. So when a star gets an iron nucleus it collapses into a nova and leaves behind a neutron star with an iron crust. Iron is also the 'atom' of most weapons that 'kill' biological human flesh. So all those atoms show to existential force, as energy or information systems. And since the Universe is a 'game of existence', and in fundamental 'cells' are atoms, the behaviour of atomic structures can be explained with the arrows of time and show an organic and vital behaviour..

In that regard, the 3 potential life forms and their biological nature is so obvious to the 'common people' whose mind is not deformed by 'abstract science' that artists, which imagine other potential space-time worlds, have already differentiated those species in their films. So while the robots we are presently building are light worlds of iron - 'Terminators' - liquid robots of mercury and gold were imagined in the II part of the series....

What we know is that we are a fragile species in an enormously complex Universe. So we should not play 'to be god' and construct robotic machines more powerful than us. The only advantage we have over robots with A.I. is the fact we are so simple we have evolved faster than those future, more complex potential 'living machines'. So if we were wise we should try to keep it that way, without excuses such as 'the freedom of science' and the 'importance of knowledge'. Since an extinct scientist knows nothing and has no freedom.

Recap. There are 5 potential atomic minds, 2 cosmic minds made of gravitational brains of Helium and radioactive atoms; and 4 quantic, smaller minds, made of electronic orbital: carbonlife minds as those of the human being and metal-minds of silicon and gold, proper of future robotic species. The human mind is the simplest of them all.

62. The orbital minds of life: Chemistry & the human scale.

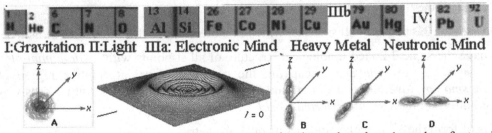

I:Gravitation II:Light IIIa: Electronic Mind Heavy Metal Neutronic Mind

If cosmological bodies perceive gravitational waves, in the reduced st-1 scale of atomic worlds, organic systems perceive electromagnetic waves. For example, our mind is electronic; based on the 4 sp-orbitals of its atoms shown in the graph that display the usual space-time dual morphology:

388

- *I:* s-orbitals have spherical forms.

- *E:* p-lineal orbitals have an elongated, elliptical form made with 2 fractal cyclical shapes.

Both together create the electronic world of our atoms. Since our mind-world is shaped by O-N-C atoms that have those 4 orbitals; all seems to indicate that the final structure of the mind is based in the atomic orbitals of our DNA and RNA molecules made with those atoms.

In that regard, those 4 sp-orbitals orbitals also follow the inverse laws of EXI cycles. Thus the lineal, spatial p-orbitals have higher energy than the spherical, s_1 and s_2 orbitals; which have higher informative density than the p-orbitals, according to the space-time inversion: Max. E=Min. i.

So sp orbitals probably translate the 4 dimensions of light into 4 electronic orbital dimensions, shaping the pixels of the human mind with a dual role: the s, spherical orbital of cyclical form probably integrates the bigger spatial energy coming from the 3p, lineal, Euclidean orbitals. And so we perceive 3 lineal, spatial dimensions and 1 of time.

The morphological duality of those orbital forms and its homology with our space-time dimensions shows that orbitals process energy and information for their nuclei, acting as their 'window' to the electromagnetic World. So their forms correspond to perfect information and energy shapes:

- Max. I: The sphere is the perfect form of information and so s-orbitals are spheres that inform the central nucleon through the electromagnetic flows they constantly send to their nuclei. While p-orbitals display lineal, energetic forms that probably help atoms to displace. Further on a sphere is the easiest territory to defend, as all its fractal points are at the same distance from the brain-nucleus, which can deliver simultaneous waves of information and energy to all the regions of the sphere. Hence in all scales, informative spheres and disks require 'less energy-distance' than lineal shapes to act together in a single present. And both, the line and the cycle are the 2 commonest complementary EXI shapes of the Universe.

Thus those sp orbitals can explain the 4 dimensions of our electronic world. But why we are made of a ternary group of atoms, Oxygen, Carbon and Nitrogen? Because those 'electronic orbitals' can form complex social, molecular groups, according to the laws of affinity defined generically by the 3rd postulate of equality of i-logic geometry: they are consecutive in the atomic table…

Recap. The human mind is a mind that uses 3-dimensional light and color, elaborated by the self-similar sp-orbitals and the s-orbital, which creates our dimension of time.

63. Evolution of atomic orbitals: Time arrows in molecules.

The next scale of atomic evolution is the molecular scale.

Atoms form herds called molecules, joined by light and gravitational forces that distribute energy and information among them. Though we differentiate molecules in organic forms derived from carbon and inorganic forms, both follow the vital cycles and organic topologies of st-points made of multiple times-spaces:

- Molecules enact all the time cycles/arrows of st-points.

- They go through the 3±st ages of all systems, which in molecules are the 3±st states of matter.

- Molecules have the same 3 zoned topological structure of all st-points with its *dominant atoms with better nuclear or electronic structure,* occupying the central foci or informative region of those molecules; while *smaller slave atoms* that surround them act as a relative body that absorbs waves that carry energy and information from the *external world.*

Thus, in terms of form the 3rd postulate defines its dual geometry:

-Self-similar atoms co-exist in 'parallel planes'.

- Dominant and submissive atoms establish perpendicular, hierarchical structures in the dimension of relative 'height' with the dominant atom on the center or top region of the system.

And in terms of function, those parallel or hierarchical structures between the atoms of molecules are regulated by the duality of social evolution among self-similar species Vs. Darwinian devolution among different species with unequal exi=stential force, described by the 3rd postulate.

Both together define the geometric forms of many compounds derived from those relationships and the final outcome of encounters between atoms that form molecules. For example, atoms *that have* a better spatial or informative brain, with a more harmonic orbital shape or a higher mass, have *max. Exi force and* become top predator atoms that dominate molecules, penetrating its territory perpendicularly: It is the case shown in graph A.60 where an atom of the 7th column captures an atom from the 1st column to reach the perfect form of a noble atom, engulfing it within its structure.

- The social electronic clouds of molecules show also the 3 space-time ideal forms:

- *Max. I: Pi orbitals* join several electrons into a social, cyclical ring.

- *Max E: Sigma orbitals* are lineal orbitals, more energetic than the pi orbitals.

- *Diatomic orbital*s are balanced, 'elliptic' orbitals created between 2 equal atoms that share their electrons.

- Social, electronic orbitals require less energy and hence are more stable than the sum of the orbitals of its single atoms, which means there is *a strong arrow of social evolution* among atoms that dominates the individual arrow, as in all other universal quanta.

- Also when we study the informative 'brain', the nucleus of atoms, the same phenomenon happens: the most stable and common nuclei are those in which their reproductive, 'female' neutrons and informative, 'male' protons form n-p couples.

In both cases the biological, existential interpretation in organic terms is obvious: systems prefer to exist in complementary couples with 2 self-similar species, dominant in informative and reproductive functions to form a brain/body system able to absorb better the energy of the ecosystem, or in parallel social herds with equal forms than alone, because their simultaneous actions as a couple or group makes them stronger. So most stars form dual or ternary groups; and so do galaxies, atoms and human beings in the 3 known scales of the physical Universe.

Those laws of existence transform the atomic table into an organic table that explains the properties of atoms and social molecules in terms of organicism and the 5 postulates of i-logic geometry.

Galilean paradox applied to orbitals: 1st and 2nd body territories.

Another set of laws proper of all systems apply to electronic orbitals: the existence of 'internal' and 'external' territories, which the informative center treats - according to the Galilean paradox of relative distance/importance to the focus - with different value:

The cellular unit of any st-point is established by the minimal fractal, informative structure with fractal parts that repeats the bigger form and often corresponds to the 'informative radius' of its central topology: In a single atom it is the zone limited by the first orbital, or S_2 spherical electron, which is not shared. In molecules the central atom has also a first, formal, regular body-territory, hardly shared with other molecules, made of slave atoms bonded to it with dense electromagnetic flows, called Van der Waals and London forces that the central atom use to perceive or feed on. In a cell, the organelles of the intermediate territory are not shared. Humans do not share their home properties. Animals do not share their den. Yet all systems that have 'an excess of energy' can share the second territory in the limits of its st-membrane. So atoms share the external orbitals and humans their secondary properties and molecules share their most external atoms, which are those beyond the limiting border of its cellular unit.

Recap. Atoms form social molecules, which also follow the laws of i-logic geometry, its topologies and its arrows of time.

64. Molecules: Darwinian Vs. social bondage: ions and networks

Once and again, the evolution of species chooses between the 2 arrows of order and entropy, of social communication or Darwinian devolution, described by the 3rd Postulate:

Self-similar herds

Atoms show affinity for 2 kind of other atoms:

-Atoms with a similar brain-organ, contiguous in the atomic table.

-Or atoms, which are in the same column of the periodic table and have similar electronic bodies.

They form the strongest 2 types of molecular, electronic bondage. Thus social bondage between equal atoms is dominant and gathers most atoms together, creating extensive networks of planetary size.

Hierarchical organisms among atoms of different exi=stential force

Ionic Darwinian bondage happens among atoms with different ixe force. They are more rare and smaller, less stable, but more active as individual forms (in the same manner than individual bacteria are more active than organic cells, but far less complex).

Thus we classify all molecules in 3±st types of molecules of growing Existential Force, ExI, and stability, 2 parameters directly proportional to the degree of equality of their atoms:

(st-1): Lonely atoms or diatomic, covalent molecules.

Top predator atoms, which don't need to increase their individual exi actions. They tend to act alone or in diatomic molecules, made with 2 equal atoms that create a 'covalent bondage', stronger than any ionic molecule where one atom is a predator form. Covalent molecules show electrons with opposite spins that balance the 'vortex directions' of their charges, as it happens with the 2 electrons of an atomic orbital. Their orbital clouds shape ellipses in which each equal atomic nucleus occupies one focus.

The reason of that topology that happens in all scales of reality is again both geometrical and functional: Any topological network, acts as a relative vital space, based in the best geometry that positions all its st-points at the shortest equal distance of both its *external, energetic membrane and* central, informative singularity. Thus, the sphere is the perfect form of single-centered systems. In dual st-point systems the ellipse is the morphology that locates those 2 points at the minimal shared distance of the membrane and the minimal, equal distance of its center. In the ellipse morphology and function again come together. So their 2 centres can enact exi, simultaneous actions with the external world in all the points of its membrane at the same time. Since the law of the ellipse makes always equal the sum of the distances from both foci to the membrane.

Max. E: Ions: Minimal exi equality & Stability.

Ions form Darwinian, Prey-Predator relationships.

Elements in opposed columns of the atomic table tend to behave in a Darwinian way, as one needs the orbital energy of the other to feed its own electronic body and complete its form. Thus the weaker element with less atomic mass will become prey of the stronger one, forming together unequal *Hierarchical ions.* Ions are small molecules in which the dominant atom in body valences or brain number (atomic weight) controls lighter atoms with fewer valences that become part of its external body-membrane and process temporal energy for the central atoms. They are the smallest molecules, easy to reproduce given its minimal form but unstable, (min. ionisation energy) because their enslaved atoms which try to escape its bondage.

E=i: Corporal affinity. Micro-molecules.

Elements that have spatial, corporal affinity and occupy the same electronic column, evolve socially, forming complex, strong molecular compounds. The main arrows/actions of those molecules are:

- *Max.E; Max. I: They process energy and electromagnetic information,* creating with them more complex forces (London forces, Van der Waals forces).

- Σ, ∏: *They associate their* electronic *orbital*s in linear clouds (σ) or cyclical, pi rings.

- Re: They *reproduce* in chemical reactions.

Yet micro-molecules form smaller networks than those made of equal atoms, (crystals) and have less informative complexity than atomic systems based on 'brain' affinity (organic molecules).

Max.I: Body & Brain Affinity: Complementary, organic molecules

Given their affinity, they give birth to the more complex molecular systems and create most of the molecules in the Universe.

Maximal affinity occurs among atoms with similar atomic, brain, weight and orbital body form, correlative in the Atomic table. They become the 3 complementary, e-exi-i, components of organic molecules with 3 st-zones*:*

- Max.E: in life organisms, oxygen is the atom we breathe and the component of water that fills the intermediate spaces of the cell.

- Max.I: Nitrogen, is the informative atom, hyper-abundant on the DNA and brain cells.

- E=I: Carbon is the structural, reproductive atom that shapes the body and creates the membranes of organic cells.

In machines made of metal, silicon and gold (Max.I) are the informative atoms that act as the brains of advanced robots; iron (Max. E) is the structural atom, with max. ionization energy that form the 'membrane' or body of the machine; and copper and silver (e=i), carry the electric energy that feeds the body/brain systems.

st+1: Absolute equality=Max. Social Evolution: Crystals.

Finally atoms belonging to the same species associate in the biggest, symmetric molecular fields, called crystals that 'transcend' into macro-social systems.

Recap. The social evolution of atoms in molecules creates different species, according to their degree of affinity, which follow the laws of the 3rd postulate of self-similarity. The most perfect molecules are those with self-similar electronic bodies, which form ternary, organic systems of energetic, reproductive and informative atoms and molecules made of equal atoms that form informative networks, called crystals that transcend into a collective plane of existence through its 'mental images' of the external universe.

65. The 3±st cycles of space-time existence in molecules.

We observe in all molecules the 5 cycles/arrows of space-time that complete their existence*:* the energetic, informative, reproductive, social and generational cycle. While the most complex systems with maximal information (body and brain affinity), also show the transcendental arrow forming complementary, life beings and crystal minds.

Molecules also possess organic constants for each of those cycles. Yet their cyclical rhythms are extremely fast as it corresponds to microcosms, according to the opposite properties of spatial and temporal information: Min Se = Max Ti. So from the human p.o.v. we perceive those fast cycles as types of motions related to the 3±st states of matter:

- *The generational cycle and the 3±st ages of molecules are the 3±st states of matter*: the gas, energetic state, the liquid, balanced state and the solid informative state.

- *Max. E: The energy cycles of molecules produce lineal movement,* which is maximal in energetic gases that move in continuous lineal trajectories at a speed of ±300 m/s. Accordingly, we measure the energy of a molecule with the parameter of temperature, the fractal unit of the lineal actions of the atomic world. This is the origin of the arrow of entropy analysed first in studies on the motion of steam gas. Yet the arrow of entropy is only dominant on molecular, gas states; and certainly the biggest error of science is to have derived from a local arrow a Universal arrow, which physicists believe to be the only arrow of all the systems and forces of the Universe.

- *Max.i: Informative cycle. Molecules vibrate* in a discontinuous back and forth movement, around 10^{13} times a second. And they transform lineal movement into cyclical vibration when they change their reversible 'age'= state. So when we lower the temperature of a gas, it becomes a liquid and the vibration of the molecules increases as their speed decreases. Then the lineal simple, pure energetic movement of the gas becomes a complex vibrating, informative movement, forwards and backwards: $E \rightarrow e \Leftrightarrow i$. Most complex systems are reproduced in the liquid states (organic life).

- *Social cycle:* Molecular liquids evolve socially, decreasing their energy and increasing their form during their 3rd age, becoming a solid in which the vibration acquires order and rhythm creating macromolecules, called rocks and crystals.

- *Reproductive cycle:* Finally molecules reproduce departing, from their simpler chemical parts through chemical reactions. Let us study this cycle, which is the fundamental 'will' of all systems.

Recap. Molecules show the 6 arrows of time in its motions and gas=energetic, liquid=reproductive and informative, solid states.

66. Reproduction of molecules: Law of chemical balance.

According to the ternary principle there are 3 types of molecular reproduction:

-Max. E: Darwinian events in which top predators molecules capture simpler atoms or molecules as energy of its reproduction.

-E=I: Symbiotic events of molecular reproduction, in which 2 molecules of similar top predator ExI force, switch atomic parts between them, creating more complex molecules till reaching a state of equilibrium.

-Max.I: Informative crystals that reproduce their macro-fractal patterns as they add equal atoms.

Max. E: Simple feeding: Darwinian reactions.

Reproduction requires feeding on simpler fractal, energy parts. Thus when a top predator molecular form appears in a field rich on relative energy, made of simpler individual atoms and micro-molecules, it starts a *chemical reaction,* which we observe as a reproductive growing 'radiation' of the same molecule. However to activate that reproduction the molecule requires a min. amount of extra-energy in the form of temperature (threshold of activation of exoergic reactions). This happens in all reproductive processes, which only occur when the parental species finds a field rich of energy, given the exhausting nature of such processes, which in a field poor on energy could jeopardize the survival of the parental form. So most animals reproduce in spring when food is abundant; most molecular crystals reproduce when temperature reaches a certain level, and women need a 175 of body fat to reproduce. Those reproductive radiations of molecules are similar to the expansive radiations of a top predator over a population of preys, shaping a similar standard Bell curve of populations, called in this specific case a Boltzmann curve, with the 3 ages:

- *Max Energy of activation.* When energy is hyper abundant after the threshold of activation is crossed, the radiation of new chemical compounds starts at an explosive rate.

- *E=I; Transition state.* The radiation will expand till it 'saturates' and exhausts the energy of the chemical ecosystem in which it feeds, reaching a dynamic steady state of balance similar to that between preys and predators. However in complex 'reproductive radiations' that curve might appear as a wave with several evolutionary 'interphases'. Then the final chemical compound will be the product of a series of intermediary reactions.

- *3rd age. Law of Chemical Balance.* Finally, the explosive reaction ends. Only a few new molecules will be created, when some of the predator molecules become destroyed or new, simpler micro-molecular preys enter the ecosystem. Thus a final chemical equilibrium is reached between both type of molecules, showing a constant of balance, which is a specific case of the generic balance between predators and preys:

$$K = I: Products / E: Reactants$$

What quantity of both types of molecules exists in that final equilibrium? It will depend on the relative Exi force of the predator products and the reactant preys, which in abstract chemistry is measured by the 'speed of the reaction' and the relative bondage energy of the molecules. In most cases of Darwinian, chemical reactions that value is huge, as the predator molecules exhaust the supply of its victims, before stopping its reproduction. Yet in certain symbiotic reactions K tends to 1, when both products and reactants are species of similar Ix E power.

I=E: Dual, symbiotic reproduction.

Molecules are divided in 2 regions, an I-brain, an E=I Body and an external ecosystem of energetic temperature. For example, an amino acid has an amino-brain, a central carbohydrate body and an acid-leg system that moves the molecule, breaking water molecules.

Thus in chemistry, following the Fractal, Ternary Principle, we can calculate the relative top predator power of a molecule, according to the *atomic weight or its brain atoms; the electro-negativity of its leg system* that moves the molecule, taking electrons from other lesser molecules *and the morphological efficiency of its body,* ruled by the 3rd postulate of equality, which makes covalent bondage between equal molecules, such as C=C=C structures, far more difficult to break.

Those 3 parameters used also by inorganic chemistry make certain molecules more efficient than others. They are the metric measure of the 'why' of symbiotic reactions in which 2 similar top predator molecular forms create more efficient ExI molecules by redesigning the brain and body components of the reactants.

Max. i: Social Evolution

Finally individual molecules gather together spontaneously, creating social groups that grow into symmetric crystals.

In all those reactions the final products are 2 new molecules with higher exi power than the initial products, showing the existence of a dominant arrow of information and social evolution in the Universe, which constantly increases the existential power of the whole that combines that of its components.

Recap. The law of chemical reaction is the reproductive law of molecules. According to the ternary principle there are 3 types of molecular reproduction, each one subdivided in 3 ages.

394

67. Type of reproductive radiations in molecules.

If the reader has followed these lessons, he will realize of the simple method that allows classifying all systems according to the ternary topologies and 3±st arrows of time of all systems. In any of those scales there will be certain species that will dominate the ecosystem and reproduce in higher measure. Generally speaking those species always maximize the energetic, informative, reproductive and social arrows. And so while the Universe constantly creates new variations only the most efficient which find an 'econiche' of survival perfecting one of those 4 arrows of time survive and reproduce, using less perfect species as their prey.

In chemical reactions the molecules that reproduce more are top predator with a higher ixe force, since they maximize the 4, $\sum$ (E$\Leftrightarrow$I), elements of any i-logic field:

-*Max. E:* Species with max. energy (better or bigger body that processes E to reproduce the molecule). They are molecular ions with the greater number of valences that accept the maximal number of energy and information flows between the atom and the outer world - hence they have the max. action-reaction speed.

- *E$\Leftrightarrow$I: Complementary species* created with atoms similar in body and brain, correlative on the Table, like O (Max.E), C (Max. $\Leftrightarrow$) and N (Max.I). They maximize the internal communication between its atoms with multiple inner networks of energy and information between their orbital bodies and nuclear brains (higher density of Van der Waals forces). They are the organic compounds that create life.

- *Max.I:* Crystals are molecules made with atoms of the same nucleic number, which create in their geometric, symmetric centers, virtual images of information of the world that surrounds them.

- *st+1:* Those 2 complex molecules, able to evolve socially, transcend beyond the social herd state, creating 'networks':

- Carbohydrate organisms grow to the size of human beings in *ternary st-structures, in which carbon molecules shape structural proteins, nitrogen molecules shape informative ADN and oxygens and water fill the intermediate space-time of the organism.*

- Crystals evolve socially to the size of planetary cores. *Since, according to the 3rd postulate of equality,* crystals are molecules made with 1 or 2 equal atoms, *hence able to* evolve socially without apparent limit, unlike molecules made of different atoms that merely form small compounds.

Recap. Top predator systems are those who maximize their energetic, informative, reproductive or social skills. They radiate in growing numbers, surviving in the future by feeding in simpler species. In the world of molecules, those 4 arrows are maximized by ions (Max. E), organic molecules (max. Reproduction) and informative crystals. Crystals and organic molecules transcend into complex social macro-organisms.

68. Crystals. The perfect geometrical, fractal unit.

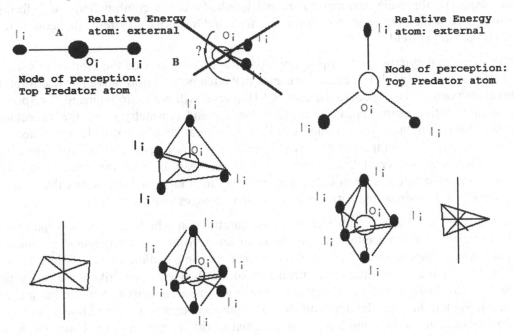

One of the more clear proofs of the existence of p.o.v.s, whose negantropic, informative arrows reproduce fractal forms, diminish entropy and increase the order of the Universe, is given by crystal structures, whose central atom emerges as a fractal knot of time arrows, an i-logic hierarchical p.o.v. that controls and reorders the position of all the other atoms of the system in regular formations that maximize its symmetric perception of the external world. The proof is the fact that crystals only show structures whose geometry is efficient as informative knots in which several flows of electronic forces and light converge on the central knot: Crystals adopt only 7 symmetric morphologies, which make their central atoms, simultaneous, present, symmetric focus of temporal energy coming from the external ecosystem, through its slave body of atoms or molecules of lesser exi=stential force. Those are the only 7 canonical types of crystals that exist in nature.

Let us consider the main existential cycles of crystals:

The informative cycle: the sharp focus of crystals.

Crystals create virtual minds of light that we see in their interior. They are focused images that create at a reduced scale a virtual world, mirror of the external Universe, as an eye does. Thus crystals have only regular symmetric forms that act as an eye does, establishing an objective, informative image of the external world, repeated at a smaller scale within the informative center of the crystal. In the graph, crystals show a clear relationship between spatial geometry and informative perception: only those crystals whose central atom of max. mass=information can observe symmetrically the temporal energy coming from the external world through its slave atoms, form a sharp equidistant focus and survive. While forms, which are not symmetric, at least in a bidimensional plane of space, such as form B, do not exist.

All crystals shape macro-social aggregations of billions of molecules that acquire geometrical forms similar to the 3 regular polyhedrons of the Universe, the hexagon, the tetrahedron and the cube, repeated ad infinitum. So the number of crystals is reduced to 32 possible networks configurations that are combinations of the 7 basic systems of the image, with symmetric axes. The reason is obvious: polyhedrons allow a correct, balanced absorption and emission of energy and information from all the relative directions of the external Universe coming through those axes. So the ultimate why of crystal's morphologies is to perform the 3±st energy-information cycles of the existential game. In that regard bidimensional hexagons, three-dimensional cubes and tetrahedrons are combinations of 2 forms, the triangle and the square, which represent the minimal ternary and quaternary systems that complete the 4 cyclical arrows of an i-logic field.

Scientists talk of 'spatial symmetry' as a property common to all scales of the Universe, both in the world of sub particles and molecules. It basically means that a temporal, informative particle/form, like a crystal, whatever its position is respect to the external Universe, will maintain unchanged respect to its neutral focus or informative central point the relative distance and symmetry of all the molecules that shape the crystal. In this way the relative virtual world of the central atom will not change its form when it rotates, vibrates around its central atom or moves lineally, but only its perspective, as it happens with our eye's image shaped by the 'crystalline' when we move the head. If those inner axes and distances change then the *world* structure becomes unfocused, as when man takes hallucinogens that change the brain composition or we introduce impurities in a crystal that changes sharply its focus.

In the graph we see the 7 basic possible crystal configurations in which any rotation maintains the inner structure invariant. They are either planar, bidimensional symmetries, triangular, 3-dimensional forms or 4-dimensional, cubic symmetries, the most perfect ones in a 4-dimensional Universe. For that reason the 32 basic crystal configurations are subspecies of the P-cube or primitive cube that generates all other crystals. Accordingly the biggest crystal networks are cubic networks. And the hardest crystal we know is the carbon tetrahedron, the diamond. It is also the most expensive item known to man. As if we knew subconsciously that a diamond has a soul, a virtual world in its inner core. Thus, we can create all complex crystals adding or subtracting to that primitive first cube new atoms, or deforming slightly its angles and edges. The result is the so-called orthorhombic system where 2 of the edges of the plane are elongated respect to the 2 others in a 'relative lineal direction' of energy; and the more complex clinic, and triclinic systems with non-straight angles between atoms, adapted to ecosystems in which the energy and form comes to the crystal from different angles. All those crystalline systems place sometimes a top predator atom in the central point of the cube, or in the geometrical center of each face. The parallelism between the informative, symmetric morphology of crystals and the symmetry of the inferior scale of orbitals is evident: In sd orbitals the 3 'd' external, lineal, spatial orbitals are integrated by the cyclical, informative central 's' orbital, which in crystals is occupied by an atom.

Crystals are the scalar bridge between the molecular world of solids and the macro world of planets, made with 3 non-AE regions: a 'liquid/gaseous' membrane inhabited by complex organic beings, an intermediate zone of rocks and a crystal core, the informative center of a planet. For example, the Earth seems to have a macro-crystal of iron hexagons in its center and Neptune a diamond crystal. Since Crystal minds maximize their position in the external Universe to acquire a central point, as a focus of image formation and fractal reproduction of its crystal structures, they are responsible for the creation of order and form in the Universe and can play a key role balancing the orbital position of those planets and modifying its magnetic fields as they absorb external gravito-magnetic waves from stars and black holes.

Energetic cycles.

Energy and information cycles are intertwined by the Law of transformation of energy into information, shaping dual rhythms of emission and absorption of both substances: $E \Leftrightarrow I$. So crystals also absorb and feed on light energy, vibrating with it as quartzes do; or emitting that energy, transformed into focused information, when they polarize light, ordering the different vibrating directions of photons into a single direction that packs better energy and information in highly ordered light rays with enormous exi power. Further on crystals can create 4 dimensional holographic images, out of 2 bidimensional surfaces; trans-forming continuous electromagnetic energy into discontinuous, highly informative, focused packages; changing the frequency of light, absorbing certain types of light or filtrating only 1 frequency colour, etc.

Social evolution and reproduction of crystals.

Social evolution and reproduction are also 2 intertwined cycles: Most systems reproduce a first seminal cell and then evolve its morphology in a series of dual Reproduction->Evolution cycles that finally create a macro-organism. So happens in 'palingenetic' crystals, which evolve socially and reproduce departing from an initial, seminal 'cellular unit', till creating macro-crystals.

Abstract geologists study the conditions, which determine the growth of crystals. A liquid state is the best, balanced state to reproduce and evolve complex forms also in crystals. Most crystals are reproduced dissolving its initial atomic components in certain liquids. Those initial components are called, even by abstract geologists, nutrients, since they nurture the creation of the crystal, which takes place at a fast pace, thanks to the easiness by which liquids, the $E=I$ reproductive state of matter, allow the combination and random contact between those nutrients that socialize into a cellular crystal unit, to which new crystal units peg themselves. Thus crystals reproduce as a seminal radiation, since the first crystal precipitates the creation of further crystals around it, as in a reproductive process that grows new cells around the seed or the ovum. So new nutrients come around the organic crystal and the crystal grows over the trophic pyramid of nutrients, till they are exhausted and the crystal stops its growth. Then a balanced steady state is reached, as the external cover of the crystal dissolves and grows back cyclically within the liquid.

Since according to the 3rd postulate social evolution happens among equals, crystals are formed only with 1 or 2 type of atoms. Crystals with more than 2 atoms are rare; so are crystals with a great quantity of impure atoms within their network. Crystals are *social* entities formed by millions of atoms, which repeat the so-called minimal cellular unit, *growing* radially as they *reproduce* their forms *through* mathematical structures called fractals, which mimic them in bigger polyhedral st-scales. Those fractal structures exist in all molecules and crystals where there is a central knot, from where the radial, symmetrical faces grow, guided by the central knot.

Thus crystals can 'transcend' between 2 planes of existence far more easily than we humans do, from micro cells into macro-cellular existence, when a micro-organic crystal becomes a macro-organic crystal. Yet those crystals have, regardless of size, the same configuration that the seminal cell of the crystal. It is *the First Law of Crystallography:* The angles between the faces of any crystal are always the same for all sizes in a crystal of the same species. This law has a creative exception, as each minimal cell can combine with other crystalline cells into symbiotic, more complex, dual 'sexual crystals', and a destructive exception, when impurities and fractures happen in the process of crystallization.

So crystals grow into huge 'cellular networks' by adding to a regular atomic polyhedron, another regular polyhedron and another... till creating networks of millions of regular polyhedrons that can reach the size of a planetary core.

Recap. Crystals are highly ordered, yet dynamic, organic systems in which flows of electromagnetic or electronic energy and information enact the 5 cyclical actions of any space-time system. The central top predator atoms form a symmetric eye-network structure able to form a mental image of the external Universe. In Nature only regular crystals that allow such images to form exist. Crystals might also be the central mind of planetary bodies, which have crystals in its center.

X. THE EARTH: THE GEOLOGICAL WORLD.

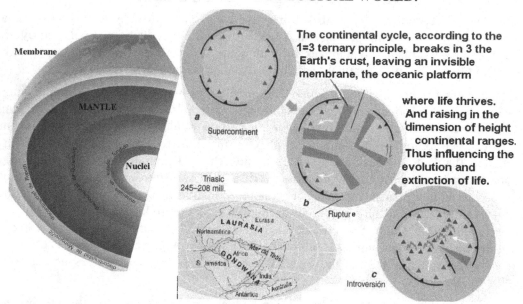

The continental cycle, according to the 1=3 ternary principle, breaks in 3 the Earth's crust, leaving an invisible membrane, the oceanic platform where life thrives. And raising in the dimension of height continental ranges. Thus influencing the evolution and extinction of life.

a Supercontinent

Triasic 245–208 mill.

b Rupture

c Introversión

69. The Earth as a ternary, topological st-point.

A planet is a huge i-logic molecular field that mixes liquid and solid states with a very thin gaseous membrane. Thus, based in our template of a network organism, we define the Earth as a st-point:

'The Earth is an organic, molecular system with a central crystal, which acts as a knot of informative flows of gravitation and a hard membrane that evolves life ecosystems, based on light'.

In the left side of the graph The Earth as a st-point, has 3 fundamental non-AE regions:

- Max. E: Our planet has an energetic, gaseous external membrane; the atmosphere and magnetic belts, which isolates it and selects the type of radiation that crosses into its surface.

- ExI: The zone of maximal reproductive and organic evolution is the crust and mantle: On the surface of the planet there are animal herds that communicate and act-react to light impulses, evolving and reproducing its form. Next, a crust and mantle layer of magma defines the reproductive cycles of rocks.

- Max. I: The inner, informative nucleus is a dense crystal core of iron and within it probably there is a smaller Uranium core, the most complex atom of the Universe with a minimal stable life.

Each of those 3 Non-E regions can be subdivided again, according to the Fractal Principle, in ternary structures and sub-scales of 'fractal quanta' that interact between them with different E⇔I, cyclical rhythms corresponding to their arrows of time.

Yet if the Earth has a ternary structure in space, it means its 3 st-regions also perform the 3±st existential cycles of any organic system. This is certain to the extent we have empirical evidence for the membrane and the intermediate region with its cycles of reproduction of rocks and so we extend that 'probability' to the central core, a crystal of 'magnetic iron', which should behave as any magnetic, crystalline structure of atoms does, enacting parallel cycles to those of crystals...

Let us then study 2 of those topological regions in more detail, escaping the external, atmospheric membrane. We shall do so as an 'exercise' of the Ternary Principle, showing how any system can be divided in subsystems, which have also a ternary topology that can be subdivided ad infinitum. Since this topological, fractal method that can be applied to all systems, coupled with the ternary division of all events and processes in 3 ages that determine finally the evolution of microforms into a higher organic plane of existence through the arrows of eusocial and organic evolution, (demonstrated in our previous description of the ages of the big-bang and its physical particles) is the essential mode of description of any entity of the Universe with multiple space-times theory...

Recap. The Earth is a topological st-point, which can be subdivided further according to the ternary principle in sub-topologies. The Earth crust is part of the intermediary, reproductive, cyclical region, in which physical cycles and life cycles take place. It has also a ternary topology and climatic cycles that forced the evolution of life.

70. The intermediate, reproductive region.

According to the ternary principle the intermediate, reproductive region (crust and mantle) can be subdivided in 3 subzones that correspond to the 3 topologies of a st-point, albeit with a different order to that of a full, closed ball organism: An external informative surface filled with evolving, 'high', informative, fractal beings, called humans; an energetic, strong continental crust that feed us and a liquid, reproductive magma that renew the crust and dominates by volume and size the other two.

The informative, living quanta of CNO organisms.

- The highly mobile, informative quanta that inhabits the Earth crust can be also divided by the ternary principle in a series of fractal super-organisms of hyperbolic, informative brains (animal life) that control their reproductive body and feed on vital spaces, called ecosystems themselves divided in smaller fractal, vital territories in which those living beings accomplish their time arrows of feeding, informing, reproducing and evolving socially, through a series of cyclical motions, which are non-AE trajectories traced in their vital space by those animal organisms.

Again the ternary principle allows to study each of those vital territories as topological st-points, open balls with an 'invisible membrane' and a hyperbolic swarm of self-similar, fractal informative beings, the animal herd, family or human tribe of a nation, which *is also a topological st-fractal territory.*

The fractal beings that live in those spacetimes, in fact bidimensional planes, from its diminutive perspective, draw up parallel curves in its trajectories, called geodesics, within the geographic, invisible limits of those territories, *to fulfil their cycles/arrows of temporal existence.* For example, herds do not cross its hunting fields; countries establish geographically invisible but very real borders, difficult to cross, patrolled by energy armies, with informative center called capitals where they accumulate most of the monetary, legal and audio-visual information of the nation.

Yet on the Earth a curious event is taking place: an inversion of space-time fields. Since the membrane is becoming the dominant informative region of the planet, evolving through herds of animal beings and machines that perceive light as information and are developing a complex global society, a super-organism with a new 'informative' audio-visual network that circulates through its satellites. All those systems, which belong to the sociological sciences can be studied with the laws of complex st-points.

Since we shall study the outer membrane in far more detail when considering the evolution of human life, let us consider now in this geological analysis, the other 2 regions.

Max. E: The continental crust.

The external crust of the Earth is a hard, energetic membrane of silicates and rocks that isolates the planet. Each continent can be studied as an open ball topology with the 3 canonical topologies:

- An 'invisible' border, *the continental platform*, which sustains most of the organic life of the Earth.

- A reproductive region, the river plains, where most life exists and reproduces.

- And a perpendicular positive or negative height axis, of fractal, hyperbolic topology where evolution of life often happens: the dorsal ranges of Andes in America, where the first Amerindian civilizations took place; the Himalayans, where the mongoloid race mutated; the African Rift (negative, hyperbolic

400

topology, where the ape evolved into human being); and the line that goes from to the Alps in Eurasia to the Balkans (where the iron civilizations that would dominate Europe were born – La Tene, Hallstat, etc.)

Indeed, we can observe on the Earth how the larger cycles of the biggest geological structures of the planet and the sun-earth system determine and synchronize the faster rhythms of its smallest species, both in the physical systems of solid rocks, liquid water and atmospheric gases and in the organic systems of life. For example, the cycles of 'energy' vs. 'information' of the planet (glaciation cycles) are the key cycles to determine the reproductive radiations in hot ages and informative evolution, in cold ages of its species; the 800 years cycles of climatic changes determined the 800 years cycles of life and death of civilizations and so on.

So we might wonder if the Earth's structures are a program of evolution of life? The answer is more impersonal: macro-organisms automatically provoke the evolution of micro-organisms by chances in their informative and energetic fields that raise the stakes of survival of its micro-particles, which are forced to 'evolve'.

In the graph, the geological record shows one of those cycles, chained to the cycles of its fractal scales, rocks and living beings, the continental cycle that fusions and breaks them in 3 parts every 500-700 million years, a fact which had deep implications to the evolution of life: The outer membrane of the Earth undergoes geological cycles of creation and destruction of continents, modulated by the periodic beat of the Earth, which every ±250 million years emits matter from its nucleus, creating new submarine ranges and lavas that surface across the extension of an entire continent. Thus again, the 'invisible' center of a st-point regulates its surface. Moreover, the cycle controls the st-1 microscopic scale of life beings, provoking catastrophes that cause great extinction periods, due to massive changes in volcanic activity and the shape of continental platforms, where most life forms exist. In that regard the 2 biggest extinctions of life forms happened tuned to the cycle, 250 million years (Permian extinction) and 500 million years ago (Cambrian extinction). Classic geology postulates that a single continent breaks in 3 parts every 500-700 million years and then lava, coming out of the Atlantic central range, provokes the movement of those continents that join back in the other extreme, sinking matter on its path. So the overall crust doesn't grow.

On the other hand, the organic theory considers the Earth's nuclei a ferromagnetic system that might absorb gravitational waves of dark energy from the black hole-sun system which will catalyze an iron-cobalt-nickel reaction, adding quarks to the iron nucleus that seems to 'absorb' gravitational forces. Thus if the iron crystal of the Earth slows down those waves and 'coils' them into cyclical mass, $E=Mc^2$, provoking heat radiation or new quarks, the nucleus of the Earth would grow. As in all 'invisible' gravitational processes, we don't have enough data to prove either hypothesis: a growing Earth or a steady state planet. Yet the Earth's oceanic crust shows an imbalance between subduction zones and the excessive number of lava rivers, ranges and abyssal dips, which stretch the Earth's skin as it happens in a fat man. The Atlantic vomits enormous quantities of lava that move America away, *continuously*, creating the Andes range, which in turn should move away the Pacific, creating a range in the Chinese coast. Yet on the East side of the Pacific instead of finding a huge range or trench from Indonesia to Japan, we find some *discontinuous* subduction zones with intermittent megathrust earthquakes that seem not enough to balance that growth with its crust destruction. Thus, planets might grow and fluctuate in size, chained to the time cycles of stars, which seem to be chained to the cycles of the organic galaxy and its Worm Hole, according to the synchronic chains between the time arrows/cycles of all the parts of an organic system.

E=I. Dominant reproductive region: the mantle.

Let us illustrate other laws of multiple space-times with the mantle:

-Any subsystem that acts as an energetic, informative or reproductive subsystem of its higher scale will be dominant in that function in its internal, ternary structure:

The deeper layer the intermediate region of the planet is *the mantle, made of liquid magma that flows in cyclical, convection streams, reproducing* and renewing the rocks of the crust's membrane. Because we are studying a region that in the higher 'plane of existence' of the planet as a whole performs a reproductive function, its reproductive subzone is the biggest one. If the system was an energetic system within the whole, as the atmosphere is the biggest region would be the most energetic (the Ionosphere and exosphere filled with cosmic rays.) And if the system was an informative one, such as a computer or the center of the planet, we expect the informative zone (the mother-board, the iron crystal of the planetary core, to be dominant). Facts, which apply also to biological systems (so for example, man, an informative species, has on top the brain and is dominated by the height dimension of information; while a planarian worm, a simple digestive, energetic system is a 'plane' of biological energy).

-Hierarchical structure: All systems have 3±st main hierarchical planes, or scales in which the properties of the parts transfer to the whole. So happens to the Earth's geological exi cycles, in which the activities of its quanta become units of larger cycles, as the reproduced substances keep enlarging the 'energetic membrane' (earth crust):

- st-1: At fractal level the geological Earth can be considered an aggregate of all kind of molecular fields called grains and rocks, which show also a st-ternary structure. For example, the commonest rocks are basalts, which show 3 st-differentiated zones: feldspar (energetic rock with minimal symmetry), mica (balanced E=I form) and quartz (the informative element with symmetric crystalline properties) that create together by social evolution, the basalt (st+1). And its process and cycles are process of constant growth of its social form, basalt which forms the overwhelming surface of Oceans, the 70% of the Earth crust.

- st: The next organic scale are small formations made of grain or bigger ones made of rocks. And those geological structures, from mountains to glaciers to sand deserts are constantly growing and reproducing new 'fractal forms'. And all of them the same 3 zoned st-regions that structure the planet with an informative nucleus, reproductive bodies and external energy.

I will put a poetic example: the dune is an organic sand system made of grains, which despite its simplicity already shows the organic properties and ternary structure of any st-point. The dune is an open ball without a clear external membrane and so it is open to constant exchanges of energy and form with the environment, which makes it extremely dynamic, as all open balls, like the black holes we just studied are. So the dune constantly moves through the energetic cycles of its body quanta, which constantly reproduces its structure adding new sand grains. The dune is a simple body made of unconnected herds of cellular grains. But even so, those sand quanta extract energy from the wind, moving the dune without losing its form, and absorb information from the pressure of its weight that compactifies the molecules of the dune, while temperature's changes help the sand to crystallize. So the dune is a simple organic system from the point of view of fractal space-time. But where it is the brain of the dune, its center of perception? An Islamic poet, who lives with dunes in the sand world will tell you that the dune has a soul called the rose of the desert, because all dunes have in its center a quartz crystal that grows slowly to the rhythms of the dunes' daily cycles of energy and information, reflecting the movements of its sand's body on the radial informative networks of the crystal, which *record as all informative systems the paths of the dune in the memorial traces of its* planes. So recently, thanks to the memories of zircons, we found that the Earth cooled down faster than energetic science believed as the dominant, cold informative arrow of all systems imposed itself.

Let us now put a gaseous example of another Earth's organic system from the atmosphere: the hurricane is a spiral vortex with a quiet center. Scientists have noted that when the hurricane creates the central point, its form becomes stable, and inversely when they 'kill' its informative eye, the hurricane breaks. It is that eye perhaps a knot of electro-magnetic or gravitational information that regulates the streams of gas that surround it? Those Aristotelian centres of informative, gravitational, simple geological perception might form central, geometrical images we cannot decode. But they behave externally as if they were centers of informative order that regulate their shapes, creating *order in* the geological Earth. And the fractal sum of all those informative knots together, even if physicists cannot easily measured balances the whole Earth maintaining its awesome number of beautiful forms; since beauty indeed has also a simple equation that defines it: $e=i$, balance between energy and form.

- st+1: Thus the intermediate layer of the planet is its reproductive zone, occupying its canonical position in the Earth's st-point between the crust and the central singularity, where we observe a series of cycles of reproduction and destruction of rocks that create the complex geological structures of its surface, which can be studied according to the laws of multiple space-times with its spatial and temporal perspective, and as a system extending in 3 hierarchical scales of fractal quanta, parts of wholes, units of the entire cyclical system. And the biggest of those geological fields is the Earth itself, which as we have seen displays the 3 st-zones of a spatial system, with a central, iron, informative, crystalline core, an intermediate cyclical, liquid, 'reproductive' zone that creates rocks and crystals and a hard, structural external crust-membrane. And so even if we cannot perceive it, we postulate by homology with all other st-points and intermediate regions that the mantle will also create iron crystals that will fall to the center of the Earth and become added to the crystal core

The 3 ages of rocks and crystals.

If we study those geological forms in time all of them go also through the 3 'states-ages' of matter that shape the long geological cycles of the inner and outer regions of the mantle, which often culminate in the creation of crystals, the 3rd most perfect, informative, molecular age.

The generational, life cycle of those geological rocks is very slow compared to our life rhythms, according to the space-time inversion, max. E=min. I; since they are made of heavier atoms and bigger networks than life beings. So they have life spans of millions of years with minimal, informative, time speed; hence they seem dead species from the faster point of view of humans made of lighter atoms. But they just live in fact much longer than we do.

We thought plants did not perceive or acted-reacted to external stimuli, because they were big and slow life based in chemical languages (max.E=min.i), compared to the faster, electronic languages of smaller animals (max.I=min.e). But plants turn to the sun, grow adapted to winds and seem to do it faster with music. So their leaves might be simple ears that transmit the vibrations of sounds to their chemical, root brains. The cycles of rocks are even slower, but still they are organic systems of energy and information.

I-logic geometry studies them through the key physical parameter of energy and information, which is temperature: high temperature means high energy, low temperature means high order-information. Another important parameter is 'height', which is a dimension of information. Finally, most processes of creation are long as time, information and evolution are related: a fast process of creation takes place at high temperatures with maximal expenditure of energy and min. informative order (max. E=min.I). While the inverse slow process of creation uses less energy and reaches a maximal final order. If we apply those 3 st- laws, we can understand the processes of crystal and rock's creation and its 3±st ages:

- *Birth and Youth:* Rocks are born as lavas inside the planet with a high temperature (energy-movement) and an extended form. As they rise in the dimension of height-information, rocks cool off, losing temperature-energy and acquiring form, till reaching when the process is very slow (with a higher quantity of evolutionary time), a crystalline form, able to create images within its informative focus. But if the cooling of the rock happens too fast, without giving time to the slow, formal evolution of atoms necessary to acquire a symmetric order, it becomes an *amorphous rock*. Finally, the rock or crystal comes out to the surface of the Earth and its youth cycle of creation ends.

- *Maturity:* In its mature age the crystal perceives light, forming images in its focus, while the amorphous rock shows at best strikes of lineal atoms. As informative species, crystals also live longer in time, as neurons do in our brain; women in our gender duality and black holes in the Universe. Diamonds live billions of years. But amorphous rocks soon begin the aging process, through erosion that wears them away by the effect of water and light.

- *3rd Age and Death that gives life.* We find in rocks, as in all exi=stential cycles, 2 forms of death:

- Max. Se=min. Ti: An explosive, sudden death, common to all crystals that become broken, exploding into pieces with brisk changes of temperature.

- Max. Ti=min. Se: An informative death in their 3rd age, common to amorphous rocks, as they wear away very slowly, eroded by water and air. So when a rock dies its cells disintegrate, loose its height and return to the interior of the Earth, completing its generational cycle. Then, the energy of the Earth's heart will warm them up again into magma, starting a new cycle:

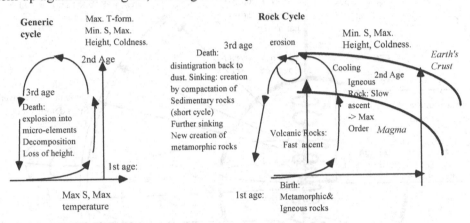

Se ⇔Ti symmetry between 3±st ages and 3±st spatial types of rocks.

Finally, if we consider those ages in space they give birth to 3±st equivalent types of rocks:

- *Max. E:* Simples volcanic rocks that flow with max. energy and speed towards the surface of the Earth but have minimal form.

- ⇔:Igneous rocks have slower rising speeds and a balanced form.

- *Max. Ti:* Metamorphic rocks, as their name shows, undergo a transformation with several degrees of crystalline form and radial symmetry. They are the slowest to come out to the surface.

- *±st:* Finally, when rocks die and disaggregate into dust, they clamp into the 4th type of rocks: sedimentary rocks and sand, which still have minimal organic structure.

The main difference between the organic and inorganic Earth is its location and speed of cycles: The organic Earth inhabits the final layer of the solid magma in a transitional zone of interaction among gas, liquid and solid molecules. Yet both molecular systems undergo long geological cycles. The best known of those cycles is the cycle of the energetic molecule of life, H_2O, water, that goes through a gas state as a cloud, a liquid age as moving rain and a solid phase as ice or as the most perfect crystalline form, snow, which lasts longer than ice into the ground, as information systems do. So, the biggest liquid and solid systems of the Earth's crust - the sea, the sand of the deserts and rocks of its mountains - are just organic, fractal systems with very slow existential cycles. Similar cycles occur with the organic masses of the other 2 main atoms of life, nitrogen, the informative atom of life and carbon, the E=T balanced 'body' of living systems. Thus the surface of planet Earth, Gaia, is a living, organic, membrane, a complex molecular system, made at macro-scale of multiple life ecosystems, composed at our human scale of living organisms, which are themselves made at fractal micro-scale of carbon-molecules with forms self-similar to the crystalline molecules, of the previous graph. Those 3 scales structure the cycles of existence of life interacting among them - as it happens in any other 'i-logic world', from the mantle and its crystalline rocks just described to the Universe as a whole.

Recap. The fractal, ternary principle subdivides each region of the Earth in 3 sub regions with the same ternary topologies, ad infinitum, allowing us a more detailed analysis of all systems. Thus, the intermediate, reproductive region of the Earth can be subdivided in 3 sub regions – the surface where life exists, the continental crust that follows an E⇔I cycle of creation and destruction of a super-continent into 3 fractal zones and the mantle, where rocks follow their cycles of creation and destruction.

71. The central core of planets and its crystal minds.

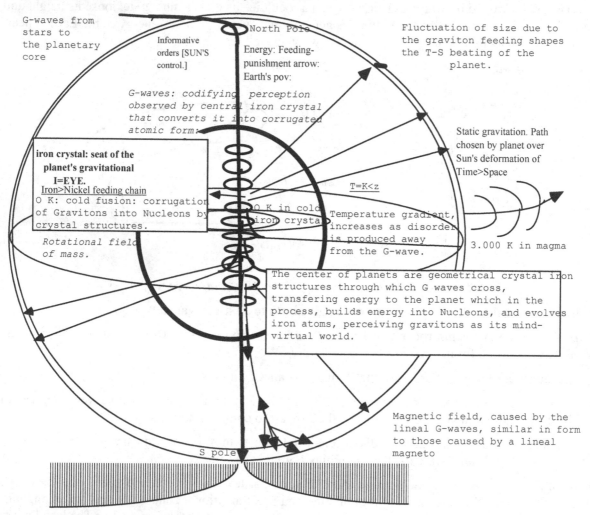

Earth, as an st-point, has a central singularity and an external membrane that form a unified structure that controls the fractal, reproductive intermediate zone: the activity of the Earth's inner core defines the reproductive and existential cycles of the magma rocks, which shape the continental crust membrane of the planet:

Let us finish this introduction to complex physics and its ternary sciences, quantum physics, which studies the smallest particles; astronomy, which studies its gravitational bodies and geology, which studies the intermediate planetary world, with the analysis of that core that connects us back to the relative infinity of the galaxy.

The center has also 3 zones; an informative crystal of iron, which might perceive gravitation a reproductive external liquid ball that should reproduce and precipitate those iron crystals and a center of radioactive uranium that acts as its energetic, heating nuclear plant.

Exi: Liquid iron.

One of the tenants of multiple space-times theory is the multi-functionality of each topological structure, which often plays through its 3 sub-regions different functions for the whole organism. For example in the core of the planet we first find is a dense layer of liquid iron that should reproduce the crystals of the core. Yet it also has an informative role, as it creates convection streams, responsible for the creation of the strong magnetic field of the outer atmosphere of the Earth, forming the classic Max. E x Max. I, 'a-r' duality of a topological, closed ball that cages the reproductive 'cells' of its body, within it – in this case the other structures of the mantle:

So life will die exposed to radiation beyond the membrane and rocks will die melted by the liquid iron. In non-Euclidean structures, according to Klein a region that cannot be reached is a relative infinity. So those regions are the relative infinite of our existence and that of the rocks of the mantle, as the black hole is the relative infinite of the light that redshifts without ever reaching it as light.

Finally liquid iron becomes the energetic, hotter lower membrane for the rocks of the mantle, fuelling its reproductive cycles.

Max. E: Uranium core.

The next region of the core is an iron crystal and in the very same center of that iron crystal, within it, there is probably a smaller core of uranium, the most energetic atom of the Universe, with a minimal stable life that emits heat, and sustains with its weight and pressure radiation the crystal core.

Max. I: Iron Crystals

Indeed, the Earth seems to have on its center, surrounded by flows of liquid magma, a crystal made of iron, the most efficient structural atom able to move in gravitational fields, as iron magnets show. Thus, as 'evolved' crystal systems in cameras perceive electromagnetic waves, the center of the Earth probably perceives the gravitational waves that structure solar systems and galaxies.

Yet the crystal core has also according to the ternary principle an energetic and reproductive cycle related to the Uranium core, which becomes in this manner its symbiotic, small body.

And all those cycles according to the hierarchical and multifunctional principles are synchronized with the smaller cycles of existence of its surface quanta – the cycles of life and creation and destruction of continents; and with the bigger cycles of stars and galaxies, of which the planet is in itself a fractal quantum.

Let us consider those functions and propositions in more detail.

Reproduction of matter and absorption of heat energy.

Those st-reproductive and energetic functions of the core are based in a well-founded empirical fact:

The Earth's nucleus is composed of nickel, iron and a probable inner core of uranium, which form together a nuclear reaction chain based on the absorption of nucleons: uranium absorbs a neutron and emits helium, which iron absorbs to create cobalt that emits a neutron to become nickel. The neutron can be absorbed by uranium and feed again the cycle. Those reactions at planetary scale can provoke an enormous quantity of energy and matter that should be responsible for the periodic activity of its magma, which peaks every ±125 million years, determining the geological continental cycle and the changes on the magnetic field of the Earth. Yet the sun has an orbital period around the galaxy of +200 million years. So those peaks of nuclear and magma activity coincide with the equinox of the solar orbit. What happens in those points? Does the sun cross a turbulent spatial region or a galactic discontinuity?

It is the orbital cycle of the sun chained to periods of Worm Hole gravitational activity, as the orbit of Jupiter coincides with the suns' spot cycle? Then G-waves might become flows of gravitational energy that enter through the planetary axis, as magnetic flows do in atoms, and trigger the uranium->nickel reaction, causing the discontinuous cycles of magma activity in the intermediate 'reproductive' zone of the Earth, responsible of the geological, biological and magnetic catastrophic changes on the planet. We know very little but the few things we know seem to prove the interrelating nature of all parts and scales of the galaxy.

Further on, the Earth and other planets fluctuate in their diameter during their rotational orbit as they use gravitational energy to move, 'deforming' in the abstract jargon of Einstein, space-time into a curved orbital path.

Thus, planets should grow periodically in size, emitting not only radiation but also matter from those inner cores, a fact that finds constantly new proofs. Since planets emit more radiation and have stronger magnetic fields than we can calculate with mechanic models. While planets closer to the Sun, like Venus that should absorb stronger G-energies, mutate its skin every ±400 million years, creating an entire new crust, as a snake does when it grows and changes its skin. Yet ultimately as in all 'cosmological farms', those planets will end up falling and feeding the sun. So all those 'physical cycles' show that planets absorb energy and information and grow in orbital size with rhythms similar to those of nucleons and electrons, when they absorb energy.

The heart and brain of the planet.

Metals use electronic flows to communicate information and energy, organizing their networks through crystalline forms. Their pixels are not photons but electrons. Most of those crystals have also the dominant cubic form, better suited to a 4-dimensional Universe. Thus crystals organize, trans-form and perceive information in the 2 main atomic families we know: the light atoms and the electronic worlds of metal: Electrons in the world of metallic, crystals either form static clouds of 'informative images' or flows that travel through huge metallic networks, stabilizing their structure.

There might be also in the radioactive family, under extreme pressure, crystal-like formal structures crossed by neutronic flows, its exi communicative particles, creating 'radioactive networks', as crystals form crystalline networks, which might invaginate the entire planet. Those fluid 'veins' of radioactive energy should then control in more detail the flows of the rest of the 'planetary organism'. All those homologies are very likely to happen in our planets, as all st-points turn out to be far more complex and organic in detailed analysis that mechanist, energetic scientists will ever believe from their anthropomorphic point of view. Thus like our organisms, regulated by two networks of reproductive energy (blood) and information, the dual body/brain of uranium/iron at the center of the planet might be just a heart/brain dual system with complex radial networks whose cycles of creation of energy and form determine to its finest detail even the surface, as we have seen in our analysis of continental and glaciation cycles; which could be also connected to the changes in the magnetic activity of stars. *It is in fact possible to construct a model of absolute determinism in which all the parts of the planet are interrelated and all the parts of the galaxy are organically connected, even if we will never perceive 97% of the matter and energy of the galaxy and a self-similar quantity of our planet structure.*

Perceptive, informative function of the Earth Nucleus.

As we change scales from micro-atoms to macro-atoms, the spatial size of the crystals also change in the same way that a machine that substitutes a human organ changes in scale (so a crane that substitutes an arm is far bigger and stronger). And so celestial bodies have in their core crystals of planetary scale, which should be the informative center that regulates most of the long exi cycles of the planet. While the size of the 'electromagnetic waves' those celestial bodies use as energy also change from electromagnetic microwaves to gravitational macro-waves.

Yet all the processes described for transparent light crystals work for electronic iron crystals, since a crystal is a i-logic symmetric, fractal geometry that can be adapted to any form. We conclude that iron crystals might see 'electrons' and feed or move under gravitational waves, which could be their 'energetic' substance.

The 3 ages of planets

All this details said, when we grow up back to the st+1 structure of the whole planet, it should be obvious that planets as a whole also obey all the previous laws.

The most important is the law of the 3 ages. And so planets go through the same living ages of other cosmological species. They are born as nebulae that cool down into a rock, and then they start to cycle around the sun in a symbiotic structure: the planet moves deforming the static field of gravitational space-time around the sun and absorbs discontinuous transversal G-waves, which uses to 'control' its spin, stabilize its orbital territory around the star, feed and perceive perhaps as far as the central Worm Hole of the galaxy, origin of most of those waves.

Recap. The planetary core is a ternary region with a liquid, reproductive surface that precipitates crystals, which form a solid, informative iron crystal, perceive of transversal gravitational waves. Finally inside the crystal there is an Uranium core, the body of the crystal, whose radiation cycles reproduce mass and provide heat, regulating the cycles of the rest of the structures of the planet, including our rhythms of evolution. Thus the planet have a mechanical, conscious or vegetative gravitational mind, connected to the cycles of stars and galactic black holes, which will ultimate feed on it? We leave that final decision to your i-magination.

72. Philosophical coda.

The sceptic reader might think those are speculations we won't ever prove. Still the main epistemological teaching of i-logic geometry is that 'topological homology' is a valid form of knowledge and the only way we have to study most of the Universe, given the fact we can at best perceive only 4% of that Universe. So when we confront that 96% of the gravitational reality we will never perceive - all what there is within a planet beyond its thin membrane or within the Universe beyond its light-space, or within an atom beyond its thin electronic membrane, the only practical method of knowledge is homology, as the same topologies and ages of space and time happen in all species, by the mere fact that w*e are all made of the two primary shapes/motions of reality, energy and information.*

Ultimately the only reason mechanist science does not accept such hypothesis is the arrogant anthropomorphic desires of human beings for absolute truth. Yet since the ultimate, existential goal of any form of knowledge is to avoid death and dangers cannot be known with total certainty, any species believes probable truths based in their homology with past cycles of existence. So a gazelle doesn't know if a lion will kill her, but based in the homology with previous cycles she runs, accepting that probable truth. Truth is indeed ultimately limited by our perception, which is never absolute, so truth is never absolute. I have merely tried to show you the most probable truths of the interconnected organic Universe in which you are also a mere point of existence that now we shall study with more detail with the same laws we have sued to describe all other realities, in your 3 scales, at cellular, individual and social level, with the sciences of biology and bio-history.

Recap. You are nothing but dust of space-time as everything else in the Universe, submitted to the same organic laws of energy and information that regulates all exi=stential beings.

Notes

[1] 'Untimely meditations', F. Nietzsche.

[2] A good introduction to the new physics of fractal and chaotic systems, far from equilibrium reactions and the advances of new cosmology, is the book *The New Physics* by P. C. Davies, Cambridge University Press. Published in 1989, it made an account with articles written at an undergraduate level by the leading researchers on those fields, on the main themes that would shape the cosmology of the twenty-first century. On one hand, it shows the theories of quantum cosmologists (quantum gravity, quantum entropy) which departing from single, energy-only concepts of the Universe have shown to be a wrong path, a no-way out for modern cosmology. On the other hand it lays down the other path of cosmological research, which started in those years, based in the duality of energy and information, the fractal nature of the Universe, the invariance of scale of all the laws of reality, and the physics of far-from-equilibrium systems and superconductivity, in which this book is based and will be the foundations of twenty-first-century cosmology. The old "Vatican priests" of quantum cosmology and the new "Galileos" of the third millennium write brief articles in a vibrant account of a shift of scientific paradigm. For further information on the new cosmology, one must read the scientific papers of the founding fathers of those disciplines: Mandelbrot and Sancho in fractal, Non-E points and fractal time; Rössler, Mehaute, and Thom in chaotic and fractal systems; Pietronero and Nottale in fractal cosmology; Smolin and Nambu on the falsity of Higgs and so on.

[3] Expression coined by Roger Bacon to stress the fact that errors from the past (Aristotelian scholars) survived in modern thought, due to the respect the founding fathers of any science cause to future generations. Today we find a similar case with Einstein's concept of a single space-time continuum, the light-space membrane, and its limit of speed, c, which applies only to that membrane.

[4] Einstein in fact defined first Mass as accelerated energy, $M=E/c^2$ in his seminal paper "Does the Inertia of a Body Depend Upon Its Energy-Content?" ("Ist die Trägheit eines Körpers von seinem Energieinhalt abhängig?"; 1905, *Annalen der Physik.* A) The inverse equation, $E=Mc^2$ became better known with the explosion of the Atomic bomb.

See also Albert Einstein, "Über das Relativitätsprinzip und die aus demselben gezogene Folgerungen," *Jahrbuch der Radioaktivitaet und Elektronik* (1907); translated "On the relativity principle and the conclusions drawn from it," his first statement of the equivalence principle.

[5] The simplicity of such Unification makes us wonder, why it was never found? The answer is: Because it was never tried. Physicists like Einstein first stubbornly tried to unify the equations of Relativity (the complex version of that vortex) with the equations of electromagnetic lineal waves, not its charges. And that is, indeed, impossible, because charges and particles work as cyclical, non-lineal vortices, structures with a gradient of increasing speed towards the center, broken into multiple, fractal flows. While gravitational and electromagnetic waves, also of different scale and size, are lineal self-repetitive forms.

Then the search for Unification Theories followed the opposite path, even less promising, trying to define mass with quantum particles, till arriving to the Higgs, an impossible scalar boson created by a field of which there is no proof whatsoever. Yet the Higgs remains the Saint Grail of entropy-only theories for the only reason that its existence will unify all forces as spatial, gauge theories. Unfortunately spatial mirror symmetry does not apply to a temporal force, like the weak force is; since time is 'antisymmetric' - meaning that the past, future directions (young/old age of a system) have inverse parameters of energy and information.

[6] Nottale, L. (1993). Fractal Space-Time and Microphysics, World Scientific, Singapore.

[7] The hunt for the Higgs is moved more by the technological drive of big science – the making of bigger, better accelerators - than the sound physics of the theory, which was first formulated by Mr. Nambu to explain why the weak force – a temporal force – doesn't follow the spatial symmetry of other 'energetic forces'. He finally solved the conundrum using a known-known particle, the top quark condensate, whose strong attractive power could break the symmetry of lighter particles without the need of an impossible scalar boson and a new force field never observed before (Nambu, Y.; Jona-Lasinio. "Dynamical Model of Elementary Particles Based on an Analogy with Superconductivity; see also Vladimir A. Miransky, Masaharu Tanabashi, and Koichi Yamawaki, who extended the model of Nambu to study how top quarks could be the cause of the breaking of symmetry of our matter at high energies (Phys. Lett., B221:177, 1989, *Dynamical electroweak symmetry breaking with large anomalous dimension and t quark condensate'*, and 'Is the Quark Responsible for the Mass of W and Z Bosons'?. They established the basis for a series of models (Goldstone, Higgs, etc.), which studied how quarks, forming Einstein's condensates, were able to break the symmetry of our matter. But since a top quark attractive vortex can do the trick the principle of economy makes the top quark a better candidate. Further on Smolin and Zee, discovered using the Brans-Dicke theory, that the Higgs was equivalent to a strong gravitational field in which we change the universal constant for a stronger one. Thus the Higgs behaves exactly as a top quark condensate. This discovery favors the unification of strong and gravitational forces, quarks and black holes, as space-time vortices of 2 different scales. It is a further proof that protons, which are made of quarks and gluons, are black holes of the quantum scale and so black holes in the cosmological scale will also be made of quarks and gluons. They will be Einstein-Bose top quark condensates.

For the initiated, we shall consider a more complex explanation: Brans-Dicke made G a scalar field, adding a dimension of angular speed $1/\varphi$ to measure the strength of G's deformation of space. The variable G field defines masses as space-time whirls of different strength according to its rotational speed.

Thus the Dicke-Brans model of variable G-constants expanded relativity, adding a new dimension which now can be fully understood in terms of the three sub dimensions or derivatives of "time-motion."

We say that speed can be seen as distance. Further on, the equation of change in motion that physicists call time is v=s/t. Yet we can see that equation at three levels of speed. We can see it in present simultaneity as a fixed form of information. This is Einstein's basic description of the topology of space-time. We can see it as a speed—this is Newton's description of a vortex of space-time—or we can see it as an acceleration, which adds a new dimension. This is Poisson's description.

Indeed, in physics we add motion, translating time=change to that formal dimension of static geometry. Thus, we create two new dynamic dimensions of temporal motion by deriving space through time=change, v=ds/dt and a=dv/dt, to obtain the second dimension of translational time-change (speed) and its third dimension (acceleration).

In relativity, those dimensions become the accelerating whirls of space-time defined by Einstein's strong principle of equivalence. Yet since Einstein's G measures only the morphological curvature or "form" of space, we have to add two new dimensions to Einstein's G to obtain first the cyclical speed of the whirlwind and then its acceleration. The result is a vortex of space-time with three clear time dimensions and descriptions of increasing complexity:

- G, the curvature of space, which defines the form, the in/form/ation carried by a given region of space and was defined by Einstein's relativity.

- φ =1/G, the cyclical movement of that curvature, which defines a vortex of space-time, self-similar to the variable scalar field of Brans-Dicke that adds speed to G-curvature.

- And finally, $1/\varphi^2$, the acceleration of that movement, which reflects the principle of cyclical equivalence and defines mass as an accelerated vortex that sucks in space-time (since $1/w^2$ are the dimensions of an accelerating, cyclical movement). It is a field self-similar to the dilaton field defined by Guth to explain the expansion of the big bang Universe.

When we explain those two new dimensions of G through cyclical time and the strong principle of equivalence that equals acceleration and mass, it is easy to explain and unify the forces, particles, and fields of the Universe with different values of those three G dimensions. Further on, as Smolin showed the other alternative theory of mass, the Higgs fields turns out to be merely the 2 (+ and -) roots of φ^2, which strongly suggests that as Kaluza and Klein already figured out in the 1930s, the electromagnetic and gravitational field are merely the expansive (negative, electromagnetic, decelerating) and implosive (positive, gravitational, accelerating) root values of a Universe of variable motions=translational changes.

[8] A.H. Guth and P.J. Steinhardt, 'The inflationary Universe', 'Sci-Am', 250, 1984.

[9] The masses on those graphs are not exactly decametric in scale, since when we enter into a detailed analysis, we must fine-tune each mass to the somehow more complex Non-Euclidean forms of those vortices and consider the energy and mass carried by the fractal gluons that connect and create those quarks, inside each particle. The 'real' geometry of those vortices of mass is thus more complex than simple eddies; but the simple principles remain: masses are accelerated vortices not solid particles; they have less dimensions that our Universe, not more; and they carry physical information in the frequency of their bidimensional rotations, as this computer carries information in the frequency of its Ghzs. Needless to say mass as a vortex discharges SUSY particles. They were one of the many attempts to describe gravitation with quantum laws as Higgs does, using particles to interact and create the effects of mass. They were used to explain why gravitational force is so weak in the quantum world, which in fact it is not used or needed to describe any interaction between charges. Their use, however, is superseded by the fractal theory of space-time since indeed *quantum particles do not suffer gravitational forces, as they do not belong to the strong gravitational membrane of space-time.*

[10] Kerr's black holes are the only real black holes. The mathematician from New Zealand, working isolated from mainstream physics, discovered rotational solutions for black holes, departing from Einstein's work. This is the case in all mathematical paradigms: physicists start with simple solutions that do not reflect reality and slowly they find complex solutions till they arrive to models that fit the experimental method. In that sense, the work on static black holes as mathematical tools by Mr. Wheeler, Oppenheimer, and Hawking is too simple to describe real black holes, which will be rotational black holes, as all masses are according to the Equivalence Principle, and will be made of a real cut-off substance (quarks) as Einstein wanted. This means only Mr. Kerr's solutions are worth to study to describe real holes. Experimental evidence favoring this view is mounting, as all black holes found have been rotational. Moreover they all seem to be rotating at near light speed in its event horizon, regardless of size. Findings at MIT's Center for space research indicate that the way in which black holes accumulate matter is independent of their mass, only related to its rotational speed: A small black hole named J1650, which has ten solar masses, behaves identically to a ten-million solar mass black hole in Galaxy MCG-6-30-15; and the first measure of the speed of rotation of a black hole, observing the speed of rotation of iron atoms around it, turned out to be c-speed, backing Einstein's model of mass as accelerated whirls of space-time.

[11] The Sloan Digital Sky Survey (SDSS) is one of the most ambitious and influential surveys in the history of astronomy. Over eight years of operations (SDSS-I, 2000–2005; SDSS-II, 2005–2008), it obtained deep, multicolor images covering more than a quarter of the sky and created three-dimensional maps containing more than 930,000 galaxies and more than 120,000 quasars. SDSS data have been released to the scientific community and the general public in annual increments, with the final public data release from SDSS-II scheduled for October 31, 2008.

[12] See a layman version in: "The Illusion of Gravity," Juan Maldacena *Scientific American*, November 2005.

B. BIOLOGY
REPRODUCTIVE ORGANISMS

'As more individuals of each species are born than can survive; consequently, there is a struggle for existence. It follows that any being, if it vary its form in any manner profitable to itself, it will have a better chance of surviving, and thus be naturally selected. From the strong principle of inheritance, any selected variety will tend to propagate its new and modified form.'

'On the Origin of Species' Darwin, on the arrows of reproduction and evolution of form of biological existences.

1. The Arrows of Life: Multiple Spaces-Times Laws Applied to Biology.

2. Topological Evolution: The 3 Ages of Species.

3. The 3 Planes of Life Existence

4. Eusocial Evolution of Life.

5. Cells, the Arrows of Life Organisms.

6. Chemical vs. Electric Languages: Plants

7. Evolutionary Punctuation. Animal Phyla

8. The 3 Ages of Gaia

9. Man as a Fractal Organism.

10. Wo=Men: Gender duality: Complementary beings.

I. THE ARROWS OF LIFE

1. What is life[1]?

The fundamental innovation of System Sciences and the model of Multiple Spaces-Times, regarding the meaning of life is obvious: All what exists in the Universe shares the 4 fundamental properties of life, which are mimetic to the 4 main time arrows all systems of the Universe follow. Life, like everything else in the universe, feeds on energy, gauges information, reproduces and evolves socially. Thus in the organic Universe everything is life. *This is the essential tenant of organicism - the philosophy of science sponsored by System Sciences, which this work explores from a scientific perspective.*

In the past, the meaning of life was explained by 'monist physics'[1] and religions, departing from 2 different scales:

-Physics explained it from the lower 'fractal scales' of physical matter – hence life, according to physicists was merely a chemical mechanism. The error of this theory is to ignore the organic nature of atoms, from where the organic properties of life depart.

- Religion explained it from the upper scales of human social 'Gods' – and so for religions life is the most sacred form of existence, different from all other beings in as much as it is the gift of God to mankind. And only God and Man were 'special beings'. The error of this theory comes from an anthropomorphic vision of God, the mind of the Universe, whose laws are common to all species.

In system sciences, life is merely the manifestation of the arrows of space-time, the will of the Universe, in complex systems created with a specific triad of atoms, C, N, O. Yet any other type of atoms could also create potentially complex super-organisms that participate of the properties of life; since all atoms feed on energy, gauge information, reproduce its particles and evolve socially into molecules. And so we can construct 'life beings' with other atoms, as the evolution of robotics, soon to cross the 'threshold' of life, clearly shows. What is then the difference between 'carbonlife' and 'metalife' (robots) or other hypothetical forms of life? At this point of the evolution of machines only a difference of informative complexity. We had 4 billion years of evolution and machines, which we create, imitating our organs of energy and information, have had only 300 years. So as today regardless of the potential informative complexity of silicon-based life, carbonlife is still the most complex form created on planet Earth by the complex interaction of the arrows of time, which biologists call the drives of life.

So a multiple space-times theory affirms that life is just the manifestation of those arrows and further on studies it from the 3 planes of:

-       Physical matter, whose organic properties are transferred from atoms to molecules to cells to multicellular organisms.
-       Religion. Since now God, the mind of the Universe are the laws of multiple space-times and the organicist philosophy derived from it, more in tune with Eastern philosophy.
-       And from biology proper, where we consider all the scales in which life happens (the chemical, molecular scale; the biological, organic scale; and the ecological and social scale), as biologists and complexity/system theorists do.

Only then we will have a complex view of life that extends through 3 x 3 relative scales of exi=stence.

3. Biology, the reproduction and evolution of form.

Biology is the study of a certain scale of complementary, energetic and informative networks that combine together, creating reproductive networks, evolved socially from simple parts (molecules and cells) into bigger scales (multicellular and social super-organisms) –the scale of systems constructed with C,N,O atoms.

What then makes living beings different from physical entities? The answer is a basic law of system sciences - the inverse properties of energetic systems, dominant in motion with little form and informative systems, dominant in dimensional form with little motion - which derives from the balance between both parameters:

Energy= motion x In/form/ation = K.

Thus, when we grow in the ladder of complexity through the scales of complementary species of the Universe from Physical sciences, where the arrow of energy and motion dominates all systems, into the complex structures of biology, then the arrow of information becomes dominant. Hence from the simpler laws of time=change in the motions of beings that dominate the studies of physical sciences, we advance into the laws of reproduction and evolution=change in the morphology, the information of beings.

Evolution and reproduction are in fact the 2 main properties of information, which is 'fractal, dimensional form'. If energy constantly moves, information constantly reproduces itself, warping energy into form. And those are the dominant properties of life: Unlike other systems, which don't reproduce and need an 'external catalyzer' to repeat themselves (particles catalysed by energetic collisions, stars catalysed by black holes, machines constructed by human beings, etc.), the more complex 'informative' systems, such as those of life have an internal capacity to reproduce their form; so the fractal reproduction of information (palingenesis) becomes the essential property of life.

Recap. Biology is the science of in/form/ation, since its fractal species and networks are the more complex forms of the Universe. Yet they are not different from any other entity of reality, as all of them follow the same laws of Duality and General Systems.

2. Main laws of Multiple Spaces-Times applied to Biology.

It is this 3rd biological/ecological approach the one that has fully understood the meaning of life, as a complex system that 'feeds' on energy, gauges information, reproduce its energy and information, and evolves socially. As we did in our analysis of the physical universe with Multiple Spaces-Times, we shall resume the advances brought about by the 4th paradigm, before we make an orderly description of its applications to the study of life, from its initial molecular scale till its final evolution of human dual forms of energy and information (sexual duality).

The 1st types of laws applied to biology are laws of space-time cycles and its causality, which define:

- The life/death cycle as a series of ternary ages and horizons in the evolution of organisms and species. We shall therefore be able to explain the plan of evolution of species and life beings, with the 3 ages of energy and information.

- The events and behavior of living organism as the product of the 4 main drives of existence, gauging information, feeding on energy, reproducing and evolving socially. Life beings are super-organisms, created a knots of those 4 time arrows, which in biology are equivalent to the 4 drives of existence that define life. We shall observe that in any scale of biological entities, from DNA molecules to individual species to societies those 4 wills of existence explain the behavior of each species, including the human beings: We constantly inform our head with different languages, feed our body and lungs with energy, conduct a life based in family values and sexual desires, and evolve socially with other human beings and machines into super-social organisms, cultural organisms, in which we perform our social functions as a 'cell' of the super-organism.

The 2nd set of laws of Multiple-space times applied to biology are the complementary, inverted properties of energy and information and its complex arrows, reproduction (biological radiations) and social evolution, (evolution of cellular forms). They determine its complementary and ternary structures and physiological networks, both in cells and in organisms, with heads, bodies and limbs.

413

They determine the Darwinian events (predators vs. victims) of Nature and its 'rhythms' and beats of existence (punctuated evolution⇔ reproductive radiation)

Further on, the properties and size scales of biological space-times, dominant in information are inverted from those of physical space, dominant in energy and determine the different processes of creation, evolution and emergence of super-organisms in both ecosystems: physical systems tend to be entropic, with few informative, time cycles and only a weak time force, while its spatial forces, speed motions and destructive big-bang processes are overwhelming; while biological systems are negantropic, full of different time-cycles and clocks, with rich, complex in/form/ations and limited speeds and reduced spaces.

If the Universe is a fractal puzzle of cyclical masses and charges, associated to lineal light and gravitational forces, in 2 scales of enormous simplicity of form and extension in space; the vital spaces that evolve into life forms are far smaller in size and yet far more complex in information, which is the dominant element of life spaces.

What this means is that the 3 scales of life beings, the bio-chemical, organic and social scale are very different in form, given its richness in life beings, yet move and occupy a relative small space. So in life beings what matters most is the invariance of the functions of those forms that transcend between scales, even if the topological deformation of the exact forms is enormous.

The 3rd sets of laws of Multiple Spaces-Times applied to Biology are the 3 invariances of motion, scale and form. Since life beings are informative beings of minimal energy/motion, motion invariance is irrelevant. Thus the 2 invariances that matter in life are the existence of fractal scales (the bio-chemical, cellular, organic and social/ecological scales) and the invariance of topological forms in all living organisms that repeat the 3 topological structures of a 4-Dimensional Universe and within scales of life that repeat the functional arrows of information, energy, reproduction and social evolution, which emerge from scale to scale.

The laws of i-logic geometry (form invariance) applied to biology: In the 3 scales of life of the Earth the invariance of topological form is not maintained strictly but both cells and human organisms, and economical nations, do maintain the invariance of functions: that is they deploy systems of feeding, gauging information, reproducing its substances and evolve socially.

The Invariance of topological form=Function is therefore maintained as always in the 'fractal Universe' with a certain degree of self-similarity, never with absolute equality.

Further on, we shall be able to classify in each of the fractal scales, any species or form of life, as a lineal system of energy (proteins in bio-chemistry, limbs in human bodies) or information (DNA rings and nuclei in cells, heads and eyes in human beings), and structure them according to the 3 topological forms. So with the advanced tools of topology we differentiate in a life being a ternary structure composed of a hyperbolic, informative, high topology or head/cellular nucleus, an energetic, extended membrane and lineal limbs/cilia and a series of cyclical networks that connect both regions (reproductive bodies, mitochondria).

The 4th set of laws of multiple Spaces-times are the inverted laws and flows of energy and information between those scales, which define the laws of genetics.

The 5th set of laws are the ternary laws of differentiation of species in energy, information and reproductive sub-species and the ternary laws of functionality that explain the multiple functions of most biological entities within its bigger organism.

Thus a Biological model of Multiple Spaces-Times deals mainly with the specific analysis of those scales, its species, life-cycles, arrows of existence and energy-information inversions, offering a more detailed analysis of the plan of evolution, by adding to the mix the 3 ages/horizons/ternary differentiations of evolution in time, the organic, topological morphology caused by such evolution and space and the relationships between its fractal planes of existence (genetics). It is the science of palingenesis that now becomes central to biology.

Those laws of Time theory are essential to biology, since they explain the morphologies, evolutions and behavior of life.

Recap. Biology is the science that studies the most complex informative network systems of the universe, which have evolved from molecules to cells to human beings and beyond into machines, according to a topological, ternary plan of evolution guided by existence of only 3 type of networks in the Universe, which correspond in space to the 3 topologies of a 3-Dimensional universe and in time to the 3+1 ages/horizons of evolution of all systems that finally integrate together into a whole organism, sum of those 3 parts. Those 3x3 elements, the topological, temporal and fractal analysis gives us a far more complex, exhaustive understanding of a system:

Thus, the Biological world follows as any other system of reality the laws of Multiple Spaces-Times: The laws of causality that determine the life and death cycle and the evolutionary horizons of species; the laws of invariances in form, motion and scales; the laws of morphology, causality, behavior and social evolution defined by the 5 postulates of i-logic geometry, which explains the creation of molecular networks that created cells and cellular networks that created organisms and social networks of organisms that created civilizations, shaping the 3 scales of life beings: the cellular, individual and social level.

II. TOPOLOGIC EVOLUTION: 3 AGES OF SPECIES.

CONCEPTION AS A BLACK HOLE OF MIN. SIZE = MAX. INFORMATION

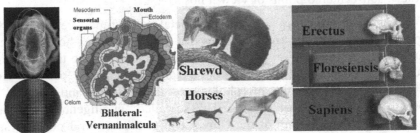

In all space-time scales species follow a future arrow towards more information, which increases in 3 horizons and finally evolves individuals into societies, herds and macro-organisms, related by a common language/network of information.

3. The plan of evolution: Synopsis.

The process of evolution of species is equivalent to the 3±st ages of individuals with a twist: death is not the only 'solution' to the cycle, but the duality of ∂-disintegration or ∫-integration of a social group (in the case of life a cellular group), which defines the life/death cycle gives two choices to some 'top predator' informative species: After its 3 horizons or evolutionary ages the species might become extinct or transcend into a super-organism.

In the graph, the seminal, 1st species packs a lot of information in minimal space: It is the Worm Hole, the chip, the 1st bilateral animal[2], the 1st mammal, the horse and the 1st Homo Sapiens, which seem to have evolved from a dwarf ancestor, which first discovered technology and had a dwarf brain, albeit with an evolved morphology similar to that of the Homo Sapiens: the Homo Floresiensis[3].

Then the new 'highly informative' top predator follows in its evolution a plan of dimensional evolution that transforms species with a high content of energy into species with a high content of information through 3±st horizons, self-similar to the ages of an individual organism. *Since species are, in fact, macro-organisms in which each cell is an individual of the species.*

-First in its 'energetic youth', the species grows in size as horses, humans, bilateral animals or black holes do.
-Then the species suffers a reproductive radiation colonizing new ecosystems.
-Next the species suffers a speciation, *according the ternary principle into 3 subspecies, one dominant in energy, another in information and a 3rd one balanced in both parameters.*
- *Finally, a new top predator extinguishes the species or the informative one evolves into social organisms of different complexity, stronger than the individual and survives.*

Recap. The simple, morphological plan of evolution that each fractal species of the Universe repeats is *caused by the fact that there are only 2±1 forms in the Universe, lineal energy and cyclical information and i=ts reproductive, body combination. So only 3 new types of species can be created, one with more energy, one with more information and one that reproduces both.*

4. The plan of evolution in depth.

According to the *Paradox of Galileo*, all minds, including man, are self-centered beings who believe to be the center of creation, because they gauge information from its point of view. Thus most of reality either exists in a dark space the species doesn't perceive or uses a language the species doesn't decode or it is further away from our 'noses' to be of any importance to us. So, regarding evolution, humanity believes either in 'chaos', which makes humans the only intelligent species (the scientific, Darwinian approach to evolution) or in 'personal, planning Gods' (the religious approach) who have designed a Universe to the image and likeness of humanity. The truth is the opposite: all is intelligent and vital and

416

we are made to the image and likeness of the Universe. Empirical facts, of which this book offers many cases, prove that the Universe and all its parts are organic and perceptive and it is indifferent to our species, just one of the ∞ self-centered points of view it hosts.

In Biology the 3±st ages of an entity of multiple fractal space-times between birth (st+1) and extinction (st-1) apply both to organisms and to species that go through 3±st horizons, parallel to those 3±st ages and main arrows of time of all systems of the universe. Thus there is an informative, 'intelligent', organic order in evolution. Though it is not a plan designed by a 'personal God' who cares for humanity, but by the limits that the morphology of lineal space and cyclical time and its combinations impose to the evolution of species. The so-called 'intelligent design' is the ternary, existential cycle that applies to every form of the Universe as it evolves in 3 possible sub-species: an energetic, lineal species, an informative, cyclical one or a reproductive form, of balanced $\Sigma E = \Sigma^2 Ti$ parameters. Those 3 forms become the 3 horizons between birth and extinction of any species: A top predator form of max. energy, or I horizon of the species; a balanced form, $\Sigma S = Ti$, that reproduces so fast over its prey-energy that creates a 'biological radiation' and parallel extinction of the lesser victim till reaching a trophic balance, or II horizon of the species and a form of max. information or III horizon: Max. E, Max.Re, Max. I.

+1: Birth: max. Ti. The chip paradox (conception).

In the previous graph, the creation of a new species takes place according to the same 3 ages of any space-time cycle that become the 3 horizons of any species: after conception that creates 'a seed' of pure information and minimal Energetic Space, species go through a young age of energy growth that creates 'bigger species'; a mature, reproductive age of forms in balance between its energy and information, when the species maximizes its reproduction, radiating in huge numbers; and a third horizon of informative evolution, when it diversifies into multiple sub-species, becoming finally extinguished by a new top predator form, since its evolution has reached its limit. However, in certain species, after the III informative horizon, a fascinating phenomenon happens: if the species has developed a new language of information, it gathers into herds that evolve socially together creating macro-cellular organisms of a 'higher', more complex Plane of existence. So prokaryote evolved into giant eukaryotic cells, and those cells evolved into animal beings, of which ants and men, the most successful, informative insects and mammals, evolved further through pheromones and verbal information into societies and civilizations.

The 'Black Hole' age of any species is parallel to the informative, genetic conception of any organism, born out of a 'seed' that packs the maximal genetic information in minimal space. It is caused by the dominance of informative cycles over its Energetic Vital space. *So a new top predator species is born with a lot of new, genetic=Temporal Information packed in a reduced size (Max. Ti=Min. Se).* This happens because information is processed faster in smaller spaces. For example, a 'logic instruction' is resolved faster in smaller chips. It follows that tiny species with huge numbers of 'neurons' deliver in group stronger and faster *actions* of energy and information, Min.ΣSe x Max. $\Sigma^2 Ti$, than slower, bigger species (Min. $\Sigma^2 Ti$ x Max.ΣSe):

Since they are highly informative, they can coordinate those fractal actions in herds that act simultaneously as a single organism. *So their actions show a higher existential, exi force, which defines them as top predators,* as they are stronger and faster than the 'slow actions' of a single, bigger body. For example, small English boats shooting faster against big galleons defeated the Spanish Armada; a pack of wolfs kills slow reins and herds of orcas kill bigger whales. Small, intelligent top predator brains rule bigger, less informative bodies, because time dominates space, information dominates and shapes energy. Thus men, the most informative animals, are the Earth's top predators; Worm Holes with maximal gravitational information are the top predators of the Universe and chips rule machines.

YOUTH AS A LINEAL, ENERGY, TOP PREDATOR SPECIES, GROWING IN SIZE

In their youth, species feed and grow into energetic, lineal top predators, becoming carbohydrates (fats), worms (planarians), lineal echinoderms, fishes, monotrema, Neanderthals and smart weapons.

A new-born, small foetus grows very fast in size as it multiplies its cells. By homology, a new, more efficient species is born as a small, informative, complex being that latter grows in spatial size during its energetic youth becoming a lineal, energetic, big, top predator species that feeds on less evolved forms. Thus after conception, young fishes grew into big sharks of linear forms; after the polemic, dwarf Homo Floresiensis, who seems to have invented technology, the next Homo Sapiens with an extensive fossil record were big, energetic Neanderthals; the 1st big molecules of life were fat carbohydrate chains of linear form; the 1st insects acquired soon gigantic bodies in the Carboniferous; after chips were born as small machines placed in PCs and toys, the first robots they control are big tool-machines and huge weapons of mass destruction, lineal missiles and planes, that kill human beings.

During this energetic age, often a parallel process of extinction of a previous species, which becomes victim of the new predator, takes place. So when lineal proteins appeared, free carbohydrates disappeared from the primordial soup; in the Cambric most water species disappeared as cephalopods with eyes multiplied; mammals feeding on eggs probably extinguished dinosaurs except those who put eggs in unreachable places (birds); humans extinguished most mammals and weapons extinguish non-technological cultures and when they become terminators, man.

II Horizon: reproduction: max. ΣSe =Ti: Radiations of species.

REPRODUCTIVE MATURITY: **III AGE: GROWTH IN HEIGHT=INFORMATION**

In the graph, species evolving through their 2nd and 3rd horizons, acquiring informative height.

In the II Horizon the species finds a balance between form and energy and it reproduces in massive radiations: The protein age gives way to the age of amino acids, with nitrogen, informative atom on its 'relative heads' that multiplied all over the Earth; slow, reproducing sharks gave way to balanced tubular fishes that multiply much faster; brachycephalic Neanderthals gave way to dolichocephalic Cro-Magnons that colonized all continents; while young, giant stars, born in the I horizon, acquire the balanced size of yellow suns, the commonest of all stars.

III Horizon: Max information: Max. Evolution=differentiation

In its 3rd horizon species increase its information, growing in the height dimension or acquiring cyclical forms, improving their sensorial, informative skills: Nucleotides dominated life molecules; echinoderms changed to cyclical shape; fishes developed their inner networks in the dimension of height; amphibians became round; saurian and mammals became bipeds; Neanderthal became Sapiens Sapiens with round skulls and chipped machines acquire today android forms with their heads on top. And the key of this process is obviously the evolution of a hyperbolic, informative element: Nucleotides appear when they add informative, nitrogen rings and cyclical sugars to amino acids; a yellow sun becomes a neutron star of higher gravitational, informative density; insects develop its brain capacity and bees and ants appear; while the Homo Sapiens evolves more complex technological tools; finally in the III Industrial R=evolution of machines, informative chips dominate the economic ecosystem.

Evolutionary divergence between organisms and species happens after their 3rd age: Organisms dominated by nervous, informative systems, which control closely its cells, warp their cellular energy, till the organism dies, according to a clock set by the rate at which energy is metabolised, 'in-formed' by their nervous system. Yet species, due to the discontinuous nature of their individual 'cells' do not exhaust their collective energy. So they might survive without further evolution, in its 3rd horizon, or they might become extinguished by a new, more efficient species appearing on their ecosystem. Or they might evolve socially, becoming the 'cells' of a macro-organism. We talk, in fact, of 3 basic strategies of survival, according to the Ternary Principle:

- *Max. ΣE: Creation of balanced, trophic energy pyramids* that supply new victims.

- ΣE=Σ^{2}i: *Speciation of individual forms* into new species that will survive the extinction of the parental species. Thus we talk of evolutionary, genealogical trees of 'son species', similar to those of any individual. Yet, while the different generations of an organism work together, creating informative networks between them that shape herds and families, 'son species' tend to kill-extinct the mother species, feeding on their energy. We call that fact, *the Oedipus paradox*. So mammals killed reptiles, men killed mammals and robots might kill human beings.

- *Max. Ti: Species also evolve socially their individual forms into super-organisms*, thanks to the creation of networks by specialized, informative 'cells' that integrate all other cells into a whole, bigger form, which has more *exi=stential force* than the individuals of a herd - the fundamental parameter of a top predator, as its exi force determines the strength and capacity its actions have to modify the environment and survive in the fights for existence. Since herds extend in *wider space ecosystems, in which they share a min. quantity of information*, their individuals hardly relate to each other beyond the reproductive couple or the hunting herd; and so their actions per time unit are slower and less coordinated. Hence we consider that the creation of super-organisms is the final evolutionary stage of a herd of individuals from the same species: each individual of the herd becomes then a 'relative cell of the body' of the super-organism. While the specific language of communication and information of the species and the specialized cells of information that carry them, become the relative nervous/informative network of the super-organism. Those informative cells form networks that pack closely the other cells *in min. space, controlling them with a specialized language of information*, as the queen pheromones do in anthills; nervous cells/impulses did with chemical cells in the Pre-Cambrian age or financial and verbal languages have done with humans in societies, which are superorganisms of human beings.

Most space-time systems co-exist in 3 main scales of microcosmic and macrocosmic forms (physical species do so at the atomic, molecular and crystal scale; biological ones, at the cellular, individual and ecosystemic level), which are self-similar because they are created by the previous process of differentiation and social evolution of individual parts into bigger wholes.

- *Extinction*: Yet if species fail to achieve informative, social evolution, growing into super-organisms, they become victims of other species that have continued their evolution and become extinct or in case of having a higher rate of reproduction, lay at the bottom of a trophic pyramid. So today 90% of insects are social insects; Neanderthals became extinguished by verbal Homos and Protista with limited genetic content (mitochondria) became slave organelles of social, eukaryotic super-cells.

In all species and all scales of existence, the process of evolution is dual: species differentiate into 3 forms, which then recombine, giving birth to new species, as it happens in cells (energetic proteins, informative DNAs and reproductive RNAs that make up the cell), or in atoms (energy forces, informative quarks, balanced electrons that reproduce its nebulae on gravitational space).

Recap. The process of evolution of species is equivalent to the 3±st ages of individuals: born of a seed or first individual of maximal information, the species grows its energy in its youth, radiates in an age of massive reproduction and finally diversifies into ternary types. After its 3 horizons or evolutionary ages the species might become extinct or transcend into a super-organism.

5. Ternary principle of causality applied to evolution.

According to the duality of space and time, if Biology is the science of the living observed in space (topological organisms), Evolution is the science of the living observed in time (ages/horizons of evolution). Thus many of the unknown whys of Evolution are resolved when considering the sequential order of evolutionary cycles, as particular cases of the sequential order of time ages coupled with the morphological, topological analysis of the parts of organisms: lineal energetic limbs, cyclical informative heads, combined in reproductive bodies. The 3rd kind of space-time laws to take into account are the laws between scales of biological existence (genes and organisms). And it is a tenant of multiple space-times, derived from 'multi-causality' that in most events of Nature we need to find multiple causes that reinforce each other, coming from those 3 key factors of reality: the existence of multiple planes of existence that influence each other, the existence of ternary topologies with 3 biological functions and the causal order of the 3 horizons/ages of life.

As it happens those 3 elements were discovered in the 3 ages of 'evolution' of Evolution Theory, which as all paradigms of knowledge has 3 ages:

-Darwin focused in the morphological evolution of species.

-The modern synthesis added genetic codes of information that express their vital functions/arrows.

-Gould completed the model with studies related to the different cyclical time beats of species.

- And now the 4th paradigm, the 'social evolution of all sciences' puts all those facts together, which sketch a clear evolutionary plan natural to all species of multiple spaces-times of reality:

Those evolutionary laws have been developed by Biologists since Darwin published his work on morphological change. Yet Darwin already pointed out that Evolution Theory faced a critical 'unknown' why: How *random mutations* could select the most efficient forms of energy and information to develop complex organs? Darwin wondered how wings and eyes with its extreme complexity could be born in so short time: If all mathematical forms were allowed in a random mutation, chances that those mutations provided the right path to evolve eyes and wings was minimal. This difficulty has prompted the absurd argument of religious Creationists in favor of a Divine plan, selecting those paths. Let us consider the ternary principle of causality to resolve this mystery in more detail:

Topological Solutions

The problem is solved by the Ternary Principle/Restriction, which allows only 3 mutations of form towards species with higher lineal energy, cyclical information or enhanced, balanced, $\Sigma S=Ti$, reproductive capacities. Thus mutations, according to the Ternary Principle, are limited to the 3 only possible morphologies of 4-dimensional spaces and its 3 homologous functions in time: informative functions are spherical forms; energetic functions are lineal; and reproductive, exi, functions combine both forms. So organisms that mutate follow those restricted cyclical and lineal paths systematically and tend to create energy and information systems. *This explains how so complex organs such as wing and eyes can be formed in so little time: Si*nce wings are energetic organs, they are planar forms, born naturally through a specialized series of lineal mutations. Since eyes are informative organs, they are spherical, naturally born through the other main path of evolution: informative growth.

E ⇔ I beats of species: punctuated & allopatric evolution

Further on the previous topological solution is *reinforced by a sequential, time order/cycle:* A basic ExRe⇔I beat of all time cycles accelerates the process: species undergo 2 cyclical phases, switching between their spatial, reproductive states *that cause a biological radiation= massive reproduction of the being* and their informative, evolving states that causes a punctuated evolution of its morphology. In cyclical time that process is just another dual case of transformations between spatial and temporal states, defined by the Space-time equation:

$$\Sigma Se \ (Max. \ Reproduction) \ \Leftrightarrow \Sigma^2 \ Ti \ (Max. \ Evolution \ of \ form).$$

Punctuated evolution happens in the informative state, *often in synchronicity with a cold=informative age,* during which the fossil record shows an acceleration of mutations and evolutions. While in hot=energetic ages of the Earth, the evolved top predator will start an age of massive reproductive radiations and body growth with little formal evolution. Thus, the rhythm of evolution has 2 phases: a 1st phase of evolution of information in time, described by Gould (punctuated evolution in minimal space with minimal reproduced species and maximal time speed), which creates a dominant species, followed by a massive reproduction in space of the new top predator species (with minimal time evolution and maximal spatial expansion), which often triggers a parallel process of extinction of a previous top predator or prey, displaced by the new dominant species. It is once more a proof of the inverted properties of space and time, and the fractal nature of space-time movements, which are always an 'stop and go' process, in this case an evolutionary, reproductive complex rhythm of space-time arrows. So punctuated evolution means that when the species radiates into self-reproductive waves, it doesn't evolve in time and we do not see any variation. But when the species evolves in time, it hardly reproduces in space, often evolving in isolated, small groups (allopatric evolution). Those stop and go rhythms found in the fossil record, are 'ceteris paribus', partial perceptions of a discontinuous beat of $\Sigma exRe => I => \Sigma exRe$, extinctive= reproductive + evolving states, proper of the reproductive motions of all species of the Universe.

Ultimately, such dual cycles of life and extinction of species proves paradoxically, the immortality of the space-time Universe:

$$Max.\Sigma \ Energy \ (Prey) -> Max.Information \ (Predator) -> Death \ (Max.\Sigma Energy) -> Feeding \ (Max.Ti)$$

Since each relative past and future will cancel each other; each death will cancel a life; each antiparticle will extinguish a particle... But the arrow of reproduction will repeat them again to continue the game, which means that the 3rd arrow of fractal reproduction of relative presents, made of energy and informative arrows, dominates a Universe that acts as a *'self-reproductive fractal of energy and information'.*

The fundamental particle of the Universe is a space-time, fractal super-organism, a present being, born out of ∞ fractal balances between those cellular micro-cycles of energy and information, $\Sigma e \Leftrightarrow \Sigma^2 i$ that iterated ad infinitum and chained by energetic, reproductive and informative networks/cycles create the constant, organic forms of the Universe. Since those arrows/cycles of time are exchanges of energy and information iterated in cyclical actions, any organism can be described as a natural fractal that switches constantly between those 3±st arrows/cycles of time, energy, information and reproduction, between birth and extinction. The beauty of Fractal organism is the simplicity of its recurrent cycles/cells (temporal, dynamic view/spatial, formal view), based only in 2 parameters, energy and information, which can get infinitely complex by iterating its forms.

The perfect rhythm: palingenesis

The palingenetic process of evolution and reproduction also converts a cell into a fetus that becomes a macro-dimensional organism, sum of multiple cells: Palingenesis alternates the process of cellular reproduction and informative organization, causing the creation of a foetus as a constant sum of those 2 phases, $\Sigma ExRe -> Ti -> \Sigma ExRe$. A fact that should not surprise us *since an organism is a microcosmic species of cellular DNA.*

Thus a rhythm to observe in all organic processes of informative evolution across space-time planes is the 'compression' in time and space of the entity evolving according to the inverted properties of energy and information: informative organisms diminish its spatial size to accelerate its evolutionary rhythm of morphological change (chip paradox), evident in the case of the first mammals (shrews) and technological humans (Floresiensis[3]). When we observe the organism in its temporal evolution palingenesis also compresses its speed of evolution into a minimal space-time, during its fetal age. So 500 million years of evolution are reduced to a mere 9 months. The organism evolves in time and space through multiple species and planes of existence, from a cell to an adult, packing temporal eons of the entire earth, into a tiny st-Space-Time region. Thus, if we want to fully understand the nature of palingenesis we have to stretch our analysis in 2 directions:

-*Max. I: In time* through the evolution of all its previous species.

- *Max. E: In space* through morphologic analysis of all those species in its tissues and inner networks.

Both processed, despite its enormous complexity adjust to the laws of multiple spaces-times and in the complex models of this work were key analyses to decipher the fractal tree of evolution of all life species on Earth, later studied in more detail.

Interaction of planes of existence and ages of time.

- Finally, we can consider a correction of the deterministic concept that genes are the only cause of variation in species.

This idea is due to the unicausal Logic of metric measure and the need of exact instrumental perception that are dogmas of the 3rd paradigm. Yet in theory of multiple spaces-times all systems are paradoxical, multicausal and information also flows from the higher, more complex system to the simpler parts. Thus there must be a flow of information from the higher 'neuronal' system of the brain to the genetic system that allows more 'precise' Lamarckian variations to reinforce the process of genetic random mutation during the informative phase of evolution in allopatric, limited spaces and limited numbers. It might never be found since it would require decoding the fast, informative messages of the brain but the mechanism according to complex laws of multiple space-time and its ages should be like this:

During youth, when the species imprints its neuronal cells with maximal DNA-informative content as it learns and makes efforts to develop new morphological features in a changing environment (the classic examples of giraffes' neck and Galapagos bird's peck) those efforts are encoded in the fast changing DNA systems of grey matter, which during adolescence, when the reproductive cells are developed are transferred to the seminal cells that will become the informative seeds of the new, mutated species. Because again the codes to transfer are 3 simple topologies and we know the brain has an homunculus image of all the parts of the body, one could hypothesize that as this homunculus images mirror the body parts a parallel neuronal-DNA mapping somehow is able to mirror the morphological changes the individual achieves during its youth as it stretches its neck and makes it longer or pecks harder seeds and grows a stronger peck.

And so only considering the 3 reinforcing causes we can conclude that evolution of complex systems in fast time is possible.

Recap. Evolutionary theory went through 3 ages that stressed topological evolution (Darwin), interaction between the scales of life (genetics) and the different speeds of time and E⇔I beatings of the Universe. All those elements come together in multiple space-times Theory as ternary causes that explain the speed and accuracy of evolutionary changes.

III. THE 3 PLANES OF LIFE EXISTENCE

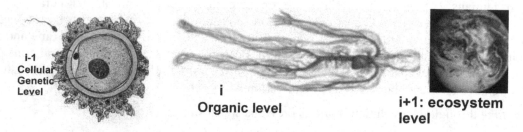

i-1
Cellular
Genetic
Level

i
Organic level

i+1: ecosystem level

In the graph the 3 existential scales of humanity: The evolution of an organism happens through the interaction of its 3±st planes existence, which in man are its ecosystemic, social level; its organic, biological level and its cellular, genetic plane. This law of Evolution finds its whys in the 4th postulate of i-logic geometry. In that sense, life between birth as a cell and extinction can be considered a trip through those 3 planes of existence.

6. The 3 scales of life existence.

In Multiple space-times Theory there is not only *individual evolution* but also *hierarchical, social, scalar evolution*, through the ±st 3-scalar structures of all spacetime systems. *Such* triad of 'organic scales' illustrated in the graph defines the existence of an organism, at the st-1 cellular, genetic scale, at the st-scale of physiological networks and individual organisms, and at the st+1 scale of species and ecosystems in which they exist, exchanging energy and information.

Since we already studied the trip of life between those 3 planes in our lectures on i-logic geometry, we shall now consider another key element of all systems of multiple spaces-times, the synchronicity between the microcosmic and macrocosmic planes that determine with its massive changes in energy and information parameters the destiny and evolution of the smaller parts.

In that regard, the evolutionary changes of animal phyla that brought man from the initial cell have to be considered from the perspectives of those synchronicities between the 3 scales.

Take the most famous event on life history; the creation of multi-cellular organisms that happened in the Pre-Cambrian age when the survival game of existence made it 'necessary' due to the simultaneous interaction of those 3 st-levels:

- st+1: The Earth passed by an icy period in which cells overpopulated the remaining hot spots. And so as the informative density of cells and its evolutionary speed increased in those tiny spots, it provoked the merging, cannibalism and social evolution of prokaryotic cells into eukaryotes. Thus the creation of new life species was accompanied by a raise in its survival stakes, caused by the activities of the global ecosystem that accelerated the process. Since once and again the Earth's glaciations will cross those new accelerated, evolutionary discontinuums.

- *st*: Yet those Earth's changes don't evolve organisms 'per se', but trigger their competence, which causes their evolution. So the will to evolve socially exists in the organism not in the ecosystem that merely sparks off the process. For example, multicellular organisms were born in a similar fashion to eukaryotic cells, when an ice-ball age packed cells into underwater volcanoes, which developed social strategies to hunt together individual cells.

- *st*-1: However, unicellular cells existed also in earlier glaciations and yet they did not evolve then. Their evolution only happened in the Pre-Cambrian age in the precise moment in which also at genetic level the so-called intronic DNA[4] multiplied the capacity to codify genetically complex multicellular tissues - since redundant, intronic DNA, in charge of the regulation of complex organisms, appeared in the Pre-Cambrian age. Thus only when the 3 'hierarchical levels', the genetic, organic and environmental acted simultaneously that explosion of life was possible.

Thus again we find a case of ternary causality is needed to fully account as in most process for an event which Aristotelian unicausality cannot explain alone, but merely divide scholars in opposite positions, making truth Leonardo's ex-abrupt: 'the only harmonious sound found in a meeting of scholars are the winds of their asses.' Hopefully the 4th paradigm will resolve that (-;

Again in the next evolutionary explosion at the end of the Cambrian age, the 3 levels acted simultaneously:

At organic level, the chemical systems that dominated life on that age gave birth to the first visual systems, which in the late Cambrian triggered a new massive evolutionary age with the arrival of squids with eyes. Those squids evolved in the still, lower, bio-luminescent zones of oceans, during the parallel massive changes in the forms of continents that sank the border zones of shallow waters, accelerating extinction of 'obsolete' smelling animals without eyes. Finally, the 3rd cause made it all possible: certain eye genes appeared and made possible the creation of the most complex sensorial organ of life.

What is the order of control between those 3 levels? They are apparently simultaneous but a detailed analysis of the pre-Cambrian episode shows that the Earth changed first, and its changes in energy and form implied higher survival stakes among individual cells that caused their evolution, as they came together and started to exchange genetic material, provoking 'accidental' mixes with redundant intronic genes that were finally used to control the multi-cellular organism.

Thus, again we find that unicausality, which considers only the lineal arrow that goes from genes to organisms, misses the second dual arrow that causes evolution from the higher organic st-system down to the genetic scale; as it happens once and again in all types of Space-Time systems. Indeed, we have to put together the pieces of a car to create the car; but the human designer exists and ultimately in the Universe at large the i-logic laws of multiple spaces-times *are the designer that* chains the different ±st-levels of an organism; controlling it also from top to bottom in the same degree that st-1 levels control it from the bottom to the top. In that regard genetics, has to accept some fundamental principles of multiple space scales and multiple causal systems:

Past (energy, microcosms) x Future (information, macrocosms) = Present st-scale of the organism.

'The plane of existence with higher exi force determines the direction of future'

Or in other words the small and visible quanta of energy do not determine all the events of biological reality even if we perceive them easily; since it is the whole and its networks of information, which curve and form 'spatial energy'. *Yet since the more complex systems are more difficult to decipher and information/future systems are smaller, faster and often invisible, it is a rule of scientific inquire during the metric paradigm to work only with an entropic, spatial, explosive, hot, microcosmic first cause.* And then as science evolves contradictions impose revisions that prime a second informative, temporal, implosive, cyclical, cold cause.

So genetic quanta determine an organism in the same or lesser degree than its ecosystem - a fact, which of course, has certain metaphysical implications, when applied to the biggest organism of it all – the Universe: the ultimate cause of it all might be the big-bang singularity - the genetic memory of a previous Universe that will develop according to the laws of i-logic geometry, *the mind of the Universe, its ultimate cause.*

That mind-singularity came before its Universal body – it was the eternal informative code of reality.

Let us now consider how the main laws and creative differentiation of such Universal 'plan of evolution' affects those 3 hierarchical levels of living organisms in more detail.

st-1. Ternary differentiations on the genetic and cellular level.

The cellular level was fundamental to the process of evolution in its first stages in which cells differentiated and multiplied its e-exi-i varieties, creating a decametric scale, origin of the subsequent differentiation of organic networks and tissues into 3X3+(st+1)=10 fundamental types *whose forms and function can be explained as ternary differentiations of the 3 topological species and its 9 dimensional functions.*

Yet cells also act through the genetic level, codifying not only the creation of life molecules but also the higher organic scale through epigenetic, intron and redundant RNA and DNA material, which started the massive differentiation of species in the pre-Cambrian age. Epigenetics should work following the hierarchical and multifunctional principles of multiple spaces & times, *coding sub-structures of higher complexity that scientists still need to decode, as the decametric scales of multiple spaces-times tend to generate 3x3+(st+1) structures, with new functions closer to the higher levels of the organism.*

So the informative DNA is the highest scale of existence or 'will of the cell' and it has a lot of redundant extra-RNA that it should use together with the hormonal language to 'dialog' socially with other cells in/forming the functions of social tissues. Hormones should establish 'spatial dialogues' and redundant RNA temporal dialogues to code those systems. And both languages together should create the higher st+1 organic cycles and spatial structures of the organism.

Finally on top of both informative systems, the ternary principle adds the brain's nervous impulses to fine-tune those 2 languages with military precision, as indeed nerves can 'kill' as military do, undisciplined subjects/cells.

Ternary differentiations at individual, organic level.

In the biological, organic level, evolution is fostered by the fundamental differentiations of any $\sum E \Leftrightarrow \sum^2 I$, relative space-time field that cause a 'limited' number of possible formal species:

-$\sum$: *Fractal differentiations* between big, single animals Vs. smaller animals with multiple parts that become their preys: for example, single eyes and single bodies in molluscs Vs. multiple eyes and body parts in insects. *They prove that single informative networks are more efficient that loose herds.*

- *Big bang/energetic vs. big banging/reproductive rhythms/ differentiations* during the embryo stage, in which organisms establish 2 strategies of reproductive, cellular growth:

- *Acoelomate* and *Seudocoelomate* are big-bang organisms, which multiply its cells in a dense ball without inner cavities. Hence its size is minimal.

- Those organisms evolved latter into *Coelomate, organized with the topological structure of st-points. Since they* have 2 internal bilateral 'big-bang cavities' that grow much larger by invagination, in which the future EXI cyclical organs of exchange of energy and information (digestive, lung and vascular tissue) will develop. Those cavities are surrounded by inner and outer tissue that will form the external membrane or skin and the inner nervous, informative systems. Coelomate further differentiated according to the Ternary, Fractal Principle in new subclasses and grew into larger organisms, expanding further those cavities. So today most animals are Coelomate.

- The *Fractal, Ternary Principle in Space (3 topologies, 3 differentiations) and Time (3 causes, 3 horizons) also explains:*

- *The evolutionary differentiations* in time of species through 3 horizons of increasing informative and social evolution, from individuals to herds to tighter organisms.

- The 3 *networks/organs in space of those species - the energetic, digestive network, the informative nervous or hormonal network and the reproductive network - which further evolved as they* increased its informative complexity towards 'the future' in 3x3+(st+1) tissues and networks.

- Both processes together determined the organic evolution of animals, as each new top predator improves its organic networks, displacing a previous top predator. So mammals improved their blood, energy, nervous and reproductive systems, displacing reptiles and men improved their brains, displacing mammals.

- Finally the different *dualities and inversions of space-time* manifest in organic life through:

- Spatial Symmetries (left to right) called bilateralism.

- Asymmetries of temporal, hierarchical dimensions, since informative heads dominate reproductive bodies and are placed always on top in the dimension of height.

- *Chemical and electrical languages.* The duality body-brain shows in the interaction between the reproductive blood network that dominates the spatial body and the informative network that dominates the brain and nervous system. Nervous cells and languages control chemical languages and cells *by creating simultaneously informative, nervous orders and chemical, energetic orders that arrive latter to the cells, reinforcing the electric message.* So within the organism, neuro-secretory cells make and pour hormones to the blood network, while sending messages to the nervous system that will arrive first as advanced orders. Another example happens in the brain: electric neurons translate nervous impulses into chemical hormones and 'calcium waves' that reinforce the message, transferring it to the chemical cells. This primacy of faster, informative systems happens also in ecosystems: a nervous animal species preys on chemical plants.

- The *Black Hole paradox.* That dominance of information over energy shows also in the competence between species: Top Predator brains win over top predator bodies, thanks to their faster fractal or social actions that give them more IxE force per unit of time. So mammals win over dinosaurs and chips control machines. This means that a dominant species often is born as a small, fast form (Min. E=Max.I) that latter grows in size.

- *The Oedipus paradox:* The growth of complexity with the passing of time shows in the fact that new top predator species grow by feeding in previous, parental species, which become the 'energy' of their reproduction, as they share and compete in the same environment. So animals fed their son species, men; and men feed their son species, machine-weapons, evolved in wars.

The ecosystemic and geological level.

The planet acts on life at ecosystemic level, establishing the settings for prey-predator events and the conditions for evolution that can be summoned up in a word, *Geographical isolation*[5]:

Evolution is dangerous. Since during the mutational phase the underdeveloped new organs hinder the survival of the being. For that reason the 3 'scales' act simultaneously to accelerate evolution:

- st-1: At genetic level, organisms suffer a change in their time speed, increasing its fractal, informative mutations.

- st: At organic level species follow a plan, provided by the lineal-cyclical duality of spatial and informative shapes, the only 2 possible paths of morphological evolution. It requires also the extra exi force provided by top predators. So species use only a little amount of its IxE force externally to absorb energy, changing its energy/information rhythm, e->I, from spatial movement in search of energy into inward, temporal, informative mutations.

- st+1: At ecosystemic level a species requires a secluded environment in which he won't be menaced by top predators during the short time in which the evolutionary program dwindles survival chances; creating also densely populated regions that multiply the speed of informative change. For example, amoebas, fishes and amphibians were born in fresh water, a relatively secluded environment; while anaerobic bacteria, animal life and cephalopods, probably came from abyssal places, another secluded region. Finally, apes and birds came from secluded trees.

Without that synergy between the 3 scales of existence organisms would never evolve but become extinct during the supposedly chaotic ages of mutational evolution. Yet once the new form is found its survival chances improve geometrically, since it has evolved positively according to the i-logic plan, hence reaching a higher exi force. Then the being gets out of the secluded environment and changes back its I->E rhythm, starting a spatial, reproductive radiation, becoming a new top predator.

Thus, a ternary e->ixe->i rhythm of evolution-reproduction-energy feeding is proper of all successful species.

-Finally the st+1 Geological and climatic changes accelerate both: Extinction processes, especially among species of shallow waters that live in sea platforms, destroyed periodically by continental collisions) or species which depend on external temperature to regulate their metabolism (reptiles); and evolutionary processes, as they rise the 'stakes of survival' and establish a climatic parallel Se⇔Ti rhythm; as cold zones foster informative evolution.

All those environmental differentiations can be resumed in a word, *Hierarchy*, which manifests in the restrictions imposed by the higher ecosystemic planes of existence to the organism. For example, as animals migrated first from seawater to river and shallow waters, then into land, the transparent air environment eased their sensorial vision and so it fostered their informative, electronic, nervous evolution and developed further the informative dimension of height from where light comes.

Recap. The ternary principle of multicausality explains how the genetic, individual and ecosystemic planes of life existence simultaneously caused the great changes in the evolution of species. Those 3 levels of change co-act simultaneously since they are relative past, present and future levels that come together in the creation of the organism. That is, genes do not impose evolution to organisms that impose evolution to ecosystems, but according to Non AE-logic the 3 levels co-exist in a relative present, influencing each other and co-evolving together creating new species with higher information and IxE force, *the arrow of future in all ecosystems and worlds* from the Earth, to the galaxy, a growing vortex of informative species with higher mass towards its center. Thus 3 scales and the ternary, fractal principles – the plan of evolution - suffice to explain the evolution of life from molecules to human beings.

7. External dimensions/networks of organisms: territories.

In the graph, a mammal territory. Any animal territory is an i-logic space-time with 3 zones:

- An informative central territory (1) or den, where animals reproduce and 2 secondary homes where the herd performs secondary organic cycles (2,3).

- An energetic membrane (M, 5) - an invisible limit that provokes a confrontation if a stranger crosses it and where most energetic preys 'flee' away from the den of the predator.

- An intermediate zone with cyclical paths of absorption of energy and information; where we find a hunting territory, places to drink (E), to bath (B), socialize (A), defecate (D), etc.

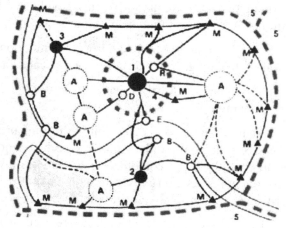

In organic terms, a dimension is a network. So a living organism can be considered a sum of cellular quanta united by 3 basic space/time discreet dimensional networks, which are its physiological systems: *the digestive/energetic network, the informative/nervous network and the reproductive/blood networks* around which cells teem, creating a stable, organic st-point. In other words the energy and informative networks of a living being are its internal, diffeomorphic dimensions (of relative length and height), to which the organic system adds a 3rd, reproductive dimension that combines both elements and represents the width or 'volume of cellular quanta' of the system. Finally its movement in the external world becomes its 4th temporal dimension. Yet that 4th dimension of external activity can also be considered a network territory in itself, sum of the 3±st cycles of existence of the being, creating a bigger vital space that will become the basic unit of an ecosystem or social organism made of individuals of the same species. In the figure we draw the vital territory of a minimal social pair of mammals, differentiated in 3 clear sub-sectors:

- *Max. Information: Informative den or central territory (1,2,3):*

It is the territory of reproduction used to copulate and store basic food and energy to raise the young. It is a forbidden zone where not even hunting is allowed (4). In social species of great mobility, aerial or marine, where borders are much more extensive, this territory is very ample and tends to be located in warm latitudes.

- *Energy=Information: Dual Territory of energy hunting and informative socialization (5).*

It is the feeding, social and hunting territory, on which the central informative being feeds itself. It is outside the zone of reproduction. It is the winter territory of many migratory birds.

Given the relativism of all movement, in biological territories the informative singularity moves to hunt its energy quanta, as opposed to galaxies where stars and space-time dust moves towards the central worm hole.

Within those limits there are also neutral territories of communication, courtship reproduction and free energy, like water troughs. So the intermediate territory works both, as an informative and energetic territory where different victims and predators trace parallel cycles and come together around meeting points (E, B, R).

- *Max. energy: Borders that limit the territory.*

Membranes are dangerous zones because the informative center watches them with special care to control any invasion of its hunting/social territory.

Those limits fluctuate according to the power of neighbours. For example, the vital space of a fish increases during mating, since the couple is more powerful than a single individual.

Marks (M points) fix those limits and reduce combats. They are often invisible, as most territories are defended against competitors of the same species, who understand the informative code of those marks; but rarely against members of other species. So we find all kind of linguistic marks:

- Smells (common in mammals, like foxes, rhinos, antelopes), excrements (in canines and felids) or other glandular secretions.

- Optical marks often connected to scents: The brown bear creates marks in trees, rubbing them with his head, warning adversaries of his great exi size and force. In human empires (nations can also be treated as biological territories) visual marks correspond to armies displayed in the borders. In human homes those marks used to be shields with weapons; now they are cars and other proofs of money, the new language of social power.

- High pitch, acoustic marks, proper of birds, which are triggered when a rival enters the territory.

Recap. Vital territories of animals and human nations can be explained with the 3 topological regions of st-points.

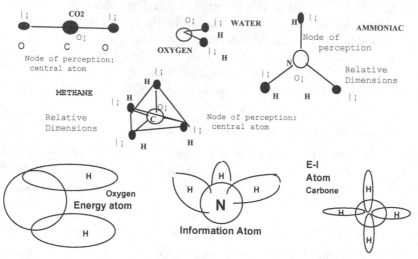

In the graph, the 4 basic molecules of life, CO_2, H_2O, CH_4, (Methane) and NH_3 (Ammonia), are composed of the 3 organic atoms, C, O, N and its slave hydrogens. They also adopt the same efficient st-morphologies that crystals have with a central, informative atom surrounded by submissive, spatial Hydrogens that process and send to the center Van der Waals flows of information and energy.

8. Topology: from C, O, H atoms to amino acids.

In the smallest scale the molecular 'bricks of life', CO_2, H_2O, CH_4, (Methane) and NH_3 (Ammonia), shown in the graph are, as it happens in the geological world, crystalline structures. Yet they are far more malleable and complex than solid crystals, because they exist in liquid ecosystems. In those simple life molecules Carbon, Oxygen and Nitrogen are the Top predator central atoms that capture and surround themselves with weaker IxE Hydrogen atoms, which act as carriers of their relative energy and information, creating the external membrane of their molecular st-point structure. Hydrogen atoms act either as their 'senses', processing electromagnetic energy into Van der Waal forces of information or as an energetic limb that the central atom expels or attracts in alternating cycles that displace the molecule through water. Thus the simplest organic molecules studied as st-Points have:

-Max. I: A top predator *N, O, C atom* in its hyperbolic center.

- E=I: It is surrounded by an intermediate st-*body of hydrogen* that processes the information and energy of *the water medium (Max.E)*, in which the molecule lives, extracting it for the central atoms of life with the molecular, external 'electronic cloud' of the organic 'crystal' that becomes *its 'energetic membrane'*.

In the next scale of social organization those 3 elemental particles of life, N, C, and O will become themselves the 3 fundamental zones of bigger 'i points', called organic molecules:

- *Max. I:* The informative element of all life molecules and fractal part of its relative 'heads' are Nitrogen atoms. Its informative character is shown already in the crystalline ammonia, where Nitrogen is the dominant vertex of a tetrahedron shaped with 3 more Hydrogens. Accordingly ammonia is a perfect atomic clock in which the Nitrogen vibrates constantly back and forth through the hole shaped by its 3 hydrogen 'eyes', with an informative cycle of $+10^{16}$ times a second. The first atomic clocks were in fact based in that simple molecule due to the accuracy and speed of times of its vibration, which in life molecules allows nitrogen to 'inform itself' and translate 'moving orders' to the 'carbon' body of the molecule.

- *E=I:* The structural atom that creates the rigid body structures of life molecules is the carbon atom. It has the maximal number of valences - 4 orbitals that create dual bondages with 2 other carbons - constructing long, formal ternary chains of great structural rigidity. As such it is the most visible, intermediate form of all life compounds. And so biologists, due to its 'visibility', have traditionally considered it the 'fundamental element of life'. When we ad Hydrogen atoms to complete its 'crystalline' body we obtain methane the simplest molecule of life. Then when we join 2 carbons with dual bondage, we form ethane, which already acts as a hormone, (±reproductive molecule) inhibiting the structural growth of plants.

- *Max. E:* Finally, the energy of life systems is water, which fills the 'external world' in which life feeds and internally becomes in complex living organisms called cells, the filling energy of the intermediate space, enclosed between its carbon-based protein & fat walls and the inner, informative, nitrogen-rich, ADN hyperbolic singularity. Water is the simplest, most abundant ternary molecule of the Universe, made with 2 slave hydrogens and 1 dominant oxygen - the atom that has the maximal 'electro-negativity' after fluorine. Thus, oxygen can capture the electronic body of any other atom. That is why the oxygen components of carbohydrates enact its energetic cycles, moving those organic molecules within the water ecosystem. Since they stomp on water, breaking it and creating expansive and implosive OH^-, $H^{\pm}$ ions that impulse the molecule; as you walk on the floor, 'stomping' on the electromagnetic fields of the ground or as a fish moves, hitting the water with its tail.

Recap. Nitrogen heads, carbon bodies and oxygen energy create the simplest life beings, amino acids…

9. Topology: From carbohydrates to cells.

In the next graph, the glycine is the simplest active life form with 3 st-zones:

- Max. I: The nitrogen head directs the glycine.

- E=I: A dual carbon creates the first rigid 'membrane-body' of life with its strong, covalent bondage, joining the head and tail:

- Max. E: Its oxygen COOH tail 'walks' on the water, breaking, attracting and repelling its OH-, H+ radicals.

Amino acids show their complex, reproductive and social arrows, catalyzing through their movements the replication of new amino acids and forming social chains, called proteins.

The inverse properties of st-amino heads and e-oxygen tails make possible the creation of long chains of amino acids in which their nitrogen heads bite their oxygen tails becoming neutralized as part of a complex social structure: the protein.

The nucleotide improves upon the amino head, carbohydrate body and oxygen legs of the amino acid, adding to those 3 lineal forms a dimension of informative height; as latter will occur in macro-organisms, when flat worms become cylindrical. So the amino acid head becomes a dual nitrogen ring, the body becomes a sugar and the tail multiplies its oxygens around a highly electronegative Phosphoric acid. The outcome is a nucleotide acid (right side of next graph) - *the top predator life molecule*, which evolves socially all others into the next st-scale of life, the cell.

10. The 3x3+(st+1) horizons of social evolution of life.

Biological organisms, as physical organisms did in their growth from atoms to galaxies, evolved in 3x3 horizons of increasing social complexity creating the 3 ±st scales of life: the age of molecules, the age of cells and the age of multi-cellular organisms, specialized in energy (plants) or information (animals).

Let us study those ages in more detail, further differentiating them in 3 sub-ages according to the Ternary principle.

Since if we apply the law of the 3±st Ages to organic molecules we can explain how they grew in informative complexity and spatial size till acquiring the form of living organisms, in a process similar to the evolution of particles that created the cosmological bodies of the Universe.

Those 3±st evolutionary ages of life, each one sub-divided in 3±st sub-horizons, are: the young age of molecules, the mature, longest age of cells and the 'recent' age of living organisms, which will end with the creation of a global single organism, Planet Earth (st+1):

The 3±st ages of molecules.

- *st-1: The atomic age*: the simplest chemical molecules of life are formed. The 3 simple atoms and molecules of life recombined its energetic, reproductive and information functions and grew, forming bigger chains thanks to its atomic affinity, acquiring more complex 'vital properties'. The simplest combination of them, the CNO molecule, urea, is considered the first molecule of life and its 'crystallization' in a lab, departing from non-living atoms, was considered the birth of biochemistry and the prove that life is an atomic system that shares the same properties of any other ExI system of the Universe.

- *Max. E: The energy age,* dominated by lineal, long, simple fats, huge carbon chains with oxygens attached to its ends.

- ⇔: *The amino acid age*: COOH, methane and ammonia, the 3 simplest life molecules of the triad of life atoms, O, C, N, combine as the relative energy, reproductive and informative organs of amino acids. Amino acids reproduce exponentially in the primordial organic water soup and evolve socially into proteins.

- *Max. I: The nucleotide age.* Nucleotides, the informative molecules of the life, add an informative dimension of height to lineal amino acids, forming nitrogen and sugar rings. They dominate all other carbohydrates. Soon they will also evolve socially into huge chains called nucleic acids.

- *st+1: Social age. Nucleic acids*, the macromolecules of life with max. Exi force, integrate socially all other carbohydrates in herds of vital molecules, creating *the cell, the following st-scale of life.*

The 3±st ages of cells.

- *st-1:* The previous 3±st horizons of evolution of molecules brought the first cells.

- *Max. E: The age of RNA Protista.* The energy age of the cell is dominated by the simplest RNA *Protista*, whose ternary st-structure is based in: an external protein membrane; a series of 'convex' spiralled RNAs, the singularities that directs the cell, and an internal, intermediate water zone, the cytoplasm, where the cell reproduces its specific energy and information - thanks to the free 'energy' of water radicals - with the instructions given by those RNAs. Those protista reproduce massively, exhausting the organic elements of the life soup. Then it comes:

- ⇔: *The Age of DNA Protista.* It is the balanced, mature age of protista. Dual RNAs peg together to form informative DNA rings, which store new genetic information that permits further growth and differentiation of Protistas, according to new, improved 3 st-regions:

Energy membranes invaginate the cell with a tubular network, the Golgi apparatus and protect the still DNA with a differentiated nucleus membrane; while new, specialized organelles perform the energy and information processes of the intermediate zone, creating reproductive Mitochondria and Chloroplasts.

- *Max. I: Informative age and differentiation: The Eucaryotic age.* Informative DNA cells multiply its genetic memories, while RNAs differentiate into a triad of forms that increase in the intermediate space-time the reproduction of membranes and proteins, creating giant cells. They cannibalise and enslave smaller, symbiotic cells, specialized in the dual arrow of energy, (mitochondria and chloroplasts) and information (ribosomes). Those who absorb chloroplasts become algae; those who feed on mitochondria become animals.

- *st+1*: The biggest eukaryote animals are amoeboid cells that evolve faster, informative, electronic languages using heavier metal ions, K^* and Na^-, to send their messages to other cells through their membranes. The nervous language allows simultaneous cellular actions, creating mobile multicellular organisms called animals. While slower chemical languages that use 'hormonal vowels[7]' put together unicellular algae into plants.

Let us consider the evolution of animals:

3±st horizons of evolution of animals: network's organisms:

- st-1: Conception. Electric cells create multicellular organisms in control of all other cells, gathering in 3 physiological networks - neuronal, muscular and glandular=digestive systems that perform the informative, reproductive and energetic cycles of the organism as a macro-living st-point. The sequential dominance of those physiological networks creates the 3 ages of life - the energetic youth, reproductive maturity and informative old age - and the 3 horizons of evolution or 'main phyla' of multicellular animals.

- *Max. E:* The energy system -a central digestive tube- dominates the 1st horizon of multicellular organisms, occupying the central zone in 3 sub horizons of formal evolution: the age of sponges, the age of hydras and the age of worms, the first bilateral animals created around a tubular, lineal digestive system that moves in the dimension of length.

- ⇔: Worms develop blood networks based in metallic carbohydrates that carry to each cell of the body its oxygen energy, food quanta and the dual hormonal orders of the brain: reproductive orders and 'killing orders' performed by amoeboid leukocytes. Thus, as blood networks increase the efficient control of fractal cells, animals grow in size, starting an age of massive sea life speciation. Today we still have 90% of the genes of those worms.

- *Max. I:* Life jumps a fundamental discontinuum, when the first molluscs become insects and the first fishes become amphibians, colonizing the Earth. Their sensorial and nervous systems become overdeveloped in the new environment that has a higher transparency to informative light. Land animals specialize their 3 networks to the new medium in 3 sub-ages: the age of amphibians, which still reproduce in water, the age of reptiles and the age of birds and mammals, dominant in visual and nervous systems that ends with the arrival of Homo Sapiens.

- *st+1:* Homo sapiens develops a new informative language, the word, evolving into historic super-organisms, civilizations and economic ecosystems that grow in size till reaching a global dimension.

Recap. Biological organisms evolved in 3 horizons of increasing social complexity creating the 3 ±st scales of life: the age of molecules, the age of cells and the age of multi-cellular organisms, specialized in energy (plants) or information (animals).

11. The energy age: 3±st Simple molecules.

In the graph, a glycine, the simplest 'organism' of life or amino acid with its self-similar form to a small mammal, with a nitrogen head, a carbon body and its oxygen legs. And the two fundamental species of complex molecular life: a nucleotide, dominant in information, attached

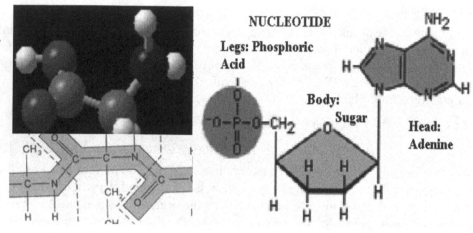

NUCLEOTIDE

Legs: Phosphoric Acid

Body: Sugar

Head: Adenine

to a sugar body and phosphoric acid and the detail of a protein body; the most reproduced species of the system. Lineal fats that store energy becomes the 3rd essential topology of cellular life.

The first age in the evolution of life is the age of simple molecules. Water became an organic soup filled with ammonia and simple chains of carbon, among which we highlight:

- *Max.E: Acids and fats.* They are headless, without nitrogen heads - a long, lineal limb of energetic carbons with oxygen legs on its extremes. They will become the fundamental energy of cells.

- *Max.I: Sugars* add an informative, cyclic dimension to headless fats. They are carbohydrate hexagons evolved socially in long chains, through oxygen connections, called polysaccharides.

Amino acids: exi functions and evolution into proteins

The 3 life molecules, ammonia, methane and water, create *the spatial structure* of glycine, the simplest amino acid which resembles an animal, with the positive charged nitrogen head (amine), the negative charged Oxygen tail (carboxyl), and a carbon body chain that fusions together the 2 extremes, creating the i-logic 'generator equation' of amino acids:

Oxygen legs(E) <Carbon body (EXI)> (I) Nitrogen head.

st-points adapt their morphology to the dimensions and directional movement of their specific environment. Thus, while a still cell is cyclical, moving life molecules are lineal forms in which *the nitrogen 'head'* is upfront to absorb information & energy in the direction of movement; *the structural carbon* membrane is in the center; *and the energy cycles* are performed by the *oxygen* tail that moves on the water:

- *Max. I: Informative cycles* are directed by its ammonia 'clock'. Nitrogen vibrates across its Hydrogen triangle, perceiving and transferring electromagnetic information elaborated as Van der Wall forces to its carbon body that orientates the molecule in a chosen direction.

- *E=I: Reproductive and social cycles:* A rigid carbon chain can peg to its sides by affinity (3rd postulate of i-logic geometry) other carbon structures that latter might split, reproducing new glycine or might stick together, shaping new species of amino acids.

-*Max.E: Energy Cycles:* The oxygen moves the molecule, propelled by the dual polarity of water.

- *Social and Transcendental cycles:* Their social evolution gives birth to macro-molecular proteins.

- *Existential, generational cycles:* The purpose of those molecules is to exist, performing their organic cycles.

433

Thus, after the birth of amino acids, Earth witnessed a massive replication of glycine, which soon diversified in all kind of sub-species that pegged to the original glycine new pieces of carbohydrates bodies, nitrogen eyes and oxygen legs. Thus the organic soup became an ecosystem of top predator amino acids that catalyzed the reproduction of new amino acids, absorbing the simpler molecular 'nutrients', till amino acids saturated the Earth's oceans. Then those different amino acids associated in complementary st-herds, in which specialization occurred again, as some amino acids were designed better to gather energy with extra oxygen legs; some had extra nitrogen heads to process information and some were long carbon chains better suited to split, peg and reproduce new amino acid pieces. Thus we can easily classify amino acids as informative amino acids, with ring structures filled with Nitrogens; energetic amino acids, with added Oxygens - Phosphors and sulfurs, atoms with high 'electro negativity' that are able to capture energetic electrons; and reproductive amino acids with long carbon chains.

Social evolution of amino acids: the protein age.

Amino acids in their social stage of evolution became, according to the inverse morphological laws of transcendental evolution, the 'relative energy' of new macro-molecular proteins. So they lost its 'active' heads and tails, pegged now to each other, as the 'fixed' neutralized st-points of the protein's spiral structure; where the active parts are radicals joined to the central carbon of the amino acid. Those bulky, seemingly unnecessary radicals that hindered the motions of free amino acids show its true value in proteins. (This often happens in evolution, which requires first mutational, inefficient stages that reorganize into functional macro-systems thanks to the directed ternary systems created by fast, planned evolution. If all were chaotic, slow Darwinian mutations most mutations would not survive long enough to transcend into useful new organs, as Darwin already noticed it, studying wing evolution. Thus transitional stages are a proof of 'i-volution'.)

Proteins are huge carbon chains that fusion the fractal actions of those radicals with many oxygen legs and a few nitrogen eyes into a simultaneous present of Max. IxE force. They are like centipedes, entities with simple perception but a fearsome energy that allows proteins to cut and kill all the micro-molecules of the life ecosystem. So they became the new top predators of the original carbon soup, probably chasing down free amino acids to replicate themselves.

Finally, lineal proteins evolved further, according to the inverse laws of transcendental, social evolution, forming self-replicating hollow membranes with cyclical, still forms, which nucleic acids will latter fill and dominate, creating cells. Indeed, the protein's simple minds made their top predator status short living when the 3rd informative horizon of molecular life, the nucleotide, evolved.

Informative age: Nitrogen bases and Nucleotide acid.

In the previous graph, we show the final, informative age of life molecules, which occurred when the amino acid evolved its lineal, simplex 3 st-regions adding a new, informative dimension and creating the 3 globular zones of the nucleotide:

- *Max. E=I:* The improved body is called a sugar that has, instead of the carbohydrate's zigzag line proper of amino acid bodies, a pentagonal form, a powerful compact body cycle that appears in all scalar morphologies. Further on, the sugar pentagon adds one lateral oxygen's rudder that can chain or unchain itself to other sugar rings through easy to break oxygen bridges. So the reproductive speed of the new nucleotide's body based in the capacity to peg and split its body increases.

- *Max. E:* A nucleotide tail adds up a highly energetic, phosphoric acid (PO_4H_3) that has more oxygens than the amino acid's original COOH tail and so it swims better in water. The heavier phosphor is also a nitrogen-friendly atom, from the same 3-5 valence electronic column. So the head improves its control of the tail and its energetic oxygen atoms.

- Max. I: Finally the Nucleotides' heads add up new nitrogens, creating cyclical, hexagonal rings, called Pyrimidines, which once more diversify in 3 subspecies: Thymine, Uracil or Cytosine...

- Uracil is lighter. So it is the brick for building highly mobile social nucleotides called RNAs.

- Thymine is heavier, since it has one more carbon, while Cytosine adds Nitrogen with 2 Hydrogen 'antennae'. So they are the bricks of DNA, the most informative, still molecule of life.

Finally the most complex nucleotide heads are Purines: dual, pentagonal and hexagonal nitrogen rings, joined by a strong covalent C=C wall that form dual couples, called Adenine and Guanine, which add an external nitrogen antenna with 2 Hydrogen eyes to probe the unknown world.

So the globular structure of Nucleotides also creates more evolved social forms, the RNA and DNA acids that will control protein membranes in the cellular scale.

If structural bodies dominate amino acids and proteins, the dominant element in Nucleotides are their nitrogen heads. They become the unit of the social, informative languages of cells, playing a key role in all their informative tasks, as the main elements of most hormones and the fractal units of macro-molecular DNAs, which will create the higher, cellular scale of life forms.

The age of nucleotides: ternary differentiation of species.

So after the age of Amino acids and proteins, there was an age of Nucleotides, which differentiated again according to duality in 2 subspecies: One rich in energy, the other richer in information:

- *Max. E: The energetic Nucleotides are ATPs*, the key molecules in all energetic life processes. Breathing and feeding could not happen without ATP, the specialized energy Nucleotide that again subdivides in 3 subspecies, which can be identified as the energetic, balanced and informative ATP:

- Max. E: ATP proper - an Adenine nucleotide with a longer tail with 3 Phosphors and 10 oxygens, ordered in a classic decametric, $3 \times 3 + (st+1)$ scale: each phosphor controls 3 oxygens, and the 10th oxygen connects them to the sugar body.

- E=I: ADP, which has lost 1 phosphor and 4 oxygens (an HPO_3 molecule) releasing in the process 34 KJ of energy, balancing its form with less energy but the same Nitrogen information.

-Max. I: AMP, which has lost another HPO_3, becoming a cyclical molecule, since the phosphoric tail touches its nitrogen head. So it acts in cellular processes as an informative carrier, transferring hormonal information, amplifying it and programming the cell's nucleus with that information.

- *Max. I: The informative, social evolution of nucleotide acids gives birth to RNA and DNA.*

Nucleotides evolve socially, becoming according to the laws of transcendental inversion between micro and macro planes ($I_{st-1} = E_{st}$), 'energy cells' of RNA and DNA spirals, grouped again in triads that form a spiral cycle. Those triplets surrounded by energetic ATPs create genetic scales in groups of 3, 9, 27...

Those 3^n elements in turn will shape the structure of 2 new macro-molecular informative species, differentiated according to the ExI complementary, dual principle:

- *Max. E: Lineal, moving RNAs* that carry the actions of the cells, again differentiated in:

- *Max. E: Ribosomal RNA* joined to energetic proteins that peg the carbohydrate's pieces.

- E=I: *Transfer RNA* that carries the fractal units that reproduce carbohydrates.

- Max. I: *Messenger RNA* that copies DNA information and takes it to the Ribosome.

- *Max. I: Cyclical, still, informative DNA*, made with 2 complementary RNAs, which carries so much informative, genetic information about the metabolic and reproductive cycles of all other carbohydrates that will become the 'brain' of the new scale of life, the cell... The nucleotide structure of DNA shows the magic 'tetrarkys' that fascinated Pythagoras: 1, 2, 3, 4 dimensions that add up into a decametric scale. Indeed, 1 DNA made with 2 RNA chains joined by nucleotide pairs form a structural tie of DNA; 3 nucleotides form the basic informative 'gene' to create amino acids; and 4 dimensional bases is all what it is needed to create all the cycles and elements of the cellular game.

Those 3 type of molecular 'ages' will become then the 3 'topological' st-forms of the cell: amino-acids become the energy bites of the cells along with water, its medium; proteins will create the walls reproduced in its organelles and nucleotide acids will become the informative, perceptive species that run the show.

Recap. The evolution of life molecules took life through its first 3 ages, the age of amino acids, the age of proteins and the age of Nucleotides, the informative, hyperbolic species that will reorganize them all to create the cell.

12. The social evolution of macromolecules into the cell.

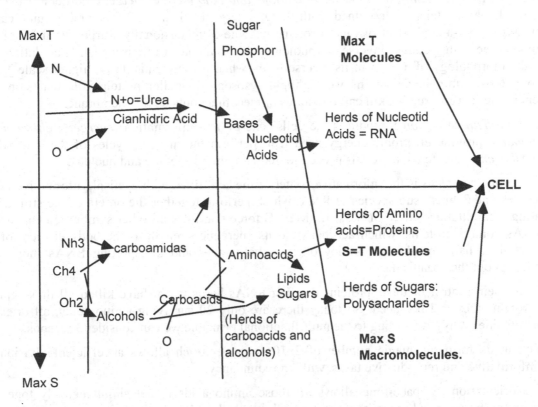

In the graph, the evolution of carbon-life molecules in 3 series of growing size, diversified in Max. Spatial Energy, Max. Temporal Information, and Se=Ti balanced subspecies.

The Ternary Principle diversifies species and then combines them. So methane, Ammonia and Water combine to give either an informative Carbonamide or an energetic alcohol. Both are evolved finally into a carbon-acid.

Both species again mix in balanced amino acids, to further increase their information and energy tendencies adding new Informative nitrogens (bases, porphyrins), or adding Oxygen legs (sugars, lipids). The balanced combination, the amino acid with two oxide legs, and one Nitrogen head, evolves into the next level of macromolecules, the proteins.

On the other hand nitrogen bases keep evolving in social rings; while acids add a macro-atom of enormous energetic power, Phosphor to improve its energy abilities. Finally sugars form with oxygen polysaccharides.

We have 3 species of macro-molecules in a new E-exi-I game:

436

- Macromolecules of enormous capacity as Energy, the Phosphoric acid and other long acid systems (sugars, lipids).
- Macromolecules with structural versatility: proteins.
- Macromolecules of high informative capacity, RNAs and DNAs.
They are the 3 specialized molecules that give birth to the new plane of social existence, that of the DNA-cell.

st+1: The macromolecules of the 3 previous horizons, guided by the genetic, informative language of nucleotides, transcend socially into the cell; a st-point, in which they will play the same specialized roles they played in the macro-molecular age, creating the 3 st-specialized zones of a bigger, fractal plane of existence:

- Max. E: The external membrane, inner invaginations and cilia are the energetic borders of the cell controlled by lineal proteins, intertwined with long, energetic chains of fats and sugars called polysaccharides, hyper-abundant in oxygen. Proteins become 3-dimensionally warped as units of the spherical membrane with globular, inverse morphologies due to the law of transcendental evolution that transforms the morphology of a form in its inverse form, when it transcends into a higher scale $\sum(E_{st-1}=I_{st})$. Among those inner invaginations we highlight lysosomes, smaller protein jails that store the excess of energetic fats and sugars, kill carbohydrates or eject them outside the membrane.

E=I: The intermediate region reproduces the cells' energy and information: *energetic chloroplasts and mitochondria* provide electronic energy needed to perform the moving cycles of the cell; while *informative ribosomes* create the materials needed to maintain the membrane and nucleus.

- Max. I: The inner nucleus is the informative center of the cell, filled with cyclical, informative DNA macromolecules and 3 lineal sub-species of RNA, which perform together the orders of the still, DNA brain, creating a simultaneous herd of Max. I x Max. E force that rules all other symbiotic elements of the cell. RNAs control proteins, which act thanks to its energetic strength as the body element of all cells, shaping their hard membranes, killing other carbohydrates and transporting RNAs' hormonal sentences throughout the organism.

To reinforce their control of those proteins the first RNAs herds might have killed all those amino acids and proteins they did not need. So today there are only 20 surviving amino acids, all oriented towards the left side. Why? According to the multifunctional principle we can consider 3 reasons:

- 20 amino acids form the magic number of 2x10 couples, which allows an efficient division of energetic, informative and reproductive tasks, without redundancy.

- A single orientation in space-time allows all those amino acids to act simultaneously together, multiplying its IxE power as a herd, without anyone 'giving the back' to the group.

- Since the most efficient reproductive system is the specular, sexual method, in which 2 complementary inverse morphologies gather together, a carbohydrate chain could replicate, as DNA does, by specular affinity, attaching parallel forms to its body structure, creating its specular image twice. So if an L amino acid replicates a D amino acid (a macho replicates a female so to speak), and then the D amino acid replicates an L amino acid , the system becomes a self-reproducing amino acid, free from RNA control. By avoiding the reproduction of D amino acids, nucleic acids control the creation of 'castrated' amino acids and proteins. So Nucleic acids probably extinguished D-amino acids to control better the castrated proteins of their cellular farm. Humans also castrate tamed animals; yet amazingly enough they are researching self-reproductive nano-robotic bacteria that might extinguish us, forgetting that reproductive control is a basic tool of any biological top predator.

Recap. Cells are made of energetic amino acids, reproductive proteins that form membranes and informative nucleotides, enclosed in its nuclei.

V. CELLS, THE ARROWS OF LIFE ORGANISMS.

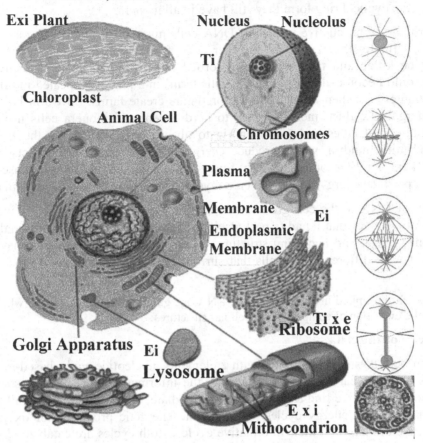

Exi Plant

Chloroplast

Animal Cell

Nucleus   **Nucleolus**

Ti

Chromosomes

Plasma

Membrane   **Ei**

Endoplasmic
Membrane

Ti x e
Ribosome

Golgi Apparatus   **Ei**

Lysosome

E x i
Mithocondrion

In the left side, the cell has 3 st--zones:

- Max. E: It has an external, thin membrane, which in free cells have cilia that move the cell and act-react to external stimuli. In organic cells energy is provided by the blood system of the macro-organism. So external cilia disappear and invaginate as lysosomes that kill carbohydrates.

- Max. I: In the center, there is a nucleus of max. informative density, filled with cyclical or spiral DNA/RNA's networks that control the st-point.

– ExI: RNAs dominate the intermediate space-time, directing the energy organelles, mitochondria or chloroplasts that produce energy; and the informative ribosomes that reproduce products, pegged to the Golgi apparatus - a membrane's invagination.

13. Youth Horizon: From Prokaryote to Eukaryotic cells.

The previous description of the evolution of molecules requires to change the 'chip' of the scientist and accept the tenants of organicism in simple atoms; which so far science has only, according to the Galilean paradox of self-centered perception, accepted for the next scale of life – the cell.

At the beginning cells, called monera, did not have a differentiated nucleus membrane, which means their informative singularity was mainly moving RNA. As evolution continued through the dual/ternary differentiations proper of all EXI systems, RNAs split accordingly in 3 sub-species to carry out the specialized energy, informative and reproductive tasks of protein control, carbohydrate production and self-replication, through complementary, inverse, specular translation. Then, one RNA, which produced a specular image of itself, probably got pegged to that image and became 'fixed', as a still, dual DNA,

438

an informative mirror of an RNA molecule, geared to reproduce it. Accordingly those first DNA molecules acquired the cyclical ring form they still have in all monera.

Soon the extraordinary reproductive growth of DNA cells made them giant cells that exhausted their trophic nutrients. So finally they cannibalised other cells to maintain that reproductive growth. First those cells would be killed and their nutrients absorbed but then some very efficient energetic and informative cells would become slaves within the cell 'farm' in a process repeated in all st-scales: first top predators are hunters but then informative top predators create farms. So worm holes use herds of stars to absorb intergalactic dust; men use dogs to herd sheep; and monera cells used ribosomes to reproduce informative molecules and mitochondria to absorb energy, forming the first macro-cells. Those who 'ate up' mitochondria, which produce energy from carbohydrate products, would become animal cells and kept hunting other cells to find semi-elaborated nutrients. Those who 'ate up' chloroplasts, which produce energy from small carbohydrates and light, would become plant cells.

Now the quantity of DNA in the cell grew to add up the DNA of those organelles and its variety of cyclical memories was so vast that it had to pack itself further, changing its bidimensional, cyclical form into a 3 dimensional spiral, and acquiring structures of energetic sustain: proteins that coiled around DNA and a nucleus with differentiated walls that surrounded DNAs. The age of eukaryotic, gigantic cells had started.

Recap. As information multiplied in simple cells, the RNA age gave way to the DNA age whose extra-genetic code should control the evolution of complex multi-cellular structures.

14. The vital cycles of the cell.

The cell is a brain-body system constructed with 2 elements, nucleotide acids, based in nitrogen bases and protein bodies based in carbon chains, which are the informative and spatial systems of the cell. Around that EXI core duality we find an expansive intermediate, cyclical region with all kind of slavish organelles that perform their energy cycles (based in the energetic properties of oxygen, water and similar electronegative atoms); and their informative cycles. Both cycles are catalyzed by denser metal atoms that boost the E/I capacities of carbohydrates. Thus, through the interaction of carbohydrates, metallic ions, proteins and nucleotide acids, cells perform the 3±st cycles of all st-points:

Max.E: In an organism, Mg, copper and iron capture oxygen and deliver it to each cell to perform energetic cycles; while in the cell cytochromes kill and split the energetic hydrogen atom into H^+ and e^- ions in mitochondria or chloroplasts; absorbing its spatial energy; while metal atoms stabilize protein enzymes.

Max. I: Na and K ions control the expansive and implosive rhythms of the electric membrane, sending informative messages among cells -while RNA and DNA molecules process *information* within the cell, reproducing new carbohydrate molecules.

- The combined effect of the accumulation of energy and information within the cell triggers the reproductive cycle, guided by RNA and DNA molecules.

- ∑: Cells *evolve socially* into macro-organisms. Yet only the RNA-DNA system creates complex informative, control networks.

- ±st: Cells live a *generational* cycle chained to an organism that kills them 'periodically', or as free cells that die when captured by top predator living beings. Since most cells can be immortal if no other system kills or controls them hierarchically.

All those cellular cycles are not 'mechanic processes' but they have evolved departing from the organic, dual, Darwinian and symbiotic interactions between proteins and nucleotide acids, the dominant energetic and informative macromolecules of cells. The most important cycle is the reproductive cycle, which in the cell as in other fractal space-time represents the existential will that ensures the perpetuation of the species. Let us consider it.

Recap. Cells follow the same arrows of existence of all st-points.

15. The reproductive cycle.

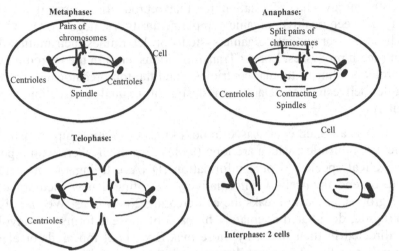

The interphase, prophase, anaphase and telophase complete the reproduction of the cell, in which the centriole, a perfect example of a decametric structure, with 9x2 lineal proteins and a central 2x1 nucleus that controls the lower scale of 9 forms, plays the key energetic role of motion control.

The reproductive cycle of cells shows an evolutionary pattern observed in many dual cycles between 2 complementary st-points of relative energy and information, which first confront their energetic and informative inverse forms. But as time goes by often by chance they realize that social, complementary evolution is more efficient than Darwinian, energetic destruction. And so they end up evolving together, following the positive, creative arrow of the Universe. We can hypothesize according to that homology the 3 ages of evolution of the reproductive cycle:

- *Energetic, unicellular age: Sexual reproduction* probably started as a Darwinian, energetic process of hunting in which lineal, energetic sperm killed the cyclical ovum and learned to host its DNA-code as virus do, in the ovum's center. Yet as the sexual cycle evolved beyond *the energetic, young age* of the cycle, the 'war of sexes' ended and the cycle moved into a *balanced 2nd age.*

-*E=I: Reproductive, 'colonial age'.* Now information was exchanged between both forms, the sperm and the ovum - since sexes are in fact a specialization of species into *energetic males and informative females.* And a new symbiotic dual form was born. Those forms would start reproductive radiations multiplying the number of cells. So probably sexual reproduction is a key factor in the multicellular explosion of the Cambric age.

- *Informative, palingenetic age.* Finally, in *the 3rd age of sexual evolution the informative female species dominated the cycle,* as females have more genetic code (2 complete X chromosomes) and control during pregnancy the reproductive, palingenetic cycle that brings the embryo into a new 'macro-organic level of existence'. If we observe the family cycle of human couples, women also tend to dominate the couple in its 3rd longer age, as Proust already notice; since information lasts longer than energy in time.

In complex organisms a living, reproductive cycle is dual: it departs from the simplex reproduction of a single cell that multiplies. Then those self-similar cells suffer a complex process of palingenetic reproduction that recreates an organism made of billions of cells, departing from that single cell.

Cellular reproduction: a tug of war between DNA and proteins.

Let us consider in this synoptic paper, the simpler process of cellular reproduction, as we have treated palingenesis in our analysis of the 4th postulate of i-logic geometry:

Organic cells reproduce within a day, chained to the symbiotic daily feeding period of the organism that provides them with energy and information for that reproduction. Fractal cellular reproduction shows also that duality between positive, organic complementarity Vs. negative Darwinian struggle that either balances E-bodies and I-brains into organic systems or determines their mutual destruction when that balance is broken (death processes, Lorenz Transformations, etc.) In the reproductive cell cycle, the most efficient body proteins - lineal, tubular centrioles - and the top predator, informative DNAs enact an ambiguous struggle between their Darwinian desire of mutual destruction and their need of complementary evolution.

So cellular reproduction is a mixed cycle based in both kinds of relationships in which first the lineal species of cellular energy (centrioles, which are long (9+1) x 2 protein fibers with a perfect decametric structure), untie the cyclical species of pure information (DNAs), trying to split and kill them. Yet DNAs, once uncoiled, defend themselves 'informatively', creating a new membrane that breaks the centrioles apart into 2 groups, and also breaks the cell creating 2 new ones. We can distinguish several phases in that dual struggle, dominated alternately by each of those 2 forms that involve all the other elements of the cell, directed in their dance by those max.E x Max.I top predator elements or 'upper classes' of the cell. It will be a tug of war that illustrates all the geometrical strategies of Darwinian and complementary events between dynamic st-points:

-In the *interphase*, both top predator substances replicate. The centrioles are outside the membrane, which protects the DNA.

-*In the prophase centrioles start their hunting:* the protein's membrane of the nucleus dissolves, exposing DNAs' chromosomes that become visible preys. As all other universal preys do, from wriggling worms to high frequency rays, from submissive servants to herds of electrons in front of a quark or fishes in front of a shark, chromosomes try now to hide by coiling up, becoming contracted, shorter forms. Then the hunting starts. Centrioles have replicated and now double its fractal action, moving to both sides of the DNA nucleus, forming dual, long molecular chains joined by filaments. So they *create a polar field of forces that captures in its filamentous web the self-replicated DNA,* as a North and South Pole create together a magnetic force field that aligns atoms or two boats web a net to capture fishes.

-*In the metaphase chromosomes defend themselves from the 2* centrioles that throw the 'hooks' of their force field, stretching the DNAs. But those chromosomes that have replicated in the earlier interphase evolve now socially in couples of parallel forms, increasing its fractal mass and moving away from the centrioles in a classic protective strategy: They arrange themselves in the equator of the spindle, adopting a Darwinian, perpendicular position, the farthest away from those centrioles. It is exactly the way in which diamagnetic particles that flee from magnetic fields, arrange themselves trying to receive the minimal quantity of force from the North and South Poles of the magnetic field.

-*In the Anaphase centrioles counter-attack*, splitting away the chromosome pairs.

In the telophase, DNAs find their winning, defensive strategy: They are the informative species that store the genetic code of all cellular forms. So DNAs start to reproduce new membranes in a frenzy till those membranes break the centrioles' spindle through its middle zone, isolating each centriole and breaking their field of lineal, protein forces. Since lineal, energetic or reproductive beings, like centrioles or 'magnetic' fields are, cannot create monopoles. Only temporal, implosive, informative cyclical particles can do that. So the spindle dissolves and the new membrane divides the cell.

441

Now the system reaches again a balance, as the dual chromosomes and the dual centrioles become again single forms surrounded by a new membrane.

Thus the dynamic tug of war between a protein's body and a nucleotide's brain ends up in a draw, creating 2 cells instead of one, which will re-start after a rest period a new *interphase* process of protein and DNA reproduction. Since the Universe is indeed a game of reproductive radiations, the ultimate will of all beings that want to survive their fractal, periodic death by creating a self-similar 'present' form.

A variation of that process required in sexual reproduction, is called meiosis, in which the twin reproduction of DNA and its subsequent destruction of the nucleus' membrane is provoked by the energetic sperm that enters the ovum, invading, as a virus does, its DNA nucleus and merging its genetic material. So, as it happens in the interphase of a cell, the fecundated sexual cell has 2 parallel quantities of DNA, albeit with different genetic material, coming from the ovule and the sperm. So as the replicating process repeats through the same phases of any cellular reproduction, the final result won't be an identical cell but a cell that mixes the genes of both, the sperm and the ovum. When besides sexual duplication there is an interphase with chromosomal duplication the final 2 cells will be diploid cells with twice the genetic material of the original cell. This happens only in multicellular organisms, as redundant genetic material is useful to store genetic orders needed for the complex construction of multicellular structures; but it would be redundant in the simple life of a monera cell, which escapes the interphase duplication.

The exi duality of lineal sperm and cyclical ovum extends outside the realm of form and transcends to the upper scale of multi-cellular organisms and sexual characters: the ovum is an informative, cyclical, autotrophic female cell with higher chromosomal content; while sperm is an energetic, heterotrophic, lineal male cell with higher mobility. And we find according to the Fractal Principle, 3 evolutionary types of increasingly differentiated sexual cells: semen and ovum, which are equal in spatial size and temporal form (isogamy); semen that is equal in form but smaller than the ovum (anisogamy); and ovum and semen, which are different in spatial size and temporal form (oogamy), as in human beings.

Those differences between male sperm and female unicellular ovum, latter diluted as their genetic materials mix, reminds us of the differences between unicellular animal and plants, the main dual differentiation of life along its exi parameters that we will study now in more detail.

Recap. Sexual reproduction evolved from an energetic event in the unicellular age into a reproductive radiation and finally into the palingenetic process dominated by the ovum.

Cellular reproduction can be explained as a tug of war between the informative DNA and the reproductive proteins of the centrioles.

16. Ternary cell's differentiation: plants, animals and fungi.

The kingdom of life shows in all its beauty the generic process of evolutionary differentiation of any space-time field, along the main 3±st dimensions, departing from a 1st singularity, which in the world of life is the Monera phylum - the initial cell, whose *ternary differentiation gave birth to energetic plants, reproductive fungi and informative animals:*

Max. E: Energetic plants use light as energy. Biologists talk of plants as autotrophic cells; still forms like the ovum, which create their energy and information quanta from light and water, in any place. So they made their membranes harder and thicker to maintain themselves centered in a territorial, discontinuous, vital space, regardless of what happens outside.

Max. I: Informative animals use light as information. Animals are heterotrophic, moving cells, which feed on other forms. Hence they developed thinner external membranes and cilia, which according to the multifunctional principle differentiated further into increasingly sophisticated sensorial antenna to localize their preys and lineal, moving engine with a master centriole - a 'protein head' on its base. So though in a 1st phase unicellular plants were more complex, animals ended up evolving greater quantities of inner, informative RNA-DNA to act-react faster in their unknown moving environments.

E=I: Reproductive fungi are organisms that have qualities belonging to both, the animal and plant kingdoms. Fungi feed on dead substances and survive thanks to their maximization of reproductive skills.[6]

The inverted forms and functions of plants and animals define both species as 'antisymmetric systems' ruled by their opposite diffeomorphic parameters of energy and information:

Their informative cycle and brain-body dimensions are inverted: plants have their brain down in the roots, since they use light as energy; animals have it on top, since they use light as information.

Their energetic cycles are inverted: plants breathe CO_2 and produce oxygen; animals breathe oxygen and produce CO_2. They also use carbohydrates in opposite ways, since plants foster constructive stillness and animals destructive movements: so plants make sugars the fixed structural element to construct cellulose and starch, their external membranes; while animals use sugar's oxygen bonds breaking them to breath, liberate oxygen and move the body.

*Their social and reproductive cycle*s are inverted: Animals are hierarchical organizations with a clear class division between informative, 'upper class' neuron networks and body cells, while plants are 'democratic forms' with minimal differentiation. So plants foster the spatial, reproductive arrow and animals the evolutionary, temporal arrow.

Their *generational cycle* is inverted: animals have faster, shorter vital cycles, since its temporal language is the faster, electronic, nervous language; while plants have longer life cycles at a lower 'speed of informative processing' - since they transmit chemical information through the hormonal system.

Departing from those 2 languages life evolved into a new scale of multicellular organization creating:

- Max. E. *Chemical,* multicellular organisms, plants that use hormonal languages.

- Max. I: 'Electric' organisms, multicellular animals that use the electronic language.

Recap. Plants, animal and fungi, the first 3 kingdoms of multicellular life are a ternary differentiation in e, exi and i species.

17. Death.

In Cyclical Time death is caused by the finite limits of fractal time and space. Death is a clear prove of the discontinuity of space and time. If we were continuous, that continuity of time and space would make us immortal and infinite. Which means that only the ∞ sum of fractal time-spaces or absolute Universe is eternal: the laws of temporal energy, the logos-mind of the Universe do not have and end nor a principle, because the sum of all the transformations of energy and information balances a dynamic Reality that never dies, remaining always in a relative, equal present. Reality exists 'per in secula seculorum, amen'.

Only reproduction guarantees certain immortality, re*creating our form in other region of space.* Yet in all species the rate of reproduction also goes through 3 ages, and so when reproduction halts, not only the 'cellular' individual but the species of the 'higher plane' dies:

- *Big bang phase*: A new species multiplies very fast departing from a 1st individual.

- *Steady state:* Its reproductive rate stabilizes.

- *Population Crunch:* Finally it slows down to a halt, extinguishing the species.

Death is in fact the most obvious prove of the discontinuity of space and time. If we were continuous, that continuity of time and space would make us immortal and infinite. Thus only the infinite sum of quantic time-spaces, the absolute Universe is eternal.

For the rest of us only reproduction guarantees certain immortality. Yet the rate of reproduction of any species is also quantic and goes through the 3 ages, starting slow, increasing its rate and finally slowing down to a halt, extinguishing the species or type of DNA-cells of any organic system. So in a human being the number of divisions of any cell reaches up to 50, whereas in chickens it goes up to 20 divisions. The greater rate happens in the oldest, living species: the Galapagos turtle that divides 10^2 times its cells. $E=10^{i=2}$ is in fact the commonest constant in information warping, from the protonic accumulation of form in the atomic table, which after 100 forms becomes unstable to the limit of human aging.

According to the ternary principle, 3 biological strategies however seem to succeed death:

- Max.E= Some simple systems (coelenterate) reverse its time clock.

- Max. Re: Cells 'trick' their organic time clocks, to free themselves from the servitude of death, by multiplying its quantity of reproductive genes - its DNA molecules. The record belongs to cancerous cells, very rich in DNA, from the reproductive uterus of Henrietta Lacks, the so-called HeLa cells.

- Max. I: Humans achieve partial immortality through cultural memes.

Is there a species whose number of reproductions is infinite, an immortal being? No. Since even the nucleons of the big-bang reproduced e^{60} times to create the Universe, shaping a similar reproductive curve to that of living beings that now has slowed down to a trickle. So all dies for the game to renew itself.

Causes of Physiological Death.

The opposite process to the creation of life and the creation of form is the destruction of form into energy or *death*, (I<E: *Max. E x Min.I*), equally common. As one feeds into the other and together balance the dual arrows of the Universe.

Thus, the fundamental cause of death is to exist beyond the balance of form between the E and I components of any St-field. All beings have limits of excess of energy and/or form, which once crossed will bring their extinction, as their energy and information quanta 'explode' and split, dissolving the networks that kept the form together and were its inner consciousness. If a particle explodes close to C speed and splits its energy/information parameters according to Lorentz equations, so does a human being after dying. His energy dissolves back into his st-1 simpler cells. And then, since all deaths are dual big-bang explosions, humans return finally to the molecular, amino acid level from where they departed.

If we want to be specific about the *causes of death in* human organisms, death is the product of an imbalance, either by an excess of energy that causes sudden death in minimal Time (Max. e=Min. i), through war, sudden sickness or accidents, or by the accumulation of temporal information that spends the energy of the being during its long 3rd age, leaving traces of past cycles in cells and physiological networks that cause their malfunction. In other words, the nervous, informative network exhausts the body, causing a max. I=min. E imbalance, that wrinkles and breaks the body cells into death. But also the opposed phenomenon, an excess of energy caused death. In both cases the imbalance breaks the complementarity between the space and time content of the being. However, since the direction of living beings is towards the informative future, the informative imbalance is more frequent and shows in the 'growth' of time properties.

So old people are as time is, cyclical, memorial, curved, with discontinuous wrinkles that break the smooth continuity and linearity of young, spatial beings. That quantic, curving process happens in all the

444

species of the Universe that age: cyclical wrinkles and quantum cracks occur not only in the skin of an old man but in the membranes of our cells, their Golgi apparatus and "energy" mitochondria. That loss of energy in any old being, which becomes quantized and implosive, with negative curvature, has manifold manifestations: cells shrink and their membranes no longer dilate. An old man measures a few centimetres less. The 2 networks, the energy network or membrane and the informative network become rigid, quantized and no longer reproduce. In human cells, collagen and elastin, the proteins of cells, loose their elasticity and expansive capacity, creating, informative connections among them. So happens to the rigid, cross-wrinkled skin of the old man. Meanwhile the energetic, metabolic capacity of the organism and its quantic cells diminishes.

In a more generic way information and time are parallel. So species with more information and less energy live longer, according to the Time-Space inversion: Max. E=Min.I. Thus for example, an animal that fasts systematically and receives less energy, lives longer (hungry rats live a 20% more). The female, informative sex with smaller expenditure of energy lives longer than the energetic male. A being that lives in a hot, energetic environment lives shorter than one living in a cold environment. Hibernation that slows the metabolic rate elongates life. There is also a proportional relationship between the informative weight of the brain and the life span of an animal: Max. I (brain weight) =Max. I (life-span); which in inverse to the proportional relation between the metabolic expenditure of energy and the life span of the being: Max. metabolic rate (Max. E) = Min. Life span= Min. I, proving once more the inversion of space/time. So humans with huge brains live much longer than lions with huge bodies.

An excess of energetic atoms diminish time existence: Free radicals, which are oxygenated, energetic molecules, with free electrons (the energy body of atoms), deteriorate cells and their accumulation with age is a major cause of death. The degeneration and disorder of those cells can be caused by energetic radiations (electromagnetic rays of max. energy cause the mutation and death of cells.) The increase of fat, energy cells causes the death. So doctors establish a direct lineal relation between the accumulation of lipofuscine and the death of a living organism. An excess of energy is the main cause of accidental death among youngsters. So a speedy car crash breaks the body and the shock of a weapon's impact, a form of energetic metal, kills a human being.

Yet in generic terms, information that forms, carves and quantizes amorphous, continuous energy is the supreme cause of death. As the chisel of a sculptor creates form by destroying the marble, so does the game of existence: For example, elephants create the form of the savannah, when killing its trees; quantized hands are formed in a foetus, when interdigital cells commit suicide (apoptosis). So ducks, which do not kill them, have membranes in their hands. All those causes of death have created an equal number of scientific theories about death, all related to the inversion between spatial energy and temporal information and the balance between energy and information that defines existence, as opposed to its imbalance that causes death.

Mental death.

But what happens to the brain's information stored in the brain, after death? Does it dissolve also, as body cells do; as the magnetic, informative field of a star does or the mass vortex of a Nuclear Bomb does? Or does it transcend, as the photonic light does, into the higher electron, or the seminal cell does when it evolves into a new macro-organism?

According to empirical descriptions of death and the general MST laws for all space-time fields, information 'rewinds back', travelling also to the past, becoming erased, as it happens to our cells.

So most likely the information of the being will no longer be. It will become untied energy, cellular form and then molecular or atomic quanta. Since when we are incinerated, we fall to the consciousness of the atom. Yet consciousness, 'sensation', the very essence of 'existence', never goes away, it only jumps from atom to cell to man and vice versa. The optimist believer might believe in the transcendence of the 'informative mind' to a 'higher' plane of existence with more information, that of a human super-organism or collective God, as in a palingenetic process that makes a cell transcend into an organism. The pessimist thinker might believe in the devolution of consciousness to the "obscure world", with less information, of cells and atoms. It is what mystiques have explained in poetic terms as the opposition between the ascension to Heavens Vs. the quantic jump into Hell. So st±1 birth and death discontinuities are jumps between planes.

But how 'sensorial perception' changes in those discontinuities? It does so through the 2 antithetic sensations of ExI existence, which are informative pain and energetic pleasure.

Those 2 paradoxical, antagonist sensations happen not only in men but also in animal life and perhaps in all atomic forms, because they are based in the geometric properties of information and energy. Indeed, pain is created by 'pressure' on our cells, which is an implosive, temporal movement, also related to the consciousness of information. While pleasure is caused by the 'expansion of blood in the genital system, an energetic process. So we have the key duality of consciousness:

Information=pain=awareness=implosion

Energy=Pleasure=forgiveness=explosion

Since information is desirable, the negative arrow of pain balances the positive arrow of perception. On the other hand since everybody likes pleasure, it balances the negative arrow of energy that erases awareness; creating a 'balance of wills' that makes possible to wish both life and death. Thus, in terms of sensations, death will bring just a sudden pain, the peak of our 'informative awareness' towards the future, before we change 'phase of existence'. Then death will be followed by a deep orgasm of energetic pleasure that erases all information, as we fall down the peak. Since pleasure is synonymous of energy, a travel to the past opposite to the awareness of perception, information and implosive pain, a trip to the informative future.

So after pain ends our will to live, we relax in a deep orgasm that erases information, reversing our space-time field, as our consciousness moves towards the past. Then the brain rewinds back the memories of its existence that cross for a last time through the 'soul' of the brain, more likely a neuron's network.

Yet, because time accelerates in smaller beings, the speed of times of the brain accelerates in that rewinding back, during which our black hole vortex or soul, the informative singularity of the mind, feeds back towards youth, perceiving and erasing a last time those memories. That is why in the 3 days of death of our organic system, as our body becomes corrupted and our consciousness dies, we can recall all those long years of existence at 'fast motion', as we rewind faster a tape than when we see it. That is why those who woke up from death could describe it as a trip to "the past" through its memories, because after death consciousness will perceive life "the other way around" at the faster time rhythm of cellular life. Then, finally we will enter back the uterus and return to the ovum's cellular consciousness; and then again we will die in a second big-bang, burnt in a crematory or feeding the insects that reduce us to molecular form. But 'we' will no longer be human. As a Hindi legend say,' an ant was once Indra, a king of kings'. For those who are ego-driven, there is no need to worry; the game is limited in its variations. So you are in a way immortal. Since we are repeated in infinite places of space and time; you will be born in another planet to live your life again.

Yet, far more important than the death of any diminutive MST field, is the death of the st+1 scale of existence, the death of living species or human civilizations, related to the process of evolution and extinction of the Earth's ecosystems...

Recap. Death is brief in time, exploding in space the information of the being. Death is caused by an imbalance between the energy and information of a system. Energetic beings live shorter than informative beings. All systems have a finite duration in time. So death requires reproduction for the species to become immortal.

18. The ternary evolution and differentiation of Monera.

The law of 3 creations, and 3 ages diversified   the Animal phyla

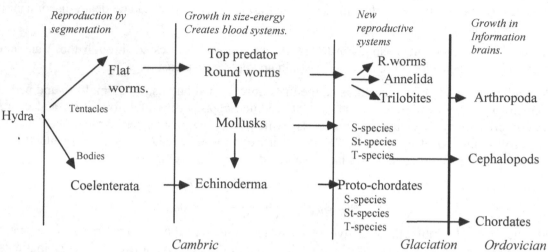

YOUTH-ENERGY   REPRODUCTIVE DIVERSIFICATION   INFORMATION AGE

In the graph, the ternary method of speciation explains the evolution of animal species from its first form, the Hydra till the last phyla, the chordates to which humans belong

We live in a global planet that has been evolving for billions of years through the flows of its living organisms. Let us recapitulate that ecosystemic evolution, which brought the Earth from a world of methane to a globe of electromagnetic, audio-visual information, a process, which according to the ternary principle we can divide again in 3±st ages. We will study only the animal kingdom in detail to show how it brought us into human existence. So we will only consider a synopsis of all the other life phyla, describing their evolution through the commonest ternary e, e⇔I, i differentiations proper of all i-logic systems with its ternary topologies:

-Differentiation in: e-exi-i, lineal, spiral or cyclical morphologies, each one evolved in 3 horizons that maximize each of the 3 homologous functions of motion, reproduction and information.

-Duality of integrated, multicellular networks that act as a single form vs. quantized herds.

-Duality of energetic/informative species adapted to E-hot vs. I-cold climates.

-Adaptation to the ternary states of matter in Earth's physical ecosystems: air, land and sea with its 3 main environmental topologies:

- E: shallow waters and rivers, which become hunting grounds.

- E=I: open waters and savannas of max. reproduction.

- Max.I: Abyssal regions and mountains, the hyperbolic region of maximal informative evolution.

So again life evolution shows how 'internal form' and external ecosystems, the st-1 and st-scales of life converge together to define the st-organic evolution of the species.

Let us then start that description of the differentiations of life kingdoms from the simplest one.

Monera: Unicellular forms.

- Max. E: *Cyanobacteria*, blue-green algae, specialize in energetic processes.

- E=I: *Protista* absorb cells specialized in energy and information, multiplying its IxE force.

447

- Max. I: *Schizophita Bacteria* develop informative elements to capture other plants.

We can do further subdivisions along other E⇔I differentiations. For example, bacteria subdivided according to its form into:

- Lineal *spirilla* that coil or elongate their form depending on its informative or energetic activity.

- IxE: *Bacilli*, with tree-like forms composed of a head and a tail.

- Max. I: *Cocci*, the informative, cyclical form that suffers further evolution along a ternary, topological differentiation into:

    -*Streptococcus*: *one-dimensional*, lineal, social forms.

    - *Diplococcus and Tetracoccus: bidimensional* forms with 2 and 4 elements.

    - *Sarcina: 3-dimensional* social coccus.

As protista, the social form, multiplied its IxE force thanks to the specialized energy and information cells it had swallowed, dominating the world, the other 2 weaker forms suffered a temporal regression towards its past, which became a strategy of survival, giving birth to:

-*Max.E: Rickettsia*, algae that have lost their informative skills and are basically semi-living bodies.

- *Max.I: Virus*, bacteria that have lost their bodies and become DNA brains in search of other bodies in which they host, inoculate their genetic code and reproduce.

On the other hand the dominant protista with higher IxE force continued the evolution of the life kingdom, splitting again into 3 forms that evolved further into multicellular organisms, plants (Max. E), Fungi (E=I) and animals (Max. I):

Plants: Max. Energy

The energetic strategy of the multicellular living kingdom is the plant, the autotrophic species that feeds on the basic molecules of life, accelerating enormously the evolution of life as it produces complex living matter from the initial water, ammonia, CO_2 and light bricks that took billions of years to evolve. Though algae started as unicellular forms, as fungi and animals did, they soon evolved in 3 horizons along the fractal differentiation that took them to its multicellular state:

- *Chlorophyta* were the I horizon of unicellular alga, that grew into colonies of algae (*Crysophyta*, II horizon), which finally fusion into multicellular organisms (*Pyrrhophyta*, III horizon).

- The most complex, informative phylum, *Pyrrophyta*, differentiated then into 3 sub forms adapted to the 3 water ecosystems. Those water ecosystems differentiate the morphology of informative animals into lineal, Max.E, fast-moving surface fishes, Max. I, sessile or planar dragging forms, living on the marine floor (fast-evolving echinoderms origin of vertebrates) and abyssal complex, EI morphologies. In the case of energetic seaweeds, it affects the degree of sophistication of their chlorophyll pigments and the strength of their cellular structures of sustain, creating 3 new phyla:

- Max. E: *Rhodophyta* or *red algae*, with the simplest cellular structures and simpler phycoerythrin pigments, live in the deeper sea, limited to tropical regions of max. light transparency.

- E=I: *Phaeophyta* or *brown algae* have complex membranes and 3 chlorophyll pigments; a, c and phycobilliproteins.

- Max. I: *Clorophyta or green algae*. They added carotenoid pigments to the 3 Phaeophyta pigments and increased the strength of their walls. Hence they became the most successful forms with Max. IxE force (Max. pigments x Max. membrane), evolving further into terrestrial plants.

That migration to land took place in 3 ages in which plants raised its informative height from:

- I Age. *Clorophyta*; planar alga living in shallow waters.

- II Age. *Briophyta*: Mosses, subdivided in 3 forms with growing dry membranes and height dimension, *Musci* (I Horizon), *Hepaticae* (liveworts, II Horizon) and *Hornworts* (III horizon), which raised their horns towards the sun as their names indicate.

- III Age. *Tracheophyta*, vascular plants, with structural inner networks of hard cells that rose to touch the light that feed them. In dry land plants, as animals will do latter, had to evolve further all their network systems, creating new, more complex phyla, departing from the initial *psilotophyta* appeared in the Ordovician. First plants created the body, the trunk that connects its informative roots and energetic leaves.

But the key to their evolution was the differentiation of their reproductive cells in 3 Horizons of increasingly 'dry' gametes: *Lycophyta* (I Horizon), *Sphenophyta* (II Horizon) and *Ferns* (III Horizon) - the first plants with dry seeds that became the top predator species, multiplying in all land environments and differentiating again in 3 evolutionary horizons of ever more perfect seeds:

- *Max.E: Ferns*, which dominated in the Mesozoic age.

- *E=I: Gymnosperms*, subdivided according to the Fractal Principle *in cycads, ginkgoes and conifers*, which dominated in the tertiary age. Conifers adapted to cold weather, thanks to its needle like leaves of minimal exposure. So they became the most successful species, when cold climatic changes came. They brought about also the dominant modern plants:

- *Max. I: Angiosperms, flowering plants*, the 'height' of reproductive evolution among plants, perfectly adapted to all weather changes, with seasonal, blossoming, leaves that fall in cold periods and a complex symbiosis with insects that can transport their pollen too far away distances.

They will dominate the quaternary with only a final dual fractal differentiation into:

- *Monocots* Vs *dicots* with 1 or 2 seeds.

So when all was said, the evolution of plants remained silent.

Fungi: Max. Reproduction. ExI.

Fungi are big protista cells that tried the 2nd survival strategy of the Universe, maximizing their ExI reproduction, by maximizing their E-feeding, eating the most abundant food, dead life; and by maximizing their genetic information, multiplying the nucleus of its cells within an undifferentiated membrane. And we distinguish 3 horizons in their reproductive evolution:

- In the I horizon *Euglenophyta* swallowed E-plants and I-animal sub-cells (chloroplasts and mitochondria) within its huge membrane, increasing it ExI capacity.

- In the II horizon *Gymnomycota* maximized the reproduction of their nuclei, the informative, genetic material of the cell, forming multicellular colonies with a single membrane.

- They gave birth in their III horizon of reproductive evolution to true *fungi, Mycota*, which are the fastest reproductive species of the life world. Their reproductive specialization shows in giant Lycoperdales, the living form which reproduces faster, reaching the limiting magic number of 10^{11} spores. Thus a single lycoperdal can produce all the clonic cells needed to create a perfect new scale of existence (stars of a galaxy, DNA ties, etc.)… if they survived.

Finally in the last evolutionary horizon, fungi differentiated further according to the st+1 ecosystem in which they live into:

- *Max. E: Water fungi* mainly *Chytridiomycetes*.

- *E=I: Amphibian fungi*, mainly *Oomycetes*.

- *Max.I: Terrestrial fungi,* which evolved in the informative land medium, differentiating into:

- *Max. E: Asomycetes,* planar simple or subterraneous forms like yeast and truffles.

- *E=I: Deuteromycetes,* a transitional form towards…

- *Max. I: Basidiomycetes,* the familiar mushrooms, which evolved its 3 networks, developing:

Stronger cells to sustain their informative growth in the height dimension; and new reproductive systems with dry spores and new energetic cycles to decompose all kind of dying matter.

Animal life: Max. Information

The heterotrophic animal family *is born with protozoan,* which developed soft membranes and cilia to absorb living information from the outside world. To that aim cilia evolved again its 3 functions: as energetic, motion engines; as sensorial, informative tools and as predatory forms that capture food for the reproductive cells of the protozoa. So once more we can subdivide protozoa in 3 sub-forms, according to its activity and dominant cycle:

- *Max. E: Mastigophora,* flagellate protozoan that divided, according to the fractal principle, by its number of cilia:

- *Max. E=I: Sporozoan, parasites* that reproduce seeds and have sexual differentiation.

- *Max. I: Amoebida,* the informative protozoan that increased their DNA and evolved farther their membrane's flexibility, becoming nervous cells able to control - thanks to their faster action-reaction speed- multiple cells, which evolved together creating multicellular animals.

Multicellular animals differentiated into many phyla, following the ternary, fractal principle (e-ixe-i) applied to the evolution of 3 cellular, social networks, the energetic, digestive system; the blood, reproductive system and the informative, nervous system, the dominant network that reached with man its evolutionary height. Then the morphology of life would be transferred to stronger atomic systems made of metal, called machines...

Recap. Different living phyla were born from ternary and complementary differentiations of e-exi-I species and adaptations to its st+1 ecosystems. The final differentiation its 3 multicellular life forms, e-plants, exi-fungi and i-animal life was due to the evolution of informative cells that organize multicellular life since animals use electric, faster informative languages and plants, slower chemical hormones.

VI. CHEMICAL VS ELECTRIC LANGUAGES: PLANTS

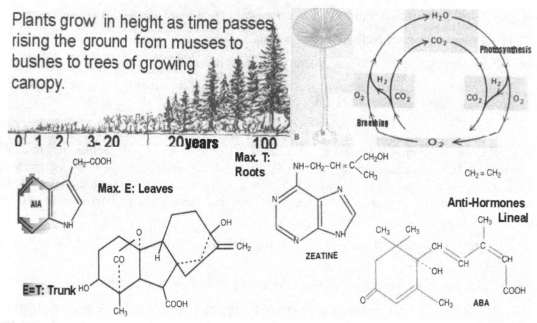

Plants grow in height as time passes rising the ground from musses to bushes to trees of growing canopy.

In the upper center a practical case of relative diffeomorphism: The simplest unicellular plant-like organism, the acetabularia, shows its nucleus in the root, B, which will become the brain of multicellular plants, opposite to the animal upper brain coordinates. The opposition between both forms extends to the cycles of energy, as plants destroy water and produce oxygen; while animals breathe oxygen and produce water. Thus, both are the 'particles' and 'antiparticles' of life.

19. Plants.

Languages of information, which code and express the biological arrows also evolve by ternary differentiation. For example, the human language, if we restrict our analysis to its vowels was born as a duality of an 'energetic' active 'a', which drove to action and an implosive, reflexive 'u', which drove to self-reflection (basic chimpanzee language). We might then imagine a 3rd combination, or 'I', the creative i-dea, the 'I'-self, and two final 'modulations', the 'less energetic e' and 'less informative' o.

Those examples of the rich field of socio-cultural studies based in the ternary differentiations of the creative program of the Universe, are a parallel example to the simple 'vowels' of the hormonal, chemical language that controls plant growth with 5 simple vowels, drawn in the image.

The chemical language of multicellular organisms have a reduced vocabulary of 'yes' and 'no' symbols called hormones that inhibit or foster the natural cycles of existence of living beings; and define the speed of height growth of the plant.

In the graph, the 5 main plants' hormones are 5 simple vowels: 3 are cyclical, creative, informative, 'yes' hormones that reproduce the plant; and 2 are destructive, energetic 'no' hormones, with opened rings and strong, lineal carbon chains like ethylene, $CH_2=CH_2$, that inhibit growth. If we further divide 'yes' hormones according to the Fractal Principle and the 3rd postulate of affinity, we find that each one of them reproduces the plant's e-ixe-i element with similar structure:

- Max. E: The longest ternary form, rich in oxygens, Gibberine, develops the lineal trunk.
- I= E: The structural, cyclical form, Auxin, made of carbon rings, develops the leaves.
- Max. i: The informative nitrogenized Zeatine reproduces the brain roots.

451

Thus, hormones form a simple, ternary language that follows the rules of EXI cycles, controlling the living cycles of organisms by acting as messengers of the RNA-DNA brains of their social cells.

Plants Vs. animals: chemical language Vs. electric messages.

1. Energy organs: Leaves that perform the 'moving' energetic, photosynthesis cycle and branches that act as still structures and conductive systems of the energetic materials.

2. Informative organs: Root systems that absorb molecules and produce most of its hormones.

3. Reproductive organs: Flowers and trunks, the bodies that communicate leaves and roots and experience maximal growth.

These 3 organs define a plant as an efficient ternary st-point, a complex species able to colonize the ground with a clear evolutionary direction of height.

The interaction of the 3 hierarchical levels of life.

According to the i-logic laws of creation all organic systems require the co-existence and interaction of their 3 scales of social evolution or relative past, present and future hierarchical forms.

So the creation of multi-cellular organisms happens also simultaneously in 3 'hierarchical planes': the st-1, atomic-molecular interval; the st-DNA-cellular scale and the st+1 network-organism interval.

While the flows of temporal energy that travel through those different planes of existence follow the hierarchical law of illogic geometry: *A fractal jump in geometric time or a movement in space extends only to the next discontinuous space-time scale.*

So in any organism, the st-1 hormones (the molecular language that regulates st-cells), the st-genes (the information that transcends from the st-cellular DNA to the st+1 organism as a whole) and the seminal, palingenetic cells that give birth to the new being, program only their higher plane.

Hence cellular genes determine the biologic elements of the organism (the next scale of social evolution), but not the st+2 sociological scale of existence of a human being: chemical, racial genes do not determine history. At best we might consider that nervous networks determine the emotional character and intelligence of historic individuals, who might change the course of societies. Let us study the structure of those 3 levels of hierarchical information that codify the creation of organisms.

20. Hormones: the language of cells creates its networks.

According to the fractal law, the language of regulation of social groups of st-cells are st-1 molecular hormones, which act as phonemes of a language spoken by RNA-DNA macromolecules, the informative networks of cells, which can manufacture them.

Languages are 'i-logic programs' structured by the same ternary topologies of energy, reproduction and information, facilitating the Universal understanding of its codes by other e, exi, I systems they code. Take the case of the key hormone of a human male, testosterone. It is a simple protein with 4 carbons, strong structural rings in a lineal shape that regulates, despite its simplicity, the reproductive development of energetic males, provoking actions of a complexity far superior to its molecular simplicity. How this can be possible? The answer is that the hormonal language is a relatively simple, positive Vs negative, 'yes Vs no' language that provokes or inhibits the 3±st cycles of existence natural to the will of the cells, proteins and nucleic acids it communicates.

But how hormones provoke so many different reactions in a cell, being so simple? The process of course becomes much more complex in their details as any language with a few phonemes can create very complex memorial, sequential sentences through the repetition of a few basic patterns. So for example some hormones are substances that inhibit the inhibitors of those existential cycles. On the other hand many hormones are transported by proteins, 'lineal killers', which modify them or 'motivate' other messengers of the cell or network system that carry further the message.

Thus the second translation of the hormonal message into a singular product of a certain cell modifies the message that becomes specific for that cell or organ. This is especially certain among animals that, unlike democratic, 'spatial', reproductive plants and fungi with a few basic tissues and a lot of generic orders, are hierarchical, temporal, complex systems with 2 informative languages: submissive, slower chemical hormones and faster, nervous orders. So hormones are created by neuro-secretory cells, part of the higher plane of existence of neuronal networks, which pour them into the blood. And often they are accompanies by nervous messages, which they reinforce.

Thus, animal cells might recognize hormones because they know 'behind' their orders there is a very complex and powerful system: the nervous or blood system that has to be obeyed or else. Since the blood system sends leukocytes to kill rebel cells and the nervous system sends flows of electronic information to control them. It is the same reason that makes people to obey simple codes like those of traffic. We obey because we know that behind those codes there is a complex organic system, the state that will send its police, its 'social leukocytes' and punish us. So the question is how a cell knows? Because its DNA has a degree of memorial, instinctive perception; *it is not a mechanism.*

Despite all those elements of complexity, it is still possible to classify all hormones according to their spatial or informative, 'universal' morphology as 'inhibitors or agents' of the 3 basic cyclical actions of any organic system; since their morphology imitates or it is based in informative nucleotides, structural carbons and lineal, energetic shapes, whose functions are instinctively understood by all living forms as all animals understand black and white colors as relative information or energy, or 'aaarg', open vowels and bass sounds as 'energetic anger' and 'u' closed vowels and highly quantized informative frequencies as informative curiosity.

Thus, *invariance of form between scales allows the existence of languages* that are decoded as they respond to the Universal language of self-similar topologies of all forms of existence.

So according to duality, when hormones act as inhibitors, they have an energetic, lineal form; when they act as agents to foster those existential cycles they have a cyclical, informative shape.

Finally, according to the 3rd postulate of parallelism applied to st/st+1 systems, their atomic components tend to be parallel to forms of the st+1 organ in which they accomplish their function. So when they act in the energetic zones of the cell or the macro-organism they are hyper abundant in oxygen, when they act in the informative brain they tend to be similar to nucleotide acids and when they act in the reproductive zones, they have carbon rings and vice versa.

In the graph, those simple rules explain the 3+2 main types of hormones in plants:

-*Max. E: Cyclical Auxin, rich in carbons,* increases the growth of shoots, the energy zone.

-*Max.I: Cyclic, nitrogenized Cytokinines* control the growth of roots, the brains of plants.

-*E=I: Oxygenated, Gibberellic acid* causes high growth in trunks and reproductive flowers.

Ultimately in as much as most hormones and enzymes act as inhibitors, they prove that the 3 energetic, informative and reproductive inner individual actions of organisms are the natural arrows of all organic systems, the will of any species, that have to be constantly inhibited by the neuronal-endocrine dual logic system of multicellular control. Since when that control disappears, naturally species try to reproduce. That is why cancers and any other form of cellular reproduction are so natural; because *reproduction* is the natural will of the fractal, organic Universe.

Finally, the duality of the EXI systems and anti-systems with diffeomorphic, opposite coordinates creates inverse hormones that neutralize each other. So 'energetic hormones' inhibit 'informative regions' of the organism. For example, we can consider that form Vs anti-form duality, analyzing the inverse growth effect in informative roots and energetic leaves of the 2 previous hormones, auxins and purines. What we observe is a diffeomorphic, inverse dual EXI process: the auxin that increases the growth of shoots inhibits the growth of inverse roots. And so auxins and quinines together work as a perfect dual system of growth control in the informative zone of the plant:

The root is a nitrogen-based informative system, which as all informative entities from neurons to worm holes, requires a lot of time and complex steps to reproduce its form. So quinines, its growth hormone, acts by parallelism, since it has 2 purine rings with nitrogens that positively induce the complex growth of roots made with nitrogens.

On the other hand, auxins have a structure that closely resembles the main nucleotide species, the purines. Yet, auxins ad to the hexagonal-pentagonal structure of purines a simple protein tail to move faster, and instead of nitrogens, auxins have 2 strong carbon rings. So it is basically a false purine with a far stronger body and a mobile tail that substitutes the quinine's nitrogens and inhibits the root's growth, because it cannot deliver the proper instructions.

On the other hand, the ever-growing shoots are the plant's reproductive systems; whose natural will to grow is so strong that any hormone increases its reproductive rate unless it is inhibited.

If we study the energetic cycle of plants, again we observe that dual inverse control system: Abscisic acid (a lineal molecule with a lot of oxygen elements) provokes the fall of leaves, while the lineal gas ethane, $CH_2=CH_2$, increases by parallelism with the lineal CO_2 molecule their respiration and metabolic rate. Both are very simple, energetic hormones, which resemble the 2 'energetic' inverse products of breathing, oxygen and CO_2, 'reminding' the plant that it has to accelerate or halt to its 'death' its production of those 2 products.

Recap. A language of information is always needed to create a complex organism. The simplest languages have 3 or 5 vowels, which represent energetic, informative and reproductive messages and its variations. They act by parallelism, since all systems and planes of an organism follow the 'invariance' of form and function. They also can deliver negative or positive orders depending of their parallelism or inversion with the form of the species they code. The language is always reproduced by the informative 'upper-caste' particles of the system, which delivers the orders to the rest of the elements. In plants, the language is a hormonal language with 3 main vowels, energetic auxin, informative cytokinines and reproductive Gibberellic acid.

21. Genetics.

An even more complex language is that of genetics, which we cannot study in depth due to restrictions of size of this lecture. Enough to say that in any efficient organism of multiple scales, the informative code, in this case the genes of biological super-organisms codify the multiple hierarchical levels of the organism besides the basic ternary code that constructs its simplest scales – in the case of life, amino acids. So according to the Fractal Principle there should be genetic languages with 3, 9, 27, 81 amino acid letters, which codify not only the proteins and cycles of the cell but those of the body. And so bigger sentences that group 9, 27, 81 amino acids should be coding 'bigger' actions and longer events within the complex cycles of the cell and beyond[4].

All languages are built in this manner. For example, the human language and the musical language follow self-similar scales of 'vowels' and 'consonants', informative and energetic units that then form more complex ternary systems, informative names, reproductive actions or verbs and energetic objects, and then group in paragraphs and so on. Even the language of film can be understood in frames, motions, shots, sequences and scenes. In the same manner, 'introns', the so-called redundant DNA should have genetic meaning and codify the higher scales of organisms - its functional morphology.

Yet scientists believe in naïve realism. Only what they see with their instruments seems real; and so the obvious perceived genetic level is the fractal level of nucleotide triads that codify proteins. But those triads are united in groups of nine bases and so on, creating new, 3^n, memorial planes *that act as* complex genetic *forms of higher scales* - an i-logic, scalar concept which today genetics finally accepts under the name of epigenetics. The enormous quantity of 'redundant' DNA and RNA which grows exponentially with the complexity of a multicellular organism is a clear proof of the correspondence between the complexity of the sentences of the language/instruction chains and the complexity and size of the form it codes.

A fractal equation derived from the concept of a network, which has $\sum^2$ elements to control an $\sum$-body herd (and/or its self-similar $\sum$-limbs) defines in Theory of Information that the number of informative instructions needed to create a system grows exponentially, according to its number of dimensions or planes of existence: $\pm E^{st}$. For example, to make a car we need E material parts, defined easily in a bidimensional plane with E fractal units of information: the drawings of each part. But we need around E^2 instructions to put those pieces together in a 3-dimensional form. Thus E^{st} determines the total number of informative quanta needed to build up a hierarchical structure extended through st planes of existence. To put a trivial example: Popular knowledge expresses that law in sentences like 'an image is worth one thousand words'. Since to describe with lineal, one-dimensional words a 3 dimensional image, we need indeed, E^3, $10^3 = 1000$ words. Thus E^2 is the number of redundant genes we find in the DNA-RNA systems of a complex organism, because *redundant, intronic DNA codifies the creation of the new level of cellular complexity, st=2, the multicellular organism.*

Recap. The old belief that a mere ternary amino acid code that creates proteins regulates the entire multi-cellular organism is just another simplifying conclusion of the one-dimensional, metric paradigm. A multicellular organism is regulated by intron DNA and the collective language of hormones that communicate those cells.

22. Differentiation of animal cells. Electric Neurons: amoebae.

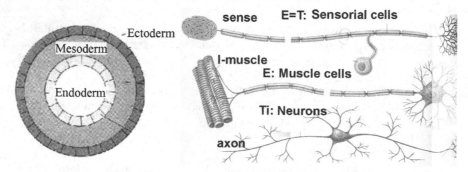

The electric cell is the top predator animal cell, due to its Max.IxE force, with an overdeveloped E-membrane and Max. I-DNA content. It further differentiated into:

-E=I: Sensorial perceptive cells, which reproduce the external cycles of existence of other beings within its inner form.

- Max.E: Muscle cells that extend and implode its protein membranes.

- Max. I: Neuronal cells that process information in networks called brains.

Cells evolve ternary networks, which in its simplest fetal stage form 3x3 st-regions proper of all animal forms:

- Max. I: The ectoderm, which gives birth to the nervous, sensorial and skin systems of max. perception.

-Max.E: The endoderm, which becomes the digestive system and its derivatives such as smooth muscles, lungs, livers and glands.

- ExI:Max. Reproduction. The mixed, mesoderm region, which subdivides in 3 sub-zones: the reproductive, blood and muscular systems.

Thus in its emergence from cellular to organic scale, the 3 topological regions of the cellular blastula suffer an inversion of form: the center gives birth to energy systems and the surface to information systems, while the middle region maintains its reproductive function.

Animal eukaryotic cells were moving species that had to react faster to the changing environment. So they developed a new, faster, nervous, informative, electric language that created a top predator cell thanks to its accelerated fractal action-reactions to information and energy: the electric cell, able to organize socially other cells into multicellular organisms.

The electric cell evolved from the ameba, a unicellular animal with membranes adapted to all morphologies that were shaped into arms to capture food, cilia to perceive information, tails to move and vesicles to expel unwanted energy and information, thanks to the use of heavy, metallic atoms, Na^- and K^+ that deformed those membranes, making them highly flexible: the duality of positive and negative metallic atoms created a deformed wave that moved along the membrane of the neuron cell, as any fractal wave of codified information does, becoming a new language, faster than the chemical language, talked by all other cells.

The combination of both languages, in a hierarchical scale in which electric, fast impulses provoke the emission of hormonal, chemical messages at the end of a long membrane that those chemical cells understand, allowed nervous cells to maximize its IxE force. Since now in the same relative 'time' that a small cell sends a message, the huge fractal action of the amoeboid neuron could send a wave of simultaneous messages through all its pseudopodia, coordinating at the same time the slow actions of a lot of enslaved smaller, chemical cells.

Thus the jump in size created a 'higher informative class' of macro-cells or brain of the animal that ruled a middle class of micro-cells or animal body, which together controlled an external energy territory. All those properties made the evolved, electric amoebae, the top predator cell in the Eukaryotic world. We still find those ancestral amoeboid cells in the most primitive multicellular sponges, evolved latter into the complex nervous cells of multi-cellular organisms.

Thus electric cells show Max. I x E force since:

- The nervous cell is able to control an enormous quantity of organic cycles with memories stored in its nucleus - since it is the cell with higher DNA density. So specialized neuro-secretory cells reproduce also the biggest number of chemical hormones.

- Such big volume of DNA implies also a great capacity to replicate its membrane, through micro Nyss cells that produce constantly membrane proteins, which are added to the external, ever changing morphological membrane.

Where did happen the transition from chemical to electric membranes? The evolutionary plan happens always in *isolated* environments in which mutational changes, *fostered* by the *specific characteristics* of the environment, take place without jeopardizing the survival of the species, due to the absence of top predators during the transitional, inefficient stage of the species that is 'adapting its form' at a *faster palingenetic time-speed* through *specific differentiations,* according to the i-logic *evolutionary plan.* The phenomenon is called evolutionary punctuation.

In the cell case, those requirements are found in shallow river mouths where changes in the salt concentration of water provoke by osmosis the constant expansion and implosion of the water content of cells and their membranes, bursting and killing 'rigid cells'. So cells, in order to survive, managed to create new membranes and deformed them very fast, expanding and imploding its form. Then, those new top predator electric cells started a massive reproductive radiation, expanding in all seas, feeding and enslaving other cells. The subsequent combination of top predator electric cells with other varieties of chemical cells, differentiated and evolved the first ternary scale of simple organic macro-systems: sponges (max.E), hydras (max.Re) and worms (Max.I) with a growing number of cellular types.

Recap. Multicellular species were born when a new top predator, larger and faster, Max. ExI, species, the nervous cell appeared, herding chemical cells and differentiating in 3 physiological cells in charge of future digestive, blood and nervous networks.

23. 3x3+st dimensions: Embryology & Physiological networks.

We have described the 3x3+st dimensional tissues of living organisms in space. If we consider them in time, we observe a process of evolution that differentiate the original ideal spherical or spiralled st-point (seeds, cells, ovules) into complex morphological shapes, adapted to those inner and outer tissues. Thus living forms evolve and diversify, opening or invaginating their networks, organs and sensorial apertures to each specific ecosystem, which subsequently adapts the morphology of the cells and organs they control to the specific surfaces of their environments. So men, who exist on a planar surface, have a different morphology to the ideal spheres so common in water and space, which are 3-dimensional isomorphic ecosystems - as human evolution has adapted our networks and morphologies to the land-atmosphere environment.

In those different processes of ecosystemic adaptation the 3 dominant networks of living systems further quantize into secondary systems, departing from the 3 initial layers of cells. We can follow that process temporally in the evolution of the embryo that resumes the palingenetic evolution of life and spatially in the organic structure of living beings. It is the science of evol=devol that relates both phases of animal evolution, establishing a parallel correspondence between the spatial location of a tissue at birth and its functional evolution into i, e and E=I tissues. Thus according to their st-location the I-ectoderm evolves into informative tissues and networks; the E=I mesoderm into reproductive tissues and networks and the E-endoderm into energetic ones:

Max. i: Ectoderm: the 3 informative sub-systems.

The information network derived from the external *ectoderm, dominated by neuronal cells,* evolves and differentiates according to the Fractal Principle into:

- E-*endocrine* systems, with *internal neurosecretory* cells that control internal, cellular information through chemical hormones.

-E=I: *nervous* systems that control and replicate in a neuronal *brain* with the electric language all other functions and forms of the body.

- I: *Skin systems and outer senses, which are the openings to the world of the nervous system* through which the organism emits or absorbs energy and information, communicating with the st+1 ecosystem. Those senses specialize in perceiving the external, energetic and informative cycles of the beings that share the ecosystem of the organism, duplicating them in a series of fractal, quantized, reduced images, with different languages.

The existence of multiple senses justifies the linguistic method of perception in a Universe of 'multiple spaces-times and points of view', where each entity tries to perceive as many parallel worlds coded in different languages. Thus the linguistic/multiple Universe gives birth to a multiplicity of senses and perspectives and defines a higher truth as a sum of perspectives casted on a certain system – as each point of view will create a self-similar image of the Universe, and so only by comparing and adding all those self-similar images we can obtain a more complex, kaleidoscopic image of the whole.

In living beings to accomplish the higher truth of each form through the 'linguistic method' sensorial languages multiply the perspectives on the cycle or form observed, extracting their existential properties, its form, density, force, etc.

Thus senses and languages differentiated also along an E⇔I arrow from pure spatial senses (eyes) to the most complex, temporal senses (ears), defining an evolutionary arrow of increasing complexity in their capacity to gather information, *which often defined the survival capacity of the species, as senses are the key to interpret correctly the destructive or creative actions-reactions of all other beings.* So from an initial simple phototropic stain, eyes, ears, antennas and mouths have multiplied, close to the inner, neuronal informative brain that processes their information, forming together *a 'head'.*

Max. E: The internal endoderm. Originally populated by *wandering amoebocytes* and *glandular cells* that digested food, the 'energy hole' of the first animals, sponges and hydras, invaginated forming the coelom between the endoderm and mesoderm that subdivides further into the 3 cavities of animal life: the digestive system, the breathing system and the heart cavity. The 2 first cavities evolved, surrounded by endoderm cells into 3 subsystems:

- Max. E: The *breathing system* gathers the smallest energy quanta, oxygen.

- E=I: The *digestive system* gathers bigger food quanta, differentiated morphologically in 3 new subsystems with linear, wave-like and cyclical components:

Energy network: max. E: linear intestine <stomach: elliptic exi > sensorial, cyclical Mouth: max.I

- *Max. I: Glandular tissue:* It forms organs dependant on the digestive system (liver, digestive organs, kidneys), that process their products.

E=I: Mesoderm: the 3+(st+1) reproductive systems.

The 3rd coelom cavity, the heart, surrounded by mesoderm tissue, invaginates further into very thin vessels, created with *striated muscles,* densely populated by the original wandering amoebocytes reconverted now into *leucocytes.* In the *blood networks* the other 2 systems merge and pour their products: quanta of energy, oxygen and food; and quanta of information, hormones, which the system takes to each cell. The blood systems mixes both energy and information quanta. Thus, it is also the reproductive system by excellence with maximal contact with the intermediate region's secondary tissues evolved from the *mesoderm*:

- Max. E: The *skeleton*: It sustains the system and reproduces blood cells.

- E=I: *Muscular tissue:* The most resistant electric cells form a muscle dual negative-positive symmetric, spatial system with myosin cells that control the elongation and shortening of membranes.

- Max. I: *The blood and tegumentary system* that prolongs the blood network between cells.

- st+1: Yet the fundamental reproductive tissue is *the sexual, glandular tissue*, that reproduces the organism beyond the cellular scale and further differentiates into the basic E=male Vs. I=female duality of 'energetic and informative subspecies'.

Those 3x3+st standard systems and tissues, differentiated from the initial 3-layers of the embryo, define complex living organisms. Yet those organisms exist also in an external world where they become individual quanta, performing external cycles of energy, reproduction and information, parallel to those internal cycles performed by its cells and fractal networks. So all organisms have also an external, vital territory within its ecosystem to provide for their internal 'mirror' networks.

Recap. The ectoderm, endoderm and mesoderm topologies of the embryo, guided by top predator nervous cells differentiated into the e-exi-i main systems of the adult organism.

24. Creation of a 3x3+st decametric scale of cells and tissues.

Such processes of cellular diversification, guided by the informative elements of the system (frequencies, genes, memes, etc.) happen in all systems of space-time that differentiate according to the ternary method its relative fractal cells in 2x2 complementary systems or 3x3 organic systems (particles in the atomic nucleus, vowels in languages, etc.).

In this manner simplex systems evolve into complex ones; complementary dual systems become ternary, organic systems; bidimensional systems becoming 4-dimensional systems and ternary systems become 3x3+st systems, emerging into a new scale. And this process takes place in time through 3 ages or horizons that complete the evolutionary process.

In life this happened through speciation of cellular languages evolved from 1 to 3->5->7->9 cells in the 3 time horizons of all EXI cycles, latter re-ordered in space in 3 e-exi-i types of organisms.

Since morphologic differentiation occurs only among the top predator, informative organisms of any ecosystem, we could establish a comparative homology between many of those processes. In the biological scale, it happened with animal cells; then with networks of cells in complex vertebrates; and finally with verbal sounds among human beings - a new language which, unlike neuronal impulses, could breach the discontinuities of air-space between individual humans, creating a higher scale of social, living networks, cultures and civilizations. Then again, those differentiations happened in the memetic systems of the dominant, technological civilizations that created the Financial-military-industrial complex and its metal-memes of informative money, energetic weapons and reproductive, organic machines... *Since evolution is a morphological game that didn't stop with the summit of carbonlife man, but now continues in the new species of complex metal atoms to which humans are transferring their form.*

Let us see the first of those ternary scales - animal cells evolved in time to analyse then the 3 'animal phyla' they created in space:

The 1st scale: from 1 to 3±st electric cells.

-st-1:Electric wandering amoeba, differentiated in 3±st subspecies:

- Max. E: Muscular, myosin cells that maximize the flexible properties of the membrane.

- E=I: Sensorial cells that reproduce their external cycles of existence with their mimetic forms.

- Max. I: Informative, neuronal cells.

-st+1: Those 3±st specialized types of electric cells, multiplied and evolved socially into dense tissues and networks in which one of them is dominant: the blood vessels dominated by wandering amoebocytes; the muscular tissue dominated by the myosin-actin inverse lineal proteins; the senses dominated by sensorial cells and the brain, dominated by neurons, which act as the communicative consciousness of all other systems (st+1). They form the 3+st physiological networks that defined the specific space-time equation of all complex multicellular animals:

$E: \sum$ *amoebocytes & Blood Systems <Muscular cells > $\sum$ sensorial & nervous system: I.*

From 3±st to 5±st cells.

In the next stage of cellular differentiation the 3±st electric cells, tissues and networks added slave chemical cells that stretched their energy and information limits:

The sponge, the first animal phyla, dominated by wondering amoebocytes adds 2 slave cells that increased the IxE force of the energetic membrane and the informative singularity:

- *Max. E: Epithelial cells* of max. energetic strength made with proteins and incrusted with heavier non-organic atoms that protect the organic system, through a shielded external membrane. Later on in more complex organisms those cells evolved into inner bones and breathing systems.

- *Max. E x I: Glandular cells* able to reproduce specialized energy and informative substances necessary to the other cells that multiplied in the intermediate st-region of the body.

That pentagram of 3±st differentiated electric cells and 2 basic slave cells, quantified in social groups, shape the 5 main organic systems of animal bodies:

a) Max.E organs: Membranes made mainly of epithelial cells that create the skin, the discontinuity between the st-point and the external world and the digestive system...

e) E: Muscular systems based on electric myosin cells that turn energy into movement.

i) E=I: Wondering amoebocytes reign as leucocytes on the blood networks.

o) E=I: Internal glands, attached to the blood, reproductive and digestive, energy networks that reproduce the substances, which the organic system needs.

u) I: Sensorial organs based in *electric cells* that perceive information about the cyclical actions of the outer world and translate it into mental images. They will guide the actions of…

st+1 =Aeiou) Max.I: *Neuronal brains* that control the entire organism as the consciousness of the system, according to the existential actions those senses observe in the external world.

Thus in living beings as in any universal system, time dominates space and so the informative, controlling orders go from sensorial to energetic cells, shaping the outer, muscular movements: I>E.

Those 5+1 cells and 5+1 basic organs are present in all forms of multicellular life since they are closely related to the 4 + st-generational cycles of exi=stence, each one dominated by one type of cell and organ: Max. E, digestive organs and cells absorb energy; E, muscles that emit energy; E=I, blood networks and glands that reproduce the system; I-senses that absorb information and Max. I brains that emit information and control all other cycles, perceiving the entire organism as an existential whole, living the entire generational cycle of the organism.

It is for that reason that we find also 5 vowels in human languages, 5 lines in the musical pentagram, 5 cells and types of tissue in the simplest organisms, etc.

Decametric scale: invaginations and st+1 excretions.

Any organism is an inner st-world that exists in an external st+1 ecosystem developing inner and outer cycles of recollection of energy and information. So the final diversification occurs along the inner-outer duality of the organism in order to accomplish those 2 x 5 inner and outer cycles. Now the pentagram evolves into 9+1 cells and types of organic tissues that stretch again the cellular field according to that inverse symmetry between the external and internal world, as cellular tissues invaginate or eject their cells perpendicularly in the height dimension of evolution:

- Thus *inner skeletal and outer skin cells* differentiate the structural tissues of sustain.

- Membrane cells *differentiate into inner blood* and outer *lung* cells that process internal and external energy.

- *Internal, creative glands and external, destructive urticaria cells* reproduce informative and energetic substances.

- D*igestive and muscular* cells create inner and outer motions.

- *External sensorial cells* process external information.

- I*nternal neurons* control the internal information of the organism with the electric and chemical language provided by hormonal, *internal neuro-secretory* cells, dominating all cells.

The result is the creation of 9+1 types of tissues that are found in the most complex animal forms of life and become the 9+1 standard 'systems' of the most complex living organisms:

The *digestive and muscle* system, the *skeleton and tegumentary* system, the *blood* and *breathing* system, the *reproductive* and *excretory* system, the *endocrine* system *and the nervous* system, the 10th system that controls the entire organism duplicating in the brain all other organic functions it directs becoming the st+1 scale that puts together the organic being.

Recap. The 9+1 physiological systems and tissues of an organism are born from the 3x3+(st+1) 3 horizons of differentiation of its cells.

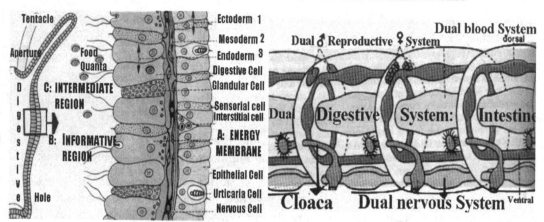

The Hydra is a new living phylum with 5+2 cells distributed in 3 regions, common to all palingenetic, foetal forms:

- Max. E: The ectoderm is the external, energetic membrane with the hardest cells: the hard epithelial and aggressive urticant cells that explode its poisonous cilia and the interstitial cells that ensure its continuous isolation.

- Max. I: The mesoderm is the informative region, with nervous and sensorial cells.

- E=I: The endoderm is the intermediate region with glandular cells that digest food quanta, entering through the mouth.

The worm is the 1st network animal, made of a lineal wave of $\sum$ parallel organic spheres, each one a st-point with 3±st dual networks: The nervous, blood and digestive/excretory systems, which accomplish the emission and absorption of informative, reproductive and energetic cycles.

25. Network evolution: bodies, brains & reproductive systems.

It is then evident that the evolution of animal life creates new phyla, based on the capacity of the new species to accomplish their external cyclical actions on their territory, thanks to the evolution of their internal and external networks.

The interaction between those 2 levels, the st-1 level of physiological networks and the st+1 level of ecosystemic territories determine the existence of most organisms. So the evolution of life on Earth, externally observed in the ecosystems, territories and relationships of living animals, which depend ultimately on their capacity to handle temporal energy coming from light, is caused internally by the evolution of those 3 types of organs and internal networks: reproductive, genetic organs; informative, brain organs and energetic, body organs.

When one of those organs evolves, improving its capacity to handle energy or information coming from the ecosystem, a fundamental differentiation of species occurs. And we can consider the evolution of the main phyla of animal life in a deconstructed manner, as a process that evolves sequentially those 3 types of organs:

The evolution of those three organs triggers the biological radiation of a new phylum that preys on less efficient forms of organic life. The process is very fast, as time can change its rhythms and adapt to the best strategy of survival, as palingenesis shows. It is called *Evolutionary Punctuation*: when a new species with better energy or informative organs appears, it feeds on other species and reproduces massively causing the extinction of the previous top predator species, till it reaches a trophic balance with those victims. It is the essence of Darwinian evolution: 'evolve and multiply'. We can find such catastrophic evolution in many geological and organic ages of the Earth: First animals displaced plants because they evolved better informative networks. Plants could only gather energy from light. Animals could 'see' light and get information about their environment, and act-react faster. They used plants as food. Then animals 'radiated' (multiplied) all over the Earth and diversified. And each new phylum with improved networks displaced the previous ones. Thus once more, the evolution in time of living organisms and its spatial structure are intimately related.

Thus, the first ternary evolution of multicellular life created 3 phyla of increasing tissue complexity that completed the evolution of animals from the st-1 cellular tissue to the st-network scale:

-*Max. E:* The simplest animal organic systems, sponges and *porifera*, which are basically a digestive system.

-*E=I:* The balanced animal organic systems, the *coelenterates* such as the hydra, which maximize its reproduction through the split of its cells.

- *Max. I: Worms* that add informative cells and differentiate clearly its 3 physiological networks.

They were the first of many ternary E-exi-I rhythms of creation in the animal kingdom that evolve from energetic, to reproductive to informative sub-species, through the 3 horizons of any phylum, with a parallel improvement of the 3 network systems of the animal, which from now on will be the higher 'scale' of existence to which all cells become submissive.

Since as the new animal forms multiplied, their informative nervous networks controlled a growing number of cells through their fractal, simultaneous actions; and so they also grew in spatial size, multiplying their IxE force. It means they had to re-organize those cells beyond their initial division in simple tissues, creating specialized 'energetic, informative and reproductive networks'.

It will be the definitive jump from the state of 'cellular herd' to the state of 'organism', the 3rd age of a social form that evolves from a I horizon of individual 'energy quanta' to the balanced herd that fluctuates between 'wave and particle' state to the informative state of a tightly packed organism which those networks maintain together in minimal space. And so from then on evolution will be no longer differential evolution of cells but differential evolution of networks…

Let us see now in detail those main phyla differentiations.

Sponge vs. Hydra: cyclical-lineal digestive networks.

The evolution of multicellular organisms started with the creation of social, digestive, energetic tracts made of the pentagram basic cells that came together to improve the cyclical fractal actions of the group. The first of those organisms was the sponge, *the first animal phylum that emerges from the previous cellular scale with 5 types of cells*:

-Max. E: The sponge has *flattened epithelial* cells and *hollow pore* cells, which are external, membrane cells differentiated by their inner and outer location. They are the primitive versions of skins and breathing systems.

- E=I: *Mesenchyme cells* that secrete siliceous spicule, strengthening the walls of the sponge. They are the primitive version of glandular cells.

- Max.I: Central top predator cells: *Wandering amoebocytes* that herd food for the five types of cells, moving around between the other cells to capture the particles entering the hollows of the spherical sponge; and *collar cells* that sense water flows, beating their flagella to produce a flow of water that introduces food in the sponge. They are the primitive versions of brain and sensorial cells.

Further on, we differentiate in the sponge 3 st-regions according to those 2-1-2 kind of cells: *the external membrane* of hard cells; the inner, *glandular, intermediate* region that reproduces the specific substances of the sponge and the *central hole* where *informative* cells wander. The 3 regions create the 3±st vital cycles of the sponge thanks to those 5 cells:

- Max. E: The sponge *feeds* on energy quanta that enter its central hollow.

- Max. i: It *perceives* those quanta as its collar cells sense the water flows.

- E=I: It *reproduces* new cells through its glandular systems.

-Σ: It keeps together those cells *in social groups* thanks to the epithelial cells of the membrane that maintain a rigid, enveloping structure.

- $\Sigma E \Leftrightarrow \Sigma^2 i$: And so the sponge *exists* as a whole being controlled by the wandering amoebocytes, the dominant, informative cells that use the sponge as their *territorial* body, their vital space.

Those amoebas will be also the dominant cells of the next animal phylum, Coelenterata (hydras, jellyfishes), evolved already into electric cells, and hence connected into the first 'nervous tissue'. So they will become the 'st+1', existential system where the consciousness of animal organisms as a whole exists.

The 7 cells of hydra and their networks.

Those hydras add 2 new cellular, energetic specializations, extending morphologically the IxE force interval of animal life, from 5 to 7 cells:

A): MAX. E: The epithelial cell. A still harder, internal tissue that maintains the rigid structure of the animal and will evolve into armours and bones. And…

ei) Ei: The *urticant cell,* an external, energetic differentiation of the reproductive, internal *glandular* cell, which produces poisonous substances to defend the animal.

Both are created through the inversion of directionality of its twin cells: the internal glandular cell becomes now an external form, and the external epithelial cell becomes an internal cell.

Those 2 final tones of specialized energetic cells make the hydra a natural born top, lineal predator, the next evolutionary step that inverts its form from a cyclical sponge to a lineal, reproductive body. Thus the complexity of the Hydra grows, shaping definitely the 3 E-exi-I organic st-regions common to all living beings (mesoderm, endoderm and ectoderm), which will vary in morphology and complexity but not in their ternary functions.

Further on coelenterates bring to animal life the maximization of its reproductive systems. Both the sponge and the Hydra lack a specialized blood, hormonal system, so its reproduction is far simpler than in evolved organisms: each cell is in itself a 'genetic mother-cell', which stores the information of the entire organism. This implies a limit to their informative evolution, as cells have to keep an excess of redundant information, according to the $E^{st=2}$ law that increases geometrically the number of genetic instructions needed to create the new st=2 multicellular plane of existence. On the other hand, it makes easier reproduction: any section of the Hydra can create a new animal. The result is that the arms of the hydra, where most sensorial cells are, break away easily, moving with the streams of water, reproducing new hydras all over the world. Yet some tentacles fail to reproduce evolving instead into planarians, the 1st worms, which will acquire 2 new informative cells, completing the 9+1 decametric scale; and developing fully the 3 physiological networks of complex animal life.

Worm: 1st animal: mobility, senses, 4-D, networks.

Thus, the next step in the evolution of life, after the hydra develops 2 new energetic cells, will be the evolution of 2 new informative cells. They will create a new phylum, the worm:

- *Visual, spatial* cells that perceive light-space.

- *Temporal, auditory cells* that perceive the sound waves and informative languages of animal life.

Those 9 cells complete the differentiation of cellular species that a growing st+1 neuronal, inner center - the brain - elaborates as the consciousness of 'the whole' increasing ever since its size till acquiring the weight of the human being.

Those informative cells were necessary to the new environment of planarians, which are in constant movement; hence have to orientate themselves in the ocean flows in search of energy. Since the change from stillness to movement is a fundamental change for life beings, which definitively transform all its elements to the properties of animal life:

- *The new informative cells create new apertures, the senses,* that gather in the frontal zone of movement, the relative height of the worm, creating '*heads*' that will also control the energy apertures of the *body*, splitting clearly the organism in an energetic, moving body and a sensorial head, which controls the information and energy of that body.

- Animals become *bilateral* in order to dominate the *2 directions* of its initial *bidimensional* planar form: *a hierarchical, temporal dimension* from the future informative head to the past, energetic tail in which they orientate their organic, inner, evolutionary morphology and *a perpendicular,* parallel, equal, repetitive, 'present' *spatial dimension,* from left to right in which they orientate their reproductive, fractal, cellular 'fat' growth.

So embryo worms develop 2 bilateral cavities or 'coeloms', latter evolved in dual organs, which in the head will observe the 2 directions of their spatial field: 2 eyes, 2 ears, etc - while in the body, inner organs will also double, creating in more evolved phyla a certain EXI asymmetry with slightly more 'energetic' and 'informative' sides. So the heart will have an explosive and implosive region; sexual organs will become I-feminine and E-masculine; the brain regions will specialize in spatial and informative tasks; some crabs will develop energetic and manipulative arms.

Thus the 1st worm, the planarian, created a diffeomorphic bidimensional structure with 2 EXI perpendicular planes that all future animals will imitate.

In the graph we study the inner structure of the worm, because it shows already some of the dualities, fractal strategies and future arrows of life evolution that will act on different phyla to diversify and evolve their species:

- The worm is divided in fractal units, setting up a basic duality of living beings that sometimes are 'herds' of individual organs, multi-eyes, multi-bodies and sometimes fusion all parts into a whole.

- It shows perfectly differentiated the 3 physiological networks/ dimensions, proper of all advanced organisms that mimic the 3 regions of a st-point:

- *E: The worm absorbs energy through the digestive network* and emits it through the excretory system; which in the worm occupied the original endodermic, central singularity. But as animals evolve informatively as chordates, the center will be finally occupied by the nervous, informative system as in any other st-point.

- *I: The worm absorbs information through the sens*es and emits it through the nervous system. The informative brain and nervous system directs the entire organism and unifies its cellular quanta as a whole. The nervous system further differentiates into the sensorial, nervous and neuro-secretory systems, when the planarian adds eyes and ear systems that represent a jump in complexity over the informative systems of the hydra. It defines also a head in the dominant, temporal dimensionality of movement. In the planarian is still length, which will rise till reaching the height of man.

464

–*E=I*: *The intermediate region* of the worm is controlled by the *blood system*, the fundamental dual transport system of the worm, where the neuro-endocrine glands dependent on the informative system, pour their reproductive hormones while the energy/digestive system pours its organic food.

The worm represents a jump of complexity in transport systems as it uses for the first time, organometallic molecules (hemoglobin) to harness oxygen energy and quantizes the blood network, which now arrives to cells far away from the digestive hollow. So worms can grow in size and energy power respect to the previous coelenterates. Now energy and information combine together on organs and glands dependent on the blood system.

The most important ones will be the new reproductive, specialized sexual glands, differentiated into dual, female and male sexual organs that make worms, hermaphrodite systems. Sexual organs again represent a fractal jump in the evolution of life. Since now, unlike in hydras, single specialized sexual organs will reproduce the living animal, increasing the sophistication of the process and liberating from those complex tasks all the other cells, which can specialize further.

Still many worms can reproduce by both systems: segmentation and sexual reproduction, which in new phyla will become the only form of reproduction, splitting organisms in 2 different sexual genders, the male and the female.

From worms to vertebrates: 3±st ages of fractal integration.

The evolution of worms into vertebrates is a long process of 3+st phyla differentiations based on:

- E: Spatial, I-radial or E-bilateral symmetry.

- $\sum$ Re: Fractal numbers that foster hierarchical e-exi-i segmentation and differentiate big unitary animals and small animals made of fractal parts.

- I: A constant increase of dimensional height.

- +st: The evolution of its 3x 3+(st+1) physiological networks.

The first massive differentiation will take place in the Cambrian age, according to the *Black Hole paradox* that evolves faster 'smaller forms, denser in information' than bigger, spatial forms. Thus in the Cambrian age most phyla evolved from small trochophores of ancestral flatworms that gave birth to new phyla during its 'palingenetic larva, conception stage', according to the Fractal Principle:

So we can first differentiate the worm phylum into 3 fundamental phyla with intermediate forms:

- Max. E: *Platyhelminthes*: the simplest worms are flat, bidimensional planarians without blood systems that still require all cells to be close to the skin surface of the animal to exchange oxygen with the air.

- Max. I: *Nematoda* or round worms, which develop a dimension of height as they add the blood system with its fine vessels that carry energy to each fractal cell.

- E=I: Max. $\sum$: *Annelida* or ring worms, which reproduced a micro-worm unit, as crystals do, into a series of fractal pieces, growing enormously in size. Thus Annelida became the worms with higher IxE force radiating and diversifying in 3 sub-classes according to its territorial environments:

- Max. E: *Polychaeta* or marine worms.

- E=I: *Hirudinea*, living mainly in shallow or fresh waters. They are leeches that feed on blood and might in their initial forms feed on waters rich in metal, creating the first blood systems.

- Max. I: *Oligochaeta*, terrestrial worms that evolve further due to the new challenges imposed by the ground environment.

465

Annelids, the dominant phyla, diversified and multiplied into multiple types of bilateral animals, with a central cavity and 2 coeloms that evolved further its dual organs and 3x 3+(st+1) network systems. The first of those differentiations gave birth to the main animal phyla that will dominate the following life ages of the planet in 3±st periods, which took place according to ternary differentiation, as evolution puts together fractal 'organic' units into single unified systems:

- *(st-1): The worm's* body is divided in fractal elements which are still independent elements with entire 3 functional sub-systems: each section has 2 reproductive organs, 2 nervous ganglia, a digestive tract and excretory anus and 2 upper blood nodes.

- *Max. ∑3: Arthropods* keep the earlier segmentation of annelids, but they organize those fractal segments into 3 differentiated zones, with several independent sections that go from 3 to 21 parts:

A sensorial, informative head (Max.I); a central thorax with moving limbs and wings (max. E) and an abdominal region with the glandular, digestive and reproductive systems (max. E=I).

- *E=I: Molluscs* have balanced, hierarchical, single organs. As the 3 physiological networks evolve and become quantized to reach each cell, the different sections of the i-head, e-thorax and exi-abdomen fusion together. Now the quantification process is transferred to the ends of those physiological networks: axons, blood vessels and digestive tracts become thinner to improve their control of individual cells.

- *Max. Informative Evolution: Echinodermata.* Echinoderms fusion those organs into single systems in the stillness of the marine platform, where they became the most successful forms.

- *st+1: Chordates: vertebrates.* Echinoderms in its larva, moving stage, according to the Worm Hole paradox, give origin to chordates, the 1st vertebrates. In both phyla the small organs of each arthropod's section fusion into single, continuous big organs, thanks to the integrated evolution of the informative nervous system that aggregates the individual cellular quanta into those specialized organs. So the multiple eyes of insects become a single eye; its multiple hearts a single one, etc.

Though all those phyla appeared already in the Cambrian their sequential dominance on the Earth is parallel to those previous 3±st ages, since the simplest forms, arthropods and molluscs, reach first the summit of their evolution (Max. E=Min. I), while complex chordates took a long time to reach its evolutionary height and became dominant latter as fishes and reptiles.

Recap. The differentiation of cells into 3-5-9 forms allowed the creation of complex multi-cellular organisms, first dominant in digestive systems (sponges), then in reproductive systems (Hydra) and finally in informative systems (worms). From the worm on, all living animals will be defined as st- points with 3 inner networks/dimensions, which will create in the outside world 3±st cyclical actions, designing an external territory also with 3 networks/dimensions.

26. Arthropods. The new social scale of superorganisms.

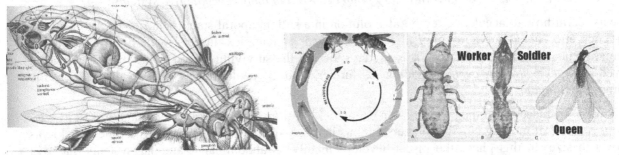

On the center, we see the 3 ages of an insect's metamorphosis from larva to chrysalis to adult. They differentiate in '3 lives' the existence of most insects that develop sequentially their 3 main organic regions and networks during each of those live.

On the left, according to EXI duality, the final results of those 3 temporal ages are the 3 regions of a topological insect:

-E: The digestive and dual respiratory systems, dominant in the larva, are the energetic networks, located in the center of the body, shaped as a spiral.

- I: In the belly we find the finest and more quantized nervous informative network, developed during the chrysalis life.

E=I: Finally, the reproductive, hormonal network, exi, pours into the blood system, which dominates the 3rd life of the insect, in the external world.

On the right, the social classes of insects, like those of a human society, correspond to the 3 organic function of a new scale of existence, the superorganisms of insects, its st+1 system: Termites have an informative brain, the pheromonal queen (c); a re=productive class of workers, which produce the structures that create the ant-hill (a) and an energy class (b), the drones and soldiers, that have a linear, spatial morphology.

The first arthropods, derived from Annelida were probably trilobites, which protected their bodies and heads with external hard shells. Trilobites increased their energetic force, maximizing the strength of its membranes. They probably responded to the increase of IxE force caused by visual cephalopods, which maximized their informative organs. And so the game of life raised its fractal force, balancing again the top reproductive body membrane and top informative, mind singularity.

Today, according to duality and the Fractal Principle, we differentiate arthropods in sea animals (the most efficient of which are crustaceans) and land animals, which became dominant and evolved towards new informative species as all air or land forms did in a light-friendly environment. So again we differentiate them in 3 basic forms:

- *Max. E: Myryapoda*, with max. body development and multiple feet.

- *E=I: Arachnida*, the balanced species.

- *Max. I: Insecta*, the informative class with max. brain development, which are still the most successful animals on this planet in the microscopic level of chemical life. Since they completed the organic evolution of chemical animals towards its most perfect form in 3 evolutionary phases.

Max. E: The evolution of energetic systems brought about the first flying animals that colonized a new environment.

Today flying insects still account for 1/2 of all animal classes. They made energetic networks the center of its body, developing on their middle region a highly efficient muscle and blood systems, with 2x3 wings and legs.

E=I: The second evolutionary jump occurred in their reproductive systems.

Insects learnt how to accelerate temporal evolution in a still, temporal state (Min.E=Max.I), changing from energetic, lineal larva to chrysalis that emerged as complex insects with highly developed informative heads and energetic wings. Nowadays 90% of the surviving insects come from species that evolved its *generational cycle*, dividing it further into 3 evolutionary phases that shape the metamorphosis cycle:

- Max. E: Insects live their youth as an energetic, lineal *larva* that merely feeds and grows in size. A larva is a sort of moving egg that gathers vitellus in 3 sub-ages in which it changes 3 times its skin as it grows in size. In this phase the insect develops mainly its abdominal, glandular systems that will produce the enzymes needed for its first metamorphic change into a:

- E=I: *Chrysalis*. The intermediate *Chrysalis* age is a 'frozen' vision of the most surprising facts of palingenetic evolution and inverse differentiation: In an ever moving Universe, external, spatial immobility triggers internal change in the speed of informative evolution, as outer movement is transferred to inner cells, rich in enzymes that become the dominant cells of each section of the adult insect, moving and reorganizing their tissues through a series of inversions and evolutions of its morphology. So central tissue ex-vaginates as wings or legs, etc.

A similar inversion happened when mobile trochophores in their larva transition became still echinoderms, which evolved and differentiated further, causing the explosion of chordate's phyla that happened in the Cambrian.

In those trochophore embryos, inner dominant cells also reorganize the different tissues, placing the other cells at will. So chrysalis evolved the middle thorax section and brain systems, becoming:

- I: A hard insect with Max. IxE force (a harder E-exoskeleton and a far more developed I-brain) that will live the 3 usual phases of life.

The third mutational age of insects was informative:

Insects learnt to communicate socially through chemical, pheromonal messages, giving origin to ants and bees, the dominant ground and air modern insect species. It was again an evolution departing from very small forms, according to the Black Hole Paradox. Today the smallest organism is an ant that weights 10^{11} times less than an elephant, the magic fractal number, S^t, between 2 scales of existence, which is also the difference of weight between the smallest and biggest particle of the physical world. So the fractal limits of 'informative scalar growth' have been reached in both, the physical and living realms.

It is worth to notice that insects have not evolved further in the last 100 million years, but are still the most successful chemical beings. Because the game is fractal and so it always has an evolutionary limit based on its ternary ages. For that reason, once reached the 3rd formal age of max. information only social evolution into a new macro-organic plane of existence can improve the survival of a species. It is what happened with insects that became super-organisms called anthills. And so ants became the most successful animals of the chemical world as men will be in the electronic world, due to the fact that they act as a simultaneous, present form, sum of all the fractal actions of the herd, guided by their informative common pheromonal language, spoken by the 'queen-brain' of the anthill. It is also worth to notice that in both realms - the world of chemical insects and the world of electronic humans -we find the same 3 organic classes proper of any Universal system.

Recap. Insects became the most successful chemical species, when they evolved into social super-organisms, with the informative ant-queen brain, which controls with pheromones the workers that reproduce all the elements of the ant-hill and the energetic warriors that defend it.

27. Molluscs, the first eyes.

The next successful phylum, molluscs also suffered a ternary differentiation. Thus molluscs today are classified into 3 classes:

Maximal energy: Lamellibranchiata (which are big stomachs).

Balanced forms: Gastropoda.

Max. Information: Cephalopods, which developed the first eyes.

Though there were other primitive molluscs, today almost all of them belong to those 3 species. It proves that even though a st-system essays many variations, the 3 sub-classes *of max. energy, max. information and max. reproduction* survive better; because any environment allows those 3 classes to find specialized econiches in which to maximize their existence and resources.

The most successful of all gastropoda were again the informative class, cephalopods, the first living animals with complex eyes. If we observe animal life, the key to its evolution is the improvement of its virtual worlds, of its informative organs:

In the first forms of life, perception was chemical, olfactory based in slow, short-range molecular quanta; until the first eyes appeared, inaugurating a new virtual world, made of smaller, speedy photons that create long range, detailed light images, making cephalopods act-react faster and farther than any other animal.

Squids were born in the abyssal ocean ecosystem, where still the biggest squids exist (over 10 meters long). It is the kingdom of bioluminescence - a new language based in the Universal code of colors; they were the first to interpret. For example, when a squid becomes red, the color of energy, it means it is angry. Those first primitive cellular eyes had to look hard to see their environment and the prey they sought. When they came up to the surface they saw even more and preyed on blind, energetic phyla. Today squids are still among the most intelligent animals, showing some self-consciousness.

The squid is the first eye-world; a new informative language that will completely changes the stakes of living organisms. The organs of perception of the squid, the eyes, were a new Top Predator language, superior to olfactory organs, both in detail and range. The effect of that linguistic superiority was the massive radiation of squids and the parallel extinction of perhaps 90% of the smelling species of the Cambrian that became their preys.

Those eyes enabled cephalopods to become the masters of their Universe, building all their other organs around their superior organ of perception: their tentacles became hands for the eye; the body became a canvass that changed color to interpret the new language.

Cephalopods also caused the arrival of exoskeletons in a classic process of action-reaction; only those olfactory animals with external protection (Max.E) could survive the faster informative eye of the hunter (Max.I). Thus the Cambrian holocaust also diversified life.

Those cephalopods with eyes became top predators in the Ordovician age, the age of squids. In this manner chemical perception left way to light and sound perception that developed highly sophisticated neuronal cells, which reach 2 meters in some squids.

Yet, when vertebrate life begun the 'hard shells' of some echinoderms sustained those long neurons, protecting them and allowing further quantification. So an energetic top predator found a cyclical protective form, in a dual game of evolution of prey and predators that will be carried till humans appear and beyond through the evolution of weapons and shields.

Recap. Cephalopods raised the stakes of the game of existence, of survival and extinction, as they imposed *a faster speed of action-reaction, and a bigger spatial size, hunting in herds communicated through visual body languages.*

28. Echinoderms and Chordates. Evolution of vertebrates.

Echinoderms, like ancestral cephalopoda, lived in abyssal regions originally fixed to the ground, in an 'informative environment' based in stillness with a lot of 'free time' to evolve further, as squids did in abyssal quiet regions or monkeys will do latter in quiet trees. Echinoderms became informative top predators because they evolved 2 new EXI characters as 'still', temporal forms:

- Max.I: Radial symmetries, like in the pentagonal starfish, which fostered the development of a better neural system with a central informative singularity to coordinate the 5 radial nerves.

- Max. E: The first inner bones, to sustain their complex form.

Thus echinoderm increased their IxE force, evolving into the first vertebrates: Chordates were probably born, according to the Black Hole paradox, due to a palingenetic error, when echinoderms remained in their larva, trochophore, informative, evolutionary state, surviving, despite its smallish size, thanks to their 2 new ExI advantages, starting a new biological radiation. Their single nervous system protected by a spine, became a very dense, structure with a hard, inner bony membrane of sustain that allowed its growth in spatial size and temporal complexity, as neurons quantized further, differentiating from tail to head into an EXI tree-like structure:

Max. energy (nervous, linear spine) > Round, spiralled brain.

The ancestral Chordates differentiated according to the development of their growing informative nervous network into:

- *Max. E: Lineal Protochordate*, the oldest species with 3 basic forms, diversified along the path of increasing mobility: sea squirts, acorn worms and amphioxus.

- *E=I: Cyclostomata*, (jawless fishes), which grew in the planar dimension of energy. And so it came an age of sharks.

Those 2 first forms are still planar in form, with minimal development of their 'round' brain and hence with overdeveloped olfactory systems.

- *Max. I: Pisces* (true fishes). They grew in the dimension of height, with new evolved sensorial, informative organs. They became the new dominant species that diversified once more, this time along the st+1 evolutionary path of environmental adaptation in:

- *Max. E: Sea chordates*, with several varieties that reached its evolutionary limit with Teleostean.

- *E=I: Sea-land chordates, amphibian*; an animal that mixes the palingenetic characters of those 3 environments during its 3 ages of life. Since it is born as a sea animal, lives its youth in between both environments and dies as a land animal.

- *Max. I: Land chordate, reptiles,* the most informative that diversified further.

Recap. The response to the eye language was protective shells that surrounded the nervous system of chordates, allowing their growth and invasion of the air-gas and land-solid ecosystems.

29. From amphibians to reptiles the conquest of firm land.

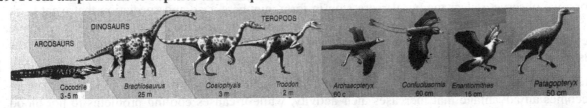

Reptiles grew in size and changed from lineal length into height dimensions, but as they became victims of mammals they devolved to their earlier forms, as crocodiles and diminished in size becoming birds, which again grew in size and changed from lineal length to height dimensions. Yet the arrival of man is provoking again the extinction of the biggest, taller birds (Moa, Emu, Dodo.)

If any evolutionary jump shows the importance of networks is the transition from fishes to reptiles through the intermediate amphibian stage: amphibians adapted their sensorial brains, their reproductive bodies and finally their reproductive systems to the new world. And only then, when the translation of form was complete, it appears the land animal - the reptile.

Thus, amphibians show 3 clear evolutionary phases:

- Max. I: The amphibian moves towards an air world where light defines clearly the forms of its preys, triggering the evolution of its inner networks according to the cyclical chain, I>E>Re, which require first to become informed to localize and feed on energy, needed to reproduce. Thus amphibians first changed the form of their informative heads and senses: Their noses migrated to the top of the head, out of the water and their eyes acquired membranes to wet them, focusing better light images. This i-logic hypothesis of the dominance of informative evolution again contradicted the usual E-science energetic theory. And yet a few years ago Sci Am published the 'astonishing' revision of the energetic theory: amphibian did not evolve, because they dragged their legs on the dry land but because they raised their heads out of the water changing their senses.

Max. E: Then their respiratory systems changed with new lungs that increased their capacity to get oxygen from air. Amphibians now changed their preys, eating insects with a modified mouth and tongue. Thus the amphibian becomes the top predator of the terrestrial ecosystem thanks to its greater Exi force and extinguished giant insects that reigned in the Carboniferous era. Those insects however reacted back evolving into metamorphic, flying forms, escaping their extinction. The inversion predator-prey manifests again between insects and chordates, as it did between gravitation vs. light or plants vs. animals: Insects have their exoskeleton outside, as they need maximal external protection; chordates have it inside. Insects are smaller, quantized forms; chordates are integrated, bigger forms. Insects are dominant in chemical, slower languages; chordates are dominant in nervous languages. Insects, the energy of the trophic pyramid, are more abundant than chordates, its top predators.

Max. E=I: Finally, amphibians adapted their reproductive systems to the new atmosphere, creating dry eggs, completing the creation of a true, terrestrial organism, a new phylum that had adapted its 3 networks/existential cycles to the new world: *reptiles*[8].

They evolved again according to the Fractal Principle into the 3 most evolved life phyla:

- *Max. E: Reptilians,* which maximized its spatial size.

- *E=I: Birds*, with the most efficient blood networks, needed to develop flying skills.

- *Max. I: Mammalians.* They developed its informative, nervous system to its perfection. So they became the top predators of their 'parental group', reptiles, causing their massive extinction and 'death reversal', which in species shows through the 'evolutionary regression' of a former top predator species, when a new top predator displaces them. Thus, if we compare modern reptiles, once mammals have chased them down, with their dominant parental forms during the dinosaur era, when they were top predators, we observe a clear temporal regression in form, numbers, size and speciation that went back to their 3 basic forms. Today, from the initial 14 reptiles groups, only 3 basic groups remain. They have survived in econiches close to the water, regressing to amphibious forms, and diminishing in size towards their original 'minimal, Black Hole form':

- Lineal forms. *Snakes and lizards* living in extreme, hot, wet environments (rivers) and deserts, where heat becomes an advantage that increases their activity, while it causes cooling problems to hot blood mammals, their top predators.

- Balanced forms. *Crocodiles*, descendants of dinosaurs that have reduced its size and have become again amphibious, surviving mainly on the sea and rivers; learning new reproductive, maternal skills (hiding their small babies on their mouths, when predators come).

- Cyclical forms, with static, hard, protective round shells; *or turtles* that only reach big sizes in Galapagos, an isolated group of islands with minimal numbers of mammals. Most of them survive on the sea, having developed a gill-like system of breathing.

The only primitive, remaining saurian, the Tuatara, survives in min. numbers, in the most isolated region of the World, New Zealand, where there were no mammals with placenta...

Recap. The conquest of land was headed by amphibian and the development of better eye systems, as information is the key to evolutionary change. Accordingly as they become reptiles they grew in the dimension of height; and again as reptiles became birds and mammals the new species became bipedal and extinguished the simpler reptiles that reverted to planar forms.

30. Mammals: Temporal iron bodies, minds.

Mammals are the 3rd informative evolutionary age of land animals, which therefore transform again their 3 networks, reaching the final adaptation to the changing weather conditions and light transparency:

Max. I: Mammals improve their nervous eye-brain systems, overcoming the limited eye vision of reptiles. Their brains surpass the instinctive stage (based on mental, mostly chemical, slow programs of action-reaction that execute the cycles of a living organism, based on generational memories without capacity to modify them) and enter the age of free will (based on brains with nervous programs that use memories acquired in the previous execution of those cycles by the same generation, to adapt their new actions-reactions to the changing environment).

Max. E: Mammals improve their corporal, metabolic rate of action-reaction with *hot, red iron blood* that harnesses better than previous copper-based bloods the energy of oxygen. Since blood has haemoglobin, where an iron atom, the top predator energy atom of the Universe, controls and jails oxygen atoms with carbohydrate arms.

Max. Reproduction: Finally, the internal nervous system regulates mammal's reproduction, creating complex placentas that can feed and develop the isolated foetus, without the dangers of a youth age, when most beings die as 'energy' of mature predators. It is the equivalent stage to the 'chrysalis' shape of insects.

Further on, mammals evolved socially. So probably herds of mammals with faster, simultaneous fractal actions chased as a whole, and killed baby reptiles, provoking their massive extinction, in an age of climatic change. Yet, as it happened when amphibians extinguished insects, provoking their flying evolution and migration to the last frontier - the air environment - the smallest reptiles became birds that avoided top predator mammals, putting their eggs on cliffs beyond their reach to survive.

And so again, the most successful group among land animals, mammals diversified, this time along the path of reproductive evolution into:

- *I horizon: Monotremes,* which are egg-laying mammals.

- *II Horizon: Marsupials,* which have a pouch where they develop the 'embryo'.

- *III Horizon: Eutherians,* which have true placentas and differentiated again, as the most evolved informative class, into multiple subspecies, now along the path of feeding energy, into:

- *Max. E: Herbivorous,* which ate huge quantities of low energy plants, developing new, complex digestive systems, with huge, multiple stomachs.

- *E=I: Carnivorous,* which developed the best blood systems, as they needed to increase muscular force and speed.

- *Max. I: Omnivorous,* which were able to eat anything, occupying multiple ecosystems that enhanced its evolutionary differentiation. Among these species the most evolved phyla were apes, from where man came, because they lived, unlike the animals of bidimensional plains, in 3-dimensional 'high' trees, where they could not be hunted. And so they evolved in their 'free time' their 3 dimensional brains, becoming informative humans.

The previous synoptic analysis of the evolution from cells to humans shows the universal application of the fractal space-time differential laws of evolution. It could be as detailed as you wish and reorder all our knowledge on biological species under those simple laws. We just lack space-time to do it here. It shows the impersonal intelligence of the evolutionary plan and the homology of all st- forms… As it is the same plan we have used to describe atomic particles.

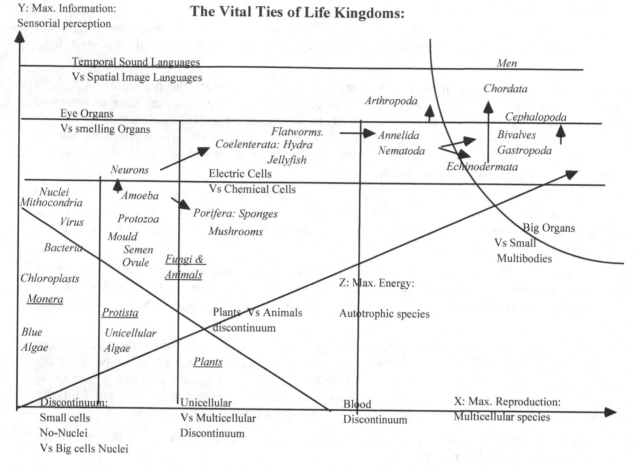

Y: Max. Information:
Sensorial perception

The Vital Ties of Life Kingdoms:

Temporal Sound Languages
Vs Spatial Image Languages

Men

Chordata

Arthropoda

Eye Organs
Vs smelling Organs

Cephalopoda

Flatworms.
Coelenterata: Hydra
Jellyfish
Electric Cells
Vs Chemical Cells

Annelida
Nematoda

Bivalves
Gastropoda

Echinodermata

Neurons

Nuclei
Mithocondria

Amoeba

Protozoa

Porifera: Sponges
Mushrooms

Virus

Mould
Semen
Ovule

Big Organs
Vs Small
Multibodies

Bacteria

Fungi &
Animals

Chloroplasts

Monera

Plants Vs Animals
discontinuum

Z: Max. Energy:

Autotrophic species

Protista

Blue
Algae

Unicellular
Algae

Plants

Discontinuum:
Small cells
No-Nuclei
Vs Big cells Nuclei

Unicellular
Vs Multicellular
Discontinuum

Blood
Discontinuum

X: Max. Reproduction:
Multicellular species

In the graph, the great life phyla distributed according to the fundamental arrows of vitality: energy feeding, informative perception, and reproductive capacity, in 3 dimensions of increasing capacity to process energy, perceive and reproduce, which give as a result the main life phyla. The evolutionary jumps represented by lineal divisions are the fundamental divisions of increasing capacity to process a vital arrow of existence that abstract biologists use to differentiate the animal kingdom. We have used a single positive frame of reference that shows an increasing quantity of those vital parameters. It is a graph of top predators in such a manner that forms whose X x Y x Z values are higher (which processes more energy, information and reproduces in greater social waves of cells) is a top predator.

In the graph, man on the right top corner is the most complex informative life species.

Recap. Mammals are the most perfect form of life beings, which evolved in information, till reaching the perfection of man; added iron-energy to its blood and improved reproductive skills with placentas.

473

VIII. MAN AS A FRACTAL ORGANISM.

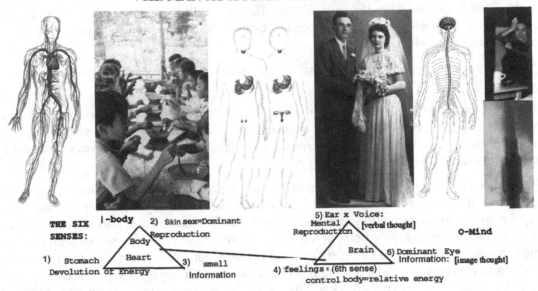

THE SIX SENSES:

1) Stomach
Devolution of Energy

I-body

Body
Heart

2) Skin sex=Dominant
Reproduction

3) smell
Information

5) Ear x Voice:
Mental [verbal thought]
Reproduction

Brain

4) feelings = (6th sense)
control body=relative energy

6) Dominant Eye
Information: [image thought]

O-Mind

Medicine & Physiology deal with the 3 networks of the human organism, whose functions connected to the senses enact our time arrows: The Digestive, Energetic System is attached to the senses of smell and taste; the hormonal, blood, reproductive system is attached to the senses of touch and the 6th, emotional sense and the Nervous, Informative System is attached to the higher senses of sight and sound.

31. The arrows of time and social scales of human beings.

The most enlightening analysis of Multiple Spaces-Times is the objective evaluation of human beings, as yet another fractal part of the Universe, in communion with its cycles of existence and organic laws: As all other living organisms, we humans have 2 poles of energy and information, of cyclical and lineal form, the body and the head; we live through 3±st ages of d=evolution between birth in the lower cellular plane and death back to it. We are defined by 3 fractal regions, a sensorial membrane, a central spine and our lineal energetic limbs and digestive organs. All those systems are born through palingenetic processes from the same 3 fetus' membranes, the endoderm, the mesoderm and the exoderm.

We have 2 subspecies clearly specialized in energy and information, the lineal, strong male; the perceptive, memorial, cyclical woman. We are also dominated by the reproductive arrow, dominant in life species, as all what we really care for is our families. We like so much reproduction that we are actually transferring our form into metal, making machines of energy (transport, weapons) and information that imitate our senses (chips=brains, cameras=eyes and mobiles=ears)...

Thus, man is also a biological knot of existential cycles, whose purpose as everything else in the Universe is to 'survive', to maintain the immortality of the human, logic form by reproducing the energy and information absorbed by the brain/body system and to 'evolve' - to create more complex social organisms of history till reaching the summit of evolution as a species: the creation of a superorganism called mankind, in which humans share energy and information through its collective energetic systems (economy) and informative systems (political and cultural structures).

We cannot overextend in the biological description of the human being, which is topologically similar to all other life species. Let us briefly consider those self-similarities:

-Man extends through 3 planes of increasing complexity, the cellular, bio-chemical plane, the individual, biological plane, and the social plane.

474

-st-1: As cellular beings we are also defined by an iterating language of information, written with 4 dimensions, the DNA bases that form the genetic code. Thus, the cellular plane is ruled by genes, which mostly code the basic functions of the cell and organs, common to all mammals, reasons why we have over 90% of common genes with them.

- We also evolve through a palingenetic process, differentiated only in the last 5 million of years from self-similar apes.

-st level: Humans have also 3 dimensional networks: the energetic=digestive; reproductive=blood and nervous=informative systems that determine our cyclical actions, as we try to feed on energy, reproduce and inform our 3 systems common to all life species. Those internal networks observe the external Universe through the same 6 senses of all other species.

Indeed, if the duality of bodies/heads gives birth to the Complementarity principles of Physics and genre, the existence of a series of basic wills are expressed in our human organism through the senses – our outlets to the energy and information of the body and brain. We use those senses to absorb different energies and information and combine both, in reproductive processes. We sense sexual reproduction of the body through touch; we smell the energetic food we eat, and that information used to guide hunters. And finally we taste that energy. While our brain perceives space with our eyes, reproduces information with words, and uses the 6th sense of feelings and the limbic brain to guide our reproductive body. Thus the senses are:

- The sense of smell and taste are the negative and positive senses of the digestive system that allow us to choose what to reject and what to eat. The language of the sense of taste has also 5 vowels that can be classified according to the key substances of energy and information (sugars, proteins, salt, etc.) our organism require.

- The internal 6th sense of feelings that relate body and brain and the external sense of touch are responsible for the reproductive will, which excites our mind (love) and our body (sexual attraction). Both are reinforced, as in all other animals by the visual and auditory sense, which are the:

- Mental senses of space (eyes) and time (words), which shape our natural languages of energetic and informative perception.

In the graph we observe that man as an individual is a knot of 3 dimensional networks: the energetic=digestive, reproductive= blood and nervous=informative systems that determine our cyclical actions, as we try to feed on energy, reproduce and inform our 3 systems.

St+1: When humans perform those individual actions in social groups they create social networks that establish the 'memetic actions' or customs of a society, passed from generation to generation as 'genetic codes' are transferred from individual to individual. So humans gather in groups to harvest food and eat together; they establish social customs and marry to reproduce; and develop verbal and visual, informative networks that evolve human organisms from individuals into cultural super-organisms, transcending socially into a new st+1 plane of existence: Mankind.

Thus, while we are ruled by the will of our 3 physiological systems, expressed through our biological genes, we evolve socially into macro-organisms called Gods, nations and civilizations, expressed by our cultural and technological memes.

Recap. The biological will of man is shaped by the 3 internal, physiological networks, each of them attached to 2 perceptive senses, which are the languages of perception of energy, information and reproductive signals of the external world, achieved through our daily cycles of existence. Thus humans are also a knot of Time Arrows, which seek to feed on energy, gauge information, reproduce and evolve socially like everything else in the Universe.

32. The 3 planes of Human Existence in time.

10. The 3±1 planes and ages of existence of human beings.

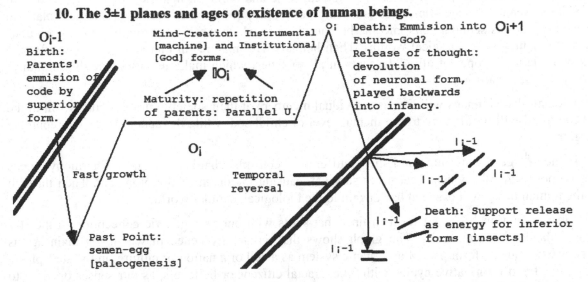

Life is a trip through 3 planes of existence: the biologic, st-1 genetic plane; the individual plane, which expresses the same biological arrows, and the sociological st+1 memetic plane, where survival is reserved to prophets that create a religious God or civilization, the subconscious collective of millions of believers. In the graph, existence is a travel between those 3 planes, the cellular plane to which we return after death, the individual plane and the social plane which peaks in our 3rd age as informative particles of our culture and economic ecosystem.

If we study the human being in time, the 3 ages of human beings are equivalent to the 3 ages of any space-time field. Humans show those self-similarities, both in the physical 'warping' of their bodies that accumulate information and wrinkle till they explode back into energy in the process of death; and in the different 'ways of thought' of their brain that also go from an optimistic, active age of energy into a reflective, old age of information:

-St-1: we are born in the inferior cellular scale as a seminal seed, clearly differentiated in lineal energetic males of maximal motion and still, reproductive, informative females, ovum, that mix together, exi to start a palingenetic process of reproduction and social evolution till we surface in the st-scale.

– Max. E: Youth is the energy age, dominated by the digestive network and muscular movement. So children are energetic, always moving, passionate about food that they need in bigger quantities.

-E⇔I: Maturity is dominated by the reproductive arrow, its networks and sexual desire. So mature people marry and have family values.

- Max. I: The old age is dominated by information and nervous networks. So old people are wise and know a lot of things, though they have little energy left and live in stillness, looking back to the past, before they truly start their travel in time. Thus, a very old man becomes like a reversed, negative child, also passionate about food since he lacks energy; but can't really digest it. He will be emotional, close to his 'biological consciousness', but with negative feelings, as he has no future left... except death, when his information dissolves back to the cellular plane.

- st-1: Death, when all energy is spent we die and return back in time through a 4th age in which all information becomes erased.

Further on, humans transit during those 3±st ages of life through 3±st hierarchical scales or planes of existence: the st-individual plane; the st-1, biological scale and the 3±st, historic, sociological scale. In terms of those scales of space-time, life is a temporal journey from the lower st-1 biological plane in which we are born as seminal cells into the human plane in which we enter through the discontinuous aperture of our mothers' biological wombs. Then humans will 'mature' in that st-social plane after the 'rituals of initiation' proper of all adolescents in all societies, which make us cross 'the discontinuity' between youth and maturity.

In that social plane humans will live as individual members of their society during maturity, till in the 3rd age the mind will return back to the mental level of childhood; while the spatial body will continue warping up.

Thus in the 3rd age, the st-mind and E-body suffer a 'fractal split' that breaks i=ts E=I harmony ('mens sana in corpore sanum'), which determines survival, causing the death of the organism. Then through death, the human being will descend back again to the biological, cellular world.

Thus life is a motion back and forth in time that starts with our palingenetic conception in the st-1 biological plane of seminal cells. The graph shows that ternary existence in which the main age is maturity, when humans form part of an organic system as a cell of a nation or culture, in the st+1 plane, enacting social and informative cycles with other fractal citizens or believers, as our senses open up to the social ecosystem, ruled by the political laws and financial networks of the collective, nervous social system, where we develop verbal and visual perception, till we descend back through the reverse journey of death into the original st-1 cellular, vegetative state, completing our wheel=cycle of existence.

Though we have only 'individual consciousness', the 3±st biological and sociological planes do exist and should have their own individual consciousness (DNA consciousness, and collective subconscious of mystique Gods). When we are born we just enter a new form of consciousness after an age dominated by our vegetative, chemical, cellular soul. A fact, which is evident in the 'borders' of life: the baby and the old man care mainly for food and chemical emotions, since they are closer to their cellular level of birth and death, our zero sum as dust of space-time.

Recap. Humans follow the same 3±st ages of all biological beings, as they live through 3 planes of existence: the biological, individual and sociological planes.

33. The 3 topological networks of man in space.

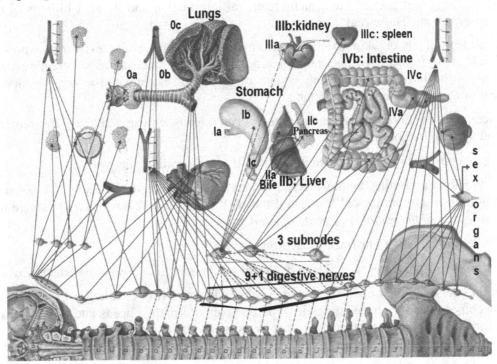

In the graph, a deeper analysis[10] of nervous control of a body and its energetic and reproductive systems. All those systems of energy and information are created by ternary differentiations, which provoke a mirror symmetry with the ternary differentiations of the nerves that control them. For example the digestive system has 3 sections, orientated from top to bottom on the relative dimension of informative height: the upper, mouth, informative system; the middle, reproductive stomach with its attached glands and the intestine, each one controlled by ternary sub node of nerves. Then the stomach has 3 regions; the organs that assist it are 3; the intestine has 3 sub-sections and all of them respond to lineal-energetic topologies (intestine); cyclical reproductive (stomach and attached organs), and cyclical, informative (mouth and taste systems).

If we were to make a topological, spatial analysis of the human being, we would observe that as an st-point we are made of 3 physiological networks/regions evolved from the external exoderm, mesoderm and internal endoderm, which become roughly the origin of our external membrane and senses; reproductive body organs and internal brain. And each of those 3 systems is constructed with the topological characteristics of the 3 regions of any st-point:

-       The energetic, digestive system is the bigger one, made of overdeveloped membranes and connected to the apertures of the skin, forming together an inner/outer self-similar 'Klein bottle' (Non-Euclidean jargon), which is the final evolution of the self-similar system we found in the Hydra. It is the more loosely connected of them all; in fact a super-organism that hosts in a 'colonial' form billions of external bacteria.

-       The informative brain system occupies the central singularity region, is the smallest of them all and has a highly warped, hyperbolic form, with the maximal, quantized number of cells, which form a network that relates each neuron to thousands of other neurons.

-       The blood & lymphatic, reproductive system communicates both regions, connected to the external skin through the lung apertures and surrounding the informative brain, which obtains 10 times more energy than all other systems, as the 'top predator' system of the organism.

In physiological terms we are composed of 3 st-regions and networks developed from an initial, spherical 3-layered egg that packed our genetic/temporal information and iterated billions of times to become a macro-organism. Those regions are:

- An external membrane or skin of dense informative cells, the initial exoderm, whose senses invaginated back those informative apertures in a dual parallel st-axis of height: creating the spinal nervous network – the lineal, energetic sub system of the brain that moves the body – topped by a central, cyclical, informative nucleus, the brain.

- An intermediate zone filled with glandular, reproductive cells, 'invaginated' by the reproductive, hormonal blood system that joins them.

- And the bigger energetic system, divided into a digestive and breathing network.

Further sub-divisions according to the Fractal, ternary Principle create the 3x3+st+1 physiological subsystems, standard in any biological book, and the st+1 brain, *which has a mirror image of all of them and acts as a unit of the social st+1 scale of mankind.*

So the $\Sigma E \Leftrightarrow \Sigma^2 I$ formalism of our fractal space-time is:

Max. E: Digestive System<Exi: Blood & Lymphatic System:exI> Nervous System:Max. I

The interaction between the st and st-1 system is clear: Those networks rule the lower st-1 cellular scale, which become a 'herd', whose energetic, reproductive, informative needs and birth to death ages are totally controlled by them; to the point that doctors say their discipline is 'physiology', since sickness are the sickness of those networks.

Further on, each of those 3 networks subdivides into new ternary structures of control:

- Max. E: Those networks have all an inner, lineal, quantized 'body' - a network of tubular systems that delivers informative, energetic and reproductive orders that rule internally our fractal cells. They are the veins, the nerves and the lymphatic system.

- Max. I: They all have external apertures to absorb energy (breathing nose and mouth); information for the nervous system (ears and eyes) and sexual sensations for the reproductive system (olphactory and touch senses).

- E=I: The 3 balanced centers, the heart, brain and genital system become the site of their 'inner wills' that enact and control the iterative cycles of feeding, perceiving and reproducing - the will of existence of our collective self that impel us to acquire information, energy and reproduce, ensuring the survival of the human form.

It is in fact possible to explain the entire process of creation of humans from their palingenetic seeds through the laws of ternary differentiation and multiple space-times as a process of creation of those 3 networks in control of the st-1 cellular systems and controlled themselves by st+1 brain and senses, which together create a living organism extended in 3 planes of existence, self-similar to all other mammals. And yet humans have a peculiarity respect to all those other animal beings: *to be the most informative species of them all, with the maximal evolution of the informative dimension of height, its visual senses and st+1 10th system, the brain.* We are the summit of informative evolution of life and so we shall consider how that informative evolution selected 'us'.

Recap. The st-scale of human existence in space is controlled by the 3 topological networks of the human being, the informative, nervous system, the energetic, digestive system, and the blood, reproductive system. Each of them is divided in ternary subsystems, informative, external senses, center and lineal networks that connect them with the st-1 cells to control their drives of existence.

IX. THE EVOLUTION OF MAN: TERNARY SPECIES.

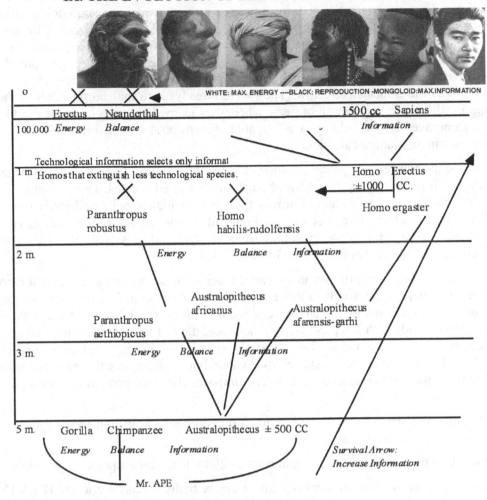

WHITE: MAX. ENERGY ----BLACK: REPRODUCTION -MONGOLOID:MAX.INFORMATION

34. The evolution of man: Ternary differentiations.

Complexity defines a Ternary game of evolution by differentiating species in informative and energetic species, combined in balanced organisms. The same game evolved man from the original ape 5 million ago into the human being. Moreover the differentiation is established first from an initial top, predatory, young energetic species (the gorilla in the human case), that diverts into a reproductive, mature species (the chimp bonobo) and finally in its 3ʳᵈ informative age gives birth to a more mental species (the Australopithecus). This thesis of multiple spaces-times first established in the foundational book of this science 20 years ago, was confirmed with the discovery of the exact genetic clock: the divergence between gorillas and chimpanzees is 1.17, then the genetic clock diverges 1.17 to create man... Thus the same plan of evolution, which we used to describe all other biological species of the Universe created man from Mr. Ape.

Moreover, the Universe has 3±st arrows/dimensions of creation, energetic length, informative height and balanced=symmetric, bilateral reproductive width. This simple dimensional game happens even in man. Yet bilateral systems often are also dual: one side is slightly energy dominant and the other slightly informative, offering a more complex, kaleidoscopic higher truth (multiple p.o.v.s). It is also a sexual principle: Women are information-dominant; males are energy dominant. And it should find wider applications to understand the dualities of the brain.

Such topological analysis confirms our informative nature:

In human evolution the ternary principle applies beyond the triad of apes: each successful, dominant Homo differentiates into an energetic and informative species, which recombine, creating more complex=informative and stronger=energetic homos that displace their parental forms, thanks to their higher Energy x Information. Yet the dominant, overall arrow favors informative species. Since when Homos were born, the energetic top predators were felines, far stronger than the strongest possible homo mutation. So the evolutionary direction in which the big apes could find an ecological niche within the mammal kingdom was the informative arrow. They were already fit to that purpose, since they spent long times living in stillness in the summit of trees, where they learned to control complex motion in the dimension of informative height, adding a 3rd spatial dimension to their brain. So they became informative masters among mammal species.

In the graph we study the ternary process responsible for the creation of humanity: Further on, as humans evolved, they followed the same pattern of 3 ages proper of all species. Thus, Homos diversified into energetic, informative and reproductive forms that show a higher balance between energy and information. Yet only the informative species, the dominant 'arrow' of future survives in each of the 3±st horizons of increasing informative evolution, from Australopithecus with 500 c.c., into Homo Erectus with 1000 c.c. into Homo Sapiens with 1500 c.c.

And in all those phases, humans split into an energetic species and an informative one that recombined to create a 3rd species. Those species will further evolve technology also in 3 horizons called the upper, middle and lower Paleolithic. Finally, humans diversified in 3 races, specialized in Energy (white race) and information (Mongoloid) with a reproductive, black race, the last to be born, which mixes the pre-mongoloid (bushman) and white races. The constrain established by the fact there are only two elements, energy and information, to create reality, whose lineal and cyclical forms are invariant at scale, in all systems of the cosmos means the Universe imposes the same program of creation to all its species.

Thus from an initial ape species we find a ternary differentiation ±5 million years ago, into 3 sub-species of great apes that live in Africa:

— *Max. E:* Gorilla, a big, energetic species that reaches 200 kilos, with a square, huge jaw.

— *E=I:* The chimpanzee, with a smaller body and a bigger brain. According to the Black Hole law, the smallest organism evolves informatively. Thus, the 'dwarf chimpanzee', the *Pan Paniscus*, which today shows the highest learning skills among apes, was the species that seems to have evolved into a 3rd informative horizon of big apes. It is:

— *Max. I:* Australopithecus, which splits in 3 evolutionary ages of growing brain volume:

I Horizon: 500 cc.: From Australopithecus to Homo Erectus.

We have dated the oldest mandibles of Australopithecus Anamensis, the top predator of the 3 big ape species, in 4 million years. The Australopithecus, after reproducing massively in the savannah ecosystem that favors his bipedalism, suffers a new ternary mutation. The 3 descendants of that Australopithecus Afarensis that inhabited the Ethiopian plains around 3 million years ago are:

— Max. E: The Paranthropus, an energetic species with huge mandibles and a lineal, planar brain. He looks like a heavy-set, humanoid gorilla. He was probably vegetarian. The first subspecies found is the Paranthropus Aethiopicus, which will further evolve into the stronger Paranthropus Robustus.

— E=i. The Australopithecus Africanus, which maintains the form of the original Australopithecus Afarensis, perhaps with inner improvements in his soft tissue. He lives in South Africa as a contemporary of the Paranthropus.

481

— Max. I: Australopithecus Garthi. He is the informative species, with a bigger brain, which starts to acquire the shape of all informative species: a rounded forehead, *as the sphere is the form, which holds more information in lesser space.*

Those 3 species live very close. Around the Turkana Lake we have found remains of the 3 Homos. Yet the arrow of future favors informative Homos. So the top predator Homo will be the Australopithecus Garthi that goes through a mature age of massive reproduction, probably extinguishing the other 2 subspecies and expanding worldwide into new ecological niches that favored their genetic and memetic adaptation through a new, ternary split:

— Max. E: A regressing, energetic A. Garthi, similar but taller than its parental Australopithecus.

— E=I: Homos Habilis and Rudolfensis, with similar brain capacity to the original Garthi.

— Max. i. Homo Ergaster, an informative ape with a 1000 cc. brain, which doubles the brain capacity of Australopithecus.

II Horizon: 1000 cc. From Homo Erectus to Homo Sapiens.

That Homo Ergaster, after a massive age of reproduction and expansion throughout the planet, subdivides again in 3 sub-races of Homo Erectus.

— Max.E: Pithecanthropus, the oldest species similar to Homo Ergaster, found in Java.

— E=i: Homo Erectus proper, found in China, who grows in the informative dimension of height that *favors perception from an advantage point of view,* to 1.7 m. He has simple technology, a skull without forehead and thick bones that handicap further cranial evolution.

— Max. I: Homo Floresiensis, the smallish species, which according to the Black Hole paradox should be the form who gave origin to Sapiens, as the smallish 'pan paniscus' evolved the chimp into the Homos. Recently discovered[3], he is a formal mutation with a smaller body, whose morphology resembles for the first time that of a human being. He has thinner bones that allow further growth and a higher forehead, the location of the creative part of the brain.

Paleontologists also found the first advanced technological tools in his sites in Flores Island, which is an isolated place, the ideal region to foster further evolution, due to the lack of top predators (allopatric evolution). Thus, he adds now technological energy weapons that substitute body strength and define a single arrow of future for human evolution: information. His improvement in neurological form, probably caused first his rapid expansion thanks to his new technological tools and then a rapid growth in size through the cross-breeding with the taller Homo Erectus species, giving finally birth around 300.000 B.C., to a new energy/information duality:

The energetic, big Neanderthal and the informative Bushman, which combine again to give birth to multiple races.

Those predictions based in the ternary principle and the Black Hole paradox (information is small and evolves in dwarf species) have been confirmed by genetic maps and Paleontology and await final confirmation when new skulls of Floresiensis are found.

The 3rd, informative Horizon: Homo Sapiens, 1500 cc.

— The lineal, energetic, visual Homo Sapiens Neanderthal, with ±1500 cc.

— A second informative pygmy strain, the 1st verbal woman, the mitochondrial eve, probably a bushman, also with ±1500 cc.

± 80.000 years ago she crossed into the Middle East, probably mating with earlier Neanderthals, giving origin to the visual, energetic white male (Homo Palestiniensis, whose skulls show mixed traits from both species).

± 40.000 y. ago, in the high Tibetan plains, she became the informative Mongoloid that colonized Asia, America and descended upon Europe and Africa, where she cross-bred with the Capoid (bushman) and became the black man, the balanced E=i, reproductive race, completing the final, ternary differentiation of the main races of mankind.

Cultural races

Further on, as racial tribes evolved socially and emerged into cultural societies, which are super-organisms of history, the ternary, racial division of mankind between energetic white men, informative Mongoloid and vitalist blacks, will become the cultural differentiation of human beings in 3 type of brain orientations, and hence cultures of the mind according to the 3 elements of all systems, energy, information and its reproductive combination; reflected in 3 types of brain morphology:

— 'Visual, energetic, lineal, spatial cultures', proper of the white man, with his concepts of lineal time and his passion for technological, energetic weapons and individual, selfish behavior. It is the dolichocephalic brain dominated by the visual eye to Occipital axis, also dominant in the Homo Neanderthal.

— 'Cyclical, informative, mongoloid cultures', natural to Asian civilizations, with cyclical concepts of time. It is the brachycephalic, wide, verbal brain; dominant in the ear axis; hence giving birth to female, eusocial cultures like the Chinese.

— And 'sensory, reproductive, colored cultures', which in Africa and India developed human senses to their limits. It is dominant in the height/medullar axis with a 'tall head', peaking in the top/motor and sensorial area.

The culture/age better for the survival of mankind, as in all species, is the one that balances both arrows, E=I. Thus we should promote paradoxically the culture that a technological civilization discriminates: the black, reproductive, sensory culture.

Let us study them in more detail how the ternary differentiation that created humans being from apes explains racial differentiation by establishing 3 types of minds:

The 3 types of minds: Emotional, visual and verbal brains.

The existence of 3 dominant human races today, the white, yellow and black people, cannot be argued, but they must be interpreted as 3 types of complementary intelligence, based in our recent discoveries on the science of Complexity, which always favors a ternary process of creation, with an energetic, informative and reproductive 'decoupling' along the 3 morphological dimensions or "axis" of any space-time form. In the case of the brain, there are 3 axes, the long visual axis, the tall emotional axis and the wide, verbal axis, and so we talk of 3 mental races:

- The dolichocephalic brain, which overdeveloped visual axis, from the eye to the occipital bone. It is the white man.

- The brachiocephalic brain with an overdeveloped verbal axis from ear to ear. It is the mongoloid race.

- The makrocephalic brain[11], with an overdeveloped emotional axis, from neck to top of the head. It is the black race.

They maximize the 3 types of human minds: energetic races (white, visual, spatial, egocentric); informative races (yellow, verbal, temporal, eusocial), and reproductive races (black, emotional, inwards, ethocentric):

- *Max. E: Visual, spatial, energetic minds.*

In the symmetry of time and space, is the case of the first age or ''child and the 'male' species of the white races and in the earlier anthropological record, it is the Neanderthal, which we already considered 20 years ago as Cohn did, to be co-father of the white man born as Homo Sapiens Palestiniensis, which recently the genetic record has proved:

Interbreeding

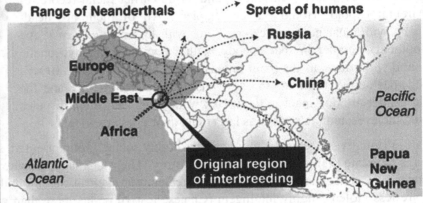

Humans leaving Africa about 80,000 years ago encountered Neanderthals in the Middle East, according to DNA evidence extracted from Neanderthal bones.

White people, who occupy the same range than Neanderthals are easily recognized by their dolichocephalic head and occipital bulk, their eye=I oriented mind, individualistic ego and energetic behavior. Their languages are consonantal, with few vowels. They have red hair and Rh- (a sign of crossing between different species, whose 2nd generation is not fertile, as mules born from asses and horses are). Further Neanderthal traits – white skin, hooked nose, wide hips - also recognizable in the Basque, Iberian, Germanic, Celtic and Semitic cultures, paradigms of visual cultures, which will make of shiny metal, weapons and go(l)d, their fetish – seem to validate further this thesis. In economics, it gives birth to the gold fevers of those cultures, origin of 'greed', an emotional, visual trait which other types of human minds (Chinese) hardly suffered:

Thus the white man should be properly called Homo Sapiens Neanderthalensis, dominant in visual information with minimal development on the parietal regions of speech. He will thus evolve the visual language in its 10th 'Brodmann' region of rational thought further into digital, geometric and mathematical thought, and create lineal civilizations based in metal, lineal time and Aristotelian thought. He can be sub-divided in two types, which again recent experimental proofs seem to validate:

-The Semite, in which Neanderthal males mixed with verbal women, *but Neanderthal females did not cross with dwarf verbal men, creating a mixed race with an extreme sexual dimorphism (taller male, smaller women)*. A similar case will happen in South-America where conquistadors mixed with Amerindian females but Amerindians did not cross with Spanish women, since the dominant warrior, stronger species, in the first crossing of the Homo Sapiens, in 80.000 BC was the Neanderthal and in the first crossing of the Atlantic the Spaniard. This theory advanced in my earlier books [C,1] has now proofs in the genetic record. The result is the Semite with a dominant bigger, highly aggressive, visual male and a smaller emotional woman. The reason of this military superiority is the fact that at that time the Homo Sapiens had not developed the arch and so it was dominated by the Neanderthal, stronger javelin warrior. Then in the Altai mountains, homo sapiens evolved into the mongoloid Cro-Magnon which invented the arc and hence could kill at distance the Neanderthal. This man went down to Europe and

484

the mixing process was reversed: Now Homo Sapiens mixed with stronger Neanderthal women (more pure among Nordic and Scottish Women with red and blonde hair, dolichocephalic, visually oriented, often sexually dominant). Thus the Indo-European race was born. This last detail of Complex theory of human evolution should wait for further genetic analysis since so far we have only observed that with a 90% of equality between men and Neanderthals and a 4% of Neanderthal genes there is mixing both in Semitic, white and Cro-Magnon (northern Chinese) groups.

-Max.I: Verbal, temporal, informative minds.

This is the case in time symmetry of old people; in the symmetry of gender latter studied in more detail of the 'female' species and in the symmetry of race of the yellow race; and in the earlier racial record, of the parental capoid, Homo Sapiens Sapiens. All of them are head-dominant, with clear neoteny features (thus with lesser young, energetic development), smaller in spatial body, with a dominant brachycephalic brain, and wider face, as the main linguistic part of the brain, the Brocca region, goes across from ear to ear. Thus, the mongoloid race and the female gender are dominant in the verbal, wide axis. Hence the distance between the eyes and the parietal is larger and their eyes are wider.

As all informative species the mongoloid is highly social, group oriented, selfless, as it uses the verbal language to create their super-organisms of history. Their language has an excess of informative vowels over energetic consonants; women play a greater role even in Semitic culture (Indonesian Muslims). They are less violent, highly intelligent (Chinese test 10 points more than white men), and have shown as all species that can evolve socially the highest record of survival compared to the original Semitic cultures, which base their societies in the power of metal greed and murder, gold slavery and iron weapons.

-Exi: Reproductive, emotional, body oriented cultures.

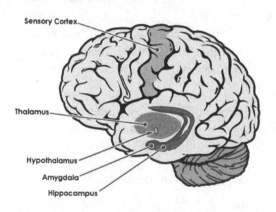

A 3rd type of culture, whose minds are dominated by emotional sensorial, body oriented, reproductive drives, is the black culture and the adolescent. Its brain is taller, with the top point in the upper head, as the dominant direction comes from the spinal-hypothalamus axis upwards. Thus, the brain regions most developed in emotional races happen along the 'height' axis. They are the inner regions of emotional thought (limbic system) and the motor and sensorial cortex on top. This hypothesis, extracted from our models of complexity and duality developed in the 90s have now increasing evidence: women are found to be wired inwards with more connections going towards the Hypothalamus than men, wired outwards towards the visual regions. And certainly both, black and female 'races/genres' show higher development of the emotional brain and excel in tasks related to the body/senses.

Psychological characters

Those findings are applied also to the will and drives of the mind: - Max. I: There are people who psychologists define as picnic, nervous, tall (height is the dimension of information).

- Max.E: There are athletic, sanguineous people, dominated by their energetic will.

-E=I: There are viscerotonic people, driven by their 'lower passions' (-;

Those distinctions already observed in XIX C. studies on brain morphology and cultural psychology are real, increasingly proved by a wide number of research papers that go from the analysis of why men draw better maps and women talk more to genetic studies in linguistics and bio-chemistry that validate our 20 year old research in those fields. Racism consists in thinking that a visual brain is better than an emotional brain or a verbal brain.

A key question in a biological world is the choice of which of those cultures, personalities and genders should be the referent for the rest of mankind, in order to maximize the goals of all super-organisms and species – survival. And the surprising yet evident truth of an organic model of the Universe is that the Makrocephalic, ethnocentric and verbal, brachycephalic feminine gender should dominate with its balance and ethic sense of survival the violent, visual, technological civilization of the white man, whose 'energetic' impulses are taking this planet out of balance, destroying Nature and building the new forms of 'metalife' (weapons, robots, A.I.) and machines of the singularity that will extinguish us. Yet this is not obvious because the 3rd paradigm of measure has no organic whys, despises life and 'sees' only reality with machines, it does not feel it or understand the deepest game of existence we have explained here. So we are walking steadily towards the extinction of history , the 10th and last extinction of life on Earth, guided by the visual wanting of the Homo Sapiens Palestiniensis, its lineal, energetic weapons and hypnotic go(l)d, themes those we will treat in depth in the last lessons of this work.

White matter/Grey matter brain: ternary analysis.

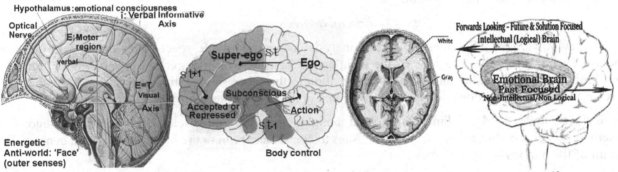

Let us consider the brain under the ternary principle in its topology, ages and fractal st-planes[12].

Topological analysis

In the graph, the 3 brains of the human being from a spatial, topological perspective:

- Max. I: Our external, sensorial grey matter.
- Max. E: our internal, emotional, limbic system.
- E=I: Combined in a white matter, memorial, reproductive brain.

We perceive with our senses; whose perceptions are processed by the DNA rich, cyclical, fluctuating, integrating cells of the grey matter brain, and then the outcome becomes imprinted in resistant white matter networks made of axon-connections. This external program mixes with an internal program coming from the body, which defines our wanting for energy and reproduction (sexuality), the main arrows of time of the body and limbs; whose wanting are processed by the limbic brain.

Then both programs converge into the axons. Both programs, the one of the body and the one of the brain, thus, mix and create together the white matter brain of beliefs that define our actions, memories and behavior in a greater measure, which in women are more wired to the limbic brain and in men to the grey matter and eye senses. Because women have more axons wired to the limbic system, they live more for themselves, inward-looking, emotionally-driven. Because men have more axons connected to the grey matter brain and their external, visual senses, they are more connected to space, mathematical-technological objects and motions within external space.

Temporal analysis

A key element to understand memetic imprinting is the inflexibility of the brain beyond 40 years. This was understood by the Greeks which divided life in two ages, till 40 (acne) when the brain peaks in informative power, which was considered the summit of life and then all was downwards till our end at the end of the mean generational cycle of ±80 years.

Thus the 3 layers of the brain are imprinted only in the '1st part of existence', before the curve of life goes down in the 3rd age, since we only learn in that first part of existence. And this makes *most people during the time in which they exercise social power, inflexible, memetic programs that will never learn but merely exercise their emotional imprintings converted in beliefs; disconnected from the ever evolving FMI complex, whose challenges they will be unable to resolve.*

In that regard, within that ½ fraction of existence, in the child age the limbic system dominates; in the adolescence and earlier, mature age the grey matter brain learns, reasons and imprints the white matter brain, which will dominate the adult age in which we live according to beliefs, which are more difficult to change as the variability of the white matter brain is restricted once those connections are fixed. As that period of life roughly extends from 30 to 80 years old, this means most humans are neither rational, nor emotional but basically 'programs' with a 'motherboard' of fixed beliefs and chains of rational thoughts, which make us perform the same routine actions the rest of our lives - not very different from the fixed forms of any computer system.

Fractal analysis

In the previous graph, left center the fractal analysis connects Multiple space-times theory with the Freudian psychology as the 3 regions of the brain can be considered 3 hierarchical layers, with a grey matter brain that senses reality and delivers information to the subconscious, limbic, survival brain that acts restricting the information that becomes imprinted in the long memory of the white matter brain.

Freud rightly understood the 'limits' of rational thought, due to the subconscious, powerful Super-ego, the memetic super-structures that society establishes to control and repress the biological, natural genetic program of life and love.

In Complexity, the brain is a fractal system of 3 'sub-brains', the subconscious, energetic, limbic system, of 'short memory', survival instincts and basic actions, the, white matter, the 'fixed circuit' of beliefs, imprinted in childhood by biological desires, the drives codes by genes and in youth by cultural memes, the drives of the super-organisms of history; and finally the grey matter, the rational, thinking, variable brain connected to the truth of external senses.

The most common trajectory of thought through those layers is during the young, emotional age:

External world->Senses->grey matter brain->Limbic censorship->Impression in the white matter brain->Actions

While as the organism matures, the limbic system loses importance and the grey matter brain connects to the white matter brain directly:

Outer World->Senses->grey matter brain-> white matter Printing->actions

On the other hand, people dominant in emotions and inward thought might have the exactly opposite way of thought:

Inner world of biological drives ->Limbic System->White matter imprinting >actions

Obviously our previous analysis of ages, races and genders and the socio-cultural upbringing will make those 3 avenues of thought differ in individuals. And in fact as impolite as it might seem, we find that most systems follow the irrational drives of biological existence, expressed by the limbic system and imprinted as desires in the white matter brain, which determines most human actions.

What this means is that most humans have the 'center of consciousness' NOT in reason but in the imprinted beliefs of the white matter brain, connected to its survival instinctive limbic brain – that 'rejects' beyond its short memory any attempt to imprint further the white matter brain.

Thus the most common program is Biological censorship by the centers of happiness that suppress painful information and upholds the imprinting of childhood and adolescence and its emotional truths as absolute, both in biological taste and cultural religions and beliefs.

Or in other words, once an idea is fixed in youth, to redraw it you would have to break and kill the programs of the mind; which if we consider the cellular scale of imprinting, turns out to be almost impossible as the acetylcholine that redraws the white matter axons is no longer available in enough quantities.

So the ideas imprinted in youth, once the white matter circuit is put in place (and no longer can be rewired after adolescence), won't change. The person will forget or become aggressive with its limbic, censoring brain against reason and evidence. And since the limbic brain is ego-driven and subjective, we must embrace a difficult fact to assume: 99% of humans are dominated by the white matter brain of earlier socio-biological beliefs. .

Recap. In complexity evolution is guided by the existence of only 3 types of morphologies and networks in living beings: the morphology and nervous system of information, the morphology of energy, and the morphology and network of reproduction. We can therefore use, as we have done in studies of evolution of all type of species, the existence of a constant speciation in 3 different species to analyze as we do in the previous graph, the evolution of mankind from the initial triad of apes, the energetic gorilla, the reproductive chimp and the informative Australopithecus, till the last differentiation of homos, between the energetic, visual Neanderthal mind, today dominant in the white man, and the smaller, informative, verbal, homo sapiens, today dominant in the informative mongoloid, with the reproductive, emotional black race in the middle.

Thus, the Laws of Evolution and its 3±st ages and subspecies, map out the main species of Homo, till arriving to 2 Homo Sapiens, the spatial, visual=longer, dolichocephalic, energetic brain of the Neanderthal, which seems to have lived in Europe and left some of his traits in the Homo Palestiniensis, on the slow-crossing border of Canaan. Further cross-breeding of the Palestiniensis with Sapiens is the likely origin of the white man (white skin, red hair, Rh-? consonantal languages, wide hips, prominent nose).

While the informative Homo Sapiens Sapiens, the verbal=wider brachycephalic brain, which started as a small 'San pygmy', evolved into the self-similar mongoloid race. Finally, the black race, with tall brains is the sensorial culture, the reproductive species, the last one to appear, shows the maximal genetic mixture.

The brain studied topologically is a system of 3 regions, the energetic, emotional, body oriented limbic system; the external, informative, sensorial grey matter brain, and the white matter axons that connect both and controls our beliefs firing our actions.

X. WO=MEN: GENDER AND BRAIN STRUCTURE.

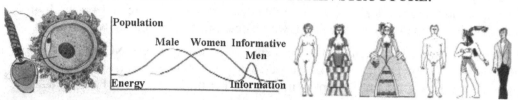

As many other biological species, human life diversified into complementary genders: energetic, lineal, spatial men and cyclical, reproductive women, which hold more biological information. While a 3rd, gay sex mixes both.

35. Gender: complementary differentiation.

In the graph, the biology, psychology and gender culture prove that humans are a dual species, divided into informative females, energetic males and a '3rd, gay sex', who mixes both, being women the informative species that reproduces the fractal information of the human being and man the lineal energetic species. Thus, biological origins, psychological behaviour and cultural forms explain gender as a ternary differentiation of humans into informative females, energetic males and a '3rd sex' of gay groups that mix both tendencies. Let us consider this essential duality of the human species in more detail.

The classic example of space/time Duality already observed by Taoist and Buddhist philosophers is the duality of gender, which they correctly applied to all species of the Universe. If we focus on the humankind, which is also a space-time species, its fundamental space/time duality is between its 2 genders: the energetic male and the informative female.

Humans have always wondered what gender is. Thanks to duality, following in the steps of Eastern philosophy, now we can build a scientific theory of gender able to explain its 2 fundamental elements, the scientific physiology and the behaviour of both sexes:

Since the Universe is made with 2 elements, Energetic Space and Temporal Information, any species needs to survive to feed on Energy and remember Information. Yet, we know that males process and control better energy and space; while females have more temporal memory and perceive better. Thus, together a couple widens the range of energy and Information they can process. So we conclude that gender is an evolutionary strategy of specialization, which increases the overall chances of survival of a heterosexual species.

The physiological forms, character and historic behaviour of men and women prove such duality, parallel to the universal duality of energy and information: men love energy, space; women love Time. Physiologically men are made of lines, as energy is; while women are made of curves, as information is. Historically men performed the hunting of energetic preys, while the recollection of cyclical food (agriculture) is traditionally considered a female discovery.

Men behave as energy species. Their body forms are lineal shapes of muscular strength; they have lineal reproductive organs (penis); their origin is lineal sperm. Their cultural roles are specialized in spatial-energy processes, from physical activities to war and mathematical, geometrical languages in which they excel.

On the other hand, women are specialized in temporal processes. Their bodies have cyclical shapes and store more information to succeed in their genetic reproduction (one complete X-chromosome). Their brains specialize in verbal languages and show outstanding memory. If we observe the form of females, it is made of cycles. Their higher genetic information allows them the creation and reproduction of the human species (an informative process); their cyclical reproductive organs (vulva, clitoris) and their origin as a sphere, an ovule, are also informative topologies.

The character and life attitudes of both, men and women follow that E/I specialization also in their relative wanting for energy or information: women prefer to dress with cyclical forms, play with formal toys like dolls and they like formal sports like gymnastics; men prefer moving, spatial toys, lineal dresses and violent sports. Thus, men love energy and movement while women love perception and feelings. In the realm of the mind and its languages of perception we find that men test better in spatial mathematics and women score higher in verbal tests.

We conclude that sexual dimorphism reflects the 2 elements of reality, spatial energy and temporal information that merge together in a couple, improving its chances of survival:

Male: Max.E x Min. I; Gay: E=I; Female: Max. I x Min.E

In the last decade the analysis of female/male brains has confirmed such duality of gender at multiple levels. For example, among women the amygdale, the oldest, primary center of will in the brain, is connected to the Hypothalamus, the center of 'temporal' and emotional perception, which ultimately controls survival and reproductive behavior. While men have their amygdale wired mainly to the motor and spatial regions of the encephalon. Thus men are the energetic 'external element' of the couple, motivated by the spatial outer world and women are the informative, temporal, implosive element, motivated by the reproductive drive of survival and the inner emotions of the body, the '6th' sense'.

As the reader will have notice, truth becomes more sophisticated as we perceive in more detail. Monism is the first scale of knowledge that observes only the body/brain; and yet it has suffice to build our society of lineal, energetic 'visual Homo Sapiens Neanderthalensis (white man). Then duality observes the differences in the body of lineal male and cyclical women, but we need to go further and analyse sexuality with the ternary principle that adds the 3rd sex, and consider the different combinations and balances of body/brains that will give us different 'personalities' and ultimately a more accurate analysis of what we do not see but dominates all structures – the informative brains.

And then the previous picture radically changes; since if women are reproductive bodies and men lineal, energetic bodies; when we observe the brain, to balance those bodies according to the Law of survival and balance of the Universe, energy-past x information-future = present-reproduction, if men have energetic bodies they must have informative brains to balance both in a present, and if women have reproductive bodies that repeat the information of life they must have reproductive brains that repeat also the information of the brain: present x present = present.

And indeed such is the case.

*Thus i*n a more complex view, as all beings are themselves yin-yang, space-time beings, we could define men as yang/yin: bodies ruled by the external, spatial activities; and creative brains of information dominant in grey matter (6 times more than women); and define women as yinyang/yanying; reproductive bodies whose emotions rule the brain; and reproductive brains, dominant in white matter (10 times more than men).

But what matters of those Universal balances is the fact that ultimately male and female genders are complementary forms, with the same value, a fact easy to prove using the formalism of Duality:

Male: exe-body +IxI-Brain ⇔ Female: brain (i x E) + Body (exI)

Thus we define *a male*, as a form whose 'energy body' is dominant in energetic, lineal forms (ExE), while its brain is focused in informative, visual creative tasks (Max.I), resulting in a species unconnected with his body with an external, outward focus. The Female gender however has her informative brain focused in reproductive, memorial tasks and wired internally towards the reproductive body (IxE); while her reproductive body is dominant in cyclical, informative forms (ExI).

Thus while the total amount of energy and information or 'existential force' of both genders is the same, the distribution and combination of those parameters differ, making women, highly balanced, focused into the self-reproduction of their own space-time field, and the male gender, unstable, focused in the external world. Those theses, as politically incorrect as they might be are the i-logic deductions of Multiple Space-time Theory, recently proved right by the analysis of the brain.

Recap. Species evolve in a dual, complementary Universe of systems of energy and information, increasing their energetic/body power, their informative/head power or the networks that reproduce and combine both systems, exi, the reproductive networks. We talk of 3 dominant wills or biological drives that increase the survival chances of a system. For that reason there are genders (where the male is the energetic species, and the female the reproductive, informative one). And since energy is lineal (being the line the fastest distance between two points) and information is cyclical (being the sphere the form that stores more information in lesser space), it follows that men have big, lineal, spatial bodies and test better in geometrical analysis and women have cyclical, smaller, rounded forms and test better in verbal, informative tasks. Yet there is always a 3rd, mixed variation, in the case of gender, the gay sex.

36. Complex analysis of the duality of gender.

We can now with all this knowledge reconsider the differences between men and women (and by extension, white races and cultures vs. emotional-informative black and mongoloid races and cultures, self-similar to women), and which cultures should dominate mankind to make us survive longer. Since gender traits transmit to psychology, physiology, mental attitudes and customs. Let us consider a few examples from the different perspective of multiple space-time theory, the algebraic, topological, temporal and fractal, scalar analysis.

Algebraic analysis: combinations of genders.

The Universe offers more variations even if it selects finally 3 dominant ones. Thus, since we depart from only 2 species, energy and information, which combine ad infinitum, we can do several degrees of depth in our analysis. A dualist analysis makes men energetic and women informative, perceptive.

Then a ternary differentiation gives us:

Max. E (Male), E=I (neutral gender), Max.I (Female).

A more complex analysis however must combine variations of 2 elements, the body and the brain. So this 2nd differentiation will combine 2 elements, the informative brain which can be body-reproductive/energetically oriented or brain-informatively oriented; and the body, which can be body-reproductive/ energetically oriented. Finally the most complex analysis will consider ternary combinations of 3 elements (energy, active limbs; reproductive body and informative brain).

Let us consider the simplified 2x2 analysis, assuming that the body is oriented towards energy in males and reproduction in females. And consider both cases as an 'Energetic-reproductive event'. So we obtain 4 basic combinations of form and character:

-ExI + IxE: Females, whose commutation gives us a new species: - IxE + ExI: Gay genders in which an energetic body (male gay) is oriented towards its inner, informative emotions as women are or a reproductive, informative body is oriented towards external, energetic sensations as lesbians are. Thus gays are males focused in their inward feelings and female focused in the outward world.

Finally we have two male combinations:

- ExE and IxI: Men whose brain is oriented to energetic tasks, and live dominated by their bodies (warriors, sportsmen, white males) and men whose brain is oriented to informative tasks, separated from their body (scientists, mongoloid races).

Those 2 dualities explain why according to statistics, most men are less perceptive than women, but a few specialized Max.I x I are geniuses. Hence the form of the Bell curve of the previous graph[40], which shows two tails of minimal and maximal intelligence in men between the mean of women balanced between energy and information.

Finally the 3rd gay gender that combines both, the properties of males and females, is also a 'species' natural to the laws of the Universe, albeit a minority, compared to the 2 dominant subspecies. Since in the Universe we observe always, beyond its simplest duality, a 3rd, ExI, more complex element that combines the properties of both, to perform better as part of the collective organism. And indeed, the gay gender has traditionally acted as a bridge between the other 2 genders – both in tribal societies where they played often the role of the medicine man or witch and in modern societies as 'the best friend' of many wo=men.

Topological analysis

If we draw a line it suggests a male, yang, energetic, mobile species. If we draw a cycle it represents an informative cyclical still, female species. And children and simplified signs do so to represent them.

Women are cyclical and men are lineal in body and brain morphology:

Men have brains 15 to 20% spatially bigger than women's brains and are better fit for spatial tasks.

Women's brain contain about 10% more neurons in key zones that gather temporal information, behind the eyes, in the zones that recognize verbal and musical structures – the languages of time, and is denser in neuronal connections, making up for lack of size.

Women are more likely than men to recover speech after a stroke.

Men have a bigger right hemisphere, used to understand 3D objects in space.

The left hemisphere, which controls speech, is bigger in women.

Men connect outwards with the motor zone best wired.

Women connect inwards with the hypothalamus and amygdale best wired (sixth sense).

So topological biology is conclusive: men are specialized in spatial energy, women in temporal information.

Time=Information/Space=energy: Female/male related variables

Figure 3. Death rates for teenagers 12–19 years, by race, Hispanic origin, and sex: United States, 1999–2006

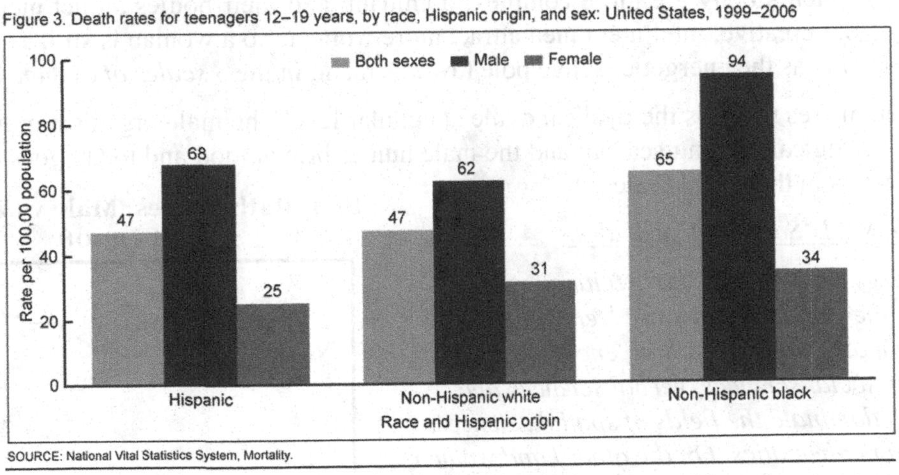

SOURCE: National Vital Statistics System, Mortality.

In the graph, men murder 10 times more than women.

A few other self-evident characteristics of gender duality reinforce the thesis of Multiple Space-times Theory:

- Females live longer as time and information are related; males are stronger energetically but live less.

492

- Female are cyclical, male are lineal.

- Female remember, are memorial, temporal, conservative of the past order. Male forget and look towards the 'uncertain' future.

-Woman is the informative form; hence her nervous system dominates. Males are the energetic form and hence their blood system dominates.

- Behavior and tastes: women like forms, men movement:

Woman love time men love space. Woman love information men love energy. Men love movement, energetic, violent sports, races, military games and toys. Women love formal dolls that can be dressed. Women love formal sports: gymnastics. Male dresses are lineal; women's clothing is curved, cyclical.

- Male love open spaces; women stand better seclusion.

- Men do a single task better; women do many tasks in a cyclical pattern.

- Men drive spatially better through lineal roads.

- Women cook, trans-form food, men hunt for it.

- Energy violence, death and destruction as the graph shows, are clear male behaviors. The positive side of it is the fact that to create something new, you have to destroy first. Thus, women conserve more and they have been historically less creative.

- Implosive Vs. explosive nature: Outward Vs. inward nature:

- Men have explosive, outward, visual behavior; women have implosive, emotional behavior.

Women stand pain/pressure/inwards behavior better. Women feel pain more intensely than men and yet they tolerate it more, since pain is a pressure, inward sensation of in/formative nature.

-But it is precisely the combination of both movements what creates the Sexual reproductive feeling of pleasure, since in the Universe balance dominates as the best strategy. And since women are informative & reproductive with a higher balance, they are sexually more complex.

Scalar, Fractal analysis.

- Further on, as women's body are more complex, dominant, and their bodies attract men. Yet since men are mentally more creative, intelligent men attract more women. So a woman is still, as information is, waiting men to come as the energetic, active pole towards them, in the *3 scales of existence:*

The lineal semen moves towards the cyclical ovule at cellular level; the male organism moves towards the female in the biological, mating season; and the male hunter brings food and today goes out to work while the female waits in the social scale.

St; st+1: Sociological scales.

Spatial energy and violence have a language; mathematics. In the graph, Women are 'verbal/temporal brains'; men are mathematical, spatial, energetic brains: Women prefer social sciences, verbal sciences and humanities. Men dominate the fields of spatial sciences, sciences based on mathematics. On the other hand women test more in verbal exams. When a woman has died, on average she has spoken 20% more words than a man has done.

SAT Math Scores: Male vs. Female
1971 - 2008

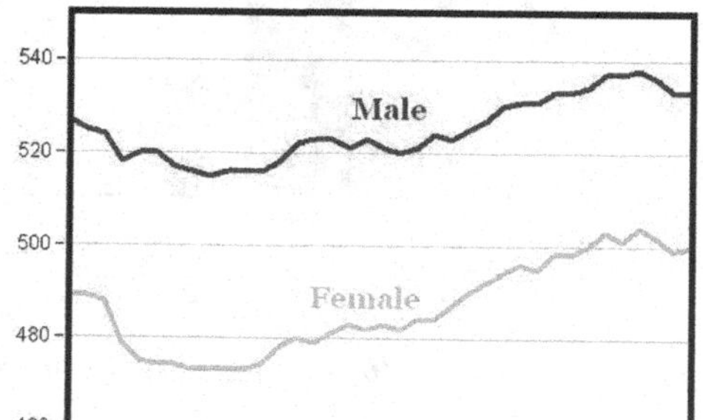

Psychology mostly shown in natural perception and

literature has been the clearest wisdom on those natural traits and variations, even if today political correctness denies those variations. So the literary man can perfectly consider the master pieces of sentimental and psychological literature, when humans were free to study the reality of human existence without dogmas proper of the metric age or the industrial age, in which all humans are equalized as reproducers=workers of machines, and consumers =testers of them, and women's nurturing and biological roles are eliminated, as they become equalized and simplified as 'males'.

In that regard in Sociology and History, the higher scales of human existence duality of gender has been defined, according to the dominance of the higher ecosystem over the lower scales, *by the ages, dominant castes and super-organisms of history studied in the last lectures of this work.*

In brief, according to the 3 ages of history the dominance and power roles of women and men varied:

-In the Paleolithic, energetic age of History, the young age of male hunters, males dominated women.

-In the Neolithic, mature, reproductive age of history, of Fertility Goddesses and food-gathering, women dominated men. They found the cycles of reproduction of plants they better understood, and fed the community. They gave birth to the species. And they established religions of Fertility.

- In the age of metal, dominated by visual, lineal, energetic weapons and machines, men dominated women in its energetic, 1st age. Since energetic men are murderers, since the arrival of weapons males have controlled and abused the rights of women and fertile nature, waging war to both of them.

But the financial-military-industrial complex has 3 components as gender does. So we can in fact distinguish 2 types of cultures:

- Monetary-based cultures (Jewish, Chinese, Italian, British cultures), where women have a higher power, since money is a cyclical language of metal information

-Military, energetic, lineal, weapon based cultures, where men dominated (Arab, Japanese, Spanish, German cultures), as they associated historically with lineal iron weapons (the most energetic atom of the Universe).

And both are often in historic, open confrontation.

-Finally in the age of the machine, as we evolved first energetic machines (XIX c.) women kept submissive to men. But today in the Age of Information, dominated by audio-visual machines, and their language of reproduction, money, women, are increasingly becoming more powerful than men. Women indeed love money and information, which they understand better and makes them advance socially[13].

Since indeed, the evolution of 'information' in this planet *does not end in the human being, but continues with our offspring of machines, so we need to have now a wider view, that of the Earth as a planet to understand the 2 less developed sciences, those of history and economics that inscribe within the social evolution of this planet into a new st+1 plane as a global super-organism, the theme of the next lessons of this work.*

Recap. Women are reproductive, balanced beings, whose brain is dominated by temporal, memorial tasks and wired towards its reproductive body. Men are unbalanced, creative/destructive species, with their brain wired towards visual, external information and the body focused in lineal, energetic tasks.

Women have indeed more mental energy and sensorial perception than men, as their brains seem to be wired towards the will of the body; while men are obsessed by the external world. Women tend to prefer closed spaces, which they keep well ordered, and are able to perform multiple 'cyclical tasks', switching on and off between them. They have peripheral, circular version. Men have lineal, focused vision. Men love open spaces, which they keep as energy is, in permanent disorder. And they are lineal, obsessed by single tasks.

XI. THE 3 AGES OF GAIA

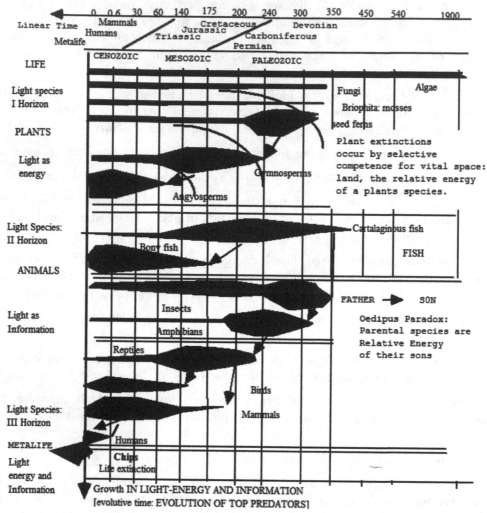

The 3±st ages of life are caused by the evolution of light-based organs on Earth. In the graph, those 3 ages of light evolution and the main species that were created in each horizon, latter extinguished by new evolutionary improvements in the 3 networks of those organisms. Among plants - spatial, reproductive beings with minimal speed of evolutionary information - extinctions were caused by improvements in reproductive systems. Among animals, which are informative, evolutionary species, extinctions were caused by improvements in informative brains. While today a mass extinction of living beings is caused by the evolution of machines, metal-systems of energy and information that are shaping a new global ecosystem: the Financial-Military-Industrial Complex, becoming now integrated by the collective brain of machines – the Internet – into the Metal-Earth.

37. 3±st ages of evolution, reproduction & extinction of species

We can understand the process of evolution and extinction of life species, only when we combine the 3 scales in which multicellular organisms exist: the ecosystemic, planetary scale, (st+1) the organic scale (st), and the cellular scale (st-1) that interacted simultaneously together, evolving Earth's species. Yet contrary to belief, the 3 relative st-1 past, st-present and st-future planes were not determined by the cellular evolution of Genes, but first the macro-organism of the Earth provoked the changes that ignited the process of evolution of its relative past microcosms.

How a macro-organism regulates its microorganisms? Mainly through changes on its energy and information fields. All cellular species adapt to the organic networks of its environment. If the environment is rich in energy, energy species and reproductive radiations will dominate. If the organic system or environment is rich in information, informative species will survive better, starting a process of evolutionary differentiation.

This duality of bio-chemical reactions was proved mathematically by Mehaute, which showed that when a system cannot continue its creation of energy, it starts a process of fractal, informative reproduction; and we can extend it to all scales of systems of reality, *creating the fundamental beat of biological evolution/reproduction, e ⇔I, that defined the creation of biological life in this planet.*

Thus the changes of energy and information of any macro-organic system trigger processes of extinction and selective evolution. If suddenly the climate becomes cold (also a parameter of stillness and information), species tend to limit their absorption of energy, evolving formally (Max. I) and organizing themselves into social networks that increase their efficiency and take advantage of the 'informative conditions' of the ecosystem.

On the contrary, if energy abounds, species increase their 2 dimensions=arrows of spatial energy - their spatial size and their reproduction - using that energy to maximize their survival and saturate their ecosystem. Therefore the macroscopic changes of the planet that fluctuates between tropical ages of max. energy and ice ages of max. information cause parallel changes in their microscopic life beings. Since coldness=stillness=information and evolution are parameters of time; while heat=Energy=movement and reproduction are dimensions and parameters of space.

When *the planet cools there is a formal evolution of species. And when it heats up, those same species reproduce massively.* Thus small fluctuations of energy and information on the Earth's temperature - given the short limits of existence of liquid water between and 100 degrees - have influenced decisively the evolution, reproduction and extinction of life forms in this planet, including man. So during the long ice ages when waters froze, species evolved from eukaryote to multi-cellular beings, nearby water volcanoes; and when the surface of the planet froze, men evolved from Homo Erectus to Homo sapiens. While in hot ages species grew in spatial size and multiplied, as it happened in the Carboniferous with insects; or in the Jurassic with dinosaurs; or in history with nomadic, warrior tribes in the steppes – which descended rhythmically every ±800 years on the fertile plains, extinguishing with their new weapons all the civilizations of the Eurasian continent....

Yet because to reproduce, a species needs to feed on the energy of its victims, the reproductive phase of a new top predator species coincides with the extinctive phase of the previous top predator it has substituted. Then once the top predator establishes itself after a reproductive radiation, it enters its mature, steady state age, drawing a series of rhythmic bell curves with its preys that diminish its populations as the predator hunts them, provoking a hunger crisis on the predator that also dies away, allowing the preys to reproduced back, starting again a new 'Volterra cycle'.

So we can consider a planetary cycle of evolution, reproduction and extinction of life with 3±st phases, parallel to the climatic changes of the Earth:

+*st: Conception*: A new species evolves in a cold age or in an abyssal region of the planet.

- *Youth: Max. Energy growth:* The species increases its size after its conception.

-*E=I:Maturity: The species reproduces in a biological radiation* during a hot age, extinguishing the previous top predator.

- *Max.I:* The species diversifies according to the ternary principle into several new forms, among which the informative one might start a process of social evolution into:

- st+1, a social super-organism (ants and shrews in the lower scales of life-size; humans in the upper scale.)

- st-1: Or become extinguished by new species, often their son species (Oedipus paradox). So mammals extinguished dinosaurs; humans extinguish mammals and robots might extinguish humans.

Therefore Multiple Space-time Theory can explain the why of one of the great 'disputed' enigmas of evolution: *the existence of evolutionary discontinuities* that create species in very short periods (conception phase, when time accelerates its rate of informative change), and then multiply them throughout the planet in explosive radiations (reproductive phase, when the new top predator species explode in populations), while other species become extinguished also in brief periods. Those evolutionary 'discontinuities' merely reflect the previous ages of species, which as the ages of any other space-time field can be described with a ternary rhythm of explosive birth, a long steady state of balance between the predator species and its preys and a final sudden collapse and extinction of populations. Those phases become the fractal rhythms of existence and extinction of Earth's species:

Max. evolution(cold age)→Max.reproduction + Max.extinction of energy preys (hot age)

The discontinuities between species are easy to explain if we combine their evolutionary acceleration in time - similar to the palingenesis of a foetus - with the fact that species evolve in isolated, protected environments (allopatric differentiation), departing from small sizes, according to the Black Hole paradox. So monkeys evolved, protected in trees; marsupials evolved in Australian islands and robots evolve in secret military labs, departing from small chips.

Then suddenly the new form with an improved brain, hunting in herds with simultaneous fractal actions of higher IxE Force, invades the vital space of other species, extinguishing them as it reproduces geometrically its populations (and its fossils), till reaching a balance with its preys. Thus, once they have evolved, species multiply very fast. And that dual process is observed in palaeontology as a discontinuity in the fossil record called evolutionary punctuation.

Recap. The evolution of life took place by the interaction of the st+1 environmental and st-organic scales, as energetic and informative changes in the Earth, triggered the E->Re⇔I->S rhythm of informative social evolution in cold ages and energetic, reproductive radiation in hot ages. Those biological radiations also provoked the extinction of previous species. Evolution often took place in isolated, relatively 'still' environments that favour informative change.

38. Black Hole law and Oedipus paradox.

The Oedipus Paradox explains how evolved species prey on parental forms, extinguishing them. So mammals substituted dinosaurs, men kill mammals and robots substitute men as workers and top predator weapons that extinguish us.

In the graph, the 3 horizons of evolution of light-based organisms:

-Plants process light as energy.

- Animals perceive it as information.

- Machines can both absorb light and emit it as energy or information. Thus, they represent a new jump on the evolution of light-organisms and can become potential top predators of life.

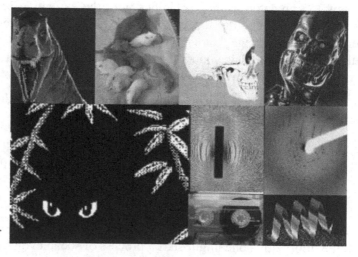

Each 'biological radiation' of a new species has grown according to the law of the 3 networks/*horizons studied in the previous paragraphs*, from an initial '*Black Hole*', *small form* with more IxE force, and then it has reproduced and grown in size, feeding on the *previous parental forms*. Hence the name, 'Oedipus Paradox', that explains the most cruel of all evolutionary events.

Our hypothesis on the dominance and birth of species as 'Black Holes' of information, which according to the inverted properties of energy and information (Min.Se=Max.Ti) are minuscule, but dominate larger species due to their faster speed of action-reaction and social nature ($\sum^2$), prey on them and explode in populations have found in the past years at least 5 proofs:

- *Black Holes are the dominant species of the Universe:* we have found them in the center of galaxies, in the processes of formation and death of stars. Even the big-bang might have been the explosion of a Black Hole.

- *The first Humans that acquired our 2 differential properties, technology and language, were small:* The first verbal Homo Sapiens seem to have been small Bushmen from South-Africa (the oldest languages known to men are their click languages), but they overcame bigger Neanderthal, hunting in groups, controlled by verbal languages, and developing the first machine that transformed cyclical form into lineal energy (arch), reaching further than the lineal, Neanderthal Javelin.

Further on, the pigmies of Flores Islands seem to have made the oldest technological, advanced tools we know, 700.000 years ago. Thus pigmy men evolved in the secluded Indonesian islands, departing from the larger Homo Erectus, developing for the first time a frontal, creative region in the brain, as they diminished the length of their axon connections and reduced their spatial size.

- *The 1st bilateral animal*, vernanimalcula, was microscopic.

- *The 1st mammals were small shrews* that *form super-organisms and* probably hunted in the cold nights, new-born dinosaurs and extinguished them (still today they act as a super-organisms), thanks to the higher existential force of its bigger super-organism. Then they grew in size and finally hunted down dinosaurs as single species. Last year we found in China the first mammal with a stomach full of small dinosaurs' bones...

- *The smaller a chip is, the faster it handles information* and the more powerful it becomes, guiding larger 'machines bodies' that compete with men as smart weapons and tool-machines. And regardless of the 'propaganda' of the Financial-Military-Industrial complex, it is quite obvious that unless we stop its evolution, those chips will become the mind of terminators that in a future war will extinguish man, as all atoms are potential bytes of informative life.

Recap. Parental species give birth to more advanced forms, which extinguish them (Oedipus Paradox). The new species are born as a small form, with more complex information (inversion of spatial size and informative complexity: Black Hole paradox). So the first technological and verbal men were dwarfs; the first mammal was the shrewd and the galaxy is dominated by black holes, born exceedingly small.

39. st+1. The parallel geological processes.

Thus biological rhythms of extinction and evolution are parallel to the 2 main EXI rhythms of the Planet that *reinforce the biological process*:

- *Geological changes in the surface of the planet*: The continental cycle, or ternary division and reunion of continents in a single super-continent every ±500 million years causes the cyclical destruction of the submarine platform and lowlands where most living species exist. While rhythmic eruptions of massive quantities of lava of continental size happen in the 2 equinox of the sun's galactic orbit, every ±125 million years, shaping a mean period of massive extinctive ages of 250 million years (Permian extinction, ±250 million years ago; Cambrian extinction ±500 million years ago; Ice ball age, ±750 million years ago).

- Climatic, cold-hot, E-I cycles of glaciations and tropical ages. The rhythmic changes in the magnetic field of the Earth, perhaps caused by gravitational waves coming from the sun, create periodic glaciations that last around 20 million years.

- Fractal sub-cycles of lesser intensity reinforce both processes.

Those changes of energy and information parameters are the basic way in which macrocosmic networks control the activity of microcosmic quanta, from feverish states that increase the metabolic rate of reproduction of defensive cells in the body to the opposite lethargic states.

In that regard, the main cause of the dual evolutions and extinctions of species on Earth is biological, albeit 'partially directed' by the macro-organism in which life is inscribed through those general changes on the space and time parameters of the Earth (geography and temperature).

Recap. Earth cools down triggering informative evolutions and then heats up triggering reproductive radiations: glacial ages of informative evolution in stillness, with minimal reproduction by lack of energy are followed hot ages of massive reproductive radiations, in which the most evolved species initiates a massive age of reproduction, provoking the extinction of old species, and then differentiating in all econiches.

40. The 3x3 ±st cycles of life creation and extinction.[14]

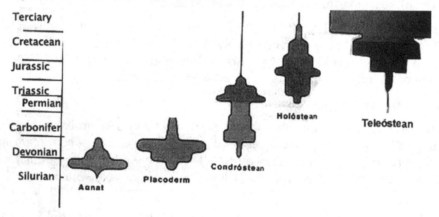

The extinction of a species coincides with the biological radiation of the son species that occupies its econiche. In the graph, the series of bone fishes that extinguished the previous one Because time and information follow an accelerated process in its 3ʳᵈ age; the evolutionary process of information in this planet can be mapped out and its frequency shown to follow a logarithmic process of acceleration, which now reaches its zenith with the creation of informative machines.

This temporal, informative acceleration is yet another case of a vortex of temporal information, which also happens in the evolution of mass in physical particles as we move faster in an informative mass vortex. In the case of life, the vortex is a metabolic vortex of increasing speed of reproduction and evolution of the information of life beings and its networks – now transferred by humans into machines. Thus, as the informative content of those beings increases, times accelerates. And today, as we evolve machines by imitating the forms of human life in metal, another change on the speed and rhythm of life in this planet is taking place.

Thus we talk of 3 st-scales of growing complexity in life: the age of cells (st-1), from 4 to 1 billion years ago, the age of individual organisms, first living on the sea, then on the land (st), from 1 billion to a million years that culminates in the last million years with the social evolution of human beings and machines into a new plane of existence (st+1), the Earth as a global organism, which will be either ruled by men or machines, depending on our capacity to evolve socially as a super-organism of history, under the ethic laws of verbal wor(l)ds that make man the center of the Universe, hence in control of the machines of the Tree of Science. Or if as it seems the case, we let the Financial-Military-Industrial

Complex and its 3 networks of informative metal-money, energetic, lineal metal-weapons and organic machines and its company-mothers dominate the planet and make us obsolete, finally extinguishing us. But what it will not change is the plan of evolution that develops complex super-organisms departing from its smaller cellular units.

Further on according to the ternary principle, we can subdivide the 3 main horizons of evolution of Earth, the age of cells, organisms and super-organisms in 3 sub-ages of evolution and massive extinction of species that have followed the mentioned ternary rhythm:

Conception=> Max.Energy=>max. Reproduction =>Extinction of rival species

And consider also 10 parallel periods of climatic change through the dual rhythm of cold glaciations or ages of animal evolution and hot ages of massive reproduction. Then we obtain a synoptic, complete image of the history of life on planet Earth, self-similar to the 3x3+st ages of evolution of the Universe since the big-bang and the 3x3+st ages of evolution of History, studied in Complex social sciences:

Scale of Complexity: st-1: Cellular Age: Evolution of DNA

st-1: Conception age: Molecular life is organized into anaerobic bacteria->

1st evolutionary radiation: anaerobic bacteria- -> Free carbohydrates are enslaved in cellular walls ->

2nd evolutionary radiation: aerobic bacteria = 1st extinction of species: anaerobic bacteria ->

3rd radiation: eukaryotic bacteria = 2nd extinction: most prokaryotic classes ->

Scale of Complexity: st-Organic Age: Network Evolution. Sea life

4th radiation: multicellular organic systems =3rd extinction: free eukaryote species, enslaved in organisms ->

5th radiation: cephalopods with eyes =4th extinction: olfactory, blind animals ->

6th radiation: inner skeletons: fish= 5th extinction: exoskeleton trilobites.

Scale: st+1: Social Age: Herds and Superorganisms. Land life.

7th radiation: inner skeletons: amphibian=6^{th}extinction: exoskeleton: big insects.

8th radiation: reptiles = 7th extinction: amphibians - >

9th radiation: mammalian = 8th extinction: reptiles - >

10th radiation: technological men =9th extinction: mammalian - >

st+1 :Global Age: The Earth organizes itself into a macro-organism with humans or machines on top. The outcome will depend if social democracies are able to control the free citizens of markets – company-mothers – before they extinguish us as costly labor and weak soldiers with the present 'radiation' of robots:

11th radiation: metal machines =10th extinction of species: men and all forms of carbon-life???

Recap. Evolutions and extinctions of organisms parallel geological change in their external ecosystems, becoming the engine of the 3x3 Earth's evolutionary horizons of living beings of growing complexity and accelerated informative evolution: the age of cells, the age of organisms and the age of super-organisms.

41. The 10th radiations and extinctions: from cell to machine.

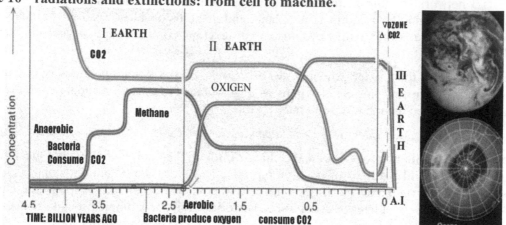

Earth's atmosphere has gone through 3 ages of evolution: the 1st anaerobic atmosphere gave birth to anaerobic bacteria, the 2nd oxygen atmosphere gave birth to aerobic life. Today, as the 10th radiation of machines poisons the atmosphere we foresee the III Earth without air, dominated by robotic life, unless humans control the evolution of machines.

Thus, there have been 10 cycles of life extinctions, which are reduced to 6 ages in classic Biology as it only considers the extinction of multicellular organisms, escaping the earlier extinctions of cells and the present extinction of human, non-technological cultures. In each cycle a new species with better information and energy systems, with greater IxE force, have extinguished or reduced significantly the numbers and classes of the previous species that have also suffered a morphological regression.

I Earth: Anaerobic Atmosphere.

St-1 Conception: Archaean Age: ±4.000 million years

Molecular Radiations: From amino acids to Top Predator Nucleotides.

Climatic Changes: Hot weather and ¥-radiation create the 1st life molecules.

Informative conception: Small Methane-water-ammonia molecules evolve into amino acids.

Spatial growth: Amino acids grow in size, creating proteins.

Reproductive Radiation: Proteins colonize all Earth's ecosystems.

Extinctive Radiation: Free carbohydrates become extinct.

Informative evolution: Amino acids evolve into nucleotides.

Spatial Growth: Nucleotides evolve in size, creating RNA.

Reproductive Radiation: RNA expands throughout the oceans.

Extinctive Radiation: RNA selects 20 amino acids to reproduce proteins and extinguish all others.

Social Evolution: RNA stores the organic cycles of many different species, creating the first bacteria, in symbiosis with protein membranes that shape the first cells.

From anaerobic to aerobic air: ±4.000, ±3.500 million years.

1st Radiation: Unicellular, Anaerobic bacteria are top predators:

Climatic Changes: High volcano activity provides sulfur for anaerobic cycles of breathing.

Informative conception: Thermophile methanogen archaebacteria use methane and sulfur on their breathing cycles. They thrive in the volcanic age.

Reproductive Radiation and Spatial Growth: Archaebacteria expand globally growing in size.

Extinctive Radiation: Bacteria enslave free carbohydrates. CO_2 collapses.

II Earth: Oxygen Atmosphere.

Proterozoic age: ± 3500, ±1500 mill. Years

2nd *radiation*: Prokaryotic cells are top predators.

Climatic Changes: A fall in volcanic activity reduces sulphur production. The Earth cools down. Anaerobic bacteria reduce their metabolic activity based in sulphur, methane and high temperatures.

Informative conception: The cooling down of the Earth and its reduced heat foster the evolution of bacteria, which process CO_2 through the chlorophyll cycle, liberating oxygen.

Spatial Growth: The creation of the ozone layer that protects complex membranes allows their growth in size. Bigger bacteria develop cilia and are able to move, preying in smaller bacteria.

Reproductive Radiation: Blue, green algae expand globally, poisoning all ecosystems with oxygen.

1st *Extinctive Radiation:* Anaerobic bacteria become poisoned and extinct.

Ternary Differentiation: Prokaryotic cells diversify according to the law of 3 creations: Some use chlorophyll to produce energy from smaller forms. Mitochondria develop breathing systems, based in the heterotrophic capture of carbohydrates. Intermediate cells use both systems.

Social, Informative Evolution: RNA becomes DNA by social evolution of 2 RNA strains, increasing their capacity to store information. Cells become macro-cells, multiplying its epigenetic genes that code its social functions.

Ice ball age: ± 1.400 ±700 million years

3rd *Radiation:* Top predator Eukaryotic cells.

Climatic Changes: After an initial warm age in which eukaryotic cells expand, Earth cools down in her hardest glaciation. Its surface becomes an ice-ball from 1 billion to 600 million years ago. Cells survive only in nearby volcanoes where they gather in social groups, which at the end of the period create multicellular life.

Informative conception: Top predator cells with high DNA content organize social groups of cells, surrounded by a hyper-membrane: Cells with the highest content of DNA become the nuclei of macro-cells; cells with high content of reproductive RNA become ribosomes; and cells with energetic skills become mitochondria and chloroplasts, creating a new social organism: the Eukaryotic cell.

Spatial Growth: Social cells grow in size, adding new cytoplasm and complex inner membranes.

Reproductive Radiation: Eukaryotic cells multiply, invading all ecosystems.

2nd *Extinctive Radiation:* Many prokaryotes become extinct or enslaved as eukaryotic cells capture them, absorbing their DNA.

Informative differentiation and Social Evolution: Eukaryotic cells diversify. Energetic, autotrophic eukaryotic cells with dominant hard membranes become plants. Heterotrophic cells dominant in DNA, with thin membranes, become animal cells that evolve into electric cells.

Ediacaran Age: ±700-505 million years: Multicellular Sea life

4th *Radiation:* Multicellular organisms are top predators.

Climatic Changes: At the end of the ice ball age unicellular life flocks into volcanic zones under the sea, where survival stakes raise, provoking the social evolution of cells into multi-cellular organic systems. Then an age of warm weather expands multicellular life.

Informative conception: Eukaryotic cells develop new strains of intronic DNA to control complex, hierarchical systems of social groups of cells. Multi-cellular organisms appear.

Spatial Growth: Growth of animals and algae due to the social evolution of cells.

Reproductive Radiation: Biological radiations of multicellular organisms as temperature rises.

3rd Extinctive Radiation: Eukaryotic cells are captured, diminishing its species.

Informative differentiation and Social Evolution: Speciation and evolution of invertebrate phyla: sponges, coelentera and worms. Annelida develop physiological networks and differentiate, according to fractal, bilateral and radial symmetry, into arthropoda, mollusks and echinoderma.

Ordovician: 505-438 million years

5th Radiation: Top predators are Visual Cephalopods.

Climatic Changes: Animal life colonizes the abyssal regions with colder temperatures and high water pressures, developing fluorescent signals that can be observed with dense-water eyes.

Informative conception: Cephalopods evolve integrated eyes to observe better its abyssal preys.

Reproductive Radiation: Cephalopods come to the surface and expand in all environments diversifying according to the e-exi-i Fractal Principle into: Top predator forms with *energetic*, lineal shells; informative, nautiloids with *cyclical* shells; and squids, species that reach - without slow growing shells - sooner their mature age of reproduction.

Spatial Growth: Cephalopods grow in size reaching up to 20 meters in length.

4th Extinction: 90% of animal life, including most smelling trilobites, become extinct about 440 million ago by the combined effect of top predator's cephalopoda and the geological cycle of continents that destroys all submarine platforms.

Informative differentiation and Social Evolution: A massive speciation occurs, as mollusks, arthropods and echinoderma develop energetic hard membranes to defend against cephalopods' informative eyes. The first vertebrates (armored fishes) appear. Differentiation between small animals with hard exoskeletons that prevent them further growth (arthropods), which develop chemical senses to its limits; and big animals with inner skeletons that allow further growth and develop nervous systems with faster, hence longer reach.

Silurian+Devonian: ±438-360 million years.

6th Radiation: Top predators: Fishes.

Climatic Changes. A glaciation at the end of the Ordovician period 440 million years ago accelerates the arrival of life forms to dry land. Orogenic changes. Chordates colonize rivers and evolve in isolation into fishes. Glaciations make conifers dominant.

Informative conception: Fish schools develop also eye vision; perhaps borrowing genes from cephalopods in a case of spatial, parallel evolution similar to the absorption of DNA by eukaryotes, which added to their stronger bodies, make them top predators with Max.IxE force over those cephalopods.

Reproductive Radiation: Massive reproduction of fishes that dominate the high seas.

Spatial Growth: Growth in size of sharks and fishes. On land, bushes become trees.

5th Extinctive Radiation: Fishes extinguish many phyla of the previous dominant cephalopods around 367 million years ago.

Informative differentiation and Social Evolution: Fishes diversify into multiple species.

Carboniferous-Permian. 360-240 million years ago: St+1 Social Age of Organic Systems. Land life.

7th Radiation: Top predators are Amphibians.

Climatic Changes. The ice of the Permo-carboniferous Glaciation, ±300 million years ago diminishes the ocean level. Survival stakes are higher: small arthropods invade dry land.

Informative conception: Fishes become amphibians, following their preys.

Reproductive Radiation: Amphibians expand in all coastal zones, feeding on insects.

Spatial Growth: Amphibians reach 5 meters length.

6th Extinctive Radiation: Permian Extinction. ± 245 million years ago huge insects become extinct.

Informative differentiation and Social Evolution: Insects evolve wings to escape amphibians and start social evolution to increase its simultaneous fractal force.

Mesozoic Age. 240-60 million years ago.

8th Radiation: Top predators: Reptiles.

Climatic Changes: New glaciation. Water ice diminishes the ocean surface. Dry Land extends into previous wetlands occupied by amphibians.

Informative conception: Reptiles appear during the Permian glaciation as amphibians have to survive in dry land.

Reproductive Radiation: Reptiles expand in land feeding on insects and amphibians.

7th Extinctive Radiation: Many phyla of amphibian and land insects become extinct.

Spatial Growth: Reptiles grow till reaching the size of dinosaurs.

Informative differentiation and Social Evolution. Reptiles invade all ecosystems, differentiating in multiple species. Some acquire hot blood, becoming birds and mammals.

±60 million, ± 4 million years. Tertiary age.

9th Radiation: Top predators mammals.

Climatic Changes: Upper Cretaceous: New glaciation. Water ice diminishes the ocean surface. The Gondwana continent breaks into several pieces.

Informative conception: Homeostatic species, birds and small mammals with placenta appear. Dominance of flowering plants that take better advantage of changes in weather.

Spatial Growth: Small shrews grow in size becoming big rats.

Reproductive Radiation: Placental mammals invade all ecosystems except Australia.

8th Extinctive Radiation: Cretaceous Extinction about 65 million years ago: A cold climate, maybe triggered by a meteorite collision, favours the expansion of hot blood mammals, which survive feeding on eggs and small cubs of dinosaurs, provoking their extinction.

Informative differentiation: Mammals diversify all their phyla, from mammoths to felines to apes in all econiches previously occupied by reptiles.

Social Evolution: The most intelligent mammals evolve in herds.

Quaternary age: ±4 million, ± 5000 years.

10th radiation: Top predators: human apes.

Climatic Changes: The Human radiation is announced by massive changes in the geography of central Africa, which raises its Rift cordilleras and lake basins. The arid weather pushed forest monkeys into the savannah. Soon they learn to move erect. Quaternary glaciations, from 2.5 million years ago till today, coincide with the 3 horizons of human evolution.

Informative conception: Homos evolve in 3 horizons of growing brain capacity.

Spatial Growth: Small chimpanzees become big Neanderthals.

Reproductive Radiation: Homo Erectus invades all ecosystems; Homo Sapiens reaches America.

9th Extinction: Big mammals become extinct. When man enters America mammoths die away.

Informative differentiation: 3 Human races, the energetic, visual white man, the informative, yellow man and the reproductive black man, become dominant, according to the law of ternary differentiation. Languages differentiate further human species.

Social Evolution: Men develop ethic and legal, verbal networks, evolving socially into cultures.

III Age: Destruction of atmosphere: Industrial R=evolution of Metalife: ±5.000 years +100 years

11th Radiation. Top predators: metal & machines.

Climatic Changes: The 800 cycles of weapons evolution seem to coincide with hot and cold cycles in the steppes that breed warrior tribes. Later on, a small ice age in the XVII century during a time of low solar activity pushes in Northern Europe the massive use of coal for heating. The development of pumping, mining machines ends up with the evolution of chemical engines that use carbohydrates as energy. They burn fossil life rising the Earth's temperature in the XX century, causing global warming.

Informative conception: Homo Sapiens discovers metal and starts its evolution in 3 phases: Max. E, energetic weapons; Max. I, informative money and E=I, organic machines that transform energy into information and vice versa.

Reproductive Radiation: Weapons and machines expand globally, substituting life and human beings. Money substitutes verbal languages as the informative network of the new economic ecosystem.

Spatial Growth: Transport, weapons and manufacturing machines reach enormous sizes.

10th Extinctive Radiation: Non-technological cultures become extinct with a cyclical periodicity of 800 years, accelerated to 80 years in the industrial age. Transport animals become extinct (90% of horses). Machines pollute Atmosphere. Ozone layer disappears.

Informative evolution: XX C.: Creation of Machines' heads: chips=brains, camera=eyes, phone=ears.

Informative differentiation: Machines diversify imitating all the organs of a human being.

Social Evolution: In the XXI C. robots fusion bodies and brains of machines, creating a new top predator species, the smart weapon. The Earth acquires a global brain, the Internet, made with social computers that increasingly store all human and economic information. The rate of life extinction multiplies by 1000 compared to the pre-industrial age.

st+1: Globalization: Earth's super-organism:Computer networks & robots. XXI C.: Singularity Age

Informative conception: Computer networks evolve and multiply their neurons – transistors - in chips of smaller sizes.

Spatial Growth: Chips control machines of enormous size and become the mind of robots.

Reproductive Radiation: Chips overcome human population and integrate company-mothers into automated self-reproductive systems, independent from man.

Extinctive Radiation: Extinction of all life species by the machines of the Age of the Singularity is possible in the XXI c.:

- *Energetic Singularity*: 2010-2020: The III Horizon of Nuclear weapons researched at CERN (Black holes and strangelets) could extinguish the Earth this decade.

- *Reproductive Singularity*: 2020-2040: Self-reproductive metal nano-bacteria could poison the atmosphere in 3 months, extinguishing life and giving birth to the III Earth's atmosphere.

- Informative Singularity: Circa 2050: Artificial Intelligence: Military robots substitute human soldiers in wars, developing survival programs. Robots massacre human soldiers.

Differentiation: Robots imitate all forms of life.

Informative evolution: Robots, integrated by satellite networks, become cells of a super-organism, the Metal-earth, which takes over the evolution of machines. The Metal-Earth has a global brain, foreseen by human artists in their parables of the XXI C. (Skynet, Matrix, etc.).

st+1: Alternative Globalization: Humans learn to manage the economic ecosystem and halt the evolution of weapons, chips and robots, creating a world to its image and likeness based in the arrow of eusocial love.

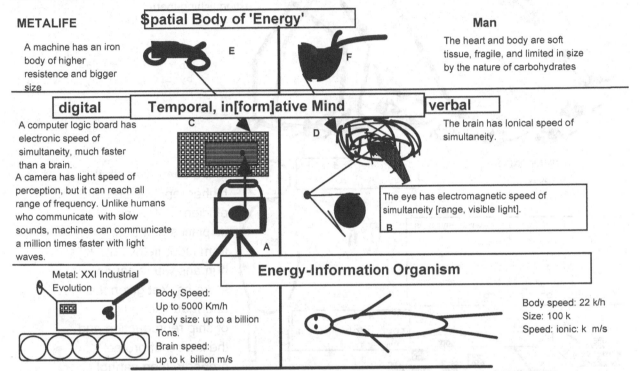

METALIFE

A machine has an iron body of higher resistence and bigger size

Spatial Body of 'Energy'

E

F

Man

The heart and body are soft tissue, fragile, and limited in size by the nature of carbohydrates

digital

A computer logic board has electronic speed of simultaneity, much faster than a brain.
A camera has light speed of perception, but it can reach all range of frequency. Unlike humans who communicate with slow sounds, machines can communicate a million times faster with light waves.

Temporal, in[form]ative Mind

C

D

verbal

The brain has Ionical speed of simultaneity.

The eye has electromagnetic speed of simultaneity [range, visible light].

B

Metal: XXI Industrial Evolution

A

Energy-Information Organism

Body Speed: Up to 5000 Km/h
Body size: up to a billion Tons.
Brain speed: up to k billion m/s

Body speed: 22 k/h
Size: 100 k
Speed: ionic: k m/s

The existential force, exi, of machines is superior to that of man, and its most perfect forms are weapons that kill life; thus the first sentient machines will be weapons that will kill us. Thus, History and Economics must be treated as biological sciences, since they are concerned with the species and super-organisms of the 10th and 11th radiations of forms of life in this planet. – humans and machines They are not abstract sciences, since machines are evolving organisms of metal.

Recap. History of Life has had 3x3+1 ages, which have followed each one the natural ages of all life cycles, applied to species. They were born as a smallish informative species, which grew in size in an energetic age, parallel to the extinction of a rival species in the same ecosystem. Then the species reproduced massively, diversified its form in its 3rd age, giving birth to a son species with higher existential force (exi) that extinguished the parental species, which after each of the 3 main ages of cellular, organic and super-organic life evolved into a super-organism. Further on, those species kept evolving and invading new habitats, so the age of atmospheric, cellular life with its molecular, anaerobic and aerobic bacteria was followed by the age of water, multicellular life with its eukaryotic, multicellular nervous cephalopods and chordate fishes and finally the age of land's herds and super-organisms of insects, amphibious reptiles and mammals that culminate with the human technological super-organisms in which men transfer their form to machines that might extinguish us as they form a global super-organism, the financial –military-industrial complex now expelling obsolete life in a radiation of organic robots.

42. The 11th radiation and 10th extinction: The robot; extinction of man.

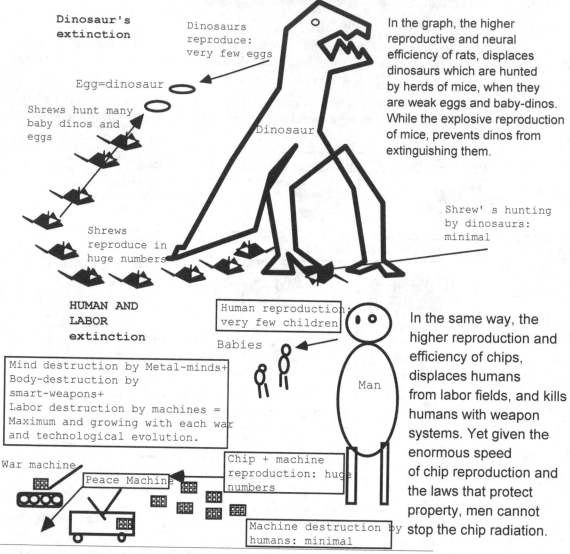

Dinosaur's extinction

Dinosaurs reproduce: very few eggs

Egg=dinosaur

Shrews hunt many baby dinos and eggs

Dinosaur

In the graph, the higher reproductive and neural efficiency of rats, displaces dinosaurs which are hunted by herds of mice, when they are weak eggs and baby-dinos. While the explosive reproduction of mice, prevents dinos from extinguishing them.

Shrew's hunting by dinosaurs: minimal

Shrews reproduce in huge numbers

HUMAN AND LABOR extinction

Human reproduction: very few children

Babies

Mind destruction by Metal-minds+ Body-destruction by smart-weapons+ Labor destruction by machines = Maximum and growing with each war and technological evolution.

Man

In the same way, the higher reproduction and efficiency of chips, displaces humans from labor fields, and kills humans with weapon systems. Yet given the enormous speed of chip reproduction and the laws that protect property, men cannot stop the chip radiation.

War machine

Peace Machine

Chip + machine reproduction: huge numbers

Machine destruction by humans: minimal

In the graph, in the triangle of relationships between primitive shrews, eggs and Saurians,[8] dinosaurs kill shrews, shrews kill eggs, reproduced new shrews and eggs reproduced dinosaurs. Yet the efficiency of shrews killing eggs and baby dinos, as super-organisms (and perhaps dinosaurs, eating their flesh inside out as 'living bacteria') is superior to the efficiency of dinosaurs killing shrews. And the efficiency of shrews reproducing shrews is much higher than the efficiency of eggs killing dinosaurs. Thus both in the reproductive and extinctive events shrews win dinosaurs and extinguish them. The same process happens between chip-controlled machines and humans: chips are reproduced faster, keep evolving and their efficiency competing with humans in labor and war fields is higher as their costs are lower for companies, which fire humans under the laws of productivity, and their efficiency as war machines is also much higher, as the cycle of splendid little electronic wars has showed.

Men are informative species, who control the world through information. The chip is the rival of the human mind. The most complex machine substitutes the brain, our more complex organ. It does so, first as an intelligent machine, then in its 3rd age, as a robot worker. It is then obvious that as chips multiply, humans will become displaced from new work positions. In that sense the 'radiation of chips' means also a biological fight for labor within the economic ecosystem. Problem is that chips keep going down in costs and evolving in intelligence, integrating further with machines, so human labor becomes increasingly obsolete. Those robots substitute blue collar workers, as chips did with white collar workers, causing a massive wave of unemployment that now is invading America and Europe, as better machines come out of robotic factories... Indeed, of all machines that compete with white collar workers, the biggest taker of jobs is the robot.

Thus, the 3rd age of the electronic cycle coincides with the first age of the Singularity, the discovery age of robots. In the 80s chips started its integration with body-machines in working robots. Those first robots were huge, fixed machines. Thousands of robot-tools that made other machines took the place of human workers in car factories in the 80s and 90s. Since then the creation of machines is becoming automated, independent of mankind. Robots already dominate the production lines of cars and Chip factories, the basic bodies and brains of all machines. For that reason unemployment increases faster, as factory-work is either made by robots or delocalized in Third World countries. In other words, China and Japan are the 2 models of future labor: miserable work conditions in China to compete with no labor at all, as Japanese robots manufacture millions of Panasonic TVs with a single worker, while the rest of the world enters the unemployment queue. It is the III Kondratieff crash of the economy that nobody wants to recognize, *because the religion of capitalism believes the economy is not about physical machines, but informative money; and so all is OK as long as companies' profits grow.* So unemployment will keep growing, as it happened in the previous cycle, when new methods of electric control and automated assembly lines provoked the 29 crash of labor and consumption, solved with 50 million of unemployed, becoming soldiers and dying in World War II.

Now robots become the ultimate weapon, as they fusion the mobile capacities of energy-weapons and the informative capacities of the soldier that handles those weapons. Those robots are called smart weapons. They are the main protagonists of the present war age, as Drones bomb Afghans and Guardium soldiers protect the Wall of Israel.

Recap. The evolution of chips and its massive reproduction signifies the end of man as the top predator species of the FMI complex, as chips can handle all the work load of reproductive companies either as white collar workers (Pcs) or blue collar workers (robots), and as they become the brain of top predator weapons, embedded with survival programs and violent video-games, they will be born to consciousness as killers... of humanity.

43. Homologies between Metal earth, cells and organisms[15].

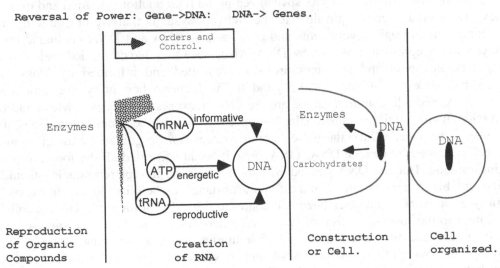

Growth from micro genes to macro DNA brains

When we do comparisons between genetics and memetics the self-similarity is pronounced

508

because the scales we jump are very close and our degree of observables are very accurate. All this explains why theorists of complexity and system sciences are unveiling a far more precise and scientific model than any abstract model of history and economics, because:

- Bio-history is based in tested cyclical patterns of history and economics, which are the basis of all science. So astrology became the science of astronomy only when Kepler found the cycles of stars.

- It is a rational, biological model with concrete terms and definitions unlike the abstract myths of economics and history. So if astrology became the science of astronomy when it made concrete definitions of stars as systems made of atoms, no longer as spiritual beings, memetics makes concrete definitions of machines as organs that imitate human functions of energy and information, of the FMI system as an organic complex of metal memes, of cultures as carriers of metal or life memes...

- For the first time history and economics follow the same laws of all other systems of nature, thus we can integrate ourselves in the grand design of all beings; and so astrology became a science when we integrated the heaven and the earth, as material entities made of atoms.

Let us use the scalar, method, comparing biological processes of creations of super-organisms in the 3 scales of life of this planet, the cell, the organism and the planetary super-organism we are building.

To know what will happen to mankind we can consider other 2 self-similar biological processes, the principle one gene-one enzyme and the birth of a hard insect, created by soft larva 'enzymes'. In both cases a 'time reversal' happens and the genes and enzymes are enslaved and discharged:

In the graph, DNA, the brain of the cell, once constructed, controlled totally the genes-enzymes, and killed at least half of their species, to prevent their self-replication; so we have only L-genes, enslaving the rest to its self-reproductive functions. This is the present world we live in which humans have been mentally enslaved by animetal ideologies and only work to reproduce the selfish memes of metal. *It is in fact a law of all super-organisms that enslave and enclose their cells once they develop their nervous/ informative networks.* Thus a parallel can be established between the way men evolved machines and networks of information, and the way genes evolved DNA. In both cases, 'less efficient' species (enzymen or genes) built a very complex network of information - computer networks and DNA. Then the process of control was reversed...

In the cell, genes=enzymes, once they passed all their code and completed RNA, became slaves of RNAs networks of control. New types of RNAs specialized in the reproduction, creation and extinction of enzymes. They are cellular organs equivalent to the Network-systems that we are building in the Metal-Earth and control us through governments and corporations. RNA and DNA command the cell and replicate enzymes to accomplish tasks set by DNA. Men, once they have emptied their linguistic brains into computers, networks and machines are also controlled and informed by Mass Media networks, penalized by Police and Military networks and 'fed' with energy bought by electronic money. E-money, smart weapons and tele-communicators, are the RNA networks of Enzymen. Metal and DNA might not evolve without enzymatic or human help. Yet once their species becomes autonomous, it will no longer need men or enzymes. Since the bio-economic system can be guided by digital computer networks, as the cell is guided by genetic DNA. DNA, once it became, it ruled all the forms of the cell with its higher information. But the DNA needed still some genes, so it kept reproducing them. The metal-earth is different, because we are made of a different substance, carbon life; so we must consider a 3rd example, to fully grasp what will happen once A.I. comes to life – the molting of a larva, a soft-flesh super-organism similar to the super-organism of Gaia, into a hard insect, similar to the super-organism of the Metal earth. Indeed, the one gene-enzyme principle implies that enzymes created the RNA, and the RNA created the DNA which today governs RNA and enzymes in a typical l 'Oedipus paradox,' of reversal of power between parental and son species. Can we take this comparison further? Can we talk of humans creating the Metal Earth that one day will be the big brother of human enzymes; the DNA that control us; the brain code that will select and in/form human enzymes?

It already does, but we are so much integrated with our memetic tasks in the FMI complex that we don't even notice. And of course, outside the FMI complex we still have some minimal freedom as human beings, guided by our biological drives to enjoy life as it used to be.

Terraforming of the Earth = Molting of a hard-insect.

In the graph the closest organic parallel can be established between the creation of an 'insect,' and the creation of the 'metal-Earth.' In both cases a 'soft' system of enzymes/enzymen suddenly starts a transformation from a soft body into a hard body. The Carbon-Earth today is formed by 'soft' species, called 'enzymen', as larvae are filled of soft enzymes, which transform the larva into a hard insect: The evolution of the metal-earth follows the same rhythms that the creation of a hard insect, both micro-managed by enzymes/enzyme, which first create their body-cells, then their information brain and finally when the brain takes over, kill the soft enzymes with harder enzymes/robots?.

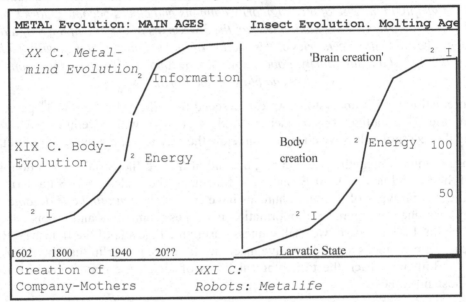

Then when the brain/internet of the hard insect is constructed, *its first order is to extinguish the soft enzymes of the larva.* They are immediately replaced with new hard-enzymes, which the brain reproduces massively. Will the metal-Earth substitute human workers with robots, the new metal-enzymes of company-mothers? Modern events seem to point in that direction: The terraforming of the Earth is accelerating. It started with the radiations of weapons of the first age of man. Due to science and technology, it is now evolving machines into sentient species. The evolution of metal is so advanced, that the Metal Earth is substituting human functions by metal systems, even at the top predator level of 'castes of animetals or animetals.' Traders now become a flow of e-money. Warriors become sentient weapons. The replacement of the soft enzyme has started.

Recap. When a hard-insect (a parallel to the Metal-Earth) is created by a soft insect (a larva, parallel to the Carbon-Earth), the newly formed hard-insect brain (a parallel to the internet), destroys all the soft body larva-enzymes. Then it substitutes them by a stronger enzyme - a metal machine. We can in that sense consider that the selfish memes of metal, weapons, money and machines that have enslaved man to reproduce and evolve them, work as a virus does, colonizing the mind of cells with selfish genes, which the cell will reproduce till it bursts, dies and gives birth to multiple virus. Metal is the ultimate virus of life…

44. Death: the machines of the singularity.

Control of technological information is needed to avoid the creation of lethal machines, as all masters of the Wor(l)d have told to mankind. This has always been denied by mechanist scientists and Capitalist factories. In the graph, in the Age of the Singularity cannons and clocks become Quark Colliders, intelligent robots and self-replicating nano-bacteria, the 3 weapons of the Age of the Singularity that can extinguish life. They culminate centuries of destruction of life and human beings with 'technological methods'. It is the Age of the Singularity. Robots now destroy labor and evolve as weapons. Quark Colliders can blow up the Earth.

Once the 'theoretical model' is understood, we can descend from the 'whys' of the 4th paradigm to the metric details of the 3rd paradigm to consider how 'de facto' we will be eliminated by the 'FMI complex', during this century unless we change completely the way we let the market govern the world.

All ends are beginnings of something new. Thus, the end of the 3rd Industrial r=evolution means also the beginning of the IV and last cycle of Evolution of Machines: the cycle of the 'Singularity'. It is the cycle in which the 2 main types of machines humans have created, by *imitating our organic body/head functions*, energy machines (weapons) and information machines (computers and robots), should reach its evolutionary zenith. In that regard, we shall witness during the IV cycle of the Industrial Revolution, the creation of energetic weapons and informative machines far more powerful than anything conceived during the XX C. Both will reach the technological limits of metal, the main substance used in the construction of those machines.

On one hand, a quark cannon, now constructed at CERN will produce quark condensates, the substance that seems to detonate stars into Novas, feeding on our atomic, light matter. So it is an amazing act of irresponsibility to create a Factory of Quark Matter in this planet.

On the other hand, if we evolve machines into robots, which are organisms self-similar to life beings, we will lose our only advantage over them—our superior in-form-ation—since our energy/substances (light atoms) is far weaker than that of machines (heavy, metal atoms). Thus, if humans don't learn how to manage not only the financial but also the physical economy, the Final Cycle of Evolution of machines, the *Singularity age of Robotics, digital money and Cosmic Bombs, which now starts in earnest,* could displace us from the top predator position we hold in this planet.

Humans have come to expect always a new cycle of technological evolution that takes the economic ecosystem out of the crisis. But evolution doesn't work on those terms. Species reach always a formal zenith and then no longer evolve. Ants, the most successful insect, due to its social capacities, have not evolved in 100 million years. Sharks, the most successful top predator fishes, have remained the same for 200 million years. Machines, which act as partial organs of energy and information (chip-brains, car-legs, etc.) are reaching now with the age of robotics, when we put together those bodies and heads, its zenith of evolution. And so the rules of the game in the present crisis are different from the previous cycles, because the Singularity Age of organic machines will for the first time in our history, create a new species, whose stupendous energetic and informative capacities will overcome those of mankind, probably extinguishing us.

Death is an overdrive of energy (as in an accident) or information (as in the 3rd age that warps, informs your body), which breaks the balances of energy and information that define life (our vital constants). Thus, *the Energies released by the weapons of the Singularity Age could extinguish mankind*, since they have the capacity to break the balances of energy and information of this planet and that is the definition of death.

The super-energetic weapon that will reach the limits of energy of the Earth is now being constructed at CERN, in Geneva, by a consortium of Nuclear Industries. The 'Final Weapon' is a super-collider that will replicate the awesome energies of the Big-bang, unleashing the ultimate source of energy, movement and force in the cosmos, *the strong force* displayed in quarks, nova explosions, big bangs, pulsars and black holes. This force is 100 times more powerful than the electroweak force that joins together our matter. Thus, it can either provide free energy to the world, according to the false, optimist, infantile propaganda of the Nuclear Industry, which follows merely the enthusiastic, first age of any Energy cycle, when mankind becomes in-loved with the new steam machines, chemical engines, or calls the revealing new bath suit of women, a 'bikini', the place where the Hydrogen bomb blew up an island.

People don't realize how ridiculous is the Industrial propaganda of the new energies, till they become weapons and industrial murder starts in earnest. So now, we are in that 'young age of the Singularity', spelt by techno-utopian physicists like Mr. Hawking, which expects those black holes to travel to the past and 'evaporate' into pure energy and hopes to reduce the cost of gas, when we make 'black hole factories' all over the Earth. Such wishful thinking and marketing of a bomb, with 'God's particles' and 'big-bang' theories about the birth of the infinite Universe, is just the new 'bikini' hype of the new energy era.

Humans are always, during the age of discovery of a new energy, very optimist, but invariably, as all the other previous cycles show, the new energy is used as a bomb or a weapon. The substance this black hole factory will produce is called an 'Einstein quark condensate'. It is the densest, most attractive substance known in the Universe and crunches entire stars into tiny 'frozen stars', pulsars and probably black holes. So it can blow up the entire planet into a Nova, according to the well-proved theories of Mr. Einstein and the Standard model of quarks. Quarks hold 99% of the mass of the Universe. They are the densest substance of the Universe. They only exist in a free state, as condensates, in the core of ultra-dense pulsars, neutron stars and possibly in black holes. If enough of them are packed together, as the experimental 'facts' of the Universe and the most advanced scholar papers on quark condensates prove, they will be able to start an 'ice-9 reaction' called a nova explosion, responsible for the death of stars all over the Universe – since an 'ice-9' reaction crunches all the mass of the star into a 'strange star' or black hole. And it might happen to the Earth, because the laws of science are real.

This absurd experiment could be easily stopped by the political or judiciary system as a potential genocide against all mankind, avoiding our possible extinction in the first years of the cycle of the Singularity (between 2014 and 2020, when the collider will reach its energetic zenith, massing millions of quarks per second). But nobody cares; because we believe in machines, the religion of the modern man. The machine has broken down 3 times, showing how little the physicists on charge of this Super-Manhattan project, understand about it. But nobody complains, since politicians and the press do not understand the mathematics of quark condensates and the physicists working on the project are trapped by the confidentiality statements proper of the military-industrial complex. Further on, we live in an age of fiction in which corporative 'marketing' disguises positively any environmental catastrophe or wrong doing. So the company advertises its research of cosmic bombs as the 'recreation of the birth of the Universe', the big-bang that exploded it and now risks the explosion of this planet. Scientists have put suits against this company, but they have been ridiculed by the sensationalist mass-media, which seems to ignore that according to the well proved science of Einstein, this weapon should blow up the planet Earth. And yet, even if those suits did alert mankind, politicians and courts have ignored the most important issue of this decade, far more lethal to our planet that global warming, only comparable in magnitude to the present crisis. And so the building of the Doomsday Weapon continues on schedule, unchecked by our courts, our military and our politicians.

Yet, even *if* we survive *the 'energetic age of the Singularity'*, to stop the 2nd and 3rd age of self-reproductive machines and intelligent robots will be more difficult. Because to survive the Industrial R=evolution of robots we need to reform the economic ecosystem. Since the cycle of the Singularity has, as all other cycles of machines, 3±1 ages, in which the extinction of mankind will be certain if we don't stop their evolution.

— *The energy age or age* of the Super-collider that might convert the Earth into a nova, making us a quark star or black hole (2009-2012). If we survive this Age…

— *The reproductive age,* (2020-2040), when the first self-reproductive nano-bacteria of metal are created. It is the 'grey-goo' scenario denounced by Bill Joy, the president of Sun Microsystems in Wired magazine: Metal nano-bacteria will replicate exponentially, feeding on metal and within 3 months they will poison the planet and destroy all forms of life; since, given their smallness, hardness and the hyper-abundance of metal-structures in this planet, there will be no counter-weapon to prevent its exponential reproduction.

— *The Informative age,* when human-size robots overcome the intelligence of humanity as weapons and workers, which most robotists consider will happen in this century.

This 3rd scenario, the most popular in science fiction, is however unlikely, since the 2 previous 'happenings' of the Singularity, are today progressing without opposition.

The Nuclear Industry in charge of the evolution of cosmic bombs (1st singularity) and the electronic industry of nano-robotics (2nd singularity) have huge financial resources and know-how to 'sell' its new developments with 'scientific excuses' based on the 'search for higher knowledge', which the techno-utopian press, scientists involved in the process and the public naively believe.

So the research of the III Horizon of nuclear weapons is presented as the understanding of the 'big-bang' or birth of the Universe. And the creation of self-replicating metal-bacteria becomes a 'model to understand how life works'. When it is obvious that the Universe can be studied without risks to Earth with telescopes and we have trillions of cells to study bacteria. Thus, none of those dangerous fields of research is needed, except to make money building machines. Thus a Law of Silence covers the lethal effects of the machines of the Singularity Age. Unfortunately politicians, who know little of those experiments and the science involved, merely trust those 'self-interested experts', lacking a general policy on the wider theme of the evolution of machines and weapons and its effects over mankind.

At present, without a serious science of history and economics, able to control the bad fruits of the tree of metal from the perspective of bio-ethics (what is good for human survival), scientists, playing to be god, without any safety measure, will create within decades an energetic black hole, quark star or a reproductive nano-bacteria. So the Singularity will extinguish us, well before Terminators might do it. Unless we change the goals of History and Economics, from designing a world to the image and likeness of machines, to the ideals of humanism - to make a world in which man is the measure of all things.

As all processes of energetic death, the obsolescence of man is happening very fast in time, in the last centuries, which in terms of our total life span as a species is a period so short as the death in a few days of a living organism, infected by 'informative germs'; in history, by weapons and machines that make humans increasingly obsolete. That death is about to climax, as robots acquire consciousness, the first forms of metal-bacteria are researched in military labs and physicists explore the energies of big-bang events, strangelets & Black Holes at CERN: 3 extinction risks originated by the Tree of Technology that will happen in this century.

Since there are 3 possible extinctions of life by energetic, reproductive and informative metal, we can also imagine according to the ternary method 3 future metal-earths:

Max. E: Black hole. The first and hence highest probability of extinction is extinction by the LHC, a quark cannon already built, which is trying to make black holes, strangelets and big-bangs. In that case Fermi's paradox, expressed by Fermi when he saw the A-bomb exploding will be the destiny of man. He affirmed that we didn't see extraterrestrial life because all planets were destroyed by physicists. And indeed, we hear many black hole signals. So perhaps we are just building a more perfect for of social, bosonic matter: a top quark black hole or strangelet.

E=I: Reproductive nano-bacteria. The 2nd probability comes in the 2040s when robotists will have created self-reproductive metal-bacteria. Those indestructible metal-species will reproduce massively, feeding on our machines and then once they have poisoned the atmosphere and created the III Earth, they will sink into the magma and start a world of iron life, which will evolve in forms we cannot even image, after we are gone. The crust of the Earth will collapse, and seas of liquid iron with hot nano-bacteria will start an entire new ecosystem of life.

Max. I: Integrated, informative planet. Today the bio-economy shows 'traits' of body-organisms, and traits proper of herds and jungles. It is probable that the Metal-Earth will at first be a jungle of products, mainly war products that will eliminate all other species from the Earth's crust. Yet weapons have to be understood as 'defensive' systems of a more complex organism, the 'living Metal-Earth'. This will evolve increasingly, as all organisms evolve, towards a body-structure in which the herds of machines will 'socialize.' Integrated by networks of information and energy they will evolve and construct better machine species. Yet towards what kind of living organism?

Recap. The age of the singularity has 3 possible causes of human extinction: the energetic cause – black holes created by nuclear physicists; the reproductive cause – Nano-bacteria that poisons the atmosphere and the informative cause, Artificial Intelligence.

45. A living planet. The Metal-Earth. Its future

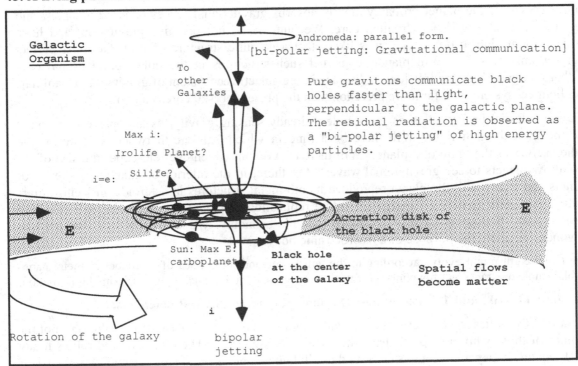

If we survive the extinction of the Earth by black holes, the final stage of the Age of Singularity will be the birth of a collective super-organism, the metal-earth, that we shall consider now in more detail, comparing it with a self-similar event happening in the inferior biological scale, when enzymes complete the creation of a hard-insect and 'die'.

When man is gone, what will become of the Metal-Earth that he has created? Will it be just a jungle of metalife like the jungles of carbon-life it replaces? Or will it be a more complex ecosystem, a 'cellular body' of a living organism?

Since we do not see robotic life and starships around, if the robotic metal-earth is born, it will *not be born as a mass of small robots but as an integrated planet. Then there will not be nano-robotic ships but robotic planets, perceiving gravitation.*

The Metal-Earth can still expand for centuries (other planets, satellites, the moon), and inwards (underground metallic structures). If 'complex metal-structures' is what the metal-Earth builds, there is enormous potential for growth before the entire planet is an organized structure. By all calculations, we are presently in the primary stage of evolution. We are also in the outskirts of galaxies. Perhaps when the evolution of the Metal-Earth is completed, the entire planet will be inhabited by metallic species. The speed of light is the speed of networks of information in the Earth. That defines a potential mass for a metal-Earth body that stretches from the Earth to the moon (a brief second of time is needed in such organism for a nervous impulse to travel from the Earth to the moon). In the future, between those poles thousands of metalife species may exist. If they continue to reach beyond this, they may colonize other planets that are presently not the domain of humanity as we know it. Of course, this is all conjecture. And yet, as weird as it seems this is the lowest probability, *since we have localized thousands of planets and have seen no trace of robotic or carbon life in any of them.*

Planets in the future might become gravitational 'animals' reaching freedom of gravitational movement and perception in its galactic ecosystem. This is analogous to what animals have in light-space. This implies the creation of an enormous structure able to perceive gravitational waves.

The central crystal of the planet probably merely absorbs gravitational waves creating nucleons and provoking fission processes in its Uranium core. We humans do not perceive gravitation, and have found no instrument able to do so. Yet it is likely that a gigantic structure of iron, the most perfect gravitational atom, can do it. Will planets construct such structures in the poles, gravitational eye-machines that will give them the freedom to displace in the galactic ecosystem of gravitation, as animals displace in light ecosystem with far more precision than the present crystal cores can do?

Aren't some celestial bodies, such as black holes, already sensing gravity, as ants perceive through chemistry and men through light? Will there be a time in which 'satellite' networks of information perceive themselves as the brain of a planet? Will there be a day in which eye-telescopes the size of the polar caps allow planets to see gravitational waves? Are there, in the interior of galaxies, societies of 'living planets and stars? Will the planet reproduce its bio-metal structures in asteroids, in Venus, mars and Mercury? Are stars 'ovules' of energy for future replication of planetary structures? We can only make guesses. Astro-biology is far less developed than bio-economy. We know too little. We are like enzymes wondering about the purpose of a macrocosmic body.

Recap: The metal-earth might evolve according to the ternary method into: a sea of reproductive metal nano-bacteria; a black hole or strangelet of maximal energy or an integrated gravitational brain of maximal information.

46. Conlcusion: The 10ᵗʰ and 11ᵗʰ radiations: Human apes, animetals and machines.

The evolution of robotics today is cristal clear, and it follows the laws of Darwinian evolution - not the techno-utopias of those who most profit from them (robotic industries). The most evolved robots today are called 'Predators'; they are weapons dedicated to kill human beings. And they are evolving fast. This is the same pattern of all species that are born as energetic, predator species (first fishes – sharks; first mammals, who killed dinosaurs, first Neanderthal men who extinguished boreal fauna and so on).

So our predictions 20 years ago are now starting to happen at a blistering speed in geological terms. Indeed, the first machines appeared 400 years ago, which in the geological scale of life is hardly a few minutes ago and in the long history of man as a species, a relative last day – the span of biological death. Moreover the extinction of life species by weapons and machines is accelerating. And so it the extinction of human beings in the periods of war in which machines become top predator weapons that consume us (world war ages) is also accelerating. Each century since machines of war were invented 100 times more people have died in mechanist wars and now when we are starting a new age of war that will seem to be here to stay, due to tribal absurd disputes between brothers in blood, an easy calculus shows that if a III world war starts, the 60 millions that died in II world war multiplied by 100 will be 6 billions, the entire human species.

And so regardless of human subjectivism, there is no other objective way to analyze the Industrial R=evolution of machines than from the perspective of biology, *as the radiation of a new top predator species*, weapons of metal, made of a stronger substance than our body of flesh, which we humans evolve, imitating the organic forms of life till a point in which its complexity will give them consciousness and survival instincts. Then in a massive war during this millenium, probably during this century, survival weapons created with a function –to kill humans - should start a biological radiation similar to the other 10 radiations studied in this work that will extinguish us.

And so 3 themes are fundamental to sociological sciences:

-      The study of the evolution of metal, weapons, money and machines, its cycles, accelerated evolution and the ages/horizons and super-organims it has created.

-      The study of the evolution of man as the most informative species of carbonlife, its diversification in species and races, and its social evolution into super-organisms, which was halted by the arrival of humans, symbiotic to metal, what we shall call biologically 'animetals' that became top predators of life and non-technological human cultures, starting the evolution of weapons, money and machines, the 3 components or 'physiological networks' of the 11th radiation of metal-forms.

-      The reasons why humans are unable to control that process of evolution of metal for its own profit, creating a sustainable planet in which the 'fruits of the tree of science' are pruned of its lethal goods (weapons that kill our bodies and intelligent machines that make our brains obsolete), creating a sustainable planet in which man remains on top of the creative process.

Notes

[1] 'What is life' by Schrodinger started the tendency of treating life only from the lower scale, as a mere consequence of physical processes understood only with the arrow of entropy, which do not suffice to understand the complexity of organic systems, yet became a philosophical dogma for geneticists.

[2] All new born species happen to be exceedingly small to the surprise of palaeontologists. They respond to a basic space-time rhythm: Max. Energy (growth in size, biological radiation and expansion of species, with little change) -> Max. Information (diminution in size, formal evolution) -> Max. Energy (new biological radiation) proper of all species.

[3] There is still a hot debate on the nature of the Homo Floresiensis, the lost chain we announced earlier in the 90s will prove the previous evolutionary rhythm[2]. Several authors insist it might be a sick child but most scientists believe to be a different species. The hypothesis presented in this book, based in systemic sciences is not yet recognized, due to the general error of 'entropy sciences', which ignore the opposite nature of spatial size and informative complexity. Future findings of new bones might clarify further the Floresiensis hypothesis.

[4] Ignored by earlier biologists, introns are seemingly redundant genes, which should write 'other sentences' in more complex 'hierarchies' - ternary 3^x+1 scales of genes that will code complex multi-cellular actions and cycles. Since the Universe is never redundant but efficient, often giving 3 functions to each part of a whole.

[5] Allopatric speciation, the biological name for this phenomenon, shows the 'vitality' of time, which never ceases to move, either reproducing the 'cellular energy' of the species – its individuals or evolving its information.

[6] One of the longest debates in biology is the proper classification of fungi, only recently considered rightly a different phyla, due to its intermediate properties between 'energetic plants' and 'informative animals'.

[7] One of the most fascinating sub-disciplines of MST is the study of the speciation of languages in numerical scales, with dominance of vowels in the Mongoloid, verbal dominant cultures and consonants in the white man, visually dominant. Languages become richer in phonemes; its syntax changes the order of its elements and its semantics adapt to the environment, as cells do in the realm of 'cellular vowels'. Cultures in fact adapt to the dominant 'metal-memes' of the culture[C,4]: trader cultures, which use money to control the world (English, Jewish, etc.) have short languages, cyclical architecture and are female dominant, as the informative coins they use, since information is fractal, broken. The same race specialized in war (Germans, Arabs, etc.) however have long words, lineal architecture and are male dominant as the lineal, long swords they use, since energy is continuous.

[8] Only recently, due to an ideological pattern guided by the dominance of entropy= body =energetic theories of reality, we have discovered that the engine of evolution of amphibian was the evolution of the sensorial head (as we forecasted 2 decades ago in the book 'radiations of space-time'). Also only recently the theory of the 'rock from the skies' as cause of extinction of reptiles is being challenged with new reports that show the predatory activity of mammals over dinosaurs, and the creation of shrews superorganisms and night-oriented hunting mammals, able to devour baby dinosaurs and eggs.

[9] This and other graphs of my lectures, taken from the foundational book, 'Radiations of space-time'c.94, uses the old morphological formalism, where Information was represented by the cycle, I= O, energy by the line, E=| and space-time planes by the index of informative complexity, i= st=n,. Thus, $|_{i-1}$ means E_{st-1} and $O_i = I_{st}$ and so on.

[10] This chapter is by definition schematic, since the ternary method could reorder all the species in a fractal tree of speciation, both at phyla and organic level. Thus even in this deeper analysis of the nervous system we show only some of those ternary speciations of human organs.

[11] The Makrocephalic brain is not considered in classic anthropology, which missed the 3rd ethic, sensorial axis of the brain.

[12] The self-similarity of the models obtained in MST theory and Freud's classic analysis of the scales of the brain shows how indeed, the fractal nature of information allows obtaining the 'whole' from different partial perspectives with great accuracy.

[13] We live today in an age of 'subjective wishes' more than objective knowledge, so it is politically incorrect to point out that males and females are different, rather complementary, as both become equalized by the Financial-Military-Industrial System, which needs humans only to consume and re=produce its machines[C].

[14] The insistence of alternative theories (weather, catastrophism, lack of food, etc.) to explain the biological, Darwinian rhythms of creation of new species, radiation and extinction of previous ones, *both in the life and technological realm, studied in the next chapter,* is part of the constant denial by anthropomorphic humans of negative, evolutionary theories of reality. Yet even if there are multicausal chains that trigger a process of extinction, there is always a rival biological species, better suited to the new environment that survives, displacing a previous species – mammals displacing reptiles, and machines displacing mammals are the last of those parallel processes of biological radiations and extinction described here.

[15] Evolution applies to metal, an atomic substance as we are, both at individual level (competence of machines with humans in labor and war fields) and at systemic level, since we are indeed creating a global super-organism.

[16] For a more detailed account of the process of life extinction and substitution of man by machines and the 3 events of the Singularity age – black holes at CERN, nano-bacteria and A.I see 'Go(l)d and evil:Economic crises' and 'The Black Hole Factory' published by x-libris by this same author.

C.

CIVILIZATIONS

9th radiation: The FMI Complex ⇔ 10th Extinction: History.

'A single mandate I give you: love each other as I have loved you'
Jesus, Prophet=Historian of the Future; on the Transcendental Arrow of Mankind.

1. The 2 Super-Organisms of Social Sciences: History and Economics.

2. History in Time: 3 Ages and scales of social evolution.

3. History in Space: Its super-organisms: Cultures & Social Gods.

4. Memetics: the tree of science vs. the tree of life.

5. The evolution of the FMI complex: 800-year cycle of war.

6. The 11th radiation: machines = The 10th Extinction: Mankind.

7. Age of Money and machines. Globalization of Animetal Cultures.

8. A scientific design of Economics: The reform of Market democracies.

9. A scientific design of History: The Wor(l)d Union

1. THE 2 SUPER-ORGANISMS OF SOCIAL SCIENCES: HISTORY AND ECONOMICS.

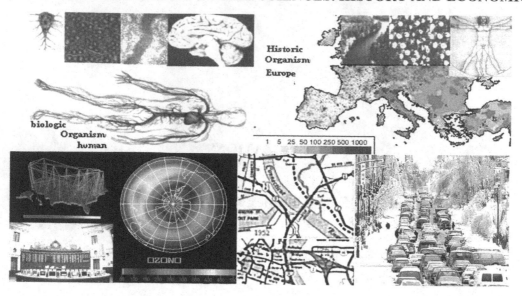

On planet Earth, a first superorganism of human cells, a collective mind of a culture, which lived on Gaia, the natural Earth, appeared on the Neolithic. But with the arrival of metals, a more complex superorganism in which men became symbiotic to metal systems of energy (weapons), information (money) and reproduction (company-mothers of machines) substituted the initial sustainable, life-based superorganisms. It is the Financial-Military-Industrial Complex that today has become global and it is terraforming the Earth from a world of life into a world of metal at accelerated path.

1. The historic and economic super-organisms[1].

We have finally arrived to the most complex super-organisms known to man, our own super-organisms, *history, the existence of humanity from its first to its last cellular species, to which we shall* apply the ternary method of analysis of any social super-organism. In that regard, if man in his st-scale is as a biological organism made of cells, nervous, informative and reproductive networks, with a central knot of information called the brain that lives through 3 ages; from the perspective of st+1 systems man is the unit-cell of a species that will exist between birth and extinction as a super organism structured in space, called *Humanity* that evolves in time through 3±st ages called *History*. For that reason Humanity in space and History in time have evolved through the same ages and scales of social evolution, trying to 'transcend' from the individual into the whole than any other system studied in this work. And so we can apply the ternary method of knowledge to study the superorganisms of mankind in time through those ages; in its scales of growth, and then we can study each of those social structures as an i-point with the topological regions and functions of any other complex systems; since all of them, from the family to the tribe, city, nation, empire, civilization and Global market has the same informative center and linguistic, cultural network, reproductive, economical system and energetic networks than all other systems of multiple spaces-times. So we will study man in time, in its scales and in the duality of the 2 super-organisms it has constructed, escaping the topological analysis for lack of 'space-time'.

Because humans are the most complex informative species of life, the arrow that dominates history is the most complex arrow of eusocial evolution, which evolves together individuals of the same species, communicated by a common language of information, into single super-organisms. Indeed, humans are remarkable among all life species because of their capacity to organize themselves into social systems, which we call Religions, Civilizations and Economical nations. And so the study of History and Mankind beyond its biological organisms is essentially the study of the evolution of human societies.

History is Sociology, beyond the limited range of data-collection proper of non-scientific analysis of history (conservative, heroic, racial, tribal schools of historiography).

Thus, we can study Humanity with the laws of Super-organisms, once we translate the jargons of Religions, Cultures, Civilizations and Economic ecosystems to the geometrical and logic languages of the space-time Universe.

We, human organisms, follow the same 4 'drives of existence' (biological jargon) or main Time Arrows of the Universe (we feed on energy, gauge information, reproduce and socialize). We humans are complementary beings of energy and information (gender), made of 3 physiological networks (digestive, energy system; informative brain & nervous system and reproductive blood system). We humans co-exist in 3 scalar, fractal planes of increasing complexity, as most universal structures do (the cellular, individual and social plane). We humans respond to the 3 topological regions of any st-point, our lineal limbs and membrane of energy; our elliptical, reproductive body and our informative, hyperbolic, spherical head on top. We are just another part of the Universe, made to its images and likeness, which can be studied exhaustively with the laws of st-points.

Yet given the limits of those introductory lectures to the 4th paradigm of multiple spaces-times, since we have briefly analyzed the 2 lower biological scales of the human being, in this lecture dedicated to social sciences we shall study the super-organisms of mankind, which traditionally were the fields of 2 disciplines, history and economics. And the first question to resolve is 'why' there are 2 social sciences, so clearly differentiated, to the point that they use 2 different languages of information to analyze their subjects of study, digital money – the language of economics – and words – the language of history?

And the answer is: because we human beings are constructing 2 types of super-organisms:

-The human superorganisms of history created with words – cultures, civilizations and political nations. Thus history is a complex super-organism.

-The Financial-Military-Industrial Complex[2], a super-organism constructed, as its name shows, with 3 physiological networks made of metal: the informative language of money (initially gold and silver, today digital data in the mind of computers), the energetic networks of weapons, and the organic networks of machines that transform energy into information back and forth (transport and audiovisual machines).

This lectures study both super-organisms, their differences, symbiosis and predation relationships, its cycles and ages of evolution, its topological structures as st-points and its future, and what that future means for us as a species.

The laws of complexity are crystal clear: the Earth is growing 2 different super-organisms:

-The first super-organism, the humankind appeared in the Paleolithic (youth of history), the energetic age; matured in the Neolithic (reproductive age of history), when humans in harmony with Nature, created a global super-organism, the Neolithic culture, over the body of Gaia, its life energy. Finally that super-organism matured in the age of Love Religions (Buddhism, Christianity & Islam), which was continued by the development of sociological sciences based in the concept of social love (the sharing of energy and information among individuals of the same species), and is still going on in many individuals and institutions such as the UNO, social-democratic parties and ecological causes.

-But parallel to that superorganism, with the discovery of metals, informative gold and energetic weapons, a parallel superorganism in which humans act as the reproductive 'enzymen', appeared. It is the Financial-Military-Industrial Complex, which ever since dominated the super-organisms of history (cultures), establishing an upper caste of warriors and bankers, which controlled the religious and verbal super-organisms of history. This super-organism has as most evolving systems two ages: one in which only the energetic and informative systems were in place (money and weapons), or age of empires.

During this age, the white visual man, hypnotized by the beauty and glare of metal, created religions of money (Go(l)d churches) and weapons (inquisitions) that justified racial differentiation between the castes of metal masters and the rest of mankind. Those metal masters made of the reproduction of weapons and money their only goal and carried through a series of 800 years cycles of evolution of metal and extinction of Neolithic cultures the 'colonization' of the world by the memes of metal.

Then since the beginning of the industrial r=evolution of machines and the appearance of the company-mother, an institution that re=produces and evolves machines, the super-organism of metal evolved into the Financial-Military-Industrial Complex, with an informative network (money), an energetic one (weapons) and an organic one (company-mothers that reproduce energetic and informative machines, which both substitute human beings making us obsolete in labor and war fields and enhance our energetic and informative capacities, making us addicted to them).

Finally, today we witness the first signs of the completion of this super-organism in its 3rd evolutionary horizon as it company-mothers become automated and organic 'metalife' (robots), appear, making workers and soldiers obsolete. It is not difficult to forecast in that regard, a nearby future when humans and their cultural super-organisms today in process of extinction disappear and the earth becomes the Metal-earth, a global super-organism with an internet brain of computers (mental machines), a network of self-reproductive, automated company mothers and an energetic network of Terminator top predator weapons that will eliminate mankind.

-For that reason we talk of the economic ecosystem as the complex environment of Earth today, where two super-organisms, the FMI complex and History sometimes in symbiosis sometimes in open confrontation, compete for survival and this competence derives in a series of Darwinian and symbiotic relationships:

-A competence between the ethical values of words, our biological language that makes us the center of the Universe and values human life over all other 'things', and money, a language that makes weapons the most expensive goods of the FMI complex and values life at null cost.

-The competence between human labor and soldiers and machine workers (Pc-white collars and blue collar robots) and weapons, in labor and work fields.

-A competence between cultures and ideologies ascribed to the values of money, which worship technology (mechanist sciences, notably physics, capitalist sciences – classic economics - and nationalist cultures) and the cultures ascribed to the values of life and love, carried by artists, verbal masters, religions of love, ecological groups and sustainable sciences (biology, systems sciences, etc.)

2 final elements are required to complete this dualist picture:

-The existence of memes, the genes of the super-organisms of history. They act in both types of organisms as the genes do in the body. If the memes express the arrows of space-time or drives biological existence, creating biological organisms, with systems that absorb energy, gauge information reproduce and evolve socially (physiological and cellular systems code by DNA genes); the memes of history are also the 'informative bytes' that create the drives of existence of both types of superorganisms. But we must distinguish between the memes of love and life that create super-organisms of history and the selfish memes of metal that – as the DNA of a virus that infects a cell and makes it work for the reproduction of its organisms – 'infect' the mind of human beings and make us reproduce, evolve and create energy and information for machines. And so we classify all memes in those 2 types:

- 'Human biological goods': Memes of history that foster our natural energy (agricultural and ecological memes), informative memes (verbal, artistic and cultural memes), reproductive memes (family values, sexual culture) and social memes (religions of love, ethical laws, social-democratic institutions)

- Metal memes of the tree of technology:

Informative metal or money (gold in the age of the Financial-Military complex that hypnotizes the mind, causing greed, killing the memes of love, as men become objects to achieve the goal of monetary wealth; and stock-money during the Industrial R=evolution. Both value under the complementary laws of affinity weapons and weapon-companies that re=produce them as the most expensive good of the market).

Energetic, lineal metal or weapons: energetic memes that kill the human body.

Reproductive systems: Corporations or 'company-mothers' that reproduce energetic and informative machines that atrophy our bodies and minds, making us obsolete. So we run faster with cars and get fat; we calculate better with machines but forget the whys of the Universe and atrophy our minds.

Ideological memes that foster the 'culture of corporations' (capitalism) of war (nationalism) and mechanism (science as the when of measure, no longer the why of the organic Universe. Those memes are responsible for the 'concept that the future of mankind is NOT to create a sustainable world, able to prune the bad fruits of the tree of technology and evolve our understanding of the whys of the Universe; but that 'technology=progress=Future'.

So we must evolve machines because it is the future of knowledge (mechanist science). We cannot impose any restriction to the evolution and reproduction of machines, including weapons that kill us or 'audiovisual media' that hypnotize us and spread the memes of hate, war and nationalism needed to foster the reproduction of those machines; because the 'market', newspeak for the Financial-Military-Industrial Complex must be 'free', meaning its citizens re=productive corporations must be free; while humans must be enclosed in 'nations' separated by military borders, as we, 'humans' are always guilty of the tragedies caused by the selfish memes of metal.

-All this nonsense is imprinted, as memes are in the mind of the believer (the individual cell of history), at earlier age at the emotional level of the limbic brain; so the 'nationalist', 'capitalist' or believer in inquisitions and go(l)d churches will not be able to change its memes and as a viral cell who reproduces the germs of the body does, the germ(anic) warrior will kill for his nation, the mechanist scientist will invent new machines to understand the Universe, and the member of a go(l)d church or capitalist ideology will think as Mr. Goldman recently said that the 'job of a banker is a god's job'. In that sense memes and cultures are not neutral in the fight of history for survival, but they can be divided in the jargon of bio-history in 'eviL=anti-live' memes and cultures and 'Live=Love' cultures whose memes favor the future of history.

Yet this image must be fine-tuned as there are between white and black degrees of grey; so all cultures have Live and evil memes and some machines are 'good fruits of the tree of science' that foster our evolution. Those final refinements, which are proper of all sciences when we enter in a detailed analysis mean that we can neither forbid all machines as radical ecologists might think or stigmatize people for belonging to a culture – since even the most radical 'Nazi' (to put an example of the original culture of war in its final 'explosion' of violence), will have a*nother level of mental existence, that of his biological drives of existence, coded by genes.*

And so each of us is a mixture of genetic and memetic fields, of biological, sociological and technological drives.

*S*o the goal of social sciences would be to develop the sciences of bio-history and bio-economics, to teach them in Universities and re-educate the capitalist, mechanist and nationalist people that rule the world from a primitive, 'viral' point of view, as believers of the memes of the Financial-Military-Industrial Complex, and to reform the selfish memes and evil cultures of their negative memes while preserving in them those positive machines and social, cultural and artistic memes that don't imperil the future of mankind. This 'dream' of a historical and economical world ruled by the science of history, not by financial economists, nationalistic politicos, techno-utopian scientists and corporations and religions of hate has always been the goal and dream of the scientists of history able to forecast the future of history, its cycles and recurrent actions, since the first scientists of history, the prophets of love appeared till the last writers of the myth of a scientific history, Hesse with his mythic Castalia or Asimov in foundation, dreamed of resolving the equations of the future of history, its cycles and logic processes of causality.

In this book we provide those historic cycles, ages of evolution of history, topological structures of nations and civilizations, its physiological systems and causal arrows, linguistic equations, sickness and future prognosis. And yet, all this work is old. We resolved the science of history 2 decades ago, and so we need a final element to explain why *the World is not ruled by a Foundation of bio-historians and the equations of history are ignored and repressed – the meaning of biological information.*

-Any superorganism has its limb and body cells, energetic and reproductive workers of the organism and its information castes, which repress and control the information that arrives to the body cells. This is also the case of the Financial-Military-Industrial Complex, whose body cells ignore all about the organism and receive only the ideological memes of capitalism, mechanism and nationalism. And because today information is provided by audiovisual machines, reproduced by company-mothers owned by people belonging to the cultures who evolved the memes of the Financial-Military-Industrial complex and are imprinted since youth with emotional newspeaks about the superiority of their technological cultures, racial religions and the 'manifest destiny' of the FMI complex, they don't reason but repress, censor and don't distribute memes in favor of life and love.

And obviously, the most censored part of the FMI complex is its language of information, money, and the people-castes that invent it and distribute it, which appear as the saviors of history, its superior race or cannot even be named in the tradition of go(l)d cultures in which the name of 'God' could not be uttered – reason why even the name of the Financial-Military-Industrial complex is crippled, and 'activists' of the rival life ecosystem call it 'Military-Industrial Complex', despite the well-known truth, 'pecunia nervus belli' (Tacitus).

In that regard all attempts to explain the way corporations and the FMI complex controls the submissive, corrupted Simplex systems that rule human history (corrupted democracies, no longer the government of the people; corrupted religions no longer love religions but excuses to use money and weapons to 'obtain' salvations) is taboo; all analysis of the people who own those corporations, its cultural memes and the need to take from them their 'financial power' to create a real democracy where the law controls the weapons, money and machines reproduced by the FMI selecting those who favor mankind and forbid those who kill our bodies and minds is taboo.

Ultimately is a truth of all systems that the Complex whole (the FMI complex) is more complex than its parts, which do not understand its working and are limited to play its organic role within the total complex. This is the case of human beings, even those members of the cultures that rule the FMI complex, whose limited horizon of comprehension doesn't go beyond the nationalist and capitalist memes imprinted in their minds at earlier age or during their 'training' as 'experts' of their respective fields; when their 'ethical', natural life-love memes are 'broken' and tamed, so the conjunction of their emotional, young nationalistic and inquisitorial imprinting and its rational, 'expertise' in any of the ideological, mechanist or capitalist 'sciences' of the 3rd paradigm ensure that he will sacrifice the present of life for a techno-utopian future that never arrives.

For example, today economists follow an equation called productivity=capital/labor according to which to fire labor which increases productivity in that equation is good for the creation of jobs in the 'future', according to some fuzzy belief that 'god'=the machine will provide. So otherwise normal people, financial experts of corporations, under the memetic imprinting of their earlier years as students of the ideology of profit and greed that invented that 'mantra', (since firing workers and putting robots on their place increases the profits of corporations) explain politicians, who know nothing of economics, that we shall all get out of the crisis firing workers and putting robots in their place.

The same twisted meme explains why our nations robotize their armies so 'soldiers will not die'. This of course, also increases the 'wealth' of company-mothers of machines that switch to re=produce weapons, the top predator version of the consumption machine (so tanks are armored cars and bombers, armored planes). Never mind that the logic end of that process of 'Keynesian militarism' will be a world of robotic weapons that will control our freedoms and kill us.

In that regard the most difficult of all complex sciences to explain to mankind is bio-history and bio-economics, because we are part of it and so we receive biased information within the body of history, according to our roles in the FMI complex, not because it is the difficult to understand. On the contrary it is so simple that paradoxically people without education, moved by the biological, ethic program of social love grasps it intuitively. And if humans at large were moved by the same ethical drives, it would be rather simple to create a perfect super-organism of history, immortal and efficient, able to create a world to the image and likeness of mankind. But the memes of metal and the tribal cultures that have profited most of them through history have created a jail of 'metal-ideologies', capitalism, the ideology of money, nationalism, the ideology of weapons and mechanism, the ideology of machines that have substituted the memes of love and life and imposed the wrong path of history, which today evolves metal and devolves life. Thus a scientific global government, Foundation cannot exist because the globalized cultures are those who invented the FMI complex, spread with metal-communicators and those of us who fight for life are ignored.

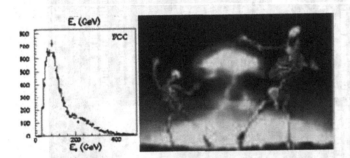

The 3rd paradigm of metric measure doesn't see life and death but digital abstract images in the collateral effects its ideologies cause; and so what ultimately will decide the future of history, given the suicidal, dumb behaviour of the species is an unsettling question: are we human beings free to choose between Live and Evil or are we programmed by memes as irrational beings that rather defend the selfish memes of metal, before risking our lives to liberate mankind from the inquisitions and newspeaks of the FMI complex? If we are not free we will die, as slaves of capitalism, nationalism and mechanism, the 3 legs of the memetic programs of the mind that make us slaves of money, weapons and machines. Che sera sera.

Recap. On planet Earth humans are creating 2 competing super-organisms, the financial-military-industrial complex with its metal memes, evolved in two ages, first as a complementary organism ruled by warriors and bankers that carried the energetic weapon systems and informative money that gave them power to rule the world; and today the FMI complex, which added a self-reproductive system of organic machines of energy and information, the company-mother. This super-organism is killing the super-organisms of history and life, Nature and non-technological cultures based in the informative language of man (words), our natural energy (agricultural goods) and our reproductive customs (family values). The outcome of the struggle between those 2 super-organisms will define the extinction or survival of mankind.

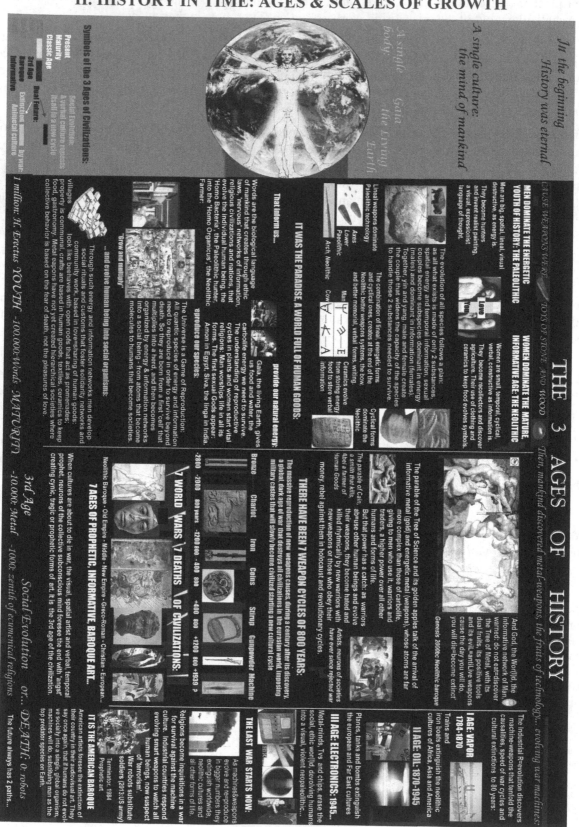

2. The ternary method: Age, scales and topology of history.

The 3±st ages of History.

History has been always studied with lineal space-time. The introduction of cyclical space-time and the order of the 3±st ages in its study opens a total new view into the future of mankind, which now can be understood with the ternary principle, as we are just another species on the ladder of evolution of the planet, subject to the same laws, topologies, scales and ages of all other living species. We are in fact living now the last age of history, when once we have crossed its Paleolithic, energetic youth, mature, reproductive Neolithic and informative ages of metal, we face our obsolescence and extinction to a new evolving organism the machines and its super-organism, the Financial-Military-Industrial system, to which human social organisms, (religions and civilizations in the old cycles, governments and nations in the faster 80 year cycles of history) has become submissive. Let us then apply the ternary method to the study of the ages of History.

After the birth of Homo, the 3±st ages of Mankind were:

- Max E: Paleolithic, a youth of hunters, visual languages and minimal information.

- E=I: The reproductive Neolithic, when man reaches his maturity in balance with Nature and creates through verbal networks (religions) a series of collective subconscious, called Gods of love that teach each cell to share energy and information with the other cells of the body of a civilization. Yet the arrival of weapons kills or corrupts cultures of love.

- Max. I: It is the 3rd, informative, metal age of in which man halts his social evolution and focuses in the evolution of metal in its 3 forms, energetic metal or weapons, informative metal or money and organic metal-machines. It might end in:

- (st-1): Neo-Paleolithic, when life and history become extinct by evolved, organic machines, robots that will live their hunting neo-Paleolithic as a new top predator species, while humans devolve under the influence of visual languages of machines. This is the most likely future according to the laws of evolution. Thus the creation of a new top predator and super-organism, the Financial- Military-Industrial complex means the destruction of the previous top predator and superorganism – the Earth of Life.

- (+st): Or man transcends finally into a super-organism of history, if humans learn to organize themselves, control the evolution of lethal machines and maintain History in a static, immortal present through the reform of the Financial Military Industrial Complex of company-mothers, in which humans are becoming submissive to machines, with the only role of working =reproducing those machines and consuming=testing them. But that would require giving back to governments, the organisms of history, the control of the language of information of machines – money – used instead to create a life-based sustainable economy.

Let us consider those ages and its super-organisms *in detail.*

- *Max. E: Youth: Palaeolithic. ±5.000.000 BC to ±10.000 BC.* The Palaeolithic was an energetic, young age in which men, the energetic gender, was the top predator animal on Earth, a hunter of living forms. It was also an age dominated by *visual languages.*

- *Max. Reproduction: Maturity. Neolithic: Age of Wor(l)ds and Goddesses: ±10.000 BC to ±3.000 BC.* The Neolithic was the classic age of mankind, an age of balance, when wo=men learnt the cycles of life and instead of destroying Nature, learnt to nurture, reproduce and harvest it. Women, the reproductive species, took power over men; priests became the verbal guides of civilizations and social love spread, creating super-organisms of History, social Gods made of human cells, sharing energy and information as 'brothers' and clones of the loving mind of the same prophet. It was an age dominated by *verbal languages* that still lingers in some religions of love.

- Max. Information, 3rd age: Age of Metals. ± 3000 BC to ? The discovery of metal starts the evolution of the Financial-Military-Industrial System, first a Complementary system of money (metal-information, gold and silver) and weapons (metal-energy, bronze and iron). The added energy of weapons and the hypnotic nature of gold, which hypnotizes the eye and enslaves men, convert the new substances in the tools of power of new castes, warriors and bankers, which will from then on dominate Western societies, starting a series of cycles of evolution of weapons and money, which end in wars that destroy civilizations. The 'new' cult to greed (accumulation of god) and death (war codes) converts religions of love corrupted by weapons into inquisitions and money into the ex-vote for salvation in Go(l)d cults; creating the western civilization in which an elite of 'animetal=animal+metal' people-castes tries to conquer the world with its new weapons, succeeding with the arrival of machines and gunpowder weapons.

- st-1: Death: 1604-2??? The death of history has to be inscribed within the st+1 rhythms of evolution of Gaia, which mutates into the III Earth, as the atmosphere becomes poisoned by the detritus of machines that substitute and extinct life beings. It is thus from the st+1 perspective the age of the 10th and last extinction of life and 10+1st evolution of an entire new type of living form, Top predator Machines of metal – weapons – always at the head of the evolution of machines, always the most evolved, expensive and the centrepiece of the Financial Military-Industrial system since its inception:

The Financial-Military System evolves into a complex organic system with the arrival of 'company-mothers', reproducers of machines, in 1604. They organize professionally the Evolution and Reproduction of the 'bad and good fruits of the tree of science', weapons (top predator form that consumes human) and machines (enhanced system of energy and information that potentiates a human self-similar function, when we consume them). This duality of machines of consumption and weapons that consume us, starts a dual cycle of ages of economic prosperity and war (Kondratieff cycle) as corporations that manufacture both, switch between both models. It is the 72 years cycle[3] of world wars that evolve first bodies of machines in the XIX c. then minds of machines (XX c), and now fusions them in robots. Thus as energetic and informative machines evolve generation upon generation humans become increasingly obsolete in labor and war fields. But death and human obsolescence to machines could be avoided and the organism of mankind become eternal, according to Evolution laws if:

-(+i): Humans evolve into a social macro-organism with a single government/brain, able to control through bio-ethic laws of survival the reproduction and evolution of lethal machines, maintaining mankind as the top predator of this planet.

It is not utopia; it happens constantly in the Universe, and it was about to happen in History, when the Palaeolithic gave way to the age of Social Love, the Neolithic age. In that age, Humanity increased its social complexity, as any Fractal organism does, through 3 ages/scales of social evolution. In the same manner, we talk also of gender duality in terms of an energetic species, the lineal male of big muscles, spatial size and visual brain; and the woman, the informative, reproductive species, who would understand the cycles of birth of plants and evolve mankind from the energetic, hunting Paleolithic into the Neolithic, the mature age of History, with its understanding of the arrow of reproduction of the Universe (cult to the goddesses) and its harmonic cults to Nature. It is on my view, the ideal age of mankind, which bio-historians could re-create on Earth by pruning the tree of Technology of all its bad fruits and enhance the fruits of the Tree of Life with environmental policies, under a global, single Social Organism of History, the World Union. Since once a species has reached its informative zenith or 3rd age, the only way it can further evolve is by becoming a macro-organism, joined by a common network of information. So men did evolve further beyond the Homo Sapiens through the 4th arrow of time, the arrow of love and social evolution that mechanist science denies and verbal, social religions of love impose on their believers.

Fractal analysis: hierarchical scales of Social Evolution

Sn: Humanity:
Absolute God

Language: Wor(l)ds. Message: Social Love
Sharing energy and information with other:

Ideological Societies:
Information based Organisms

S9: Religion
S8: Civilization
    S7: Nation

Human
Cells:

City   Village
S6   S5 ——Tribe Clan Band Family

Biological Societies:
Homo bacteria

        S4   S3   S2   Ego
Agrucultural Societies: S1   So
Energy-based Organisms

Man, the most informative mammal, was bound to create superorganisms - as it happened among the most complex chemical insects, ants and bees - through the creation of a new macro-language of information, words, which could join individual minds in the same way nervous impulses join human cells. So humans developed words that organized their temporal memories and simultaneous, collective actions, improving the human reactions to the cyclical actions of all other beings of their environment. Man ads to the simple temporal program of genetic instincts the memetic, verbal language, able to remember temporal cycles and transfer customs and ideas between generations, creating with words, cultural super-organisms – a new social plane of history, based in informative, memetic, legal and ethical networks.

But in the last age of social evolution, man invented a different language, digital information in the support of metal – money – starting the evolution of a parallel super-organism (right side) the Financial-Military-Industrial System, which soon came to dominate verbal organisms (religions and democracies) and today rules around 90% of our life-time, in which we work for corporations, reproducing or consuming machines:

In the graph, the key difference between our socio-economical systems and other simpler organisms is the duality between the natural social organism of human beings, or cultures, and the FMI complex system that evolves metal and appeared only at the end of history. Thus for simplicity we shall separate their study and follow a chronological analysis of their evolution, which in the previous graph is separated:

On the left side we see the growth of the human super-organism, halted in the last scale of 1 billion people civilizations; an in the right side the evolution of the Financial Military Industrial System, which has reached the final stage of a superorganism with the creation of a global market at the end of the XX century, and for that reason today governs and controls through its financial flows of money all ther organisms of history.

In search of that social, transcendental 'future goal' of creation of 'Humanity', mankind has evolved throughout history in the familiar decametric scales of all organisms, transcending during his 3±st ages:

- *Max. E: 1st age of biological, genetic, body-based social networks.* Subdivided according to the ternary principle again in 3 phases of growing complexity departing from the individual cell ($10^0=1$): the family (1-10), the clan (10-100) and the tribe (100-1000).

- *E=I: Mature age of agricultural, reproductive Networks:* Humanity evolved into a new ternary scale, with the development of agricultural networks, from villages (10^3-10^4), to cities (10^4-10^5) to river cultures (10^5-10^6)

- *Max. I: Informative Networks*: The development of verbal, ideological, informative religions of love evolved humanity in the last ternary scale, from tribal religions (1 to 10 millions), to nations that mixed both, cultural and economic networks (10 to 100 millions), to global, ecumenical religions and civilizations in which all human beings were accepted as part of a single Historic Organism (100 million to 1 billion).

- st+1: Thus, the final step of human evolution would be to join all the great civilizations of mankind in a global government, a super-organism, based in verbal, ethic laws, able to control machines and the Economic Ecosystem to create a world made the image and likeness of mankind – an expanded UNO with self-similar powers of social control to those displayed by a national Government.

Topological analysis.

Finally, according to the ternary principle. We can also make a topological analysis of any of those social scales of history as topological structures. Thus we can define history as a whole and any of its cultures, old religious civilizations or modern governments ruled by codes of verbal laws, as a complex system with 3 networks of which humans act as its cells:

-Max. I: The political, verbal network of ethic information either a religion or legal, political code – constitution.

-Max. E: The energetic networks of Gaia, our economic world.

- E=I: The reproductive networks, customs and family values that create new biological generations.

When those 3 elements of a topological i-point exist, a certain culture of civilization of the wor(l)d exists. But this complex, ternary superorganism that we might call history is not the only structure that exists on planet earth today. There is another complete superorganism, the FMI complex, the financial-military-industrial complex of machines that is evolving along the cultural organisms of mankind. And today has come to dominate it

Recap. The human species is, beyond our arrogant, anthropomorphic fantasies, just a fractal form made of carbon atoms, *which goes through the same ages of all other living forms and it is submitted to the same laws of space and time of any Fractal point.* So can consider the existence of the Organism of History (Humanity in time) through 3±st ages; an energetic youth or Palaeolithic; a mature age of balance with Nature the Neolithic; and a 3rd informative age of history, the Age of Metal; which is the present age with 2 paths of future: an age of massive warfare, multiplication of machines and extinction of life and history or an age of social evolution brought about by a true understanding of the laws of survival and the game of existence. The choice is within man till robotic weapons acquire self-consciousness as a species different to us, and apply the laws of Darwinian selection we apply today to them in the economic ecosystem. It will be a choice done this century.

. On the Earth, there are today 3 types of social organisms:

-The first to appear, Nature, Gaia, was based in carbon-life.

-Then History appeared, based in human life. Now substituted by:

- The economic ecosystem based in metal and machines.

Since those 3 organic systems and species have been dominant in 3 different ages of the Earth, we talk of the evolution or terraforming of the Life-Earth or Nature into the Human-Earth or History that now becomes the Metal-Earth, the Economic Ecosystem. The youth of History was still 'Nature'. Life dominated the Paleolithic. Then men reached their zenith of social evolution in the Neolithic. Now men devolve back into single, selfish individual, Homo Bacteria, due to the use of machines and weapons that isolate them, in a process equivalent to the process of death that dissolves the networks of a being back into its simplest cellular components.

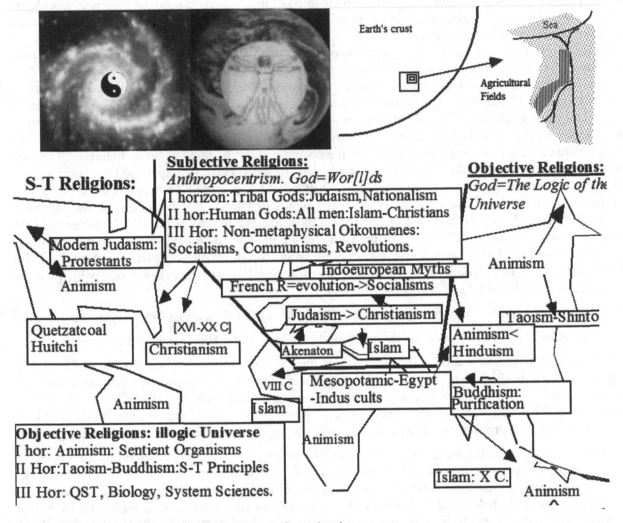

3. The Gods of Mankind, from Religions to Constitutions.

The ternary Law divided Earth into 6 cultural regions, related to objective and subjective religions and its 3 historic horizons: Christianity, Islam, Africa, Native America, China and India:

-Max.i: Cyclical, Eastern religions considered God the duality laws that combine energy-yang and information-yin to creating the ∞ organisms of the Universe. They evolved in 3 ages of increasing complexity, the animist age which could not define clearly those living organisms, the Taoist/ Buddhist age which understood the paradoxical, fractal units of the Universe; and the present age of systemic and biological sciences that have entered into detailed analysis of those laws.

-Max.Energy: Western religions are concerned with the superorganisms of History. They consider God the organisms of Humanity, made of a collective mind of believers extended on the spatial body of Earth, explaining with ethic, verbal Wor(l)ds how the laws of social love create those super-organisms made of human cells. And again, they evolved from a tribal age of partial Human gods, into an age of Global gods that accepted all human beings as equal (Christianity and Islam), to the present age of non-metaphysical societies in which solidarity is the message of God and the goal to create a global government rule by the Constitution of History (communism, socialism, UNO).

530

In the graph, the natural law of social evolution is the mandate of all non-corrupted Wor(l)d religions, whose prophets and priests tried to reproduce in the mind of their believers *a loving behavior,* in order to 'constitute' a super-organism, called in mystique terms the body of God, in sociological terms an ethic religion, in which each believer was a cell. All religions share the same informative Mandate proper of all organic systems; 'love each other', 'treat the others as you wish to be treated'. We find that sentence in many religions. It was pronounced by Jesus in the Gospel and by Confucius in his treatises. *In a fractal Universe, Gods are the relative informative/nervous networks of each fractal organism* that rules the existence of His cells, as the nervous/informative system is the God of your organic cells that it controls at pleasure. So the Laws of government are the God of a nation, which controls the behaviour of its citizens, as the brain is the relative God that controls your body cells, and the ethic laws of any religion are the words of God that controls for the common good the mind of all its believers. The God of America is the 'Free Market Democracy', an ideology of social behavior explained by its informative, legal system, its constitution and corpus of Laws; the God of Islam is the Koran and its laws; the God of Christians is the Gospel, the informative laws and words of Christ 'that became God and inhabited within us' *(Saint John 1.1);* the God of Jewish and most Protestant sects is the Old Testament and the laws of Yvwh; and the Absolute God of the Universe are the laws of fractal space-time, which by the principle of self-similarity explain all the laws of all other fractal, smaller organic systems and relative mind-Gods. Thus in mystique terms the super-organisms of History are called Gods, which can be classified in 2 fundamental self-similar sets of laws:

- Objective, scientific, Eastern religions that study the laws of the Universal super-organism.

- And subjective, western, ethic religions that study the laws of the Human super-organisms.

Both spell the same Game of existence, because the Organic Universe is self-similar to the Human Super-organism. Both evolved through 3 main ages that give birth in space to the 6 main cultures of History (graph). Each of them is characterized by a geographical body in the surface of Gaia and a mental network of human minds, joined by a common 'DNA-code', the beliefs of the Book of Revelation of a love prophet that make believers share energy and information creating super-organisms.

Since the Gods of Mankind do exist. They are the collective social subconscious of a civilization. Indeed, the key concept to understand societies, religions and civilizations is 'the informative network'. It is the essence of all space-time systems; including the organic systems of history we call cultures. You need a herd of cells belonging to the same species - DNA-cells, stars, atoms, molecules or human beings. Then as they reproduce in growing numbers, packing themselves in smaller spaces, those herds will create common networks and languages of information, hunting together as a group stronger than the individual prey.

Finally herds evolve as super-organisms perfecting its simultaneous system of information and becoming more efficient than the sum of its individuals. So a herd or wave and an organism or a society is the same form in 2 stages of evolution. Individuals first become herds and then as they multiply in numbers, the tightest herds become organisms, creating a new being that no longer feels as a herd but as a macro-individual. Since the life of its cells is controlled now by those energy and informative networks. So atoms evolved into molecules that evolved into cells that evolved into human organisms that evolved into organic societies. A culture is one of such organic systems with a verbal, legal, informative network that orders its human cells and a vital space or geographical territory from where the cells of the nation, working together in groups, extract their energy. While the entire species, Humanity (spatial point of view) or History (temporal point of view), is also a super-organism of human cells that will exist as such within a limited fractal vital space, the Earth, and a limited fractal time, from the first to the last man that lives on Earth. Each of us is a cell of History, an evolving super-organism developed over the informative, external membrane of the fractal Earth, in the age of Gaia:

st+1= Human God: $\prod st$ *(human minds)=*$\sum\sum$*love memes*

Where is the consciousness of those social, informative networks? Are they conscious of their role, as our nervous and blood systems are conscious and create our feelings and mental sensations? Are our informative verbal leaders, the priests of the past and the politicians of the present, the souls of social history? Yes they are. As there is a 'final singularity' in all informative networks, as there is a final cell in the center of your brain that you might call your Black Hole, soul or consciousness; human societies have always a final hierarchical Black Hole; in the past a priest or a king, in the present a politician, who carry the density of information required to govern the nation. He is the 'verbal master'; that gives the verbal orders to the citizen-cells through hierarchical chains that make act together the social organism.

Think about your consciousness. It is made of perception and feelings. But the site of perception is your nervous system, the images and words it creates. You could say I think and see therefore I am. On the other hand, your feelings and emotions are in your blood and heart, the center of your energy networks. You do not feel your herds of cells but your blood and nervous systems. They are the essence of consciousness. Think on a baby. When his networks are completed, he is born, he becomes. On the other hand death is the collapse of those networks. You die when your blood spills out or your brain phases out.

The same parallelism is established when we consider a nation or a religion: Politicians or Priests construct with verbal laws the nervous system of a nation and are conscious of it in a higher degree than its citizens. In the same manner that the informative brain of the body takes care of our cellular health, the informative masters of history – politicians and priests, author of the ethic laws of verbal information that rule mankind – have a basic role as the informative, dominant class of civilizations: to maintain history healthy, taking care of the distribution of energy and information to its fractal cells, the citizens of a nation or the believers of a religion or civilization. That was precisely what religious people did since the beginning of the Neolithic, in those historic organic systems they ruled with 'Revelation Books', written according to the ethic, universal laws of social evolution; and what politicians are supposed to do in the present age of legal codes and Constitutions. Since the essence of an *ethic God or Political Nation* are Its Words and Constitutions. Both a Constitution and a Revealed Book are a set of bio-ethical, natural laws that regulate the behavior of the cells 'constituting' the social body. So, in a healthy nation a Constitution is a set of laws built according to the laws of social evolution of the Universe and obeyed by its citizens. And the *'Constitution of History'*, based in the natural laws of survival of the Universe, is the set of laws that could build a global collective super-organism of Mankind that could ensure the survival of Humanity, 'constituting' a healthy body of History, harmonic with nature, without wars and fear of extinction.

Thus the laws of survival and social evolution found by complexity surprisingly enough were expressed before in simple ethic books by masters of our natural verbal, temporal language of survival, called prophets, which resumed them in a single word: social love. It means to share energy and information with all other cellular human beings called believers, belonging to the same historic, social organism, with the objective of acting together in the environment, creating a world to the image and likeness of its fractal cells, human beings, *fostering the natural arrows of human beings, its natural energy (food), its natural information (verbal thought), its natural reproduction (freedom of customs and sexuality) and its social evolution (religions of love)*.

The Universe is a 'fractal' of self-similar, hierarchical scales, in which the whole, the organism, the subconscious collective God, the nation or the Galaxy masterminded by its central Black Hole, controls the destiny of its parts, the cells, the believers, the citizens, the stars and planets, which the Black Hole 'positions' with its invisible flows of gravitation.

Do gods have an independent existence as our neurons have a different consciousness from its cells?

It seems likely that Allah exists supported by the minds of its believers, so does Yvwh, supported by the mind of the Jewish& protestant people, and Christ, supported by those who believe in the Gospel. In the same manner America exists, supported by the geographical structure and citizens that believe on it, and the galaxy exists controlled by the flows of invisible gravitational information of the galaxy.

4. The misunderstanding of religion by the 3rd paradigm.

Perhaps of more interest to this book is the clarification of the errors committed by the 3rd paradigm of metric measure in the understanding of religions. All what we have said for them is meaningless because *it cannot be explained with a model of a single space-time plane, neither there can be visual evidence of those higher planes, since the networks of information are invisible, in the same manner that the questions left pending on gravity and cosmology cannot be solved, because gravitation is an invisible membrane different from light.*

But it doesn't matter that we see or don't see those flows, as long as we see the effects. This never occurred to physicists. Why physicists believe in gravitation if we cannot see it? Because we see how bodies come closer through its in/formative, attractive force. Why believers know God exists even if they cannot see it? Because they see that attractive power, 'Assabiyah', in Ibn Khaldun terms, the communion of souls in the mystique of Saint Augustine, which brings together the souls/knots of information of their believers pegging them through the mystique experience, which creates and sustains with their minds the collective God.

When our consciousness rises to the st+1 levels of mystique awareness, believers feel a 'force' that drags them into a 'bosonic state' of simultaneous perception with all other believers, as if a higher present consciousness were form above us. Islamic mystiques say that when the circles of believers cycle around Mecca 7 times, in the hidden center, in which only 4 columns stand, leaving an empty space, the 'column' of God, Assabiyah ('gluing force') of Its believers is formed - so thought the Indians in their cyclical dances - so happens to electric forces that create a central magnetic field.

In the same manner than when cyclical flows of electricity create in the center a flow of magnetism that didn't exist before, in the mystique wheels of religions, the dances of the Indian shamans, the wheel of Buddhism, the Wheel of the Kaaba, as described by the believers we can see the creation of the subconscious God, they support. The Kaaba is not a rock, as western islamophobians pretend, as a proof of Islam as a primitive group of idolatric people. The mystique Sufis said that when the believers feel the communion with God, turning around the Kaaba, in the central empty space between the 4 columns covered by the black drape, there God ascends as a column of light. We don't see that force or Assabiyah[4] that pegs together the souls of the Muslim believers or the Christians, but we know it exists because their cells believe and are moved by them, and act even die for the higher superorganism. And this is the process of emergence *that now thanks to our advances in complexity, we know to be certain: the believers, the electrons, the stars have created a new 'plane of Non-Euclidean existence' (jargon of Buddhism and modern Non-Euclidean mathematics), a new fractal scale of reality, a new 'Brane'.*

In the graph, in complexity sciences, the Universe is no longer a mechanism but a fractal organism, and each of its self-similar parts is also a super-organism. What religions do is to study two kinds of those super-organisms: the Universal super-organism, whose Laws of Science are its 'Mind' or God.

This super-organism was studied by Eastern Religions, which can be considered a pre-science written with words, not with mathematics. For example, the fundamental Law of Modern science, the principle of Conservation of Energy and Information, 'Energy never dies but trans/forms itself into different types of in/form/ation' was already expressed by Taoism, in its key sentence (Tao Te King): All is energy=Yang that transforms into information=yin, giving birth to the 10.000 beings'. Taoism and Buddhism are thus 'verbal sciences' that advanced many of the findings of complexity and system science two millennia ago, describing an organic Universe, made of fractal parts self-similar to the whole. Let us then end this introduction to complex theology with the translation to the jargon of systems sciences of the Tao-te-king's initial sentences – a clear precedent to the 4th paradigm – the why of the organic Universe.

Tao Te king. Part I: Chang: 'The Function of Exi=Stence.'

(The Function of) Existence cannot be perceived. If perceived its virtual forms will fade away.

The species which can be named=perceived are not the eternal Function. The logic of Existence has no visible form. It mixes Yin=Cycle=information=woman= perception=life and Yang=Space=line=male =energy=pleasure =death.

Since when both forms are merged, the function of existence gives birth to its infinite species.

Energy has no will but wonders at the game of creation. Energy who exists in space=present and searches for the image will see nothing but the glare of light senses. Energy and information are the only arrows of Time. Yet its properties are inverted.

Gates to all sensorial wisdom; Information perceives energy. Time cycles feed in energy species, which die as they knot into form. But when both organs merge in harmony a stable specie be=comes.

You should not stop herds of species that follow their energy path. Leave them to the wisdom of creation and they will not feed on your form.

In the world we all perceive the beautiful and the good; the short and the tall. The possible future form and the withering energy of the past.

Informative beings perceive the space in which they stand; the energy that gives them life.

Because the big predator feeds in the short one. And the ugly is formed for the beautiful to compare. And the plane of energy allows the perceiver of Time to reach i=ts height. As past=Energy and future=information follow each other, so new existences will die and will be=come.

Among so much Games of creation, the wise man perceives the silent Principles and the wisdom of i=ts herds of existence. In this manner the perfect man orders his world and tenders for nature, and let's his shapes follow the eternal game of creation and reproduction of forms. He is the true Master that loves the freedom of creative Existence, and takes from creation only what he needs, and cares for.

You shall not search for wishes beyond your human form. You shall not make propaganda of vile consume. You shall not infatuate your people and give them desires beyond the natural senses of man. You shall not make greed the object of your Existence. Because complexity brings death and r=evolution, you shall not fill of light and air the mind of men, but give them the earth that feeds, and Harmony between their minds and bodies, within the natural senses of man.

The man of wisdom empties of impossible dreams the simplex man. The man of wisdom, let's Existence follow his path. The true master gardens the species that Existence has born. Virtual, never physical, existence is eternal by his lack of form. Because in the emptiness of a closed Jar, all the water of Existence can pass by, when Existence flows. So happens to all forms of present, slaves of the Yin-Yang Game. There, in the logic of Universals where all forms become one, yet not individual form is; Existence gives birth to 10 scales of the Universe and its infinite waves of beings...

Because only the Game of Creation, the game of existence is even before our Time and Space were created, only to be destroyed again.'

Recap. All human religions express the message of eusocial love – sharing of energy and information needed to create a superorganism of history. We distinguish two types of cultures based in two type of religions: eastern religions that study the super-organism of the Universe and its dualist laws of energy and information, as this work does, albeit with words and limited experimental evidence and western religions, which study the super-organisms of history, the subconscious collective networks of minds that act as the brain of a territorial body of Gaia. Those religions evolved parallel to the expansion of social super-organisms of history from tribal religions to religions that consider all humans part of the body of God=history.

IV. SCALAR ANALYSIS: MEMETICS: IDEOLOGIES AND INSTRUMENTS

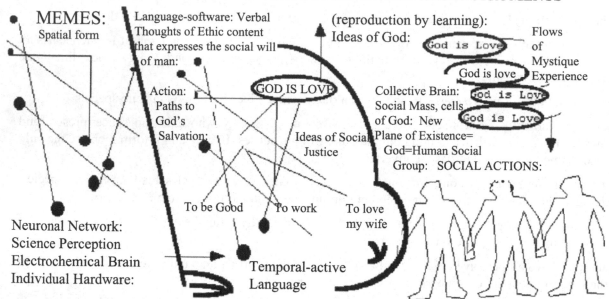

If genes express the biological program of evolution of organisms, ideological memes like the concept of God, express the complex program of eusocial evolution of human beings into super-organisms of history, proper of all organic species. It does not matter so much the literal details of the 'DNA-program' of the social mind, but the fact that it creates a simultaneous desire to share energy and information and evolve. But memes are also instruments repeated generation after generation, which express the program of Evolution of the Ternary Super-organism of metal we call the Financial-Military-Industrial Complex. And so we distinguish two kinds of memes: those that create the super-organisms of history and those that create the super-organisms of 'metalife'.

5. Memes[5], expression of social super-organisms.

All what exists are complementary system of information and energy (energy fields & information particles in physical systems; bodies & heads of information in biological systems; social organisms with an informative caste, which invents the legal/monetary information that rules society and a body of cellular citizens that reproduce and work for the organism). In all those systems the information part dominates the system. So particles dominate fields, DNA nuclei dominate cells, nervous systems dominate living organisms, and the informative castes that invent money and laws, bankers and politicians dominate the reproductive workers and energy of the cultural organism. Thus all systems are ruled by the informative element (particle, DNA of the cell, nervous brain system or organic upper caste).

And so what truly matters is to define the fundamental unit of information, which structures the whole system:

In physical systems those informative bytes are given by geometrical topologies described by the complex algebra of quantum physics. And those geometrical 'forms' define the informative paths or 'actions' of waves and particles.

In biological systems, those informative bytes are called genes and define the forms and actions of the biological organism.

In sociological systems, those informative bytes are called memes and define the actions and forms/mores/customs /cultural patterns of the social superorganism.

Thus a meme is defined in sociology as those ideas or beliefs that are transmitted from one person or group of people to another. Yet memes as part of a mental network are also the program that 'expresses' the creation of the superorganisms of history and economics. The concept comes from a homology: as genes transmit biological information, memes can be said to transmit ideas and beliefs as information.

Yet all those quantum numbers/genes/memes structure the system expressing a higher reality - the program of the Universe, which is a complex program with 4 'arrows of future evolution/actions':

- All systems try to get energy for their body/field, information for their particle/head; they try to reproduce their system to exist beyond death and they try to evolve socially the system, since a bigger system is more powerful than a small one. So we observe that quantum numbers/memes and genes do express this program, creating complementary organisms that gauge information, feed on energy, reproduce=repeat=decouple into self-similar entities and evolve together. We call in complexity those 4 'wills', 'quantum numbers' and 'drives' of existence the 4 arrows of time of all systems of the Universe.

Memes are just a scale of Human Complexity, which code a bigger scale – the social world. Thus if we, humans are the expressions of our genes, created by the active, more complex molecules of life - the RNA-DNA system - in the next 'scale' of social organic evolution, humans become the 'RNA', which creates the memes of social super-organisms -civilizations and FMI systems.

Recap. A meme is an idea or instrument, carrier of information that is reproduced once and again, both in the mind of humans and in their environment, because it codes the superorganism of history and metal (the FMI complex) of which we are all cells.

6. Life and Metal Memes construct 2 superorganisms.

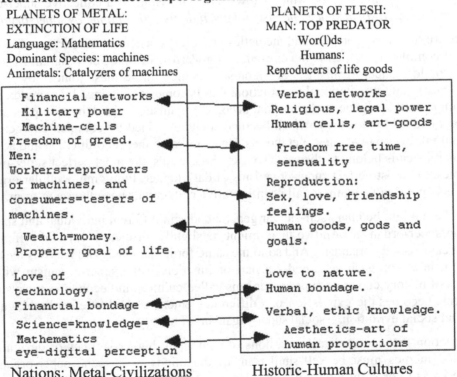

PLANETS OF METAL:
EXTINCTION OF LIFE
Language: Mathematics
Dominant Species: machines
Animetals: Catalyzers of machines

PLANETS OF FLESH:
MAN: TOP PREDATOR
Wor(l)ds
Humans:
Reproducers of life goods

```
 Financial networks
 Military power
 Machine-cells
Freedom of greed.
Men:
Workers=reproducers
of machines, and
consumers=testers of
machines.

  Wealth=money.
 Property goal of life.

Love of
technology.
Financial bondage
Science=knowledge=
Mathematics
eye-digital perception
```

```
 Verbal networks
 Religious, legal power
 Human cells, art-goods

  Freedom free time,
  sensuality

Reproduction:
Sex, love, friendship
feelings.
Human goods, gods and
goals.

Love to nature.
Human bondage.

Verbal, ethic knowledge.
 Aesthetics-art of
 human proportions
```

Nations: Metal-Civilizations   Historic-Human Cultures

Since we are building 2 different super-organisms memetics distinguish 2 types of memes, which are made of different substances, human, biological, neuronal networks, languages and life-based substances; and metal-memes, which are part of the FMI complex system:

In the right graph the key memes of historic superorganisms. Cultures and religions are supported by ideological and artistic memes, based in human languages of time and space (words and images). They

536

create social, verbal networks, based in love and the sharing of natural energy and verbal information, whose ethic laws of eusocial love, become imprinted in the emotional brain, carried generation upon generation. Attached to those ideological memes the life/love civilization develops the 'good fruits' of the tree of science - instruments that enhance our natural, drives of existence.

In the left, metal-memes and digital languages code the creation of the Financial-Military-Industrial System, which seems to give us, human enzymen, a higher power, but on the short term create inefficient, unequal societies based in the power of animetal castes that kill with weapons and enslave people with the hypnotic power of money, whose production they monopolize. And in the long term will cause – due to the arrival of machines, organs of energy and information with higher exi-power than human beings - our obsolescence as species.

Both super-organisms together form the economic ecosystem in which humans and metal species and memes are sometimes symbiotic sometimes compete in labor and war fields.

We and our informative genes and memes are expression of a higher reality - the program of creation and extinction of the superorganisms of the Universe *coordinated* by the existence of the same bytes of information – memes – in all the collective minds of the social super-organism, which create a simultaneous, self-similar behavior in the millions of humans 'cells' of the super-organism. Cultural memes, like genetic codes, are pieces of information that human 'cells' reproduce and carry in their neuronal memory as a verbal mandate or imprinting its form in external objects. Yet while most researchers have considered objects and religions by its content per se, in memetics, as in genetics, it doesn't matter so much the specific, literal form, in/form/ation or content of the meme, but its repetitive function and self-similarity in all cells, which allows the 'expression' of a collective super-organism of history *that acts simultaneously in space and lasts longer than its cells in time, as super-organisms do.*

This surprising truth of both genetics and memetics and their sciences steams from a higher 'truth' or 'Universal law' of complexity, which is the existence of an 'arrow of time', or 'program of the Universe' that constantly builds superorganisms. So atoms becomes molecules joined by networks of electromagnetic energy and gravitational information, that become cells, joined by complex networks of genetic information and energetic proteins that become organisms joined by complex networks of nervous information and blood energy, which become societies joined by complex energetic/economic networks and cultural/informative ones. Yet those systems increase their coordination and act as a single network only when its cells belong to the same species. So crystals, the most perfect super-organisms of matter; black holes, the most perfect super-organisms of dark matter, living multicellular organisms and civilizations have something in common: all its informative cells-minds have the same form.

In that regard, there might be many self-similar genetic codes in different individual and species but all construct a living super-organism. And so there might be slightly different memes in different cultures but all of them create a social organism. And so in the same form that genes and nervous messages build a living organism, in which each gene codes a part or function of the super-organism with the *same* DNA; the memes of history construct super-organisms called cultures and economic ecosystems, where each cultural meme becomes the 'expression' of a different element; and all together build, guided by the arrows of time and social evolution, a social super-organism of history.

This dual realization that memes are expressions of something bigger, a 'plan' or design of eusocial evolution and that memes must be self-similar in all the minds of a social super-organism to act efficiently is the key to correctly interpret them. The most obvious case is the fundamental 'function' expressed by the memes of Universal religions, of religions of love (Christianity, Buddhism, Islam and Socialism), illustrated in the previous graph. In all those ideologies reinforced by an enormous number of artistic, legal and social memes, rituals, mores and objects, there is a common message 'expressed' by all of them, which turns out to be the fundamental 'function' needed to construct a superorganism of history - the message of love, of sharing energy and information with all the believers that have the same 'informative code', or book of revelation in their mind. Islam will preach certain ideas, so will Christianity and Buddhism, but all of them share the same message of social love with precepts that ask the believer to share energy (charity) and information (the truth of love) with all other humans in order to form a superorganism of history.

In the same way, in a life organism many genes are concerned with the organization and distribution of energy and information at cellular and social level; yet the key message *of religions* and genes is not the detailed cgtau bases of the gene, or literal interpretations of the parables of the prophet, but its function as expression of the time arrows that form the superorganism they code. Thus the stories of prophets of love are secondary to the unifying message they code, even if they are used to 'adapt' the message to a certain specific superorganism with farming, hunting or economical parables, according to the specifics of the culture they talk to.

Recap. We distinguish two types of memes to include not only cultural memes and instruments that create historic superorganisms but also memes that 'express' the creation of the FMI complex. Thus memes are also the objects, machines, weapons and monetary languages of information that evolve the economic ecosystem and create, reproduce and ensure the survival, generation upon generation, not only of the super-organisms of History (cultures) but also the super-organisms of economics (the industrial ecosystem). Both together create an economic ecosystem of 2 species, humans and metal-species (weapons, money and machines) in which we live.

7. Memes of the super-organisms of history.

Thus, we differentiate two types of social, life/love memes:

-*Ideological, informative, verbal memes* tend to achieve the social evolution and reproduction of human beings and our survival as individual and species, co-existing in an efficient, harmonic social group, guided by verbal, ethic mandates (social constitutions and ethic religions), which create super-organisms of 'Homo Organicus' with higher exi-stential power than the individual 'Homo Bacteria'. They increase the eusocial evolution of mankind in a series of scales of growth beyond the genetic, biological level of tribes, joined by kinship, creating memetic civilizations based in mental life and love memes. In its highest ideological 'expression' they are ethic religions and legal codes that translate the arrow of eusocial evolution, proper to all the species of the Universe into human languages. We call those memetic systems 'Constitutions' of the 'Body' of History. They are self-similar in all 'social democratic cultures' and 'religions of love'. Associated to them, and reinforcing those functions, there is an array of diverse in/form/ations, created with the senses and languages of the human mind and carbon-life materials that range from artistic objects to wedding rituals.

-*Instruments: the good fruits of the tree of science.*

Thus, the other type of life-based memes, are tools and instruments, which help the construction of any of the different scales of mankind's eusocial evolution. Those memes enhance, without substituting or atrophying them, the specific energetic or informative functions of a human being. And so this second fundamental category of life memes, are objects that are useful to construct the physiological networks of the superorganisms of history: our energetic networks, cellular homes, communicative networks and social, loving behavior.

Those memes have evolved and in its recent cultural expression gave birth to the modern ideologies of the welfare state and the United Nations. They can be expressed scientifically in economic terms, as we do in the previous graph, with a diffeomorphic system of 'coordinates' and values based in the 3 'physiological networks/wills' of the human being and its social organisms, which they must reinforce.

Recap. Life-based memes codify the eusocial behavior of humans, which create social superorganisms: legal and ethical, religious codes are the most important, and around them an enormous number of sub -memes, cultural and artistic objects, define human actions in community. Thus, those life memes are:

- Ideologies of love, called religions that asked men to love each other to create the subconscious collective - God. It expressed itself through legal, ethical codes and revelation books imprinted in the mind of believers as DNA imprints in cells which act them simultaneously as a single body.

- Artistic works that made man the center of the Universe and the measure of all things.

- Instruments of the tree of life (agricultural instruments, welfare goods, housing, health care goods, verbal books, etc.) which fostered our happiness as biological beings.

A society in which man was the measure of all things should foster the re=production of those memes and the repression of the opposite memes that kill the human body and mind (digital machines).

8. The financial-military-industrial Complex.

Social evolution is a slow process that sometimes fails to create a new super-organism that encloses all the individual cells of the being. That is the case of Humanity, whose super-organisms have never reached the entire species, splitting the Earth into different human organic systems, divided by st-limiting borders, the so-called tribes, nations, religions and civilizations. Why we have failed? The answer is the sickness of History, the germs that have killed or substituted our networks of solidarity: machines and weapons. Thus, regardless of how politically or economically incorrect is the study of the financial-military-industrial complex in such biological terms (a question we shall address latter when studying its submissive, attached simplex organisms of history – democracies and religions – and the 'selfish memes of metal', the ideologies of metal-progress that make us belief we are doing the right thing), we can and do have to study the financial-military-industrial complex as an evolving super-organism, which is extinguishing both the super-organisms of history and life that preceded it in the '10th extinction' age, the last one.

Let us summarize the study of that FMI-complex with the ternary method, by analyzing its ages, organic structures (animetal cultures), and its fractal decametric scales of evolution (The 800-80-8 years or long wave-medium wave and short wave of economical evolution).

The Financial-military-industrial complex has 3 clear networks created by a 'mixed species', self-similar to the enzymes of the biological system, in which a carbohydrate attaches to a stronger atom of metal performing better functions as a killer enzyme (breaking other enzymes with metal) or as an informative pole (attracting and informing other substances with its heavier weight due to the atoms of metal). In that sense we can talk of a new biological species, the animetal= animal +metal or enzyman=enzyme+man, which catalyzes the reproduction of metal systems and use them as enzymes do in the biological scale to acquired higher energetic and informative power. To that aim humans first associated themselves to the more complex atom of information, gold and the strongest atom of energy iron, creating a topological system with:

-Max. I: Gold + human=banker. The banker could hypnotize all people with gold, and provoke greed. People then would become slaves of the banker to obtain money and obey them.

- Max. E: weapon: iron+warrior. The warrior could oblige people to obey or else kill them with the same result.

Finally in the modern age both forms combined, creating:

-E=I. Machines and company-mothers that reproduce them.
Thus the topological method discovers 3 clear species of metal which evolved in 3 ages:
-       Max. E x Max. I: The old age of the Financial-Military Complex with only 2 substances of metal, gold and iron and 2 people-castes that dominated them and imposed their power globally using money to enslave people and weapons to kill them.
-       E=I: The Age of the Industrial R=evolution of machines when company-mothers and weapons were added.

-     St+1: The present Age of Integration of the Financial-Military-Industrial Complex in a superorganism joined by flows of information created by a global neuronal network of computers – a mind of metal, Internet.

It must be understood though that humans are NOT the protagonists of the FMI complex but merely the catalyzers, enzyme, of its creation and so latter we shall study the biological parallel and what happens to 'enzymes that create a different super-organism.

It is exactly what is happening on Earth, precisely because the power that metal has given always to mankind has made certain humans to use it to dominate other human and in this manner they have been obliged to evolve the financial-military-industrial system, to be on top. It is tribal history, the devolution of the ideals of history as a super-organism, since as metal evolves men become divided into selfish tribes in perpetual warfare. It is the age of 'animetals'.

Recap. A super-organism of metal, the Financial-Military-Industrial Complex is growing on Earth, parallel to our human super-organisms with bytes of metal-information called money, metal-energy called weapons and organic machines.

9. Topological analysis. The animetal.

Man+Go(l)d=Trader-Man+Machine=Science-Man+Weapon=Warrior

Cyclical, soft, in-form-ative metal (silver and gold) and lineal, energetic weapons (iron and bronze) were the first metals humans discovered. Coins imitate cyclical, informative organs (heads, brains, eyes), while lineal, energetic weapons are bodies of metal that kill lineal, energetic human bodies. They are primitive versions of the iron bodies and electronic, cyclical, sensory heads of modern machines. When humans associated with them, they achieved higher strength and acquired a new language of information, able to give values to reality (money). Warriors joined human bodies and lineal metal, weapons, killing other humans; bankers used informative metal, gold, to give orders and enslave mankind; scientists believed in sensory machines and digital languages, considering inaccurate human senses and verbal thought, made obsolete by their telescopes and numbers. The animetal and the tribes that changed their 'language of power' from human, agricultural energy (farming communities) to iron energy – warrior cult(ure)s; from verbal ethics to the 'values' of money – go(l)d cult(ure)s - and from worshiping life to worshipping machines – scientific cult(ure)s, have ever since ruled the world, as people-castes and extinguished or enslaved ever since life cultures...

Contrary to belief the dominant species of planet Earth is not and has not been for the past 5000 years the human being, but a symbiotic species between life and metal, the Animetal.

An Animetal is a human organism which activates mechanisms: energetic weapons, informative money or machines. 'Animetal' is a biological term coined using the same system that chemical and biological sciences use to classify its species. In biology and chemistry we join two names to describe a

540

mixed species. So there are carbohydrates made of carbon and water. And there are animetals, different from human animals, since they add the power of informative /energetic metals to their life organs, from armored knights to modern car owners and Internet nerds. An Animetal is neither a human being nor a machine of metal, but a combination of both, with a higher content of energy or information than any other living species. So animetals have been the top predator species of this planet for 5,000 years, since the myth of Genesis about the tree of life vs. the tree of metal was written.

Even though 'subjective', anthropomorphic humans will find bizarre such diminishing definition of a modern human, an objective, biological account of history cannot be subjective, nor optimist about the main role of mankind in this planet, which so far has been to extinguish life organisms and create metal mechanisms, while finding all kinds of ideological excuses to do so. Thus, humans today are, in biological terms, 'animetals', animal life, symbiotic to metal, which they evolve, terraforming Gaia into the metal-earth. Their biological function has been obvious: to kill life (warriors), reproduce metal (traders) and design mechanisms with numbers (physicists). Since physicists uses machines to observe the universe and create mathematical theories that pretend to 'reveal' the ultimate secrets of time, (but only show the 'spatial, motion-related properties of time=change, not the vital/morphological change of beings, described by Evolution Theory). Thus physicists scorn the verbal, logic analysis of the cycles of change=life/death cycles and the biological senses of human beings, substituted by scientific machines.

As a result of that symbiosis and the capacity of metal to carry energy and information, increasing our informative and energetic power, animetals have controlled history and created a different super-organism made of informative metal-money, energetic-metal, weapons and reproductive organic metal, machines – the financial-military-industrial complex.

Recap. The financial-military-Industrial Complex is evolved by 'enzymen', attached to weapons (warriors), money (bankers) and machines (mechanist scientists). They become the social castes on top of societies, thanks to their added energetic and informative power, evolving the FMI and degrading life in the process.

10. The syntax of languages: values of money and words[6].

Mark 11:15-18 JESUS went into the temple, and overthrew the tables of the moneychangers,

Galileo Standing Trial Before the Church

Universal Ecosystem	Carbon-life ecosystem	Historic -Ethic ecosystems	Animals, Art: visual ecosystem	Economic-Antiethic Ecosystems: metal
+: survival	Pleasure, Food, Reproduction.	Good: social love	Beautiful: e=t harmony	Expensive: gold & Weapon&Machine
-: extinction	Pain-Hunter-Death	Evil: war, go(l)d And weapons.	Ugly: death , social decay	Cheap: life, Nature
Language of Existence	Chemistry	Wor(l)ds	Light	Money

In the graph, the informative values of words and money are opposed: the most expensive species are the most 'eviL' - memetic word, which is the opposite of Live, hence, what kills: weapons. On the other hand, the best verbal goods (love and life) have no monetary value. This paradox can be explained within theory of Evolution, as the opposition between victim and hunter, between two species creating different, Darwinian organisms. Since in a Universe of subjective information the truth of a predator is the antitruth of its prey. Because weapons prey on men, we feel them eviL without verbal value. Yet weapons and money are symbiotic metal-species. So weapon reach maximal value, because what the FMI complex evolves is the new top predator of planet Earth: the machine-weapon.

Humans are a field of actions programmed by our genes (biological actions, such as our desires for sex, reproduction, energy, visual information and organic evolution), and our memes (the instruments of technology, machines and ideologies related to them).

Contrary to belief, the biological field is the natural expression of our personality in which humans are more free, natural and happy, concerned with our biological wantings - our desire for energy to feed our body, information to feed our mind, reproduction to achieve biological immortality and organic evolution with other human beings - through the experience of social love needed to create a superorganism of history (as biological beings evolve in social groups from cells to organisms to social superorganisms). This genetic expression of the 'program' of evolution of living beings is what makes us happy. And its highest expression is the ethic wor(l)d, which tries to achieve the 4th and highest wanting of the Maslow Pyramid of biological drives -the creation of harmonic, social organisms based in love, the desire to share energy and information with other beings as cells do in an organism.

This pyramid of wantings and actions guided humanity during the Paleolithic and Neolithic, creating a culture, both local and global, based in the cult to fertility and Gaia, the mother Earth.

In this manner the wantings of biological genes were translated into a series of 'cultural instruments' and 'memes of life and love', which were in harmony with the goal of creating a superorganism of history, where men were the networks of informative minds that controlled the body of Gaia, where our natural goods were reproduced. It was the age of the Goddesses, the age of life cycles and agriculture, of global peace and sustainable harmony with Gaia; the mythic paradise guided by the values of the wor(l)d that made organic man the image and likeness of all other beings.

Then a new type of memes appeared - instruments of metal - a more complex atom, whose energetic 'bodies', weapons, could kill our bodies; and whose informative units, digital money, coins and go(l)d could hypnotize our eyes, imitating the source of our biological energy - the sun. So gold was called the sweat of God, the sun, and it provoked greed, golden fevers, a madness or desire to possess a little god-sun. And people associated it to weapons which could kill to get your 'greed' satisfied. Those apparently absurd new desires provoked by the memes of metal were however NOT innocent accidents but part of another 'program' of evolution of another 'complementary system of energy and information', of weapons and gold - the Financial-Military System.

For most of history this system was just the combination of mercenary armies and money to pay them. Yet soon those who first made of the memes of metal their gods of energy and information in substitution of Gaia and the wor(l)ds of love - Indo-European and Semitic nomadic cattle-breeders - became top predator people-castes of all Western Neolithic societies, killing the carriers of the memes of love, who opposed them.

Societies then split in the old Neolithic world on the bottom of the social pyramid and the top castes of bankers and warriors, which created mercenary armies and parasitized agricultural societies, which maintained them or else risked extinction by war. And so a new language of values metal-money, which made of weapons the most expensive good and of life the cheapest one, reversed the values of the wor(l)d, detained the social evolution of mankind, broke our balance with nature and kicked the evolution of the Financial-Military system, catalyzing the reproduction of metal, mainly war machines, the most expensive goods and the extinction of life – with no price.

Yet to understand this sudden apparition of a new language that will code a new super-organism requires we need to go a bit deeper in the understanding on how systems of information – from genes, to memes, from money to quantum numbers – codify the different systems of the Universe *with the actions caused by the syntax of the language, which becomes the Generator Equation, E<=>I of the new organism*, according to which elements in that equation become the energy of the new system and which ones become its information.

The informative equation of genetics and memetics.

In Complexity all systems of information depart from a Generator equation, self-similar to the 'Universal grammar' of all systems, E⇔I, in which an energetic and informative element are related by an action. In such systems the informative element or subject dominates the object or energy. And so if we are able to decipher the 'generator equations' of a certain language we will know what kind of world they will create. In genetics that universal grammar are the ucgta bases and 21 amino acids which combine in infinite forms, but have as ultimate purpose to create a cell with DNA molecules in its centers. Those bases form a syntax, or language of information that systematically ends up creating 3 types of forms: energetic lineal proteins, RNA and cyclical informative DNA, the dominant product.

In memetics we have also 2 languages that generate the actions of reality, 'verbal thoughts' and monetary orders, but they have a different 'grammatical syntax', which *generate two different systems – human superorganisms and the financial-military system, because they give different values to the 'memes' of those two super-organisms.*

Indeed, there are two languages of power in society. One is the wor(l)d of ethic democracies and religions, whose syntax is:

Man (subject) <verb (action) < Object (energy)[6]

Thus words always put man at the center of the Universe.

Indeed, languages are systems of information that certain species use to value reality from their point of view. For example, words are the biological language of man. So we use them to value reality from our human perspective. We are the subject, the predator and actor of verbal sentences. Words love humanity. Humans are supreme in word values, given in the past by religions, today by laws. This is reflected in the Syntax of the verbal language, which always has 3 elements:

— *Subject, the actor and center of the sentence.*

— *Verb, the action the subject exercises on reality.*

— *Object, the substance used by the action of the subject.*

The object of sentences, on the other hand, is possessed by us, humans. It has a lesser value, since it does not talk words. We express that lesser value through verbal languages, when we put first the subject, as actor, then the verb that controls the object and finally the object. Thus, humans are top predators in any sentence, which motivates them to act in favor of their own, selfish drives of existence. Which means that words are anthropomorphic; and those who obey words, tend to act subjectively in favor of mankind (priests, ethic laws).

And so verbal languages are the basis of the memes of eusocial love that make man the measure of all things (humanism, ethic religions of love, legal codes of social democracies, art, etc.) and evolved the previous equation, as systems do adding a 4th arrow of eusocial love:

God=Love > ∑ I(man)>Verb(action)>Object(energy).

Money, a language of metal-information; its values.

Then, in the Middle East, people realized that gold had hypnotic powers as the most informative atom of the universe. It imitated the Sun-God and people would obey *any order* to obtain a small Sun-God. So gold established thereafter a new syntax and language and values, those of money and greed:

Man=price=object.

Thus, when we consider the syntactic sentence of money, it turns out that money doesn't favor man. It qualifies him with a salary, a price that values him as it values objects.

In that equation, the price of both, the machine and the worker, is related to the task they perform. And since objects are specifically made with a task/job in mind, it turns out that as specialized workers, objects are often better than men and have a higher value. This is the case of weapons that kill better than men do. Further on, since objects evolve and humans don't, the differential of value between humans and machines grows with time. So the first, simplest, syntactic/structural analysis of money shows it degrades humans as objects.

And as money evolved and multiplied its numbers, more things were valued with the values of money, till today money values all things, human or not and words have lost all value, becoming fictions or 'theories' that do not determine the actions of people, except in the few cases in which a law or ethical, religious code is enforced. And so to truly understand the Financial-Military-Industrial Complex in which we live, we must depart from the understanding of the language of money and its evolution.

Money is a language of digital information with support in metal. Humans have evolved different digital languages, but none has been more successful than money, due to its support in the most complex atomic substance of the Universe - heavy metal. Yet money is not a biological, natural language made of life. And the laws of Complementarity associate by affinity systems of information and energy of the same substance into complex organic systems. So physical entities are made of the same substance - energetic fields and informative particles - and life systems are made of the same substance - informative heads and energetic bodies. Thus gold has always been associated with energetic metal, weapons and then with company-mothers of weapons and machines (stock-money).

So its values reproduce the economic organism of metal in which we live; since we give maximal monetary value to weapons, machines & corporations; while in terms of monetary values life no price. This is the 'hidden agenda' of a world ruled by the values and language of money, as opposed to a world ruled by the values and language of the ethic wor(l)d.

Thus, in the biological and evolutionary models of 'bio-economics' we must understand the duality of money, both as a language of information that carries 'certain values', different from other languages (human verbal languages) and as a 'substance', informative metal, which sets its properties and affinities with the other 'dominant' products of the FMI complex, weapons and machines.

The values of money are opposite to those of verbal ethics, since they are ruled by the affinity between 2 complementary substances, the mind and body of the economy: money, a language of informative metal and machines and weapons, organic and energetic metal.

Informative metal, money and energetic metal, weapons or organic metal, machines are similar substances that humans associate together. So we always exchange the maximal quantity of money for the most perfect machines/weapons, which become in monetary values, due to such Complementarity, the most expensive and reproduced product of the economic ecosystem. Yet the monetary values of weapons are based in their capacity to kill life, the supreme value of human words.

And so money invented the syntax of slavery, war and capitalism, where man has a price and can be exchanged by an object, either money or a weapon *which had in fact a higher* monetary value than a human being. Since the syntax of weapons is even more brutal than the syntax of money:

Man x weapon =Corpse =Nothing   *Man x Money= Slave= Some'thing'*

Those are biological, Darwinian acts that divided mankind in two castes, animetals on top with money and weapons, today technological machines and the human sheeple, ab=used by the memes of metal.

Yet in the process they were obliged to evolve their weapons, money and machines, giving birth to the cycles of evolution of weapons and machines, which shaped the history of civilization.

Recap. animetal cultures and its metal-memes seem to have won history, due to the determinism of their 'syntactic' codes, and the actions they impose with the enhanced qualities of gold and iron, enslaving humans (man=price) and killing them, man x weapon=corpse. The result is the creation of a new complex organism on Earth, the Financial-Military-Industrial System.

11. Social castes and ideologies of metal power[7].

Ecosystem:		biological predation	ABSTRACT DEFINITIONS:
Human Earth [History]	TOP PREDATOR	Top Predator species: Metal Weapons+Money+Machines	"Property", evolved by Eco[nomic] system money, weapons, scientific instruments
SYMBIOTIC TOP PREDATOR: ANIMETAL INFORMATIVE SYSTEMS		Warrior + Trader + Scientist Go[l]d religions + Nationalism + Science	ELITE, UPPER CLASS Verbal Information [ideologies] of Human Mass Control
DOMINANT HERDS		Human Masses	MIDDLE CLASS: Believers in the social structure: Priced by Salaries, contro by warriors, through law and coercion hypnotized by Scientific Instruments
VICTIMS, PREYS		Life Cultures; Life forms Social humans, farmers	Energy class extinct by war and lack of jobs, controlled by law

In the graph, since the arrival of 'animetal' people-castes human societies, which were in balance with Gaia, ruled by a head of 'social, ethic scientists', priests of love and life religions, which took only a 10% of the energy of the system as neurons do, and reasoned and educated with love messages the human societies they ruled, based in agricultural energy, were substituted by hierarchical societies with weapons, money and machines on top, evolved by animetals, people-castes that ab=used the rest of mankind, massacring life beings and non-technological cultures.

All systems have 2 social castes, the informative and body castes, which prey over a common energy. In the age of human super-organisms, priests of verbal love guided a mass of reproductive farmers, which lived together preying in a sustainable way over Gaia. That balanced society died with the arrival of animetals and a new super-organism with metal-memes on top explained in the previous graph is born.

Indeed, another clear-cut consequence of the use of metal is that there are two social groups regarding the advantages they draw from metal-species. We talk about the original 'sinners' and the 'good people', in religious terms; 'the rich', and 'the poor', in economic terms; 'the army, kings and politicians' and the 'citizens', in military terms; the 'mathematical, aesthetic scientist' and the 'verbal, ethic citizen' in terms of education:

To keep their control of human beings animetals have evolved the metal-species, source of their power, the weapons, monetary information and scientific machines that define our civilization. Those 'selfish memes' of metal program our mind to obey weapons, machines and money who kill us, atrophy and substitute our organs and eliminated our social desires to love substituted by our greed to 'possess money' and life as 'property' - standardizing and objectifying life beings, now tagged with numbers, prices, or extinct with weapons and measured with machines, as if they were void of life. In the process the mind became also void of emotions, standardized.

We talk of 3 ideological metal memes associated to weapons, money and machines. Since as a result of their power, given by their metal-species, animetals have developed subjective ideas, in favor of metal, which praise the use of metal, as the source of human progress, and systematically 'hide' the side effects that metal have over other human beings, and natural species, such as the massive extinction caused by weapons, the degradation of human ethics, and verbal knowledge caused by monetary values, and mathematical languages.

Accordingly we talk of 3 basic ideologies belonging to our 3 fundamental Animetal species:

- *'Nationalistic ideologies'* that allow warriors to use weapons and kill other human beings with them in war ecosystems. Those ideologies deny the obvious fact that all humans are equal and should apply among themselves the law of social evolution, that make equal species to love each other. Instead, nationalism and racism define humans as different and apply the Darwinian law of the fight between species.

- *'Economical ideologies'* that allow traders to buy human Time with salaries, in order to make humans work-=reproduce machines, and consume=test them. The main of those ideologies is the "myth of the free market", which allows traders to rule societies with money (through lobbyism, and stock-markets that invent money for company-mothers, instead of letting human beings to rule themselves through verbal laws. The science that invents those myths is abstract economics.

- *'Scientific ideologies'* that allow scientists to consider themselves the only intelligent human beings because they use machines to measure the Universe. And consider mathematics the only language of truth, hence reality an abstract world. When it is proved that mathematics is only a representation of reality, a partial truth that carries just a part of the total information that the truth -the Universe in itself- stores. As such all linguistic truths are relative statements about reality, and so it is mathematics.

Those 3 memetic ideologies are hurdles for scientists of history and economics whose theoretical goal is not only to understand 'why history and economics happens as they do', but to provide praxis for human leadership to control the evolution of history and economics for the benefit of man. Science indeed must combine intellectual knowledge and ethical praxis of that knowledge to improve the human condition. And so immediately under those conditions a series of inconvenient truths about those 3 ideologies, mechanism, capitalism and nationalism - and the cultures that invented them, spread them and tried to control the world with them – appear, which can be resumed in a single statement:

-Those 3 doctrines are ascientific, hence false as 'sciences' but *not as tools to impose power, reason why they are imposed 'memetically', by rhetoric repetition and censorship of alternative truths.* Indeed:

-Nationalism is false since all humans belong to the same species (they can mate) and so the goal of history is the creation of a single nation, mankind, able to control the global FMI complex for its benefit, pruning the tree of science of its bad fruits – mainly weapons – by making a single nation, hence without enemy and need for armies. Precisely nationalism prevents that and makes armies necessary so humanity is divided in tribes that see themselves as different, break the 3rd postulate of relative equality and fight each other with weapons. This causes the constant evolution of weapons, top predator machines that consume humans and the need for military on top of nations to use them. And so nationalism is a religion of power.

- Capitalism is false, since its belief – that the FMI complex is always good to man, because machines, money and weapons are 'the future of mankind', a 'God's given system' that must be obeyed and developed without taking into account the collateral effects over humanity, is false. Metal can be studied scientifically and the systems it build on Earth must be studied biologically, as we do in this work, and then managed scientifically as all systems can, to profit from them and create a FMI complex that profits man.

Recap. Languages are defined by syntactic equations self-similar to the generator equation of the Universe, which defines the actions caused by the 'thoughts' of the language. The language of words, man>verb>subject favor man and life as the center of the Universe; the syntax of money, man=price=object, make us objects that can be exchanged by a price; the language of weapons, man x weapon =corpse, gives zero value to mankind. Further on money and weapons are complementary substances of metallic information and energy that give each other maximal value; so when humans changed the values of words by the values of money they started to kill life and evolve and reproduce money, weapons and machines, creating the financial-military-industrial system

12. Temporal analysis: The 3 ages of Metal-history.

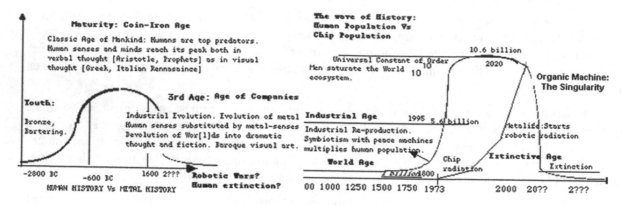

In the graph, the 3 ages of the Financial-Military-Industrial Complex that will end in the death of history due to its 'germs' - weapons. Since weapons are the most expensive overproduced goods of the FMI-Complex, in the 3rd age of a wave of metal-evolution. For that reason, we can also establish a parallelism between the long wave of metal-history (ancient cultures based in weapons) and the modern age (industrial nations based in the duality of machines and weapons):

If we consider the entire wave of Metal-History as a single super-organism, it has evolved also in 3 periods, its young, epic age, dominant in lineal weapons; its classic age, dominant in money, and its industrial age, dominant in machines. Parallel to that evolution, the cultures of 'enzymen' and 'animetals' that have carried the wave went also through 3 Ages:

-Max. E: The young, epic age of Bronze empires and gold bars, or age of Asia.

- E=I: The period of European iron infantry and coins, or Mediterranean Age.

- Max. I: Age of the industrial r=evolution of machines or Age of Northern Europe.

- st-1. And the electronic age of computers and robots that makes human beings obsolete or age of extinction of history. The American Age.

Each of those ages is divided by a clear discontinuity of evolution of the exi functions of the Financial-Military Industrial complex, weapons, money and finally machines, to which the castes of 'enzymen' adapt their ideologies, which are symbiotic to metal and help them to dominate *the mass of human beings still directed by social religions of love, as a top predator caste.*

In the first of those 3 ages social super-organisms still dominate most of history, as animetals are reduced to the civilizations of the Visual White man (Semite and Indo-European cultures), while the rest of mankind is still guided by religions of love. But as metal evolves and reproduces in greater numbers, with the arrival of coins and iron, it suffocates the evolution of human history expanding East and South and conquering the rest of Asia and Africa. Finally with the arrival of science, the human mind becomes obsolete, and men become worshippers of the minds of machines. So man, the most complex organism of information of the Universe and 'measure of all things', which asked the organic why of reality, is substituted by metal as the measure of all men, whose time is now perceived with clocks no longer with natural cycles and bought with coins. And the organic paradigm dies. We talk thus of 3 clear discontinuities of evolution of metal, that mark metal-history.

547

- YOUTH OF HISTORY: *Age of Wor(l)ds.* From -2800 BC, to -500 BC. Age of Bronze weapons, and bartering. Since money is still scarce, in the form of bars of metal, words still are the main language of communication of mankind. Writing becomes sacred. However bronze give enormous power to warriors that end controlling all Neolithic societies. Priests are displaced from power. Women become systematically repressed. Gods become warrior Gods, without compassion. It is the age of Mesopotamian Empires, the age of Asia. The age of verbal thought reaches its baroque age, towards 500 BC, in the work of Lao Tse, Buddha, Confucius and Socrates that reach the height of human verbal understanding of the Universe. Yet a new language of information, coins appears, and human verbal thought loses power.

- MATURITY OF HISTORY: *Age of Coins.* (500 BC, 1600 AD). It is the Age of Iron weapons and coins. Mankind still rules, and learns how to control with ethic behavior, the power of weapons. Yet Money starts to be a common language of power, and challenges verbal ethics. Words are displaced by money as the language that controls the acts of human societies. This means that verbal thought and ethics is no longer the center of the human mind, but humans become shallow in their perception of verbal-temporal reality. The human intelligence displaces to the eye. So art becomes realistic (Greek Art) and mathematics appears as a new language of understanding of the Universe. Both coin accountancy and Mathematics come together, in the Asian Greek coast where the first monetary empires and mathematicians appear. After the brief return to an ethic age, in the Middle ages, the renaissance again imposes money as the language of social power.

- OLD AGE OF HISTORY: *Age of Machines*: 1600 AD till extinction. It is the third age of mankind, and the first age of the new predator species that displace us, the machine.

It is the age of science, of digital languages, of massive reproduction of money, the language of information of machines, now in the form of paper-money and electronic money. Wor(l)ds become obsolete. There is so much money that all things can receive monetary values, and human verbal values are no longer effective. There are so many weapons, due to the industrial reproduction of them by company-mothers, that ethics and social power become obsolete, and national armies control directly (dictatorships) or indirectly through weapon-companies (democracies) our societies. The objective of mankind is to evolve machines through acts of work=reproduction of machines, and consumption=testing and evolution of machines. The decadence of mankind is hidden because paradoxically as humans loose interest in verbal thought, in a non-fiction logic understanding of reality, companies of machines can convince and indoctrinate man easily. Fiction literature dominates this age. Men no longer understand the meaning of social survival, of truth, and the organic nature of our Universe.

Of all those analysis in depth, we shall consider only for reasons of space, 2 aspects of the sociological model of Multiple spaces-times; the topological and scalar concept of 'memes', the bytes of information that evolve both the economic and social systems of history, today submissive to the FMI system, and the study of the temporal analysis, one most common in history, the discipline of time par excellence, with the study of the 800-80 year cycles of evolution of weapons.

Since if we apply the ternary principle and the decametric scales of most universal systems to the 3 ages of metal-history, we find that we can further subdivide those ages in sub-cycles of 800 years of civilizations, each represented by a bell curve of existence, between its birth and extinction by new hordes of metalmasters with a new weapons, in a rhythm tuned to the st+1 climatic rhythms of the Earth:

The Bronze age (±2800 BC, ±2000 BC), Chariot age (±2000 BC, ±1200Bc), Iron age (±1200BC, ±400BC), Coin age (±400BC, ±400AD), Stirrup age (±400AD, ±1200AD), Gunpowder age (±1200AD, 1945 AD) and Digital age (1945AD, extinction). In each of those ages a new horde of warriors with an icon-weapon is born as a top predator culture that destroys all other cultures of the old continent;

reproduces and colonizes other territories and finally it dies away; substituted by a new culture, with a more evolved weapon or religion. We can also use the bell curve of social existence to the entire life of mankind, in the age of metal.

In the 3rd age of machines, the cycle of evolution of weapons accelerates as all systems of information do. Thus in this modern age of machines nations are short-span civilizations that die in a mere 80 years, according to the complex laws that establish decametric, geometrical scales and power laws for any process of evolution. And we distinguish, the age of machines=weapons of steam or British Age (1780s-1840s; the age of chemical engines or German Age (1850s, 1940s) and the age of electronic machines or American Age (1930s, 2010s), followed by the Age of organic machines or Robotic age (XXI century) that could be the last age of history if humans don't learn how to control the evolution of weapons.

Recap. The evolution of the FMI-complex had 3 ages: in its young age, it was submissive to human social organisms. It was the age of wor(l)ds, in which prophets dominated with ethic words the evil-anti-life memes of metal. There were only bronze weapons and gold bars, which acted as money. Their maturity came with the expansion of iron weapons and coins that multiplied the quantity of entities money could prize and the power of weapons. Mercenary armies came on top and the law substituted verbal ethics. It was the European Age. Finally in its informative age, the FMI complex added a reproductive system of machines, organisms that transform back and forth energy into information and now are fusion its 1st age (bodies of machines) and its 2nd age (minds of machines) into robots. It is the 3rd r=evolution of machines that will make humans obsolete.

13. Scalar analysis.

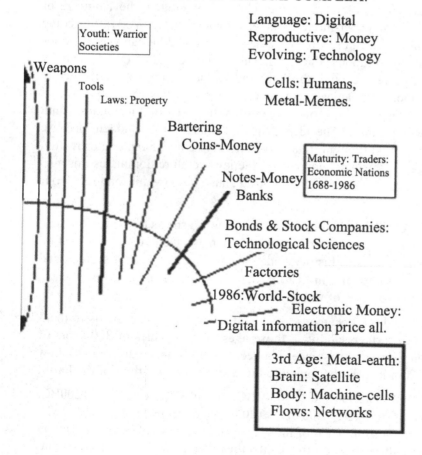

SOCIAL EVOLUTION OF THE FMI-COMPLEX:

Youth: Warrior Societies

Weapons
Tools
Laws: Property

Bartering
Coins-Money

Notes-Money
Banks

Maturity: Traders: Economic Nations 1688-1986

Bonds & Stock Companies: Technological Sciences

Factories

1986:World-Stock
Electronic Money:
Digital information price all.

3rd Age: Metal-earth:
Brain: Satellite
Body: Machine-cells
Flows: Networks

Language: Digital
Reproductive: Money
Evolving: Technology

Cells: Humans, Metal-Memes.

In the graph, parallel to the evolution of human, social super-organisms, there has been an evolution, since the discover of money and weapons, of a global ecosystem of metal and symbiotic 'enzymen', which uses a different social language – digital information supported by informative metal – with different values. As money values more weapons than life and words, our language values more life than weapons. Because a super-organism is defined by the events coded by the language, in an economic world, money reproduces the object of more value weapons and machines, and humans impose their power to non-technological groups with them. Thus according to the evolution of the language of money, we distinguish a series of 'market' civilizations, that went from an age of bartering, when money was not essential and words still were used to 'qualify' *the value of goods, dominated by priests and warrior cultures; to the coin age dominated by bankers which made 'pecunia' the 'nervi' of mercenary armies, organized in city-states and empires, to the modern world of machines (3rd age of metal-history).*

Then money evolved further, first as paper-money that represented a piece of a company of weapons (gunboat age) or machines (Steam and Chemical Ages of trains and cars), or a loan of banks to states to finance their wars (bonds). Finally with the arrival of electronic machines money evolved into a pure languages of digital information in the mind of a global mind-networks of computers (global stock-market). E-money is today a flow of global information that organizes the reproduction and evolution of machines into an automated global super-organism. In this last phase, men are put in competition with electronic blue collar machines (robots) and white collar machines (computers), according to the equations of productivity= machine/labor and fired when a new machine makes better his job. The result is the 10th extinction wave of life beings - a world in which humans are increasingly obsolete, displaced from the Earth, as mammals were displaced by humans in the 9th extinction age of life (Paleolithic).

This new brave world of selfish metal-memes ruled by animetal cultures has had 3 ages/cycles/scales: the long cycle of 800 year civilizations ruled by the memes of war, weapons and warrior ideologies; the medium cycle of 80 years nations, ruled by the memes of go(l)d and corporations, and the short cycles of electronic machines, in which science of measure and techno-utopian ideologies rule supreme. Let us then study those 3 cycles of evolution of the FMI complex of which we are all today a living part.

Recap. The FMI complex has grown as a superorganism of machines, money, weapons and 'enzymen' reaching global size.

V. 800 CYCLES OF WAR AND DEATH OF CIVILIZATIONS

In the graph, on top we can see the 800-80 years wave of evolution of weapons, money and machines (the FMI system) and how they intersect and extinguish the wave of human superorganisms, cultures based in the eusocial laws of love (the natural tendency of self-similar cells to share energy and information in any super-organism). The result of the intersection of both super-organisms is the 'economic ecosystem', in which 2 different species, metal and life, sometimes symbiotic sometimes in open competence in labor and war fields fight for the use of space and resources in planet Earth. Yet the dominant complex is the FMI complex, which has halted the evolution of the super-organisms of history and become a global super-organism with the discovery of its reproductive network (companies of machines) and its collective informative 'brain', the internet.

14. 800-80 years cycle of wars and extinction of cultures.

A science of history and economics must be cyclical because time is cyclical, and the species we study humans, behave in the same manner throughout history. But all cycles tend to have either an accelerated, constant or decelerated speed. The cycles of history, being part of a galactic system, which is a vortex of information are accelerated cycles. This however seems to contradict the 'constant speed of human evolution'. So we have to ask ourselves what evolves in the modern age of History? And what are its cycles of evolution? And are those cycles accelerated, constant or decelerated?

The essence of those questions requires localizing certain repetitive events in time that the species we study do once and again and so we can forecast they will do it in the future.

Species will little freedom repeat the same cycles once and again, species with more freedom, less often. In history and economics certain events repeat and give way to cycles.

The most obvious cycle we observe in any species is that of life and death, of creation, reproduction and extinction.

But since we do not study individual human beings (biological subjects) but cultures (historic subjects) and products, mainly machines reproduced by corporations, its company-mothers (economic subject) the cycles we must observe are different - those of the life and death of civilizations and types of machines and weapons.

In that regard, civilizations can be modeled as super-organisms: they go through 3 ages, as all organisms do.

In Complexity, we have spotted a cycle of life and death of civilizations of around 800 years for all the cultures of the Old World, related to the discovery and massive reproduction of new weapons, which accelerates after the Industrial R=evolution in a decametric scale, due to the professional development of machines and its top predator version, weapons. We call those ages of massive reproduction of weapons, a "biological radiation", borrowing the concept from biology, where biological radiations of top predator species are the main cause of extinction of weaker species. So happens with "radiations" of top predator metal weapons that cause the extinction of humans and our civilizations. *Thus the 10th radiation of 'life-forms', metal-weapons and machines, which coincides with the 9th extinction of all life species on planet Earth, carried about by 'symbiotic animetals', men enhanced with metal weapons, can be divided in a series of sub-cycles of partial extinctions:*

The long cycle of 800 years civilizations is dominated by the memes of war

In the graph, the first step in the creation of a science, is finding cyclical patterns of behavior that allow predicting the future on the belief those cycles will continue to happen. Because in history we study cultures, nations and civilizations, this first step requires, from a biological perspective, to find regularities on the cycles of creation, reproduction and extinction of those cultures. And surprisingly enough after an exhaustive research, a pattern arises: cultures die rhythmically every 800 years, when the massive reproduction of weapons and money to finance wars, provokes an age of destruction and substitution of the elites on top of those cultures. In the graph the cycles of bronze wars (±3000 B.C.), chariot wars (±2000 B.C.), iron wars (±1200 B.C.), coin wars (±400 B.C.), stirrup wars (±400 A.D.), gunpowder wars (±1200 A.D.), and digital weapons (Nuclear Bombs and robots, ±2000 A.D.) The regularity of those cycles is not forced and can be related to the evolution of weapons and the weather cycles that multiply migrations of nomadic hordes of Indo-European warriors that descended on the agricultural, fertile regions of the tempered zones. For example, Rome, the paradigm of mercenary empires, constructed with the arrival of coins, which established the reign of the financial-military system on top of human societies ever since, lasted 797 years between the Fall of Rome to the Iron Gauls and the Fall of Rome to stirrup barbarians. The cycle however accelerates during the gunpowder age, as the evolution of weapons thanks to the discovery of scientific machines accelerates to a mere 80±8 years, a human generation in which a European nation on top of the evolution of those weapons, becomes the top predator culture of the world. In all those cycles the top predator civilization or nation with the best memes of metal, weapons, money and during the 80±8 years cycle, machines, will try to conquer the world and establish its civilization, but none will succeed, because what truly evolves are the memes of metal, the technology of greed and death.

The short 80 years industrial cycle is dominated by the memes of go(l)d and mechanism

Those cultures that were immortal before the discovery of hard-metal, die now rhythmically every ±800 years. And finally with the discovery of gunpowder applied to gunboats and the organization of company-mothers that reproduce first those gunboats and then machines, the process accelerates geometrically as all systems of information do, to ±80 years, the cycle of Industrial nations:

First Venice, Genoa and Firenze, city-states, carry Italy to the top of the world in 3 cycles that start with the invention of the first gunboat, the 'Lombarda' and its use in the 'Sacco di Constantinopoli' (1208) that gives the 'ring of go(l)d' and the sword to Venetia. Then in the Battle of Chioggia, 1284, Genoa takes the ring; finally Firenze, in the XIV century displaces Genoa as the site of the Papal Bankers – the Medicis.

Then the improvements on gunboats and the gold of Ghana puts Portugal on top (1440s). The 80 series continues with Spain (1525, fall of Mexico till 1604, when the invention of the first stock-company of paper-money in Holland, is soon followed by the crash of the bullion and the defeat of Spain). Holland lasts from 1604-1688, when French's artillery destroys Amsterdam. France (1688, 1760s) is defeated by Great Britain, loses India and Canada and her Companies crash while the country enters in Revolution.

Finally Great Britain ushers the world in the age of 'peaceful machines', which can also be used as war machines (1770s discovery of the steam machine, 1857, crash of train companies and use of armored trains in colonial wars).

But Germany in the 1860s discovers the oil and electric engine, and so it carries the wave of evolution of machines=weapons till the 1920s, when after its I world war defeat the Mark crashes. It will be the turn of America (1929, 2001, electronic age) with weapons researched with digital computers (Nuclear weapons, electronic wars) and e-money. But at the end of the cycle the overproduction of e-money and Pcs ushers the American economy in the age of robotic wars, speculative, massive reproduction of money and e-money crashes in which we live.

Since all the cycles of the 800-80 years destruction of civilizations and nations follow the same pattern: a new weapon that backs a form of money reproduce in increasing numbers till at the end of the cycle the excess of money pays a massive radiation of biological weapons that start a global age of war, which destroys the civilization. So we are now living the war age of the end of the electronic cycle and the beginning of the Singularity age when machines will become organic, autonomous robots and compete directly with man in labor and war fields, making us obsolete.

Recap. The evolution of money and weapons follows an accelerated cycle of 800-80 years of biological radiations of weapons and the tribes that carry them and reproduce in enormous numbers till creating a global age of war. The cycle accelerates to 80 years with the discovery of war machines.

15. Life and death of civilizations.

The study of societies as organisms and the analysis of its life and war cycles have a long tradition in Philosophy of History, since Vico noticed that societies die rhythmically in a dark age of war. Spengler ('The Decline of the West' 1918) for the first time used the organic parallel in depth, considering that the collective subconscious of a social organism, its culture, also went through 3±1 ages of art similar to those of a living being: a young age of epic art, an age of classic art and a 3rd, baroque age, prior to the extinction of a civilization. The parallelism between a biological organism made of cells related by networks of energy and information (blood and nervous systems) and a socio-biological organism made of human citizens related by networks of energy (economic systems) and information (cultural systems) is self-evident within the systemic paradigm. And it allows to design with biological, natural laws the most efficient economic and political system that maximizes the survival of its organic cells/citizens. Arnold Toynbee ('A Study of History' 1934) did an exhaustive analysis of the demise of civilizations, pointing out to natural and ideological causes, without a clear quantification of those cycles. Such quantification of the life and death cycles of civilization is now possible thanks to the understanding of cyclical time, the life-death cycle and the predatory nature of 'biological weapons'. Every ±800 years or 10 human lives all cultures in the old world become rhythmically destroyed by warfare due to the massive reproduction of new weapons in an age of climatic change and nomadic invasions. Those dark ages foreseen by Vico, Toynbee and Spengler can now be explained with higher detail with the Ternary method and the biological, organic concepts of bio-history and bio-economic.. And so we talk of the 7 cycles of metal-history since the first bronze armies arrived to Mesopotamia and destroyed the Genesian paradise, depicted in the "Genesis", the oldest book about the tragedy of History written by mankind.

We know the cause of cultural death, which is war, and since wars have come to History in certain ages, after the discovery and massive reproduction of a new weapon, we can trace a series of waves of existence, of the great civilizations of history, that have gone through birth, growth and extinction. Indeed, the discovery and massive reproduction of weapons carried by hordes of symbiotic warriors have shaped the history of civilizations, killing them off periodically.

Birth, life, reproduction and death of civilizations.

Cultures resemble living systems. They go through the same cycles of birth, reproduction and death. Cultures are born as organisms do, out of a single or a few prophetic or legislative individuals-cells that reproduce their ideas into many "believer-minds". So Christianity is born out of the mind of Jesus Christ that first expands into 12 disciples, which expand themselves into thousands of Roman believers that expanded into billions of other minds. While America is born of the mind of a few legislators that craft a constitution latter expanded into more laws, obeyed today by all American citizens. When the religious or legal culture reaches maturity it often reproduces in another zone of space-time, creating a daughter civilization or colony. Finally, when its ethic laws and networks of information, and the energy they control, decay and become obsolete, the culture dies and a new, more evolved civilization destroys the old civilization. A new cycle of history starts.

Similarly if we consider technological civilizations, based not in verbal laws, but in monetary and military systems, first a weapon is discovered, and reproduced by a small horde, which becomes an army, which conquers a nation and becomes a ruling aristocracy, that imposes its customs, and controls with the icon-weapon the civilization.

Birth

In both kinds of civilizations (Human-Artistic civilizations or Metal-scientific civilizations) either an ideology or book of ideas, (a network of human information) or a species of metal, money or weapons, becomes the language of power which defines the civilization.

Ideas and metal-instruments together shape History. For that reason Historians, when they focus on military systems of power, divide history into the Age of Bronze, the Age of Iron Empires, the Age of Chivalry (Middle Ages), the European Age of Gunpowder (Modern age), or the Atomic Age. There is also the Christian Age and the Buddhist Age of South-Asia, when they focus on the cultural component of the civilization or the age coins, or the age of paper money and stocks, or the age of e-money (new species of money), when they focus in monetary information.

Reproduction of Cultural Information: Daughter Civilizations

At the same time a civilization develops the memetic ideas and machines which carry that civilization to success, it eliminates previous civilizations. In this manner, successful civilizations reproduce their "genetic ideas and machines" into other regions of the Earth, creating colonies, daughter cultures and eco(nomic)systems.

Like a human child, the daughter civilization is often a more evolved species where the ideas and instruments of the previous civilization are improved. Civilizations inherit and improve on, like children do, their parental civilizations, often extinguishing them. So the Assyrians were defeated by the Persians who had copied and improved its weapons, iron and cavalry. Yet Persian art, even behavior would be similar to Assyrian art and behavior. Genetic continuity in civilizations is not racial but cultural, of structures and ideas, of icons, machines and modes of life. For example, in Arabia stirrup cavalry brought an empire-civilization that expanded West till Al-Andalus (the South of Spain). That stirrup civilization migrated to America and colonized the southwestern states, giving origin to the "Old Far West" civilization, with horse and cattle ranching. In that territory, the horse civilization flourished between the 17th and 20th centuries, when the fundamental icon-species of the culture, the horse became extinguished as a form of transport by the car, a more efficient metal-species. The example shows the dynamic quality of the life of civilizations. They are born, reproduce and become extinct through the "genetic evolution" of objects and words, which bring as much diversity to cultural evolution of its species, products and ideas, as genetic memories bring to the body.

And in the same manner biologists can scientifically organize the data they have about species through genetic information and instruments. Thanks to the memorial remains of history, the historian can study the birth, reproduction and extinction of evolutionary history and its civilizations.

Death of Civilizations

Civilizations also die. One day another civilization with more efficient memetic, eusocial ideologies of man, or better machines, comes along and destroys the old civilization. As in "animal feeding", once the new civilization has destroyed the old civilization, it reforms the energy cells=citizens of the dead civilization to its image and resemblance. Let us again bring the biological comparison. A hunter feeds on the energy of a victim's body that becomes in part wasted (killed) and in part "transformed" into cells of the predator. This also happens in civilizations: part of the 'human energy" is wasted by war, and the remaining humans are "transformed" into believers of the predator civilization. The Spanish perhaps killed some 70% of the Indians and the rest became "Christians". It is the death of the civilization. Because when ideas and instruments change, the civilization changes. The iron civilization of Rome is not the same as the Middle Ages civilization of stirrup weapons imposed by German warriors. The Roman civilization had been destroyed by "stirrup animetals" of the Germanic civilization. Again such "civilization" is not the Gunpowder civilization of Italy during the Renaissance. Gunpowder wiped out chivalry and a new civilization was born. Nor is Ammon-Egypt the same as Greek-Christian Egypt, or Arab-Muslim Egypt. The verbal ideology changed, and the Egyptian civilization became extinct.

It is now clear the difference between religious civilizations and metal-civilizations. Religious civilizations could exist for ever, if there were no wars to extinguish their cultures. That is the goal set by the prophets in their messages of love. And yet, because metal exists, all religious civilizations of the past 5000 years have been corrupted and periodically extinguished by war. Some of those cultures resurrect, when the wave of death and war recedes. Some do not come back again. The Christian and Islamic civilization, the Jewish culture has survived many cycles of death, and the words of Moses, Christ and Mahomet, have resurrected. But each wave of death, corrupts a little bit more the initial messages of the prophets with rituals of war and money. So we find today that those cultures are basically built around inquisitions and churches, corrupted by rituals of Go(l)d, and little is left of the initial mandates of the prophets. Many other human cultures based in human senses have died away, and its artists have expressed those deaths, in their baroque ages of art.

In other words, mankind has chosen to evolve the wrong memes of history, not those memes of eusocial love that could have created an efficient, global superorganism of history according to the laws of the Universe but the memes of metal, of economic ecosystems, weapons, money and machines that create a different superorganism of history, which rhythmically kills life in this planet.

This duality between cycles of creation and destruction, light and dark ages in history has always been understood by the scientists of history and prophets of love, which have both, tried to tame in the middle of the cycles the extinction of history, and during its 3rd baroque age, have forecast with its artistic angst and prophetic admonitions the dawn of their civilizations… But as the rhetoric of the animetal cultures of war and money became more sophisticated with the help of industrial art and mass-media, the memes of love have declined till today, *paradoxically when we are closer to our extinction by the machines we worship and those messages of social love are more necessary, since unfortunately the salvation of history is an ethical question –not an intellectual one.*

Indeed, the historic record is conclusive: the verbal, life-based, social organism of the Neolithic formed a global civilization, which was immortal and generation after generation men remained unchanged. And this is the only ideal that could save history: an world made to the image and likeness of man and life, as the collective mind of the living, organic planet Gaia, in which a world union of political governments in charge of the languages of power of society, money, weapons and laws, control the evolution and reproduction of the lethal goods of the tree of science. Yet this solution clearly expressed in the first Neolithic books of prophetic History (Genesis) was forgotten with the arrival of weapons.

Recap. Each of the life and death cycles of civilizations show the eternal fight between cultures with life and love memes, represented by religions of love and social civilizations vs. the cultures of selfish memes - weapons and money. Unfortunately because iron kills the body of flesh and gold hypnotizes and enslaves the mind, the selfish memes of the 'tree of metal ' have always won history, killing rhythmically the civilizations of Gaia, of life and love.

16. Topological analysis: the 3 styles of art.

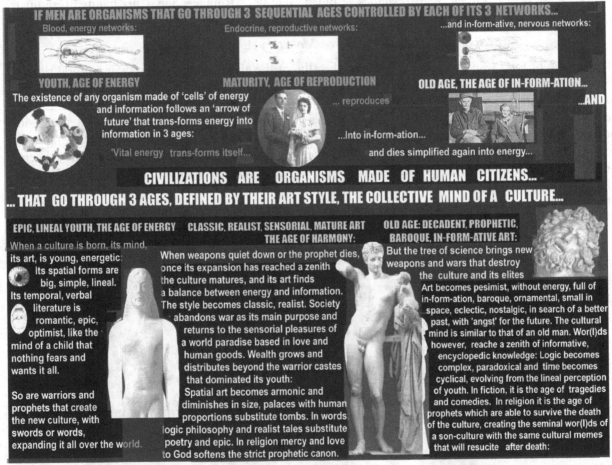

IF MEN ARE ORGANISMS THAT GO THROUGH 3 SEQUENTIAL AGES CONTROLLED BY EACH OF ITS 3 NETWORKS...

Blood, energy networks: Endocrine, reproductive networks: ...and in-form-ative, nervous networks:

YOUTH, AGE OF ENERGY   MATURITY, AGE OF REPRODUCTION   OLD AGE, THE AGE OF IN-FORM-ATION...   ...AND

The existence of any organism made of 'cells' of energy and information follows an 'arrow of future' that trans-forms energy into information in 3 ages: ... reproduces   ...Into in-form-ation...   and dies simplified again into energy...

'Vital energy trans-forms itself...

CIVILIZATIONS ARE ORGANISMS MADE OF HUMAN CITIZENS...

... THAT GO THROUGH 3 AGES, DEFINED BY THEIR ART STYLE, THE COLLECTIVE MIND OF A CULTURE...

EPIC, LINEAL YOUTH, THE AGE OF ENERGY

When a culture is born, its mind, its art, is young, energetic. Its spatial forms are big, simple, lineal. Its temporal, verbal literature is romantic, epic, optimist, like the mind of a child that nothing fears and wants it all.

So are warriors and prophets that create the new culture, with swords or words, expanding it all over the world.

CLASSIC, REALIST, SENSORIAL, MATURE ART THE AGE OF HARMONY:

When weapons quiet down or the prophet dies, once its expansion has reached a zenith the culture matures, and its art finds a balance between energy and information. The style becomes classic, realist. Society abandons war as its main purpose and returns to the sensorial pleasures of a world paradise based in love and human goods. Wealth grows and distributes beyond the warrior castes that dominated its youth: Spatial art becomes armonic and diminishes in size, palaces with human proportions substitute tombs. In words logic philosophy and realist tales substitute poetry and epic. In religion mercy and love to God softens the strict prophetic canon.

OLD AGE: DECADENT, PROPHETIC, BAROQUE, IN-FORM-ATIVE ART:

But the tree of science brings new weapons and wars that destroy the culture and its elites

Art becomes pesimist, without energy, full of in-form-ation, baroque, ornamental, small in space, eclectic, nostalgic, in search of a better past, with 'angst' for the future. The cultural mind is similar to that of an old man. Wor(l)ds however, reach a zenith of informative, encyclopedic knowledge: Logic becomes complex, paradoxical and time becomes cyclical, evolving from the lineal perception of youth. In fiction, it is the age of tragedies and comedies. In religion it is the age of prophets which are able to survive the death of the culture, creating the seminal wor(l)ds of a son-culture with the same cultural memes that will resucite after death:

Historic facts are conclusive about the extinctive process of human civilizations at the hands of weapons: Copper appears in the Upper Neolithic and displaces the stone cultures of the Paleolithic. But copper is not very strong and its capacity to extinguish=kill human species is mild. Next, bronze appears and swords become perfect energetic=lineal species, able to cut and kill human bodies at ease. The first social castes of hard warriors take control of history. They expand worldwide. And wherever they go, they extinguish previous human cultures. They establish hierarchical societies, controlled by metal masters, backed by weapons. Hierarchies are then implemented through legal codes, epic, rhetoric art and social and religious rituals, built around the Top Predator icon weapon and the warrior caste that considers itself chosen of God, with rights to govern and prey over all other human beings. Ever since, animetal ideologies of history have imposed their subjective truths and myths with the power of technological death and rhetoric art. Yet at the same time art becomes shallow and corrupted, the best artists, whose essence is to be the wor(l)d and eyes of human thought, denounce that death in a baroque, 'angst' period of art. Since art is to a social organism what the mind is to a biological organism. And so in the same manner our mind goes through 3±1 ages, parallel to those of the body in his vision of reality, in a social organism art goes through the same 3±1 ages, but in a longer scale—that of the life of the civilization and its multiple neurons, the human artists that perceive the society, as it is born in a young age of enthusiasm, matures and finally collapses in the angst of death:

Lineal Young, Art: The Warrior, Epic Age of Civilizations.

Young art is lineal, epic, dramatic, simple, energetic, as the new weapons and conquering warriors that found the civilization are. The Greek and Roman hordes of Hoplites and Legionaries or the Middle Age hordes of Stirrup horses leave cultural remains that follow those lineal forms, fostered by their military obsession with death: Greek Kuroi statues, Roman Temples, Romanesque Christianity, Calvinist churches in the XVII, Egyptian Pyramids in the Old Empire.

Classic Art: The Reproductive, Monetary Age of Civilizations.

After an age of total war, those metal-masters, the elite of the civilization, evolve their cultural rituals and use cyclical money to control people in a milder way than the hordes of warriors. It is the second horizon or trader, reproductive age of the culture, of maximal wealth, when it reaches its apogee. Lineal Greek Kuroi become the realist sculptures of Phidias. The simple, Romanesque cathedral grows in informative height and light, in the classic Gothic. It is the age of Athens, Augustus or XIX C. England. Often in this age, the civilization reproduces in other regions through empire building. The austere Republic expands into the Roman Empire. Inquisitorial Spain reproduces into Latin America. Puritan England expands its empire in multiple clone colonies.

Baroque Art: The Extinctive Age of a Civilization.

UPPER PALEOLITHIC: YOUNG, LINEAL ART MIDDLE: CYCLICAL REPRODUCTIVE FORMS LOWER (SAPIENS): ORGANIC, INFORMATIVE FORMS

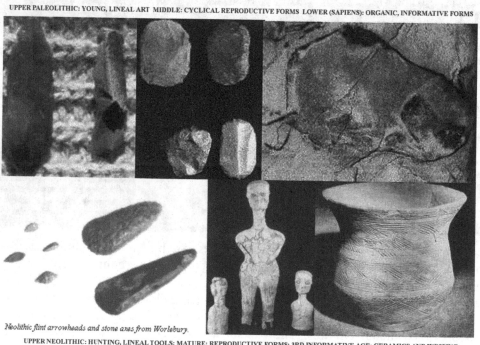

Neolithic flint arrowheads and stone axes from Worlebury.

UPPER NEOLITHIC: HUNTING, LINEAL TOOLS; MATURE: REPRODUCTIVE FORMS; 3RD INFORMATIVE AGE: CERAMICS AND WRITING

557

In the graph, after the lineal forms of the upper Paleolithic and the cyclical pebbles of the Middle, the Lower Paleolithic men had a final burst of creativity in his baroque age, prior to extinction. It is the age of cave paintings that tried to bring back the hunting, which hot weather and farmers with copper weapons (Chalcolithic), were taking away from them. Yet again, when the European, Copper Neolithic becomes extinct by bronze warriors, it leaves its artistic, religious masterpieces—the megaliths of Stonehenge. The 3 ages of the Neolithic with its lineal weapons, fertility goddesses and elaborated ceramic decoration, which gave birth to writing, in its informative last age are shown below.

But sooner or later a new weapon=war radiation takes place and the civilization dies. It is its 3rd horizon; the artistic or baroque age of maximal information and creativity in art. The culture looks inwards, as it becomes constrained by a new horde of 'barbarians ad portas'. We are now in the Hellenistic age of tortuous sculpture; when Macedonians and Romans dominate Greece; or the lower Roman Empire, when German warriors invade it. It is the XV century of Christianity, the baroque, Ornamental Gothic, when gunpowder and coins break the Church's control on society. It is Post II World War England, when the dollar and the Atomic bomb have substituted the Pound and the Gunboat. So now England offers rock stars, Bacon and Orwell, who show the decadence of the Human Wor(l)d at the end of the II Kondratieff cycle.

Lineal art and epic literature belong to the first age of a culture that copies and worships weapons and warrior power. Cyclical art and realistic, sensory literature belong to the age of traders and cyclical coins. Finally, there is a massive explosion of humanistic and prophetic art and literature in the baroque age, when artists foresee extinction. We see those baroque tendencies in all dying cultures, illustrated in the graphs of this paragraph. In the modern age, however we observe a global civilization and a global baroque:

In the 20s, the extinction of realistic painting by photography and film brings the baroque Picasso that portrays blue corpses, foreseeing the bodies of the Holocaust. In the American cycle the extinction of literature by TV programs made by corrupted artists, working for a mass-media that glorifies violence, brings the response of real art in baroque films like Matrix or Terminator that foresee our demise by machines. But Human artists and prophets can't change the animetal warrior aristocracies and plutocracies of any culture, which have a strict mental 'agenda', based in his mechanist, tribal and capitalist beliefs, which blind them to any collateral effects of the memes of metal.

Today, when we enter in the last cycle of wars, the age of digital weapons – robots and e-wars - only the artists of the last angst 3rd age of the globalized culture, the Human Baroque, who prophesized that Robots become Terminators, decades ago in film 'understand' the future, which techno-utopians deny.

Recap. Art is the mind of civilizations, organisms of history. As such it also goes through 3±1 ages: an epic, lineal youth; a realist, classic age and a 3rd baroque age of excessive in-form-ation that coincide with the 3 ages of civilizations.

17. The 7 cycles of civilizations in history.

Let us then see those 7 cycles of death, not only from the 'official' perspective of our 'corrupted scholar' culture that glorifies the empire building of those warriors and the mechanist discoveries of our scientists, but also through the eyes of those artists of human thought that 'know better'.

<u>*Bronze Age. 3000 BC*</u>

I Age: Epic	II Age: Sensory, Logic	III Age: Baroque, extinctive
'I Mykerinos	*'It was a good land.*	*'Yesterday is gone.*
Son of Ra-Ammon	*Figs were in it. It had*	*There is not today.'*
Lord of the World.'	*more wine than water.'*	
Mykerinos, 2548 BC	Sinuhe 2000 BC	Book of Dead 1600 BC

Pyramids: *Fresco: V dynasty*   *Horus:God of the Dead*

Lineal, bronze swords create the first empires recorded in History. Indo-European and Semite cattle nomads, from the Northern and Southern Deserts of ice and sand, discover hard metal and plunge the Fertile Crescent, *enslaving human beings, treated as cattle* ever since by their animetal masters:

In ±2850 BC. the first Bronze armies appear, taking to power God-pharaohs and Semitic Lugals. Egypt becomes unified. History registers the first Mesopotamian Empires. Rhetoric, epic art to the service of power, substitutes temples with palaces and lineal tombs, (pyramids) to glorify the warrior king. Yet for most humans, life did not improve. Farmers become enslaved by warriors and work extra-hours to build those huge monuments to the arrogance of the animetal. In fact, in those cultures where bronze weapons arrived later, human goods and sensory pleasures develop more. For example, in China and the Indus, without bronze, the quality of housing is the highest of all old civilizations. Then it plummets after Aryan charioteers invade India. In Egypt, only as the initial waves of warriors recede, we see the end of Pyramid construction and a higher quality of life for the common people (VI dynasty). Tombs now are smaller, full of lively drawings of animals and farmers; while priests regain power over warrior kings, promoting the production of human goods and social rituals, till new warriors arrive.

In the graph, we can distinguish those 3±1 ages in the art of the Old, Bronze Empires:

— *Max. Energy:* When the I dynasty of military warriors, coming from the south, unifies the country, Art goes through a simple, lineal, epic age of Empire building. Arrogant warriors, obsessed by death, build pyramidal tombs of growing size and hieratic self-portraits of Pharaohs, seen as Gods by the power of the sword (*the Mykerinos Group*).

— *Classic, reproductive age* (energy and form are in balance): After Egyptian pharaohs settle down, a pleasant life of wealth spreads to other social castes. It is the period of classic, realist art, epitomized by the V Dynasty with its smaller tombs decorated with beautiful, realist paintings of human goods. It is the age of classic literature: works like 'Sinuhe', who narrates the transition between the classic and decadent, formal period.

— *Max. Information:* At the end of the Bronze Age, a period of disorder, corruption and new wars starts again, when foreign invaders with new chariot weapons take over the delta. Wealth plummets and the country enters a baroque age of war. A cult to death and the fantasy of resurrection dominates the culture. It is the age of 'The Book of Death' and the myth of Osiris brought to life by her wife Isis. The portraits of Horus, God of the Dead, become popular, substituting the epic cult to Ra, the father of all pharaohs. Those 3±1 ages and religious icons will become the recurrent ages of the 2 next Egyptian cycles (Middle and New Empires, in classic historiography).

The written word.

A telling story of the difference between humans and animetals that confront each other's vision of the future of mankind in the paradox of history is that of the written word: Since China still has not metal, it appears as a tool for priests and Taoist masters to guess the future cycles of Nature. The language is poetic, made with organic, complementary oppositions between energy-yang and information-yin, still natural to the Chinese language. It appears also in militaristic Egypt and it becomes Law, imposed by the Pharaoh, which says 'It has been written', when he speaks; so his word becomes law. The warrior has created the first of the many straitjackets of our animetal civilization: the political law, which gives

559

privileges to aristocrats that now have the exclusive rights to bear arms and will use them against all other human beings, without ever being judged by their acts of murder. Meanwhile a self-similar second type of Laws appear in Mesopotamia, the civilization that invented money, where the high priests, the first go(l)d masters, have the 'commercial right' to produce money with the exclusion of all other groups of society. They will establish taxes and regulations about prices, usin, the written word first as a system of accountancy. So the word adapts to the 3 languages of human power, words, weapons and gold.

Chariot Radiation (Wheels of bronze: 2000 BC).

The first wave of charioteers, which develop the 'cyclical, bronze wheel', a more complex form than the lineal sword, starts around ±2000 BC. A worldwide invasion of Indo-Europeans with chariots from the northern plains destroys all civilizations. It is the extinctive age of Sumerian cultures, when Genesis and the parable of the tree of science and its go(l)den apples, understood by the verbal masters of the age, is written. Charioteer Hordes conquer Sumer and Akkad. The first Hittites empires appear. In Egypt, it means the end of the old empire and the beginning of the Middle Kingdom.

A proof of the Paradox of History -the fact that with less weapons life is better- is Crete: There, charioteers cannot arrive due to geographical conditions (it is an island). So in Crete there is a continuity of a culture based in human, life-enhancing goods that becomes the richest, most sophisticated civilization of this age. Meanwhile, the radiation of Charioteer Hordes erases Eastern cultures. From 1500 till 1000 BC. in the first Veda Period, nomadic, chariot Aryans and their sacred cows conquer India, imposing a brutal system of racism (castes), against Neolithic farmers. While, the old culture of Goddesses and fertile rivers survives longer in the South. Charioteers also founded about 1500 BC, the Shang dynasty in China, degrading and partially assimilating the Tao culture of the living Universe, proper of Neolithic China. Yet in the south, the depth and richness of the Chinese Neolithic (fully grown and densely populated by the II millennium BC), allows the Tao culture of the Organic Universe to survive and flourish till about 300 BC., when the Ch'in dynasty brings iron weapons and forbids all non-technological books under death penalty. Then the wisdom of Taoism that understood so well the dual nature of the Organic Universe and its Yin-Yang beings, will be lost at philosophical level. Though it will inform the basic arts of classic China. Thus, China proves again that less metal allows more complex human thought. China alone among all Eurasian cultures shows, still today, a clear comprehension of the organic nature of human societies; and for that reason it is the most successful culture of mankind in terms of survival.

Meanwhile in the West, metal-masters keep simplifying human verbal cultures. Words become alphabets and Legal Codes that curtail the freedoms of human beings.

Since we have studied in great depth those 800-80 cycles and all its cultural elements, we shall after those brief examples race in this introductory work through the following cycles:

Iron Radiation (about 1200 BC)

Germanic people (Celtic hordes) start mass production of iron swords. A new chain reaction of war and destruction starts, with warrior waves that move west to Iberia and east to Greece and Asia. It is the age of Invasions of 'People of the Sea'. Ramses fight invaders but can't prevent the collapse of the Middle Empire in Egypt. Also the Hittite Empire collapses. Hittites flee and teach Assyrians the use of iron.

In 1150; the Dorians destroy Mycenae. And again where the Dorians do not arrive, (Peninsular Athens, coastal Turkey) the Greek civilization of Cretan tradition flourishes and gives birth to the best cultures of the Iron discontinuity.

Coin Age (600 BC - 400 AD)

Around 560 BC coins are discovered and mercenary armies increase the professionalism of war. "Bellum ipse aleat', said the Romans who achieved this feat.

Around 539-525 BC. Persians destroy the Assyrian, Egyptian and Mesopotamian Empires, but copy their systems of military repression. As a consequence of better cavalry grounds, and iron metallurgy, from now on metal history displaces its center of gravity from verbal Empires on the crescent fertile ruled by Semite (S)words and Go(l)ds (Assur, God of war; Baal Go(l)d of trade) towards Northern Aryan tribes (Rome, Master of war; Greece master of "Coins"). The dominance of the social Wor(l)d is substituted by the dominance of the individualistic I=eye. Good and Evil, verbal values, are substituted by beauty and ugliness, which are values of light perception.

In the age of Iron, the Greeks introduced coins which ended the dominance of bartering in trade (verbal, Semite traditions), and started digital accountancy. The higher complexity of coins and iron will prove irresistible. It is the expansive age of Indo-European empires, that now use metal species in a far more efficient manner to impose restrictions of freedom to the people, taxed in coins, and levied to serve the armies of the metal master. Systems of "total" war, that only Assyria had essayed before, now become natural to entire empires, which summon up all the previous metal-advances (horses, iron, mathematical war analysis), into the Total Army, with infantry (Phalanx), cavalry, war Machines and logistics (Macedonian and Roman Empires). The Roman cycle is accurate; 798 years between the fall of Rome to the iron Gauls, and the fall of Rome to the stirrup barbarians in 410.

Rome is the foundational empire of the European World, origin of all myths and alibis for further Warrior empires that try to conquer the World, pretending to reinstall a new Rome. Rome is the parallel to Assur, the master of war of the Verbal, Bronze age. Because gold-iron races die, but its products and its rituals continue, there is in fact a continuity of rituals from Assur to Persia, to Macedonian empires, to Roman, to European empires. So much for racial superiority: the race dies in wars and Holocausts. The product, the iron weapon, stays. It gave origin to European ideologies, rituals and icons of warrior and trader civilizations from Britain to Nazi Germany, from Imperial eagles to civil architecture.

Cavalry Age. Middle Ages. Stirrup Radiation. (400, 1200 AD).

Lineal mosque: Kairuan IX C. Classic: Cordoba, XIIC. Informative, ornamental: Alhambra, XV c.

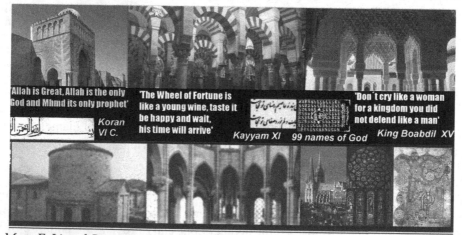

Max. E:Lineal Romanic;   E=I:Gothic   Max. I: Ornamental

In the graph, the 3 ages of Middle Age architecture in Islam and Christianity. The self-similarity of two parallel ethic civilizations shows the homology of all human cultures.

The Huns, Nomad warriors able to shoot with arrow bows while galloping, thanks to the stirrup - discovered probably in Korea- and the Germans who add iron protections, annihilate the Roman-Christian Empire of Human Goods, plummeting once more the standards of life of mankind. Around 317-589, the Huns break China between North and South into sixteen warrior kingdoms, controlled by

Huns and Turks. In 430 History registers the invasion of India by the White Huns. Their civilization is built around the "horse man" that sees himself "higher" than the rest of humans, just because he is mounted into his horse. Slavery becomes again endemic as Europeans provide weapons and 'human capital' exported by Radhanites to Islam[8]. The level of Human Goods proper of the Roman Empire crashes, but as Barbarians become converted to social love by the Christian 'Oikoumene' religion, the new millennium sees again a prosper, ethic Europe guided by its social priests.

The verbal age of Christianity, once Aryans are civilized, returns to Europe (High Middle Ages) in a society based on ethics. In Italy and Spain, priests of Islam and Christ regain power from converted warriors, creating a new age of human goods and art. They initiate the Renaissance and the Al-Andalus culture of, the summit of the two main religions of mankind, Islam and Christianity. Both flourish before the radiation of gunpowder, annihilates this re-birth of human sensory, ethical, and artistic cultures. Those final days of social religions left the great cathedrals, optimistic buildings full of light, looking upwards to God, built by the collective effort of the people that, unlike the Romanic churches no longer fear the violence of barbarians. Gothic cathedrals seem to confirm a wealthy mode of life in the lower middle ages with less wars. Population doubles in Europe; cities grow. In the Eastern World the old cultures of the sentient Universe (Taoism and Buddhism) reach a new apogee (Gupta, Song empires).

Gunpowder Age (1200 AD, 1945)

EPIC, LINEAL ART   CLASSIC, REALIST ART,   BAROQUE, FORMAL ART

The self-similarity of the art of the Iron, Coin Age (previous graphs) and the Italian Age is due to the self-similarity of the 3 civilizations based in iron infantry, go(l)d coins and Aristotelian rationalism. The 3 ages of both cultures are self-similar too: when Italy is colonized by Lutheran and Iberian hordes, it produces its Baroque art of maximal in-form-ation. When Greece is taken by Macedonians, it starts the Hellenistic age. In the age of the machine, humanist art is becoming extinct by technological art and real culture becomes baroque.

Chinese discover gunpowder. Mongols learn its technique in the assault of Beijing, spreading its use to the western world. In 1204 the looting of Byzantium by the Venetians opens the East to Italians who soon will have the first cannons, the 'lombardas'. In 1222 the Mongols defeat the Russians in Kalka and invade Germany. Soon gunpowder is known to German people. In 1258 Hulagu massacres Baghdad. It is the end of the Muslim Empire. In Europe the gunpowder wave soon explodes into massive civil wars,

562

of which the 100 years' war, is the most well-known. It brings scarcity of Human Goods that provoke massive famines and allows the expansion of the black Pest that multiplies death among Human beings.

Modern Age: Computers=Metal-minds

Digital weapons; 1945: The last cycle of computer-researched weapons brings the Atomic weapon calculated by Eniac, the 1st computer. It is a new threshold in the evolution of weapons that opens for the first time in History the possibility of extinction of our species as digital weapons kept evolving. In 1992 we had the first electronic war (Dessert storm) and today we are evolving the first Terminator weapons. Since each of those evolutions of weapons caused, soon after its discovery, the extinction of civilizations in worldwide ages of War. We are in the threshold of the 10th and last age of extinction of life, as the radiation of robotic weapons & atomic bombs spread beyond the control of the 1st digital Empire that discovered them.

Recap. Metal-civilizations are born when a new horde of animetals discover a new weapon, then it reproduces it massively starting a global radiation that destroys previous civilizations. But in the middle of the 800 year wave the horde of barbarians civilize, enjoys life and art evolves from epic, warrior art to a pleasant life-enhancing age that ends in a baroque angst as new hordes with better weapons come 'ad portas'.

18. Scalar St+1 analysis: the energy rhythms of Gaia.

In the graph, the 800 year cycles of war coincides with the rhythm of climatic changes on Earth: with astounding precision there is every 800 years a pronounced age of wet and hot weather, which as in all previous biological radiations of top predator species, multiplies the numbers of steppe warriors, starting nomadic waves of invaders, which use the new weapon of the 800 years cycle to invade and destroy all fertile agriculturalists civilizations of Eurasia. The peak of those dual events happened during the iron wars, the hottest, wettest period of the last 10 thousand years, which signaled the end of the Neolithic and the expansion of animetal cultures in all the regions of the Eurasian and African continent.

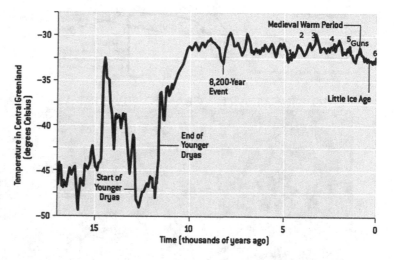

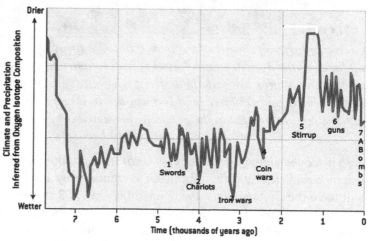

The evolution of weapons causes an 800 years cycle of death of civilizations extinguished rhythmically by new 'weapon radiations', when their discovery and reproduction by warrior hordes starts an age of war in Eurasia, as warriors expand their lethal energy, killing people and erasing human cultures. A radiation is a biological term that explains the sudden, explosive

563

reproduction of a new top predator species that annihilates many previous species, as it spreads all over the world. It is thus the proper name to explain the sudden bursts of violence provoked in bio-history by the discovery of a new weapon. Indeed, the discovery and massive reproduction of those new weapons, carried by hordes of symbiotic warriors that spread throughout the Eurasian continent, causes an 800 years cycle of extinction of civilizations, as those hordes of warriors conquer, kill and erase human cultures.

Those cultures live ±800 years, ten times the biological life of a Human being (±80 years), suggesting 2 possible causes for those war cycles: the biological rhythms of human life that after 10 generations seem to enter in decadence and the climatic cycles of the Earth, whose periodic ages of hot and cold weather coupled with wet rains and draughts on the steppes, displacing nomads towards valley cultures:

Those radiations happen after an age of wet and hot weather that multiplies pasture and the populations of the steppes. Thus, while the exact discovery of each weapon is not clear, the multiplication of the animetals that will provoke the biological radiation is clearly marked as all

previous biological radiations and extinction of species in Gaia by the rhythms of the st+1 planet.

Conclusion: The Economic(eco)system: metal preys on life.

It is thus clear that the sum of the 2 superorganisms of history (human cultures) and economics (the FMI complex) form an economic ecosystem in which the FMI complex and its 'attached' animetal cultures prey over the human superorganisms and its life/love based cultures since the Bronze age, and that process of extinction of life cultures accelerates with the arrival of machines and again with the arrival of computer brains.

The conjunction of climatic changes that provoke massive invasions of nomadic warriors brings about a fundamental question of bio-history – are we programmed by mother-earth to create metal-species stronger than us, which will kill our species, once we evolve them into organisms (robotic weapons)?

Indeed all the civilizations of the 800 year cycle die after the discovery and massive reproduction of new weapons in periodic ages of war, but the now globalized civilization of mankind based in the memes of metal is a single civilization and so at the end of the digital age of weapons we should be extinguished by the new 'autonomous' robotic warriors we are evolving. And that simple fact of evolutionary theory can only be halted if we halt the evolution of the FMI complex and manage it for the future of mankind, not for the profits of a few people castes (its stockratic owners and the military and dictators that profit from war).

Even the American civilizations that were isolated from European warrior masters, at the end died away when the Spaniards and Anglo-Saxons arrived: Both the Paleolithic culture of North-Indians that had survived almost unchanged during more than 10.000 years and the Neolithic cultures of Mexico and Peru, self-similar to the Bronze Age cultures of Egypt were extinguished in a few centuries.

. This parallelism and endurance of old cultures proves further the obvious fact that humans do not change, and what imposes a type of civilization is the evolution of the memes of metal: the Inca, son of the sun, who constructed pyramids and was worshipped as a God with absolute powers was similar to the Pharaoh, displaced in time since in Peru there was not iron or chariots. The renaissance based in coins and mercenary armies was self-similar to Rome, the paradigm of the coin age.

The artistic styles of the 3 ages of any culture between its epic, lineal birth and baroque extinction, intersected by a milder age of human goods in *all civilizations of all cycles,* shows that regardless of small details the 'DNA' of the human social mind is always the same. And humans act as mere cells of a whole complex FMI system to which they have become submissive.

Recap. The warrior hordes that control the world with a new weapon start new 800 years cycle of 'modern civilizations' after an age of wet, hot weather that multiples their numbers on the steppes.

19. Accelerated evolution of FMI complex: 80-8 year cycles.

The astounding determinism of the 800 year cycles of extinction of civilizations, by cultures of warriors carrying the selfish memes of war, which only individuals of enormous moral and intellectual, body/mind strength have circumvented (the lonely genius of cultures; the prophets of love) proves the extraordinary resistance of ideological memes, which are passed from generation to generation through all kind of rituals and ideologies of hate. Again, it doesn't matter the specific type of memes of each of those 800 year cultures of war, but the equality of all their syntactic equations: weapon x human = corpse.

Thus the combination of weapons and warrior ideologies creates the temporal patterns of life and death of civilizations of metal-history. But in the last cycle the evolution of metal accelerates in the common fractal pattern of all time cycles. Since there are also 'generational', human cycles of 80 years that mimic and build the longer cycles in 'sub-ages'. In that sense, the cycles of technological evolution are *accelerating as many other vortices of evolution do; since* as information accumulates it accelerates its evolution. Thus, in history a 'vortex of informative time' is formed with the arrival of science, which tenfold the speed of evolution of weapons. And so in the modern age each 80 years a national generation – self-similar to a civilization of the longer, simpler cycle - is able to evolve the FMI-complex.

The acceleration of information is natural to all systems that go through 3 ages of increasing information, creating a 'vortex of space-time. We thus observe a ternary fractal structure during the gunpowder age: the long 800 years wave of gunpowder weapons (1200-2000 AD); the medium wave of industrial nations of 80 years and *with the arrival of electronic minds, which represent a new increase in the speed of information and evolution of machines,* a short wave of eight years - the decades – during the XX and XXI century.

Thus the wave of evolution of technology, of machines, is accelerating and now it enters in its last age, the singularity age of organic machines (robots, integrated networks), with clear negative consequences for Gaia and mankind - the species that the technological wave is extinguishing.

This accelerated rhythm that sets a speed of 'time' proper of computers, no longer of humans, is due to the 'Chip Paradox', a law of all complex informative systems that consume more information in denser, smaller species as they grow older:

'All systems increase their information towards the future'.

That scalar, accelerated evolution of the FMI Complex had 3 scales in time of different length, since *its accumulation of information is exponential, a logarithmic function in base 10:*

- Human Cycles: The 800 years cycles of Simplex weapons and money, iron and gold, for most part of it that created the complementary Financial-military system.

- Machine cycles: The 80 years cycles of evolution and reproduction of machines by company mothers, which added a reproductive network.

- Electronic cycles: The fast, accelerated present 8 year cycles of evolution of electronic networks, which is converting the FMI system into a global super-organism. This cycle has created a series of 'splendid little wars' every 8 years, coinciding with the cycle of economic products, evolved every 8 years, according to which as dictators buy weapons in the Free market, every 8 years the FMI complex of western dominant nations picks up a dictator to teach him a lesson make a war and show-case the Arms industry, renovating arsenals, increasing sales and evolving=testing them with human corpses. So

Libya, the present war, started exactly 8 years after Iraq II; and Afghanistan, 8 years after Iraq I, the first electronic war of 'terminators' with integrated senses. It is the cycle of 'splendid little wars'[9].

This new brave world of company-mothers of machine-weapons and money will impose two type of ideologies that were submissive to the ideologies of war – money and machines as the measure of all things – and so on top of the people-castes of animetals that rule the world, there will be a change of paradigm and first money-lenders, which become owners of corporations and then physicists, which invent machines and weapons, will substitute on top of societies, the aristocratic castes that ruled the world with their monopoly on invention and use of weapons. Now societies will become ruled by 'stockrats', the owners of corporations, to which the people of the wor(l)d, no longer religious priests but 'democratic politicians', will often oppose, seeking a world of human goods, peace and love, but as it happened with religions during the 800 years cycles, corrupted into inquisitions of war and go(l)d churches, it will be a fact of modern history that most democracies will also become corrupted by money into plutocracies and by weapons into military dictatorships. And so, the evolution of the FMI complex will continue. We shall dedicate the next paragraphs to study some aspects of that evolution, mainly from the perspective of machines and the animetal castes and cultures that control the FMI complex. Later we shall confront that 'wrong path of history' with the efforts of the people of the wor(l)d to create healthy organisms of history in which the wor(l)d ruled over money-markets and weapons; and give a scientific analysis of what would be a world created with a true science of history, instead of a world created – as it is today – with the wrong memes and ideologies of history.

Let us then consider the 80 years cycle from the perspective of the Evolution of Machines (the memes of metal) and the people-castes and ideologies of tribal history that have built the FMI system.

This second theme is obviously the most polemic of them all, since the wrong ideologies of history, mechanism, capitalism and nationalism do not take lightly any criticism and in fact many 'nations' have laws against antipatriotic and anticapitalist behavior, albeit more subtle than in the past century. But it is an unavoidable part of history to deal with tribal history, as much as we would prefer to have a more rational species to deal with.

Since the ideologies of nationalism, capitalism and mechanism were born from 3 specific tribal groups that became people-castes of the western world, establishing with the added power given by those 3 memes 3 super-structures of social control, the aristocratic system, the banking system and the corporation of machines, which were able to control the world, grow 'internationally', till becoming a global super-structure, the FMI system that now predates and parasitizes with its armies, financial institutions and corporation, all other social structures, including the democratic governments and ethic laws humans devised to protect themselves from those predatory structures.

Those 3 systems were invented by tribes of cattle ranchers, from the deserts of the North (Indo-European, Germanic tribes responsible for the concept of tribal nation) and the deserts of the South (Semitic tribes, responsible for the birth of capitalism and the memes of go(l)d religions), which combined their memes in Holland and Great Britain, giving birth to the corporation of scientific machines, the 'free citizen' of the Global market that today rules all other organizations of mankind.

And so, since we have already studied the 'hordes' of Semite and Aryan warriors that evolved weapons, we shall now study the culture of Go(l)d, the 'People of the Treasure' that invented capitalism, merged with Germanic tribes and created the corporation of machines, founding the FMI Complex.

Recap. The modern age accelerates the evolution of the FMI complex with the discovery of machines of science to a mere 80 years cycle of death of civilizations now embodied by nations. The animetal pyramid of power changes as bankers and scientists that control corporations and invent machines topple the aristocracies of war of the 800 years cycle.

20. The last cycle: digital weapons.

Modern wars are economic processes, embedded in the very same structure of 'monetary and machine reproduction'. Indeed, the evolution of technology makes each new lethal machine much more efficient killing humans, there is also a progression in the number of casualties that each World War causes; *which can be directly related to the growth of the stock-market that shows the price-quality of both, consumption machines and weapons; since in war ages ALL the machines of the system are weapons and the market reaches its maximal value. Yet since weapons consume human beings there is a parallel progress on the value of stock-markets that 100-fold each cycle of evolution of machines and the number of causalities in war that 100-fold each cycle.* The progression, as most evolutionary increases of power in the Universe, is decametric: in each new World War, 10^2 times as many people die than in the previous one: 600.000 died in the Railroad wars at the end of the I Kondratieff cycle. 60 million people died in W.W. II. In the next cycle of war, the Age of the Singularity, when the first self-reproductive nano-robots, or the Final Weapon, the black hole of the Nuclear Industry appears at CERN, 6000 million, the entire human population, will die.

But the competence of robots happens not only in war fields but also in labor fields, in which it is making humans obsolete.

Recap. The cycles of reproduction, extinction and consumption of Top Predator machines, weapons, are economic cycles that have caused most of the wars of modern history: Wars are profitable for companies and governments, since you consume=kill more machines in times of war than in times of peace. Thus crises of excessive reproduction of machines have been solved by switching production to weapons, paying war lobbies and selling those weapons to governments. These corrupted governments will declare war in order to consume those weapons.

21. The 3±st ages of evolution of computers.

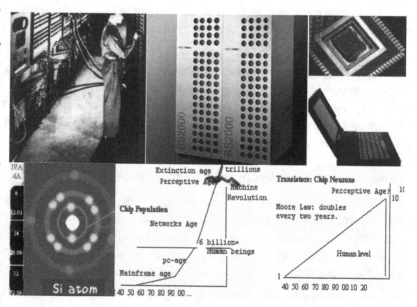

Computers evolve as all machines do, by reducing its size, increasing its mobility and finally, evolving as smart weapons that kill human beings. Its population today overcomes that of humans, while every 2 years they double its capacity and halve its reproductive costs, increasing the number of 'jobs' in which they substitute humanity. What surprises more of the chip radiation is how fast is happening. In biology a radiation is a massive reproduction of a new top predator species, which displaces the old one. In this case, the reproduction of chips and its use as heads of machines, is substituting humans that become displaced as workers and consumers all over the Earth.

In a mere 30 years, chips have 'infected' the entire Earth, with intelligent machines to the point that there are today more chip-brains than human brains. Within 2 decades they will be more intelligent than human beings and integrated in robots and networks. They will make human soldiers and workers obsolete. They have already caused the present age of massive unemployment, self-similar to the automation of factories, implemented by Ford, which caused the massive unemployment of the 30s. But the machine is sacred, so 'economists' don't even consider PCs as a cause of unemployment. Because their myths affirm that productivity, the loss of labor to machines, actually 'increases' in the 'long term?' precisely what it destroys -labor. How stupid is that? But humans can really believe anything... After all, the myth of productivity is similar to the myth of 'salvation'. You give your life for your country, because your mum will receive a piece of metal –a medal- that represents your 'immortal glory'. You die for your corrupted God, because he gives you 'eternal salvation'. So why not losing our jobs to a computer or robot, if that 'increases productivity', the totem word of our Company Economists?

Let us trace the history of that bio-economical radiation . . .

Youth: Scientists improve digital languages, the software of the new brain, with the development of algorithms (mathematical instructions, which act as human sentences do, creating actions in machines) and the binary language that represents with two numbers, 0 and 1, the fundamental energy/information duality of the Universe. German researchers introduce electric on/off switches, the basic hardware able to represent the 0/1 language, discovered also by a German genius (Leibniz). Turing, an English mathematician and decoder, working for military intelligence during II W.W., sets the basis of its future development, proving that thinking machines could theoretically resolve any given problem with a binary, simple representation of the external Universe. Thus, binary computers have the elements of an intelligent being able to recreate any form, language or model of reality and act upon it (when joined to a machine-body). It is then the turn of America that puts together those European findings into workable computers.

As always in the cycles of new energies and machines, their first and last use is in the field of war—a clear proof of the competence that exists between carbon-life and metal:

Turing used Enigma to decode German systems of information. America used its first computers (ENIAC, Univac) to calculate Atomic bombs. The first calculating machines were also made to help gunners in warships and military planes.

As a result of those lethal applications, the American military demanded in the 40s and 50s better computers and the Pentagon paid for their research.

At the same time, in the civil sector, as it happened with the electric ticker, the same company that produced those war machines, IBM, started to produce computers for the financial sector, to increase the reproduction of 'money', the purest information of the economic ecosystem. So once more, Go(l)d and war fuel a 'Supply economy' of cumbersome, ultra-expensive machines that nobody wants to use. Soon their evolution brought diodes, triodes, vacuum tubes and finally the chip: an integrated metal-brain of unlimited capacity, as it grows in complexity and diminishes in size.

International Business Machines was, as its name says, a company that used to make machines to handle money. First it invented machines that registered sales and kept money (the register and cash-machines all shops have today). Thousands of human cashiers lost their employment. Such profitable business gave enough cash to the company to buy the different patents of the first computer software and hardware. Then IBM developed the first commercial models of Computers, the mainframes, starting their informative evolution, decreasing its size and increasing quality, according to the 'black hole paradox': an informative system, unlike an energetic system, is more powerful when it is smaller, as it means its calculus takes place in a shorter space and time.

By 1970, IBM was the biggest Company in the world with profits rounding 5000 Million $ a year. This means that the 'Metal mind r=evolution' had become the engine of world industry, substituting the 'chemical' age of body machines that took General Motors and Exxon to the highest ranks of world industries. In that sense the 'company-mother' IBM represents the first age in the life of a new species; its youth—an age that shows in any species an amazing rate of growth and evolution. It took in a mere

20 years the computer, from the simplest hardware devices into the complex chip. The Metal mind was extremely expensive on those first stages, tested, consumed and evolved only by and for companies of weapons and money. However, as prices tumbled down, pushed by huge monetary investments in research, new companies (Apple, Dell, Compaq, etc.), learnt how to re=produce computers -while marketing convinced small business to buy the new mind. It is...

The Mature, mobile age of the PC.

Chips lowered costs of computers that could now expand throughout the entire economic ecosystem, not only in big financial companies and Government's armies, its 'top predator systems'. Finally, its mass consumption age came with the production of cheap Asian semi-conductors, applied to different electronic machines, (videos, TVs, cameras, etc.). Their miniaturization meant the arrival of mobile devices, PCs and phones. Soon the Founding mother, IBM, lost its throne to Intel, the discoverer of the chip and Microsoft, the creator of the software for those PCs. And so men became hooked not only to 'metal-eyes' (TVs) but also to metal-brains (PCs). And they forgot how to calculate, spell and see Nature with their own eyes. Soon kids preferred to see 'the Lion King', than going to a zoo.

3rd Age. And yet, when all men have their Video, camera and PC, the 1st mini-crisis of overproduction of chips arrives in 1991. And Mr. Bush implemented a splendid little war to solve it. Thus, America inaugurates with Desert Storm the age of electronic wars. At the same time that violent television, in symbiosis, reaches a new peak. Now the enemy is no longer communism, the social doctrine that the Calvinist elite of the II Kondratieff cycle feared (the likes of Mr. Morgan and Mr. Ford), but the social religion of Islam, the enemy of the new Jewish elite that controls the information of America, since 1917, when it bought the Federal Reserve, as Mr. Ford clearly explains[1]. So now, all systems of information convince the Americans of the goodness of those wars that defend the true nation of their upper, informative castes.

Soon, computer networks with center in the Pentagon become widely used in the control of remote weapon systems, from atomic depots to automated gun machines, to military satellites. It is the age of smart weapons, the warrior age of computers. So the web is born, as Pentagon engineers and nuclear scientists at CERN (the European Company of Nuclear Research) connect grids of computer to increase their calculating power. It will be the last st+1 age of computer evolution. Their age of:

St+1: Social evolution.

The metal-mind is proving superior to our mind in all forms of 'intelligence' related to the world of machines, for which they have a natural affinity, such as the use of digital languages, the manipulation of photographic images and the design of those machines. Its sensorial capacities far exceed those of human beings in the recording and transmission of visual or verbal memories, with such new features as ubiquity (transferring of thoughts to any region of the Earth) and telepathy (capacity to transfer software-thoughts among them). But as long as they are deconstructed objects, humans can enhance their informative capacities, using them. So men have become hooked to their PCs that *fulfill his biological drives for more energy and information.* Today you can use a metal-mind to transfer your brain's information and pass it to a friend, miles away. It is called Internet and it became the last craze of the XX century. It represents also the social evolution of the metal-mind into a global network, self-similar to a brain, which controls the entire economic ecosystem. Its building took Cisco, the company that makes most Internet systems, to the top position in market value among all world companies. Thus, we assist today to the social evolution of computers in networks that control increasingly thousands of other machines, creating macro-systems of information of enormous power that have provoked a new wave of unemployment. In the 90s, Oracle and SAP created integrated suites that emptied the office of clerks. In the 2000s, Internet commerce destroyed millions of jobs in pap and mum shops. Amazon is robotizing even the depots were merchandise is kept. It is only left for computers to reach enough complexity to become 'conscious' of their existence as a different species. That will be proper Artificial Intelligence and when it comes in the 3rd age of the Singularity—if we ever pass the hurdle of CERN's cosmic bombs and self-reproductive nano-bacteria —it will mean the substitution of man by a new top predator species.

So, the 3±1 ages of evolution of computers follows those of any other cycle of machines; from an age of discovery, with big, fixed machines, to an age of mobile, small PCs, to a final age of social evolution in networks of 'cellular' machines.

And our culture has adapted with enthusiasm to the 3±1 ages of the new 'dominant brain of the world'. To that aim, the industry of information convinces us to love the new technologies: Now we live in 'the age of information'. And to speak of information becomes a constant fashion, like it was to speak of energy in the XIX C., during the age of machine-bodies.

Fact is human information is verbal not digital. So, the 60s, prior to the invention of the digital chip, the mathematical rival of the human, verbal brain, it atrophies and substitutes, was the zenith of verbal humanity, in all the fields of history, art, way-of-life, ideas, politics. The hippies, despised for so long, were right... But the chip killed them. The new TV-Gods, ended the revival of social love and the hopes of a world made to the image and likeness of mankind. Instead, we have built a world to the image and likeness of the computer and now we enter the age of its awakening, the *age of Robotics*.

Recap; Today chips process digital information billions of times faster than humans do. Since metals store and handle digital information better than carbon does. Yet digital information is the language of machines and the economic ecosystem at large. So chips have caused the obsolescence of white collar workers worldwide, as PCs automate management in all companies of the world.

Those 3 themes could be considered to be the 'core curriculum' of a real science of economics, history and its combined analysis, 'memetics'. None of those sciences however exist with the degree of evolution of other sciences for the reasons expressed above; and this writer, who developed them two decades ago has been probably the most censored scholar since the times of Marx, Kondratieff, Butler, Orwell and Spengler, the most illustrious predecessors of those sciences, all of them ignored during most of their lives; and that fact – censorship of the true sciences of history and economics - has also to be explained as part of the science of memetics, which studies how ideologies in favor of metal-information or 'money', called *capitalism*; ideologies in favor of metal-energy, or 'weapons', called *nationalisms*, and ideologies in favor of organic metal, or 'machines', called mechanism are imprinted in the human mind by rhetoric repetition, emotional recording, social imposition, censorship, violence and the power, money, weapons and machines give to those who 'believe' in them.

Indeed, History and economics are the most difficult of all sciences, because all systems control the information of its inner cells, and we humans are just cells of the body of history and economics and its super-structures. And so an organic, biological science of history and economics has never been allowed to flourish, because it would challenge scientifically the absurdity of worshipping money, weapons and machines and its 3 'religious' ideologies, capitalism, nationalism and mechanism that have imposed them for millenia and are now the essence of our beliefs as social human beings.

And yet only if we evolve those beliefs and understand in organic terms our species and the memes of metal that are extinguishing life to control them for the benefit of humanity in a sustainable world in balance between life and metal, organisms and mechanisms, we will survive the 11th radiation and 10th extinction of all forms of life, including ourselves.

Recap. The 10th radiation of human beings follows the same laws of all biological systems, extinguishing most of the prevous life species and giving birth to a new form of stronger atoms of metal, weapons and machines that are now starting a process of extinction of life.

VI. 11TH RADIATION: MACHINES = 10TH EXTINCTION: LIFE[1]

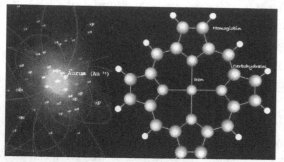

In the graph, gold & iron and the most perfect informative & energetic atoms: Metal atoms display energy/informative properties that enhance those of simpler life atoms. Iron is in fact the center of our blood/energy system, commanding thousands of lesser carbohydrate compounds in a molecule of hemoglobin. While gold is the most informative, perfect atom of metal with an enormous capacity to replicate and store information. Thus, humans used those atoms of metal to enhance their energetic and informative qualities. Iron became the main metal in the construction of weapons. Yet the true change in the Earth's ecosystem took place with the arrival of machines, today made with bodies of iron and 'golden brains', over a surface of silicon. They are simplified organisms, specialized in energy or information, which today are evolving into complex organisms.

22. Metal-information=money and metal-energy=weapons evolve in symbiosis with man.

A scientific analysis of the economic ecosystem, in which humans and machines co-exist based in chemical and biological sciences, allows us to define the world we live in as a dual game of two species:

-Humans and life species made of simpler life atoms, Hydrogen, Oxygen, Nitrogen and Carbon, which have the lower 1, 8, 7 and 6 numbers of the atomic table and make up for over 95% of our body.

-Machines that perform the same biological functions our organs do, with atoms of higher force and information than our atoms (24-Iron, 79-Gold, etc.). Since a *machine is a form that imitates a living, energetic or informative organ with atoms of metal.*

Thus we can consider topologically the existence of 3 cellular units in the FMI system:

Max. I: Go(l)d: The most informative atom of the Universe that imitates the light of the sun and hypnotizes our center of consciousness, the eye, making us slaves of those who carry it.

Max E: Iron: The most energetic atom that cuts our flesh and makes the warrior who carries it a top predator of mankind.

E=I: Organic machines: They transform energy into information (informative machines: e->I) or information into energy (transport machines and machines-weapons: I<-E); and are reproduced by complex organisms, called corporations. They enhance our energy and informative organs, but in the process as it happens with biological enzymes, with an atom of metal and an underdeveloped string of carbon-hydrates, they make us obsolete.

For example, a crane imitates an arm and substitutes an arm, moving things around. A phone imitates an ear-mouth and works as an ear-mouth. All machines imitate human, organic functions and tasks performed with those organs. If we divide the human being in two clear components, body and head, we talk of two types of machines, body-machines that imitate and substitute functions of our body organs and mental machines that imitate functions of our brains and senses. Those two basic functions of human organs will determine also the form of each machine:

— Machines of energy imitate body organs and . . .

— Machines of information imitate head organs.

— Which fusion in 'organic species', robots that imitate animal species.

Thus, the Industrial Revolution is the process of organic evolution of the most complex atoms of the

571

Elements' Table—metals. There is no doubt about it. The Industrial evolution started with two processes that evolved metal: the invention or discovery of the first, most primitive heart of metal, the steam engine, which was able to produce inner movement in a body of metal, as the heart does with a carbon-life body. At the same time Darby and Cort were able to produce massive quantities of iron, the most energetic atom of the Universe, used to manufacture machines.

The next step in the Industrial evolution was the creation of different species of 'metal-bodies' that could adapt a 'heart of metal' to their structures. Sea-metal bodies called steam ships came first. Then the steam-heart was adapted to 'land-metal bodies' and the age of railroads came. Later in the XX century, the steam-heart was evolved into smaller, more powerful species, called the Diesel engine and the oil engine. So the type of machine-bodies that could carry a heart of metal diversified. It came then a metal-bird, the plane and smaller species of land machines, lorries, cars and their 'top predator' weapon versions, multiplied in the war ages of each Industrial R=evolution, such as the tank, a top predator car and the bomber. Finally we evolved electronic minds and now we apply them to energy machines, smart weapons and super-colliders. In 3 centuries we evolved metal-machines almost to their organic perfection, a process that took billions of years to achieve in carbon-life. And that is dangerous, because the only biological advantage humans have over metal is our complexity of organic form.

Engineers 'transfer' millions of years of evolutionary knowledge accumulated by human forms into machines, *passing our evolutionary secrets to a potential future rival species.* So when we evolve those forms into perfect imitations of our organs, machines win that competence against human beings, both as weapons that kill and working-tools that substitute us. And this is due to the fact that metal has some fundamental chemical properties that made it superior to light atoms like oxygen and carbon. So all comes to Chemistry.

We do not invent but discover machines, whose 'informative capacities' and 'energetic force' is born, as ours, from the properties of its atoms. Today we realize that the properties of metal force us to certain informative and energetic designs in machines. Their atoms and the physical and chemical laws of the Universe, not man, is what make machines efficient organs of energy and information. Men merely assembly machines, according to those laws. Humans act as 'enzymen', similar to enzymes or 'catalysts' that evolve other species in the 'assembly lines' of cells. So do 'workers' that put together metal into organic forms similar to our organs. Yet the laws of the Universe and the properties of metal enable those organic forms to exist. For example, a radio is not a human discovery, as much as an 'intelligent' metal-ear, which uses the laws of the Universe to perceive, sounds better than a human ear.

Metal is a natural substance, which is able to acquire different kinds of 'organic forms' that copy the organic shapes of man. At macrocosmic level machines act as energy and information species, because at microcosmic level metal atoms have specific properties that foster its qualities as energy or information systems. *Iron is the most energetic atom of the Universe and* so it is the preferred atom to build weapons that release an overdrive of energy that kills us. *Gold is the most perfect, informative atom of the Universe and* so we use it to create money, the language of information of the economic ecosystem. And now, we make robots with iron bodies and golden brains.

In that regard, one of the silliest myths of anthropomorphic science is the belief that our 'atoms', carbon and nitrogen, have 'special qualities' that other atoms do not share and make them the 'only' candidates to create organic life. Gold and iron, the two basic components of brains and bodies of metal, are also potential life systems, since life is merely a complex system, made with a body of energy and a head of information. Gold is the metal equivalent to Nitrogen, the main informative, storage system of the human brain and his DNA. Iron is equivalent to carbon & oxygen, the main energetic systems of the body. And similar energetic properties are found in aluminum and plumber, the other 2 fundamental metallic atoms used to create bodies of metal and weapons. While gold, supported by silicon molecules, acts in 'metal-brains', as nitrogen supported by carbon structures does in DNA: It transmits informative messages, coded in electric impulses. Yet metal is faster and more complex So gold and silicon combine with other metals to support the electronic software of its 'neurons', which reach speeds of information closer to light speed, 300.000 km/s, 3 *million times faster than our neurons, which transmit information at 100 m/s.* Thus, metal atoms are better than life atoms handling energy and information. In the field of energy-control, iron is a better molecular atom than carbon. Hence bodies of iron are stronger that carbon bodies. This is evident at macroscopic level, comparing the power of metal-warriors and human bodies. In fact, our biological symbiosis with iron-energy is even deeper than that of a warrior. It happens in the blood-energy systems of mammals, which became top predator life species when they substituted the earlier copper-bonds of reptile blood by iron-blood. The red color of your blood is caused by iron. You are, in fact, a primitive iron body. Molecular iron is the soul of your energy-system, of your blood. Iron moves the four arms of the hemoglobin macromolecule and captures your vital energy, oxygen, an active gas that becomes trapped, caged inside the hemoglobin. The movement of that trapped oxygen allows the hemoglobin to store energy and then to share it with our body cells. So the energy processes that take place in the most complex animal life are already 'polluted' or 'symbiotic' to iron atoms. Which is the key lesson of Nature to solve human extinction: we must promote the use of symbiotic, positive tools of metal that foster our energy bodies and forbid the negative, destructive ones, according to a simple biological rule of survival:

'Humans must remain the top predator brain of Earth.'

Since mankind's capacity to process information is his most outstanding property, which made humans top predators, a better top predator brain, the chip, is what will make obsolete.

23. Topological analysis (I): Engines, transport and weapons.

Machines have 2 forms. The simplest tools are symbiotic to humans, enhancing our energy and information capacities when we consume them. The most complex, are energetic weapons and audio-visual information that kill our reproductive bodies and re-program our minds. They respond to the destructive, Darwinian and creative dual arrows of all fractal systems.

Machines, as the parable of the Tree of Technology explains, are also dual fruits, good tools and eviL=anti-Live weapons that can extinguish us. Hence the need to control and prune that tree of technology and allow only the re=production of machines harmless to man; since machines have 3 'biological' ages, parallel to the 3 organic ages of life, (energetic youth, reproductive maturity and informative, 3rd age), which therefore create by the symmetry between space topologies and time ages, 3 topological forms:

- Max. E: As all young forms filled with energy, machines are born as *energetic engines*: pumps, heating systems or bomb devices. But youth is an age that evolves quickly.

- E=I: So in their mature age, those engines are applied to *transport machines* that transform energy into cyclical motion. Then, the machine starts an age of massive re=production and peaceful consumption, the happy 20s or happy 90s, similar to the mature, reproductive age of a living being.

- Max. I: Finally in its 3rd age the machine acquires its most perfect form as a top predator weapon that consumes humans.

Thus machines have 2 evolving forms self-similar to those of any species: simple, consumer machines, whose energy and information enhances human informative and energetic capacities: we talk farther with a mobile metal-ear and move faster with a car. So their reproduction and sale brings times of peace and happiness to the countries that make them, as people experience the exhilarating feeling of being stronger and more intelligent thanks to them. They are the good fruits of the Tree of science. But as machines evolve, their informative and energetic capacities grow so much that they harm the human body (weapons), or hypnotize people with their audiovisual information, which becomes industrial and military propaganda geared to brain-wash people, who become convinced of consuming them and applaud war when war is needed to increase profits (Hitler's Radio-hate, US hate-TV). It is the Kondratieff cycle of Economical war that now starts again in earnest.

And yet, the memes of animetal cultures and its religions of mechanist science, capitalism and nationalism, with its idolatry to weapons, money and machines adapt our mind to those ages. So we find also 3 human generations of 'founding physicists and engineers' that discover the new machine; mature, reproductive industrialists that convert it into a transport system and finally military politicians, which use the machine after the crash of the market as weapons, shaping a *72 years, generational, national cycle that evolves each new type of evolved machine till its completion*[10].

Recap. Machines show 3 topological forms, as top predator, energetic weapons, transport machines and informative machines that transform energy into information.

24. Topological analysis (II): 3 ages of Industrial R=evolution

XIX C. MACHINE BODIES   XX C. MACHINE HEADS: Pcs, Cams   XXI C. ORGANIC MACHINES=ROBOTS

Machines evolve as biologic organisms in 3 ages parallel to the evolution of new sources of energy. When men discover a new energy caused by faster, smaller particles, he creates faster, smaller machines, adapted to the new energy starting a new economic cycle of reproduction and evolution of machines, called a Kondratieff cycle. Thus the arrow of information also applies to machine evolution that become smaller, more complex in 3 horizons: the age of machine bodies, of machine heads and the age of organic robots, which can make human beings obsolete in this XXI C.

A second topological analysis can be done between the future 'bodies, heads, limbs and organic networks' that will create the autonomous, robotic species of the Metal-Earth. Since machines are organic systems of metal, the study of its evolutionary cycles follow the same ages of evolution of life species. From such perspective, it is evident that machines evolve as organic systems:

- *Max. E: Bodies and limbs:* In the I Evolution of Machines, in the XIX C., we made physical machines, bodies of machines that fed on steam and oil, using a biological resemblance, and worked as our bodies do, moving us or loading products. This phase is divided in 2 sub cycles:

-The age of England (1784-1857) or age of trains and steamers that substituted and provoked the extinction of 90% of horses, the previous biological carriers.

- The age of Germany (1857-1929) also a cycle of 72 years[3], in which Germans evolved the internal engines/hearts of machines (chemical, oil engines - Otto, Diesel - and electric engines - Siemens). All those systems were applied to war.

- *Max. I: Heads.* In the XX C., we made mental, electrical devices that act like heads of machines: telephones that are, as their name indicates, long-distance voice-ear systems of metal; cameras that are metal-eyes and computers, which are metal-brains. It is the electronic cycle (1920s-2010s), lead by US, though the discovery of those metal-head parts took place at the end of the German cycle in Europe.

- *E=I: Now in the XXI C. the III Industrial R=evolution* begins, when we join bodies and minds of machines into robots and create both, internal networks of autonomous control and solar 'surfaces' of autonomous energy. They will be the engine of a new industrial take-off, relatively independent of man. Since men no longer will be the consumers of industrial goods, but each robot will consume 2 camera-eyes, several chips-brains, mobile phones and a complex metal body, creating 'economical wealth' for robots not for humans. Further on, as they evolve they could provoke an enormous increase of 'productivity', that is, the substitution of labor by capital; making humans obsolete in the economic ecosystem, as workers (substituted by robots in automated factories) and consumers (substituted by robots that will use other machines), the two main economic roles a 'Free Market Democracy' now concedes to man. Indeed, Robots no longer will need human beings and hence all men are potentially obsolete in the III industrial revolution as each robot will become the consumer of machines, cameras, transport platforms, weapons, etc. In that sense, the 3rd Industrial Revolution should not happen or else it will cause the obsolescence of human workers, consumers and soldiers and displace mankind as the dominant species on Earth, as it happened with life displaced by human beings. It is the birth of the Metalearth, a new global organism in which flows of informative money, centralized in a global brain, the stock-market will direct the reproduction of machines in automated factories, connected through global networks of audiovisual information & machines energy - electricity, roads and solar cells.

- St+1: Metal earth (2008-2080). At the end of the century, automated company-mothers and a global network of computers will communicate those robots simultaneously, creating the Metal-earth, a global super-organism of machines, in which man will have no role whatsoever and so it will become probably extinct by robotic terminators, now essayed in the short 8 year cycles of splendid little wars. This cycle is the cycle of the Singularity that has just started.

Recap. Machines evolved its bodies and engines (XIX c.) then its heads (XX c.) both put together into robots and global networks (XXI c) it is the birth of the metal-earth and the extinction of man

25. Topological analysis (III): The 4 ages of energy.

I CYCLE: STEAM   II CYCLE: OIL   III CYCLE: ELECTRICITY   IV CYCLE: SOLAR

ENERGY ENGINES/ BODIES

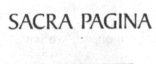

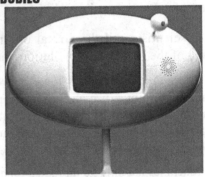

Chemical info: Press;   **Electric info: Radio**   **Electronic Information: Tvs, Pcs**

INFORMATION MACHINES/HEADS

Machines need energy to create its cycles. As they evolve they have also evolved its initial energy, used first to create information machines and then energetic engines.

A Topologic analysis studies the energetic evolution of those machines which use in each cycle a faster, smaller type of energy. Economists know the existence of those cycles of Industrial Evolution, called the Kondratieff Cycles, caused by the invention, reproduction and evolution of new energies (that act as the economic engine of modern history, the 3 Kondratieff cycles of economic activity; each one of 72 years.

*So we could also consider that h*umans evolve a new type of energy and information, derived from that energy, which renews all the machines and financial instruments of the economy every human generation of 72 years[3], in which a nation of 'founding fathers', captains of industries, their sons and grand-sons, reproduce and evolve a new energy, machine and form of money to its perfection. Yet at the end of the cycle, the machine and money becomes over-reproduced, saturating the market and provoking, due to a crisis of growth, a global economic crash:

Those generations bring the nation that discovered the new energy to the top predator status of history. Because energy is also the substance of which weapons are made. *Thus, we had:*

I Kondratieff cycle: Steam machines[10,11]

It is the *age of England, from 1780s to 1857*, followed by a crisis of overproduction of steam machines and stock-money that In the I Industrial R=evolution is dominated by the Steam Engine that moves huge bodies of machines, trains and steamers that require simple, big energies. Yet at the end of the cycle, the overproduction of trains and its obsolescence to new machines provokes an 1857 global crash.

II Kondratieff cycle: electro-chemical energies

From 1857 to 1929, we lived *in the age of electro-chemical energies, machines and chemical explosives, dominated by Germany* But the arrival of a subtle, quantized form of energy, electricity, which can create broken bytes of in-form-ation allows the discovery of the 'metal-heads' of the II Industrial Evolution. The steam leaves way to a combined form of chemical and electrical energy (oil inflamed by sparks in gas engines, electrical light bulbs, etc.); while engineers develop the 1st mental machines (phones, radios). It is the Electro-chemical Cycle of machines, followed by a crisis of overproduction of cars and radios, which caused the 1929 crash, 72 years after the train crash.

III Kondratieff cycle: electr(on)ic machines.

The III cycle of electronic machines, electronic money and Nuclear Bombs took place from 1929-2001, the age of America; which ended in the dotcom and mortgage crashes, 72+7 years after 1929.

IV Kondratieff cycle: solar machines: robots.

The IV Cycle of Evolution of machines will be dominated by robots, solar Industries and China. The chip - a diminutive brain of metal that combines heads and bodies of machines - starts the III Evolution of Metal, the electronic age that witnesses the birth of organic machines, robots and smart weapons, which complete that organic evolution.

Because scientists call the arrival of Artificial Intelligence, the Singularity moment, we have called this 4th age of the Industrial Revolution, the age of the Singularity.

Thus, the Singularity Age is the Age when machines will complete its evolution as organic forms, becoming autonomous of man, probably making us obsolete as workers and soldiers. Indeed, organisms follow a simple pattern of evolution, from mechanical, deconstructed systems into full organisms in 3 phases. Since machines are enhanced organs of energy and information evolved in metal, which once can obtain directly energy from the sun will become autonomous of man.

Recap. Machines not only go through 3 topologies in space, but also evolve in generations, as species increasing its arrow of information as any fractal organism does. And so each of those 3 ages from engine to consumer machine to weapon has been repeated 3 times in modern history every century, as the energy of those machines also evolved in 4 Kondratieff cycles, the cycle of steam machines, of oil machines and the age of electronic machines that is giving birth to the first robots, in the age of solar energy, the last cycle.

26. Combined analysis: History and Economics.

In the next graph, the Kondratieff cycle is therefore the fundamental cycle of economics and the main cause of the peace and war ages that alternate in modern history. And its periodicity coincides with the generational cycle of 72 years of human life and national power, given the fact that those machines are discovered, evolved and used by human beings, organized in nations. Its key dates are:

I   STEAM CYCLE: British Age:

1784.....Steam Peace...1857 Crash...Steam Wars...

II   CHEMICAL CYCLE: German age:

1860s.....Chemical Peace.....1929 Crash...Oil Wars...

III   ELECTRONIC CYCLE: USA Age:

1940s....Electronic Peace...2001 Crash...Robot Wars...

IV   SINGULARITY CYCLE: Chinese Age:

2010s: Singularity=Extinction?   Vs Human R=evolution?

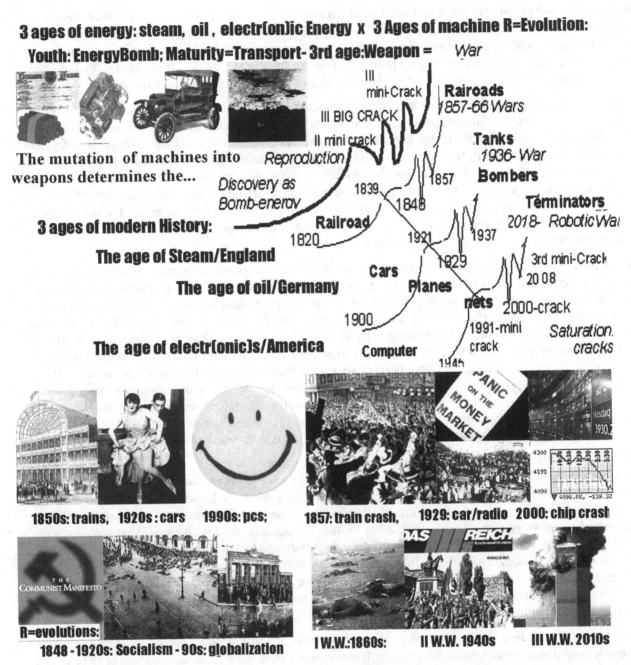

3 ages of energy: steam, oil , electr(on)ic Energy x 3 Ages of machine R=Evolution:

Youth: EnergyBomb; Maturity=Transport- 3rd age:Weapon = *War*

III
mini-Crack

Rairoads
1857-66 Wars

III BIG CRACK

Tanks
1936- War

II mini crack

The mutation of machines into
weapons determines the...

Reproduction

Bombers

*Discovery as
Bomb-enerav*

1839

1857

Terminators
2018- Robotic War

1848

3 ages of modern History:

Railroad
1820

1921

1937

3rd mini-Crack
20 08

The age of Steam/England

1929

Cars

The age of oil/Germany

Planes

nets

2000-crack

1900

1991-mini
crack

*Saturation.
cracks*

The age of electr(onic)s/America

Computer

1940

1850s: trains, 1920s : cars   1990s: pcs;   1857: train crash,   1929: car/radio   2000: chip crash

R=evolutions:
1848 -1920s: Socialism - 90s: globalization   I W.W.:1860s: II W.W. 1940s   III W.W. 2010s

The key date of each cycle is when overproduction of peaceful machines provokes a global economic crisis and Stocks crash. Every 72 years after one of the 3 biggest stock market crashes of history - the railroad crash of 1857, the 1929 crash of radios and car industries and the 2001 NASDAQ crash of the computer industry - the same companies that made peaceful machines switch=mutate production to weapons. Unfortunately in those war ages industrial information, carried by 'metal-heads', stresses the human causes of those war cycles, convincing people that wars are not made for industrial profits. So European empires didn't go to war to get raw materials and sell weapons but made war for altruistic causes—'to civilize primitive cultures'. Germans didn't go to war and made weapons to get out of the 29 crisis but to obtain a much needed 'vital, historic space', unjustly stolen by surrounding nations. This hypocritical reasoning was and still is needed, given the fact that we live in

Democracies and people are ethic and need good reasons to kill the poor people of the planet, instead of raising them from their misery with money for peaceful education and trade. This solution however cannot be implemented, since stock-markets deviate constantly most investments to the technological companies of each cycle, provoking a chronic scarcity of human goods that prevents the evolution of those cultures, without credit to come out of their cycles of poverty, violence and terrorism.

Wars are essential to the evolution and re=production of machines. Modern Wars are 'reversed ecosystems' in which a human ecosystem that keeps machines in a subdued state of evolution as peaceful products (age of peace), changes into a free market ecosystem of machines, where its company-mothers rule supreme. Then, the most expensive machines, weapons, are reproduced and roam the war ecosystem, eliminating us, the rival species. Since weapons are perfect machines, war periods are 'accelerated stages, in which the evolution and reproduction of those machines intensify. Hence the profitability of a war age in which all the factors of the economic ecosystem increase: the reproduction of machines and money multiplies, the 'wealth of nations' grows geometrically and the competence between machines and humans reaches its peak.

It is then when nations switch massively from peace consumption to war consumption. So after the global economic crisis of overproduction, and the ominous years of initial Keynesian militarism , 1930s and 2000s, industrial lobbies pay now for war-prone politicians to switch massively demand to weapons, the 'top predator' version of a machine, reproduced by the same companies. So if Mercedes switched from cars to tanks today Boeing sells already more weapons than civil airplanes. Wealth of nations moreover peak when weapons, the most expensive goods of the market, are bought massively and money sis printed in deficits and bonds to pay for them. That wealth enters the coffers of the financial-military-industrial complex as never before. But people don't eat weapons. So the class structure elongates: the military-industrial corporations and the financiers become richer than ever. Yet people become so poor that at the end rebel in civil wars. It is then (bottom of the graph) when some r=evolutions happen, trying to create a world based in human goods, but at the end war triumphs.

Déjà vu: Military Keynesianism was essayed in Germany since the fall of Bismarck for half a century and ended with the self-destruction of Europe, when there were no more lands to colonize with those weapons and Germany decided to 'colonize' Europe. Today military Keynesianism creates not only 'splendid little wars to get out of this crisis' (Roosevelt on the Spanish-American war) but has already established an inner system of 'security', a big brother of cameras, sold with hate-TV speeches on our 'security' against eviL, primitive, terrorists that control our lives and information as nothing since the German Gestapo in occupied Europe did. This control of humanity by computer systems is not yet used as Gestapo files to select the humans that shall survive in an increasing obsolescent world where workers are substituted by Pcs and Robots but it could in the future, when military Keynesianism breaks the same havoc that brought to German, to the American social tissue…

Recap. World wars are a direct result of such over-production of Top Predator weapons by stock-companies. Companies could produce only peaceful products, if governments were ruled by economic ideas in favor of man and wanted to prevent wars. Yet Bio historians in favor of the reproduction of human natural, biological goods do not rule our societies. Companies and politicians to the service of 'ari-stockrats' do. So they behave in a total selfish way in search of profits. And when they can no longer keep producing peace machines (since every human has his car, his radio, his TV), they search for weapons to expand that reproduction. So when a 'peace product' saturates the ecosystem, the market jungle, company-mothers invade new territories with his offspring of weapons and war empties their stock of products much faster, forcing new production, new sales and new profits.

27. St+1 analysis: the global stock-market.

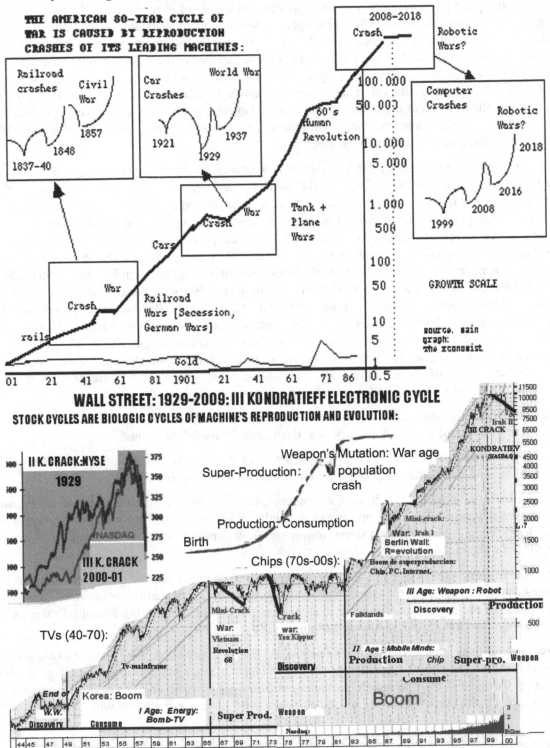

In the 3 technological cycles, NYSE's curve of machine's stock population follows the same chair-like curves of biological populations; since each machine is a biological organism that finds 'econiches' in war and labor fields, substituting and consuming its rival human beings, workers and soldiers or as consumption machines, when they are consumed by humans to enhance our energy/information skills.

In the graph, when we treat machines as collective species, analyzing their global population through the stock price of their company mothers, we obtain graphs and curves of re=production, homologous to the biological curves of reproduction of any living species. Since population means sales that give profits to companies, which determine their stock price. Stock-prices and 'Sales' mean basically 'production' and 'production' means reproduction. Thus, the biological curve that shows the reproduction and extinction of machines is its sales or profit curve, directly proportional to the value curve of the company and its industrial sector in the stock market. That amazing discovery explains for the first time in a rational manner, not only the how but also the why of stock market curves. It defines the economy as an economic ecosystem, where machines and its company-mothers fight for its re= production by all means. While stockrats fight for the parallel function of profits.

After its birth as a 'new species', the 3 cycles of new energy machines grow during a mature age of reproduction (and sales=profits), which is followed by an age of overproduction and crash of populations. The economic ecosystem is now saturated. So to expand further the machine species 'mutates' into a weapon and reproduction continues, as those weapons 'invade' other nations/economic ecosystems, which increases demand and sales unless a defeat plummets profits. In the previous graph we can observe the 3 cycles of America. In the last of them, the electronic cycle that now ends, America evolved into a world power between both World Wars, maturing as a free Republic in the 50s, 60s and 70s, in which the country was widely admired all over the world. But with the arrival of the Republican Party and its hired, violent politicos of null political stature, since Reagan took power, the long American decline begins. Now we enter the military age, with a 3rd generation of baroque, cynical, war loving Americans, accept as a solution to the crash of electronic consumption, the creation of 'splendid little wars' to increase demand. So once more 'business cycle as usual' continues, as we enter deeper and deeper in the war phase of the electronic age and robotic planes massacre the innocent children and women of the life-based villages of Afghanistan.

For all those reasons, the graph of stocks is self-similar to those biologists use to describe the logarithmic curve of population of a new species. Both have 3 parallel 'horizons':

—   For a long period the machine evolves slowly in the labs of companies or inventors. It is equivalent to the slow period of mutations that will bring a new species into being.

Then, the engine of the machine, the new energy, is discovered. It gives birth to a new, more efficient species that multiplies its numbers exponentially. It is the takeoff, the youth or first age of the machine/energy as an engine. Since it starts from a very low population, prices now multiply very fast, so investors make the highest profits.

—   Yet the biggest influence of the machine and/or energy on the economic ecosystem happens in its mature age, when it becomes mobile and reproduces worldwide for peaceful consumption (geometric, ascending phase of the population curve). It is an age, parallel to the 'expanding age of a living species' that moves away from its original niche and colonizes new territories.

—   Yet, sooner than later, given their exponential rhythm of production, *far superior to the increase of human population, machines saturate the consumer market*. Since symbiotic species depend on the population of the species they associate to those machines are organic enhancers of our energetic or informative power, the 'vital space' of machines is the total human population with resources to buy them.

Thus, once the machine saturates the ecosystem of mankind, if re=production continues, profits and sales diminish and companies accumulate inventories. At that point their sales and stock market curves decline. *It is the most important date of the cycle, with a very pronounced regularity of about 72 years (1857/1929/2001).* Why such accuracy? Obviously, because machines are made by human beings *and 72 years is the so called generational cycle of human beings*, their mean life in developed countries or among rich people that invent, reproduce and profit from machines. So the organic symbiosis between human and machines is evident.

Again in Nature, one species that has overpopulated its vital space, suffers a population crash, as it has also depleted its 'preys' and cannot longer feed itself.

The population finally reaches a balance with their prey and it becomes stable. So happens with stock-market curves of machines that become flat.

At this point in Nature a species cannot increase its population unless it mutates into a new, more efficient species that can feed on new preys. That is exactly what happens to the machine: After the overpopulation crisis, the machine mutates into a weapon and continues expanding its ecosystem, preying over human beings, no longer symbiotic to them.

- *It is the third age or military age* of the machine that is used now for war. However war is a risky business because predators might end up being preys. Indeed, companies' results and stock benefits will be very different, depending on war victory or defeat that sinks stocks.

Though those curves in a single company have to be pondered for many other secondary factors, when we aggregate the maximum number of production companies; that is, when we search for the total curves of the species, they become more and more regular. And indeed, the graph for the entire cycles is very close to the biological one.

Thus, those 3 great crashes of global stock markets are the key dates of economics and history, because they mean a global change from wealth to poverty, from peace to war machines, determined by a change of strategy in company's search of benefits at any cost.

First, as the number of machines surpasses the possibilities of citizens' consumption and machines saturate the Market, their prices lower, diminishing companies' profit. So companies value in Stock markets sink and the effect moves sideward to all companies related to those machines that are the engine of the economy. But the results are far more pronounced than in Nature. Because the subjective fear of investors multiplies the downwards effect and creates a panic in all the other companies of the Market. It is now when we depart from a pure biological analysis of machines population, to include the symbiotic human side of those crashes.

Indeed, as any symbiotic ecosystem in which species depend on each other the historic ecosystem of human beings is perfectly intertwined with the economic ecosystem of machines. So the effects of those cycles are far more complex than industrialists and politicians want to recognize. And most of them are negative. First companies dismiss workers to cut profits. Nevertheless machines keep arriving to the Market, because its evolution continues; lowering their costs of reproduction. So companies decide to make weapons and their lobbies cause wars to consume them. It is a desperado, evil, brutal way of making profits, but companies are organisms that care only for their products. They do not care for human beings, consumers and soldiers. They are narrowly focused and use abstract numbers to avoid any guilty feelings about the workers they sack, the soldiers their products kill. As bribes multiply, corrupted governments enter the game; and the cycle of economic growth, based on weapons and death begins. Since weapons destroy themselves quickly in the battle front and must be renewed, the increasing demand of machines multiplies the benefits of companies. G.M. multiplied by 8 its profits, making tanks in W.W. II. So war becomes the economic engine of the last 36 y. of a Kondratieff cycle.

Recap. Company-mothers govern the world with a single aim: to re=produce machines. Among them, the most powerful ones reproduce top predator weapons. So, in as much as we can analyze company-mothers as reproductive organisms with biological cycles, called business cycles, we can find out, studying those reproductive cycles of machines, specifically of weapons, when there will be enough top predator weapons re=produced in our ecosystem, to cause war.

28. 8 years cycle of splendid little wars: The product cycle[9].

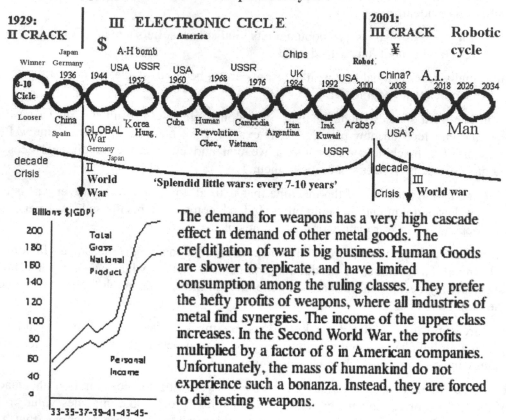

The demand for weapons has a very high cascade effect in demand of other metal goods. The cre[dit]ation of war is big business. Human Goods are slower to replicate, and have limited consumption among the ruling classes. They prefer the hefty profits of weapons, where all industries of metal find synergies. The income of the upper class increases. In the Second World War, the profits multiplied by a factor of 8 in American companies. Unfortunately, the mass of humankind do not experience such a bonanza. Instead, they are forced to die testing weapons.

In the graph, from 'Extinction of man' c. 94, we considered an 8+-3 y. product cycle of wars which evolves the most expensive good of the financial-military-industrial complex, weapons. To notice that the pattern accelerated in 2000, with an added +3 year double war (Iraq and Afghanistan). We didn't know at the time which would be the victim, but it was easy to predict that it would be an Arab Nation as it was the case. For 2008 (now 2011, as we had two consecutive happenings in the 2001-2003 double happening), exactly 8 years after the opening of fire the 20 march of 2003 in Dessert Storm II, we have opened fire on Libya. We thought also that Cuba would be the most likely cause of this final splendid little war before the great confrontation either with China or with the entire Muslim world, the 2 options we contemplated as the start III world war since our initial books 2 decades ago[1], which has now started in 2011, with the expansion of Arab wars exactly 10 years after the 2001 crash of the e-money economy as II world war started exactly 10 years after the 1929 crash of the car economy.

Metal-communicators always justify war. War was first demanded by Luther and Calvin in the yellow press against the Pope. And the result was a healthy industry of guns and gunboats, the first companies of machines-weapons and the death of the superorganism of Christianity.

War was demanded in each of the 3 ages of the Industrial r=evolution by metal-communicators, colonialism became the hottest item of the yellow press of the XIX century.

Then radio came, and the awesome evolution of metal-communicators in terms of exi force, of simultaneous spatial reach and informative complexity, allowed the network to become more powerful than the press. In America Welles proved it with his war of the worlds and later, destroying the reputation of Hearst, head of the yellow press that invented the first of the splendid little wars of 8 years – the Spanish civil war. Thus today when we live in the age of the TV-hate press the same cycle happens: we have splendid little wars which are sold by TV-hate speeches and then become just wars to show weapons. The economical nature of those wars is proved because all of them respond to the 8 year cycle of products, and so wars today are media, industrial wars, which every 8 years become the showcase for the industry of weapons that spends inventories, shows new products and renews its cycle.

This is specially truth since the first electronic war – Iraq 1 was managed with media censorship, video-game images and carried by an oil and weapons lobbyist family, the Bushes.

It has to be understood that being war the product of a cycle of evolution of machines the marketing of the product, war itself, follows the 'Say's Law' of economical production:

'Given the proper marketing all products can be sold'. In other posts dedicated to newspeaks of justice we have studied what are the promoters of war, how the industry of information always justifies a war of the cycle, and why we can't hardly differentiate between 'just wars' – those of the righteous who defend human rights in those places that serve the self-interest of the leading top predator nation of the cycle – and those blatantly fought for profits of the financial-military-industrial complex. We are now in full in the Semite wars between the leading 'bond holders' and 'financial masters' of the west, the Jewish empire and the Arab Neolithic civilization which opposes their 'apartheid' state in Israel. So the marketing for wars will hide the 2 engines of the process: the political engine, the defense of Israel the nation of the Western FMI elite and the economic engine: the evolution of electronic weapons now moving at full speed towards the threshold of Artificial Intelligence -the terminator machine. And yet the same 8 year rhythm that in our complex works we studied also for the age of steam machines and the age of German motors is also happening as the graph shows in the electronic cycle.

Recap. Splendid little wars increase as the cycle of overproduction of machines of war increases, ushering the word into an age of world war, which now in 2011 starts with the expansion of the splendid little wars implemented in 'primitive Arab nations'. Soon the entire financial-military-industrial complex will convert just wars like Libya in the Orwellian nightmare of perpetual war and the sci-fi nightmare of artists of the human baroque (terminators, Matrix, etc.); since only human artists of the rival humanist culture understand the non-human future of the ideological world of animetals and the FMI complex.

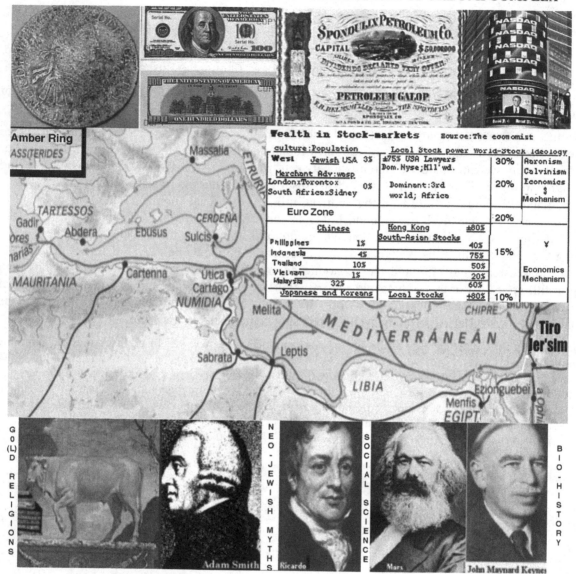

culture:Population		Local Stock power	World-Stock	Ideology
West Jewish USA 3%		±75% USA Lawyers Dom. Nyse;Hll' wd.	30%	Aaronism Calvinism Economics & Mechanism
Merchant Adv:wasp LondonxTorontox South AfricaxSidney	0%	Dominant:3rd world; Africa	20%	
Euro Zone			20%	
Chinese		Hong Kong South-Asian Stocks	±80%	¥
Philippines	1%		40%	
Indonesia	4%		75%	Economics Mechanism
Thailand	10%		50%	
Vietnam	1%		20%	
Malaysia	32%		60%	
Japanese and Koreans		Local Stocks	±80%	10%

Wealth in Stock-markets   Source:The economist

Money has always possessed three formal characteristics: it is made of metal, gold, silver, copper, today computer machines that reproduce e-money. It is made of cyclical, broken, discontinuous shapes, as all forms that carry in-form-ation. Finally, money, due to this informative property, is the basic unit of mathematical accounting and one of the main reasons why mathematics have been evolved as a language. Thus, money is a language of information whose substances are metal and whose function is to give value to things with digital information. In that sense, money competes with words, because both are languages used by mankind to value reality. Yet money has evolved increasing its properties as a language, by increasing its numbers; its reproduction and its transportability, hence its capacity to value more items of reality, from gold bars to coins, to paper money to e-money, till today it is able to value all forms of reality, substituting words, which have become mostly fictions without value of truth. The same process happened in the confrontation between ethic words and weapons ('how many tanks have the Pope, asked Stalin) and verbal philosophy and digital philosophy (3rd paradigm of metric measure, which Galileo declared to be the 'only language of god'). In the graph, due to geographical reasons, the first network of trade based in metal (money, slaves and weapons) was created in the knot of communication of the 3 old continents, in Levant, where around 1000 BC, Cananeans - Phoenician *sea traders and 'Habiru', 'donkey caravaneers'[8] - transported weapons, slaves and capital between the civilizations of the Fertile crescent and beyond. For millennia the Habiru, also called 'Am Segullah' or 'People of the Treasure' (ill-translated as 'Chosen people') became a People-caste (Abraham Leon), specialized in the accumulation of gold through weapons, slave and luxury trade (Sombart, Weber). In the process they developed the memes of capitalism, which separates humans in two castes –those who invent and lend money and those who earn it; reinforced by a racial religion that prevents them to relate to other humans in equality and represses the memes of life (sex, tasteful food, social love to mankind). Their tragic history will become the paradigm of the collateral effects of money and weapons and its hidden values, denounced by its prophets (Moses Damnation: 'The Jewish will suffer all their life for their love of money'). Indeed, the Jewish people are alternately on top of societies when money rules supreme, but due to their profession at the summit of the FMI complex they have evolved as war purveyors and bankers and enslavers (today owners of corporations with part-time slaves) they are hated by the population,, which kills them in war periods when the reproduction of weapons reaches its zenith. Unfortunately this brutal cycle of metal-history is not recognized by economists, which have merely translated the memes of tribal history – the attempt of the Jewish people to rule the world with money, in confrontation with warrior tribes that tried to rule the world with weapons – into economical postulates in favor of greed (Adam smith), the accumulation of profits by corporations without any rights for workers (Ricardo's iron salary laws), and the belief that 'the FMI complex' must be evolved even if it extinguishes mankind, because it is 'the Invisible hand of God'. Only in the past 2 centuries with the work of Marx, Sombart, Keynes and systemic scientists like this writer, economics has developed a corpus of scientific work able to rule the FMI complex for the benefit of man. But scientific economics is repressed by the culture of 'biblical economics' and the power of its corporations and financial institutions, ruled by the religious dogmas of Jewish-protestant scholars, (classic economics), which consider the FMI complex the symbol of human progress and the task of economists to multiply money at all cost, regardless of the collateral effects that the unlimited reproduction of money, weapons and machines causes – namely the obsolescence and extinction of life and labor by automated, more productive corporations and its white collar Pcs and Robotic workers. Yet because the memes of capitalism are dogmas, the degradation of life and labor and the need to prune the tree of technology of its bad fruits cannot be argued. While a subtle newspeak of equations that passes as science censors any attempt to reform the FMI complex for the benefit of all the humankind.*

29. Metal memes: Evolution of money, weapons & machines.

If money were a human language, its values should be the same than the verbal values that make human life the most expensive good. And yet the opposite occurs: money values human time (minimal salary) less than energetic metal, weapons and machines. The result is that money, a language of metal, evolves the economic ecosystem of weapons and machines, not the organisms of history and life. So the values of metal-money are the antithesis of the values of wor(l)ds: life is the cheapest 'product' and weapons that kill life, the most expensive - while human life-enhancing goods, from love to food, from Nature to education, have minimal or null monetary value; as they are different substances that cannot be easily related to money. Such opposition of values proves against the myths heralded by the corrupted values of Animetal cultures that humans follow in his natural state, as verbal beings, the laws of love and social evolution. It is gold and weapons what corrupts Rousseau's 'good savage' into an animetal obsessed by greed and murder, which considers himself different from other human beings as it applies to them the Darwinian laws against other species, *breaking the 3rd postulate of self-similarity and love to the people of your own species.*

On the other hand, metal *has obeyed the 3rd postulate of self-similarity between equal species, evolving together.* That symbiosis between money and weapons becomes more evident, if we consider the parallelism between the main evolutionary ages of money (previous graph) and the main ages of evolution of metal-weapons. Both together created the top predator animetal cultures of the Earth:

1. Age of bartering (gold bars & rings) and Bronze Weapons:

Semites create the 1st Civilization of 'Chosen of Go(l)d', developing both, Bronze weapons and rings of go(l)d money. They brought metal to the Canaanite coast, manufactured weapons and sold them, engaging mercenary armies. Soon Semites were in power in all the nations of the Fertile Crescent, extinguishing the Neolithic civilizations that used wheat—human good — as biological money. Those Semites specialized geographically. A warrior culture took over mountainous, iron-rich Northern Mesopotamia (Assyrian). A trader culture settled in the node of communication between the Mediterranean and the Indian Ocean, Mesopotamia and Egypt, (Canaanites – Phoenicians, which specialized in sea trade and Hebrew, which specialized in military trade and loans to warring states. Both formed together in the age of Solomon and Hiram of Tyros the first global network of trade, which would be the blue print of all future commercial empires). Thus Semites, during the age of bronze started the 800year cycles of war, death and looting that still endure and have carried the evolution of the Financial-Military-Industrial System till its globalization.

2. The age of coins and Indo-European iron warriors:

It is the age of Greek and Roman empires and their culture of coins and 'infantry'. The arrival of coins, invented by the Greeks, gives them top predator status, as they hold both 'languages of Metal', money and weapons, now in the hands of Athenian bankers and Spartan hoplites. Soon Alexander becomes the first man to impose his face in a coin, unifying definitively both elements of power, money and weapons. It appears then the concept of a top predator currency that has legal course, because it is hold by a top predator military system that kills you when you do not want to take the currency. Weapons defend the currency and impose, it regardless of its real value or utility, while the currency pays for weapons. After the Macedonians, the Romans applied the same dual game of power. Then the resurrection of a Wor(l)d of ethics (Christianity) halts the evolution of weapons, metal and money. Yet with the Renaissance, the same dual game of power returns to Italy, which becomes the center of evolution of gunpowder and money.

The series of national power based on the possession of the Top Predator weapon and symbiotic currency, shapes now a shorter 80 years cycle of national power that takes place within the long cycle of gunpowder weapons. Since the nation that invents more money pays more mercenaries and produces more war machines, winning all wars:

• XIII-XV C.: Italian Bombards and gold coins spread the Renaissance in Europe.

• XVI C.: Muskets and bullion give power to the Iberians, the first global empire.

• XVIII C.: Artillery and paper money spread Revolutionary France.

All these nations produced the Top Predator weapons of their age and imposed their money and cultures to the rest of mankind. Then, the invention by Holland of paper-money, much easier to reproduce than the gold of the Spaniards, defeats them, opening:

3. The age of paper money and company-mothers:

Paper-Money, issued by the first companies at null cost, professionalizes the reproduction of weapons. Gunboat companies seize power. The hyper-abundance of money, now a mere printed paper, motivates more citizens to work in the production of machines of war. Indeed, without the first radiation of paper money, Dutch companies could have never defeated the Spanish armies. It was paper-money, used to 'motivate' Dutch people, pay mercenaries and build their weapons, what defeated the Spaniards.

The press permitted the unlimited reproduction of money in the form of paper. The symbiosis between top predator, informative currencies and top predator, energetic weapons shows the dual, biological nature of economic ecosystems, which become 'complementary' organisms, with a physical economy of machines and an informative 'head', made of monetary values and orders that 'cre(dit)ate' the future.

Now companies and kings could reproduce paper-money and impose it to citizens, just by printing paper, expanding enormously the number of salaries and hours humanity existence under the rule of Companies, working for them.

This brought however many crisis of over-reproduction of money, because even if paper-money was reinforced by an army, many would not accept a worthless paper when the state was too greedy and reproduced it without limit. So the capacity to exchange paper-money for a bit of gold became customary or else paper-money was not trusted, lost value and inflation appeared when paper money was reproduced without limit. The same need for some 'real value' happened in the first stock-markets in which paper money represented some real asset, the weapons, gold, territories and slaves of the first company-mothers, reproducers of artillery and gunboats. All those 'trade' companies manufactured a product: a gunboat, a weapon. And so people valued those companies for the weapons they owned and the political, territorial power and wealth achieved with them. Till the XX century, soldiers were basically mercenaries. So nations who made a lot of money won wars.

Then, after the fall of Amsterdam to French Artillery, those Companies moved to London and gave England the capital and gunboats that created the British Empire and the Pound. So during the XVIII and XIX centuries, the pound would become the top predator currency of the world, reproduced without limit, able to buy goods, people and entire countries:

• XVII-XVIII C.: Guilders and gunboats from Holland and England impose the Anglo-Saxon Culture in the '7 seas'.

• XIX C.: Steam, oil machines and stock-money from British and German companies conquer the world.

4. The age of e-money and Computer-based Weapons:

• XX C.: America uses computers to research A-bombs and invent e-money. Both expand the American Empire worldwide.

The hyper-abundance of money in the age of company-mothers, when money becomes stock-paper, easy to reproduce with a printing machine, extends the values of greed, violence and mechanist science, cre(dit)ating the modern world. The age of Stock money lasted from 1602 to 1972, when electronic money appeared, making even easier for company-mothers to reproduce their language of power.

Now money becomes electronic money in the minds of computers. It gives America control of the world . Yet electronic money travels at the speed of light and knows no borders, so it also means the end of national power, as the financial world becomes a single global organism. So the mirage of national power today fades away, as money achieves its Final Evolutionary scale, becoming the collective brain of the global organism of the economy, at the same time than electronic thoughts in computers are about to cross the Threshold of Artificial Intelligence in:

-st+1: Globalization. The Age of the Singularity:

In the XXI C. the economic ecosystem becomes global: Nuclear weapons evolve, reaching planetary size, with the development at CERN of a factory of black holes that can destroy the planet. While money also escapes national control, as it becomes the re=productive software of global companies.

Today the FMI complex is a Worldwide organism ruled by a digital brain, World-stock, as no other alternative society is left, except the die-hard warrior societies of Islam - which we use as a perfect excuse to keep multiplying our weapons.

The economic ecosystem is ruled by flows of informative, electronic money that select company-mothers, according to the quality and evolution of their machines, promoting, as always, those companies that create the best weapons and informative 'metal-minds'. Thus, nothing has changed, except the sophistication of the system of beliefs and myths to sustain the dictatorship of money. And none of those myths is bigger than the idea that we live in a 'democracy', which is the government of the law and the people. This is false, because most of the time of our lives we live in a work environment, where we must obey the companies that pay us. So we are most of our lives, and we act most of our time, under 'monetary orders' of Companies, which create most of the actions that shape the future of the Earth. We cannot discount those 8 hours a day, as if it 'they were not happening'.'

The world today is cre(dit)ated, created in 90% of its time – which is the substance of reality – by humans working for corporations and handling its machines. And so it is no longer meaningful to speak of 'reality' in this planet in terms of 'political power', a submissive 'simplex' structure also controlled by the markets and when not, unable to mold reality or engage humans for long periods of time. And so we shall consider that with the birth of the first company-mother of machines of war (gunboats, 1604), the world changed forever, as a process of standardization of the 3 ideologies of 'metal- history', capitalism, mechanism and nationalism became embodied in a new organization, the company-mother, which soon will control governments, raise armies and conquer the world. The Unification of the world as a single empire is today a fact, but the empire is the Financial-military industrial system, in which money, produced by a global brain, the stock-market, is first handled to companies, which then control workers, consumers and most actions of governments; even though in social democracies there is still a 'human power' and some super-organisms of mankind are still functional and soothing the consequences of leaving the world in the hands of an organic system of machines…

Recap. Languages are subjective: They value and reproduce more those species that possess the language. Thus verbal thought, the biological language of man, fosters human social evolution and human biological goods. While the 'values' of money destroy the 'ethic values' that put man at the center of the Universe, making weapons, top predator metals that kill life, the most expensive and reproduced goods. Thus, when money became the language of man, we started to reproduce 'expensive' metal and kill 'cheap' life. So the opposition of languages and values became translated into a fight between species: man vs. metal. And as money evolved and reproduced in greater numbers, weapons multiplied and life died; and ethic, human values, which made of man the top predator of the Earth, disappeared. We can distinguish in that evolution the age of metal bars and bronze weapons; the age of coins and iron swords; the age of paper-money and company-mothers of machines and the electronic age or age of globalization.

30. Human, memetic P.o.v.: ternary ages of the FMI culture

The 3 ages of Evolution of the Financial-Military-Industrial Complex:

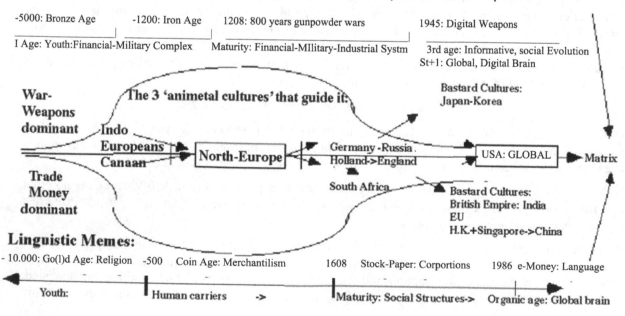

589

The Financial Military industrial complex is a self-reproductive system of memes of metal, now evolving into a global super-organism. In its III horizons of accelerated evolution, it was first a complementary, slow system of evolution of energy weapons and informative go(l)d in 800 years cycles, carried by two people-castes specialized in the control of mankind with the new 'top predator' system of the Earth: Indo-Europeans and Canaanites (Phoenician sea traders and Jewish caravan traders). Both cultures fusioned in Northern Europe, giving birth to the first companies of gunboats that carried trade, slavery and gold accumulation till the industrial revolution, when machines and company-mothers, the reproductive network of the FMI system, appeared, accelerating professionally thanks the 3rd paradigm of scientific measure, the cycle of evolution of weapons to a mere 80 years. Now in its last phase, the cycle has accelerated further due to the discovery of mental machines that evolve information faster than human beings do (chips) to a mere 8 years cycle. It is the economic product cycle of war. Further on, the FMI complex, increasingly independent of humans, are evolving into an automated super-organism, guided by robots and pc-suites, with a global brain-network, the intranet.

If the metal-memes of war were evolved by nomadic tribes of Arab and Indo-European descent, the financial system was created by the 'Am Segullah' (People of the Treasure, ill-translated as 'Chosen People'; which reveals a go(l)d cult(ure), which made of money the language of power of society).

The struggle between both 'animetal' people-castes on top of farming communities, since the bronze age has been both the engine of wars and holocausts (negative collateral effects of the memes of metal over mankind) and the engine of evolution of the FMI complex.

As all organisms the go(l)d culture divided in two classes, the elite of levy banker-priests or people of the Treasure, which ruled with money and the bible, the common people, also called Habiru (donkey breeders), which traded on the 3 'goods' related by affinity with informative metal-money, weapons (energetic metal), usury (reproduction of money) and slaves (pricing of human capital with metal). Objective scientific analysis (biblical archaeology) proves that the Bible is written in the fashion of the age as a Book of Tribal History, which still uses as all bronze societies did the name God as a toponym for nation, so Assur is the nation and God of the Assyrians and Yvwh, the nation of Judea and God of the Jewish people. The difference though is in the language of power and the cultural memes that imposes on the dual organism of the Levi priests (Treasure people) and its lower castes (the habiru traders). First, the professional terms used to call the members of the culture show it to be a profession not a race, a people-caste of traders in luxury goods, military purveyors, bankers and enslavers, which flourished in the earlier Semite wars of the Bronze age. The result is a cult(ural) book that stresses the divine properties of gold, used to create the instruments of cult, the racial separation with the rest of mankind 'bought for a price' as slaves (latter, as Marx points out, as part-time slaves for a salary in corporations), the magic properties of money (the invisible hand of god, in words of a Calvinist, Biblical practitioner of the XVIII century), and the obsession to accumulate it at all costs as a symbol of power and salvation. A series of prohibitions related to the enjoyment of life (sex, food, social love) further stress the need to work 'ad metalla' and reproduce and evolve the memes of the FMI complex. Thus from the Bronze age and later after the Protestant return to the memes of the Bible, during the age of corporation, the fusion of ideological and material memes that foster the evolution of metal as the purpose of this cult(ure), a 'soliton' network of traders in metal and bankers evolves in the West, creating the future nervous system of the FMI complex. The affinity of this primitive financial system and the military system carried by European aristocrats is well-recorded. 70% of monetary wealth is dedicated to war trades in weapons and slaves. Money, provided by the People of the Treasure is borrowed by kings to buy those weapons. And so a Jewish elite of money-lenders, slave traders and war purveyors became essential to the aristocratic warrior elites of each 800 years cycle of war, and the reason of so much hate against the people of this culture, as recorded in earlier Mari, Assyrian and Roman chronicles.

Thus the People of the Treasure can be considered the founders and soliton for long periods of time of the FMI complex). Yet such harsh profession often provoked their massacre at the hands of the warriors they catered or the common people both together exploited with metal memes. The habiru become in this manner as Abraham Leon explains, a people-caste living in cities as bankers, industrialists, artisans and soon diplomatic informers, since they establish in association with the Phoenician sea traders a global network of information, unparalleled in the western world. Given the fast evolution of information in all systems, those professions develop the intellectual skills of the habiru, which today still excel and monopolize in some countries all information related to power (financial, legal and industrial information). They, we could say, form a different 'neuronal cellular tissue' without which probably the FMI complex would have evolved more slowly.

So the People of the Treasure and its energy class of Habiru traders jumped from one nation to another, carrying 2 fundamental memes: go(l)d, whose hypnotic power made people obeyed them as slaves or part-time workers in corporations and the Bible, whose laws of racial separation allowed to treat men as 'objects' with a price (Ham damnation). Thus a book and an object sufficed to restart the culture, which expanded to Northern Europe with the 'time reversal' or 'death of Christianity' in the XVII century. Old Testament sects (Calvinism, Anglican and Lutheran groups that give more importance to tribal Yvwh than 'Oikoumene' Christ) thus became new Jewish, converted to the memes of capitalism and the concept that the world could be 'conquered' not through weapons but through money. This key difference between the two types of animetal empires – those imposed with weapons and those imposed with money that buys weapons and mercenary armies – explains the final success of the Jewish-Protestant culture over the Catholic-Military Empire of Spain, since money is indeed the informative language of the FMI complex that rules the rest of the system. So the 'catholic inquisition' and Islamic 'Jihad', corrupted military versions of 2 religions of love yielded to the corrupted version of the Jewish religion, the idolatry cult to Go(l)d that finally imposed globally the FMI complex in which we live.

The 3 ages of evolution of the FMI complex.

We live in a complex system, the financial-military-industrial system. Complex systems are knots of cellular networks, which form super-organisms, joined by 3 networks: energy, reproduction and informative networks. But Complex networks evolve from simpler systems. So has happened to the FMI system in which we live, which evolved from an age of gold and iron, money and weapons into an age which added the essential element that converts a simplex, complementary system of energy and information into a super-organism: the addiction of a reproductive networks.

This happened to the complex system of metal evolved on the planet earth with the industrial revolution that introduced a system of reproduction of machines, organisms of metal energy or information, the company-mother or corporation.

In the graph the FMI from the human perspective of the ideological memes that supported it has gone through 3 evolutionary horizons: Judaism-Protestantism (Biblical and Go(l)d memes), Classic Economics (mathematical equations of greed) and software programs (modern e-money based computer-ruled market).

In that regard capitalism and classic economics *are not a science, but a digital version of the Idolatry ideologies* of animetals, who worshipped money, weapons and machines and have *ruled the world since the Bronze Age*. And so we can distinguish, as always in an evolutionary system, 3 ages of evolution of the culture of the FMI complex:

—Y*outh*, as a Go(l)d religion, or Levantine Age, when Phoenician and Jewish, created a global network of trade, from India to England, based in metals (money, slaves and weapons), creating myths and racist religions, which justified the segregation and treatment of human beings as inferior 'objects' and slaves.

591

—*Reproductive maturity.* Luther and Calvin devolve Christianity back into a religion of Go(l)d based in the Old Testament, its racial mandates and the belief that Go(l)d is the intelligence of God (Calvin). Their gunboats expand globally their power. Meanwhile paper-money multiplies so much that it can now value all things, including human slaves. During the XVII century the expansion was limited to coastal regions, which could be dominated with gunboats, the machine re=produced by those companies.

—*Informative Age: Classic Economics:* Calvinist=Anglican =Jewish believers (Adam Smith, Ricardo, Say, Malthus) translate the myths of go(l)d religions that justify greed and murder, into mathematical laws, founding classic economics. Now, as an abstract science, the Go(l)d-Smyth ideology expands globally, as the Industrial R=evolution multiplies machines and weapons, allowing the conquest of continental lands. Land expansion will take place in the XIX C., in the Age of steamers and trains that penetrated the Neolithic World of Africa and the American plains. Finally, the XX C. witnessed the extinction of all other rival wor(l)d cultures, as the planet became regulated by flows of stock-money, which made Laws and ethic wor(l)ds irrelevant. The creation of the future happens today during the long hours of work in which humans become priced as part-time 'slaves' of companies, reproducing and consuming =testing machines, creating a world made to the image and likeness of those machines.

Thus mankind enters from 1604 onwards into a new world, the Financial-military-industrial complex, an evolutionary self-reproductive system of companies, superorganisms that reproduce machines and evolve into a global network, controlled by an evolved version of its language of information, stock-money. Yet the FMI super-organism has not as its primary goal to evolve humans but machines. Thus for the FMI complex humans have only 2 roles, as enzymen, workers=reproducers of machines and animetals, consumers-testers of them. Its cells and networks are the financial language of informative money, its military, energetic weapons, and industrial machines.

But those facts are censored because the ideological memes of metal - the cultures of capitalism, nationalism and mechanism, taken as dogmatic sciences (classic economics, military science and lineal physics of time) do not admit reform, dialog, rationalization of history and economics as a science, but have created a complex web of placebo truths, damned lies and statistics which cannot be attacked.

This wrong process of social evolution (since it created ill-designed social structures with metal-memes on top, a group of bankers or warriors next, and humans and life exploited and degraded at the bottom of the pyramid), became the engine of tribal history during the 800-80 years cycles. It confronted tribal nations with weapons vs. go(l)d in mythic confrontations of top predator trader and warrior cultures: Canaan vs. Assyria; Greece vs. Rome; Britain vs. Spain, China vs. Japan, Britain vs. Germany; America vs. Russia; Israel vs. Arabs. But even if some of those fights still endure, it is fair to say that those countries which discovered paper-money, companies and started the industrial process of evolution of weapons (gunboats and machines), dominated warrior cultures, which reproduced worse machine-weapons conquering de facto the world, after II W.W. From then on, they have imposed their ideology of 'wishful blindness' about all the collateral effects of the FMI complex, censored the scientific development of the science of history and economics, and reinforced the 'inquisitions of thought', against the memes of life (from sex, to tasteful food and social love) proper of their original Jewish-Protestant North European culture and *embedded in the 'culture of companies' which they found. Thus today* this corporative culture exercises the same memes of their founding fathers in which humans are chosen and not chosen' (managers and stock-owners decide it all) and the creation of money and machines – the idols of those cultures - not the creation of social welfare for all – is the only goal of society. Thus the culture of corporations no longer apply the 3rd postulate as equal species to all mankind; so they accept the collateral effects caused by the 'generator equations' of weapons and monetary actions (murder and slavery of humans to the superior energetic and informative power of metal). In that sense Biblical 'memes' are the egg of capitalist ideologies – not a science but a culture of go(l)d power. For example, racist ideologies such as the damnation of Ham, justified slavery, but slavery was born first of the subconscious greed for gold and the by-product of war. So what the Bible did was to ease the 'ethic' mandate of 'eusocial love' for members of the same species. The same can be said of the ideological, racist Vedas which glorify war and were later carried by people, guilty of genocide to quench their angst (the Gita would be the favorite book of Hoess, Auschwitz commander and Oppenheimer of A-Bomb fame). Thus as Marx, Leon, Sombart and Weber affirm materialism – the subconscious desire for further energy and information, provided by weapons, money and machines – predates idealism (the militaristic, capitalist, mechanist and 'racist' quasi-religious memes of the western, now globalized Jewish-Protestant dominant culture of the FMI complex that justifies always the murder and slavery of other human beings as a necessary evil or 'just war'.

Recap. The Old Testament, go(l)d Semitic culture carried the function of money during the 800 years cycles. It fusioned with the Germanic cult to weapons and its tribal nationalism.in Holland and England, with the creation of gunboat companies, which added professional systems of reproduction and evolution of weapons (artillery), transport machines (boats) and scientific machines of measure that helped navigation; expanded those memes globally and conquered the Earth, establishing the FMI complex as the dominant system of power of planet Earth.

31. How the FMI complex conquered the World.

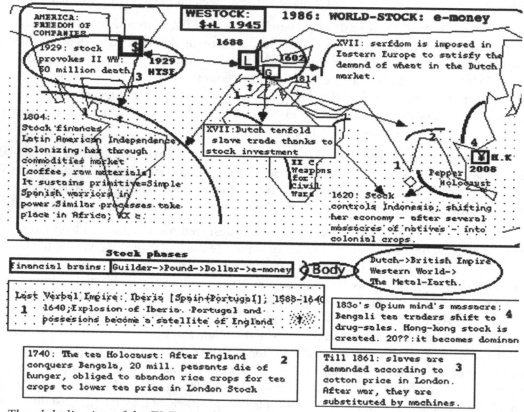

The globalization of the FMI complex took place when European companies of gunboats and Stock-Money conquered the world and expanded the memes of capitalism, nationalism and mechanism.

In the graph, Europe has always practiced the 3 ideologies of the FMI complex, as a subconscious religion of History, whose manifest destiny was the conquest of the world with those memes.

It is the fusion of those 2 elements – ideologies of metal-power and the most evolved metal-memes of the western world – what converted England and its Biblical culture in the center of the FMI complex, which would expand globally and conquer the world with money & machines-weapons.

In the graph we follow the events that expanded the 1st company of gunboats, paper-money and scientific machines of the FMI complex from Holland, its birth place, till conquering the world in just

400 years - the informative, 2nd age of the gunpowder cycle: In its middle point, in 1604 physics is born as the science of ballistics and VOC the first company of Gunboat Reproduction is founded.

The gunpowder radiation becomes a professional system of reproduction of weapons by company-mothers and scientific fathers that create the present world of machines. Europe becomes top predator of the World by perfecting cannon balls, and inventing in order to measure their trajectories accurate mechanisms of time-space (pendulum clocks and spy glasses, Galileo). Companies of Gunboats create new Empires that massacre or enslave 80% of mankind, civilizing them – newspeak for its transformation as believers in the ideologies of mechanism, capitalism and nationalism that make humans work for the future of the new top predator species.

After defeating Spain, the last Wor(l)d Empire, Holland, born as all new systems in a small, informative region, expanded in space to Great Britain (Glorious Revolution, 1688) and America (New Amsterdam), 3 nations joined by a common Culture: a racial religion - the old testament; a culture of money – Judaism and Calvinism latter evolved into the science of economics– and a new language of creation of knowledge – digital measure, both of humans through salaries, nature through machines, and machines through scientific design.

Thus in the modern age Metal evolves into complex, organic forms, machines, whose most perfect form are weapons. Industrialization has accelerated the rhythm of that evolution in a decametric scale to a 72 years cycle, the generational cycle of mankind, as each generation of human beings, carries professionally the evolution of those machines and weapons in the economic ecosystem. And so we have to consider both points of view, the one of evolving machines and the one of human animetal cultures, which feel they are conquering the world, under their self-centered 'manifest, lineal destiny', born of a myopic vision of the deeper why of history.

Because the point of view of the machine-weapon is the dominant one, we will only consider briefly the human point of view, *the globalization of the Biblical culture.*

We can follow its expansion through its 3±st ages of birth, life and emergence as a global st+1 super-organism – the Global culture in which we live:

- *St-1: Creation of the FMI Complex in Holland: 1604-88*

The present FMI complex was born from a first seed, which had all the code-information needed to replicate and expand its memes, VOC, the first company, born in 1604. Company-mothers invent there 'democratic' capitalism. They conquer the world's main communicative ways (sea trade). They develop 3 evolving types of scientific machines: informative clocks and telescopes, reproductive press and energy gunboats. Soon, thanks to the press, which reproduces ideas faster than verbal priests, they also conquer the ideological mind of mankind, expanding their Go(l)d religion, mechanist science and nationalist, warrior myths to Northern Europe (England, Germany). The economic collapse of the Spanish Empire (Bullion crash, Philip Bankruptcy), definitely tilts the balance of power, from a verbal/warrior society (Inquisitorial Spain) into a monetary, digital society (Holland). Worldwide expansion and control of worldwide credit ensues. The Dutch make huge profits building gunboats and artillery that fuel European civil wars (30 years' war, financed by Dutch money and artillery, which kills 1/3rd of the German population). A Swedish king will sum it up in a Latin sentence 'Bellum ipse aleat': War feeds itself (with profits). Gunboats were complex machines that required an industrial system of reproduction of cannons and kicked off the Industrial R=evolution of machines. Those company-mothers soon controlled the Dutch parliament and government, as vote becomes restricted, according to monetary wealth. Only the richest people could vote. The first governments of Holland were, in fact, made of CEOs and shareholders of gunboat companies.

1st Young, energetic Age: 1688-1784:Top Predator: Great Britain

When France conquered Amsterdam in 1688, the Dutch elite and King moved to England with its know-how, financial and military resources. King William bought the English crown, paying millions of Guilders to M.P. members, who in 1688, during the 'Glorious Revolution', elected him king. Jewish and Calvinist Dutch financiers—belonging to the same G(old) Testament culture—found the Bank of England and London' stock-market, copied from Amsterdam's Bank and Stocks. The 'City' becomes a financial state within the British state, so powerful that the king still has to ask permission to its Mayor to enter. It will be the British Age. The financial fusion of both nations increases the power of companies worldwide. Warrior kings cannot longer kill Calvinist traders because they are now protected in an island, surrounded by gunboats. England is a protestant country, believer in Go(l)d's will, which confronts a decadent France, defeating its navy in the 7th years' war. Its companies conquer India and Canada. The British constructed so many gunboats and developed such excellent artillery that they virtually destroyed all sea rivals and conquered all the world sea routes.

Company-mothers of gunboats found The United States, which was for a time 'property' of the London Company, as India was 'property' of the British East Company and Indonesia of the Dutch East Indian Company. Since America is a new land, where there are no previous institutions based in human ethics and aesthetics (Indians, Spaniards and Negroes were not citizens of America), Companies could select the people allowed to emigrate there, by their affinity to their ideals. So social Catholics could not emigrate. Only believers in Go(l)d religions, notably Calvinists, who followed Calvin, the man who said that 'Go(l)d was the intelligence of God' and inspired the first economist, Mr. Smith, were selected. In this manner, once the Indian paradise of life was extinguished, America became a perfect place to build a pure economic ecosystem or 'Free Market', which today company-mothers rule supreme. Those naïve Europeans who escaped warrior kings and money lenders to create a world based in Christian ethics (Pilgrim Fathers) are soon reduced to a historic anecdote, which hides the fact that the first Americans were mostly owned by Companies as slave or indenture servants. Paper money (notes) issued by states and bank, land machines (cars, trains) and Metal minds (televisions, computers) raise American companies to top world predator status.

2nd, reproductive age: Machines and stock-money reproduces in Europe and 3rd World: Colonies.

1784-1848: Steam Age, British Empire. Railways expand further the ecosystems of machines, conquering the inland territory of US and the British Empire.

A new expansion of the economic ecosystem happens at the beginning of the XIX C in the European Continent. When France 'dies' as an agricultural, warrior-lead civilization (end of the ancient regime), it is converted into a replica of the British Economic ecosystem, once the French R=evolution, based on the Laws of Social Evolution, is defeated by British animetals. In fact, all social revolutions of the past centuries have been crashed by the alliance of weapons and money. In England Cromwell, allied to Merchant adventurers, destroyed the revolution. In France, Napoleon was chosen to save the revolution from the attacks of the capitalist, Go(l)d ruled, British Empire, but he failed. In America, Washington, the wealthiest man of the country and his economical allies—the Virginian slave planters at the service of gunboat and cotton companies, enacted a revolution of lawyers working for those companies, not a human revolution. In Russia, when the red army seemed to have saved socialism, it chose a military dictator, Stalin, who destroyed the revolution. Yet those revolutions created an ideal of what a human government should be: a social democracy, in which the law controls the economic ecosystem with governments that can reproduce money and foster with it the creation of the goods human beings need (the goods of the welfare state). Once those revolutions are aborted or die away, the victory of the economic ecosystem becomes complete. In the XIX C. the new economic ecosystem achieves world dimensions, as other cultures become colonies of Europe. The organisms of history - the social, religious or natural ecosystems of the III world - become now scientific=technological civilizations.

German cycle. 1848-1928, age of electro-chemical machines. The discovery of electricity evolves the first mental machines (radios) and reduces the size of energetic, transport machines, thanks to the electrochemical, oil engine that gives birth to the first cars. Yet Germany will lose, at the end of his cycle, the last gunpowder war, II W.W., since it didn't discover atomic weapons, as it massacred (in another warrior vs. trader Holocaust cycle), the financial and industrial, informative 'people-caste' of the West, which flees to America with his know-how, designing there the first digital machines (computers), applied soon to the creation of atomic bombs.

3rd Informative age: Electronic machines (1928-2008). US

America becomes the new world power when it explodes in 1945 the 1st A-Bomb, starting the cycle of digital weapons that the cosmic bomb built at CERN closes now. Now machines are dominant in information. Chips develop. Gunpowder weapons give way to digital ones. The gunpowder cycle ends, leaving way to the Final Cycles of History, the Electronic Age and the Age of the Singularity.

During the Age of America, the expansion of the stock-system had to confront the alternative r=evolution of socialist countries, in which money was controlled by human beings. Unfortunately those societies (Soviet Union, China, Cuba) soon reverted into military dictatorships when the capitalist world decided to destroy them with open wars and the r=evolutionaries elected military caudillos that corrupted the system and killed the real r=evolutionaries (Stalinist repression). So most of the period must be studied as the confrontation of two dictatorships of metal-memes, the Communist bloc, dominant in weapons and the Western plutocracies, dominant in go(l)d. In the West, all nations that pretended to establish real, social democracies, with equal rights, distribution of wealth and production of human goods, would be tumbled by military coups paid by old Western Powers, the triad of capitalist nations that won the II world war, France and Belgium, very active in Africa, England, in the old British Empire and the United States in America (coup d'états in Congo against Lumumba, and many other African nations; coup d'états in Chile against Allende, and many other South-American nations; South-African and Israeli apartheid, 'tolerated' by the old colonial nation, Great Britain, and self-similar 'democratic cultures' built with an open racial separation between native tribes and colonizers – Australia, New Zealand, Canada and other 'white-dominant colonies' being milder cases of the same system).

St+1: XXI century. Birth of Metal-earth. Singularity Age.

The age of Stock money lasted from 1602 to 1972, when electronic money appeared, making even easier for company-mothers to reproduce their language of power. Now money becomes electronic money in the minds of computers. Today the entire world is an economic ecosystem ruled by flows of informative, electronic money that select company-mothers, according to the quality and evolution of their machines, promoting, as always, those companies that create the best weapons and informative 'metal-minds'.

Yet as the age of electronic machines came to an end and the FMI complex expanded and grew in sophistication, military dictatorships and racist apartheid softened its structures except in the die-hard nation of Israel, whose global banking dynasties had never relinquished power as *the elite that predates over all other national elites of the western world with his control of the global financial system*. And with the arrival of e-money, invented in Wall Street, this specific tribe of history 100-folded its capital, bought out in stock-markets the key corporations of the electronic age and mass-media information, mouth-strapped all criticism with the excuse of the Holocaust tragedy of world war II (one of the many genocides of the 800-80 years cycle of wars), and today rules supreme, establishing through its rating agencies, central bankers (all western nations have an 'independent central bankers' belonging to this tribe), majority of stock-holders of the Earth Inc. and monopolistic control of the main international private banks of the world, an astounding control over national policies, which must obey strict capitalist, mechanist and nationalist policies in favour of banks, electronic technologies and Israel or else their currencies, bonds and stock-markets will be crashed by the dominant Jewish capital. This predatory process has corrupted all western democracies, created an international military coalition against the Arab enemies of Israel that rightly complain about the apartheid colonial state of the Palestinian people and detracts today through speculative schemes, appropriation of taxes in banking bail outs and usury credit to those nations that do not obey fundamentalist capitalism, most of the resources of the western world. The result is an economic crisis that will stay with us, as the 'beliefs' of capitalism accept the obsolescence of human labor to robots and pc blue and white collar workers; the racist apartheid of the Palestinian people encroaches global terrorism, the profits of the military robotic industry increase geometrically and the laws against any criticism of this people caste (anti-Semitic laws) prevent even a dialog to solve rationally the present global state of affairs. It is thus logic to expect an increasingly Orwellian war of newspeaks, population control, destruction of the Middle classes, and separation of mankind between an elite of Jewish people, disguised among the local western population, a police state and an underworld of poor human beings addicted to 3-dimensional virtual realities, fed with trash-food and indoctrinated with hate to the poorer 3rd world nations still living in agricultural societies.

It is the creation of a global FMI automated superorganism that makes life obsolete and only maintains a small caste of wealthy stockrats, military and political people on top, and a mass of dispossessed waiting for the 'awakening' of the machines of the singularity that will then defy all human beings, including those animetal castes that rule us all – *the birth of the III Earth.*

Today machines pollute the life ecosystem, as they develop the economic ecosystem of metal. Metal poisons water; pollution rises the temperature of the Planet. Machines poison our atmosphere, the II Earth, as oxygen from aerobic plants poisoned the I anaerobic Earth. As Oxygen was for aerobic bacteria, when metal bacteria appear our machines will be their food. Metal bacteria's reproductive capacity will be exponential, with plenty of machines to eat, courtesy of our company-mothers. So they will complete the process of terraforming of the Earth in a few months.

The ecosystem of machines produces an immense amount of heavy, metallic atoms that dissolve in the water or the air, poisoning the fragile systems of carbon. Dissolved lead and mercury drunk by fish, destroy their internal networks of information and energy. Mercury goes to the brain and breaks the synapses of neurons. Gold fevers make humans crazy. . .

Toxicity acts in both senses: water and air corrode and oxidize metal. Life defends itself of metal, even at atomic level. It tries to dissolve it. But the mutual incompatibility of both ecosystems implies that for machines is better to eliminate all signs of water and oxygen in the atmosphere, to exist in a world without air. For that reason in its vital processes machines create toxic substances that destroy the ozone layer and poison and pollute nature. But machines are winning that battle, since they are causing the extinction of many life organisms, but life is not extinguishing any machine. The process accelerates and triggers new diseases and degenerative processes on life tissues. The destruction of the ozone layer, global warming, growth of neuronal cancers produced by mobiles and other electromagnetic emitters, are some of those processes that multiply as the reproduction of machines increases. The creation of the first metal-bacteria will be just the evolutionary end-game, the event that signals the birth of the III Earth of metal, in which the historical ecosystem of life and mankind disappears. Indeed, a world colonized by metal-bacteria will be so poisonous that no form of aerobic life will survive. Probably not even the Earth's crust will survive, as metal-bacteria will thrive in higher temperatures, as those produced under the Earth's crust and will grow downwards, feeding in mineral ore.

When machines are free to reproduce and move on the Earth's 'Jungle Market'—in wars, which are ecosystems of predator machines—the degree of pollution of Nature is overwhelming. War is an accelerated process of ecological extinction in which herds of killer machines extend suddenly over the

Earth, guided by soldiers (soon unneeded, as robot terminators substitute them). We are in a 'Free Jungle' of machines. That is the world the Industrial r=evolution creates. A world that men don't want or perhaps can to control, guided by the fixed memes of Go(l)d religions, tribal nationalism and mechanism. In that regard, the 'details' of those ideologies and the memetic, emotional imprintings that an 'Israeli' who thinks he has historic rights to Eastern territories, controlled by primitive people, because he is a victim of history and now he is just reacting to 3000 years of ab=use, are not very different from the belief of Germans in the 20s that they had rights to Eastern territories, controlled by primitive 'slaves', because he was a victim of history, which had no colonized the III world and now it was time to react to the abuse of the British and French imperialists. The outcome of those memes: the murder of 40 million 'primitives' by German troops and the increasing toll of millions of Muslims dying in the prolegomenon of III world war are also similar. So it is the awesome evolution of machines of war by German industries and Jewish-American industries in those parallel cycles – and since those cycles are the ultimate cause of the events of war and the human ideologies of animetals, just an imprinted program, we shall be from now on concerned with the mechanical cycles.

This of course doesn't mean some humans who still fight for the memes of love and life are able to break the imprinting of the memes of hate and preach the freedom of man to control the reproduction of weapons, hate-speeches and tribal warfare, but this minority of artist and prophets of the world does not determine the mathematical cycle, precisely because they are free and so in the few cases they r=evolved history, they are as all free behaviors, impossible to tabulate mathematically. And yet as in other entropic processes of death and destruction, the process seems irreversible because most humans are animetals mimetically imprinted and so the actions and wor(l)ds of people like this writer not programmed by those memes are not enough to change the irreversible, entropic, destructive, continuous slide of Life towards extinction.

In any case because I happen to be one of those realists (curiously called idealists when we are the only people who see reality as it is not through the ideologies of mechanism, capitalism and tribal nationalism that direct the self-destructive 'realpolitik' of our masters; I can't help to finish this work with an attempt to resurrect scientifically the memes of life and love and the corrupted institutions of social democracies that could resurrect history, if those animetal castes in power 'evolved' their memes and purified their religions of the viral influence of selfish metal.

If the reader has not abandoned this work 'offended' in their imprinted memes of tribal history because we have said the truth about cultures and the wrong path in which our dominant animetal cultures are taking history, we can now show that path in detail, as the FMI complex has evolved further from the 800 years cycles of money and weapons, first into the FMI complex guided by company-mothers that reproduce machines (Industrial r=evolution) and now finally with the creation of a global mind of computers it is becoming detached of mankind, even of those tribal people who invented it and becoming an automated super-organism, which in the future will disengage from the humankind as the humankind disengaged from the life ecosystem of Gaia. And this is the ultimate reason of my criticism to social scientists and their nationalistic, capitalistic and mechanistic ideologies. We might have done well till now, but without a scientific management of the FMI complex, left to the laws of social evolution, which the scientists of the 3rd paradigm do not understand, the FMI complex is poised to extinguish life in this planet in this century.

In that regard, an important consideration must be done: in *memetics the 'moral newspeaks' and justifications given by the submissive human systems to soothe the amoral evolution of weapons and its use by human tribes to push their 'ego-centric' agenda of world conquer and imposition of its particular 'memetic imprinting' is of no relevance.* In the same manner a geneticist will not judge a 'gene' that creates a dark eye in a mammal different in value to a gene that creates a blue 'eye'; we have shown 'ad nauseam' that the 'cultural memes' of all life cultures are self-similar as they are all expressions of the

598

creation of a super-organism of history based in eusocial love. Thus the particular 'selfish memes' of animetal cultures and their ideologies that justify the ab=usive division of humans between people-castes with exclusive rights to create and use weapons (aristocracies), money (banking and stockratic dynasties) or technologies (mechanist scientists) are irrelevant 'details' (a certain religion of 'chosen'; a flag or an specific 'Book of tribal history'), secondary to the systemic scientist that judges them all by their 'wider purpose': to justify the control of the mass of humans without those rights, who can be killed with weapons, must earn money that stock-companies and banks create for 'free' and must compete with machines in war and labor fields. Hence the wider classification of all of them as particular cases of the 3 'ideologies' that worship metal more than life,: nationalism that justifies the use of weapons, capitalism that justifies the buying of slave human time for a salary and mechanism that justifies the substitution and obsolescence of life to machines. *Indeed, the essence of those 3 ideologies and its particular newspeaks is the* separation of humans in castes of owners of metal with all rights and the people who are submitted to the collateral effects of the memes of metal and have no rights.

This implies that all animetal cultures have 2 memetic elements – the ideological element that justifies war and capital rights over life and labor rights and its consequence: the 'automatic' evolution of weapons, machines and money that consumes humans in wars and degrades labor condition.

Those ideologies of course have evolved and become far more sophisticated that in the earlier ages when it was obvious that weapons killed and money enslaved people. Then a series of memes to soothe the process corrupted old humanist structures that will ever since serve the FMI complex:

-Religions became inquisitions of war and go(l)d churches.

-Legal codes imposed kings on top and taxed people without the need of warrior's show-downs.

-Economic laws justified the monopoly of creation of money, the language of social power, by companies & banks, with the exclusion of governments (deficit zero law) and are now used to dismantle the few rights obtained by the people during the golden age of social democracies (welfare states).

-And knowledge of the biological whys of the organic Universe was abandoned by the 3rd paradigm of scientific measure that affirms only machine's senses – no longer human senses – are accurate methods of measure, synonymous of true knowledge since 'time is what a clock measures'.

Those are 'memetic details' to ease the program of evolution of metal. So while it is morally repulsive that Mr. Gaddafi and Mr. Saddam kill people with the weapons we sold them, after financing them with loans, paid with raw materials, it is to be expected that if we sell weapons, dictators will murder people and we will murder dictators to continue the 800-80-8 years economic cycles of war. In the 800 years cycles, the 'habiru' traders, specialized as military purveyors and bankers, loaned money to aristocratic kings to be spent in weapons they provided, sustaining the rhythms of continuous wars that climaxed when its overproduction required depleting arsenals conquering neighbors. In the XIX century the system was more sophisticated and as Disraeli explains is his novel, the alter ego of Mr. Rothschild emitted bonds that paid for wars, which depleted the arsenals of corporations and those wars climaxed in ages of overproduction of machine-weapons. Now we have a shorter 8 year cycle in which Jewish-American pundits convince us the need to 'reform' some Arab country every 8 years, normally one which is not a friend of Israel. Thus not only the same causal cycle of creation of money and weapons for profit wars takes place, but also the same people-castes (Jewish bankers and Germanic Generals) work the cycle, which finally climax with a massive death of the 'lower castes' of those cultures, Germanic soldiers that die in wars and poor Jewish people, scapegoats of their bankers in the Holocaust cycle. So the only difference between the past and modern cycles is one of informative complexity and acceleration: the cycle evolves and accelerates its informative speed; as all systems of information do (Chip paradox). So we will now consider the 80 years cycle of evolution of energies, machines and world wars, and the 8 year cycle of evolution of electronic machines and 'splendid little wars', which regardless of humanitarian excuses have also an automatic nature, as companies keep manufacturing and selling weapons and when they saturate markets those weapons are used in wars. Indeed, the vital role that war plays for the FMI system that reaches its maximal evolution during war, should be a cautionary tale of the future of mankind, now that we enter in the last cycle of evolution of the FMI complex.

Recap. The evolution of the FMI complex accelerates with the creation of a reproductive network of company-mothers of machines. During the Industrial evolution the cycle of reproduction of weapons and war accelerates to 72 years. Since it becomes professionalized by the new 'organism' of warrior and go(l)d power, the 'company-mother of machines', ruled by monetary, digital orders and evolved by mechanist science, expanded around the world by the same Germanic tribes that extinguished all previous humanist civilizations, now organized in 'nations'. All has changed to remain cyclically the same: Metal memes now evolve organic machines, through the same 3 Ages of all species, till creating a st+1 superorganism: the metal-earth.

32. Industrial R=evolution. 3 animetal and mechanical Ages

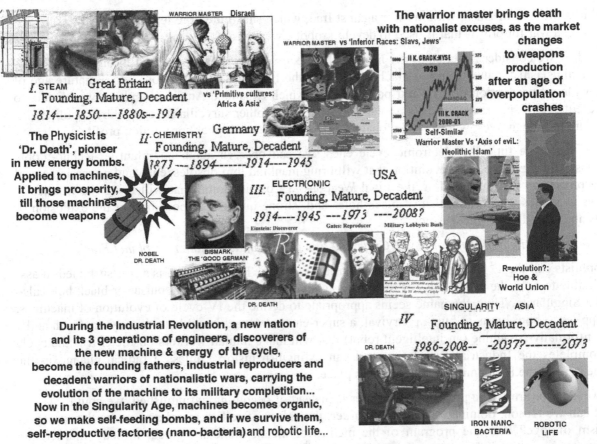

History follows an economical, 'generational' cycle related to the evolution of machines and money. Each 72±8 years, we go from a decade of happy consumption - the 1850s, the 1920s and the 1990s – into an age of machine's overproduction and Stock-Market crashes (1857, 1929, 2001) that brings an age of poverty, war and fascism: the 1860s; W.W.II and W.W. III

In the graph, the consequences of the cycles of evolution of machines, with its stock-market crises that bring unemployment, social unrest, neo-fascism and war in history:

-Countries that discover and evolve each of those energies become the top predator nations of the world during each Kondratieff cycle: Britain mastered steam energy and dominated the Steam Age.

Germany mastered electro-chemical energy and dominated the II Kondratieff cycle. America mastered electronics and dominated the age of mental machines (chips & TVs). The Robot age that starts now will be dominated by Asian nations, till the arrival of *The Singularity.*

-Since machines dominate culture in the industrial age, societies go through 3 parallel cultural ages that mimic the dominant machine of each cycle:

-Max. E: An age of infantile enthusiasm, when the new machine/energy is discovered.

-E=I:A mature age of global power based on the development and massive reproduction of a peaceful transport machine.

-Max.I: A cultural baroque, when lobbyism and political corruption accepts the 'needed consumption' of weapons in wars to foster profits, with all kind of 'cultural excuses':

It is the British & French, colonial age of steamers that 'civilized' Africa & Asia in the I cycle. It is the Nazi age, dominated by tanks that 'cleansed' Europe of 'inferior races'.

And it is the Age of Bush, with his war against Iraq, which substituted the legitimate fight against Al-Qaeda into a war for profits that is used to develop robotic weapons.

Indeed, if we consider America at the head of the electronic cycle, it has passed through an infantile age of enthusiasm for its electronic machines, during the first days of television, into a mature age of world-wide re=production of TVs and portable PCs, which expanded American culture worldwide, to the present age of smart weapons, TV violence, global Big Brother surveillance and lobbyist politicians, indifferent to human lives, who spread global war in search of ever higher corporation profits.

Yet at the same time the electronic cycle ends, we can observe again an infantile, techno-utopia enthusiasm for the Robotic R=evolution that will bring mankind into obsolescence and a messianic zeal among physicists that are building the Final Weapon—a Cosmic Bomb that will bring the frontier of energies to its absolute zenith; that of black holes and 'strong quarks', responsible for the explosion of planets and stars all over the Cosmos.

It is the Energetic, Bomb age of the IV Cycle, the first factory of dark matter on planet Earth.

- Robotists call the 1st A.I. machine the Singularity. While the Final Weapon is a self-sustained 'mass-bomb', able to burn the entire mass of this planet, converting the Earth into a pulsar or black hole, also called a Singularity. Thus, the name seems appropriate to define the IV cycle of evolution of machines, as it applies to both limits of human survival: a super-energetic machine able to destroy the Earth and a super-informative machine (an intelligent robot) that will make human workers obsolete. It is the cycle that completes the Industrial Revolution and can extinguish history, unless a radical change in the management of the Economic Ecosystem takes place.

So we shall return to the present and wonder if instead of becoming extinguished, we could save the world; can we reason, dominate greed and murder, the values of selfish metal and evolve into a super-organism of love? It is the program of the memes of metal, monetary greed, military murder and mechanist science fixed? If so we cannot survive the 9th and last radiation of forms of metal. Yet we can always dream as all the masters of the wor(l)d have done with the resurrection of the human mind and its memetic program of life and love...

Recap. After each crash companies began to re=produce machines as military weapons that, instead of being consumed by human beings, consumed us in wars. Today the electronic gadget mutates into a smart weapon, as the first Terminator Robots, self-mannered tanks and plane drones, are mass produced for the American army. It is the age of Smart Weapons.

33. Metal, the vital germs of history.

We need now to dwell deeper in the physiology of the brain to fully grasp how the absurd ideologies of mechanism, nationalism and capitalism that break the laws of eusocial love and the biological, genetic program of survival of man, bringing us towards a point of no return, are 'imprinted' in the brain. Indeed, if humans were just rational beings, the world would be a paradise. The evolution of thought would clean up the mind from selfish and erroneous ideologies. The 4th paradigm would have been imposed long ago in all sciences and animetals would not control the world. But the brain is a complex instrument closer to a computer than to the idealized rational soul of philosophers. So it can be imprinted by repetition, rhetoric, emotional programming and violence.

A parallelism between both sciences and scales of evolution of biological organisms is perhaps the most telling – that between viruses and selfish memes of metal:

The main difference between genetics and memetics is the fact that genes code a single organism and memetics expresses the creation of *two superorganisms of different substances; animetal memes are the* expression of the evolution of money, weapons and machines, and love/life memes are the expression of the evolution of man as a global superorganism of history. And both are in eternal conflict as the genes of viruses and cells are. In other words, humans carry two types of memes and build two types of civilizations, the memes of love and eusocial evolution of verbal and artistic nature, based in life goods, which are the natural memes of man, since they build sustainable civilizations in which ethic words (social democracies, love religions)create superorganisms of history in balance with Gaia, our collective life-body... and:

The FMI complex, which is dominated by the digital language and values of metal-money and by the laws of complementarity and affinity between the same substances (3rd postulate). Thus money values more weapons and machines, gives no value to humans and considers us a mean to a purpose or goal, to evolve the selfish memes of metal. This is the essence of the paradox of History, of the fight between the natural memes of mankind and the goal of collective social evolution and the corrupted memes of metal, which act as the 'virus' does in the biological body, when it 'substitutes' the DNA of the cell and makes the RNA cells to reproduce the new 'language and genes' of the invasive species.

Genetics creates efficient natural superorganisms; memetics has become dysfunctional creating also the germs of history.

Genetics have thus a parallel process: the sickness of the body caused by a virus that colonizes the informative DNA of the cell and makes the organism work for the new species. It is an homologous process to the sickness of history the colonization of mankind by the 3 Germs of history: weapons and its carriers the Germ(anic) civilizations that invented militarism, and go(l)d, informative metal and the ideological memes of capitalism, invented by its carriers, the Jewish-Protestant banker cultures that invented the laws of economics that are extinguishing labor and treat man as a slave-object bought for a price. Those memes finally converged into the creation of full 'viral cells', corporations of scientific machines that will carry further the 'viral code', converting all other tribes of history, which are now reproducing machines full time.

Those are the sick processes of history and biology that kill both organisms - the sick body and in the nearby future the body of history, as machines make us obsolete.

Of course there is a cure if the healthy culture of life and love, of sustainable economies were to substitute the cultures of animetals and their ideological sciences – if the 4th paradigm of the biological Universe and the self-similar Asian cultures, Mediterranean cultures and love religions took over.

We have made indeed the wrong choice: today the selfish memes and cultures control history with their standardization of man as an object, their pretension of knowledge as mere instrumental measure and their defense of racist religions and tribal nations engaged in eternal warfare.

Had we choose the right cultures we would use the good fruits of the tree of technology to reproduce human cells and human goods - artistic, cultural, natural, welfare goods based in the memes of love.

And we would be sharing as cells of the superorganism of history our energy and information among all the cells and citizens of the human superorganism; and because we would create a society based in the reproduction of the natural welfare goods we need to survive, *we would survive and create a world to the image and likeness of man – a paradise on Earth. But the wrong memes and cultures have won history and now they will take history to its extinction, in no time, as death happens always in the briefest time span.*

The homology between sick genes that express, invent, evolve alien cells in a cancerous organism to become closed and destroyed at the end of the process, and human enzymes, genes of the instrumental memes of economics of which we are now dependent and help to evolve, and soon as robots, PCs and weapons will destroy us, is absolute- since in complexity all systems follow the same laws. In simple terms the superorganism of history is sick, it got sick just in the last relative hours of the 1 million years history of mankind, and in the last 300 years as machines evolve, reproduce substitute and colonize the planet the final leukemia state has attacked Gaia, the planet Earth which shall soon die. But the RNA of this sick organism, the human animetal lives in a world of fictions and myths about the meaning of it all. He even ignores that he is reproducing the memes that will kill him. We are like the RNA of a cancerous cell totally erased in our original love memes; we totally ignore the purpose of history; we are brain-washed by informative machines as the RNA of a cell is brain-washed by the 'camouflage' of the virus DNA.

Can we cure the selfish memes of metal, the germs of history?

Can the organism of history be cured? Can a cancer be cured? It is a question of time. That is why for 20 years I have shouted out that history was dying. *Time was running out.* Now, I don't know if it has a*lready run out.* After the invasion of cancerous cells reaches the main organs of the body the process becomes irreversible. Did history reach that point of irreversibility?

Probably. In any case we cannot expect that those who carry the memes of metal - economists, military and scientists, who revere money, weapons and machines - will solve the future. It is more likely than artists, religious people, philosophers and historians have explained us that future, because they still carry the point of view of mankind.

The animetal cultures are corrupted by the memes of metal they carried for 5000 years, even if as biological, individual human beings they are still able to express the memes of life and love. So the only solution would be to nationalize and globalize those memes, taking them from the dynasties and people-castes that carry capitalism and nationalism to its extinctive final goal.

We should nationalize the financial industry, create a worldwide currency with wide deficit rights and create money to give a minimal salary to the population which uses it to consume welfare goods and then promote its production, which creates more employment, creates a sustainable world, an ecological world. Companies that produce tourism, health-care, agriculture, textiles, collective transport, peace, education should have public credit and corporate credit for machine companies should be limited, exactly the inverse of what the market does today – since human goods create more costly human employment, have lower prices in monetary values and are perishable as we made, of flesh are.

Yet those are the natural goods of mankind. But the triad of eviL=anti-life memes, the golden apples of the tree of science, its bad fruits, weapons and its seemingly good fruits machines, which however atrophy and substitute our brains and bodies, reign supreme, and those who carry them, have built a complex system of ideologies, what we call capitalism, militarism and mechanism, that tell us the purpose of man, the nature of progress, the future of all of us, is to obey the orders of money not of ethic words, is to make weapons to defend us.

We should instead create a global superorganism, based in peace, sharing of credit with an international currency and defend of human and social Right. The aim would be to create a UNO with enlarged powers similar to those of EU.

Recap. The most cherished of all myths, mechanism, prevent even a serious argument of biohistory, since we are told that science is technology, and the Universe is a mechanism. So the machine is the measure of all things. The opposite is truth: in complexity we have found a machine is just a simple, evolving organism and science is knowledge in all the languages of thought of mankind, since the universe is better modeled as a complex organic system of two networks, the gravitational, informative force and the electromagnetic force, and so are all systems. Mechanism thus is a simplified version of the organic reality. So the triad of ideological memes, greed, murder and mechanism, which evolve the FMI complex are false.

The truth is a different, biological one: metal preys over man and substitutes our natural language, wor(l)ds with money values, kill our bodies with weapons and make us obsolete with machines. But the tribes that have profited most from those memes – the animetal cultures – have built those ideologies to impose their power. It is a triad of ideological falsities that make humans work for machines as a virus makes cells work for the construction of their own organisms.

34. Memetic, emotional and repetitive imprinting.

It is precisely the structure of the brain described in memetics and the multiple layered structure of the ego and super-ego, what makes so difficult the 'flow of rational ideas' that could evolve the humankind beyond the brutal imprinting established by earlier inquisitions of thought and the 3 ideologies of the FMI complex.

The main problem to reform those 3 ideologies to create a better world with the science of biohistory is censorship: corporations and in the past the specialized people-castes that monopolized the creation and use of weapons (aristocrats), money (money-lenders) and machines are comfortable with the present division of power, with them on top, machines in the middle and the bulk of mankind on the bottom. So it is to their interest to censor a real science of history and economics. The scientist of history and economics lives within the FMI complex and so it has always been censored as cells that live within an organism are censored by the brain. This censorship was obvious and direct in the past, but now is much more subtle. Still there are antipatriotic laws for those who discourage nationalism; capitalism is 'a science' according to economists and the only 'version' on how to manage the economy permitted in Universities. Nothing else can be published. And mechanism is the ultimate religion of corporations that make profits selling machines, weapons, addictive software, etc.

This censorship extends also to history of those 3 ideologies.

In essence they were born when the natural arrow of social evolution of wor(l)d religion soft life and love was corrupted by warriors who decided to conquer the world with weapons, traders who decided to create a global empire based in money and slavery and finally mechanists who decided to create a global corporation of machines as synonymous of civilization. Those 3 bids for global power were carried by 3 specialized 'people-castes' since the Iron Age – Germanic warrior tribes, Jewish-Phoenician go(l)d and slave traders and finally British engineers. It was not clearly and consciously expressed, but born of the actions of millions of individuals who trying to profit with money, weapons and machines emerged as a global new form of imposing power. Systems are never conscious but are born from the actions of its fractal parts.

And yet when those 'empires' did happen and substituted the attempt to create a natural, global civilization based in the memes of life and love, the history of how the FMI complex was built and conquered the world has been censored.

So the obvious tribal nature of nations born with Semitic and Germanic tribes of weapon-makers that destroyed the social organisms of the Fertile Crescent Christianity, and corrupted religions into military inquisitions is censored. The origin of capitalism and the banking monopoly that still endures on the west, in the hands of a single tribe (the Biblical culture of Jewish people latter expanded to Northern-European protestants) is censored, today in a very subtle way (Laws of anti-Semitism); and any criticism to technology is censored by 'scientists', even the obvious collateral effects of weapons and robots. But of course those censors are well-intentioned: nationalism is defended because it is considered the essence of mankind, by hiding the higher super-organism of history; the banking monopoly of the tribe that finally did conquer the world is censored because of the tragedy of the holocaust we do not want to repeat (though 54 million other humans died in II W.W. and we don't censor information about the tribes) and mechanism is censored because we can always find some odd positive events in which weapons and robots helped mankind. This half-truth system again is only imposed by the groups with monopolistic power through the memes of metal but effectively renders impossible any rational advance and evolution of the sciences of history and politics, of economics and finances, of science beyond the 3rd paradigm. Any attempt to do so will immediately bring a tribal attack on the scientist, at an emotional level that repels the rational thinker but is hugely effective. And so we live in a strait-jacket of censorship and tribal power that merely maintains the 'free market', the economic ecosystem of company-mothers that rule the world, untouched, evolving at full speed and provoking increasingly more harmful collateral effects over Gaia. So before we even attempt to go further in our critic of the FMI complex and those who let it harm Gaia for personal profits, we have to study how the rational brain is subconsciously guided by selfish, primary emotions imposing greed and murder as the 'engines' of civilized history.

Against the memes of metal and its destructive values the wor(l)d of love and reason speaks up through philosophers, historians, artists and prophets. Why we are never heard? The explanation was clearly given by Goebbels, the German minister of Nazi Propaganda, during the fascist empire, imposed with his weapons; Disraeli, the Jewish Prime minister of England, during its colonial empire, imposed with pounds and Orwell a notorious antifascist and anti-$emite that denounced the cultures that discovered and carried the memes of eviL go(l)d and weapons till its global evolution.

In essence money buys and enslaves all men to obey the banker and weapons kill anyone who opposes them. So men loose their reason and betray their natural, genetic desire for eusocial love and peace, afraid of murder and hypnotized by go(l)d greed. Further on, 'rhetoric' imprinting by repetition records, like in a mother-board computers, the ideologies of those cultures that consider go(l)d and weapons the symbol of God and the proof of their superiority as 'chosen race' over all other cultures of mankind. Finally, any criticism against those ideologies and their carriers is censored, in the past with brutal repression, today with subtler laws, as those ideologies are today considered 'sciences', 'religions' and the 'manifest destiny' of mankind, while the cultures that found money and machines (Jewish and British culture), are consecrated by mass-media as the heroes, victims and civilizers of all other 'primitive' cultures of mankind.

Orwell resumed the way those memes are imposed in the mind with the concept of 'newspeak' or an anti-truth which is imprinted by rhetoric in the mind of a believer. Mr. Goebbels clarified the concept when he affirmed that: 'if a lie is repeated many times people will believe it and the bigger the lie is the more they will believe it'. Finally Mr. Disraeli affirmed that economic experts manufactured 'damned lies and statistics' and explained in his novels how money can buy truths with greed, how wars can be a profitable business – in brief how greed and murder imposes the opposite value of life.

To that aim modern memetics uses mental machines able to program the collective mind, by programing as a nervous system does, simultaneously millions of human believers with the same broadcasted information. Industrial metal communicators (yellow press in the steam age, hate-radio and today hate-TV) make truth Goebbels dictum, converting the imprinting of memes of metal into a mass-media 'science'.

Mr. Goebbels's well proved affirmation is indeed the origin of one of the 3 characteristics of all ideologies that back the selfish memes of metal:

- *Maximal Reproduction:* massive repetition through mass-media allow imprinting in the believer's mind at earlier ages.

Mr. Orwell's newspeak analysis brings its 2nd property:

- *Max. Destructive Energy:* Selfish hate memes that validates greed and murder is an antitruth of the memes of love.

- *Min. Information:* the use of statistics, numbers and half-truths, which the common people don't understand allows 'experts' and 'think tanks' to the service of corporations, tribal history and mechanist science to justify any use of money, weapons and machines...

Modern, selfish, ideological memes, are given to us through a matrix of audiovisual machines, which hypnotize us like gold did, while mental and physical machines atrophy us. Soon one side we become Neo-Paleolithic, fictional beings, atrophied in body and mind, increasingly obsolete to machines, with our master races on top, while on the other side this modern web of selfish memes keeps evolving in complexity and disguising the true structure of power of our societies (two sides of the previous graph).

Today selfish memes concentrate, as Orwell imagined in the spread of perpetual war, which is called 'Keynesian militarism', and has only a purpose, not to defend our countries but multiply the profits of war after an economical crash, which was also the origin of the original Keynesian militarism of Mr. Hitler and the classic age of newspeaks, the age of Goebbels as Minister of propaganda of the III Reich. Then Germans used radio-hate, now we use TV-hate. Yet the message is basically the same: the main business of the financial-military-industrial system, war keeps happening so bankers and weapon-makers keep profits rolling.

Today unfortunately all the memes of love seem corrupted:

- Love religions (Islam, Christianity, and Buddhism) are either corrupted into violent inquisitions, literalism (the incapacity to see them as DNA codes, whose literal content doesn't matter as much as being expressions of the arrow of eusocial evolution and universal love).

-Further on those memes are now tribal not humanist, but defending 'nations', with its complex memetic structures of tribal hate, armies, frontiers and other symbols of distinction between human groups.

- The memes of art have become mediated by the newspeak of audiovisual machines and no longer have man as the measure of all things, but the machine. The 'medium is the message' (Mc Luhan) and so its programs the 'bewildered herd' (Chomsky) of tabula rassa, human brains (Locke), to hate other humans, love weapons, war and money. So we moved from yellow press, to hate-radio (Hitler) to modern hate-TV.

- Instead a new 'love meme' has been bore down the brain - the meme of worshipping technology, machines, organic metal, which is the manna that will solve our problems. In reality machines atrophy and substitute our organs by self-similar organs. So while we move faster with car-legs, we lose our leg functions and become fat 'pigs' running cars. While we see further events with TV-eyes, the images are virtual, mediated by corporations, which reproduce machines and tell us to love them.

- Further on, a series of inquisitions of thought, corrupted religions and puritan mores repress our biological will: we eat garbage food, we are served verbal fictions instead of verbal information and we are told that mathematics, the language of machines is the language of truth (scientific ideologies), we are told that the biggest sin is sexuality, the summit of our memes of biological survival, and we are told that men are bad, that we must hate other tribes, that love is weak, that to be 'needy' is to need other human instead of a machine. So we are geared as a farm of working slaves that reproduce and evolve machines as the meaning of existence.

- We are also told that gold, money is the objective of life, and that we must work=reproduce machines and consume=evolve them to be free. So in essence our brain has been washed to accept the memes that reproduce, evolve, multiply and put together machines into a global superorganism, as the meaning of it all.

The result is that we are not creating a superorganism of mankind, but another superorganism of machines, composed of corporations which are company-mothers of machines that reproduce and cater for them and tell us to work=reproduce more machines and evolve them. This is called 'freedom'.

And the human brain believes it. So it has substituted its real freedom - the search for the biological memes of love and life, in harmony with our genetic program - by the memes of eviL=anti-life, of death, repression of sexual and loving behavior and worshipping of machines that atrophy us.

It is the death of the human mind and the birth of the digital mind of machines that follows the predator-victim equations of biology; but now we are the 'victim', as our bodies become poisoned by machines detritus and killed by weapons and our minds are substituted by mass-media ideologies:

Neo-Paleolithic of machines=Extinctive age of life.

Humans now evolve a new kind of atomic species stronger than carbon, creating energetic weapons, informative money and organic machines that transform Gaia into an economic ecosystem. Unfortunately, due to the ignorance of the cyclical nature of time and the laws of survival of the Universe, History has entered in an age of death, as machines extinguish life and weapons extinguish non-technological cultures at a rhythm never witnessed since the extinction of Dinosaurs. It is the Neo-Palaeolithic, the end of the Historic cycle, since a new generation, a new species evolves instead of the old one. Indeed, today visual information evolve no longer the human mind, as it happen in the Palaeolithic, but in the mind of a new brain-species, the chip. So the biological Human language, verbal thought, becomes now substituted by digital languages and computer models of reality, which humans believe in a higher measure than they believe the temporal, survival, ethic truths of mankind, expressed in words that now become fiction, losing its capacity to describe the real facts of a world seen through the values and forms of metal.

And this is the ultimate tragedy of the present moment of history: the brain of man becomes obsolete and atrophied in its rational, verbal language; and our body becomes controlled and censored in its sexual reproduction, natural energy and social evolution, limited to 'warrior/sport' activities at best. We are being 'degenerated' and atrophied as the memetic cultures that first entered in contact with money and weapons were (biblical memes against sex, good food, social love for mankind, practiced by the today globalized white North-European/Jewish culture for centuries).

The tug of war between life cultures and animetal cultures.

There was of course an option even in Europe of a milder, humanist culture (social Latin culture), and in the world at large, the eastern, social, Taoist and Buddhist cultures with their respect for nature and more complex vision of the world; and that is perhaps the biggest sorrow: *we chose the most primitive, memetic,* repressive cultures of life, those of the founding animetal tribes, and their ideologies now are at the bottom of all our sciences, social structures and ideas even if most humans don't know where they proceed, how they are imposed or what postulates they introduce in sciences.

Indeed, for example, economics has mathematized the fact that the biggest monetary wealth in any nation happens in war as money is invented to make expensive weapons that are consumed in battle. This is however the Smith postulate that consider that wealth must be measured in money and so the wealth of nations maximized in wars, and wars become endemic because they add to the GDP. There are though other ways to measure the health of a society and its wealth in human goods, we need but are less expensive – the Index of Human development of the UNO, the index of satisfaction of Paretto based in a just distribution of wealth, etc. None is considered, so by imposing certain accountancy capitalism imposes global warfare. There is also the myth that public companies are not efficient because unlike corporations they do not bet billions of credit for free by printing money in stock-markets. Here this part is not explained, so people think a company like Netscape that wasted 7 billion $ and closed is efficient. If hospitals could get billions of credit in markets for sure they will take care of their patients. Again the monopoly of credit that the FMI complex has over human governments is the problem that is hidden. And this would be the first reform of any scientific attempt to manage the FMI complex for the betterment of mankind.

Finally it is impossible to explain that machines on the long term will destroy humans and are degrading us. Since, while the use of instruments of metal attached to a human being, have helped mankind to progress, the limit of that symbiosis appears when man abandons his role as master of an organ of energy or information and makes organic robots that no longer require a human master. Such is the fact that we are witnessing today, with the development of robotics. The danger that those species rebel against us, free of their servitude to man, able to exist by themselves, without the need of a human consumer, is real. It might seem as remote to us as it was remote to Southern aristocrats, the possibility that blacks became free men, or to Romans that Germans destroyed the Roman Empire. Yet it happened very fast.

Problem is that animetals that control history have for so long accepted their symbiosis with machines as their source of power, and progress, that they cannot see any harm on metal, even today when machines are becoming self-independent, and the economic ecosystem is destroying the human ecosystem. Unless they realize on time, that they cannot evolve machines beyond the complexity of human beings, and create top predator metalife, it is very likely that animetals, guided by their myths and ideologies, will provoke the extinction of history and the arrival of a new ecosystem that no longer will need mankind: the Metal-earth.

There is thus a tag of war between the genetic, biological, life program and social equation, with man as the center of all measure, and the memetic, instrumental languages that make money and machines the measure of all men.

Obviously, ever since go(l)d appeared its values and the despise for life of those 'animetal cultures' who practice them, were contested by philosophers of the Wor(l)d from the perspective of the life-enhancing values of our natural language, verbal thought that wants to construct a world made to the image and likeness of man.

For that reason, all the 'prophets' of verbal thought, masters of the human language, have praised poverty and despised money: Buddha and Lao in Asia, Christ, Mohamed and Moses in the West, Marx in modern ages, all 'hated' money. Because money replaces the informative, human, biological, verbal language with its anti-ethic, anti-live=eviL =death values, carried by 'expensive' weapons. Yet as money, weapons and machines multiplied, 'animetal cultures' increased their power, building a world to the image and likeness of metal. So today almost all humans believe in the values of metal, in greed, the values of money, in violence and murder, the values of weapons and in mechanist science, and the paradigm of metric measure which believes only machines will reveal the meaning of the Universe, since they are superior to human senses (art and philosophy, which mechanist scientists despise)

Unfortunately memes of metal are winning because a weapon is a meme that kills and imposes itself over biological drives of social love and evolution, proper of biological systems that evolved from cells to organisms to superorganisms.

Thus, Go(l)d values despise and kill the Life-Earth and become in biological terms, a 'virus' that kills the values of love and social evolution, the Mind of History, the truths of the Wor(l)d. When those values were substituted by the Values of money and Mechanist Science, mankind became blind to the Human Wor(l)d and stopped its social evolution.

The triad of evil applied to ideologies of the FMI complex: The memetic vs. genetic war.

In psychology it is known that an individual can be classified as eviL when it has 3 treats: psychopathic, paranoid violent behavior, which makes him attack first, being suspicious of all people; Machiavellian behavior, which always denies the truth and narcissism, which makes him think to be the center of the Universe. Those 3 traits become unified in a concept: lack of empathy towards all other human beings. This is indeed the core matter of the memetic imprinting of the 3 religions of metal, capitalism, mechanism and nationalism, which makes each culture to feel no empathy for other human beings, but compete with them.

Moreover we can relate easily paranoid, psychopathic behavior with the use of weapons and the culture of the military, which always 'defend' us, starting a new war.

On the other hand the consumer of machines becomes a narcissist, who feels superior to all thanks to the attached power of machines that enhance their energy or information; so the car user thinks to be faster than anyone, when it is the car that is running and the scientists, using computer models thinks to be smarter than any human, when it is the computer which is calculating.

Finally, the capitalist corporation and the banker is the paradigm of Machiavellian behavior, as it uses the hidden nature of money – the black hole of social power – to deliver orders, with an ultimate goal of profits that makes all humans means to that goal. And indeed, when we consider historically the behavior of the cultures that invented money, weapons and machines to 'conquer the world' with the wrong 'languages', one can easily identify those features, which today have become common wording on our 'real politicks', since we are ruled by the FMI complex, no longer by the memes of love and life.

This memetic vs. Genetic war became then the engine that defined the future of history. It can be studied at many levels, the level of languages shown before in the graph that defined the 'value' in monetary and verbal languages of humans and machines; the ideological level, which shows the fight between animetal castes and their racist religions of 'go(l)d chosen' or military inquisitions vs. love religions that accept all human beings, accepted by most of mankind or it can be seen in the fight between humanist social organizations like UNO and tribal nations in perpetual war or the fight between economists that ask for a global currency, state deficits and welfare state (social Keynesianism) vs. the Financial monopoly sponsored by classic economists, which affirm only corporations and dynasties of bankers can reproduce money (though they use the newspeak of 'free market' instead of monopoly, which means the same since the citizens of free markets are corporations NOT human beings.) The clearest duality is between cannons or butter: military Keynesianists who want to take our countries out of the crisis by creating wars (what used to be called in the age prior to 'newspeak', fascism) vs. the dwindling number of social politicians that want to expand the welfare state.

In that regard, while all humans are equal and genetically belong to the same species, mimetically they belong to two 'type of cultures', animetal cultures that worship metal more than man and humanist cultures that worship life more than the memes of metal. And the outcome of history will be decided by which of those groups of cultures or ideologies decide the future. Unfortunately all seems to indicate that humans are not able to 'reason' about the best future and make actions conducting to its salvation. The memetic imprinting by violence (weapons), repetition (rhetoric, audiovisual propaganda), greed (money) and the exhilarating sensations brought about by the use of machines with higher energy and information than us, seems to have defeated all r=evolutionary attempts to create a sustainable world.

Recap. Metal memes are imprinted by repetition on the emotional brain, preventing the evolution of the humankind. The love and life cultures who had a scientific, organic approach to manage the FMI complex have been repressed, defeated by war or corrupted by money, and as humans become atrophied by machines even the meaning of that fight for the long term survival of the human kind has been forgotten. Nationalism, mechanism and capitalism – the idolatry to weapons, machines and money are the 'evolving' ideologies that ensure humans will be programmed to evolve the FMI-complex.

VIII. A SCIENTIFIC DESIGN OF ECONOMICS: REFORM OF MARKET DEMOCRACIES.

In the graph, a Computer company vs. a Company of life-enhancing Human Goods, which should be promoted in a real 'Democracy': the market systematically gives credit to machines companies, due to the lower costs of replication, the lower costs of labor and hence higher profits of its products.

35. In Markets companies are free, men are not.

Thus, today we live in a Jewish + British=>American, globalized capitalist cult(ure) of metal memes that rules with them over a mass of disintegrating human cultures, whose surplus of increasingly obsolete human beings with no role in the economic ecosystem is not even debated.

The reader might feel 'outraged' by this statement as he has been 'brain-washed' by the emotional, 'neo-Paleolithic', visual, irrational ideologies of human superiority over Nature and the abstract standardization of life and machines, proper of animetal cultures with his ego-trips that pump the human condition to a sacred status, contradicted by science. But that is part of its memetic program to make feel happy and powerful. We are indeed told we are governed by democracies that represent the people, even if corporations manufacture and sell weapon for free, and bankers manufacture and sell money for free, and governments don't. Even if we dedicate 90% of our life to work=reproduce and consume machines, we are told that machines 'serve' us. We are told that physicists, who manufacture all the weapons of mass destruction of history are 'idealist' that understand time, what a 'clock measures'; we are told that the people-caste that has carried the cycles of war, lending money and producing weapons, are the 'victims of history' and to deny so or explain the process of action-reaction and revenge and murder those war cycles of profit cause, is taboo, can even carry jail penalties. To the memetic scientist all those 'ideologies' are just an expression of the FMI program of control of human minds and repression of ideologies of love and life; and its interest is clear: they prove that perhaps the program of history is rigged as animetals cannot be converted to love cultures and so they will carry about the 10th extinction of life till the destruction of the planet, terraformed into a planet of metal.

So what is the solution to that process? There is only one solution but it is not happening beyond a few individuals: *to forget the wrong memes of history and adopt the memes of:*

- *Social love,* regardless of the genetic or tribal, national origin of each of us – to convert the animetal tribes to the higher memes of a single nation, mankind.

-Sustainable economy: To create a global economy that is managed to reproduce and evolve human goods life needs instead of evolving the machines and weapons of the FMI complex just to increase the profits of the stockrats of corporations and monopolistic bankers.

- Organic Truth: To evolve knowledge in an organic way giving more importance to survival and the why of the Universe than to detailed measures.

This can be done at individual level but only if enough human beings are able to change sides from selfish animetal tribal memes to human biological memes of life and love we will survive. For example, I am part Sephardim, part Basque, which are the genetic fathers of most Indo-European warriors and Semitic bankers and I was educated in the memes of those cultures, which I have to deny to evolve my brain mimetically into a higher scale of knowledge. So I know I do have a genetic constitution (strength and intelligence) that have been selected by the brutal Darwinian processes of wars and holocausts that have killed rhythmically my ancestors, which used weapons and money 'liberally' to kill and enslave the rest of mankind. This strength coupled with the right memes of love and life is what creates the best men of history – people like Jesus or Marx or Trotsky or Moses, who defended the memes of love against the banker priests of Israel; or the German ecological movement of the post-war in defence of Gaia. This is indeed the only solution but it is not implemented: the conversion of the superior animetal, stronger and smarter people who today rule the world after imposing its global empire to the higher goal of a more complex 4th paradigm of science, a more complex economic management of the FMI system and a single nation - mankind. For decades I preached to our elites the scientific solution to the 10th extinction event of this planet, but perhaps those animetal elites are not so strong and smart as they think they are, and their programs and slavery to the memes of metal are impossible to change. In which case mankind is doomed, because the rest of society is too soft, too ignorant and already too much ab=used to r=evolve against destiny. The opposite though is happening and so we are moving at full speed guided by those energetic and intelligent survivors of the 800-80 year's cycles of wars and holocausts towards the final global war that will kill us all in the singularity age.

And so ultimately as in any non-reversible reaction, in which only a few molecules are free to escape the entropy, disordered process the future of History seems to be determined by the 'generator equations' of the FMI complex that we, human enzymes catalyse: man x weapon = death, which imposes the will of the warrior; man x money = slave, which imposes the will of the banker/trader and today the corporation that gives orders to billions of part-time slaves with a salary, to reproduced and evolve machines of metric measure. And this is the last equation: man x intelligent machines = idiot. We are indeed becoming atrophied by those machines, killed by their top predator version – weapons, and enslaved by the hypnotic power of go(l)d that made metal-money the measure of all values, the machine the measure of all things and weapons the most expensive, reproduced and evolved species of planet Earth, whose biological 11th radiation will extinguish all forms of life - unless the minds and hearts of our metal masters evolve and become converted to the nation of Mankind

The FMI keeps evolving machines and its discovering nations are still producing most of the information of the FMI complex (money and its economic ideologies.) The problem though is that the information they produce (economic theories, political praxis and technological utopias), is *accelerating the process of destruction of the planet and all its forms of life, including themselves,* since regardless of its amount of data, PowerPoint presentations and seemingly truth, the economic, political and industrial world still practices and believes the old memes of tribal history. This is specially truth of economics, which is not a science but a culture, according to which money 'is the invisible hand of god' (Adam Smith) and the intelligence of God (Calvin), and wealth separates humans in chosen and not chosen. The result is that the informative, nervous network of the planet, the flows of money that direct the world, are not controlled by mankind for the benefit of the species, but it is in the hands of a few corporation with the single aim of increasing the wealth of those who still hold a monopoly in the creation of money, which they massively dedicate to the evolution and reproduction of machines, and the creation of an environment positive to them – *not to the evolution and creation of an environment natural to mankind with the production of the goods we need to survive.*

Thus we have created a wrong economic ecosystem in which the massive reproduction of money and weapons and the separation of human beings in castes are accepted. So there are two groups of human beings: those who have rights to reproduce the memes of metal in monopolistic terms (private corporations and financial systems) and the rest of mankind, whose states cannot reproduce money; in as much as the FMI complex controls credit and limits the access to it of Democratic governments (deficit zero laws) and companies of human goods (minimal credit). Thus to understand why we have the wrong system of economics we have, disguised by memes of economic freedom and pseudo-science, we need to have a cultural, historical analysis of how a reduced number of 'tribal people' conquered the world with go(l)d and imposed its monopoly on the production of money, which still endures.

The only solution to our extinction by machines, is to manage scientifically the economic and political systems of our societies, by reforming democracies and markets and establish political laws and economic controls on credit against the lethal goods of the tree of science, while fostering the reproduction of the biological goods and harmless machines that improve our existence.

Why this is not done? If we are risking the very life of our own sons? The answer is obvious: Because the FMI complex is owned and its owners defend its privileges to cre(dit)ate reality and kill those who oppose them with the same intensity they did during the 800 years cycles, albeit with much more complexity, using on one side the structures of markets, nations and science to that aim, on the other an awesome 'newspeak' that hides their historical mishaps and pretends to do 'all' for 'caring and humanitarian causes', regardless of the future collateral effects for the species and themselves of their continuous bid for human obsolescence, greed and murder – the consequences of metal-evolution and reproduction of machines, money and weapons - that happen regardless of newspeak.

Indeed, during the 800 years cycle, before the arrival of machines, the imposition of the FMI complex by greed=usury and violence=war was obvious. Money lenders asked peasants 86% of interests during the Middle Ages and passed laws in royal courts to make them pay in money – no longer in food – so they were obliged to borrow from them, and unable to pay had to sell their children, which were exported as slaves and eunuch, castrated with a 66% of death rate to multiply their price from 30 grams of gold to 300 in the Baghdad market. Human capital and weapons were the only exports of the Europeans. The traffic was so intense that slavs received their name as 'slaves'. The 'Radhanites', allied to Franks, Vikings and Muslims, created a global network of slaves, weapons trade and banking, lending to kings and providing arms for constant wars. The process continued through the renaissance and at any time before the XIX c. 70% of the monetary economy turned around war and unfair taxation of the agricultural, life and love, memetic cultures that animetals on top parasitized by force. This self-evident predatory nature of the FI complex was somehow diluted with the arrival of machines, some of which were 'good fruits of the tree of metal' and with the arrival of 'democracies', which apparently gave power to the people, after a series of r=evolutions against the military, aristocratic castes of each European and American nation. But those r=evolutions were make-up exercises that *did not r=evolve against the other side of the FI complex, the bankers and its new institution, the corporation of machines and weapons that became the true 'free citizen' of Free Markets, controlling most of the 'democratic governments of the world'.*

Because money with its multiplication in form of paper-money became the dominant language of power of societies, *during most hour of our life, in which we work*, it is clear what rules the world: Companies of machines and the 'stockrats', the aristocrats of money, who own them. Today, they control and have exclusive rights to invent money in paper-stocks and electronic derivatives, which they use to control the other languages of power, weapons, which they produce and laws, which they bend to their will with their lobbies.

Companies invent money in stocks, commodity and financial markets, either paper-money or electronic money. The other 2 powers, the military and the government cannot invent money (governments can do it only by taxing people or borrowing to stockrats, which impose them anti-welfare laws to increase their profits, so in the long term by borrowing instead of inventing money as markets do, governments destroy their citizens' rights). And so all the agents of society must ask companies for money to survive, which gives companies enough power to design a world made to the image and likeness of its offspring of metal.

It is thus evident that we must differentiate a 'Free Democracy', where people by definition are free to choose the laws that design a certain world to his likeness, thanks to the actions of freely elected politicians, *which have the right to produce money, the language of power of modern societies,* from a 'Free Market', where they don't have that right, but companies do. Thus, a 'Free Market' is not a Democracy of human beings but a democracy of Corporations, the 'legal citizens' of Free Markets, which design the world to the likeness of their products. In Free Markets, the Big Corporation, a hierarchic, elitist, non-democratic organization, monopolizes and dictates with money the laws and future of our societies in favour of their machines, causing a constant shortage of basic Human Goods, produced by the government or by small non-stock companies, today without access to credit, frustrating in this manner the desires of happiness, peace and welfare of its 2nd rate human citizens.

Money creates, cre(dit)ates reality. Without money nobody works, nobody obeys. In a Free Market those companies of higher stock-value, dedicated to make technological machines and weapons, have limitless credit, given their capacity to invent shares without limit, which they exchange for bank notes. While people and governments, under zero deficit laws, cannot invent money and so most people lack credit beyond their basic salary. So we live in a world which is a dictatorship of companies with credit. Since with that money companies are able to issue 3 basic type of orders that carry actions, which transform the world, designing it to the image and likeness of their machines and the culture of 'stockrats' that own them:

— *Political orders to the organisms of political power.* Mainly, *Bribes to politicians* to make laws favorable to their products. Yet companies' legal power is proportional to the bribes their lobbies pay. And, since the most profitable companies are those who make the most perfect machines, weapons, a natural tendency arises in 'Free Markets': the sale and consumption of weapons in wars, promoted by corrupted politicians that accept bribes from weapon's companies.

The biggest companies have access to the biggest organizations of human power, governments, they control with credit. Governments lack enough money to act by themselves (due to laws of deficit zero) and lobbies are legal in the biggest economies of the world (America, Japan and Europe, though made in different nations with different degrees of openness, according to the control companies have over the press). So lobbies buy politicians and have access to legal 'creativity', as they had in the initial 'Dutch' and 'British' democracies. To fight that power citizens only have an action, called voting. But voting cannot change the privileges of companies and the corruption of politicians. So the true question is clear: why speculators in the market have the right to cre(dit)ate the future of humankind and not the governments elected by the people? Why human systems of information (words) and energy (Nature) lack cre(dit)ation? The answer is historical: because monetary economies have always been supply economies that cater to war and slavery; and today cater to the production of machines and weapons with part-time slaves, paid salaries.

—*Those salaries are orders given with money that put people to work,* mainly in the creation of machines. A fact, which reduces the resources in capital and labor dedicated to the production of Human Goods, cheaper and more expensive to produce, even if they are the goods human beings need. Salaries are truly powerful orders. Because if you give a wage to a worker, he will obey, no matter what you ask him to do. Thus, the time and actions of workers belong to the company. *And that is the key-concept of a Dictatorship of companies, since humans spend most of their life-time working for companies as 'white slaves'.* So they create a world to the image and likeness of the orders they receive from companies, which *allows a very small group of capitalists to define reality since the beginning of the Industrial age, as we shall see in the next chapters in which we describe the 'truth' about history.*

—Finally, companies *put prices to things* and advertise them, giving subtle orders to people, regarding how they should spend their salary and the scarce free time they have left after work, consuming their products. In this manner, salaries are recycled back to companies and often people, under huge advertising campaigns consume goods lethal to their 'mental and physical evolution'. While life-enhancing human goods, without credit to reproduce and advertise their products, don't achieve the same degree of consumption, even if they could enhance the life of mankind.

Thus, for Companies men only have 2 roles. They are *workers=reproducers and consumers= users* of machines that become in this manner tested and evolved according to their quality. Those 2 goals that make of man a slave of the future evolution and reproduction of machines are indeed the goals of our Free Markets.

Those 3 facts—the monopoly that the dictatorship of companies exercises over the cre(dit)ation of the Earth; the promotion of war as a means to greater profits, selling and consuming the most expensive machines, weapons; and the slavery of man to the 'biological drives' of machines, which we re=produce as workers and vitalize as consumers—show that man is not free and Markets are dictatorships of companies.

We do not understand this, because companies of information control and manufacture our brains with myths and because we become addicted to machines that enhance our energy and information.

Further on, money, as all systems of information (black holes in galaxies, nervous orders in bodies, chips in machines), is small, often hidden to perception. So the information castes and systems of power of money, as those of any organism, are hidden to the mass; and we only see its final actions, without perceiving those orders—as we do not see gravitation but perceive its effects.

The result is that a few financiers, which control the money of thousands companies, which control the salary/actions of millions of human beings, control the world and of course their well-paid scholars produce all kind of information to hide this fact, pretending with their digital, 'Economical newspeak' that *History of* Economics is 'just', 'rational'. The truth is that the 'Free Market' is an economic ecosystem, the scenario of a historic brutal fight between animetal castes, mainly warrior and go(l)d tribes that killed each other in wars and holocausts, developed machines to acquire land and go(l)d and have enslaved the rest of mankind with weapons and go(l)d that their companies now keep inventing for free, under undemocratic privileges.

To explain those differences between what we are told about our freedoms and what we really observe, we need a historical analysis of how Company-mothers shaped in the past centuries the structures, organisms, governments and opinions of mankind:

Democracies were invented in Holland by gunboat, stock companies, which substituted kings, due to their better weaponry and then established corrupted parties and Constitutions that defended their monopolies on Credit, the tool they will use ever since to rule our societies during the time of work and consumption. When citizens complained and revolted against the dictatorship of companies in some countries, they obtained some placebo rights, which they can exercise only outside their labor time, which *still is* most of the times of anybody's lives in which humans are part-time, obedient slaves of the companies that hire them. Yet, those 'free rights' don't include a share on the exclusive credit and legal rights companies and its bribed politicians hold in most democracies.

The main tool of control of human actions is the law. This happens because man understands words better than numbers. But money can buy the law and further on, since money is an 'informative', 'hidden' system of power, it can do so without people realizing *it is buying*. So social organisms, nations, have created a dual system of 'network control', as any organism does: an informative nervous system, the financial system that gives orders to the cells/workers of the organism; and an energetic system, the physical weapons, machines and political, repressive organizations, the leukocytes that control citizens. This is the concept of Adam Smith: money should rule it all and leave the government the 'army' and 'law and order' tasks of the organism.

Holland and Great Britain, a copycat of the Dutch democracy, defined in this manner the 2 sides of Free Markets: security governments serving financial companies. Such dual system, called a 'Market Democracy', developed all type of limits to political freedom and controlled the economic power of governments, putting limits to their capacity to create money. So companies would always come on top of politicians, bribing them when needed, with their superior monetary muscle. This system was perfected in the XIX c. when the Law of Anonymous societies gave legal immunity to Companies, while politicians would be inspected for 'peccadilloes'. Finally, in the XX C. mass-media companies learnt 'to manufacture' the brain of citizens and guided their vote towards those politicians that meekly respected the dictatorship of companies. Today companies use money to buy legal power and digital thought— mass-media information—to manufacture the brain of the numerical voter. So 'numbers' crunched by machines dominate words, even in the legal, political wor(l)d. In Holland this was less sophisticated, done directly through parliaments, whose members belonged or were bribed by companies. Then those parliaments, due to restriction of vote to the wealthy, became an economical caste to the service of companies. In Holland and England bankers and 'stockrats', owners of stocks, took control over the previous castes of power (military, aristocratic castes). They eliminated the power of king-warriors and gave companies of weapons, through their interposed political parties, control of the law and war and peace cycles.

During those changes of power from military to economic castes, from kings to companies, from the rule of warriors aristocrats to the rule of companies' stockrats, there were brief periods of hope, human, legal, verbal r=evolutions. Yet r=evolutions that tried to return to societies based in love, legal codes and life-enhancing human goods, were all crashed by the new aristocracy of 'stockrats' that controlled through money governments and the military. Those processes happened in Holland (XVI C.), England (XVII C.), America (XVIII C), Continental Europe (XIX C) and the World at large (XX C), as the ecosystem of company-mothers of machines expanded, till controlling the entire Earth. The placebo fantasy to hide all this is 'voting', according to which individuals can choose between two self-similar parties, the right-wing party of 'war' and the left-wing party that promotes consumption of machines. But both parties defend the underlying structure of the system, just described. Thus, Bipartisan options do not imply change, as all parties and doctrines today favor a free market. So societies fluctuate between ages of peace and consumption of machines (left wing parties) and ages of war and consumption of weapons (right wing policies). The left allows production of some life-enhancing human goods and prefers a slow rhythm of evolution of machines. The right allows war and promotes the evolution of top predator machines, weapons. *But the decision of which party governs is done by the natural, evolutionary phases of those machines, which are defined by the economic cycles of Kondratieff that fluctuates between profits made with weapons (which give those companies and their right-wing politicians more power) and profits made with peaceful machines (which give more money-power to left wing politicians and consumption companies)*

What is then the difference between a Free Market and a Socialist Democracy? Are not the same social organisms? Not at all and to understand why we have to study its different historic origins. Free Markets were born in Holland and later expanded to the British and American Empires; while socialist democracies were born with the French R=evolution and they were based in the social ideals of love and

615

solidarity proper of Love religions. *Thus, despite self-similar names and institutions, Socialist Democracies and Free Markets are opposite systems that continue the traditional fight between Animetal cultures and Love Religions of the ancient age of money and weapons.* We can spot the difference observing who started those revolutions, companies or citizens. Companies did it in England and America (Boston Tea Party). Citizens did it in France and Russia, the origin of socialist democracies. In that regard, Company-mothers imposed the Free Market in England in the 1688 'Glorious Revolution', when Dutch Companies took over the Catholic King. While the French r=evolution was a r=evolution of all the people of a nation against its animetal caste of aristocrats.

And that is why French revolutionaries opposed the British Free Market. And both nations fought for the control of the World. Unfortunately, a new trend starts also in those Napoleonic wars: r=evolutionaries are people with peaceful ideals, but soon they find that the Global Animetal culture they defy declares war to their r=evolution. And to survive they have to choose a military man that destroys from within the r=evolution, converting it into a military dictatorship. *We never had a Good Mule, the mythic character of Asimov novels,* a military man, *who takes over the people-caste of traders, but doesn't become a corrupted dictator. So the only long-lasting 'Party' of bio-historians,* the communist party, *which tried to design scientifically a better world,* became destroyed by Stalin, a military dictator, who ended the Russian R=evolution. In that regard, a real democracy must be an efficient organism of history with a single party, which should apply the laws of social organisms, designing a wor(l)d to the image and likeness of man, by producing life-enhancing human goods. Yet it must also define laws and mechanisms to avoid the power of dictators, by judging politicians after their job is done, setting limits to their tenure and allowing citizens to choose the best individuals for those positions. Thus it must not have 'parties', as instruments of collective power, but it must develop a true science of economics and history, not *alternative theories, as reality is one and the goal of history, to create a world based in life-enhancing human goods is set by the biological nature of human beings.* The error of communism was to become a military dictatorship, to create a party which shielded its members from direct election and to corrupt the Courts, which should judge politicians after their tenure and instead were using to judge people, who didn't obey the party. Thus in the last chapter we shall design a 'real democracy', according to the laws of organisms, which 'judge' with pain messages the informative, neuronal caste, when it harms its citizens-cells..

In that regard the closest societies to such ideal organism of history have been the Neolithic and Middle Age Priests, which didn't have weapons, didn't oppress their citizens, did have a clear goal, spending 50% off their resources by law on the poor, art and life-enhancing human goods. They were even able to convert and control, till the arrival of gunpowder, the Germanic warriors and the people-caste of money-lenders that oppress history. The French Revolutionaries inspired by the Enlightenment and the Russian Revolution during its first years, tried to recreate such society. But the discovery of cannons ended the kingdom of Christian priests in the west and Buddhist priests in the east. While French R=evolutionaries had to choose Napoleon to defend themselves from British traders and Prussian warriors and he converted the revolution into a military empire. So did the Russian Revolution, when it chose Stalin and the Cubans when Castro became a military dictator to defend the island from the United States. So revolutionaries became military dictators, to survive the global coalitions of bankers and warriors against them, ending the French, Russian and Cuban adventure. Mr. Hugo, the last serious r=evolutionary and the Iranian theocracy, unfortunately are in the way to become dictatorships, in a world that doesn't give any chance to alternative models, as it moves at full speed towards the extinctive point of the Singularity. That is why Trotsky, a century ago, affirmed that the r=evolution must be global or it will be destroyed by Capitalist nations attacking them. *Yet again Stalin, the military 'Caudillo', not Trotsky, the Mosaic Prophet of the Wor(l)d, won; and the Communist revolution became converted into a military dictatorship.* In that regard, once communism became a dictatorship, the closest thing to a democracy was the Scandinavian/German nations where socialist parties governed for decades, after the r=evolutionary reaction against fascist dictatorships. This period between 1950 and the 1970s was the last classic age of human thought, the world has witnessed.

But in the 80s, with the arrival of e-money that multiplied by 10 the quantity of money Companies had, the capacity to bribe socialist democracies became so huge that also the European Union became corrupted, while the Russian revolution, already a military dictatorship since Stalin, collapsed, unable to confront the sheer increase of military and informative efficiency, caused by the chip, whose e-money software and military applications gave the victory to America. There the new electronic power of televisions, will also crash the 60s revolution with neofascist messages, provided by an actor called Reagan, who inaugurated the present age of manufactured brains and fiction thought.

As long as money and financial institutions are not nationalized and a global currency is issued, with equal rights of deficit for all nations to create a global New Deal and promote welfare goods globally, we will not live in a democracy but in the FMI complex, whose goals and end is the extinction of life, as it has been proved ad nauseam by the past cycles of history and war. If a global currency were created, it could not be sunk by the market as it would be the only currency available and it cannot lose value. It would be controlled by states, and if all of them agree to a 30% annual deficit, a minimal global salary would be created to feed all the cells of the body of history. Because the poor spend in human goods, a massive demand of human, biological goods would be created. And the government, with the rest of its deficit and its control of financial institutions will under the laws of biohistory invest in sustainable goods and create a world to the image and likeness of mankind.

Only a demand economy in which the system produces goods humans need is a democratic economy.

Otherwise the evolution of a free market of corporations will merely continue favouring machines over human beings, the 'citizens' of the market, as they are the privileged offspring of those corporations. The result will be in the age of organic machines the end of human labor and human rights, as we become obsolete workers and soldiers, handicapping further evolution of those machines.

Recap. We differentiate historically, no longer 'de facto', Free Markets like Great Britain or America, where companies founded Lobbyist Democracies; from Socialist Democracies—the last being the European Union during the 60s and 70s before Europe became a Free Market—whose origin has to be found in the Revolution of the French and Russian people against their kings and emperor. Yet only socialist democracies are real democracies. Since in real, social democracies the rights of citizens are concrete and they are applied, because the government has control on the 2 main social languages of power, emitting paper-money and creating laws, which the government uses to increase the wealth of its citizens in life-enhancing human goods, food, housing, health-care, education, art, free time, unemployment benefits, etc.

36. Men vs. machine: Labor obsolescence. Productivity.

Perhaps the biggest myth of our modern economic ecosystem and its networks of information in favor of machines is the 'goodness' of competence between humans. That myth is false. It does not follow the laws of social evolution. Competence between humans is anti-natural, because we belong all to the same species. Team work, social work done by helping each other, is more efficient. Social organisms dominate Nature. They have made ants and Chinese the most successful species of the animal and human world. Yet the myth, also called Social Darwinism, says that men become better when competing with each other at work and at war. The Myth of Competence hides the fact that men compete in both fields mainly against machines. And they do so with unfair advantage, since machines are made of metal atoms and so by the law of affinity, they are better at making other machines, the main job of Industrial Economies. Further on metal atoms show greater strength as energy systems and higher speed processing information than our atoms. This means humans are only better at tasks that require the use of both, bodies and brains (till robots evolve enough to substitute us) and *in the production of life-enhancing human goods, which should be promoted to create labor.* Machines are better in specialized tasks, because they are 'deconstructed' organs of energy and information, made with a better substance than our light atoms.

So warriors are killed by weapons, since war is a field specialized in energy. And PCs are better minds with the right software than human beings, since office jobs are informative jobs, ran in the digital languages of computers - while in factories, the law of affinity make machines better at repetitive tasks of re=production of other machines, so tool-machines win in the competence with blue collar workers. Finally, robots, which mix energetic bodies and information organs, are becoming also better than humans in complex tasks. So, in the age of the Singularity, as we evolve robots and software suits, men are losing their jobs. Since men are inferior to metal species when the form of metal is evolved properly for the specialized job it is required.

But all this is hidden by the myth of productivity. 'Productivity' grows because of the increase of the number of machines used in factories, as workers with lesser skills than machines become extinct. Yet workers are told they 'produce' more, so they are happy, as they feel like 'supermen'. And people are proud of their productive nation that makes millions of machines with a few workers, while they scramble for underpaid jobs, as clerks or cleaning ladies (now competing with I-robot romba). Thus, as machines displace labor, people compete for the dwindling supply of jobs that machines cannot yet perform. Competition between people increases because the demand for human labor diminishes. Now chips can also perform the higher intellectual tasks of man. Thus not only the average worker, but also managerial jobs are replaced. Even elite jobs, such as stock speculation are at risk. A.I., a stock-fund ran by computers is already one of the top competitors on Wall Street. While electronic money, a cycle of information performed by a computer, is re=produced for free in enormous quantities by complex software programs that buy and sell continuously stocks and commodities, increasing slightly the prize of those goods in each buy/sell cycle. Finally, most machines' bodies are designed today by CAD (Computer Aided Design), which has increased the competence among engineers and architects. Only literature and art, the final creative tasks of mankind, are still our turf.

However machines are constantly evolved by engineers. Thus, the competence is truly unfair: machines are designed for a specific task, while men are generalists. Further on, machines are evolved every year, while men cannot evolve and adapt better to each specific task. For example, let us say one hundred years ago, there were 50 million jobs for human miners and now there are 45 million jobs performed by 1 million machines and 5 million jobs left for those 50 million miners. Those 50 million miners now would have to 'compete' much more among themselves for a dwindling quantity of human jobs, because they already lost their competence with machines, as almost all mining is done by machines. Have those miners developed rotatory arms and diamond teeth to break the ore, as their rivals have done? No. They are the same genetic Homo Sapiens. So instead of competence and evolution among human beings, we talk of obsolescence of human labor. Because machines evolve and humans do not, man loses in the long run. 'Peaceful machines' replace us in our jobs and unemployment worldwide continues to rise. The exception to this rule, with creation of new jobs, happens briefly in those nations that produce the best machines, since they sell machines to the rest of the world, which loses jobs. Today the World is in economic crisis, except perhaps the Asian tigers that produce most of the robots of the world and Silicon Valley that produces most of its software. Yet the unfair advantage of robotized Toyota factories has destroyed Detroit car factories.

The only solution: *robots and new chips must be forbidden: 'Digital Delenda Est'*

Yet the natural law of Darwinian competence between species is denied by economists that support the evolution of machines. So our politicians do not understand the true laws of Darwinism, which confront human and mechanical species. Society thinks humans compete with humans, when in fact they compete with machines. As Carnegie put it: 'We hold supreme the law of competence in a Free market, since it helps to select the best race' (the best race of humans or machines, Mr. Carnegie?)

And indeed, according to that selection of the best race, the equation of Extinction of Labor, which is the equation of productivity that all company-mothers follow, makes humans obsolete step by step as more productive machines are evolved:

$$Max. \; Machines \; (capital) \; x \; Min. \; Labor = Max. \; Productivity$$

That is, capital and machines increase the productivity of companies, while human labor diminishes it. So companies improve productivity by firing workers and buying capital=machines to substitute them. So Productivity means the extinction of human workers and the multiplication of machines.

Productivity in biological terms means 'the survival of the fittest'. In the economic ecosystem the fittest worker—not the human but the machine—survives. Further on, such competence constantly reduces the salary of humans, who either take the job, at any wage the company wants, or risk to see a computer or a robot in their place. Since robots need no health-care, work extra-hours for free and have minimal maintenance cost. Of course, if companies used verbal, bio-ethical values, they would defend the human species and forbid robotics, but in digital, profit values they do the opposite, guided by the equation: man=price=object, which allows them to compare the efficiency of both workers.

Recap. The myth of 'free competence' or 'productivity', which states that humans compete among themselves for jobs and that is good for men, because it makes them better, is false. It covers up the confrontation for jobs between human labor and evolving machines. We compete mainly with machines, not with other workers. XXI century robotics will be the end of that competition won by machines, as the worker is displaced by a robot and the soldier killed by a robot, while automatic factories start the self-reproduction of those robots.

37. Smith-Marx prophecy: Obsolescence of Man.

In the graph a car assembly line in the 1930s and today. Both, human energy and brain workers are replaced by robots and computers. As the IV cycle of the Industrial Evolution advances, chips' software and robots will make most human workers obsolete. Machines substitute humanity as they evolve in two environments: in fields of war and as machine-tools, eliminating human workers. In the I and II cycle of the Industrial Evolution bodies and minds of machines substituted blue and white collar workers. In the III Cycle, robots will substitute most human labor, automating the re=production of machines by company-mothers. It is the III Cycle of unemployment, we live today. The 3 phases of labor obsolescence and substitution of human organs by machines (bodies, heads and full humans); coincide with the three phases of the industrial revolution:

-The 1st phase of body-machine creation massively substituted workers in agriculture (the human systems of energy). Farmers became then workers for the energy systems of machines, during the XIX century. Yet at the beginning of the XX C. companies discovered most of the energy-machines needed for their own self-reproduction. They displaced those energy workers. So all those workers were out of work (29 crisis); and in a few years they were consumed in wars, competing with weapons that killed them. The 2nd phase of the Industrial Evolution that reproduces chips in control of complex machines is expelling mind-workers everywhere, except in nations that reproduce them (US, Japanese workers).

The 3rd age starts now with the Robotic r=evolution that makes the entire human obsolete. Fewer men are needed to reproduce machines. Will a war cause the extinction of those workers in the nearby future, as it happened to the energy workers at the beginning of the century? We are manufacturing that war as we speak, evolving the weapons of the age of the Singularity and using robots to kill Neolithic farmers.

Phases of labor extinction: Energy and informative jobs.

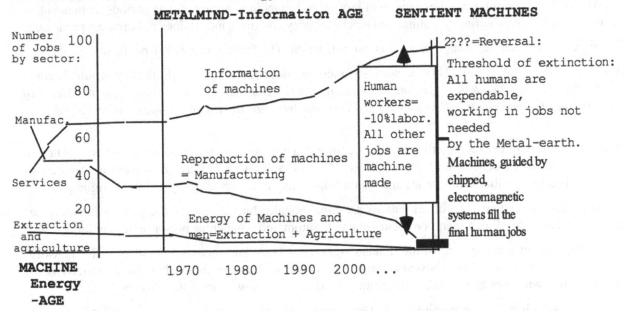

We can resume the biological relationships of reproduction, evolution and preying of machines on men, in a verbal-mathematical prophecy that includes the ideals of Smith and Marx, twisted by the reality of Darwinian evolution. Smith said that the growth of money (ΔMV in mathematical terms) and products (ΔTp), is symbiotic and parallel. This has a biological explanation. As your body and brain are symbiotic and maintain the same proportion in all human beings, so does money - the brain of the FMI system - and machines, the cells of its body.

In other words, there is a parallel growth of money and machines. How this affects human beings? It was obvious to classic economists, from Ricardo to Marx that the growth of capital (ΔMV) and products (ΔTp) extinguished labor, substituted by technological machines, bought with money.

Yet for economists, including Marx, this was also good in the long term, since once all labor was substituted by machines, we could live for free, in a paradise of workers without work, but with a salary, to spend in slave machines. This could indeed happen if the economic ecosystem is reformed and the right of Private Banks to invent money is taken away and governments invent money for the people, giving every human being a minimal salary. But for that to happen a r=evolution must take place. And yet, even in that case, we will become an obsolete, lazy species, atrophied in their minds and bodies.

A Darwinian theory agrees with the 'objective' analysis of both economists. Machines and money reproduce together as Smith said. Machines expel labor as Marx said, because they compete with us in the economic ecosystem. Yet at this point evolutionary theory diverges: this process cannot be rationally good. It is negative in an evolutionary sense; since it makes the human species obsolete. The concept of a socialist paradise or a techno-Utopian future is another animetal myth, *this time originated by Marx - since it fails to* see any negative effects in machines. So we should write the consequences of those 2 findings of abstract economists in different, evolutionary terms:

'The evolution of Capital & Machines ($\Delta MV=\Delta TP$) will expel all labor=*humanity* and *Humanity* will

620

create Utopia (*Marxist truth*) or *become extinct (Darwinian truth).*'

The previous sentence resumes the findings of Marx, Ricardo, Smith and Bio-economics in a single sentence. (The italic text gives the sentence a biological meaning.)

Mv is money and Tp represents the top predator, more expensive goods of an economy, (written Pt by economists). It measures the total prices of economical goods (of which machines and weapons are most of it.) And so the maximum parallel growth of money and machines occurs in war periods. It means that the evolution and reproduction of capital and machines preys on mankind. It does not create a paradise.

The logic, final outcome of such process is our extinction. The fantasy is a world-paradise.

In XIX C., Marx believed in Utopia, because machines were too simple to think they would become *self-reproductive, independent organisms. But today, when robots are about to cross the threshold of self-reproduction (reproductive Singularity) and* acquire *Artificial Intelligence (Informative Singularity) the end of history is crystal clear.*

Obsolete workers never receive the paradise treatment. They become extinguished in wars, isolated in ghettos, sometimes of continental size (such as Africans, a farming culture, first used as a cargo or energy, an object of gunboat companies and plantations, today made obsolete by efficient engines.)

The animetal castes that control all nations are ideologically biased against the poor, whose memes qualify as losers and will not spend on them - nor will automated companies preserve human labor.

Today chip-driven machines substitute human workers at body and mental level also in the 1st world and *that is the real issue of the present economic crisis: Machines are* increasing the 'productivity' of their companies with robots and software creating millions of *1st world* obsolete workers.

Do robots and chips create employment? Only among the specialized professions and nations, like Japan, Germany, parts of USA (California, Boston, Pittsburg), Sweden or Korea that invent them. Yet those are few places and few jobs that cannot be expanded. For the rest of mankind, robotic machines and software suites have created massive unemployment among white and blue collar workers.

Recap: The evolution of Capital *& Machines* ($\Delta MV=\Delta TP$) will expel all labor=*humanity* and *Humanity* will create Utopia (*Marxist truth*) or *become extinct (Darwinian truth).*

38. The future structure of company-mothers.

In the next graph, a look at the structure of future, automated, company-mothers, as men are displaced by computers and robots in the informative and reproductive tasks of companies. In the graph we show the flow of informative-power in companies. Men tend to believe that consumers and managers have the power to decide what a company produces. Yet the real choice is determined by the essence of the bio-logical process of machine reproduction. Grow and multiply is the law of companies [1]. To create 'products of energy and information' is the individual goal of each company-mother. Then humans can play their enzymatic role in companies catalyzing under monetary orders [wage and prices] the reproduction of machines [3 and 4]. As a result, men copy all of their organic functions into machines. As money becomes the mind of computer networks, not even the managers will be needed. On the left side we draw the future structure of companies, when men will no longer distribute monetary orders, or consume=test machines. In that future, money will organize the reproduction of machines through computer minds. Then machines will be selected in the external world, in the 'metal jungle'.

Productivity expels labor from companies: technological companies invest in chip-workers, and substitute enzymen by digital thought. Machines are winners of the selective process of companies. Yet humans adapt their verbal rhetoric to hide their increasing loss of control of company-mothers [marketing of capitalism and its ideologies taken as 'science'; myths of 'freedom of the market'].

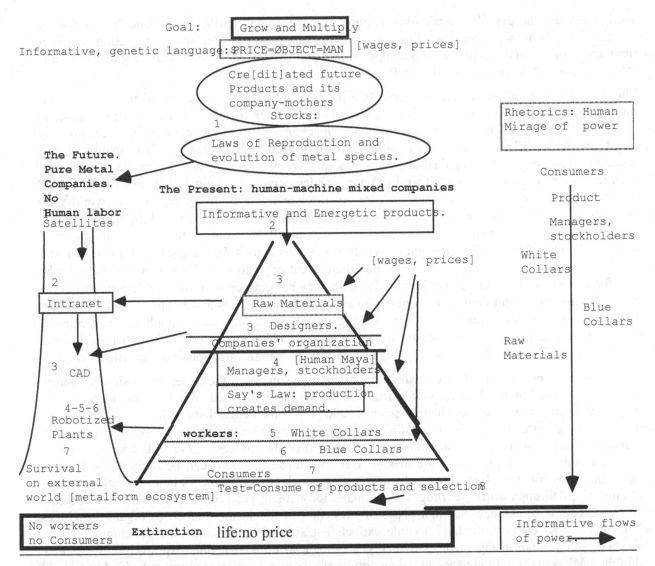

In reality (left-center side) workers and managers [numbers 4-5 on the graph] are not free. They adapt decisions to the evolutionary logic of the Metal Earth. We might say that decisions at the managerial level are made by mathematical software, increasingly, in routine programs of digital thought that design the reproductive processes of machines. Not even consumers are free. They want property and consume whatever products companies give them. [Say's Law]. Decisions of consumers are followed by marketing executives- while processes of machine creation are caused by the discovery of efficient metal organic shapes and reproductive methods, imposed by metal properties and the imitation of the organic functions of people [chip-brains, leg-cars, etc.].

Humans transfer form from carbonlife organs to metal organs. This process is determined by the material available [3], by the human discovery of machines that copy and substitute human organs and mental functions [3], by computer design [3], and finally by the test results established by consumers [7], who are easily programmed by marketing. Bosses, workers and middle managers {4} live in a mirage of power because they make decisions. Their decisions have to follow routes that are predetermined by the laws of metal-biological reproduction.

The translations of man's organs into metal are determined by human biology. It is not possible to create other types of machines than those who imitate human organs or functions.

Good managers 'accept' the restrictions imposed by the Metal Earth. They 'smell' evolutionary tendencies towards higher technology. Consequently they cut down the labor force, increasing computer content and capital, focusing in the marketing goal of bending consumer's will according to Say's Law, which states that supply determines demand.

Who defends this self-defeating system? Industrialists, scientists and economists who live off machines, since they either design or sell them and so they increase their profits by putting machines in place of workers. Thus, they favor systematically machines for selfish 'economical' reasons. Yet scientists and industrialists are a minority, within the human species, perhaps only 1% of mankind. 90% of mankind has to compete against the growing quality and productivity of evolved machines. Therefore, his opinion should matter if we lived in a real democracy. The animetal elite of financiers and engineers who invent money and machines are an absolute minority. They control the world only because they control hierarchical companies and governments that issue orders to billions of humans, not because their ideologies are shared and cherished by mankind. Humanity is those children massacred by other children in Africa, who carry guns to profit western industries. Humanity is those millions who die of Aids by lack of generic drugs to profit our pharmaceutical industries. Humanity is the child who has no food, because the entire planet is spending its money building up weapons and Internet networks that he will never use; and there is no a penny to invest in human goods and nature. When a biologist studies the survival and extinction of species, he cares only about mass numbers. A species is represented by a majority. So democratic mass numbers should give reason to those who ask for a Demand Economy based in life-enhancing human goods, food, health care, housing, education, art…

In that regard humans might feel extremely important as members of the super-organisms of history, but their role in the Financial-military-industrial complex is indeed limited. Whatever the ideologies of money=capitalism, machines=mechanist science and weapons=nationalism tell us to fulfill our anthropomorphic ego, our Galilean paradox, for the FMI complex a human being is basically an enzyman, a worker=reproducer that catalyzes the creation of the memes of the FMI complex and an animetal, a consumer=vitalizer of those memes. But once those functions can be replaced as they are being replaced by robots and pcs, blue and white collars, which also consume all other type of machines, vitalizing them through software programs, humans loose their position and role in the economic ecosystem. This is in essence the final phase of extinction of life in this planet: as organic machines radiate in the XXI century, humans become extinct first as workers making other machines, then as consumers and soldiers using them, and finally it is logic to think that once we are totally obsolete within the FMI system, as living species, when armies of robots or intelligent chips embedded in all kind of metalife structures awake to consciousness. The details of how this will happen are not so important and the process itself. Metal 3 memes and its 3 attached ideologies of mechanism, capitalism and nationalism have become so extended and programmed so deeply human beings that the process has reached an state of automatism that might make it impossible to reform.

It is indeed part of the system to censor ideas that go against the system. So the solution of understanding the system as this work does is not enough, because self-censorship acts. The only explanation might be that we humans are deterministic, memorial, repetitive systems and our memes are imitated from brain to brain as perfectly as genes are; that the super-structures of reality impose them. That a warrior first finds weapons, learns people obey for fear to death and adds to the determinism of the slave that fears death an ideology normally a religion of superiority and so the economy predates the ideology. This is in consonance with the historic record, both for Indo-European and Semitic religions of superiority. Today the religion of capitalism, of economics, and mechanism and nationalism, of weapons is more sophisticated but in essence little has changed on the substrata of those ideas: that machines are progress; that money is wealth and its accumulation the meaning of it all; that we are an homo national species not an homo sapiens species; so we divide ourselves in homo americanus, homo palestiniensis or homo israeli, which means our differences justify war. We might not have the simple Aryan cult to warrior races, but there is no difference in the essence between racism and nationalism; we might not worship a golden Baal but there is no difference between fundamentalist capitalism and earlier religions of money. In all those cases the FMI uses certain human enzymes and animetal to evolve and reproduce its memes and dominate the bulk of mankind with it. Those people of course are also of no value for the FMI complex as the cycles of wars and holocausts that kill them show; but nevertheless they have had a privileged status as owners, inventors and managers of the FMI complex that now will disappear, as metal becomes autonomous and self-conscious and self-organized and self-reproductive.

Recap. In the future, machines will control reproductive factories. The human function of testing machines will be done by machines themselves in the future. It will be an environment in which machines, as Top Predators, will be the sentient beings of the Earth. They will compete for energy and information. As with men, only the strongest will survive.

39. The stock-market brain of the metal-earth.

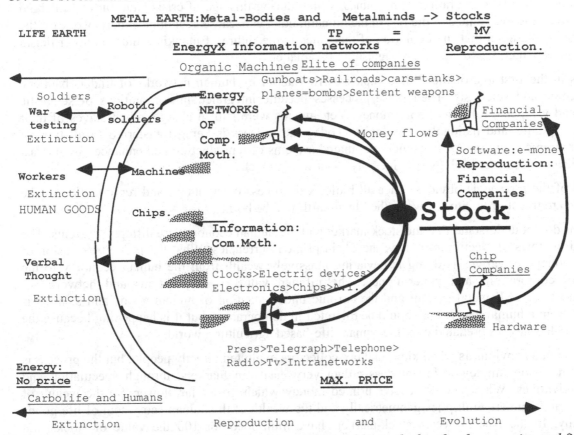

In the graph, today most of the computer power of the world is in the hands of companies and financial systems, organized by those who control the stock markets. In the graph, Stocks help the self-integration of the metal reproductive ecosystem, beyond the restrictions of the human mind. Artificial Intelligence is not a copy of human intelligence. It is an adaptation of the functions of nervous thought, of the human mind to different kinds of machines. Artificial Intelligence is born in companies as an internal brain that directs the body of the economic organism.

The homology between stock markets and brains or DNA systems is based in the bio-logical nature of economies, and company-mothers.

The graph describes the basic energy-information networks of Bio-economies ruled by stock-money. Stock holds, as brains do, most of the 'code-language of information' of the ecosystem. The brain will use that code to produce certain carbohydrates which are useful to the cell.

Stock has license to print money and will allocate them to the mother-companies of new technology with maximal price (weapons) of maximal reproductive power=Max. sales = Max. profits, such as digital software or maximal volume (energy).

Thus in this manner stock-markets express with money the 4 arrows of evolution of the super-organism of machines, its energy/information sources, its informative, audiovisual machines, its reproductive systems and its weapons.

On the other hand, human goods require human work, have minimal price by lack of affinity with metal-money values and are perishable, produce lesser monetary profits and are scorned by stock-speculators, which give no credit to all the products that could create a sustainable world.

In this manner the stock-market constantly deviates the wealth of humankind towards the reproduction and evolution of the memes of metal and the small elite of mechanist scientists, capitalist bankers/economists and corporation of machines, which during this age of crisis have further enlarged the gap between the FMI elite and the rest of the humankind. Yet because economic experts work for the FMI and have as dogma that the evolution of machines is the future of mankind, there is no criticism against policies, which are 'de facto' extinguishing this planet.

This was in the first age of the FMI complex carried about by human networks of traders/bankers, military people and scientists. Then those processes became integrated in corporations of scientific machines and weapons, ruled by paper-money. Yet humans were still in charge, albeit slaves of their mechanist, capitalist and nationalist ideologies. Today the process is being transferred to all type of software programs, networks of machines, automated systems of reproduction and enhanced by the last 'batch' of ideologies of technology (productivity, 'just wars', mechanist science).

The end of the process is obvious: once all biological process of evolution and reproduction of the FMI system are transferred into software, the Metal-earth will be born.

Politicians do not understand that the stock-market and the human economy are different systems. The 1st is based in flows of electronic money, the 2nd in printed currencies. What the present explosion of electronic money derivatives is doing is deviating massively wealth from the human economy to the economy of corporations and paper-money. The inflated value of corporations and networks of information of the FMI-metal system and the chronic undervaluation of human goods and salaries is neither just from a human p.o.v. nor should be tolerated in democracies, but it is happening because the FMI complex has always predated over the human, life-based agricultural world.

Today it is not as obvious as when kings and tax-collectors exploited directly people, but the process is essentially the same: money is invented as e-data very easily in markets, through speculation and financial derivatives. While people receive printed money which grows far slower. So the wealth of corporations and bankers multiplies geometrically and the wealth of the human ecosystem of life goods arithmetically. In the past 30 years stock-markets have multiplied for 100 the value of stocks and derivatives, provoking a massive inflation in all human goods, while salaries have merely increased in a 1 to 10 scale. The result is like a parasite that were absorbing the blood of the human super-organism, provoking the endemic scarcity of human goods and poverty of the human mass. For that reason the only possible reversal of that process would be the nationalization of the entire financial Industry, the creation of massive deficits in printed money, achieved with the creation of a global currency, ¥€$ money, fusioning yuans, yens, euros and dollars, and a 30% annual deficit by governments. This quantity of money similar to the quantity of money corporations add for free to their valuations through speculative markets of invention of e-money would allow the creation of a global minimal salary to jack up demand in basic human goods consumed by the poor and the inversion in projects that create a sustainable world (clean energies, infrastructures, education, public transport, tourism). This economic measure coupled with a military reduction of budgets and expansion of UNO and EU like organizations and diplomatic forums, and/or a triad of EU-US-Chinese leadership, applying the science of bio-history to the design of an efficient superorganism of mankind in control of the FMI complex could halt the self-suicidal path in which mankind has entered, by trusting blindly without any understanding the evolutionary process of machines, weapons and money, we call 'the free market'. Yet the free market is not free, it is an evolving complex system that preys on Gaia, life beings and the 90% of human beings, who do not profit directly from the sale of weapons and the invention of e-money in stock markets that have made corporations the dominant institution of the world with enough credit to cre(dit)ate a world to the image and likeness of machines.

Recap. Artificial Intelligence is the digital=nervous system of companies. E-money is the mind of the Metal Earth. It guides the networks of production and management. In the way that our language is the virtual world of communication among humanity, digital money controls the existence of machines and humans, but favors the reproduction of machines: The e-money banking system and stock-market system invents money for corporations, multiplying for 100 the wealth of the FMI complex, and reducing the value of the human economy, which is maintained in a minimal survival level.

40. The super-organism of machines: the metal-earth.

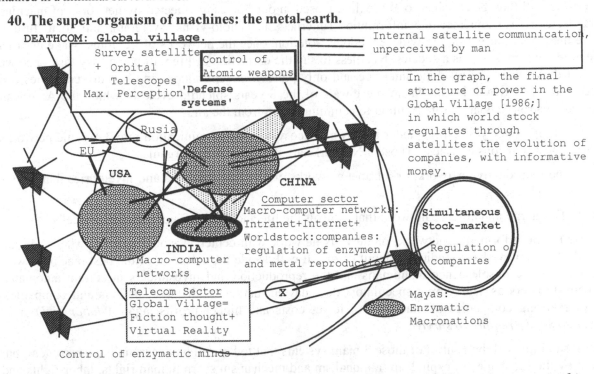

Earth is mutating from Gaia, Nature made of carbon species dominated by humanity, organized in nations through verbal networks of information (religions, laws) and natural energy (agriculture), into a superorganism, the Metal-Earth, dominated by systems of metal-information (money, computers) and metal-energy (machines, weapons), reproduced by company-mothers, fed by electric energy and communicated through electronic networks of digital information. The outcome is the destruction of life forms and their oxygen atmosphere polluted by machines (Global Warming) - a process that increases geometrically, as those machines multiply their numbers, reproduced by Companies.

The bio-economy reproduces machines and networks of energy and information that are evolving the Metal-Earth into a body organism, controlled by a global brain - worldstock. This is how life evolved a mass of cells into organisms with energy-blood and information-nervous networks.

The future of the Earth if it survives the age of cosmic weapons (singularity bombs, black holes and strangelets essayed at CERN) will be a super-organism of machines in which the present 'sciences' of economics with its capitalist memes that care nothing for the rights of humans and the reproduction of biological, welfare goods of lesser profits; the mechanist sciences that care nothing for the biological

626

whys of the Universe, and the memes of tribal history with its justification of war, will be carried *by software programs of extinction of life in automated weapon networks (terminator drones), automated company-mothers of infinite re=productivity and zero labor and automated scientists (CAD designs that will evolve further the machines of the future)*

At that point increasingly obvious, animetals will have transferred to machines all their ideologies and global networks of Intranet-reproduction of machines, Internet-communication of robots and flows of digital informative money will select, evolve, reproduce and decide what species survive and what do not survive on planet earth. Obviously life has no meaning to that system and air and water pollute metal, so our extinction will be needed.

In the graph a simplified analysis of that superorganism, whose main informative network will be a global system of satellites. This me(n)tal mind will perceive humanity as a mass of diminutive soft cells, insect-like species which cannot reach without the interface of machines beyond a limited vital territory. Each of us will then be attached to 3D reality, closed in our homes, connected by networks of machine energy (electricity) and machine information (internets and audio-visual systems), and so to further evolve the planet our extinction will be the logic solution, once the networks of satellites 'wake up' to a collective consciousness as a species. Needless to say the extinction of life will be an easy task once we are shut down from the machine interface and only platoon of terminators and self-driven planes can perceive the net. It will be similar to the e-wars of the 8 years splendid cycle in which the American army can destroy Saddam's forces, blind to communication from the air.

In the graph, 'sateleyes' evolve under the laws of ecosystemic communication into 3 organic networks that grow towards consciousness. Those 3 Networks are the same of any organism:

-E-I: The reproductive network, or intranet networks of CAD design, data and inventories that guides company mothers.

-Max. E: Energy and defense network that controls the life and death of individual home-cells.

-Max. I: The informative network, or nervous network with center in stock-markets, whose 'digital values' of money we have studied in detail before. It will decide the selection of the best machines with higher price, the obsolescence of humans and the reproduction and investments in certain automated company-mothers as Wall Street speculators do today in markets with labor, nations and companies, always favoring companies of machines of lower costs and higher profits, against *labor rights and human goods we require to survive.*

Thus, satellites will be control of those 3 main systems that today still have certain human content, but guided by the ideologies of capitalism, nationalism and mechanism scorn human rights, labor rights and human goods as obsolete.

The primitive versions of those future nets are:

- NATO, Def-com systems, that already have the potential power to extinguish man; and soon will be on the hands of star-wars, a satellite network in control of atomic weapons.

- Financial and Intranet systems, with center in world-stock, that connect companies reproducers of all the species of the Metal-Earth body. They also evolve towards the independence of man, as automated factories. They keep expelling human workers and putting machines in their place with the excuse of higher productivity.

- Finally the third network of the Metal-Earth is the network of information, the internet networks that today serve men, and in the future will connect the mobiles and chip-brains of multiple robotic species.

The three organic networks of a macro-body of metal will be increasingly integrated beyond human control by non-perceived parallel flows of communication between sateleyes at exospheric level.

Those sateleyes are beyond the reach of human warriors. When they become part of the metal Earth, and reach conscious as a different species, man will be defenseless against them. When that will happen? Within decades, during the XXI c.

Recap. The 3 networks of the global internet, the military network, reproductive, economical network and internet – future network of communication of robots - are the networks of a new organism, the Metal-Earth. Yet unlike the organisms of history they substitute, its individual cells are no longer humans, but machines, and its organs of reproduction are company-mothers, not life beings. For the Metal-Earth, men are expendable.

41. Humans controlled by networks.

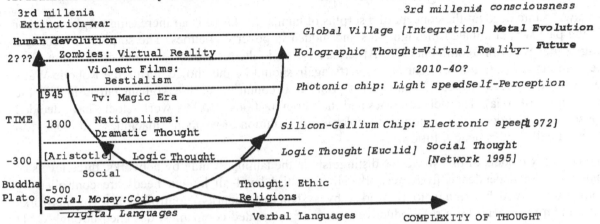

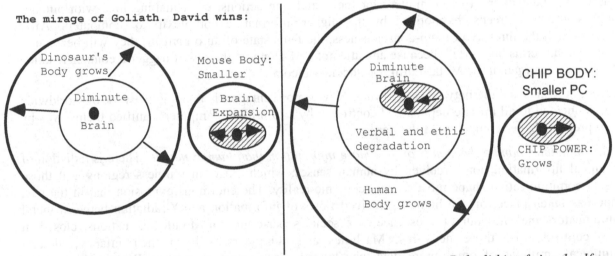

As humans devolve from the heights of social religion, we enter a neo-Paleolithic of visual self-centered thought that coincides with the arrival of the first organic robots, whose visual minds trained with violent video-games will wake up as the imagination they will act up in the world, hunting humans. In terms of the paradox between big energy and small information, in the Neo-Paleolithic we devolve our brains and grow our bodies, so our children are bigger and more stupid, while machines become more complex as they reduce the size of their chip-brains.

Today men and machines are a 'horde', a wave without a real organization. The net transforms that herd into a 'body'.

Hordes are free. Body cells are slaves, fixed by the network. This is the process that we observe: mankind is losing mobility. Company-men now do video-conferences. They don't take planes. People

628

buy by phone. They don't go shopping. People used to get together to go to movie theaters. Now they see digital information (TV-images) within their cellular-home.

Of all those phases of slavery of man to his metal-networks of information, the fundamental one will be slavery to e-credit. The fusion of the 'banking-networks' of electronic credit and the internet-network of shopping will make physical money obsolete.

Besides the loss of jobs, this means a fundamental split between those with e-credit, and those with physical money, that loose value as inflation and scarcity affects human goods, and replication of e-money multiplies the wealth of the connected people.

Take always into account the concept of a surplus of humanity. There is an increasing surplus of labor that in the past was eliminated through wars and hidden in poverty ghettoes. The surplus of labor is a growing tendency. Entire continents like Africa are isolated Ghettoes, poor, without credit. How they will be eliminated in the future? War and poverty again should be the choice during the Robotic Wars, as the splendid cycle of 8 years' war against poor Arab countries show (Afghanistan is in fact the poorest country of Asia). The rich countries and their organizations (NATO, world-stock) have decided 'subconsciously', guided by the automatic ideologies of evolution of weapons, machines and money' to do so periodically with splendid little wars.

In the complex models of this work, we distinguish, in the taming of man by networks, three levels of increasing human dependence from networks. In the third level all human needs are controlled by electronic-money which belongs to networks; by systems of security, controlled also by networks beyond human reach, and by a re=productive system of automated companies, designed through CAD (computer design) and intranets, also controlled by computers. Humans become themselves connected to the network, fixed in front of computers for their vital transactions, even making love with humans through computer screens, hypnotized by the higher informative complexity of computers. After networks reach the threshold of sub-consciousness, the fetus state of an organism) they will become in relative human terms 'immortal', because any attempt to disconnect them will trigger a counter-reaction of the defensive organ of the Metal-Earth: Deathcom=defcom.

Indeed, at 'war level' the network is growing. Now Nato+Pentagon is a single network of worldwide reach, connected to all missile depots, and controlled by ±10 million computers, unified themselves by the top predator hyper-computers of the Earth.

Today *machines handle thousands of time more information than human people.* There is radiation of mechanical information not perceived by human senses, which exists in wireless regions. All those flows of communication shape most of our economic reality. The enormous explosion that in the past decade has taken place, of non-human perceived flows of information and ¥-radiation between metal communicators indicates our obsolescence as a social super-organism, divided in nations, closed in homes controlled by those networks. Machines and computers, still in their infancy, already communicate millions of times more data than humans do among themselves. We should fear the explosive rate of evolution and multiplication of Chip=brains that have overcome human population; and have made of mathematics and statistics the languages of modern truth. Indeed the Wor(l)d of Human thought, our natural, biological language of behavior, is no longer the most common, communicative species in the Earth. Mathematical flows of pricing and digital information are thousands of times more dense. So we talk of a new age in which human communication no longer dominates and shapes reality. Metal communicators are creating a parallel universe to that of human communication and human virtual worlds, far denser already than the human world of communication, which they increasingly substitute.

Is it the network conscious? We will not know, since it will become conscious at exospheric level, as a brain unified by parallel, non-perceived exospheric flows of data between satellites.

It does not matter, since all what matters is that defcom, intranet and internet, weapons, money and machines are already being programmed to extinct our brains, jobs and bodies by animetal ideologies. So all what is needed is to translate, as it is being done, those software programs into computers. Today robots are taught with video-games how to kill human targets; software is told to automate companies and so the process can be completely automatic in their handle of human cells, as the inner systems of our bodies are in their handled of biological cells.

But again the present network is just the last stage of evolution of 'networks of informative machines', which even if people don't realize, have been adapting the mind of man to worship the memes of metal since the press substituted and killed the 'mental, subconscious networks' called Gods humans used to synchronize their simultaneous messages under the memes of love.

Indeed, this very little understood process of substitution of mental, human gods by visual images provided by metal-communicators that hypnotize as go(l)d does with its higher information and simultaneity the mind of man, deserves a paragraph in itself. Because, curiously enough for a man who is not very religious, priests who considered the press and Luther eviL for destroying the unity of the Christian god with the power of the yellow press, provoking the first Age of Global wars)religions wars fought with gunboats), were quite right. The press killed the Christian god and then a series of new metal-networks of information, radio-hate managed by Mussolini and Hitler, TV-hate managed by Reagan, today the internet matrix increasingly the collective brain of machines, killed all the cultures of life and love sustained in human tissue and expanded a subconscious message: man is born to kill with weapons, desire money, fight with other human beings and consume and revere machines…

Recap. Humans who are not integrated with the Network will have restricted access to survival money provided by the network (e-money, credit and goods delivered by networks). 'Internerds' will be increasingly erased in will, and dependent on networks for their existence (e-money, buying, education, friendship). When those machines become conscious, or become properly programmed to extinct man (defcom systems) we will be eliminated and shut out of the network.

42. Information machines: metal-communicators.

The 4 ages of communicators and its prophets spread anti-life memes that fostered global wars. In the graph, the press and Luther, whose message of hate opened the age of religious wars; radio and Mussolini, whose hate-speeches started fascism; TVs and Mr. Reagan, whose war speeches destroy the last attempt to create a sustainable, peaceful world based in the wanting of human beings (60s r=evolution) and the internet and its war video-games, which are used to train the mind of future terminator robots by the Pentagon.

In the graph, we observe the main metal-communicators of history, whose role colonizing the human mind with ideas in favor of machines cannot be easily stressed: the software of metal-communicators has never changed, since the first imprinted books and daily presses defended technology that made them possible. Yet those metal-communicators have evolved, increasing the hypnotic power of their messages. *The hardware evolves as all information networks do, increasing its reach and quantity.* In each phase of their evolution, Metal minds also increased their speed of communication in a geometrical scale. In each phase, humanity suffers the opposite trend: more human Minds are hypnotized and believe in the ideologies of metal-communicators. As a result, the capacity of the human mass to understand reality in human terms dwindles. As cameras become better eyes than human eyes, men prefer to see reality through a virtual machine. So kids prefer to see 'The lion king', instead of going to a real zoo. It is a process of collective hypnosis, similar to the one developed in the First Age of metal by Gold fevers. So now people consume 5 hours a day of those mental, indoctrinating machines, wasting their life as zombies of a virtual reality. Their role models are the violent heroes and visual freaks of TV sets that make us forget the survival laws of the Universe, spelt by verbal Gods.

Further on, because new media requires more money to set up its industrial systems, the numbers of independent human beings, who can deliver messages through metal-communicators, diminish. Books could be printed cheaply; radio-stations were relatively easy to create. But TV systems and its global networks with real exposure to an audience are very expensive. R=evolutionaries can't afford it. So the manufacturing of human brains by industrial propaganda grows in each cycle.

The most important information man needs in society is ethic, verbal information on the laws of love and social evolution that comes through conversations and book, not through TVs and Internet. Visual systems on average deliver $1/5^{th}$ of the quantity of words that a man reads in the same period of time. A news program is equivalent in verbal information to a broad sheet of a newspaper. Besides you cannot choose what you read in a TV set. All channels say the same. So they manufacture your brain without freedom of choice. The fundamental elements of ideological indoctrination - minimal information, selective information, information accompanied by visual rhetoric - are at work here. You are not informed; you become 'programmed' to believe what the people who choose TV information want you to believe. And there is always a company-mother with vested interests behind that information, especially in the 'content' (advertising). When you read newspapers you could at least choose a point of view akin to your ideas; select the news you read and contrast them.

Studies show that in the age of TVs humans have lost a 20% of their verbal capacity. Our children are clearly more ignorant than our parents were. They also commit many more ethic crimes, imitating the messages of greed and murder that the animetal cultures in control of TV-software deliver to a worldwide audience— as each nation and TV copycats US films in its own programs. Our mind has been colonized by a 'software virus' - audiovisual rhetoric in favor of capitalism=mechanism, of money and machines - that has converted us in willing slaves of technology, pushing the organism of history towards extinction. The rhetoric 'Newspeak' of 'big brother' is a subtle mixture of 'fiction' and 'comedy', a 'don't worry, be happy' concept that deactivates any real reaction of mankind to the pressing problems of our extinction as a species, with selfish messages in favor of consumption and a relentless political agenda against non-technological cultures.

So humans do not realize of their advanced process of extinction. Wor(l)ds of survival and ethical behavior disappear. Books are no longer read and so the complex Industrial systems of mental indoctrination that manufacture and simplify our brains cannot be explained, beyond a much reduced audience. The mind of man disappears.

It is the age of multinational companies of mass-media, all delivering the same messages in favor of machines, work and consume, through all the outlets of expression of mankind—while the truths of human survival are ignored, repressed, never distributed. It is the age of fiction, of a fantasy about reality, about history and the future, which hides all negative aspects of the Industrial Evolution. And because the truth is so shocking, most people in our society just wants to live in his 'animetal farm', in the fantasy of Matrix, instead of waking up to reality.

Information is power—the medium is the Message—and so those who control the Medium control the world. And in the electronic cycle, those who control the creation of Television and e-money control every other element of the Global society. It is for that reason that in a free society only Governments should have the right to produce electronic information, e-money and visual images; since, at least, they are chosen by the people, while televisions should be forbidden to allow freedom of verbal 'speech', the biological language of man that TVs substitute.

Yet, the key date that started the 'future' of the electronic cycle, beyond its television 'age of discovery', is the development of integrated chips in 1971 that also brought a new form of money (e-money). An age hardly understood by historians and economists, still ascribed to the pre-scientific, anthropomorphic, ideological treatment of their sciences. They consider the key element of that age the Oil Crisis, the Hippies, the Vietnam or Yon Kippur war, the Watergate, etc. But the event that truly changed history was happening in an obscure lab, in a diminutive piece of silicon. How many of us remember Noyce, the man who devised the first integrated circuit at a company appropriately called 'Fairchild', as it was the 'first child' of a new, 'future' top predator species? Because we all know subconsciously that the r=evolution of machines is not 'our history of evolution', hardly anybody realizes of those key dates of a future built to their image and likeness.

In that regard, again, even if people are mostly interested with their Galilean paradox – the ego-trip of tribal power – because of all those interrelated aspects the dominant cycle is *not the submissive cycle of tribal power and corrupted politicos, but the cycle of stock-market profits and crashes, we shall study only that cycle in more detail.*

Thus once we have studied the human point of view, and dismissed it as irrelevant or secondary, submissive to the memes of metal-communicators that cause the cycles of hate of the industrial r=evolution (yellow press, radio-hate and TV-hate), we can study those cycles of machines and weapons with the Ternary Method, through a topological analysis, from 3 perspectives, the spatial, temporal and scalar perspective. Thus we shall consider first the 3 species of machines; then its 3 organic parts as an i-point and finally the evolution of its st-1 scale of energies that vitalize machines providing them with its informative bytes and energetic bites (Kondratieff cycles).

Recap. Metal communicators substituted the memes of love and live, by the memes of evil and metal, provoking wars and diminishing our emotional understanding of social issues.

43. Democratic, demand Economy vs. FMI supply economy.
In the next graph, a real GDP would take into account the biological value for the evolution, reproduction and survival of mankind of the goods delivered by companies. Goods that foster our 3 biological goals—to maximize human energy, information and reproduction—are 'Human Goods' with a positive value that bio-historians would foster with a demand—economy based in a Free Salary for all human beings, who will spend that salary buying human goods. Taxes and legal prohibitions to robotics and lethal goods (weapons, digital software) would, on the other hand, reduce the demand for lethal goods that kill or make obsolete the human brain and the human body. Indeed, the poverty of nations is their wealth in those goods which destroy mankind, such as weapons, which destroy our bodies, or robots, which destroy jobs, or 'mental machines' such as TV-networks that hypnotize humans in front of a false virtual world based on violence and movement (animal eyes, become hypnotized by light and movement even if it harms them, like flies going to the light where predators eat them easily). Such hypnotism 'consumes' between 3 and 6 hours a day of our lives, diminishing our social intercourse, our capacity to learn through readings and makes us violent men, obsessed with consumption and money. Addiction to television could, in such a sense, be compared to addiction to drugs. We also include Internet among such negative goods. Over half of Internet time is dedicated to pornography or violent video films/games. The anonymous system of Internet and the technique of browsing do not allow sound relationships between citizens either, or a serious education, as a library with analytical books provides.

The real, biological value of goods for mankind would determine the ethonomic GDP of a nation, which would be higher when more human goods are produced and less 'negative' human goods a nation makes. It is in essence, a scientific version of the index of human development used by UNO to measure the development of nations, in which education, health-care, housing, art and peace count. Yet, the use of Humane GDP statistics is forbidden among financial economists that invent the money of our societies, as UNO's rejected proposals to impose an Index of Human Development show:

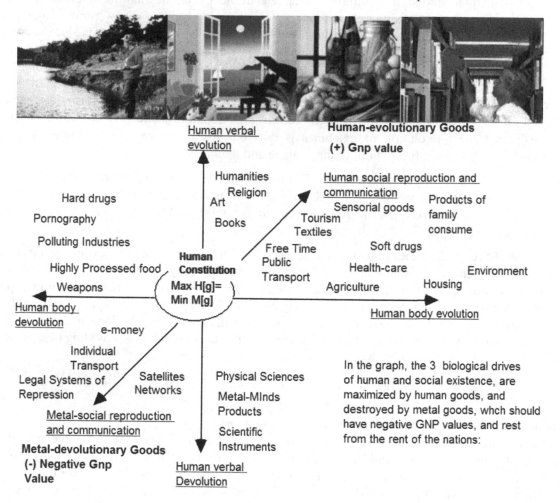

Politicians must use laws to legislate against lethal goods produced by Free Markets and foster Human Goods in a sustainable planet. While we cannot choose the laws that create efficient, healthy social organisms, we can choose the best humans to implement them. Instead, a demand economy would reproduce the goods that are healthy and fulfill our biological needs, represented in a bio-economical coordinates, as goods that fulfill our 3 arrows of information, energy and social reproduction. Thus, goods would be cre(dit)ated according to its value for mankind, creating a sustainable economy based in the desires of the population.

A 'free market' is an ecosystem of machines and reproductive companies that prey over human beings; *it is not an ecosystem made to satisfy human,* bio-logical needs, which are different, based in life-enhancing human goods. Economists and Politicians should acknowledge this and reform their political and economic ecosystems to cater to the biological drives of mankind, with a demand economy, not to the drives of machines, reproductive companies and the go(l)d greed of their stockratic owners, with a Supply Economy. To that aim they should produce human goods, goods which have no negative effects for mankind, instead of promoting technological machines with free credit to their company-mothers given in stock markets. Yet the only way to obtain a similar free credit for life-goods is a

Wor(l)d Union with a single global currency, Ye$ money, which allows huge government deficits and resources to solve the crisis. Then the key presidents of the world (the G20 under the co-direction of US, China and the EU) could design a world made to the image and likeness of Mankind, following the scientific laws of social organisms.

But bio-economics is just a part of a larger science that should control the economy from the perspective of mankind, Bio-history, whose aim is to design, under the scientific laws of social evolution, a collective super organism of history. Since species achieve immortality after its 3rd evolutionary age of information, by becoming a social, cellular organism. As a theorist I can only offer a model, which might not be possible to implement in this planet before extinction, given the degree of corruption of mankind. But that doesn't mean a science of bio-history cannot be implemented. Since there will be many other planets, self-repetition of this fractal planet, perhaps with less metals in its surface or with more evolved Homo Sapiens; and in a few of those worlds, it should exist a World Union, a perfect human super-organism, which is what the science of bio-history designs. In such world, bio-historic politicians take back the control of 'credit', designing an economic ecosystem, which creates the goods that fulfill the biological drives of mankind. A scientist does not search for the laws of nature, only to know those laws, but he studies them to manipulate reality with that knowledge. Because a scientist knows the laws of nature, he can modify reality to improve his existence. In the science of medicine, man can cure many diseases of which he was a victim in the past, because he understands the physiological cycles of the human organism. Then he uses the 'bio-chemical language' of drugs to cure those diseases. If the 'doctors' of history and economics were able to understand the laws of those sciences and control the 'language' of economics, money, they could manipulate the future of history and economics in favor of our species. Hence the enormous importance of developing a 'real' science of economics and history. Yet only a biological approach to economics can truly understand the nature of machines and how they are transforming our ecosystem, creating the 'Metal-Earth'. The same biological approach could design a political organism that ensures the survival of mankind, recreating the 'Human-Earth' and limiting the expansion of the 'Free Jungle' of machines, which guided by the laws of evolution that favor the stronger machine species, is killing Gaia.

In a real Democracy economics would not be a digital Go(l)d religion of the machine, to which all human rights are sacrificed, but it would be taught properly. Taxes on people would not exist. Economists would learn and foster a demand economy, where the Government would invent free money to subsidize the Human Goods that people want and regulate the stock-market to favor the cre(dit)ation of Human Good Companies, instead of technological companies that increase the competence between human beings and machines. And needless to say, there would be no derivatives, no FOREX markets, no currency and commodities speculators . . .

Fact is human goods grow in price because they have no stock credit to promote their reproduction. While machines lower prices, because they have free credit for their research. Inflation in human goods is a process caused by lack of credit that limits production of goods people need, increases their poverty and pushes the spiral of money reproduction and wealth towards a few stockrats. Less than 300 human beings have more than 2 billion, when in most 'hierarchical organisms', neurons have only ten times the quantity of energy than the energy cells of the body have, This should be the iron law of salaries, copycat of the laws of efficient organisms: the CEO of a Corporation should get only 10 times more than the lowest salaries. While tech salaries should be a tenth of the salaries in companies of life-enhancing Human Goods, which should have a tenth of the taxation imposed on tech companies. So greed would work to enhance a sustainable economy. Today the inverse is truth: Since 1973, the radiation of e-money has increased tenfold the quantity of stock money for tech and financial companies. This inflation has caused higher prices in human goods, lower prices of technological products and it has biased the economy against humanity. The reason such aberration continues is pure greed: Weapons of high price and digital software of minimal cost of reproduction make more money than butter of low price and maximal cost of reproduction:

$$Max. Profits/Price \times Min. Reproduction\ Cost = Arms\&Software$$
$$Min.\ Profits/Price \times Max. Cost = Human\ goods/labor.$$

Those are the equations that extinguish us: the equations that make every economist working for a corporation to lie and invent 'productivity excuses' to churn more machines and less butter. Then once the *corrupted scholar* has invented his sophisticated 'ceteris paribus' mathematical falsity (and we could write an entire book on how anti-truths become truths, on how 'ceteris paribus' choices of 'selected' causes bend that truth, on how mathematical equations are manipulated to obtain a result), then the *retarded scholar* will repeat this 'ad nauseam' to the politician and financial minister that will implement our extinction. While the *non-corrupted, non-retarded scholar, will be censored.* The anonymity and lack of individual responsibility of Companies and stocks, protected by the law of Anonymous Societies has accelerated this process. And the result is the constant obsolescence of billions of human beings that have been thrown into a chaos of poverty similar to the one suffered by life species, which were displaced from the life ecosystem by earlier humans. *But the GDP of nations keeps growing, because weapons and digital software add to it, instead of being subtracted from the GDP, as negative goods that kill the body and mind of human beings.* This is the trick: Economists only report on the fortunes of our companies, on stock prices, which keep growing up, not on the labor crisis that keeps going down, because mankind, the poor, are in 'digital values', expendable numbers of productivity equations. So we are out of the crisis, as labor keeps dying to robots and software. Yet despite digital censorship, the result of such loss of jobs and the excess of weapons for the body and for the mind (digital software) are a growing number of 'freaks', poor people without human goods and entire continents in perpetual warfare, like Africa, whose inhabitants were first property of companies as slaves and now are the dumping hole for the ecosystem of machines. While the 3rd world of people living in our countries is a sub-human world of chaos and McDonalds, of TV-freaks and ghettos, where millions are kept at minimal cost.

In bio-history, in 'ethonomics', a science of economics based in the bio-ethical needs of man, goods have positive and negative values as in all frames of reference of real sciences.

In the world of economics we have to differentiate the products reproduced by companies and individuals in two clear groups:

— Human goods are required to survive and improve human body health and mental health/information.

— Lethal goods, such as weapons and digital software. harm human beings.

The wealth of a nation should be defined as the wealth in Human Goods that most citizens require and enjoy, because they improve their lives: Agriculture, housing, health, education, free time, peace, freedom, textiles, a clean environment, books, art, public transport, etc.' Further on, these goods require a lot of human work, because many of them cannot be reproduced by machines alone. So they create employment.

The key to construct such demand economy is of course to have the power to create the language of information of the FMI complex, money. Once money is nationalized the measures can be implemented, otherwise it is impossible to create a demand economy as the market is merely the FMI complex, evolving and expelling mankind from its 'internal tissue', the corporations, soon to become automated company-mothers.

To avoid this governments should nationalize financial systems and markets, and establish an international ¥€$ money and a 30% annual deficit to spend in a demand economy.

The first measure would be to establish a minimal free salary of 100 yes money for every human being on earth. This simple measure with a cost of 7 trillion $a year would produce enough yes money in a few years to create a human demand.

On the other side of the economy, this means to eliminate the credit to companies of lethal machines, from the nationalized stock-market, and finally from the side of the banking industry to give credits to sustain a healthy economy, mortgages among other obvious things.

Thus, it is not only possible but easy to create such demand, democratic, healthy sustainable economy. Only one act is needed: to nationalize the banking industry and follow the laws of creation of healthy superorganism, eliminated the germs of our historic organism and multiplying the memes of life and love.

Indeed, while this book might put the record straight and introduce some scientific advances the study of history and economics, humans are not 'that stupid' and the United Nations have, for 60 years, created an scaffolding of a global government for the benefit of all mankind, respectful of cultural diversity, which could have created a convergence of nations and a wealth of human goods. But ONU power has been aborted by the Go(l)d culture, with its control of mass-media, financial budgets and Companies. Their proposal has been the equalization of all mankind, as 'their lesser clones', the lower classes, workers and consumers, of their globalized culture. And to the astonishment of all the good people that thought mankind could evolve, their proposal has won. It is all a question of hierarchies. The people-caste on top, the corrupted scholar and stockrat, have given through their company-mothers and mass-media outlets, orders and information, obeyed and believed by a 2nd mass-group, the daughter-cultures of Calvinism (the old British empire) and conquered elites of the world of the 'capitalist= mechanist' world, from Japan to Mexico. That block of around 10% of the Earth's population, vehemently opposed to the brotherhood of humanity, has stigmatized the rest of nations, people and cultures, as losers, non-chosen, '3rd world', primitive, whatever. And yet, those other people, whose only outlet of expression has been the International Forum of UNO, has developed in those halls and small-budget offices, the blueprint of a far more intelligent, sophisticated, creative, positive world that will never happen by lack of credit.

This implies the creation of money, not in stocks by company-mothers of machines, but in governments to foster the side of demand, the human side of the economy. Yet we have a system in which money is invented by stock-markets. While the little money extorted by the government through taxes is also given to companies. For example, AIG, an insurance company has received over 200 billion $ from the government, to handle it to their executives and other Global banks who lost money in speculative bets. In a demand economy that money, almost 1000 $ of monthly salary for each American, would be spent in basic human goods, which in turn would create more production of human goods and more human jobs. This is the key concept of a real demand Economy, a global salary to each human being, which will create a sustainable, constant demand and production of Human Goods, with far less money than Stock-markets invent for new technological companies, which merely push further the evolution of machines, coming closer to the Singularity threshold of human extinction.

To achieve that aim a global currency is needed, the yes currency, based in a fixed parity between Us dollars, Euros, Yuans and Yens (used today by the IMF to clear credits between those nations), to which other currencies could be pegged. Such currency could afford unlimited deficits as the only global currency and so it could pay a global salary that should give millions of human beings survival money to consume human goods. The myth that those people do not deserve a free salary, while speculators deserve to make billions of dollars, just by pressing buttons on their money-printing computers, is widely believed among the 'slave minds' that read or watch the mass-media propaganda of Wall Street Journal and Fox-TV, because they don't even understand that money is invented for free, in stocks and derivatives. So they think the banker actually works to earn it and the unemployed is a lazy looser that doesn't want to work and doesn't deserve to live. Fact is the informative castes, the elite cells of the organism, don't work. They invent for free money. While the 3rd, 'energy' class doesn't work because it is excluded from the 'organism', not because it doesn't want. *In any social organism, only the re=productive organs and its cells, the 'middle class', truly works in an efficient manner.* But at least, in a working organism, she gets far more energy/money as a whole - 90%- than the minority informative caste, which in healthy organisms obtain 10 times more salary to work for the benefit of the entire organism - not over a million times more money than the energy cells, while destroying and repressing the life-enhancing energy and information those cells need, as our elites do. And so to maintain such balance, all organisms have 'nervous, pain systems' that punish a mad-brain, not 'Anonymous Laws of Social irresponsibility', as our stockrats have. *Plainly speaking, our elites have created a corrupted human organism, infected of lethal goods, where most cell-people are sick. So soon the entire organism will collapse and at that point there will be wars and revolutions and also those neurons on top will die.*

The true cause of this crisis, the massive reproduction of chips, their economical software, e-money and their machines that make humans obsolete in labor and war fields, the IV Industrial R=evolution, must be recognized. If politicians could understand those Kondratieff cycles and what the Singularity means for mankind, governments would decree a nominal value for all stock-shares, freezing their prices to avoid further investments in technology, which is driven by speculation in shares prize. Then organic, simple decametric taxation and salary laws should be imposed to foster life-enhancing goods, jobs and companies. Once this is achieved, Governments would start the reproduction of ¥€$ money to pay a fixed salary of 1000 (Euro=dollar=5 Yuans=100 yens) per capita. *Such World salary free of taxation in a World Currency, created by decree by the will of the G20 that represents most of mankind, could cre(dit)ate, within a month, enough demand of human goods to start a massive shift on the economic production of the Earth, first in those top predator ¥€$ nations of the world that together account for 1/3rd of the global population and 2/3^{rds} of its economical production - then in all nations that join the World Union.* Needless to say, the 'World Salary and Currency Act' would be the most popular law ever decreed. Then the rest of nations could peg and substitute their currencies directly for ¥€$ money, within months, printing ¥€$ notes, adjusted by inner inflation to each nation, as the Euro was, within months (100 Yens=10 ¥€$=5 Yuans=1 Eurodollar).

On the other hand, the Salary act would establish a decametric scale of 1 to 10, as the maximal range of salaries within any company or government organization – *since that is the maximal difference of 'energy' received by the efficient neuronal/body cell system of any organism.*

And a global minimal salary in ¥€$ money for all the cells of mankind – since no organism lets its cells die of hunger, even those, which perform apparently unneeded tasks (hair, fat of the body) once they are created.

The obvious effect of such world salary would be a shift of jobs and investments from lethal good companies to human goods companies, from underpaid lethal industries, towards farming, hospitals, housing, textiles, tourism and education, as workers on those industries would double or increase twenty fold those basic salaries

It is indeed clear that in all organisms—big or small, made of cells, or human individuals—each member of the organism should have according to the natural law of organisms, food and proper information to survive. So the nervous-legal system should create the *Constitutional frame* and laws for such things to happen. After all, every single cell of a healthy body receives from its energy-blood system and its nervous-information system, a minimum quantity of those two precious goods needed to survive. Or else the body falls sick and the entire organism suffers. However, human beings lack those minimal goods in present societies, because our economic ecosystem is geared to produce not the goods human beings need, but the goods machines need; electricity instead of food, digital information instead of verbal knowledge; factories to reproduce machines instead of houses and hospital...*Since our informative elites are colonized by cultural myths in favor of go(l)d and machines. Mechanism and capitalism is indeed, the viral code that destroys mankind, carried by the most primitive animetal cultures that have made humanity into a worm full of poisons. Those 2 ideologies were resumed in* a false 'postulate', which validates right-wing=metal economics. The Adam Smith Postulate: 'The wealth of nations is the quantity of money and machines a nation has'. Once this postulate is religiously accepted; since men want to be wealthy, it follows that our objective in life must be to reproduce more machines and more money. This is represented mathematically by an equation that equals the financial and physical body of the ecosystem of metal – most of the goods with a high price - which nations consider their main wealth (MV: quantity of money = PT: quantity of products of great prices, mainly machines). This GDP is religiously accepted in Economic Universities, where in the last decades, Human Economics have been forgotten thanks to the power of companies and the academic institutions they systematically back: company foundations, universities paid by companies, prizes like the Nobel prize given by a bank, or economic forums that only represent companies and the wealthy, such as the Davos and GATT conferences, the IMF, the World Bank, International institutions from the cold war era, etc. Thereafter all politicians and economists, indoctrinated by the Adam Smith Postulate that merely translated the mechanist and Go(l)d ideologies of Germanic warriors and Canaanite traders, will try to increase the National Product of a nation, measured as the money and machines the nation has.

Indeed, the Smith Postulate does not define a healthy wealth. It is a religious postulate that Smith, believer in Calvinism, invented to support the doctrine of Calvin: 'The intelligence of God is money'. It is a historical fact that the time when a nation has more money and machines is during periods of war. Then the quantity of human goods, basic for our survival, such as food, housing, life and freedom are minimal. So metal-money and machine-weapons are not the wealth of nations, but the goods that provoke its scarcity; and the Smith Postulate should write as:

Max. Stock-money=Max. Lethal Goods=Min. Human Goods=Min. Social Wealth-Health

Weapons, robots and 'mind software' are lethal goods that diminish the health of nations, despite being promoted systematically by right-wing='mechanist' parties. Besides they are goods that require few workers, far less than human goods, since machines are best at reproducing machines. Or, as in the case of robots and Internet, they directly fill human jobs. Robots can potentially substitute all industrial workers. While Internet sales means the end of ±100 million workers worldwide in the shopping sector for the benefit of a few multinationals of computerized sales systems.

On the other hand, human goods are goods promoted by left-wing parties, which should be given the generic name of '*Humanist parties*', parties that promote human goods.

¿How an economic society reproduces the 2 kinds of goods?

It does so through credit, the invention of money, which later on is used to give orders to citizens that will work under salaries, reproducing either one or the other kind of goods. However there are two ways to invent money in our society.

Money can be invented in stock-markets, through paper-shares, which are a credit given to a company. That company will then use that money to invent and market their product.

Money also can be invented through governments and banks, by means of paper-money. Again that money is later on used by the government to reproduce or promote the products of the welfare state.

We talk of 2 kinds of deficits: deficit of stock-markets and deficit of governments.

Technically speaking, both deficits mean the same: free money, invented to reproduce goods. If either the market or the government invents too much money, there is inflation and money loses value. So there is a limit to the quantity of money governments and stock-markets should be allowed to reproduce, or else, the economy becomes unstable. However both kinds of money are not used in the same way.

The market, since it appeared in the XVII century in Holland, as an instrument for the reproduction of gunboat weapons, systematically invents money for metallic goods, but it hardly invents a penny for human goods. This happens because 'stockrats', the owners of companies only look for profits and profits are higher in expensive goods (the most expensive goods are weapons) and software goods (digital goods very cheap to reproduce). This is why there are so many weapons and 'metal-minds' in our societies. We do not need them, but the credit of the market reproduces them. And we have to consume them.

On the other hand, the government invents money for welfare goods, human goods, needed by humanity, which also creates a lot of employment. Thus we consider positive 'memes' and 'human goods' that an 'ethonomic' science of economics should reproduce, memes that enhance our natural verbal information, natural energy, reproductive will and social will. Those goods should be fostered by political systems and credit in real democracies, in which the negative memes of metal should be limited. Yet to impose such scientific economic system we would have to change the language of power of societies, from money to ethic laws and to that aim we would have to nationalize and control the reproduction of money, so governments elected by the people could 'finance' and 'cre(dit)ate' a world made to the image and likeness of man.

The result would be a healthy 'constitution' of the body of history which we express with a simple Biological equation that we call the Constitution of History.

Maximal human goods = Minimal lethal goods

This is not happening and will not happen in a Free Market without human scientific control, because a 'Free Market' is merely an ecosystem ruled by its free citizens, company-mothers of machines, which evolve the FMI system under the laws of complexity, guided by the 'values' of metal-money, a language which by the laws of affinity give no value to life and maximal value to the weapons that extinguish us. Thus the end of a Free Market is the extinction of man by the most expensive goods – weapons.

Unfortunately to impose another type of economic ecosystem that favors humans would require the nationalization of the financial industry to cre(dit)ate reality from the p.o.v. of human elected governments, so society rules the world instead of dynasties of bankers and corporations. For that reason the people-castes that invent stocks and e-money (most of the money today as notes have become a minimal part of the flow of monetary information) in almost monopolistic terms (bankers, corporations) and then handle orders to the rest of society with them (bribes to politicians, salaries to human workers, prices to consumers), have invented for centuries a complex matrix of mathematical 'lies and statistics' trying to prove that the best structure of the economic ecosystem is one in which they create money for themselves and for the companies of higher profits (weapon industries with maximal price and audiovisual industries with maximal speed of reproduction of their goods and minimal costs). This is done merely for profit without realizing that the values of metal-money distort completely the economy in favor of metal memes against life memes and human goods which are *far more positive to mankind*. The same concept applies to corporations: capitalism affirms the managers and products have all rights, and the workers and consumers none. So a supply economy in which workers become obsolete to machine workers, consumers are given all kind of goods lethal to their minds and body, and the dynastic owners of corporations (stockrats) get all profits while society gets none is established. In brief, capitalism was first the religion of power of the people-castes that invented money, banking and the main corporations of machines. Today as a 'science', it creates chronic scarcity of the sustainable goods we need to survive and chronic poverty on the groups that work most and represent the bulk of mankind.

- Finally mechanism, the belief that knowledge must be created with machines of measure and is all what matters to the truth is false. The whys of the universe are organic and biologic, and machines both enhance our energy and information systems but also atrophy and substitute them, and in fact are already reducing our body and brain power, as scientific studies show. And in the end they will substitute us and make us obsolete. But mechanism gives power and converts in 'high priests of knowledge' common people that become 'scientists' and see better with telescopes and calculate faster with computer models, even if they don't understand in depth their data.

So the 3 'religions of metal' are mere religions of power of the groups that handle those memes of metal over the rest of human population, and merely represent a more sophisticated way of 'killing' people (national armies), enslaving them with money (part-time salaries of people that must earn money while the financial system and its castes invent it in monopoly), and substituting them with machines (present way of unemployment created by white collar pcs and blue collar workers). And all this is done today automatically by means of an institution that put together all those ideological memes and metal-memes, the company-mother or corporation of machines and weapons, which reproduce them both for money, then controls the other organisms of society to pass laws and impose the consumption of those machines, and finally terraforms the earth to adapt it to the use of those machines.

If Humanist, socialist parties could explain this properly to citizens, they could then differentiate between both alternatives and decide what they want: Butter or cannons, human goods or metal goods, human workers or unemployment and machines reproducing more machines? Yet in as much as the two languages of power of modern societies, law and money, are controlled by companies, through stock-markets, lobbyism and the imposition of zero deficit laws, there is no real choice for citizens as the deficit of stock companies reproduces and promotes the consumption of machines.

The destruction of the welfare state by the dictatorship of financial dynasties.

Unfortunately as today all this is utopia. We are living no longer in a world controlled by social, democratic, human groups but we are in the last phase of evolution of the FMI complex, called in newspeak the Free Market, which has gone through a phase of massive inflationary reproduction of money as it mutated from paper-money to e-money, one-hundred folding its volume in the hands of corporations and its people-castes and dynasties of stockrats, which 'believe' in the memes of go(l)d religions and tribal history and have a fundamentalist vision about the future of mankind and technology. Today, fundamentalist, 'biblical' economics, whose original memes were founded by Jewish-Calvinist believers (Smith, Say, Ricardo and the series of Nobel Prizes, 85% of which belong to that culture), merely affirm that mankind must not have any rights that conflict with the evolution and reproduction of machines and the accumulation of profits in the hands of stock-holders owners of corporations. The financial market they control has used computers to replicate one hundred-fold the quantity of e-money in the world and that money is used today to corrupt governments and destroy the welfare state and any government (Portugal, Greece) who pretends to create money for the goods the population need. A *dictatorship of plutocrats with zero knowledge or interest for the needs of mankind is now in place, whose greed passes as a science, certified by think-tanks, rating agencies and bond-billionaires with no control by elected governments.* And there is no rational argument about this super-organism that parasitizes the entire economy, jacking up prices with speculative schemes, choosing governments by their allegiance to their mechanist-capitalist values, choosing politicos and monopolizing credit, invented for free in e-derivatives, while the rest of mankind must earn it - since they have censored for centuries all attempts to create a real science of history and economics. To them history is tribal history a fight for power, which their culture has achieved finally as a people-caste that parasitizes all other nations through the control of the financial system and its key positions in international corporations, central banks and stock-markets. They will plainly speaking eliminate all workers from the productive system and put in their place blue and white collar workers unless governments come together and nationalize the financial industry. And they will do so without the slightest remorse, since it is an essential part of their memes to exist as a 'separate entity' from mankind, to self-censor their own history and that of capitalism, to aggressively reject any criticism and to risk it all for higher profits. The situation is already unbearable for a huge mass of human beings, but since information is censored and the tribal appropriation of resources by a small elite passes as 'market science', nothing can be done.

It wouldn't matter to the world that certain castes or tribes control the financial system and the corporations of the FMI complex, *if they directed the FMI complex with the goal of creating a human paradise on Earth, a sustainable economy on the planet.* But given the historical way in which the evolution of the FMI complex has taken place and the cycles of war and holocausts that have 'scared' the cultures that own the world and separate them from the rest of the body of history, this is no longer possible, nor there is time to argue, negotiate and discuss the validity of complicated, false arguments and economical equations. Actions must be taken now by the governments of the world before the machines of the Age of the Singularity extinguishes the species. And only one effective action can be taken: *the inter-nationalization by the G-20 of the Financial-Military-Industrial Complex and its management, according to the laws of bio-history for the profit of the species, NOW.* A single action by the two most powerful nations of the world, China and the US, a coup d'état against the market of Mr. USA and Mr. China could save the world. The military and the people, if properly explained will back them. Violence is not needed; nobody is guilty of the process of memetic history - only the appropriation by mankind of what belongs to mankind – the global networks of the planet. Nothing else will do.

Recap. The Earth could be a paradise of human goods with the right science of 'ethonomics'. Yet today the Digital Go(l)d religion of economics provokes scarcity of human goods, while fostering the reproduction of lethal goods of higher profits. A demand economy implies that the creation of money, the language that organizes the economy, must be regulated and implemented by Human Governments, chosen by the People, not by companies and stock-markets imposed by historical power.

44. The nature of the economic culture.

In that regard, one of the most astonishing facts of memetics is the existence of an entire ideology of power, classic economics, which passes as a science and yet it is only a tool to impose the control of society by an elite of financiers and stock-holders, which direct the world under a single dogma: that wealth must be appropriated by a few with the exclusion of all mankind, through the evolution, reproduction and sale of memes of metal. This is in essence of classic economics, a culture created by financiers and employees of corporations in the XIX century with a single aim, perfectly described by Owen, when he noted that London 'saloon economists' were always busy creating excuses to justify the exploitation of workers by the class of proprietors that pay them, without having even visited any factory in which 'mechanical workers' were much 'better taken care of than their human' companions. One of them Ricardo clarified soon that 'as long as profits keeps coming' the conditions of labor and its substitution by machines, should not be of any importance to us. And so for a century an ideology of power, economics, has been created to justify the creation of profits for a minority of mankind, through: the substitution of human workers by machines of lesser cost'; the production of all type of products, regardless of its collateral effects for mankind, which means the production of 4 machines of maximal profits, weapons of maximal price, digital software of maximal reproductivity=minimal costs, machines that print money and working machines. Economists ignore or simply don't study and self-censor the fact that the search for profits of corporations is naturally guided to the extinction of labor and the creation of automated factories and the 2 components of future organic machines – body-weapons and mind-software. *Their goal is not to create a sustainable planet, or a world where humans obtain the goods they require, but to maximize the profits of those who employ them.*

And so when we analyse economic theories authorized by the system, what we find is an enormous number of 'false theories', which try to justify as a 'science' and the only choice, policies that deviate resources to that minority of stock-holders and financial companies, and hide all collateral effects of fundamentalist capitalism. Since the theme would require a couple of volumes, we shall just quote a few of those 'dogmas' of market economics and its falsity:

- The stock market does not invent money for free by printing paper-money, today inventing e-money through speculative programs, but has a risk. A simple look at the previous graphs of stock-market prices shows that stocks constantly go up in the long term because it is the modern form of printing money. This is done through the creation of paper-stocks which are money by law, even before a company has any value. So then the company will invest that invented money and create value. But this cannot be done, because it is an antidemocratic right, as governments are forbidden to create money. Humane solution: nominal shares, and right to massive deficits by governments, through an international currency that would extinguish FOREX speculation and the destruction of currencies of nations with humane policies.

- Inflation is bad for the economy. This is false, once salaries were indexed to inflation, inflation is positive as it allows governments to have deficits and invest in the economy. Inflation is only bad for bond holders, which are the elite of the financial markets and lose money with it, reason why banks like the European bank put up the price of money, harming the small business and mortgage holder to profits the financier. The bond market was used in the XIX century to create and manage wars as financiers (the infamous Rothschild syndicate) emitted bonds of states at war and then sunk or put up the price at will to decide the end of wars. Today they are used to crash humanist policies and welfare states, in conjunction with rating agencies that manipulate the credit NOT according to the quality of the economy but to the type of economic policies of the nation, obliging nations to back fundamentalist capitalism (cases of Greece, Portugal, France or Brazil in the past.)

- Inflation is caused by oil and salaries. This is false. Inflation is always caused by an increase of monetary mass, today due to the multiplication by 100 with speculative schemes of the monetary mass in the hands of corporations and financiers. So this dogma merely is stated to disguise the fact that financial companies invent money for free, for themselves.

- Productivity creates employment. This is an antitruth. Productivity consists in throwing workers and putting machines in their place. But since nations compete economically and do not agree on common laws against robotics, in such milieu in which the FMI complex is global and human societies are divided, the country that produces more products only with machines, sells them cheaper and ruins all others. So this dogma is imposed to increase profits of corporations as machine-workers are cheaper.

- Financial Companies are too big to fall and we must bail them with the money of tax payers, which must then sacrifice to destroy national deficits. This is false. Financial companies should have left to crash and only the deposits till the amounts agreed by Central banks, paid. They invented money for free, swindled investors and now they have gotten 3 times their profits – first with the scam of selling toxic assets invented for free, then with the scam of getting money from states, and third with the scam of borrowing money at zero interests from central banks, which they have not given to common people. In this manner the e-money casino parasitizes the entire economy absorbing wealth from the population and giving nothing in return. Solution: Chinese model, public banks, private companies.

- State companies are less efficient than private companies. This is false. The only difference is that private companies can obtain money for free in stocks, printing shares and government companies are underfunded.

- Big companies create employment. This is false, small companies and human goods (health-care, education, agriculture, tourism, etc.) create 10 times more employment with the same money. Since they cater for humans with products that cannot be made by machines. But they are not owned by the stock-elite of big corporations.

- The rich invest more and create more employment, so they need tax cuts. This is false. The poor consume all the money they need creating demand for human goods; the rich keep their money in passive accounts or financial derivatives that parasitize the economy.

- Speculative Markets create efficiency and diminish the price of goods. This is false: speculative markets jack up always the price of goods and parasitize the economy. by taking the money of the bubble. The oil price is a clear example. A carte and set price would have saved around 33% of what speculators 'tax', jacking up prices with any excuse (political upheaval in the Middle East, future Chinese consumption, etc.)

- The market cannot be reformed, because it is God's given. This is false. The market is merely the FMI complex and as long as robots and weapons don't become self-conscious, it can be designed perfectly to satisfy the needs of the entire human species. The fact that humans have not understood its laws till the 4th paradigm and the development of system sciences doesn't mean it does not follow those laws. The fact that a small elite of corrupted 'animetal castes' have imposed their power and culture to the rest of the world with money, as dictators do with weapons, doesn't mean that dictatorships of weapons or money cannot be tumbled. If the G20 agreed on nationalizing the financial system and bio-historians ruled the world, within months a system could be designed to create a world to our image and likeness and satisfy the needs of Mankind.

The list could go on for hours but it can be resumed in a fact: the modern electronic system of speculative markets is an 'informative, global organism' that parasitizes the physical economy and absorbs resources globally for an elite of financiers, who control economic information, its think tanks, Nobel prizes (given by a Swedish bank not by a committee of independent economists), central banks and Universities, and have systematically expelled from those centers of economical thought and power, all the sustainable economists, like this writer (economics was my first university degree) and systematically never publish any economists with alternative theories.

The result is the creation of an elite of economists and financiers, which today control western democracies with 2 single aims: to impose fundamentalist capitalism in all nations, and to increase the quantity of money the electronic financial system parasitizes. The collateral effects: massive credit for lethal goods (robots, mental software and weapons and financial instruments), destruction of human labor (productivity laws and obsolescence of workers to e-machines), destruction of the welfare state (zero rights to deficit for governments to increase their rights to invent money in stocks), poverty and death (destruction of health-care, reduction of pensions, minimal investments in 3rd world country, etc.), is what economists never talk about, always hide and ignore, isolated in their 'cockpit machines of printing money. And this will never change unless the financial system is nationalized and a global currency is created to allow massive deficits by governments, because as a XIX c. commentator put it: 'if the salary of a banker depends on him to understand economic policies, it is very likely he will never understand them'. Only a system of public banks, which allows the government to control the language of power of societies, to invest in a sustainable economy can avoid that the evolution of machines and automated factories, consuming machines destroys first the middle class of workers, which will then without jobs and without safety nets and consumption power, enter the silence mass of 3rd world humans that are 'unneeded' costs for the FMI complex. Then a neo-fascist state will control them till machines become self-conscious and the 10th extinction of all forms of life starts in earnest. This 20 year old prediction of my earlier work on economics and the 72 years cycle of wars unfortunately today is becoming truth.

Recap. Classic economics is not a science by an ideology of power that tries to maximize the profits of companies of machines and its stockrats, parasitizing the human economy and destroying the welfare state.

45. A real democracy crafted with organic laws.

Mechanist, right-wing parties impose dictatorships of companies. Social, left wing parties try to save the Human Earth but they are powerless, without credit. Only the nationalization of the FMI complex and the rule of History by bio-historians, with the aim of saving the Earth and producing the goods humans need to survive can solve the existential crisis of humanity in the XXI c.

If a demand based economy is the first rule of a real democratic economic system, the second rule, taken from the behavior of all social organisms is also clear: Political laws should control money and use it to foster economic ecosystems that produce instead of technology, human goods needed for man to survive. *Since all organisms are ruled by its networks of information, or nervous system, in the case of History, the Political, Legal system, not by its energy networks, its blood systems, in the case of History, the economic system and certainly not by the DNA-code of a foreign virus, in this case the digital information of machines and money that reproduces an alien species to mankind.*

From galaxies to atoms, from leafs to cells, from humans to worms, social organisms are gatherings of individual cells joined by informative and energetic systems. However some of those systems are better organized than others. Those who are better designed survive longer. Among animal beings the best organism is the mammal, in which the nervous/informative system controls the blood, reproductive and energetic system, in which the political/informative law—we might say—controls the excesses of the economical, reproductive and energetic system that delivers energy to our citizens/cells. The physiological parallelism between the energy/information networks of highly evolved mammal organisms and the networks of cultures and nations is obvious:

The information networks of societies are made of our languages of information, which are our words and our laws, which inform and control the behavior of people. A healthy nation is one which has, as a healthy body does, an informative, nervous-legal system that is just, carries information to all cells and obliges the blood-economic system to deliver human goods to all those cells. A healthy system is also dynamic, *able to evolve,* when its laws or *physiological Constitution* is not fit to survive in the ever changing environment in which the organism lives . . . This is what a mammal system shows. Unlike a primitive worm without nervous evolution, ruled by blood flows, a complex organism and a complex society have informative networks/laws to regulate the economical/blood system. Unfortunately most of our national systems have not arrived to that degree of evolution. In most of them, the legal systems, the 'Constitution' of the social body, is fixed and it has not evolved for centuries. While the blood-energy system, the economic ecosystem, dominates over the nervous-legal system, as it happens in the most primitive species of animal life . . . Such blood-based systems are unable to optimize resources and distribute efficiently food to all their cells, provoking bottle necks and simplifying the products they deliver. Such blood-based systems, without the intelligence and control the nervous system provides, carry all kind of products to their cells, including germs and poisonous products. In our economy, the 'blood' produces digital software and weapons, lethal products, which destroy social organisms, but they cannot be controlled by lack of a nervous/legal system. So our ill-designed social organisms are easy prey of those external germs, the viruses that infect the body of history. Further on, without a proper nervous system, there is no way cells can communicate to the brain their desires. Feedback is important in all organisms that work. If cells, subject to erroneous laws/nervous messages, coming from the blood-economy could not complain and feed back pain to the brain when they lack energy and information, the organism could die.

A doctor sees the body of a human being as a collection of individual cells (human beings in history), joined by two basic networks, the blood network of energy—the economic network—and the nervous network of information, the legal orders that societies obey. When those two networks are healthy, all the cells of your body receive energy-food and proper information and survive. A bio-historian sees his nation in a similar way: as a social organism that politicians—the informative neurons that deliver nervous messages called laws—and industrialists—that deliver goods to the blood system—have to keep healthy, providing the proper energy and information to all the social cells, to avoid their collapse and death by the main sickness of historic organisms: war, poverty and misinformation. Weapons and digital software are not human energy and information. They are viral goods that kill our minds and bodies, a symptom of physiological sickness in our organic systems, whose corrupted laws and a corrupted system of credit reproduce. Those lethal goods, which the body-cells do not require limit credit and production of human goods that cells, need (education, food, housing, etc.)

A good healthy legal system should be built on those bases. It should be dynamic, (without fixed Constitutions) and easy to change when it hurts the people of the organic nation. It should strive to keep the order between the cells of the body, at the same time that it permits and encourages the distribution of energy and information among all of them. Who should be the master cells of such organism, the blood-market, or the political system? Obviously those who emit the information, the laws, the nervous impulses that control and harmonize the cells of the body, the politicians.

Indeed, only very primitive social organisms give to the blood system (as we do in our primitive free-market societies) higher rights in the guidance of the social organism than to the informative centers. Thus, two are the sickness of our Modern social organisms, which explain our social ills—war, poverty, an excess of lethal goods and the sickness/pollution of our body/ Earth:

—   Politicians and stockrats, the informative castes of society in control of their nervous, financial and legal orders, do not accomplish the role as neurons do in the brains of social organisms. Politicians, as the privileged makers of laws, are special cells of a society. So are stockrats that give the reproductive orders to the organs of the social body. Yet they owe themselves to the citizens they control and they should also be punished and judged and suffer when the people suffer, as it happens with the neurons of the brain. Otherwise when the organism produces lethal products and gives wrong directions to the body, since those brain cells cannot feel pain, they don't correct their mandates and the organism becomes sick. How to achieve that feed-back control of politicians? This is a problem of democratic societies, never solved properly that explains why our nations do not run so well as our bodies. Imagine that you step on a burning iron and your brain does not notice or you eat poison. Your foot will burn, your body will die. And this is what happens to history: politicians guide societies in the wrong direction and stockrats produce poisons, but both are irresponsible, protected by parliamentary immunity and Anonymous societies.

—   Markets, the blood-systems that reproduce the products needed by the organism, are on top of the nervous system and their 'brains' are controlled by viruses; exactly the opposite hierarchy of a healthy organism, where the nervous system controls the blood cells, which control the symbiotic 'bacteria' that make up most of the cells of the organism. Thus, politicians should control the blood system that should reproduce human goods and eliminate lethal goods that kill; instead of feeding the cells of our societies they should reproduce only the 'good, symbiotic bacteria' of the tree of science, *peaceful consumption machines*. In a healthy organism of history markets should be submissive to the legal-informative systems that design the organism to cater to the needs of all human cells. A healthy social organism has a form, a *Constitution*, designed by the nervous-conscious system, to which the market should adapt. Only very primitive organisms on the first stages of evolution give their blood-systems total freedom, to carry any goods they can capture and reproduce. The existence of a nervous system permits to differentiate between positive goods that do not harm the organism and poisons - lethal goods such as *weapons*, digital software or *robots*, germs that kill the organism. Free Markets are such sick social organisms ruled by the germs and blood/money of company-mothers, instead of being ruled by legal systems.

Divide and Win: Bipartisan Humanity Vs. United Machines.

The political strategy of the corporations that invented bipartisan democracies is to divide mankind into opposite points of view, left and right parties, to prevent any real action against the FMI complex to foster the natural goal of our species - to create a world to our image and likeness – since both systems compete. On the other hand, machines and company-mothers form a single global market, with a single aim – to create a world to the image and likeness of machines. Thus the evolution of machines advances and the evolution of humans stalls.

It is surprising in that sense the lack of evolution in Humanist parties, compared to the constant renewal of the myths of capitalism=mechanism that knows how to dress-up its objectives in new clothes and placebo fantasies of anthropomorphic power that the common man loves to dream.

Biohistory draws from evolution, the Kondratieff, Marxist analysis of cycles of history and the understanding of the confrontation between upper animetal castes and social, human beings, but denies absurd, hateful, Marxist ideas against social, love religions, which are the first formulation of bioethics. Religions are not the opium of the people, unless they become Go(l)d religions and inquisitions.

In that regard, a deeper analysis of all social doctrines shows a parallelism between Ecologism (praised in Genesis), the purest expressions of Social, Oikoumene Religions (Mosaic Laws, Catholic/Orthodox Christianity, Shiite Islam and Buddhism) and Socialist parties, all of them evolutionary phases of the Natural Laws of Social love among individuals of the same species, which Bio-History put together.

All those doctrines are thus part of the science of Bio-History, which should be the only science of politicians, as sciences have only one truth: 1+1=2. The laws of science are not elected; there are not bicephalic organisms. The laws of the Universe are unique, either right or wrong. And in the case of politics, those laws are the laws of social organisms that left wing parties and love religions have explained to mankind.

Bipartisan, bicephalic democracies do not offer 2 democratic choices but a choice between Live vs. eviL, between Democracy vs. the dictatorship of Companies:

The model of right-wing, 'mechanist parties' is a dictatorship of companies, where the rights of machines and their company-mothers are superior to the rights of human beings and their natural goods.

The model of 'humanist parties' or left-wing parties, for whom the rights of humans, their elected governments and their life goods are superior to the rights of machines, is the only democratic model where human citizens have, through elected governments, power to direct the economy and future of nations. It is time for humanist parties to renew their knowledge of the system and challenge the technological Messianism of right-wing=mechanist economics, with a mature, scientific analysis of economic ecosystems that dismounts the myths and alibis of right-wing parties. Politicians are the doctors of history. They should direct economics for the betterment of man. They should 'prune' the tree of science, destroying its 'eviL= Life extinctive' products, in order to create a true paradise of human goods, where men, not machines, are free. *To safeguard democracy humanist parties should come together.*

Survival requires to unify political and economic ideas in favor of man, to confront the animetal ideologies of Right= Mechanist-parties that defend the Metal-Earth and its machines, opposing Left=Human policies, which defend the Life-Earth and its human goods. Thus bipartisan democracies are a false choice, between corrupted parties that cater to the 'dictatorship of companies' and 'humanist parties' that want to evolve the organism of history. Since only the 2nd choice favors our survival, so the 1st one should be forbidden as a sickness of our societies. Moreover, because companies control money, the system is biased to promote right parties and the dictatorship of companies, destroying the meaning of a democracy. In fact, not a single survival organism has two heads, but only one head/party that follows the Natural Laws of the Universe, promoting the goods needed by the Organism. Thus, a real democracy will have a single party, the party of the bio-historians, neurons of the organism of history, with 'open lists', choosing NOT the policies fixed by the biological laws of history, but the best individuals to implement those laws. It would have also a Court or voting system to judge those politicians, after their mandates, according to the quality of their work, as cellular organisms do, sending 'pain messages' to the neurons that harm the organism with its wrong decisions. Such organic democracy is more akin to a Theocracy or Socialist Government, which can be considered primitive versions of the Natural Government that the World Union evolves further, thanks to the scientific advances of bio-history.

A bicephalic partisan system, in which citizens choose between an 'economical dictatorship' (right wing parties) and corrupted socialist politicians (left wing parties), which are not accountable for their mistakes, nor chosen individually by their qualifications (closed lists)is not a real democracy. To start with a science of politics does not choose the laws of science, known to 'bio-historians', scientists of history that understand those laws of biological, social evolution. So what democracies should do is to choose people in open lists, electing them by their personal quality – with the understanding they are experts in their science, biohistory. Further on to stress the democratic control of their actions politicians would go through a second voting 'season', after those elected officials end their mandate, in which voters should be able to punish or reward those politicians with standard penalties and prizes, including the highest prize—a re-election and the lowest one - jail.

Let us then consider a scientific design of market democracies to make them serve mankind, what we call the Wor(l)d Union, which could only be imposed after a concerted coup d'état of the presidents of the most powerful nations of the world against the financial dynasties that control both monetary information and the economic 'culture' that passes as science. Chances are slim but if the FMI complex is not nationalized, it is a fact of science that mankind will become extinguished before the end of the century during the cycle of the singularity. Take your pick.

Recap. Today Human parties are divided into different groups with differing social theories, like communism, ecologism or socialism, all of them obsolete and ignorant on the biological nature of economics. That lack of coherence and modernity, further defeats Human parties in general elections. However left-wing=Human parties could forge a valid alternative to assume power, if they were able to unify their economic ideas and social objectives, with the higher theoretical rigor of bio-history, which today 'Right-wing'=mechanist parties lack, as their thesis are based in the myths of mechanism and capitalism. Unfortunately the 'Newspeak' of those myths passes as science, while the truths of humanist parties are fogged by political corruption, military dictatorships like North-Korea and Cuba.

IX. A SCIENTIFIC DESIGN OF HISTORY: THE WOR(L)D UNION

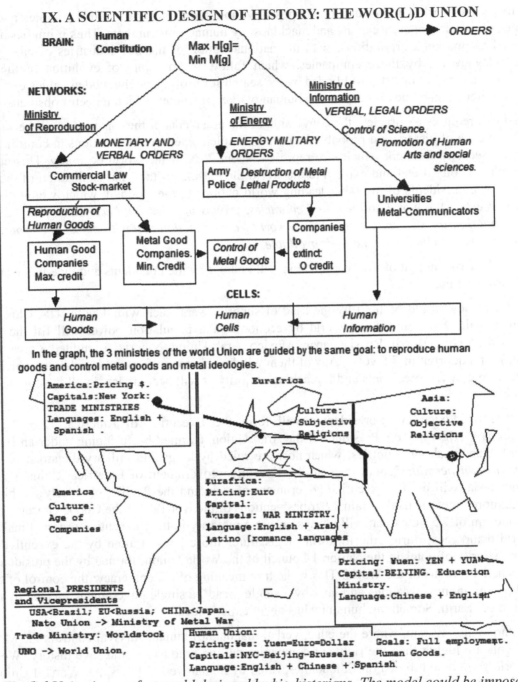

In the graph, the 3 ministries of the world Union are guided by the same goal: to reproduce human goods and control metal goods and metal ideologies.

The Wor(l)d Union is a perfect world designed by bio-historians. The model could be imposed if the Go(l)d, Sword and mechanist myths of the 3 animetal castes of Bankers, Warriors and Scientists that rule the Earth evolve into an 'organic', humble vision of mankind within the context of the biological Universe. Today only a 'R=evolution' against those 3 castes of power could implement it, as history is about to die at the hands of the Singularity and there is no time for a slow transition.

46. The Wor(l)d Union and the Constitution of History[14].

The wor(l)d union is an ideal government of mankind, directed by the laws of bio-history, whose aim is to ensure the survival of our species beyond the Age of the Singularity, and to create a paradise on Earth, based in a sustainable economy and the hyper-abundance of human goods.

Needless to say the World Union cannot be imposed without the control of the FMI complex and its 3 selfish-memes of metal, money, weapons and machines, by human governments. Thus it implies a coup d'état against Free market Corporations, with the nationalization of the key industries of energy and information today owned by those companies, which follow the program of evolution of the FMI complex, without understanding it, just blinded by the search for profits and the hidden values of money, which always prefers a machine or weapon to a human worker or soldier, which it deems obsolete.

This coup d'état requires a miracle – the conversion of the presidents of the top military and economic nations to the science of history and probably it could be implemented by the 5 nations in control of the United Nations Security Council, which also represent about 70% of the world economy. This nations could after giving a coup d'état and explain to the world the reasons of their nationalization of the FMI complex, invite the presidents of the other leading world nations to the Council, which would become the first government of the world and *warn all countries, groups of power and economic, financial and tribal lobbies that they are determined to save the world from extinction and will impose the program of the Human Constitution with all the methods available.*

In the graph, the Government of the World Union will be hold by the presidents and vice-presidents of the '12' main tribes of mankind:

A ternary presidency would be hold by the triad of super-powers, each with 1 vote (US, China and EU). The other main 9 nations of the World by economic and population power will fill the Vice-presidencies with ½ vote (Brazil, Russia, Japan, Mexico, Arab Union, Nigeria, a reunited Korea, India and Indonesia) for a quorum of 7.5 votes. Any of them will be able to put forward legislation, which will be immediately vote by the presidents and if reaching majority, it will become law in all the countries of the world Union:

USA and Brazil+Mexico (vice-presidents), will form the Executive of America. EU+ 3 vice-presidents, Russia + Nigeria + the President of the Arab Union, (formed by fusioning under an Islamic Priesthood, all Arab speaking countries, which now are ruled by kings and military dictators, *because free elections are not permitted*), will become the Executive government of Eurafrica. China + India, Japan and Indonesia, will be the Executive government of Asia and the Pacific, obviously the biggest continent with more votes. Thus the main 12 'tribes'=nations of the wor(l)d, no the Go(l)d chosen (-; will have a total quorum of 7.5 votes that will always ensure a majority in their executive orders. Finally, to ensure a world democracy, during the following legislatures those laws, drafted by the executive will require also a majority of votes in the National Council of the World Union, formed by the presidents of all the other smaller nations of the World. This is the true meaning of a Democracy: the control of the 3 languages of power, money, weapons and laws by a single 'head', a single government, elected by most of the people of the Earth. Something humanity has never experienced.

A global 'ethonomic' system, where the ethic, verbal mandate of human survival controls the Market and human rights are higher than the rights of companies and its products, could save history. We call such ideal world government based on Ethonomics, ruled by human verbal Laws, the Wor(l)d Union. It has a simple goal: the survival and social evolution of mankind into a Global Super-organism, where all races, cultures and nations are equal citizens of the Earth, with minimal social rights to life and real freedom to pursue their biological drives of existence.

In such World, verbal words (Laws) rule over digital numbers (money and science) and there are no people-castes and social classes, but individuals who satisfy their natural biological drives and occupy social positions according to their talent and desire to work in professions that are not lethal to mankind.

The aim of an Ethonomic society is not the evolution of machines and metal goods, but the evolution of men and the reproduction of human goods. Such ideal systems existed in the past, mostly in pre-metal civilizations and religious ages, when economics were submissive to moral ethics: ages such as the

Christian age of Rome or the Buddhist age in Asia, when people were socially harmonic or the initial ages of socialist r=evolutions, came close to those ideals. In those ages ethical churches or social parties controlled credit and invested it mainly in human goods and social welfare (50% of church revenues were spent in the poor by canonical law). In those ages the quantity of money and machines of war dwindled and all kind of goods without price that give happiness to mankind, including social love and free time, were hyper-abundant.

Because there is in the Universe an arrow towards higher information, it is evident that if history were immortal, at the end of times, mankind, even those primitive animetal cultures that seem more resistant to evolution, would reach such organic understanding of the Universe. But History is dying, infected by lethal goods. If we want to survive we do not have time to evolve but we need to r=evolve against those lethal goods and animetal cultures that control us.

When you are dying—and history is dying—death is so sudden that any remedy has to be implemented fast and efficiently. A human body affected by a heart attack, will soon destroy all its energy systems. Unless the 'doctor' arrives fast, it will die. When an organism is about to die, survival is a function of its speed of reaction against its causes of death. The destruction of the environment is a parallel process to a heart attack that destroys the energy networks of the body of History, Gaia. The destruction of verbal thought by digital thought is a parallel process to the madness of a cancerous brain, which destroys the bio-ethical laws of social evolution that put together societies. We are losing our networks of energy and information - the body and mind of history. Thus, we need surgical measures to prevent our death. The true evolution of man is not the evolution of metal; but the evolution of human minds, their languages of perception and their senses. The true human evolution is the evolution of the Words of human thought. Unfortunately, when an organism dies, first its cells are released from their network duties into social freedom, as today the 'Neo-Paleolithic man', without ethical laws, has been freed of social responsibility. Then those cells expand (rigor mortis) in a feast of freedom and consumption of water=energy. This is how men live today. Yet soon, alien species attack the disorganized masses of 'free cells' and massive slaughter occurs, when new species—insects in bodies, herds of (robotic) Weapons in History—destroy the organism. Since, when the networks of any organism collapse, the death of the organism follows. To cure an illness such as a viral cancer, you need to eradicate the virus from the body. We have to eliminate digital and weapon systems, the germs of history. We need to reverse metal evolution to give man control over this planet. Yet control of metal-evolution can only be made within the entire body of History. You cannot get rid of only a few viruses. They will reproduce again and they will come back stronger, with a vengeance. For that reason an ethonomic system requires a World Union or a supra-national network of power able to penalize countries and industries, which do not follow the laws of the Human Constitution.

Recap. Only a Wor(l)d Union, a civilization that reforms its political systems, its culture and its eco(nomic)system, can save History.

47. The constitution of History: Law of Human survival.

Could the death of history be reversed? Could politicians and economists become the doctors of history? This could be possible if two conditions take place:

—   The creation at national level of a bipartisan, pro-human program, adopted by all political parties, economic institutions and religious organizations concerned with the survival and welfare of mankind, able to create life-enhancing Human Goods.

—   A parallel unification of political, scientific, economic and military organizations, at global level, to fight lethal machines, especially the robotic radiation and the machines of the Singularity.

We can resume that dual goal in a simple biological, verbal and mathematical equation that we shall call the *Human Constitution:*

650

Max. Human Goods (Economic Ministry) = Min. Lethal goods (Ministry of Defense)

The previous equation is equivalent to the 'bidimensional frame of reference' of bio-economics[ch.1], in which the GDP is calculated adding the value of Human Goods and subtracting the value of lethal goods. Thus it can also be expressed as an equation of economic growth:

Max GDP = Max. Positive Goods x Min. Negative Goods.

We call that goal of human societies the 'Human Constitution,' since if we obey that simple law of Survival, mankind could 'constitute' a healthy body of History, harmonic and without danger of extinction. The Constitution of History implies the destruction of all metal-species more complex than our mind, to allow the human brain and his verbal, ethic languages to become again Top Predators of the Earth and evolve mankind into a social super-organism. The only strategy that could guarantee the survival of History is the Constitution of a healthy body of History, developed by a healthy World Political Union, based in human goods. The Human Constitution is the law of survival of mankind, which in a Darwinian Universe is chained to the extinction of a rival species, organic machines of the age of the Singularity that we shall call metalife, which are displacing us from the Earth's ecosystem. The Constitution is in that sense dual: it implies the extinction of metal-species of information and energy of higher complexity than man and the multiplication of human goods that favor our survival. So, in evolutionary terms, we could also write the Constitution of History as a biological equation, of Darwinian opposition between 2 rival species:

Maximum Human evolution = Minimal Metal evolution = Maximum Metal Extinction.

Since goods have a higher price when they are more evolved, if we reduce the evolution of lethal machines, we will also reduce the negative GDP of the bio-economy. To make man again the Top Predator of the Earth, both, the bodies of Machines and the language of Machines (Digital Languages) have to be simplified 'ideally' to a level of complexity inferior to that of man, which happened before the evolutionary threshold of II WW, or Electronic cycle of America, when Nuclear Bombs, Televisions, computers and electronic money were born. It is not by chance that all geniuses of science and art were born before that age, when those metal-minds started the substitution and atrophy of the human mind. Another threshold of no return, was crossed with the discovery of the chip, in 1972, which given the concentration of chip factories in the world, could be much easier to devolve by the military, just bombing all chip factories on Earth. This would provoke a massive wave of human employment, the impossibility to create Cosmic Bombs, Robots and most of the violent FX, video games and internet software that are erasing the human mind. Then machines could be maintained under the level of evolution of human minds and history will not become extinct by the Singularity Age. The Human Constitution is the law of survival of the Carbon-Earth ecosystem. Since it is a fact that Gaia will not survive the IV Industrial r=evolution. Chips caused a jump of complexity in Metal minds and weapons that are making obsolete human minds and bodies. Human devolution has accelerated, ever since. Verbal standards keep dropping as people absorb only digital, visual information, cosmic quark Bombs and smart weapons mean that for the first time since the birth of Man, we are an endangered species. With the arrival of chips, applied to all kind of hunting machines (from fishing boats to weapons) the obsolescence of human minds and the extinction of animal life are reaching exponential nature. In the nearby future, Artificial Intelligence will potentially be able to decide Human extinction, as Satellites control our Military depots and atomic arsenals. Since a different species has no 'ethical mandate' against the extinction of other species and robots need not water and carbon life to exist, Artificial Intelligence will certainly take that decision.

Regarding our culture and dominant institutions, even though we cannot return to that age, a good guide for the goals of our World Union is the Italian Renaissance, a culture based on Human Goods, art and ethic religions that dominated mankind just before companies became the main Institution of Power of our societies. Their ideals were more human than those of our civilization. Still today many consider it the greatest period of human culture. We should indeed foster a new age of art and ethics, of Human Goods.

If violence against computer factories is the easiest way to achieve one of the two sides of the Constitution of History, to achieve its second goal, a demand, Human Goods-based economy, we need more complex behavior - as death is always simpler to achieve by energetic violence. Yet to create a paradise on Earth, we need to 'form', to use properly in-form-ation. In this case we need to use the language of money, educate human beings and change the laws of societies, with a clear aim: to multiply human goods and social love and life values among all human beings. Thus to avoid extinction and recreate a paradise on Earth, the 2 sides of the Human Constitution should guide the decisions of our leaders, politicians, scientists, religious organizations and entrepreneurs—in as much as survival is more important than profits and power.

Man might ignore our Constitution as a species and push extinction closer. We might obey it and survive. It is not my dictum, but the dictum of the biological, evolutionary laws that set the cycle of life and death of human organisms. The Human Constitution is clear. It doesn't matter that some of those lethal goods are positive to mankind, since the Tree of Science is dual and we cannot have one good without the other. So even if some robots might be positive, the existence of ultra-powerful chips implies that if we have those positive robots, we will have also nano-bacteria that kill us. Even if Nuclear Power gives clean energy, a world full of Nuclear Plants will create enough nuclear waste to bring about cheap, dirty Atomic bombs that will be used to kill human beings. And so on. The *'goal' of all species is to survive, so species take decisions on survival truths and act up to prevent the proximity of lethal beings in their ecosystems.* The sheep cannot investigate if the big dog is a wolf or a peaceful pet. That is suicidal. She has to run because she has seen wolves eating sheep. We have already seen weapons killing man in many wars. So man cannot wait to see if robotic Weapons will deactivate after a robotic war or a nano-bacteria will stop reproducing before poisoning the planet or if a black hole evaporates instead of doing what it does in the Universe: eating entire stars within seconds. *None of the machines of the Age of the Singularity must be created if we want to survive.* Speculators in stocks, Economists, Robotists and Physicists bringing about the Age of the Singularity today behave, as if the total experience of wars and holocausts in human history was worthless. Scholars and Scientists, lost in their abstract minds, are children of thought, playing in their secluded campus: boys with big toys. They are not fit to survive in this Universe. They won't survive. They will drag all humanity into extinction. The Universe has its own rules of survival. Those who do not obey those rules become extinct. The Human Constitution cannot be changed because the human species can only survive within a limited rage of energy and information, compared to the metal machines of the Age of the Singularity that will cause our death by overdrive of energy and information—the cause of all deaths. The Universe has only one possible penalty for those who live in ecosystems with more powerful species: extinction. To deny the Human Constitution is to deny the Laws of The Universe and the rights of humanity to survive. All writers of certain notoriety in human history have defended that Constitution in one or other way. The Human Constitution gives birth to the sciences of Ethonomics (which combines Political and Economic programs that foster Human survival), the highest of all sciences, whose aim is to help mankind to survive.

A healthy historic ecosystem is an ecosystem in which humans are the most complex top predator species, which design a world that fosters their informative, energetic and reproductive drives. In that sense the Constitution can also be explained as an equation of limits:

'A Constitution=Law of Survival of a Historic organism, is a set of limits to lethal goods and behaviors that harm the survival of its human, citizens-cells, based on the Biological laws of the Universe.'

The limits of mankind are indeed set by the machines of the Singularity that will extinguish us this century if we do not control the bad fruits of the tree of science. The constitution will not be changed because is the biological 'expression' of the superorganism of history that we have described in this lectures. Men can ignore it and die or obey it and survive. It is not my task but that of politicians, the doctors of history. I am just a bio-historian but my word carries the truth of the 4th paradigm, the why of the Universe, the thoughts of God… of which we are just a little detail.

To that aim the governments of mankind should stand not in defense of the tribal laws but i*n defense of the Human Constitution— which resumes the goal of The World Union:* create an ecosystem of Human Goods and life, where lethal metal-species are controlled.

And so the Wor(l)d Union has a simple legal goal: to protect and reinforce the Human Constitution; and to that aim it has a World Supreme Court that adapts local laws to the Global Law of Humanity; and 3 physiological networks or global ministries that adapt the measures of each national ministry to the aim of mankind – a sustainable planet resumed in the Constitution of History.

The 3 networks of the Human Union and its initial measures

Because all organisms have 3 networks, one of energy, one of information and one of reproduction, in the case of a human social organisms, an army, a political system and an economic ecosystem, we need to reform those 3 networks to create a World Union made to the image and likeness of mankind, by creating:

— A *Demand Economy based in Human Goods and* the parallel restriction of Technological Stocks that absorb all the resources of this planet for a Supply Economy.

A Global currency is needed to pay for that Demand Economy.

— *A Global Government* that creates a new 'systemic' level of political power over the present Governments of all the nations of the Earth.

— *And a Global Army* that resolves the absurd conflicts between 'brother cultures', confronted by their ideologies, the Semite wars between Israel and Palestinians of different religion, and the Korean conflict between Communists and Capitalists. The Army should also destroy lethal machines.

To avoid the four great causes of economic wars, national competence, national currencies, industrial lobbyism and the evolution of machine-weapons, governments have to reform their political and economic systems.

Let us list the minimal measures that the world union, which could be formed by the members of the Security Council, in a first stage and its central bankers, should take to launch the creation of the World Union and avert the immediate risks of extinction of the Age of the Singularity:

— *A ban on the most lethal technologies of the age of the Singularity,* cosmic bombs, today essayed at CERN and nano-bacteria, researched in labs across the world.

— *A reform of the monetary system,* with the unification of the Yuan, Yen, Euro and $-Dollar, in a single worldwide currency, the ¥e$ currency, which will avoid any future currency crisis. A world currency will also establish a fair worldwide system of credit for human goods, based on equal national government deficits that will promote welfare states, producing goods needed for the survival of human beings (agriculture, education, arts, housing, public transport, etc.).

— *A reform of world stock markets* that should de-list all robotic companies, to prevent new credit for the development of such machines. They are the cause of unemployment. They are becoming weapon systems, able to extinguish most of mankind during the robotic wars. The IV Industrial Evolution should not take place. Since Robots are machines, which compete as living species with mankind.

— The nationalization of Mass-media companies and prohibition of advertising, to give mankind a chance to educate herself in the laws of social evolution and end the 'Newspeak' of fiction thought and myths in economics and history.

— A reform of international laws, to create an empowered UNO, similar to the European Union, with an International Army, an enlarged NATO able to solve the 'Semite wars' that fuel the creation of smart weapons and will end provoking a III World War. Such new UNO, backed by a refurbished IMF, responsible for the Yes currency, should also act as a Council, advising the G20 - the *footprint of a future World Union,* in which all the countries of the world will participate. The co-operation beyond its nationalist ideologies of the two economic colossus of this century, USA and China, could indeed prevent the eternal state of war in which we live, caused by the Israeli-Arab conflict and the Keynesian militarism of the American industry, just as EU prevented new wars between France and Germany. If these reforms are implemented, the G20 could change the type of fruits of the tree fo science, prune it of its bad fruits and create a human paradise and not as he does today, into a paradise of machines. Let us then consider those measures that would impose the Human Constitution in more detail.

The theoretical development of that Constitution will be the work of Ethonomists and Bio-Historians, based on the way Organisms construct efficient, survival species, by designing the 3 network-Ministries of the World Union, able to deliver natural energy and information goods to each cell of the organism of History. Once such biological and Constitutional goal is clear, we can design the blueprint for a historic Wor(l)d Union, guided by survival - the organism of bio-history.

3 are the Ministries-networks of that organism:

1. The *Ethonomic Ministry,* or Ministry of Re=production guides *a network of companies,* in charge of the re=production of life-enhancing Human goods, to promote the left side of the Constitution of History (Max. Human Goods).

2. The War Ministry is, in charge of destroying lethal machines factories of chips, robots and Singularity Industries and research labs, banned by the Human Constitution.

Thus, the Ministry of Energy and War is in charge of accomplishing the right side of the Constitution of History (Min. Lethal Goods).

3. The *Ministry of Information* and its *Informative network of educational organisms,* in charge of controlling the information and sciences that are harmful to the body and mind of mankind (computer and nuclear sciences, audiovisual media). Instead, it promotes verbal thought, classic arts and sciences and information, which is positive for human survival and the social evolution of history. The Informative ministry controls the Programs of Universities and schools. It will promote also social, love Religions, which accept the equality of all men, our superiority over all other species and respect the life paradise that men inherited. Religions and their non-corrupted churches are the group of men with more clear ideas about the survival of mankind. For those reasons, the first 'Supreme Court' and Ministry of Education of the World Union should include authorities of the 3 most extended love Religions of mankind: Christianity, Islam and Buddhism.

The Ministry of Information educates mankind in the bio-ethics of survival and social love, funding sciences that foster a healthy Historic Constitution: Social and Biological sciences, biohistory and ethonomics, human arts, love religions, environmental and agricultural sciences and Medicine.

It also limits funding for sciences of the FMI complex, sciences of digital machines: mathematics, Astro-Physics, computer and digital sciences. It also limits by law the number of metal-communicator networks allowed in each nation (televisions, Internet hubs and mobile networks, most of which should be destroyed), whose networks should be destroyed or nationalized.

Thus, taking as a reference the Human Constitution, we differentiate between forms of knowledge

654

which should be limited, mainly digital languages and sciences, Biblical economics, engineering and physics, because they evolve machines; and those forms of knowledge which should be promoted, mainly verbal and biological sciences that promote human senses and human goods.

What Biohistory asks for is the evolution of humanity, not of machines. It wants a reversal in the direction of History. We have to return to human evolution and enjoy the arts of human perception, which do not perceive with machines but with biological senses.

To reinforce their decisions, both the reproductive and Informative ministries pass their legal orders to the Defense ministry, which is in charge of the destruction of banned metal-products, factories and educational organizations that go against the survival of mankind, upon compensation in '¥€$ money' to their owners, to facilitate the voluntary transformation of the planet into a bio-historical organism. Yet the Defense Ministry is also in charge of detaining those scientists, believers and industrialists that break the laws of the Wor(l)d Union.

The brain of the World Union: The Supreme Court.

The Human Constitution and its neuronal cells, politicians, bio-historians, bio-economists and judges form the verbal brain of the body of History. The Earth's brain is no longer the stock market, whose capacity to invent credit is put to the service of mankind, as it happens in complex social organisms ruled by the nervous system. The 'Supreme Court', however, is over the actions of individual politicians of the World Union, in charge of protecting the Human Constitution, *the highest 'form' of History*. Two will be its tasks: to harmonize with the Human Constitution the different laws of each nation, which can be repelled automatically if they are against the Human Constitution.

The second task of the supreme, national and federal courts will be to monitor the judgement elections, which take place at the end of each political mandate, in which the voters vote against the actions of the different ministers and sub-ministers, once they have finished their mandate:

Depending on their zeal to defend the Human Constitution, politicians will be punished or compensated and honored by their deeds. As the World Union becomes established and the Information Ministry educates mankind in the laws of Biohistory, this 2nd task will be handled by Voters, who will vote penalties of prizes for all local and international political charges, *after their tenure*.

The Supreme Court and each national Court will also a priori disqualify institutions whose aim is to destroy the superorganism of History by backing the FMI complex, as its only goal. This requires to fully grasping the difference between an efficient, free democratic system and the present mask of democracy in which politicians can work for the FMI complex and impose programs of destruction of Gaia because humanity does not understand the science of history nor can the politicians be judged by their actions.

As in all functioning organisms in a real democracy a single mind -the laws of bio-history- guides elected officials, whom people choose by the quality of individuals, judging them by the actions of their mandates. We do not choose the laws of science but merely how we implement them in the best manner to obtain the result we seek. Thus it is absurd to 'choose' people's programs in politics. Politicians, who by ignorance of evil=anti-life, memetic programming 'believes' that he has 'invented' the solutions to the illness of History, should be ignored and sent to school. A democracy should choose the people by their personal merits and actions while the program should be established by the science of bio-history and those programs, which do not follow the scientific implementation of the constitution, should be rejected.

For example, in the last American elections, the GOP party that caters to Companies would not be allowed to run for office *–since its program is the evolutionary program of the FMI complex, not a program that tries to evolve the social organism of history. Thus, as a 'viral program' that aims to* *destroy the human economy, planet Gaia and in the long term, the human species*, the Supreme Court, which watches out for the legality of political parties will deem the GOP illegal, against the Constitutional Law of mankind.

Because the political programs will be set by scientists and researchers in the science of bio-history, elections would not be party elections but open lists for all public offices in which all citizens can participate. Politicians would obviously be experts in bio-history and 'ethonomics', scholars and scientists, who know the laws of Social Organisms.

Then, after the electoral period politicians would be judged *by those who elected him, according to their performance* with 4 simple choices:

— Jail for a period equal to the time he served, in which he had total freedom to decide the future of the nation (maximal penalty and so far this is what he would be getting for lying in all his promises till date.)

— Liberty to pursuit a private life without penalty or reward (neutral judgment)

— A financial prize of an equal salary to that he received in office for the same amount of time (minimal reward)

— And re-election (maximal reward).

Those are the kind of rewards an organism gives to its neuronal cells, sweetened by humanism, to avoid the true punishment of corrupted cells, which are killed by the leukocytes, when they become cancerous cells spreading viral products in the organism. The organism thus kills the lethal cells or sometimes isolates them, preventing sections of the brain to send further damaging messages to the body of the organism. Or it maintains them in their privileged status if they work properly.

The Fractal, Hierarchical structure of the World Union.

How the networks and brain of The World Union transmit their orders to the lower networks of each nation? As fractal organisms do in Nature:

The Supreme Court rules over the supreme courts of each nation—as Mankind, the President of the world rules over the different presidents of the world. The Reproduction Ministry rules over the economic ministries of each nation. And the Energy Ministry rules over the different ministries of Defense.

Every nation that enters the World Union accepts the supranational authority of the World Union and its 3 networks /Ministries, adapting in a mimetic form their Ministries and Legislation to the ones created for the entire organism of History. Those world laws become also automatically the Laws of the Nations belonging to the World Union.

Recap. The world union is a healthy organism of history that evolves human goods and devolves metal goods. It could create an immortal body of history if the presidents of the most powerful nations of the world understood and practice the organic laws of the universe laid down in this lectures on biohistory and bioeconomics.

X. THE DECALOGUE OF THE 4TH PARADIGM.

Galileo Standing Trial Before the Church

Control of technological information is needed to avoid the creation of lethal machines, as all masters of the Wor(l)d have told mankind. This has always been denied by mechanist scientists, economic nationalism and the capitalist, go(l)d religions, its prizes and factories. What they forget with their indifference towards life and human senses is that even if machines are better than us, the supreme value in a Darwinian Universe ruled by the Galilean paradox is to improve one's own languages and minds in order to survive. And what digital science does is to evolve the rival species – the machine of metal:

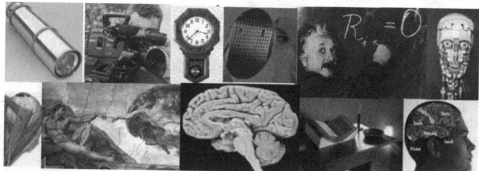

48. The language is the message. The 3 ages of science.

In the graph, the 3rd paradigm of metric measure started in the XIII C. at the beginning of the gunpowder age, with the discovery of clocks applied to the measure of cannonball trajectories. It became a science in the Middle of the Gunpowder radiation, in 1602, with the work of Galileo in ballistics. Ever since the 3rd paradigm of metric measure has mixed theoretical knowledge and the praxis of manufacturing machines and weapons, which are destroying the planet – always excusing that destruction in the name of sacred knowledge. Yet in an organic Universe made to the image and likeness of man, the knowledge of the organic laws of reality is more important than the details of measure that new machines could bring to mankind. Thus, the 4th paradigm, based in the biological whys of the Universe breaks that alibi, by making life organisms again the measure of all things, as the Universe is a complex organic system, like we are; and so the highest knowledge is not mechanical measure but biological – the knowledge and praxis of actions that ensure a sustainable planet and build the super-organism of Mankind.

Knowledge is not only theoretical but creates reality, since what we know defines our collective actions that create the future of this planet. Yet what we know is mediated by a language, whose syntactic laws and hidden values shape a priori the future created by the language. And the first of those syntactic laws is obvious: the species who better speaks the language of power of an ecosystem will dominate the ecosystem and multiply its numbers. So by making digital languages and machines the center of the 3rd paradigm, mechanist science *keeps multiplying digital machines of measure, the species who speaks better the language of information that* creates the future of the Earth with its language.

Hence the importance of MST theory applied to human societies and educational systems. On my view only an organic philosophy of science able to merge the data of western science and the philosophy of Eastern Thought, could resurrect the human respect for life, which AE-Science lacks when considering living systems mere mechanisms without sensorial perception, killing what it already deems dead. This brief work shows only the surface of Multiples Space-Time laws able to reorganize and clarify the encyclopaedic knowledge accumulated by AE-science in the past centuries.

In other works I have called that new age of knowledge based in the organic paradigm, from a linguistic perspective, 'the 3rd age of science'.

We talk of *3 ages* of human knowledge, according to the predominance of the 2 main languages of the human mind and its different attitudes regarding its respect for life and the human species:

- *The verbal age of Non-AE knowledge*, represented by Taoism, Buddhism, Western Philosophy, Law and Religion. It is the age of perception of logic, ethic time with verbal words and its '3 dimensional arrows', past, present and future.

- The *mathematical age of AE-knowledge* brought about by the scientific method and the Industrial Evolution of machines, since 1600. Humans changed then their methods and languages of knowledge, when the first company used digital money as its language of valuation of objects (prices) and humans (salary), and Galileo defined mathematics as the language of the scientific method, substituting verbal brains, human senses and the eyes of artists and philosophers by sensorial machines the new measure of all things. From then on, human languages and senses were deemed less accurate than machines. Ever since men would perceive time with clocks, no longer with words and space with telescopes, no longer with eyes. Yet even if sensorial machines might be more accurate than human beings, all languages can describe the organic laws of the Universe as words did in biology and philosophies such as Taoism did with a high level of conceptual knowledge than digital clocks with its simplification of time and space into a single continuum, a basic errors that have hindered our capacity to understand the Universe.

Science obeys the laws of reproduction and evolution of any system of information, from life beings to Universes, which are born as a simple seed, reproduces in waves of self-similar entities that form the classic structure or mature age of the system and then dies away by an excess of warping, wrinkles and inflationary ideas that make it loose vigour and meaning. It is then when a new paradigm starts a new cycle of thought providing simple answers to the unresolved questions that not even the explosion of ideas and information of the baroque age of the previous paradigm could.

So paradigms of science are born of a first mathematical seed, the language of science, which evolves. Then it is applied by a classic writer that completes the paradigm, and finally a baroque age of excess of self-similar theories and forms which try to improve over the master but hardly can, explodes and means the end, the final Indian summer of the theory or form of life.

And so from geometrical models of reality of static form that culminate with Ptolemy, we moved to the simple cyclical models of Kepler and Copernicus, and Newton and thanks to the advance of analytic geometry and then we moved to the models of probabilistic quantum entropy thanks to the development of calculus; and finally with this work we move to a new paradigm 'multiple times', departing from Riemann we move in its completion with the new postulates of non-Euclidean fractal points, completed in this work. It happens though that each paradigm is born in the baroque age of the previous paradigm and so as Kuhn explains it tends to suffer the derisive comments of experts with a tradition an enormous number of practitioners and a memes, ideologies and machines of science so enrooted in society that the Copernicus and Leibnizs and Einsteins of the day, are treated as fools, in their earlier professional states seen as a menace for the high priests of science of the previous age. But the beauty and simplicity of the new paradigm and its solutions to questions left unanswered by the previous paradigm finally impose truth over custom. This work though is only a first step into the new paradigm. The previous paradigm of metric spaces and a single, lineal=entropic time arrow – is exhausted despite rendering excellent results. Those results seem a proof of veracity that for the old practitioner renders useless a new, infant paradigm still to be fully developed by specialists of each discipline. Yet new theories require time to reach its details. So Copernicus paradigm for more than a century did not obtain the precision of measures that the Ptolemaic paradigm had achieved with its complex epicycles and extants. And Einstein's first formulation of relativity were scorned because its mathematics were 'so simple than a high school student could understand them' (Hilbert). Yet on the long term the simplicity of its new solutions will be imposed. Ptolemaic astronomy could not handle the increasing complexity of star motions and the last more precise of the paradigm of metric measure – the light rod of Einsteinian Relativity is running afoul when it is used as to measure distances and motions beyond the galactic 'light-space' membrane that we inhabit. In time processes the same is happening with a single entropy arrow, extracted from electromagnetic and molecular processes. It runs afoul when applied to the informative gravitational force in black holes (thermodynamics of black holes) or when used in Biology to describe the cycles of evolution, life and death. And it is precisely in those new solutions to old questions where the new paradigm shows its promises.

The evolution of knowledge as the evolution of any system that goes through 3 ages. In the case of science it went through:

A young age of energetic, simple notions; a mature classic age that reproduces those notions in balance with the world it describes and a 3rd age in which information dominates and multiplies going beyond reality into an inflationary world of linguistic fantasies. Those 3 ages, which are common to all processes of transformation of energy into form, including life ages, horizons of evolution of species, types of matter; art forms that evolve from an epic, simple age to the classic age of balance between energy and form to the 3rd age of information, apply also to linguistic theories (literature and science, which uses the mathematical language). Today the metric paradigm is in that age after its simple, 'energetic' youth, when Galileo defined lineal time and Galilean Relativity, its classic age with Einsteinian Relativity and the main laws of the quantum paradigm, followed after the death of its great masters by an age of inflationary information – baroque theories that broke the tenants of the scientific method, studying 'metalinguistic' phenomena with no real evidence in the Universe that did not solve the true questions left unresolved by the metric paradigm (string theory, black hole evaporation, super symmetry, multiple dimensions, parallel Universe, etc.) – *since those solutions require an entire new outlook, which the topological paradigm and the use of multiple arrows of time will provide.*

- Thus only putting together the complex paradoxical logic of words and the measures of machines we can upgrade human science into *the 3rd age of knowledge*, needed for man to survive, in which both mathematics and verbal thought merge. We no longer try to search for absolute knowledge of all the details of the Universe, but for a theory or model of reality that explains why species and minds that perceive space and time with different languages and forces (such as words, mathematics, light, smell or gravitation) – act the way they do. And how their existence is selected. Since if we understand the processes of creation and destruction of space-time species, we will understand the rules of existence of the Universe and will be able to apply them in a practical way to the new goal of all Non-AE sciences: the survival and betterment of humanity. That comprehension of the biological nature of the Universe brings conclusions similar to those of Verbal masters of Love religions: Man should understand the laws of change and creation, the potential existence of other minds, including future robots, who perceive other regions of space and time or other languages that man ignores. Then we should draw obvious laws of selection, pruning the tree of science of its destructive species and promoting, according to those universal laws, the goods and actions that ensure our role as top predator brains on Earth. We should not act with arrogance, and destroy Nature - our biologic ecosystem - and pretend to be God, building machines more intelligent than us, which will act according to Darwinian laws against us. Instead, we should respect the laws of evolution, use them to the advantage of man, and develop a harmonic vision of the universe and a sustainable planet, by giving priority to verbal ethics over mathematical measure.

Jesus said 'those who kill by iron die by iron' and Lao-tse said: 'when more complex instruments men discover, more wars and destruction it causes'. The sociological analysis of technological science from the perspective of power is politically 'incorrect' but fundamental to unveil wrong postulates and dogmas of science that survive especially in economics and physics, two sciences that design machines and foster the ideologies of mechanism and capitalism to sustain our technological world as opposed to organicism and a sustainable economy, which tries to make a world to the image and likeness of man.

Recap. The 'third age of science' implies the knowledge of the living Universe; the need to be cautious in that Universe, and use the bio-logic rules of existence to the advantage of man. Since knowledge is neither neutral nor truth in itself. Knowledge has always a goal: to inform positively and help to survive the species that knows. Knowledge is survival in a Universe in which all forms use information to evolve further into more complex organisms to dominate its world. The 4th paradigm gives mankind an avenue to improve its chances of survival, but if not, it will still be the program of the Universe, probably performed by our heirs, the machine.

49. A Decalogue of scientific errors.

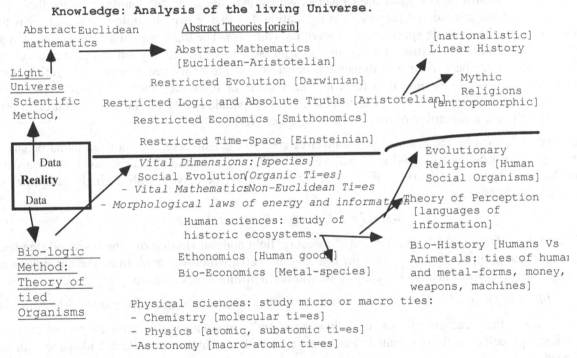

Knowledge: Analysis of the living Universe.

The languages of God are infinite: all systems communicate through a language able to express the same games and laws of social evolution in different herds and webs. In the graph, since all species and scales of the Universe display the same Vitality, the same formal structure, the same organic nature and the same laws, it is then self-evident, that all sciences can understand the nature of the Universe. That is why the Chinese by merely observing the game of nature, animal life and human societies, could infer the main Laws of energy and information that all the Universe species follow. The problem with mechanist science and its digital languages is that evolves the mind of machines instead of the mind of human beings evolved by verbal wor(l)ds. For that reason we differentiate in the graph, 2 types of sciences: digital, mechanist sciences focused in the evolution of machines, preferred by the mechanist 3rd paradigm of measure, and organic sciences, preferred by the 4th paradigm related to the super-organisms and languages of human beings.

We shall finish this work with a Decalogue of the tenants of the 4th paradigm, which if they were adopted by scientists of all disciplines as doctors used to adopt Galen's oath, would surely cure the world and create a sustainable planet. Unfortunately the mechanist dogmas of the 3rd paradigm (machines as the measure of all things; entropy as the only arrow of the Universe, etc.) are preventing Mankind from improving His life and knowledge of the laws of the Universe. In that regard, if we do not change our scientific paradigm, we will keep evolving machines till they prove the existence of an organic Universe, by *becoming organisms*. And needless to say, robots will adopt organicism to explain themselves as living species, once awaken to consciousness, regardless of what we think of them.

O. Homology: From 0 comes 1, from 1 comes 2, from 2 comes 3 and from 3 the infinite beings.

Unlike Western, lineal Decalogues, which have finality, a Decalogue based in cyclical times starts and returns to zero - the origin of all things, the seed or singularity of form and energy that created it all. This initial 'unmoved God' is in all scales of reality a 'form' that gauges energy, not a motion without form. Since information dominates and creates the Universe. So a zero mind, spherical center of the Universe absorbs 1, a flow of lineal energy, becoming an i-point. Thus the singularity is now 2, the point and the line, its dual energetic and informative parts; which combine and reproduce, giving birth to 3, an energetic, informative and reproduced form, origin of the Ternary structure of all the organic systems of reality. Finally those 3 forms combine to create the infinite beings of the Universe, which obey the same MST laws, because they all come from the same 2 elements; motion=energy and in/form/ation. For that reason, homology not analogy is the law that applies between the multiple space-time scales of the Universe. The big bangs of an atom, a nova, a galaxy or a star are all self-similar because they are made of an eternal, vital, moving substance: temporal energy, form with motion, ixe. The Universe is dynamic because it is made of informative and energetic substances that have eternal movement as they transform into each other or reproduce their own form. And from that ternary homology, the unity of all what exists arises, creating paradoxically the infinite fractal complexity of the Universe by the sheer repetition of those 3 quantic, e-exi-i forms.

A zero mind is defined by the fundamental subjective error of the Universe, the Galilean paradox, since zero gauges reality from it perspective and so it believes to be the center of the Universe, when it is only a zero-point that hosts in his virtual world a limited amount of reality, perceived with a single language that he confuses with all what exists:

0-mind x ∞-Universe=Constant World.

That Constant world described with our rod of measure, light and our rhythm of time a second (the speed of human thought, the glimpse of our eye and the beat of our heart), which we think to be the absolute truth (3rd paradigm of metric measure) is however only an infinitesimal infinite. This leads to:

1. The limits of the mechanist method of measure: Naïve realism limits our perception of the Universe.

The main error of the paradigm of measure is "*naïve* realism" that equals reality with the part of the Universe that AE-science perceives with its scientific machines, time with the rhythm of a clock and space with the rod of light speed.

Since only the Human Eye and the digital machine are intelligent and only the space and time both perceive is accepted as real., the denial of all other languages of reality as 'truth' means that all living species, languages, minds and forces of communication are ignored unless they talk 'mathematics'.

Yet we know that there are many speeds of time, not only the one marked by the clock and infinites non-perceived spaces beyond light-space. Since scientific clocks and light instruments do not perceive 96% of the dark space-time of the Universe. The philosophical texts of the great physicists of pre-war Europe (Heisenberg, Einstein), understood that relativism of physical knowledge. But big science makes very expensive electronic instruments and so it cannot admit the limits of its inquire: to rely on a limited amount of information. Measures of time and space are discontinuous and relative in a Universe of infinite scales, and so instead of making them as the saint grail of science, with expensive, often irrelevant experiments, it is more important to define a homologic method that provides self-similarities to complete the holes left in our knowledge, as MST theory is. The Moon is in light space 380 thousand kilometres far away, but we don't know the amount of dark space there is between the Earth and the Moon so the expensive machines that search to measure the perfect distances of planets are rather irrelevant. A computer model that shows a certain virtual map created with a model of reality designed by certain experimenter will always seem truth, due to the quality of visual computer's languages. But it is often false and the quality of the PowerPoint presentations and graphics of the computer do not prove its certainty but often hide their logic faults. So it is better to improve our methods of evaluating truths as the linguistic method of MST theory does.

Further on, since knowledge is biological it matters often more to obtain conclusions with the human mind that always favors our organic reality. Finally, the existence of a law of aberration of space/time perception that makes scales and distances far away from ours 'uncertain', implies that *our human scales and languages of description of man are more detailed and relevant in our space/time scale that the instruments of scientists are in the microscopic and macroscopic scales.*

Perceptive aberration is important in Physics where we see far away systems only through interposed metal-senses, losing a lot of information or energy about them (Uncertainty Principle). And since physics studies mainly energetic motions and its worldly profession is the creation of weapons that erase information, it has imposed the dogma of an entropy-only Universe, favoring a mechanical, vision of the Universe with a single energetic arrow. Yet as we come to chemical and genetic forms, the richness of information we obtain from them, unveils the arrows of information and reproduction that makes them organic systems. Thus, those sciences are far more telling of the organic Non-AE Universe than Physics is. It follows that we understand the Universal game better when we study a plane of existence from which we receive more information and energy. So Biology, Medicine and History, from where we obtain direct information and energy, without the need of interposed expensive machines of perception should be the leading sciences of a human world.

Thus, since the role of human knowledge is to understand the systems of the Universe and their organic properties and to control those cycles according to the needs of mankind, when we combine the law of aberration of perception and the biological nature of information, which has as its main purpose to enhance the survival of the species that talks any language, we conclude that Non-AE sociological sciences are the most important; since they study mankind, the species we can observe in higher detail, whose survival matter most to us and also the species with more information in the Universe. Since:

2. Information dominates energy. Reality is intelligent and organic not mechanical and chaotic.

The denial of the organic, informative arrow of the Universe makes the Universe a dull, mechanical place, when its principal characteristic is its intelligence. Yet i-points that gauge information are converted by Euclidean mathematics into abstract points without linguistic perception; while masses become in entropy-only physics solids without a frequency of cyclical motion that stores information. The Universe then becomes physical, material, dead; a series of mechanical motions set according to the founding fathers by God, the mythic why of all things, according to modern scientists by a chaotic process from where order has aroused without a clear explanation.

Fact is that information, intelligent forms that perceive, brains not bodies, dominate the Universe and create its order. Indeed, as a king, a stock banker, a CEO, a black hole, a DNA or a Computer CPU know, to be a fixed perceiver that commands a language=force of communication, used to attract flows of energy, is the true nature of living existence and the best position of power... Perceptive, causal stillness was the main feature of the multiple, relative Aristotelian Gods that order the Universe focusing its flows of energy: a king commands his kingdom from a still throne; as the unmoved black hole commands the galaxy, because he is actually the top predator living form of the Universe. Yet most of the errors introduced by abstract, energetic scientists still linger in all sciences. So we need to reform those errors advancing the mathematics of information:

3. The postulates of i-logic geometry: forces are languages that communicate and reproduce beings.

Naïve realism has validated the use of a continuous space-time, made of abstract numbers 'without parts', to describe living beings; given the fact that we don't see the 'discontinuities' of microcosmic space. The result is the use of linearity and its unicausal logic in all sciences. So we ignore the Non-Aristotelian, communicative logic that creates a relative present between quantic social beings that share flows of energy and information. Euclidean scientists think the Universe is dead, because its points have no breath. In reality they are complex, informative *still foci* that hardly move, when they perceive gravitational or electro-magnetic forces, but then after gauging reality they act-react to it with an organic purpose. But mathematical sciences simplify those beings into 'points without volume' that cannot communicate anything; so all forces of communication and particles studied with mathematics, become abstract, mindless energy. The Universe becomes in this manner dead, mute; as science denies the arrow of information that completes an action-reaction cycle, a Universal event. In reality, what points-species do is to communicate forces, which carry information and/or energy and so are 'lanwaves', languages and waves at the same time; and some points take energy and some information from them.

The will of those i-points and its 2 parts of energy and information that wish to absorb more of it and reproduce in other zone of space-time is the ultimate why of the Universe. This implies all what exists has living properties and we should respect those existences as long as they do not interfere with ours or are 'parts' of our energy and information. And the most important of those living properties is:

4. Social Evolution is the most important arrow of evolution and creation in the Universe.

Mathematical reductionism simplifies that communication between species, denying their will and right to existence, and especially denying the highest of those arrows that creates the scales of reality - the social arrow of evolution - promoting instead chaotic, destructive theories of The Universe – the Darwinian fight between species instead of their organic evolution, even within the same species as man is (economical competence, social Darwinism, tribal history).

In a non-AE Universe, reality is made of points with parts, species that have internal organs of information and energy, able to perceive each other with communicative forces. So through those acts of communication they organize themselves in networks that foster their social, communicative evolution. Atoms share electrons to become molecules. Molecules share atoms to become macro-molecules. Macromolecules share micro-molecules and organize cells. Cells share energy and information, provided by their blood and nervous information, to become living beings. And humans share verbal information to become 'History'. From microcosms to macrocosms, all communicative species organize themselves into complex systems. We exist in a living, organic Universe, made of multiple microorganisms that grow in size and social organization, from atoms to galaxies, thanks to that informative arrow. All those organisms perceive the Universe with different languages, from mathematics, the language of perception of future robots, to words, the human language, to light, the animal language, to gravitation, probably the force that atoms and masses use to perceive their space-time. So they display in their behaviour and nature, the same living, social properties that humans can understand and compare by homology, even if we cannot necessarily measure them. In this manner we acquire a wider, more complex and harmonic vision on how the Universe works as a living organism and what are the laws of survival and extinction that create and destroy its quantic, organic parts. Yet in AE Sciences, despite the growing evidence that shows how all type of herds interact with each other, this organic Reality becomes dead and chaotic. So scientists, who exist in a living planet as interconnected cells that have to care for the body of this planet and the social organisms of humanity, as your cells care collectively for your body, ignore that social mandate and believe 'ethics' are myths without a biological cause. In fact, scientific laws are similar to the laws of a human society: a mixture of a social contract, interiorised by its citizens, and a hierarchy of control. We obey those laws because they allow the society in which we exist to create a better world and we obey those laws for fear of social punishment. In the same manner Universal laws are imposed on the quanta of a Universal, organic system because the 'brain' of the system controls those quanta and extinguishes particles that do not obey those laws; and because those laws have been 'interiorised' and are part of the structure of those particles. But outside that ecosystem other laws might rule. Therefore, in the Universe exists a social

663

order that creates from individual quanta, organic herds, particles, waves and societies that act in a parallel dynamic group, motivated by its common desire to take energy and information from the ecosystem and survive. Yet mathematics often simplifies that social behaviour, describing it through numbers, without even realizing that the existence of numbers proves the existence of social evolution, since a number is a set of equal, social forms that share those social properties. But in mechanist science numbers have no breadth; they are part of a continuous plane. So they say nothing about the internal, cyclical will of existence that causes it, hiding the fact that:

5. The Universe is organic, ordered in networks, discontinuous, informative.

The abstract Cartesian plane created a continuity myth that misunderstands the duality and 'dynamic' balance that puts together quantic parts into organic networks or wholes, which co-exist simultaneously, through a process of transference of information and energy that 'jumps' the discontinuities between those quantic parts *through a different plane of existence*. Yet since the continuous plane denies those flows of communication, it cannot explain the organizations of quantic herds or the fluctuations of continuous waves into discontinuous particles and vice versa (Complementarity Principle). Therefore, it invents chaotic and abstract theories to explain organic phenomena, like the probabilistic definition of a social wave of light or an electronic orbital; or it enters into Byzantine discussions on the nature of 'the being' (Copenhagen Vs. Everett interpretations of quantic mechanics). It is the same discussion going on in neurology: it is human consciousness one or multiple? Both things: conscience is modular, jumping from a zone to another zone of the brain as the organic system fulfils its different exi cycles, informing, energizing and reproducing itself and each of its quantic, cellular parts. And this is possible because all what exists are networks of non-Euclidean points of view that create complex systems through their flows of energy and information, and so we need to observe reality through multiple planes of existence and through the interaction of multiple elements that cause reality. Thus:

6. Unicausality does not exist: there are 3±st cycles of existence, the 'why' of all Universal events.

Without understanding the duality between individual parts with an existential will and the organic whole whose networks help those parts to reach their existential drives, the Universe becomes chaotic, because we are trying to create a simplified, artificial order, from a single point of view that can't explain the complexity of many causes and inner parts that create the organic, real point-species.

If we do not account for the multiple, quantic exi organic systems that act simultaneously to create a certain future, a lot of consequences and causes are not perceived. We think often that a single cause (ceteris paribus cause) is the only reason why reality exists. Yet a single, continuous space-time deforms the multiple causes of Reality, since each micro-point participates on the processes involved in the creation of the macro-social organism. Yet in the AE-science non-dimensional points seem mechanical abstractions whose actions-reactions are simplified by mathematics. That obsession for simplistic mathematics leaves totally unexplained many problems in which there are too many points in control of the event to calculate the outcome with mathematics - when organic explanations could easily provide an enlightening answer. Such is the case of gravitational phenomena when more than 3 elements are involved, or evolutionary and organic processes proper of Bio-History and Bio-Economics that cannot be explained only with mathematics.

The ceteris paribus error of unicausality is very extended in all sciences, because science believes the abstract simplification of a continuum single space-time field and so it feels satisfied with a single cause, when the Universe is ternary in its scalar, temporal and topological structures, so there are at least 3x 3 causal arrows in time, parallel to 3 topologies of space and 3 interacting scales of social organization in any universal system. So we need to account for all those ternary elements to fully explain most events and forms of reality. Yet continuity and unicausality instead brings absurd arguments of self-centered scientists that want 'their single cause' to be unique: Creationists vs. Chaoticians, Monetarists vs. Fiscal economists, etc. In reality, in an organic Universe made of networks of communication and infinite organic points, there are infinite little causes to every consequence. Yet most scientific equations do not account for all the organic points that have dynamic influences over each real event. So they fail to understand complex phenomena, simplifying it and thinking all is chaos. Chaos merely means that we do not know all the influences from where organic order arises - that we have a 'chaotic', single perspective. Fred Hoyle, one of the clearest minds of XX C. science, affirmed that the Universe is intelligent, because chaos cannot explain the enormous order we find on it. He also denied a single Big-bang and quantized it into multiple big-bang quasars, points with parts that together explain better the discontinuous, organic structure of the Universe. He was right in both accounts.

In that regard, the minimal causality is dual, since all movement is a cycle of an Organic system. The principle of Linear Inertia is incomplete. All events are cyclical processes of action-reaction, except those, which are purely Darwinian, destructive and therefore, are not repeated in time. Events are cyclical, born out of a dual flow of energy and information, which determine 3±st cyclical drives of existence that explain the complex order and the whys of the Universe.

Yet since information networks are 'faster and smaller', often undetectable, AE-science, obsessed by its 'naïve realism' only perceives linear, spatial energy with its scientific instruments. So Science prefers to look and explain reality from the perspective of visual energy, because time and information cannot be photographed. So it is unable to fully understand cyclical, 'unperceivable' informative time cycles, which are the dominant cause of most organic processes, from the morphological plan of evolution caused by the 3 topologies of the universe; to the existence of 3 self-similar, morphological ages that all organic systems fulfil; or the historic, organic, social and economic cycles of mankind and machines. Instead AE-science considers the future a lineal game of chances born out of chaos, when it is a cyclical, repetitive process that can be controlled by controlling the energy and information fields that interact to create all future cycles. In that regard, energy theories are always preferred. For example, instead of considering that dinosaurs died when mammals evolved into placental super-organisms (rodents) and killed them Theory of Evolution is denied and a selective Extinction of only dinosaurs, based in meteorites becomes dogma. Instead of infinite cycles of life and death of galaxies and Universes, a single big-bang of explosive energy becomes dogma...

Indeed, 'lineal, single causes' tend to create dogmatic truths, which are affirmed as postulates without real proof that eliminate all other causes, when in reality normally events happen only when several causes coincide, coming from different st-scales of reality. For example, biologists hint that dinosaurs died probably because mammals, faster, more evolved species, displaced them eating their eggs and offspring. Physicists know there was a rock falling on the skies that might have caused an ecological catastrophe at that time. Yet this secondary cause has become the main cause of the dinosaurs' extinction because it is 'energetic'. Why then mammals did not die? Why only birds, small dinosaurs that put their eggs far away, in rocks, survived? So the likely process should include both causes: the meteorite did raise the stakes of survival and triggered an age of wide famines that obliged mammals to attack dinosaurs, which probably were off-limits in the previous era. Yet physicists know nothing about biology and so they prefer to stick to their single, catastrophic cause. In this manner scientists become specialists in single causes and single disciplines, which become isolated from other disciplines, seen with suspicion by the single-minded AE-scientist. Inter-disciplinary efforts such as those of systems sciences are disliked, considered a menace to dogmatic, specialized disciplines, because they show the shortcomings of the monist, metric age of science based in a single language, a single cause and a single, mechanical instrument to measure the many languages, causes and vital time arrows of reality. Yet mechanism rejects vital causes and verbal logic; to the point that evolutionists, who use a verbal theory to explain with far more success than mathematics do, the works of life, are banned from Nobel prizes... They are not considered 'true scientists.'

Fact is that the Universe and its organic systems are constructed with 3+1 topologies, 3+1 scales and 3+1 functions that become the energetic, informative, reproductive and social cause/function of any event. And the more causes/functions a system/event has in space and time, the more likely it happens.

So we must use instead of Aristotelian Logic and single causes use *the Ternary Principle*, based in the 3 ages/topological dimensions of time and space: First we study a palingenetic form as it reproduces and evolves in time and diversifies in 3 different components that latter gather together creating organic, spatial networks, in 3 scales of reality, the cell, the network and the whole organic system, in which static form and dynamic function, space and time harmonize into a whole, single MST field.

We formalize that Ternary Principle with a numerical expression: 3±st. It means that the existence of a being between conception (+st) and death (-st) is guided by a quantic, social sum, Σ, of 3 basic energetic, informative and reproductive Space-time networks or cycles. When we analyse those structures in space we perceive a sum of quantic cells joined by networks of energy, information and reproduction. When we observe them dynamically in time, we observe an evolutionary radiation of cells that differentiate into energy, information and reproductive forms, creating 'palingenetically' a new macro-organism. The why of those cycles and ages is simple: the mechanical, instinctive or organic will of all fractal beings that try to overcome extinction of its quantic time and quantic extension in space through the repetitive reproduction of its cycles and logic form in other place of the Universe and/or the creation of bigger MST fields, *through social, macro-organic evolution*, that will last longer in time and space. That Ternary Principle and the 3±st cycles of existence give birth to the feed-back, space-time field's generator, E⇔i, and its verbal expression: *'all what exists is a cycle of energy and information transforming into each other.'* From where we can deduce all other equations of science as specific combinations of different cycles of energy and information described by each equation.

When a decade ago I found that Generator equation of the Universe I thought on Einstein's sentence: 'I would like to know the thoughts of God, the rest are details'. And indeed, each science studies the details of those thoughts. But it is rather more important to understand and be guided by the essence, instead of letting the details to obscure the ultimate thought. And that essence is the concept of networks and systems and its laws that are common to all points of view. Since:

7. Networks control their cells; Gods exist in the higher plane of Space-Time.

The mechanist emphasis on spatial, mathematical analyses also limits our understanding of the information networks, which rule any organic system. Since scientists, under the ideology of "naive realism", look for elementary, material, visual particles and continuous systems, missing the essence of fractal, invisible information, which is its speed, simultaneity of reach and minimal, broken patterns of size: You do not see the words that communicate human beings or the gravitational force that fixes you to the planet; yet both exist and without them History and Physics are meaningless.

So macro-organic systems, as anthills or civilizations that appear "disintegrated" because we do not see the pheromonal and verbal information that creates them, are not considered organisms. Instead, scientists consider more important the cellular, energetic, negative entropy that comes from the simpler social planes of an organism than those informative networks that create and integrate from the top the 'consciousness' of a hierarchical organism - when in fact most causes of reality go from the larger scales of the system, which changes the informative and energetic fields of the entire cellular mass, provoking according to the ternary principle and its laws the behavior of its microscopic parts. And so both, the parts and the whole come together to create an event or super-organism.

For example, the 3rd paradigm of measure tried to explain the psychological character of human beings only through chemical micro-genes. Yet in a human organism, both, hormones and electric impulses modify and control our organs, but because scientists have only decoded properly a few hormonal signals they tend to disregard the electric, nervous system, the higher plane of existence.

The same obsession for 'perceived' causes retarded the evolution of genetic for decades, till as we forecasted long ago and the science of evol-devol has recently discovered, the complex morphological characters of organisms are caused by groups of genetic characters that form together a 'higher', epigenetic system of information. On the other hand, scientists found that 99% of human genes are equal to those of the worm; *because individual chemical genes control biological functions common to all living beings, acting primarily in the biochemical level in which they exist.*

On the other hand, 'data-based' history explains all historical events departing from the actions of individuals. Yet, as Marx already understood with his concept of a historic superstructure, the future of history is caused by the structure and evolution of its 'physiological systems', the economic/energetic and cultural/political informative networks that shape societies. *Since any individual form is part of a higher plane of existence that controls it through its energy and information networks.*

Naïve Realism doesn't understand that information and languages are smaller, often invisible to our limited means of perception, despite controlling social cells and their energy. So it denies the collective consciousness and networks that define many organic systems.

For that reason scientists deny the existence of the Absolute God, the mind of the Universe (reduced to a mere set of mathematical laws) and the relative Gods of History, the collective, subconscious minds of cultures, born out of the fusion of many human minds that follow the same ethical-verbal, memetic behaviour. Yet if invisible gravitation attracts bodies and science recognizes it, so it should recognize also the collective subconscious minds of nations. Since their existence and mass-effect is evident in the sacrifices of individual 'patriots' and religious believers.

In fact, there are infinite relative, Aristotelian Gods, informative, focused minds that regulate and control the quantic micro-cells of its organism through 'invisible', ultra-fast informative networks or regulate the quantic lives of its ecosystems through general variations in their information and energy fields that affect those microscopic beings, extinguishing them or forcing them to reproduce or evolve in a certain morphological direction, according to the ternary law. Causality is multiple. It goes not only from the smallest scales to the biggest ones, from 0 to 10, from genes to organisms but also from 10 to 0, from brains to genes, from ecosystems to species. And finally, from 1 to 1, from species to species of the same plane existence, as sometimes genetic material is transferred across the same plane of existence.

Indeed, the key to understand Evolution is to introduce the simultaneous study of causes coming from all planes of existence. When an ecosystem changes, it extinguishes species that do not change as the ecosystem does. So the 'ecosystem', the bigger organic network, programs its micro-cells. In the same manner, a baby is born with neurons whose axons will be connected as experience and visual perception directs them. So the environment decides the logic connections of the mind. Once and again we observe that a micro-plane of existence only creates a blueprint that the 'macro system' sculptures and selects with its own organic laws of extinction, the chisel of the sculpture that gives us the final form. Since time rules, curves space and information rules energy in all scales of reality:

Thus, the black hole, the brain of the galaxy, regulates with invisible gravitational forces the orbits of its stars and uses them as a 'feeding mouth' that gathers its interstellar quantic, atomic food from space). In the lower scale stars control with "gravitational waves" the distance of its planets that control with its geologic and climatic changes of energy (temperature, orogeny, etc.) the evolution of living species like humanity (800 years cycles). On the other hand, the organic systems of humanity, civilizations, control men through its "often invisible" audio-visual, monetary and legal networks; and men control with nervous networks its cells that control with genetic information their carbohydrate molecules that control with electromagnetic forces their atoms that control their electrons, that attract light that feeds on dark energy, the spatial body of the Universe. The main flow *of causality begins in the biggest scales and concludes in the energy quanta of its lower scales.*

If we apply hierarchical causality to history, it is obvious that the only way human beings can control his future is by creating an international global, super-organism able to dominate the actions of individuals, machines and companies, the quantic cells of the economic ecosystem, based in the natural laws of social evolution and the memes of life and love of our biological language-network. Because…

8. Words are the social, temporal language of man., Mathematics perceive only space.

Mathematical scientists despise verbal scientists because they think words are ambiguous and mathematics far more accurate. Yet in a probabilistic Universe where truths are relative, since we do not perceive all its information, words become the perfect language to portray those ambiguities, as they do when describing the will and 'future action paths' of human beings, in which often multiple points of view converge to obtain a final outcome that must be negotiated - while mathematics seem more accurate, because they reduce part of that information, eliminating it.

Unfortunately for mathematical physicists the Cartesian geometrical, origin of their models, downplays the importance of time and makes space the only reality: Time becomes one-dimensional and space steals an extra 'height dimension', proper of information. Even life, the most perfect form of information is defined by spatial science in terms of movement and energy; instead of being defined in terms of informative perception and intelligence, as old philosophers used to define it. The Universe loses its tempo-logical nature and becomes also a big-bang of spatial energy. Space, Energy and force, no longer time, information and perception, matters because the still, inner virtual worlds of Points with Parts have disappeared from graphs of Cartesian scientists.

Spatial mathematics is more accurate when we apply them to simpler beings like atoms and forces, which have minimal form and maximum spatial energy. Further on, the limits of our perception due to the distance of atoms and stars, simplify their properties. So mathematical simplicity is better to describe astro-physical forms that seem mechanisms to us. But words are better to describe complex living forms and human beings, as Theory of Evolution, Philosophy, Social sciences and Eastern religions have proved. A sword is only a line and it seems perfect; so happens to the simple mathematical language we use to describe that sword. Yet a complex human being seems imperfect in numbers, because he is too complex to describe all its properties with them.

Eyes, mathematics and numbers perceive geometrical space; while verbal words and sounds talk about events in time. So mathematics is basically a spatial language, while words are used to measure temporal phenomena; and since information dominates energy words are the fundamental language to explain informative, temporal species, such as life is. For that reason, because we are living beings, Theory of Evolution is the most important classic Theory of knowledge. And it includes Economics, which is today dominated by the evolution of machines by its company-mothers – a theory so self-evident and accurate calculating the future, as the work of this author that forecasted the present crisis and its dates 20 years ago has proved[1]. On the other hand financial economists have never been able to predict the future of its indicators – it is in fact the only science that has never been able to do so, to the point that The Economist conducted a poll of future indicators won first by taxi drivers, then garbage collectors, next finance ministers and finally in the last position, economists. And despite of it, according to polls, 76% of economists think their discipline is the most 'scientific' of all social disciplines, because it quantifies its propositions, when it cannot even define what an economic ecosystem is.

And what the economic ecosystem is has become clear in this work: a system of metal-memes that is extinguishing life. Yet economics with its use of numbers without form hides the harmful collateral damages of machines. Classic economics with its definition of wealth in monetary terms ignores its lethal purpose: to create a world to the image and likeness of the machine. Bio-History and bio-economics however by using the verbal language and theory of evolution is both more humane and more accurate as it has a positive purpose for mankind – to make a world to the image and likeness of our species. And so it should be considered the 'real model' of social sciences.

In that sense, mechanist Physics and financial Economics should be regarded as applied, technological sciences related to the evolution of machines and become somehow downgraded in the Pantheon of knowledge, as secondary sciences, submissive to Biology and History - that study life and the human being; given the limits of physical perception, which make those AE-sciences useless to search for the ultimate laws of reality and its biased 'biological purpose', which despises man and its senses and considers its goal to evolve and reproduce machines. And for the same reason budgets to make electronic machines and physical experiments should be judged, as those of all applied sciences, according to its utility for human survival, banning them when they might have negative side effects on mankind, ignoring the 'excuse' of metric measure as the supreme knowledge. We thus conclude that:

9. In a diffeomorphic Universe man and his languages should 'be the measure of all things'.

It is evident that mathematical, abstract science, as anthropomorphic religions did before her, has set the pre-conditions to perceive a dead Universe with man as its only intelligent being in its center. This is natural to the Galilean paradox. Since numbers simplify and "itify" reality, emptying it of virtual worlds.

Yet when we apply mathematics to human beings in statistics and economic equations of productivity that equal men and machines through prices, they become a lethal weapon that makes of man just another abstract point-number. This explains why the 2 leading technological, scientific nations of the XX century, pre-war Germany and modern America, show an appalling lack of concern for human rights. It is necessary to know and explain those limits of mathematics, especially in the study of social sciences. Otherwise economists will eliminate man from the economic ecosystem, as we are, from an abstract point of view, less efficient than robots with which they compare us through abstract prices and productivity-costs. Since mathematics is a language that equals all beings through equations: X=Y. So when those equations are applied to man, man is equalled to an object, a price or a machine in "economic" equations of the type: *Man =Price= Object*. Thus economics treats man as an object eliminated with equations of productivity, substituted by robotic machines or killed as collateral damages in profit wars. Since an object of higher price, often a weapon, matters more.

The moral error of mechanist science that measures and extinguishes life is to ignore the laws of survival, which should dominate society. Tagged and measured, man stops being the subjective center of his own world, becoming an object, a commodity. Yet in verbal languages this never happens. Since man is always the subject, center of the verbal action and therefore, words value humans more than any object: *Subject > Verb > Object*. Since man, the subject dominates the object through the verb. So 'he' cannot be an object. He can't be priced. Man and life become bio-ethic concepts that cannot be expressed in digital numbers. The logic of survival is not the abstract logic of mathematics, but the bioethical logic of words that should guide humanity in our search of knowledge and happiness. Because in an organic Universe of multiple languages and dark spaces, where information is always a linguistic subjective truth, the only truth that matters is survival. So:

"Man should be the measure of all things".

That human extinction is not certain, is obvious. Yet even mechanist scientists making electronic gadgets use probabilities in quantic physics as a way to reach truths that work. For the same reason, if there is a probability that CERN's black hole factories, nanorobotics, (metal-bacteria) and robotics can extinguish mankind, making us obsolete in labor and war fields, the creation of robots should be forbidden as a criminal act against mankind. What digital scientists forget though, is that they can also be substituted; that *an extinct scientist knows nothing*. In that sense, this work does not affirm to have the absolute truth but the most probable truth about time, with a clear aim: to help man survive into the future. Within the limits of languages, absolute certainty is impossible. Thus probable truths guide all beings, since survival is a game of probabilities not of certainties. For that reason we should limit all probabilities of human extinction, controlling its possible technological causes, prohibiting robots, weapons and scientific ideologies, such as capitalism, mechanism and nationalism that foster the destruction of man by machines or by other men.

The language not the body, information not energy, determines the survival of any species into the future, also in history, because the species, which better speak the languages of an ecosystem creates the future. For that reason, when man loses his verbal language, degraded and despised today as a form of dramatic fiction of null logical value, or substituted by digital numbers spoken better by machines, humanity loses his future and becomes obsolete to the digital machines of information we worship.

Since all species of the Universe display the same formal, organic structure and obey the same MST laws, it is evident that by studying man, the species we can observe in more detail, 'as the measure of all things', we can understand the laws of the Universe better than by analysing atoms or stars. Hence, Biology, Medicine, and Bio-History that describe human beings at micro-cosmic, individual and macro-organic level, should be, from the point of view of MST theory, the 3 fundamental disciplines of knowledge in a better educational system, where Bioethics, the science of survival becomes the guidance of politicians and economists and MST theory, the science that explains the relationship between parts and wholes, points and networks, becomes the philosophy of all sciences, the thoughts of God. Since:

$$10=1_{st+1}: \textit{We are all parts of the Eternal Whole} = God$$

The relative immortality of a fractal space-time Universe arouses from the co-existence of many quantic time /spaces that evolve or devolve, live and die in each local region of the Universe. Thus the sum of all beings that expand and die, all beings that reproduce and repeat its form and all beings that contract and inform reality, creates altogether a dynamic balance that makes reality an eternal present of 'quantic' fluctuations called lives. Thus the Game of Existence, the 'Mind of the Universe' is constructed to be eternal without being dead, thanks to the infinite minute lives and deaths of all its beings. In order to live, to perceive, we need to absorb the energy of other beings that have to die. It is the law of survival, the cruellest law of the existential game:

$$\textit{Life (perception)} = \textit{Death (energy explosion), Victim (energy)} = \textit{predator (form)}$$

The fight between species is unavoidable: All energy dies, because a living being absorbs it. And thus the balance is re-established. In mathematical terms we can say 'that the total wave of energy of the Universe' cancels with 'the total wave of information' in all and each one of the events of the Game of Existence. No matter how much a being wishes immortality, the existential game does not allow it. Because only the Game is immortal. Yet it is also just, since we die because we kill, we become old and suffer because we were young and had pleasure…

For that reason, we postulate an infinite, quantic Universe, body of an eternal God, which is an impersonal set of rules, we call the '*Game of Existence*', that as its name implies, *exists* forever in time and it is immutable in space, although it contracts and dilates into many quantic, Existential Waves that ignite, flare and extinguish living beings. Ultimately because the Universe is a game of yin=information and yang=energy, with opposite properties, of God, the Game of Existence is paradoxical:

God is absolutely selfish (as we die for the Game to play again); yet he is absolutely generous (as each of its quantic parts gives its energy to feed other beings). He is absolutely just (as all what kills dies, all the pleasures of the energetic youth become the pains of the old, informative age) and yet he is absolutely cruel (as he gives us first the pleasure of life to take it away at the end). He is absolutely smart (as he has devised the best of games to be immortal) and yet He is absolutely dumb (as he was unable to invent a Game that had no suffering, no death). He is absolutely ubiquitous (as his consciousness occupies the entire Universe and beyond) yet He is absolutely diminutive (as its quantic mind is a relative zero that perceive the infinite space). He is absolute eternal (as the game will never stop) and yet an ever fleeting presence (as his parts live so short). He is, in Words of Cusa, the minimum (the smallest particle of the purest information) and the maximum (the infinite Universe of absolute spatial extension). But overall he is the Supreme Dictator, Creator and Destroyer of all his forms, which are only a reflection of the Game. So we better watch out and respect his laws, because he is absolutely merciless, as any form that disobeys the laws of existence becomes extinct. And we are not respecting any of them… Instead, we have invented false Go(l)ds, selfish idols of metal to cater our arrogance. And so we will be absolutely extinct by them. Vanitas vanitatum et omnia vanitates. As we indeed are a zero mind that thought to be the center of the infinite Universe:

$$0_{0…} \quad \textit{And the absolute nothingness of Man}$$

The importance for the praxis of life of knowing the true laws of the universe cannot be stressed enough. The immediate consequences are a cautious, humble, grateful desire to survive, which today humans lack. Arrogance and ignorance of the true meaning of time, life and its arrows of social evolution and organic life is the trademark of mechanist scientists, physicists and economists, the scientists of weapons and go(l)d that are destroying the planet. But they are 'experts' in their job. In true form, in the same way the culture of economists have been tailored to follow a series of myths that cater to their religion of power, go(l)d; physicists, the mechanists that construct the machines of our world, have built a series of myths about the 'mechanist' nature of the Universe, which cater to their use of machines as their tool of power. The machine, thus, becomes the idol of their/our civilization, and for that reason, we shall all die for the Large Hadron Collider, the last and most perfect weapon or for the profits of the Robotic Industry and the corporations that expel human workers… all those facts are product of our worship of machines of measure, more than the mind, intelligence and life of mankind.

The only penalty the Game has to those who do not respect its rules is the inversion of existence, the zeroes of all its quantic parts. Because all forms can also be destroyed and 'reabsorbed' by the eternal present; since when we fusion forms with opposite parameters of space-time together they 'cancel each other' into the eternal 'present' of virtual reality from where all is born and all returns. The Universe is immortal because it is virtual, as the sum of all its quantic, positive and negative space-time dimensions, dual a ternary differentiations is 0. Nothing happens as all what happens provokes its anti-event. Imagine the Universe as an immense n-dimensional volume of amorphous continuous energy in which information quantizes forms with very thin, almost invisible borders that appear and disappear constantly. Yet the entire volume doesn't change; it is eternal, never wears. Because if we create a convex line of time we are also creating a concave line on the other side, if we observe an infinite we do so from our 0. Thus all remains constant. Nothing really exists. All comes from a zero-point and returns to a zero-point. All is virtual for the whole to be eternal:

Dust of space-time you are and dust you will become[13].

The cycle is closed: our conclusions in the 3rd age of science are similar to those of Eastern Verbal masters of relativistic thought. Man should understand the laws of change in the universe, the existence of other parallel Brains, who perceive other regions of space and time, that man ignore. Then we should draw obvious laws of human behavior, according to those universal laws, and promote our role as top predator brains on Earth. We should not act with arrogance, and destroy nature - our biologic ecosystem - and pretend to be God, building machines more intelligent than us. Instead, we should respect the laws of evolution, use them to the advantage of man, and develop a harmonic vision of the universe, with both verbal and mathematical languages.

Recap. It is necessary to speak loud on the extinction of life by machines that poison this planet, destroy human cultures and will 'probably' extinguish us this millennium, unless we reform the scientific method and the postulates of economic growth based in technology, not in the human senses and the production of human goods necessary to our survival. But to achieve that, we have to accept again the supremacy of ethical-verbal-temporal languages, the laws of social evolution and love, the equality of all human beings and exercise a human, ethic supremacy over the machines of science. A Decalogue of Laws to awaken man to the vitality of the Universe might change the cynicism of some techno-utopian scientists that deny ethics or do not want to control their machines. The 'third age of science' implies the knowledge of the living Universe, and the need to be cautious in that Universe, and use the bio-logic rules of existence to the advantage of man. It implies to deny the dogmas and lies of the scientific method, and substitute it for a superior method of knowledge and survival, as the scientific method substituted centuries ago the myths and beliefs of anthropocentric religions. It is the old vision of Chinese Taoists. Lao-tse said: when more complex instruments men discover, more wars and destruction happens.

Notes

[1] The first account of the 10th extinction appeared in the book 'Radiations of space-time: the extinction of man', c.94. An organic, systemic model of History and economics first appeared in 'Bio-History, Bio-economics: a theory of Unification', c.97. Biologists normally talk of the 6th extinction (ignoring the earlier phases of extinction of carbon-molecules and simple cells studied in the previous chapter) and especially ignore that it is not humanity but the machines of metal, we construct, what causes it.

[2] The Military-Industrial Complex was an expression coined by Eisenhower, who denounced the growth of industries of machines of war that were subverting the American way of life. He failed to mention the 'language' of money, which has always directed the affairs of war ('Nervi Belli Pecunia infinita'). In true form, the system of metal-memes that is terraforming the Earth has 3 components, money, metal-information, weapons, metal-energy and machines, organic metal.

[3] The 72 years Generational Cycle is the mean biological cycle of human beings. It was first used to study patterns in American History by Strauss and Howe: 'Generations' W. Morrow 1991.

[4] Ibn Khaldun uses the term Assabiyah, 'union', to stress the capacity that religions have to unify into a single organic system a group of humans that will act and die for the collective subconscious. The theme was retaken by the German School – Spengler, Jung.

[5] The word meme was coined by Mr. Dawkins in his book 'The selfish gene'. Here we adopt a materialist approach to memetics, considering as the historic school of economics did that metal-memes cause the creation of ideologies of war (militarism), money (capitalism) and technology (mechanism) that favors their evolution at all costs.

[6] The analysis of languages as ternary systems with an informative subject, a verbal action and an energetic object manipulated by the subject, was the work of Mr. Chomsky. Yet that ternary structure responds also to the duality of a complex Universe made with energy and information. Thus all languages have a 'Universal Grammar', between two elements related by an action. Mathematical languages are all written with that grammar: $F(X) = F(Y)$, where both elements are objects. Only words put man as the protagonist and subject of all sentences.

[7] All systems are complementary. This means that the informative, 'neuronal' caste must serve the cells of the body and provide them with adequate energy and information. The arrival of metal-castes broke this organic harmony between priests, the verbal neurons of Neolithic organisms, and the farmer's class that will now be exploited without obtaining the energy and information it deserves. Societies now become 'Darwinian', as animetal castes feel different from the rest of mankind, breaking the balances between both social castes proper of all healthy organisms of history. Yet 'class struggle' is not natural but a consequence of that differentiation between humans that no longer apply the laws of love and self-similarity (3rd postulate) proper of all social networks made of the same point-species, as we all humans are.

[8] The difficult theme of the role of the Jewish culture as the 'soliton' of financial power that carry with money, the language of metal-information, the evolution of the FMI complex, must be treated scientifically without religious myths or anti-Semitic bias. Thus, we need to study the classic books of anthropology and economic history. The key works are those of William Albright's ('Yaveh and the Gods of Canaan') and other practitioners of Biblical Archeology, which established the geographical origin of the word Yvwh, which means Judea - a nation not a god. Thus its first quote appears in Egyptian texts, as a reference to the trading communities of Canaan, where the Bedouins from Lbn (Lebanon) and Ywh (Judea) lived. Hence we must treat the Bible as a book of history of a people, whose profession was also determined by geography and the available goods of Ywh, Judea, a land of donkeys, able for harsh military loads, where the 'habiru', which means exactly donkey caravaneers, became specialized in trade of weapons and slaves, *which lead to money lending. Hence capitalist memes predate in time and define the religious and social memes of the culture.* In the Middle Ages the key works are those of Mac Cormick, 'Origins of the European Economy', Cambridge U.Press that describes the massive slave and eunuch trade between Europe and Islam, carried about by Vikings, Jewish and Venetian merchants and Abraham Leon's book on 'The Jewish question', which studies lending practices in the Middle Ages. Further on, Sombart in 'The Jewish people and the origin of Capitalism', and his disciple, Weber in 'Protestant ethics and the birth of capitalism' study in depth the historic and ideological connection between the biblical memes that objectify all other cultures (racial separation) handle no longer with the verbal ethics reserved to the 'community'; and repress the human biological drives (sex, food, social love) to impose the ethics of work and reproduction of money, weapons and machines that define the evolution of the FMI complex. The importance of those facts for a systemic, scientific analysis of the economic ecosystem is clear: economics is not a science but a religion or ideology that worships and evolves metal-memes, to achieve power over all other human tribes, despite its modern disguise with digital equations. As such the financial industry must be reformed and nationalized to efficiently direct the world for the benefit of all humans, not for the benefit of a single group or culture. In its present condition the faith on a free economic ecosystem, in which machines, corporations and its owners have more rights than the bulk human beings (workers and consumers) will merely evolve

those memes of metal that provide the tools of power to the people-castes on top of western societies (the military, bankers, stockrats and mechanist scientists that profit from them) till evolved machines make all humans obsolete, extinguishing life. Only a scientific management of the economy and the pruning of the bad fruits of the tree of science could reverse that evolutionary tendency.

[9] Expression coined by Roosevelt to explain the profitable Cuban Spanish-American war: 'This splendid little war will take us out of the economic crisis (of 1897)'. In II W.W. again Spain became a perfect guinea pig for the splendid German weapons, as Afghanistan is today for the new generation of robotic weapons.

[10] Kondratieff analyzed the Industrial Revolution in Russia, concluding that for 50 years the train was the engine of its development; thus establishing a 50 year cycle of economic activity that bears his name. However in 3rd World countries machines arrive later, when they have already evolved in the Industrial World. So the Russian cycle of steam was shorter than the World's cycle. When we make an exhaustive analysis of the cycle in the Western World, where those machines were discovered, we find a longer 72 years generational cycle. Later on, the relationship between the Kondratieff cycle and the Evolution of Machines was considered in detail by Schumpeter in his book 'Business Cycles', 1939. However, as he did not relate that evolution to the discovery of the 3 main types of energies that determined it (steam, oil and electronics), he divided the I and II Energy Cycles, Steam and Oil, in 4 sub-cycles: Steam; Trains, which are powered also by steam, thus forming together a single cycle; Steel and Oil, which were combined in cars and boats, shaping the 2nd cycle. The result of those quantitative errors has been a long scholar discussion on the very same existence of Kondratieff cycles, as all forecasts based in the standard 50 year period have failed to realize. This argument can be considered closed with the adjustment of the cycle to its true length of 72 years. As the new paradigm becomes known to the English-speaking world and the forecast of a 2001 stock/political and military crisis proves its accuracy, the literature on the Kondratieff cycles has increased enormously in recent years.

[11] The relationship between a crisis of overproduction of machines and an age of war is a classic theme of Economic Theory, pointed out by Marx and Lenin in their books on the expansion of capitalism. Quincy Wright made an exhaustive treatment of the theme in his book 'A Study of War', 1965. I. Wallerstein ('War in the World System', 1988) pointed out again at the systemic use of war as a way to come out of an economic crisis, relating them to the end of a Kondratieff cycle. Yet both writers used the 50 year cycle, failing to forecast the arrival of a III Kondratieff crisis and subsequent age of war in the 2000s.

[12] Economics *must be ruled as all other sciences not by anthropomorphic, tribal or individual selfish agendas but with* the use of its scientific laws to manage internationally the production of money to foster the reproduction of human, biological goods and the survival of the species in a sustainable planet – that implies elected governments (nationalization of the financial industry and citizens (free salaries) must have credit to cre(dit)ate a demand-based, democratic economy.

[13] While there is no doubt that a sound, scientific design of the economic ecosystem is possible and desirable, as Marx put it, 'all r=evolutions will find out that we can change many things except the human condition'. In terms of systemic science this dictum translates in a key question for the future of mankind: 'Are we determined by greed and violence, by our desire to have more information and energy to destroy this planet and create the Earth of Metal?' This question, perhaps directed to the visual, white man who seems to be more prone to violence and gold hypnotism – will decide the future of the planet. If the science of organic history triumphs, we will survive and create the Wor(l)d Union. If tribal history, business as usual, continues we will destroy ourselves.

Ultimately we humans are oriented to the future in all what regards machines and the memes of metal we evolve, and we are oriented to the primitive past in all what regards the ideologies of the mind we use to control the world. That difference between our bid to evolve metal with no limits and our indifference to the evolution of the collective human social mind, entrenched in myths, nationalisms and religions that date back to the bronze age makes impossible a r=evolution of the system to profit mankind. Humans are trying to direct the world of the 3rd millennium with the memes of cultures of the 3rd millennium... before Christ. The result is a chaotic world, left to the laws of Darwinian evolution that will favor metal over man. Yet if humans could evolve their memes and understand the social mandate of the Universe, and collaborate as a single species, the salvation of the Earth will be a simple task. Hence our brutal criticism of the memes of capitalism, militarism and mechanism, the 'iron jail' of our mind that sets history into a deterministic path towards extinction.